Blood Science

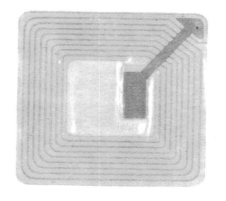

Blood Science
Principles and Pathology

Andrew Blann
University of Birmingham Centre
for Cardiovascular Sciences
Department of Medicine
City Hospital
Birmingham, United Kingdom

Nessar Ahmed
School of Healthcare Science
Manchester Metropolitan University
Manchester, United Kingdom

WILEY Blackwell

Library of Congress Cataloging-in-Publication Data

Blann, Andrew D., author.
  Blood science : principles and pathology / Andrew Blann and Nessar Ahmed.
    p. ; cm.
  Includes index.
  ISBN 978-1-118-35138-3 (cloth) – ISBN 978-1-118-35146-8 (paper)
  I. Ahmed, Nessar, author. II. Title.
[DNLM: 1. Blood Physiological Phenomena. 2. Blood Chemical Analysis. 3. Pathology, Clinical–methods. WH 100]
  RB145
  616.07′561–dc23
                                                                    2013019467

A catalogue record for this book is available from the British Library.

Wiley also publishes its books in a variety of electronic formats. Some content that appears in print may not be available in electronic books.

Set in 9.25/11.5pt Minion by Thomson Digital, Noida, India.
Printed and bound in Singapore by Markono Print Media Pte Ltd

1 2014

For Edward, Eleanor and Rosie
Dr Andrew Blann

For Neha, Aryan and Saif
Dr Nessar Ahmed

Contents

Preface

Blood science is a relatively new discipline driven by changes in the pathology service and is the merger of traditional disciplines of clinical biochemistry, haematology, immunology and aspects of transfusion science and clinical genetics. Blood science departments are growing in number in UK National Health Service laboratories and have been established not only to cut costs, but also to bring together common aspects of laboratory practice, reduction of overlap and following review of roles in clinical diagnosis and management. This new super-discipline will require a special breed of scientist not only to lead the laboratory management, but also to ensure the success of this initiative within pathology and the wider hospital community. In addition, changes in education and training for healthcare scientists have resulted in the development of practice orientated degree programmes, many of which offer modules in blood science. *Blood Science: Principles and Pathology* has been written to meet the needs of undergraduate students taking such modules on BSc programmes in biomedical and healthcare sciences. In addition, this book will provide suitable initial reading for those students embarking on blood science modules on MSc programmes. It will also be of value to new graduates entering the profession and starting their career in blood science departments and will supplement their practice-based training with the required theoretical underpinning.

Our book consists of 22 chapters, where the first chapter introduces the subject and the second chapter provides an overview of the techniques used in blood science. The order of the remaining chapters is based on groups of analytes investigated in blood; namely, red blood cells, white blood cells, platelets and then to the constituents of plasma such as water, waste products, electrolytes, glucose, lipids, enzymes, hormones, nutrients, drugs and poisons, and so on. Each chapter starts with a series of learning objectives; this is followed by the main text and ends with a summary and further reading, the latter consisting of a list of selected journal articles and web sites. Most chapters include their own chapter-specific case studies with interpretations to demonstrate how laboratory data in conjunction with clinical details are utilized when investigating patients with actual or suspected disease. There is a separate final chapter devoted to more detailed case reports, the purpose of which is to integrate different aspects of blood science. Throughout the book we will present examples of how blood science integrates the five different disciplines of biochemistry, haematology, blood transfusion, immunology and genetics into a single discipline. These forty-eight 'Blood Science Angles', and their locations, are given below. The units used for measurement of analytes are typical of those used in hospital practice within the UK. We have also provided a list of abbreviations, reference ranges for common analytes and a glossary of key terms to aid the reader.

Blood science angles

Dr Andrew Blann
Dr Nessar Ahmed

Acknowledgements

A number of individuals have assisted us in the writing of *Blood Science: Principles and Pathology* by reviewing chapters, donating figures and providing feedback, for which we are eternally grateful. They include: Joanne Adaway (Manchester), Lakhvir Assi (Birmingham), Gwendolen Ayers (Manchester), Wilma Barcellini (Milan), Hani Bibawi (Oxford), John Bleby (Milton Keynes), David Bloxham (Cambridge), David Briggs (Birmingham), Paul Collinson (London), Adrian Ebbs (Kings Lynn), Angela Hall (London), Mark Hill (Birmingham), Tim James (Oxford), Colm Keane (Brisbane), Sukhjinder Mahwah (Birmingham), Stephen McDonald (Altnagelvin), Garry McDowell (Ormskirk), Paul Moss (Birmingham), Mohammed Pervaz (Birmingham), Drew Provan (London), Walter Reid (Manchester), Susan Rides (Dudley), Doreen Shanks (Manchester), Roy Sherwood (London), James Taylor (Birmingham), Tony Traynor (Birmingham), Pat Twomey (Ipswich), James Vickers (Wolverhampton), David Wilson (Aberdeen), Ulrich Woermann (Bern) and Allen Yates (Manchester).

We are also grateful to Lucy Sayer (Commissioning Editor) and to Fiona Seymour (Senior Project Editor), both from Wiley-Blackwell, for all their encouragement, support and guidance. Although we have taken steps to reduce any errors in the text, we take full responsibility for any that do arise. Please feel free to contact us should you detect any errors or have any suggestions for changes and we will endeavour to rectify these for any future editions.

Dr Andrew Blann PhD CSci FIBMS FRCPath
Consultant Clinical Scientist and Senior
Lecturer in Medicine
University of Birmingham Centre for Cardiovascular
Sciences
Department of Medicine
City Hospital
Birmingham
United Kingdom

Dr Nessar Ahmed PhD CSci FIBMS
Reader in Clinical Biochemistry
School of Healthcare Science
Manchester Metropolitan University
Manchester
United Kingdom

List of Abbreviations

AAE	acquired angioedema	CABG	coronary artery by-pass graft
AAT	alpha-1-antitrypsin	CAH	congenital adrenal hyperplasia
Ab	antibody	CAT	computerized axial tomography
ACEI	angiotensin converting enzyme inhibitor	CE	capillary electrophoresis
ACR	albumin/creatinine ratio	CEA	carcinoembryonic antigen
ACTH	adrenocorticotrophic hormone	CEL	chronic eosinophilic leukaemia
ADCC	antibody-dependent cell-mediated cytotoxicity	CETP	cholesterol ester transfer protein
		CFU	colony forming unit
ADP	adenosine diphosphate	CGD	chronic granulomatous disease
A&E	accident & emergency	CGL	chronic granulocytic leukaemia
AF	atrial fibrillation	CHF	congestive heart failure
AFP	alpha-fetoprotein	CK	creatine kinase
AGE	advanced glycation endproduct	CKD	chronic kidney disease
AIDS	acquired immunodeficiency syndrome	CLIA	chemiluminescence immunoassay
AIHA	autoimmune haemolytic anaemia	CLL	chronic lymphocytic leukaemia
ALA	aminolevulinic acid	CML	chronic myeloid leukaemia
ALL	acute lymphoblastic leukaemia	CMML	chronic myelomonocytic leukaemia
ALP	alkaline phosphatase	CoA	coenzyme A
ALT	alanine transaminase	COSHH	Control of Substances Hazardous to Health
AMH	anti-Mullerian hormone		
AML	acute myeloid leukaemia	CPD	continuing professional development
ANA	antinuclear antibody	CRP	c-reactive peptide
ANCA	anti-neutrophil cytoplasmic antibody	CSF	colony stimulating factor
ANP	A-type natriuretic peptide	CV	coefficient of variation
APA	antiphospholipid antibodies	CVD	cardiovascular disease
APS	antiphospholipid syndrome	DAT	direct antiglobulin test
APTT	activated partial thromboplastin time	1,25-DHCC	1,25-dihydroxycholecalciferol
ARB	angiotensin receptor blocker	DHD	dihydropyrimidine dehydrogenase
AST	aspartate aminotransferase	DHEA	dehydroepiandrosterone
ATP	adenosine triphosphate	DHEAS	dehydroepiandrosterone sulphate
ATPase	adenosine triphosphatase	DHPD	dihydropyrimidine dehydrogenase
AVP	arginine vasopressin	DIC	disseminated intravascular coagulation
BALT	bronchus associated lymphoid tissue	DKA	diabetic ketoacidosis
BcR	B cell receptor	DNA	deoxyribonucleic acid
BCSH	British Committee for Standards in Haematology	DOAs	drugs of abuse
		2,3-DPG	2,3-diphosphoglycerate
BJP	Bence–Jones protein	dsDNA	double-stranded DNA
BMI	body mass index	DVT	deep vein thrombosis
BNP	B-type natriuretic peptide	EBV	Epstein-Barr virus
BPH	benign prostatic hypertrophy	ECF	extracellular fluid
CA	carbohydrate antigen	ECG	electrocardiogram

EDTA	ethylenediaminetetraacetic acid	HE	hereditary elliptocytosis
eGFR	estimated glomerular filtration rate	HELLP	haemolysis, elevated liver enzymes, low platelet count
ELISA	enzyme-linked immunosorbent assay		
ENA	extractable nuclear antigen	hFABP	heart-type fatty acid binding protein
ESI	electrospray ionization	HH	haemochromatosis
ESR	erythrocyte sedimentation rate	HIT	heparin-induced thrombocytopenia
ET	essential thrombocythaemia	HIV	human immunodeficiency virus
FACS	fluorescence activated cell scanning	HLA	human leukocyte antigen
FAI	free androgen index	HNA	human neutrophil antigen
FBC	full blood count	HOMA	homeostasis model assessment
Fc	crystallizable fraction (of immunoglobulin)	HPC	Health Professions Council
		HPFH	hereditary persistence of foetal haemoglobin
FcR	crystallizable fraction receptor		
FCH	familial combined hyperlipidaemia	HPLC	high performance liquid chromatography
FFP	fresh frozen plasma		
FH	familial hypercholestrolaemia	HRT	hormone replacement therapy
FIA	fluorescence immunoassay	HS	hereditary spherocytosis
FISH	fluorescence in-situ hybridization	H&S	health & safety
FITC	fluorescein isothiocyanate	HSST	Higher Specialist Scientific Training
fL	femtolitre 10^{-15}	IAT	indirect antiglobulin test
FN	false negative	IBD	inflammatory bowel disease
FP	false positive	IBMS	Institute of Biomedical Science
FSH	follicle stimulating hormone	IBS	irritable bowel syndrome
G6PD	glucose-6-phosphate dehydrogenase	ICU	intensive care unit
GC	gas chromatography	IDL	intermediate density lipoprotein
G-CSF	granulocyte colony-stimulating factor	IEF	iso-electric focussing
GFR	glomerular filtration rate	IF	intrinsic factor
GGT	gamma-glutamyl transpeptidase	IFG	impaired fasting glucose
GH	growth hormone	Ig	immunoglobulin
GHRH	growth hormone releasing hormone	IGF-1	insulin-like growth factor-1
GIT	gastrointestinal tract	IGT	impaired glucose tolerance
GLC	gas–liquid chromatography	IHD	ischaemic heart disease
GLP-1	glucagon-like peptide 1	IL	interleukin
GM-CSF	granulocyte-macrophage colony stimulating factor	IM	infectious mononucleosis
		IMD	inherited metabolic disorder
GnRH	gonadotrophin releasing hormone	INR	international normalized ratio
GP	general practitioner	IRMA	immunoradiometric assay
GSD	glycogen storage disease	ISI	international sensitivity index
GSH	glutathione	ITP	immune thrombocytopenia purpura
HAE	hereditary angioedema	IV	intravenous
HASAWA	Health and Safety at Work Act	kDa	kiloDaltons
HbA$_{1c}$	glycated haemoglobin	LA	lupus anticoagulant
Hb	haemoglobin	LCAT	lecithin cholesterol acyltransferase
HCC	hepatocellular carcinoma	LDH	lactate dehydrogenase
25-HCC	25-hydroxycholecalciferol	LDL	low density lipoprotein
hCG	human chorionic gonadotrophin	LFT	liver function test
HCL	hairy cell leukaemia	LH	luteinising hormone
Hct	haematocrit	LMWH	low molecular weight heparin
HDL	high density lipoprotein	LPL	lymphoplasmacytoid lymphoma
HDN	haemolytic disease of the newborn	LR	likelihood ratio

LTA	light transmission aggregometry
MALDI	matrix-assisted laser desorption ionization
MALT	mucosa-associated lymphoid tissue
MBP	myelin basic protein
MCH	mean cell haemoglobin
MCHC	mean cell haemoglobin concentration
MCV	mean cell volume; mutated citrullinated vimentin
MEN	multiple endocrine neoplasia
MGUS	monoclonal gammopathy of undetermined significance
MHC	major histocompatibility complex
MODY	maturity onset diabetes of the young
MPV	mean platelet volume
MRI	magnetic resonance imaging
mRNA	messenger RNA
MS	mass spectrometry
MTHFR	methylenetetrahydrofolate reductase
NAD	nicotinamide adenine dinucleotide
NADH	nicotinamide adenine dinucleotide hydrogen
NADP	nicotinamide adenine dinucleotide phosphate
NADPH	nicotinamide adenine dinucleotide phosphate hydrogen
NALT	nasopharynx-associated lymphoid tissue
NAPQI	N-acetyl-parabenzoquinone imine
NASH	non-alcoholic steatohepatitis
NBS	National Blood Service
NBT	nitro-blue tetrazolium
NEQAS	National External Quality Assurance Scheme
NHL	non-Hodgkin lymphoma
NHS	National Health Service
NICE	National Institute of Health and Clinical Excellence
NK	natural killer (cell)
NPT	near patient testing
NPV	negative predictive value
NSAID	nonsteroidal anti-inflammatory drug
OA	osteoarthritis
OGTT	oral glucose tolerance test
Pa	pascal
PCOS	polycystic ovary syndrome
PCR	polymerase chain reaction
PE	pulmonary embolism
PEG	polyethylene glycol
pg	picogram 10^{-12}
PK	pyruvate kinase
PNH	paroxysmal nocturnal haemoglobinuria
POCT	point of care testing
PPV	positive predictive value
PRCA	pure red cell aplasia
PRP	platelet rich plasma
PRV	polycythaemia rubra vera
PSA	prostate specific antigen
PT	prothrombin time
PTH	parathyroid hormone
PTHrP	parathyroid hormone related protein
PTP	Practitioner Training Programme
RA	rheumatoid arthritis
RCC	red cell count
RDW	red cell distribution width
Rh	rhesus
RhF	rheumatoid factor
RIA	radioimmunoassay
RNA	ribonucleic acid
SCID	severe combined immunodeficiency
SCIT	subcutaneous injection immunotherapy
SD	standard deviation
SHBG	sex-hormone-binding globulin
SHOT	Serious Hazards of Transfusion group
SIADH	syndrome of inappropriate antidiuretic hormone
SLE	systemic lupus erythematosus
SLIT	sub-lingual immunotherapy
SNP	single nucleotide polymorphism
SOP	standard operating procedure
STP	Scientist Training Programme
sTfR	soluble transferrin receptor
T3	tri-iodothyronine
T4	thyroxine
TAFI	thrombin-activatable fibrinolysis inhibitor
TcR	T cell receptor
TDM	therapeutic drug monitoring
TFPI	tissue factor pathway inhibitor
TfR	transferrin receptor
TIA	transient ischaemic attack
TKI	tyrosine kinase inhibitor
TN	true negative
TOF	time-of-flight
TP	true positive
TPMT	thiopurine methyltransferase
TPN	total parenteral nutrition
TPP	thrombotic thrombocytopenia purpura
TRH	thyrotrophin releasing hormone
TSH	thyroid stimulating hormone
tTGA	tissue transglutaminase

uACR	urinary albumin/creatinine ratio	VTE	venous thromboembolism
UFH	unfractionated heparin	vWd	von Willebrand disease
U&E	urea & electrolyte	vWf	von Willebrand factor
VKA	vitamin K antagonist	WHO	World Health Organization
VLDL	very low density lipoprotein		

About the Companion Website

This book is accompanied by a companion website:

www.wiley.com/go/blann/bloodscienceprinciples

The website includes:

- Powerpoints of all figures from the book for downloading
- PDFs of all tables from the book for downloading

1 Introduction to Blood Science

Learning objectives

After studying this chapter, you should be able to:

- explain key aspects of blood science;
- understand the role of blood science in modern pathology;
- describe the role of blood science in the wider provision of healthcare;
- outline the overlap between different areas of blood science.

In this chapter, we will introduce you to blood science – not only the study of blood, but also how the subject relates with other disciplines in pathology. You will also get a feel for blood science in the wider aspect of healthcare.

1.1 What is blood science?

Put simply, it is the study of blood. However, as with many questions, a short answer is often inadequate, and this is no exception. Blood itself is a dynamic and crucial fluid providing transport and many regulatory functions and that interfaces with all organs and tissues. As such, it has a very important role in ensuring adequate whole-body physiology and homeostasis. It follows that adverse changes to the blood will have numerous consequences, many of which are serious and life-threatening.

Blood itself is water which carries certain cells and in which are dissolved many ions and molecules. These cells are required for the transport of oxygen, in defence against microbial attack and in regulating the balance between clotting (thrombosis) and bleeding (haemorrhage). The blood is also an important distributor of body heat. The blood also carries nutrients from the intestines to the cells and tissues of the body. Once these nutrients (and oxygen) have been consumed, the blood transports the waste products of metabolism to the lungs and kidneys, from where they are removed (i.e. they are excreted). In some particular diseases and conditions (such as diabetes, myeloma and renal disease), the investigation of urine can be valuable. Although clearly not blood, blood scientists will perform and comment on the analysis of this fluid.

An historical perspective

From the early 19th century, little was known about the make-up of blood, and blood cells in particular, until a way could be found of stopping it clotting once outside the body. Thus, the development of anticoagulants was an important breakthrough. Once this was achieved, it became possible to separate intact blood cells from plasma. This led to the discovery of the differences between serum and plasma, the former obtained from clotted blood.

As the Victorian age progressed, chemists were refining old tests and discovering new ones, and so initiated the development of modern biochemistry. However, the most well-developed disciplines were (what we now call) microbiology and histology. The former was built on study of diseases such as cholera and tuberculosis, and the germ and antiseptic theories of Koch, Lister and others. Histology was benefiting from the development of dyes, enabling the identification of different substances within tissues of the body.

The first organization dedicated to non-medical laboratory workers, the Pathological and Bacteriological Laboratory Assistants Association, the forerunner of today's Institute of Biomedical Science (IBMS), was founded in 1912. Members of this group include biomedical scientists and clinical scientists. Other professional bodies for

Blood Science: Principles and Pathology, First Edition. Andrew Blann and Nessar Ahmed.
© 2014 John Wiley & Sons, Ltd. Published 2014 by John Wiley & Sons, Ltd.

laboratory workers include the Association for Clinical Biochemistry, founded in 1953.

Further developments in biomedical science during the remainder of the last century saw the emergence of four disciplines within pathology: biochemistry (also known as clinical chemistry), haematology, histology and microbiology (the latter having evolved from bacteriology in recognition of the role of viruses in human disease). Immunology appeared as a discipline in its own right in the 1970s, followed in the last decades by genetics (possibly also known as molecular biology, or more correctly as molecular genetics).

Therefore, biomedical science has been evolving over the last 200 years, driven by advances in science and technology. This evolution has seen the merging of several of these disciplines (bacteriology and virology into microbiology) and the development of new ones. This principle has also been rolled out for other scientists, such as cardiac physiologists, audiologists and medical physicists. Biomedical sciences (which may also be known as the life sciences), encompassing all those working in modern pathology laboratories, may be classified in to three groups: infection sciences, cellular sciences and blood science (Table 1.1).

Therefore, blood science (in common with infection science and cellular science) is simply another step in the development of a particular part of pathology. However, blood science is not simply a group of disciplines thrown together. Haematology and blood transfusion are sisters, and have historically grown up together over the decades. Being based around the functions of certain cells in the blood (leukocytes), immunology is effectively a 'daughter' subdivision of haematology.

The merger of biochemistry with haematology, blood transfusion and immunology at first seems strange. However, all take as their source material blood in special blood tubes called vacutainers, some of which have anticoagulants to stop the blood from clotting. Furthermore, to some extent, many tests in each of the four disciplines are amenable to measurement in batches by autoanalysers. As we shall see, many diseases call on both haematology and biochemistry, and often immunology. The serious consequences of many diseases may call for the transfusion of red blood cells or proteins to help the blood to clot. The inclusion of genetics in blood sciences comes from the fact that many diseases have a genetic component, such as the bleeding condition haemophilia, and the cancers leukaemia and lymphoma.

The reference range

The function of the laboratory is to provide the practitioner investigating or treating the patient with useful information. This information is almost always numerical; and if so, the particular numbers need to be compared with a range of other numbers to provide the practitioner with an idea of the extent to which a particular result is of concern. This is the set of numbers that we refer to, and hence the term 'reference range'. We prefer this name to alternatives such as 'normal range' or 'target range'.

The expression 'normal range' is inadequate simply because a result that is normal (i.e. is present in a lot of individuals) in a population does not necessarily imply it is desirable. A good example of this is low haemoglobin that may be endemic in some parts of the world, possibly because of malnutrition, genetics and parasites – none of which we would consider healthy. In addition, merely because someone appears healthy (i.e. are asymptomatic), it does not automatically follow that their blood result is satisfactory, and vice versa. Similarly, 'target range' is not fully appropriate as it implies a level of a result that we are trying to achieve – this may never be possible in some individuals, resulting in disappointment and a sense of failure. However, there are cases where a target is a useful objective.

It is also worthwhile discussing where 'normal values' come from. Who is normal? Many people have unsuspected asymptomatic disease that may well impact on blood science. In the past, results from blood donors were considered to be representative of being 'normal', but we now recognize the shortcoming in this definition as blood donors are in fact highly motivated and healthy individuals who are, therefore, on the whole, 'healthier' than the general population.

Individuals who are free from disease are often described as being 'normal'. In a medical setting, someone who is not complaining about any particular condition

Table 1.1 The biomedical or life sciences

Infection sciences
Bacteriology, epidemiology and public health, molecular pathology, virology

Cellular sciences
Cytopathology, genetics, histopathology, reproductive sciences

Blood science
Biochemistry, haematology, blood transfusion, immunology, molecular genetics

(such as chest pain) is said to be asymptomatic. This is not to say that person is free of disease, simply that it is not so bad that it impacts on their lifestyle. It is important to recognize that normality does not always indicate health, but is merely an indication of the frequency of a given condition in a defined population. Some diseases occur with such frequency in the population that they might be considered to be 'normal', such as dental caries. A further example of this is a high level of serum cholesterol, which is asymptomatic, but which predicts, and is a contributor to, cardiovascular disease.

The normal distribution If we examine the distribution of an indicator of health, for example the level of serum cholesterol in a population (Figure 1.1), we can see that it follows a symmetrical bell-shaped curve. Most of the data are in the middle; the result that is present at the greatest frequency (the tallest 'tower', which represents the number of people with that result) is about 4.9 mmol/L. Indeed, the average value (known as the *mean* by statisticians) is 4.9066, which we happily round down to 4.9. This shape has several names, one being the normal distribution. It represents a way we can visualize the spread of values of a particular index in a population. The distribution is called 'normal' because most sets of data (height, weight, serum sodium, haemoglobin, numbers of hairs on the head) take this distribution – it does not make any statement about what is normal or abnormal about the data itself. This type of distribution is also often described as 'bell shaped'.

A close inspection of the distribution in Figure 1.1 shows a small number of low values (on the left, about 2 mmol/L) and an equal number of high values (on the right, about 8 mmol/L). So although the data set ranges from 2 to 8 mmol/L, most values are in the middle, between say 4.2 and 5.6 mmol/L. We can use a mathematical expression to be more precise about this spread, which we call the standard deviation (SD), which is 1 mmol/L. The importance of the SD is that the mean plus or minus two SDs should include 95% of the results, which is 2.9–6.9 mmol/L. It follows that 5% of the results are outside this range.

A common error is to assume that just because someone's result is above or below this 'magical' range of mean plus or minus two SDs this automatically implies this result is abnormal. This assumption is incorrect. The definition of normal or abnormal is generally made independently of the data, usually by a panel of experts. Indeed, the concept of abnormality varies from place to place and over time. Decades ago, before we knew that cholesterol is a risk factor for cardiovascular disease, a cholesterol result of, say, 7.0 mmol/L in a 65-year-old man would probably not have been regarded as abnormal, and so would not have been treated. However, we are now fully aware of the danger of raised serum cholesterol, and so need guidance as to what is normal, and what is abnormal, and so treatable. These days the same serum cholesterol would be regarded as abnormal, and so would be treated.

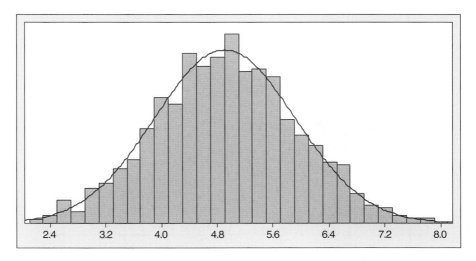

Figure 1.1 The normal distribution.

The non-normal distribution   This is the most common alternative to the 'normal' pattern of distribution. A good example of such a distribution is that of another type of fat (lipid) in the blood, the triacylglycerols (also known as triglycerides). In this pattern, the individual data points are not equally spread around the centre of the data set, with the same number of points on either side of the most frequent result (the mean). Instead, the data are skewed, or shifted, to the left or right, and the bell-shaped curve in a normal distribution is not present. This skewed distribution is called non-normal; in itself, it makes no statement of which individual data point is normal or abnormal.

In a typical set of triacylglycerol results from a large population of people, the most frequent result may be, for example, 1.7 mmol/L. This result is called the *median*, and is the middle point of all the individual data points. The smallest value may be, say, 0.5 mmol/L, but the highest may be perhaps 11.0 mmol/L (Figure 1.2). Clearly, therefore, the median of 1.7 is not in the middle of 0.5 and 11.0, unlike the mean cholesterol result from Figure 1.1, where 4.9 is not far from the middle of the full range of 2.2–8.0, that being 5.1 mmol/L.

The point is that that the criteria of what is normal, and therefore acceptable, cannot be used when the particular index (the level of serum triacylglycerol) has a non-normal or skewed distribution. We therefore have to consider a different set of rules, but these are still based on the middle 95% of people. In this case the results from 95% of the people lie between 0.8 and 4.7, so that 2.5% of people have a result less than 0.8 and 2.5% of people have

a result greater than 4.7. This does not mean that someone with a result of 4.8 is abnormal. However, the further away from the median we get, then the more likely that a result is indeed abnormal. Accordingly, we may suspect pathology in the patient with the highest result; that is, 11 mmol/L. They may have a metabolic disease.

Variation in reference ranges   It should be noted that reference ranges vary both from hospital to hospital and over time. The former is often because different auto-analysers give a slightly different result on the same sample of blood. Furthermore, the reference range should serve the local population that the hospital serves, and local populations can vary a great deal.

As we improve our knowledge of biomedical science, it becomes clear that some reference ranges need to be changed. In the 1975 edition of a major haematology textbook, the middle of the 'normal' range for the average volume of a red cell in the adult is given as 85 fL. However, in the 2001 edition, in the 'reference range and normal values' table, the mean volume is given as 92 fL. In 1975 the reference range for a type of white blood cell called a neutrophil was $(2.0–7.5) \times 10^9$/L, but in 2001 the range is $(2.0–7.0) \times 10^9$/L. It follows that in 1975 a result of 7.25×10^9/L was considered to be within the 'normal' range, whereas 26 years later the same result is outside this range, and so may be described as a mild neutrophil leukocytosis (full explanation given in Chapter 5). Whether or not this is actionable is another question.

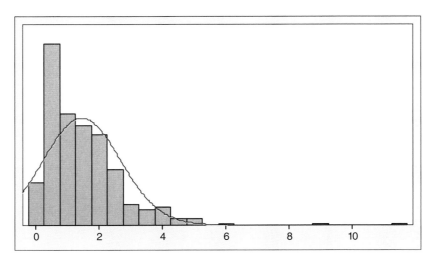

Figure 1.2 The non-normal distribution.

A note on units. Not only do the units of blood tests vary around the world (such as total cholesterol being reported in mmol/L in the UK and as mg/L in the USA), but they also vary in time. Until recently, haemoglobin was reported as g/dL. However, the unit (dL, decilitre) is not fully part of the international system, which reports in terms of the litre (L). Hence, the unit for haemoglobin has transformed into g/L, so that the result of 13.9 g/dL simply becomes 139 g/L.

Interpretation

All routine blood science results sent out (from whichever laboratory) are accompanied by a reference range, which is often a set of numbers enclosed by brackets (see Figure 1.3). In addition, the laboratory will often draw the attention of the reader to those results that are considered to be out of range and, therefore, worthy of attention. There may be an asterisk or other flag alongside these results. Indeed, for this reason the reference range may also be considered a 'concern range'. This is because a result fractionally outside the reference range does not always carry a serious health hazard. However, the further a particular result is outside the reference range then the more seriously we must address the result, as it may well be the consequence of actual disease and so should be acted upon.

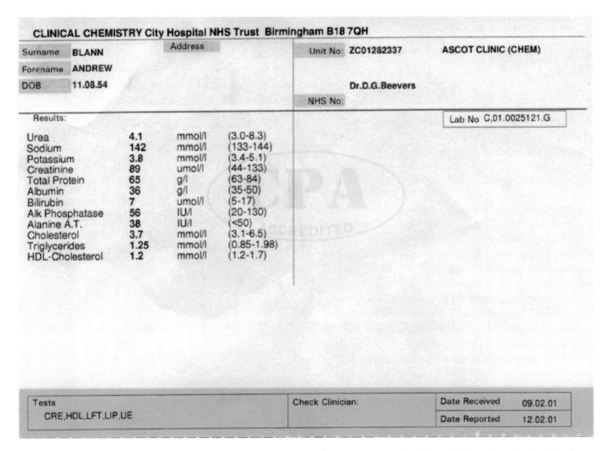

Figure 1.3 Common biochemistry tests. Selected biochemistry results on a presumed healthy middle-aged male. The blood tests themselves are printed out in the first column on the left (headed by urea). The next column is the actual result (in this case, 4.1), and then the units (mmol/L), and finally the reference range on the right (3.0–8.3). Results which are outside the reference range are generally highlighted by an asterisk. The fact that there are no asterisks present means that all results are acceptable and no further testing is required.

The concept of reference range/normal range/target range is common over all pathology tests, regardless of the particular laboratory producing the data. We will now briefly summarize the different components of blood science.

1.2 Biochemistry

Biochemistry is the study of the chemistry of the molecules, ions and atoms of various elements in the serum or plasma. To do this, the serum or plasma must be separated from the blood cells by the process of centrifugation. Different tests need to be performed on serum or on plasma; and if plasma, then the type of anticoagulant may be important. If in doubt, always consult the laboratory.

The results themselves are generally printed out on to a standard form that finds its way back to the patient's medical notes (Figure 1.3). The information is also held on a computer which can be accessed remotely from the ward or the clinic. It can also be e-mailed to a general practitioner.

Many biochemistry tests can be grouped together. The major biochemistry tests are grouped together according to certain types of physiology or pathology. For example, 'urea and electrolytes' (U&Es) together tell of the function of the kidney, whilst the liver function tests (LFTs) are self-evident. A lipid profile will generally include cholesterol and triacylglycerols; and in confirming and treating a suspected heart attack, several blood tests may be needed.

Urea and electrolytes

These tests include urea, creatinine, sodium and potassium and are described in detail in Chapter 12. The first two are the body's way of getting rid of the waste products of excess nitrogen, urea being synthesized in the liver. The final point of removal of these molecules is the kidney, so that high concentrations of urea and creatinine are a sign of renal damage. Sodium and potassium are part of our diet, and concentrations in the blood need to be regulated within a fairly tight range. Another function of the kidney is to regulate concentrations of these ions, and also the pH of the blood (the focus of Chapter 13), so a great deal of complex biochemistry is required.

An additional aspect of the concentration of these substances is the amount of water in the blood, which the kidney regulates and, of course, one's drinking of water can vary a great deal during the day. Consequently, changes in sodium and potassium also provide information about renal function. These tests are used to help diagnosis and then monitor the effect of treatment in diseases such as renal cell carcinoma, renal artery stenosis, acute renal failure, chronic renal failure and nephrotic syndrome. If the cause of high concentrations of U&Es is suspected glomerulonephritis, this may be due to autoimmune attack on the kidney, which can be determined by the presence of an autoantibody, calling for the help of the immunology section. The latter is discussed in Chapter 9.

Liver function tests

These tests (detailed in Chapter 17) include bilirubin (a breakdown product of red blood cells), and four enzymes: alkaline phosphatase, gamma glutamyltransferase, alanine aminotransferase and aspartate aminotransferase. Increased concentrations of all five of the LFTs imply malfunction of this organ. However, a problem with these enzymes is that they are also found in other cells and tissues (such as of the heart and bones) and that concentrations are influenced by other factors, such as drugs. Consequently, only one LFT by itself is difficult to interpret.

The liver also makes many proteins, including albumin, so that low concentrations of certain proteins are also indicative of liver disease. If the autoimmune disease primary biliary cirrhosis is suspected, testing for antimicrosomal antibodies will be useful. This may be undertaken by the immunology laboratory (Chapter 9).

Lipids, glucose, diabetes and heart disease

The major risk factors for atherosclerosis, the disease process causing most heart disease, are smoking, hypertension, dyslipidaemia and diabetes. There is a blood test for smoking – cotinine – but the laboratory is most unlikely to offer it or even to agree to send it to a reference laboratory. There are no blood tests for predicting or monitoring hypertension – this is by the clinical measurement of blood pressure with an automatic sphygmomanometer. This therefore leaves lipids and glucose, which are detailed in Chapter 14.

Dyslipidaemia Total cholesterol is actually made of two components: low-density lipoprotein (LDL) cholesterol and high-density lipoprotein (HDL) cholesterol. We know that high concentrations of LDL are a risk factor for cardiovascular disease. We also know that low concentrations of HDL are a risk factor for cardiovascular disease. Therefore, to use the word 'hypercholesterolaemia' (i.e. high total cholesterol) as a risk factor for atherosclerosis is to fail to understand the value of measuring LDL and HDL. This is why we now prefer the word 'dyslipidaemia', as it does not focus on LDL or HDL.

It is still a matter of debate as to whether or not isolated high concentrations of triacylglycerols are an independent risk factor of atherosclerosis in the general population. As we shall see, there are abnormalities in lipid metabolism when disease is associated with high triacylglycerols, but these are uncommon. Triacylglycerols show a strong diurnal variation (i.e. up or down at different times of the day), so that the best sample is first thing in the morning, before breakfast, and is called a fasting sample.

Diabetes This risk factor is probably the most prevalent treatable and preventable risk factor for atherosclerosis. It is characterized by high blood glucose (hyperglycaemia). The trouble is that glucose, like triacylglycerols, also has a diurnal variation, and so calls for a fasting sample. Diabetes mellitus is not simply about too much glucose in the blood, but how this glucose is handled by insulin, the hormone required (amongst its other roles) for moving glucose out of the blood and into cells. A good way of looking at this is to take a sample of blood before and 2 h after having had a drink that contains 75 g of glucose. This test is called the oral glucose tolerance test.

Glucose can easily enter the red blood cell, and stick to the haemoglobin, leading to glycated haemoglobin (HbA_{1c}). This stickiness is irreversible, so that the haemoglobin stays sugary for its entire lifetime; that is, approximately 120 days. This means that a one-off measurement of HbA_{1c} effectively provides a long-term view of the degree of hyperglycaemia.

Cardiovascular disease The serious, acute, clinical aspects of cardiovascular disease are myocardial infarction (heart attack) and stroke, where the arteries of the heart and brain (respectively) are attacked by the combined effects of the risk factors. There are no major acute consequences of atherosclerosis of arteries of the groin and the leg – but these are long-term and chronic diseases.

There are no blood tests to help diagnose the consequences of disease of the arteries of the groin, legs and brain. However, the consequence of disease of coronary arteries (a heart attack) is damage to the muscle cells of the heart. This causes the release of various enzymes and other molecules, including creatine kinase and troponin. Measurement of these molecules is important in differentiation of chest pain caused by a myocardial infarction from chest pain caused by other factors, such as muscle strain or gastrointestinal problems.

Calcium, phosphate, magnesium and bone disease

Bone is made almost exclusively of calcium and phosphates, so that abnormal plasma concentrations of these ions can be indicative of bone disease. A key enzyme in bone metabolism is alkaline phosphatase, but this enzyme is also part of the LFT panel. Vitamin D and concentrations of parathyroid hormone are also important in bone health, where the major diseases are osteoporosis, osteomalacia (called rickets in children), Paget's disease and osteomyelitis. In some cases the blood protein albumin may be part of a bone panel as this molecule can carry calcium. The importance of these indices, and their related diseases, are explained in Chapter 15.

Hormones and endocrine disorders

Although diabetes is the most common endocrine disease, there are several others that can be tested for in the laboratory, and these are detailed in Chapter 18. These include disease of the thyroid, where thyroxine, triiodothyronine and thyroid-stimulating hormone can be measured. Several metabolic conditions (Addison's disease, Cushing's disease, diabetes insipidus) focus on the adrenal glands, and the measurement of cortisol, adrenocorticotrophic hormone and vasopressin may be informative.

Other blood tests in this area include those for growth hormone, testosterone, estradiol (oestrogen), progesterone, follicle-stimulating hormone and luteinizing hormone. Some of these are released from the pituitary, others from the gonads (ovary and testes). Measurement of these hormones is undertaken when assessing reproductive disorders and subfertility.

Many endocrine conditions have an autoimmune basis, so that the immunology laboratory will be needed in order to find presumed autoantibodies, as are listed in Chapter 9. Indeed, the red blood cell disease pernicious anaemia is also part-diagnosed with an autoantibody.

Other tests

Naturally, there are dozens of other tests of undoubted value in the diagnosis and management of human disease. Unfortunately, the pathology laboratory does not have an infinite budget, and it can only offer those tests most commonly requested. However, regional or specialist laboratories may offer rare tests if there is a sufficiently large demand and if critical mass is achieved. The most common biochemistry tests are listed in Table 1.2.

The laboratory can also test for drugs. This is clearly important in cases of accidental or deliberate overdose, where concentrations of the drug (such as paracetamol) can be crucial in treatment. We often need to assess concentrations of prescribed drugs in the plasma. These

---

Table 1.2 Common biochemistry tests

Urea and electrolytes
Urea, creatinine, sodium, potassium

Liver function tests
Bilirubin, alkaline phosphatase, gamma
    glutamyltransferase, alanine aminotransferase, aspartate
    aminotransferase

Bone panel
Calcium, phosphate, vitamin D, parathyroid hormone

Atherosclerosis and its risk factors
Total cholesterol, HDL, LDL, triacylglycerols, glucose,
    HbA$_{1c}$, creatine kinase, troponin

Endocrine and metabolic disease
Thyroxine, tri-iodothyronine, thyroid-stimulating
    hormone, adrenocorticotrophic hormone, cortisol,
    vasopressin

The pituitary and reproduction
Testosterone, estradiol, progesterone, follicle-stimulating
    hormone, luteinizing hormone, growth hormone

---

include levels of lithium (a treatment of bipolar disease), digoxin (to treat certain types of heart conditions) and methotrexate (in cancer and rheumatoid arthritis). Collectively, this is described as therapeutic drug monitoring, and is the subject of Chapter 21.

1.3 Blood transfusion

Blood transfusion is not a blood disease, or a test, but it is blood science and a form of therapy. Its objective has changed over the years, from being a crude instrument to maintain haemoglobin at a certain level, to a therapy that saves lives. Additional changes have been the realization that a blood transfusion is far from a simple and trouble-free treatment, and that it can promote damage, possibly permanently, to the health of the recipient. A further development has been the provision of blood products such as platelets, albumin and coagulation factors, provided by the National Health Service (NHS) Blood and Transplant (NHSBT) service (formerly the National Blood Transfusion Service).

Blood groups

All blood group systems consist of two parts. First, certain molecules present at the surface of the red blood cell, also known as 'antigens' (an antigen is a structure that invokes an antibody response). The second aspect is a series of corresponding antibodies that recognize these antigenic molecules. The ABO system is based on the presence of two molecules, A and B, that may be present on the surface of red blood cells. Some people have one or the other (group A or B), some have both (group AB) and some have none (group O). There are also plasma antibodies that recognize blood group structures, but these are the reverse of your blood group. So if you are group A, you will have antibodies that will recognize group B (i.e. anti-B). Likewise, group B people have antibodies that recognize group A (i.e. anti-A). Group AB people have no anti-A or anti-B antibodies in their plasma, but group O people have both anti-A and anti-B antibodies. This is summarized in Table 1.3.

The second most important blood group system is the rhesus (Rh) system. It is very much more complicated, being composed of over 40 recognized glycoproteins, although on a day-to-day basis five different structures on the surface of the blood cell are commonly dealt with in the blood bank. In practice, we tend to focus on the molecule known as D (i.e. RhD). The main distinction

Table 1.3 ABO blood group factors.

Antigenic structures on the red cell surface	Antibodies in the plasma
A	Anti-B
B	Anti-A
AB	No antibodies
O	Anti-A and Anti-B

between the ABO and Rh systems is that, in the normal person, there are always anti-A and anti-B antibodies in the absence of reciprocal molecules on the red blood cell surface (which makes ABO incompatibility potentially fatal), but people with the D molecule on the red cell surface do not have a corresponding antibody in the plasma. There are also hundreds of other blood groups of diminishing frequency and importance, such as those of Lewis, Kidd and Duffy.

Blood groups present a serious challenge to the transfusion scientist. An incorrect transfusion may well precipitate a major clinical crisis in the recipient, and is termed an incompatible transfusion, which could prove fatal.

The practice of blood transfusion

Basic training in blood transfusion demands competency in several techniques: the determination of blood group, antibody screening and the cross-match.

Blood group determination The request to 'Group and Save' (G&S) is made by a practitioner with the implication that a blood transfusion may be needed in the near future. The request is to find out the patient's blood group (*Group*), but then keep the blood handy (*Save*, generally in a refrigerator). Most blood banks determine the ABO and Rh groups.

Antibody screening Many of us have natural antibodies to blood groups A and B, but exposure to other people's tissues, perhaps by a previous blood transfusion or pregnancy, may induce the formation of antibodies to other blood groups. If present in the recipient, such antibodies may cause serious problems, and so it is prudent to test the recipient's blood for antibodies to other groups.

Cross-match Not every G&S is translated into a request to cross-match. But when the call comes, scientists will mix red cells and plasma (that potentially contain antibodies) from the patient with a sample

from different packs of donor blood. A good match is where the red cells are unaltered by this mixing, and therefore should not react together when in the patient.

However, blood that does not match will aggregate, forming small clots, indicating an incompatibility. This is inevitably because the molecules on the red cells and the antibodies in the serum or plasma recognize each other, and react together, causing blood to clump. It is presumed that the same reaction may happen in the blood vessels; hence the danger. The same principle of incompatibility also occurs in other systems, such as antibodies to Rh molecules (such as D).

Blood components (previously blood products) The blood bank can provide not only red cells, but also platelets and coagulation proteins such as fibrinogen, factor VIII, factor VII, fresh frozen plasma and cryoprecipitate. The latter will be needed by people at risk of, or with actual haemorrhage (uncontrolled bleeding). Albumin is available for people with critically low concentrations, have had heavy burns, or have ascites (fluid in the abdomen).

Clinical aspects of blood transfusion

No treatment is 100% safe or effective. Therefore, does the patient *really* need a transfusion? Alternatives may be possible, such as transfusion of the patient's own blood, which is possible using cell salvage systems. Therefore, transfusion should be reserved only for those in danger of losing their life, or when it will show a measurable improvement not achievable by other means.

Sources of error Errors can and do occur in all laboratories, regardless of discipline, and so also in the 'journey' of blood from the donor to the recipient. However, it is generally recognized that most errors happen in the laboratory and/or once the blood has left the blood bank for its destination. However, with increasingly rigorous safety checks, errors are becoming increasingly rare.

Packs of blood arrive from the NHSBT service already typed for ABO and Rh, and screened also for major infective agents, mostly viruses. However, the blood sample from the recipient may be labelled incorrectly. The next source of error may be the incorrect labelling of the same portion of each potential donor pack. Further errors are possible during the cross-match itself. Fortunately, these are rare because the laboratory invests heavily in the technology and reagents to ensure that

Table 1.4 Some signs and symptoms of a transfusion reaction.

Symptoms	Signs
Cough	Fever (temp. spike $>40\,^{\circ}C$)
Flushing/rash	Hypotension
Anxiety	Oozing from wounds
Chills	Haemoglobinaemia
Nausea and vomiting	Haemoglobinuria
Tremble/shakes	Tachycardia

if an adverse reaction is happening it is detected as rapidly as possible. However, if the cross-match goes wrong, which is a false negative, a possible incompatible unit or units of blood may be issued.

Responses to an incompatible transfusion   Symptoms and signs of a transfusion reaction vary enormously (Table 1.4), but if suspected, it should be immediately stopped. All good hospitals will have a defined protocol that must be followed. Treatment will depend on the severity of the reaction, which, if not too bad, can be rapidly reversed. Problems may occur perhaps days after the transfusion, and include renal failure and jaundice, the latter being an indication of destruction of red cells.

Post-transfusion purpura (bruising) is characterized by a severe thrombocytopenia (low platelets, which can last from 2 weeks to 2 months) and is caused by antibodies to molecules on the surface of platelets. If heavily haemorrhaging, then the transfusion of platelets may be required.

Repercussions   Naturally, there are many steps designed to prevent a transfusion error, such as the use of laser bar-coding so the sample can be traced from the requesting blood sample all the way back to the patient. Many hospitals have a policy of at least two members of staff checking that the blood they are about to transfuse into one of their patients is the right type. This approach has proven to reduce mistakes and serious hazards of transfusion. Chapter 11 has more complete details about blood transfusion.

1.4   Genetics

Genetic diseases are surprisingly common. Perhaps 0.5% of all newborn infants have a chromosomal abnormality, a figure that rises to 7% if the infant is stillborn; and 20–30% of all infant deaths are related to abnormal genetics. Many 'adult' diseases have a genetic component, and this proportion is highest in the cancers, where inherited susceptibility accounts for 15% of cases.

The broad title 'genetics' includes several other names, such as molecular biology or molecular genetics. The newest recruit to the pathology laboratory, genetics has yet to find its place alongside the other disciplines, but doubtless will eventually do so. One of the reasons that molecular genetics has not been rapidly and extensively adopted is that in many cases the techniques, such as Southern blotting and the polymerase chain reaction, are very sophisticated and demand highly skilled scientists and complex equipment, which together generate a high unit cost. Accordingly, this service is often offered at regional level, and is frequently attached to a university teaching hospital. However, with advances in technology, many methods are becoming simplified, so that they can be offered by a district general hospital.

A further aspect of a genetics service is its function. Our present healthcare system starts with the patient complaining about a particular problem, or in response to a coincidental finding. The next step will be investigations, ideally a diagnosis, and then management (cycles of treatment and testing to ensure the treatment is effective). This model works well for well over 90% of problems, and so does not call for a genetics input. It follows that genetics is often called upon to help in confirming or refuting a diagnosis. There are dozens of instances where a particular disease has a genetic component, but this need not always be confirmed.

Genetic disease in families

However, a valuable service offered by the genetics service is to confirm or refute a potential diagnosis in someone who is asymptomatic, but in whom occult (hidden) disease may be suspected, such as in a strong family history of a particular disease. Of those cancers caused by gene mutations, multiple endocrine neoplasia (lesion present on chromosome 10), von Hipple–Lindau syndrome (chromosome 3) and Wilm's tumour (chromosome 11) are the most common. However, one of the strongest family 'cancer' genes *BRCA-1* (found on chromosome 17) is closely linked to the development of breast cancer. A second mutation also linked to breast cancer is *BRCA-2*, present on chromosome 13, and both *BRCA-1* and *BRCA-2* are linked to ovarian cancer. Some gene effects are so strong that one can say with a firm level of confidence that the individual has a likelihood of actually developing a particular disease at some time in the future. This is called penetrance.

Penetrance

Some particular gene mutations (the genotype) always produce a physical problem such as a disease (the phenotype). If so, we say there is complete (100%) penetrance. A good example of this is haemophilia, caused by one of several possible mutations in the gene for coagulation factor VIII so that low levels or even none of the molecule is produced. The mutation is always present and active: you cannot have 'partial' haemophilia. A second, allied, haemorrhagic condition is von Willebrand's disease, caused by one of several mutations in the gene for von Willebrand factor. These mutated genes have varied penetration into the disease. In some, penetrance is high and the disease is severe (complete absence of von Willebrand factor). However, in others, the mutation causes a partial reduction in concentrations of the protein, and so minor bleeding, or the disease may even be largely asymptomatic, in which case penetrance is low.

BRCA-1 has a penetrance of over 50% for breast cancer, and over 30% for ovarian cancer, by the age of 70. The predictive power of this marker was demonstrated over a decade ago by the case of a middle-aged woman whose female relatives all suffered breast cancer and who were positive for *BRCA-1*. Although her breasts were entirely normal by established tests, she opted for bilateral mastectomy. Early neoplastic changes were subsequently found in both breasts. More recently, chemotherapy with tamoxifen and surgical removal of the ovaries are options.

Genes, chromosomes and DNA

All the information regarding the working of the body is carried by an individual's deoxyribonucleic acid (DNA). An individual piece of information, such as an instruction to synthesize haemoglobin, is carried by a particular section of DNA: a gene. The sum total of an individual's gene and DNA make-up is their genome. The DNA itself is a long chain of nucleic acids, wrapped about certain proteins (histones), and which form into chromosomes. Normally, the information that is carried in genes is tightly controlled, but if it goes wrong, and a section of a particular gene is altered, the resulting product may be abnormal, and we say the gene is mutated.

A good example of this is the case of sickle cell disease. Haemoglobin has evolved as a specialized protein to carry oxygen, but a certain mutation in a gene for haemoglobin means that there is a different amino acid sequence of the abnormal haemoglobin, which means it is less able to carry oxygen. This can lead to the symptoms of anaemia, and so is called sickle cell anaemia. However, the sickle gene has variable penetrance: in some it causes considerable discomfort, but in others the consequences of the same mutation can be mild. The mutation can develop 'naturally', but in most cases it is inherited from one or both parents.

Chromosomal disorders

Some abnormalities are present at the level of the whole chromosome. A normal individual has 46 chromosomes in their nucleus, two pairs of 22 (autosomes) and two sex chromosomes (an X and a Y in males and two X chromosomes in females). This collection of chromosomes within the nucleus is called a karyotype.

Perhaps the best-known chromosomal disease is Down syndrome, which is caused by an extra copy of chromosome 21, so that the karyotype is 47 chromosomes. An extra copy of chromosome 13 defines Patau syndrome. Kleinfelter syndrome is characterized by an extra X chromosome in males (i.e. XXY), whilst Turner's syndrome is characterized by the presence of only one X chromosome in females (i.e. XO), so that the karyotype is 45 chromosomes. Disorders of entire chromosomes almost always have complete penetrance into the genotype (it is not possible to have 'partial' Down's syndrome).

There can also be sections of a chromosome missing (a deletion), an extra part fused on to a chromosome (a duplication), sections of DNA swopped between chromosomes (a translocation), and a section of DNA the wrong way round (an inversion). Examples are provided in Table 1.5.

Table 1.5 Chromosome abnormalities.

Abnormality	Example of condition (location of lesion)
Deletion	Prader–Willi syndrome (chromosome 15)
	Wilm's tumour (chromosome 11)
	DiGeorge syndrome (chromosome 22)
Duplication	Charcot–Marie–Tooth disease (chromosome 17)
Translocation	Chronic myeloid leukaemia (chromosomes 9 and 22) (the Philadelphia chromosome)
Inversion	Acute myelomonocytic leukaemia (chromosome 16)

KARYOTYPE

Cytogenetics Laboratory
Department of Pediatrics/Genetics
University of Utah School of Medicine

Name:
Lab No:
Results:

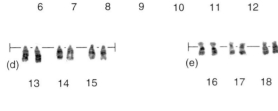

Markers

Figure 1.4 A normal karyotype: 46 chromosomes arranged in 23 pairs, dependent on the length and the banding patterns, of 22 pairs of autosomes and two sex chromosomes. (Iles & Doherty (eds) 2012, Fig. 3.14, p. 104. Reproduced with permission of John Wiley & Sons, Ltd.).

Determination of a subject's karyotype requires growing some of their cells (often lymphocytes) in tissue culture, and then arresting their growth cycle at a key point in mitosis. At this point the chromosomes are not bunched together in the nucleus but are spread out and so can be stained and identified (Figure 1.4).

Gene disorders

The sequence of nucleotides in a particular section of a chromosome that have a defined function (such as ultimately generating a protein) make up a gene. Changes in this sequence can give rise to genes that generate a different protein; an example of this is normal haemoglobin and sickle haemoglobin. The ability to detect differences in genes has been a crucial step forward in biomedical science, and those pioneers developing the techniques were justly rewarded with Nobel Prizes. With these tools, geneticists have identified hundreds of thousands of alternative forms of different genes, and these give rise to the desired variation in the human condition. However, other variations are unwelcome as they cause disease, and most of these can be classified as deletions or as translocations. Some of these mutations cause cancer, and so are called oncogenes.

Deletions   These mutations are simply sections of DNA that are missing, so that genes fail to generate a molecule that functions correctly. For example, a normal gene may produce a protein whose amino acid sequence 'spells' a sensible word such as

* * * * * * * * understandable * * * * * * * * * * * *

A deletion mutation in the gene may produce a protein that is missing some amino acids, so that the word now 'spells' something unrecognizable, such as

* * * * * * * * * * undeandable * * * * * * * * **

Although some mutations are beneficial, many are associated with disease. A good example of the former is a deletion of a section of a gene for a lymphocyte molecule called CCR5, which confers resistance to the human immunodeficiency virus (Chapter 11). However, the list of gene deletions leading to harmful mutations is considerable. One of the best known is a deletion in the dystrophin gene on the X chromosome that causes high levels of calcium in muscle cells, and that is the basis of Duchenne muscular dystrophy.

Cystic fibrosis is caused by deletion of three DNA bases in a gene on chromosome 7 that codes for a structure on the cell membrane with a role in regulating chloride and sodium ion transport in and out of the cell. The consequences of the abnormal gene are increased mucus secretion that ultimately causes the disease, mostly of the lungs and digestive tract.

Translocations   Diseases caused by this process develop as a result of two different sections of DNA being brought together, and so a new section of DNA is created that codes for a new gene. For example, on one chromosome a 'nonsense' DNA sequence may 'spell'

* * * * *unfortu

whilst another section of DNA on a different chromosome may have a sequence that 'spells' an equally obscure word such as

nate * * * * * * * * * *

If a translocation happens, the two sections of gene could merge and produce a new protein, which may be unfortunate. Translocations commonly cause cancer, an example of which is a gene called *PAX8* on chromosome 2 becoming adjacent to a gene called *PPAR-gamma* on chromosome 3. This often results in follicular thyroid cancer. Translocations of a part of chromosome 1 to chromosome 6, chromosome 14 or chromosome 19 are found in many cases of malignant melanoma.

Oncogenes An important series of deletions involves oncogenes – genes that cause cancer – and in many cases this is to do with the rate of growth of the cell and its differentiation. One such gene is *MYC* (pronounced 'mick') located on chromosome 8, which codes for a protein involved in the regulation of many other genes. Inappropriate activation of *MYC*, therefore, leads to abnormal gene expression, which in turn can lead to cancers. A good example of this is the translocation of *MYC*, so that it becomes adjacent to a gene on chromosome 14 that codes for part of an immunoglobulin molecule. This translocation is often found in Burkitt's lymphoma. Another oncogene is *SRC* (pronounced 'sarc'), acting on internal cell enzymes also involved in signalling. Fascinatingly, the gene sequence of *SRC* resembles a virus that in chickens causes a tumour of muscle tissues.

An important group of genes are the tumour suppressors, whose function is clear. Many cases of retinoblastoma are caused by a mutation in the *RB1* gene on chromosome 13. A function of the protein product of the normal *RB* gene is to regulate the cell cycle and cell growth. The mutated *RB1* gene produces an abnormal protein that fails to regulate cell growth and that can lead to cancer. However, the pathology of this type of retinoblastoma is an example of a two-hit disease, as a second factor in addition to *RB1* is needed to cause the disease. Both *BRCA* genes are tumour suppressors: the normal genes code for proteins that help repair broken and damaged DNA. The abnormal protein produced by the mutated gene fails to repair damaged DNA, which may then precipitate a cancer.

Genetics as an independent blood science

Genetics has the power to unequivocally define a disease, and also, by the absence of a mutation, to deny a disease.

However, current research can provide a clue as to which treatments, such as a particular combination of cytotoxic drugs, are likely to be more successful, as different chemotherapy can target the products of different mutations. Indeed, knowledge gained from the molecular genetics of a certain type of chronic leukaemia has led to the development of a completely new and very successful class of drugs that treat only this type of cancer.

The preceding section has introduced the value of genetics in many different types of diseases, including cancer. Indeed, it could be argued that clinical genetics is in fact a collection of methods called on by pathologists to help with diagnosis. Accordingly, this textbook will not have a separate chapter on genetics. Instead, each individual chapter will emphasize genetics where and when it is appropriate. Examples of the impact of genetics in the other blood sciences include:

- haemoglobinopathy, leukaemia, haemophilia, von Willebrand's disease, Factor V Leiden, and hereditary haemochromatosis in haematology;
- alpha-1-trypsin deficiency, LDL cholesterol receptor, cystic fibrosis and 21-hydroxylase deficiency in biochemistry;
- ABO and rhesus blood groups in transfusion science;
- the heritability of autoimmune disease, X-linked agammaglobuminaemia, variation in human leukocyte antigens (HLAs), and DiGeorge syndrome in immunology.

All these conditions, and others where genetics impacts into a particular blood science, will be explained in their own chapter. Other examples of molecular genetics used by non-blood science disciplines are shown in Table 1.6.

Table 1.6 Molecular genetics across pathology.

Test	Discipline	Disease
Hepatitis C virus	Virology	Chronic hepatitis
CK19, with or without mammaglobin	Histology	Breast cancer intra-operation lymph node investigation
HER2 (human epidermal growth factor receptor 2)	Histology	Breast cancer
Chlamydia	Microbiology	Population screening
KRAS (Kirsten rat sarcoma viral oncogene homolog)	Histology	Colorectal cancer

Blood science angle: Haematology and molecular genetics

Certain haematology malignancies, including polycythaemia rubra vera, essential thrombocytopenia and myelofibrosis, are often characterized by increased numbers of red blood cells and platelets. In the past, these diseases have been diagnosed by a combination of blood tests and examination of the bone marrow, and difficulties in specificity and sensitivity of these tests often brought false positives and false negatives.

Janus kinase 2 (JAK2) is a protein involved in a cell's response to growth factors. In 2005, several groups linked a mutation in the gene for this molecule (*JAK2 V617F*) that effectively conferred a hyperreactivity of cell to growth factors, and which went a long way in explaining the high cell counts in these diseases.

This link is now so powerful that the presence of *JAK2 V617F* now makes the diagnosis of polycythaemia. This is a classic example of molecular genetics providing a new diagnostic tool. Moving to populations, a study of almost 50 000 Danes demonstrated that this mutation is also a risk factor for ischaemic heart disease and deep vein thrombosis. Further details of JAK2 are presented in Chapter 4.

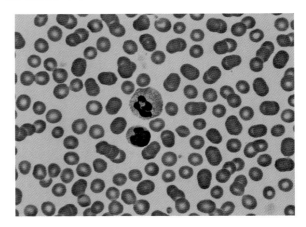

Figure 1.5 A blood film. (Courtesy of D. Bloxham, Cambridge University Hospitals NHS Foundation Trust).

1.5 Haematology

Haematology can be summarized in four areas: the blood film, the full blood count, coagulation and haematinics. These are used in the diagnosis and management of a broad range of diseases of the blood.

The blood film

The blood film provides the opportunity to view the components of blood under a microscope. A drop of blood, generally from the same tube that provides the full blood count, is smeared on to a glass slide and is allowed to dry in air at room temperature. It is then fixed, stained, allowed to air dry and examined under a light microscope. We can also use the shape of the nucleus to help classify the different types of white blood cells; this is called morphology.

Figure 1.5 shows a typical blood film. The principal features in the centre are two white blood cells, characterized by a nucleus that has taken up stains so that it appears purple. Note that in these two cells the structure of the nucleus is different. We will return to the importance of this in Chapter 5. A second feature is the large number of roughly round cells that are all a single colour; these are red blood cells. The third feature is the single

small purple body to the lower right of the white blood cells; this is a platelet.

Autoanalysers now provide an excellent breakdown of the different types of blood cells, so the blood film has become less important and is now not routinely examined. However, there is still a place for examination of a blood film with a microscope that may be necessary to confirm the autoanalyser's profile or to perform other assessments of the blood. The same principle is also used to look at bone marrow, which is where blood cells are generated.

The full blood count

This is the most requested and valued blood test in haematology. It provides a package of information on red blood cells, white blood cells and platelets (Figure 1.6).

Red blood cells   These are unusual among cells of the body as they lack a nucleus. Consequently, they are relatively easy to identify in a blood film (Figure 1.5). Red cells carry haemoglobin (Hb), an iron-containing protein that absorbs oxygen from areas of high oxygen content (i.e. at the lungs) and then releases it in areas were oxygen levels are low (e.g. in the tissues). The haematocrit (Hct) expresses that proportion of whole blood that is taken up by all the blood cells as a decimal (e.g. 0.42) or as a percentage (e.g. 42%). The reference range for haemoglobin, red cell count and haematocrit varies between the sexes, with lower levels in women.

The red blood cell indices comprise three measures of different aspects of the red cell, its size and the amount of

Sandwell and West Birmingham Hospitals NHS Trust.   Haematology, City Hospital

Surname	Address	Unit No.	OPD – CITY
Forename		Consultant/GP	
D.O.B.			Prof G.Y.H.Lip
Sex			
NHS No:		Lab No. C.13.4250374.P	

Ethnic group: White – British   **Clinical Details:**

HB	145	g/L	(125–180)	ESR	2	mm/h	(1–14)
MCV	94.6	fL	(79.0–99.0)				
WBC	5.8	10*9/L	(4.0–11.0)				
PLT	241	10*9/L	(150–450)				
RBC	4.62	10*12/L	(3.50–6.50)				
HCT	0.44	L/L	(0.38–0.54)				
MCH	31.4	pg	(27.0–34.5)				
MCHC	332	g/L	(316–365)				
NEUT	3.38	10*9/L	(1.70–7.50)				
LYMPH	1.85	10*9/L	(1.00–4.50)				
MONO	0.49	10*9/L	(0.20–0.80)				
EOS	0.06	10*9/L	(0.00–0.50)				
BASO	0.01	10*9/L	(0.00–0.10)				

RANDOM SAMPLE

F: Please note new units for Hb & MCHC (Now reported in g/L).
   e.g a Hb previously reported as 12.5g/dL will from 27/03/13
   be reported as 125g/L

Tests: ESR, FBC		Check Clinician	Date collected 13.05.13 14:30
Report Run: 73	Specimen: Blood		Date received 13.05.13 16:19
			Date reported 13.05.13

Figure 1.6 The full blood count.

haemoglobin it contains. These indices are: the mean cell volume (MCV), the volume of the average (mean) red blood cell; the mean cell haemoglobin (MCH), which reports the average amount (mass) of haemoglobin in the average cell; and the mean cell haemoglobin concentration (MCHC), the average concentration of haemoglobin in a given volume of red cells. The red cell is discussed in detail in Chapters 3 and 4.

White blood cells   White blood cells, or leukocytes, defend us from attack by microorganisms (viruses, bacteria and parasites), when raised levels of these cells can be expected. However, increased numbers (i.e. a leukocytosis) may also be present in a number of diseases. There are five different types of white cells: the neutrophil, lymphocyte, monocyte, eosinophil and basophil. Each can be defined on morphological grounds, but also by their function, as is explained in Chapters 5 and 6.

Neutrophils are the most common leukocytes, and also the most common polymorphonuclear leukocyte, so

named because they have an irregular nucleus (the upper cell in Figure 1.5). They are called neutrophils because they take up dyes at a neutral pH. The second most frequent group of leukocytes are the lymphocytes. Their nucleus is round and regular and takes up almost the entire cell (the lower cell in Figure 1.5). Monocytes, the largest white cell, also have a regular nucleus, but the nucleus takes up perhaps two-thirds or three-quarters of the cell. Eosinophils are so-called because of the reddish colour, owing to the chemical make-up of their granules. Basophils are the least frequent leukocytes; they contain numerous granules that take up different dyes, and so appear as black or dark blue.

Platelets   These are not true cells, but are small fragments of the cytoplasm of a larger cell found only in the bone marrow (the megakaryocyte). They form a clot, or thrombus, when aggregated together with the help of the blood protein fibrin and so reduce blood loss. This process is focused upon in Chapters 7 and 8.

A low platelet count (thrombocytopenia) may be caused by drugs, poor production (as may be present in disease of the bone marrow) or by excessive consumption. This condition can lead to an increased risk of bruising and bleeding. The converse, a raised count, is thrombocytosis and is often present in many physiological and pathological situations. These include infections, after surgery, some autoimmune diseases and after short but intense bouts of physical activity. A high platelet count may lead to thrombosis.

Erythrocyte sedimentation rate   The erythrocyte sedimentation rate (ESR) is a global score of physical aspects of the whole blood. The result is obtained by measuring a band of plasma on top of a thin column of blood that has settled after standing for an hour. An ESR can be abnormal in a large number of conditions, including inflammation, infection, the acute-phase response, after surgery, anaemia, leukaemia and almost all forms of cancer. Indeed, it follows that an abnormal ESR is present in most patients in hospital.

Haemostasis

Haemostasis is the balanced orchestration of interactions between blood vessels, blood cells and plasma proteins. Together they keep the blood in a fluid state, and also limit and stop bleeding upon damage to the blood vessel. We have already mentioned platelets are part of the full blood count.

The clot (or thrombus) is generated by platelets and fibrin, which together form a net that traps red cells; but clots can occur without red cells. The laboratory offers tests on the ability of the blood to generate these clots. The inappropriate formation of a clot can lead to a disabling deep vein thrombosis or, more seriously, a stroke or even, in the lung, death. Conversely, inability to form a clot can also lead to serious disease due to excessive bleeding (haemorrhage).

Prothrombin time   The prothrombin time (PT) assesses the ability of plasma to form a clot based on certain components of the coagulation pathway. It employs a reagent called thromboplastin that activates part of the coagulation system. The time taken from the addition of thromboplastin to the patient's plasma to generation of a fibrin clot is the PT itself, which is recorded in seconds.

Activated partial thromboplastin time   This test assesses the ability of a different series of coagulation proteins from those of the prothrombin to form a clot. Patient's plasma is again incubated with a complex collection of reagents, and the time taken to clot from the addition of the calcium ions is also recorded in seconds.

Fibrinogen   An adequate level of fibrinogen is crucial if coagulation factors such as prothombin and thrombin are to have their desired effect. The laboratory measurement of fibrinogen activity is performed using a modified version of the thrombin time where the patient's plasma is induced to clot.

Pathology of thrombosis and haemostasis   Thrombosis is the most common ultimate cause of death, such as is caused by a clot in an artery of the heart or the brain. However, a great deal of disease is also caused by clots in veins (venous thromboembolism), mostly of the leg (deep vein thrombosis) and the lung (pulmonary embolism). High numbers of platelets in the blood can also predispose to thrombosis. Consequently, the ability of the laboratory to assess these disease processes is at a high premium.

We can intervene in the process of thrombosis with drugs such as warfarin, heparin and aspirin. However, in many cases, the activity of the former two drugs needs to be checked with an appropriate blood test, such as the prothrombin time and activated partial thromboplastin time. A serious problem with these drugs is that too much warfarin or heparin can grossly reduce the ability to form a clot, which therefore results in bleeding. There are several examples of diseases where coagulation factors are 'naturally' lacking or are ineffective, haemophilia being a good example, where spontaneous or accidental haemorrhage is a constant concern.

Haematinics

A small number of vitamins and minerals are needed for blood cells to be generated effectively, and all must be provided by a healthy diet. These include iron, vitamin B_{12}, vitamin B_6 and folate. Deficiencies in any one of these micronutrients may result in impaired production of red blood cells by the bone marrow and therefore result in anaemia.

Iron   This micronutrient is the key atom in the middle of the haemoglobin molecule where the oxygen is carried.

It is placed within a complex molecule, haem, by an enzyme called ferrochelatase, a process that happens within stem cells in the bone marrow. If there is not enough iron being placed in haem, an iron-deficient anaemia will result. However, the root of the problem may be not enough iron in the diet (malnutrition), or perhaps the inability of this iron to cross the gut wall and enter the blood (malabsorption, which may be related to, for example, inflammatory bowel disease).

The polar extreme of too little iron is too much iron. This can cause the disease haemochromatosis, and in many cases the ultimate cause is a gene mutation; another is too many blood transfusions.

Vitamin B$_{12}$ The second most common type of deficiency is of vitamin B$_{12}$, which is required by key enzymes in the synthesis of haem, a process that happens in the cytoplasm and mitochondria of bone marrow stem cells. Only a tiny amount of the vitamin is needed, so that malnutrition rarely causes the anaemia. Instead, the major cause of the anaemia is not being able to absorb enough of the vitamin across the gut wall, so is a form of malabsorption, the most frequent type being due to autoimmune disease of the stomach, and if present this is called pernicious anaemia. There are no known examples of problems caused by too much vitamin B$_{12}$.

The laboratory in micronutrient deficiency The opposing features of microcytes (with a low MCV, often due to iron deficiency) and macrocytes (with a high MCV, often due to vitamin B$_{12}$ deficiency) are key steps in the diagnosis and understanding of anaemia. Treatment of these deficiencies, in many cases, is simply by replacing the missing micronutrient. However, if there is malabsorption, the oral route is not appropriate, and the iron and vitamin will need to be given by injection or infusion. The laboratory is then needed to ensure the treatment is effective – which is that the size of the red cell becomes normal and that the haemoglobin rises.

Haematological disease

These tests and processes are used to diagnose and manage a large number of diseases, such as cancers (leukaemia, lymphomas and myeloma), different types of anaemia (caused by lack of micronutrients or perhaps by the destruction of red cells (haemolysis)) and by excessive bleeding (haemorrhage) or clotting (thrombosis).

Table 1.7 Common routine haematology blood tests.

Tests performed on anticoagulated whole blood
Full blood count: provides information on red blood cells, white blood cells and platelets
ESR: a physical property of whole blood

Tests performed on plasma
Coagulation tests: prothrombin time, activated partial thromboplastin time, fibrinogen (important in thrombosis and haemorrhage)

Tests performed on serum or plasma
Micronutrients iron, vitamin B$_{12}$ and folate
Proteins transferrin and ferritin (used to investigate different types of anaemia)

Table 1.7 summarizes key features of laboratory haematology.

1.6 Immunology

The laboratory and clinical aspects of immunology are certainly the most recent developments in mainstream biomedical science. The British Society for Immunology was founded as recently as in the mid 1970s. Indeed, the subspecialty of immunology can be said to have grown out of haematology (since the function of white blood cells is immunological defence), and in many hospitals the immunology laboratory is physically linked to the haematology laboratory. Furthermore, immunology is the basis of almost all reactions in the blood transfusion laboratory.

Laboratory aspects of this subject can be simply classified into serology (literally the study of serum) and cell biology.

Serology

From an immunological perspective, serology focuses on the assessment of levels of antibodies to particular pathogens or tissues, total levels of the immunoglobulins, complement, and other proteins.

The key molecule in immunology is the antibody, and all antibodies are immunoglobulins (Igs). They are classified into one of five groups, dependent on their protein make-up – IgG, IgM, IgA, IgD, IgE – and all have a common structure, which is often presented as a 'Y' shape (Figure 1.7). All have two 'heavy chains', of which there are five types (A, D, E, G and M) which define their

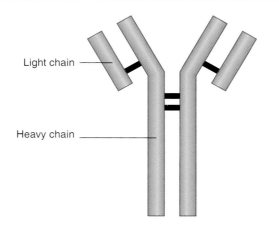

Figure 1.7 An antibody molecule.

class, but also two 'light' chains', which can be one of two types: kappa or lambda. All antibodies are made by B lymphocytes, most in the lymph nodes and (to a lesser degree) in the spleen, and to do this they need the cooperation of T lymphocytes and perhaps some specialized monocytes. Antibodies are found in the plasma and also as part of the membrane of B lymphocytes. These important cells and molecules are revisited several times, in Chapters 5, 6, 9, 10 and 11.

As with different roles of the various white blood cells, the different antibody classes have both specific and common functions, but all are designed to bind to structures believed to be foreign. These foreign structures are called antigens. However, some antigens are not foreign – such as the AB blood group molecules – and this is the basis of the incompatible blood transfusion that scientists in the blood bank seek to avoid.

Complement is a family of nine molecules (C1–C9) that have a number of functions. Some are broken-down fragments that influence the cells of the blood vessel wall (endothelial cells) and attract certain white blood cells, whilst others promote the ingestion of bacteria by those white blood cells. Others come together to form a complex that can punch holes in the walls of certain cells and bacteria, resulting in their destruction. Details of complement are found in Chapter 9.

An alternative aspect of serology is the ability to detect plasma antibodies to defined pathogens, and this effectively proves the presence of an infection with the particular bacteria or virus. This is of interest to microbiologists, who seek evidence of a possible pathogen and so determine the most appropriate form of antibiotic or antiviral chemotherapy. Many inflammatory states, with

or without an infection, are characterized by increased concentrations of a c-reactive protein (CRP). Consequently, in many cases, this molecule defines immunological activity within the body. However, CRP may be measured not in an immunology laboratory but in biochemistry.

Cells

Cellular aspects of immunology focus on two areas: the enumeration and function of neutrophils and of lymphocytes, which can be achieved morphologically on a blood film, or by identifying cell-specific molecules on the surface (the CD family of molecules). Neutrophil function can be assessed by the ability of the cell to attack and digest common pathogens such as yeast (the process of phagocytosis), and by their ability to switch on certain biochemical pathways designed to destroy bacteria. There are no commonly used tests of monocyte function.

Perhaps the most common lymphocyte investigation is to determine the proportions of the two major classes of lymphocytes; that is, T cells and B cells. An extension of this is the T lymphocyte subgroup, mostly the number of the CD4 subset, as this is important in those infected with the human immunodeficiency virus (HIV). Almost all of this work is performed with a machine called a fluorescence-activated cell scanner; this technique is also of interest to haematologists, as there are many other leukocyte groups that can be assessed and that are important in leukaemia. The fluorescence-activated cell scanner can also help diagnose diseases of the membrane of the red cell and of platelets.

Immunopathology

The most successful immune responses are characterized by the serological aspects working together with the white cells to defeat the infection. However, this can be defective and result in disease. Like many different pathological states, immunological disease can be classified according to two extremes: where there is an excessive or inappropriate activity (too much), and where an immune response is weak, or even absent altogether (too little). These topics are examined in Chapter 9.

An inappropriately excessive immune response
Ideally, an infection is countered in physiology by the immunological system with minimal adverse side effects for the body, the most common being a fever and perhaps some aches and pain. However, if the response to an

infective agent is strong, these side effects can also be strong, and may actually cause illness. An example of this is where an acute infection (such as of the lungs) transforms into a chronic inflammation that requires immunosuppressive chemotherapy. An overactive immune response can also cause allergic reactions, hypersensitivity and asthma.

However, there are many inappropriate responses where, instead of attacking a foreign invader, the immune system attacks the body. These conditions are collectively called 'autoimmune diseases' perhaps the most well-known being rheumatoid arthritis. All autoimmune diseases are characterized by an abnormal antibody; that is, an autoantibody, generally to the cell or tissue that is presumed to be abnormal and so is being attacked. Thus, an autoantibody to the thyroid causes thyroid disease, and we have already noted the autoimmune attack of certain cells of the stomach that causes pernicious anaemia.

A weak or absent immune response Failure to mount an effective immune response to an infectious agent inevitably results in the success of the pathogen. This immunodeficiency may be due to problems with the cells, with proteins or with both.

A common cause of cellular immunodeficiency is the effect of cytotoxic chemotherapy as in the treatment of many cancers. This happens because these sophisticated poisons damage the bone marrow as well as the tumour, resulting in fewer blood cells, and so a lack of the white blood cells that are needed to attack microbes. HIV preferentially attacks and destroys T lymphocytes, which are required to attack cells infected with viruses and are also needed to help B lymphocytes make antibodies. As mentioned in Section 1.4, DiGeorge syndrome is characterized by lack of a thymus and so lack of T lymphocytes. In the case of HIV, this eventually leads to the complete destruction of the immune system and the acquired immunodeficiency syndrome (AIDS). Other cellular deficiency diseases include chronic granulomatous disease, where neutrophils can ingest but are unable to kill bacteria.

A consequence of HIV/AIDS, and contributor to the heavy burden of infections, is falling levels of antibodies (hypogammaglobulinaemia), but this can also be seen in myeloma and leukaemia. However, a complete lack of antibodies, agammaglobulinaemia, is also known. This lesion is a defective gene on the X chromosome that leads to inactive B lymphocytes. Specific deficiencies in individual antibody classes are known, such as lack of IgA.

Table 1.8 Immunology

Tests performed on serum/plasma
Total immunoglobulin levels
Levels of specific antibody classes (IgG, IgM)
Complement components
Antibodies to defined pathogens, e.g. bacteria
Autoantibodies, CRP

Tests performed on cells
Neutrophil function
T and B lymphocytes
T cell subsets (CD4, CD8)

Immunological disease
Overactivity:
   hypersensitivity, allergy autoimmune disease, chronic
   inflammation
Underactivity (immunodeficiency):
   hypogammaglobulinaemia, lack of, and/or failure
   of function of lymphocytes and neutrophils

Total deficiencies of complement components that lead to major disease are exceedingly rare, but some diseases, such as systemic lupus erythematosus, are often associated with low levels of some complement components. Table 1.8 presents a summary of immunological matters.

1.7 The role of blood science in modern healthcare

It has been estimated that, in hospitals, 75% of the information needed to make a clinical judgment, be it a diagnosis or a decision to initiate, change or stop treatment, comes from the laboratory. Put around the other way, the laboratory provides three times as much information as do all other sources (history, signs, symptoms, imaging, etc.) combined. Whether or not this staggering statistic is different in primary care is not known. Therefore, healthcare professionals provide a huge resource for the practitioner facing the potential or actual patient.

Blood science in human disease

There are many ways to classify human disease. One is to consider three broad areas of pathology: cancer, connective tissue disease (such as rheumatoid arthritis, osteoarthritis and their allied conditions) and cardiovascular disease (principally heart attack and stroke, to include their risk factors of diabetes and hyperlipidaemia).

Together, these constitute 70–80% of the healthcare burden of the developed world. The remaining conditions include, for example, infections, endocrine diseases and psychiatric illness.

One reason for bringing together the major disciplines into blood sciences is the fact that 'pure' diseases of haematology, biochemistry or immunology are exceptionally rare. The pathological basis of many diseases is multifactorial, and so demands a more comprehensive understanding of pathology than any one discipline in itself can provide.

Almost all congenital disease (present at birth or noted in the immediate neonatal period) is genetic, and in some cases will need to be formally confirmed with molecular genetics. Several have been mentioned in Section 1.4. Non-genetic congenital diseases will generally have been acquired from the mother, and these are generally metabolic and infectious diseases, such as some types of diabetes, HIV infection and syphilis. Others may be caused by drugs taken during pregnancy, such as thalidomide and warfarin, or perhaps by lack of adequate nutrition, such as the relationship between folic acid and neural tube defects.

Cancer We can consider cancer in two aspects: genetics and environment. In some cancers, these aspects are separate, and in some they are combined. Some are apparent at birth or shortly after, whereas others develop later in life. The disease may manifest as a lump or growth (breast, lymph node) or as unexplained weight loss, pain, an unexplained blood clot in the leg and excessive tiredness.

Investigation of cancer in an adult will inevitably call for many blood tests and genetic analyses, as well as imaging such as X-ray. In many cases these may also indicate cause, although for some the likely cause may be clear, such as tobacco smoking and lung cancer. Many cancer markers can be measured in the blood, such as CA-125 and breast cancer, and carcinoembryonic antigen (CEA) in gastrointestinal cancer (Chapter 19). Many of these will be measured in the biochemistry laboratory. However, numerous cancers are associated with non-specific changes, such as an increased ESR and a normocytic anaemia.

Connective tissue disease These include diseases of bone, joints, muscle, tendon, ligaments, collagen and skin. Epidemiologists tell us that this group has a high prevalence, with osteoarthritis and rheumatoid arthritis at the top of the list. The former is more common, and is

likely to trouble over half of the population at some stage of their life. The biggest risk factor for osteoarthritis is obesity. Rheumatoid arthritis affects approximately 3% of the population, affecting three times as many women as men, and rises with age. Rheumatoid arthritis brings an increased risk of cardiovascular disease and death.

Rheumatoid arthritis certainly has a genetic component, linked with certain HLA molecules, and as a typical autoimmune disease will attract the interest of the immunology laboratory and so the measurement of rheumatoid factor. The chronic inflammation will also be noted by haematologists, with a raised ESR and (often) a low haemoglobin leading to a normocytic anaemia. Rheumatoid arthritis is a systemic disease, and spreads out of joints to attack other organs such as the liver and kidney, in which case the biochemistry laboratory can offer LFTs and U&Es respectively. Osteoarthritis is restricted to joints, so is unlikely to be linked with grossly abnormal blood tests. These issues are discussed in Chapter 9.

Cardiovascular disease This complex and multifactorial condition is dominated by atherosclerosis, which has four major risk factors. Two of these (dyslipidaemia and diabetes) and a consequence of atherosclerosis (such as myocardial infarction) can be tested for in the biochemistry laboratory (Section 1.2). In many cases, and certainly in middle age, atherosclerosis is acquired from lifestyle, but in later life it seems an inevitable consequence of ageing. Indeed, it has been argued that, as the pathophysiology of this disease is so complex, genomics will never be useful in arterial thrombosis. However, despite this, there are several clear examples of the influence of gene mutations on the disease process of atherosclerosis, such as polygenic and familial hypercholesterolaemia.

Table 1.9 shows some examples of diseases where the pathology impacts into each of the blood science disciplines.

Blood scientists: who are they?

These are simply scientists working in blood science, and they can be employed in one or more of the major disciplines we have been looking at. There are several different types of scientists, the most numerous being biomedical scientists, with smaller numbers of clinical scientists and biochemists.

The major professional body for scientists in the pathology laboratory (whether NHS or private) is the IBMS, being the natural home of biomedical scientists,

Table 1.9 Blood science and human disease.

	Haematology	Biochemistry	Immunology
Cancer (e.g. malignant myeloma)	Anaemia[a] and thrombocytopenia	Increased calcium resulting from bone resorption, increased U&Es reflecting renal damage	Acute-phase response, increased cytokines, paraproteinaemia
Connective tissue disease (e.g. SLE)	Anaemia[a]	Increased U&Es, reflecting renal damage	Acute-phase responses, autoantibodies
Cardiovascular disease (e.g. atherosclerosis)	Thrombosis	Increase lipids and glucose	Low-grade inflammation

SLE: systemic lupus erythematosus.
[a]Blood transfusion: a treatment for profound anaemia and blood loss regardless of aetiology (rarely called upon in cardiovascular disease, except perhaps ruptured abdominal aortic aneurysm).

although clinical scientists are also members. Many clinical scientists with a background in biochemistry are members of the Association for Clinical Biochemistry. However, the diversity in the biomedical and blood sciences and the requirement to specialize at higher staff grades (such as Agenda for Change level 7 and upwards) often demands membership of additional professional bodies. These include, for example, the British Society for Haematology, the British Blood Transfusion Society and the British Society for Immunology, although a host of other professional and quasi-professional bodies exist, such as the Health Professions Council (HPC), to which practicing biomedical and clinical scientists are required to register.

The IBMS provides a formal structure of examinations and diplomas, from which follow levels of membership (Licentiate, Member, Fellow), position statements on professional practice, and standards of proficiency. It also provides the basis of keeping its members up to date, via a series of education modules summarized as 'continuing professional development' (CPD). One of the requirements of both the IBMS and HPC is that members complete sufficient CPD to warrant registration.

Clinical scientists may also be members of the Royal College of Pathologists. The award of Fellow may be obtained by examination or by the submission of a thesis of research publications. Whilst an MSc is mandatory for those at higher grades, scientists of all disciplines may well have a PhD.

However, the individual professional groups are being drawn together into one, as healthcare scientists, with a defined career pathway and specific training.

The role of the higher education institutions

Historically, many universities provided, and continue to provide, formal education in biomedical science, and so in blood science. Many do this by merging their existing modules in the subdisciplines. However, in order for these to be of value to the student practitioner they must be validated and approved by external bodies that include the IBMS. In this way, the particular university benefits from ratification from the profession, and the IBMS ensures that its new recruits have an appropriate background and are well on the way to formal competency, and so state registration.

More recently, the government has licensed certain higher education institutions to provide formal courses on blood science at undergraduate and postgraduate levels. These are linked to formal training courses with placements in hospitals. Foremost among these are those administered by the National School for Healthcare Science.

Training in blood science

Whilst the terms 'biomedical scientist' and 'clinical scientist' were made protected titles in law during 2003, there is as yet no clear definition of a blood scientist. However, in their document 'Modernising Scientific Careers', the UK Government has determined the existence of this speciality and that a formal training system is set up to deliver trained blood scientists to NHS hospitals. This training is in conjunction with selected higher education institutions, and operates at various levels:

- Associates and assistants – this includes National Vocational Qualification and foundation degrees, underpinned by an awards and qualifications framework;
- Practitioner Training Programme (PTP) – undergraduate level;
- Scientist Training Programme (STP) – postgraduate entry, pre-registration training;
- Higher Specialist Scientific Training (HSST) – at doctoral level.

Figure 1.8 shows the scientist training programme.

These initiatives lead to the following grades of employment, generally with increasing responsibility, and increasingly focusing on management. This process therefore provides a uniform career structure that can be accessed at different levels, dependent on the qualifications and experience of the individual. The grades of healthcare scientist are as follows:

- Healthcare science assistant
- Healthcare science associate
- Healthcare science practitioner
- Healthcare scientist

- Senior healthcare scientists
- Consultant healthcare scientist.

This pathway is also present in the other NHS scientist groups, such as audiologists and respiratory physiologists.

1.8   What this book will achieve

The objective of this book is to provide a firm foundation in the new discipline of blood science. It will also provide an informed view of each particular discipline that will be attractive for those experienced in another (i.e. haematology for biochemists, and biochemistry for haematologists). Naturally, in order to produce a volume of manageable size, not all aspects of all disciplines can be addressed. For further details of particular subdisciplines, a reading list is provided.

All these tests and procedures felt to be of major value in blood science will be discussed in more detail in the chapters that follow, and can be grouped together as follows:

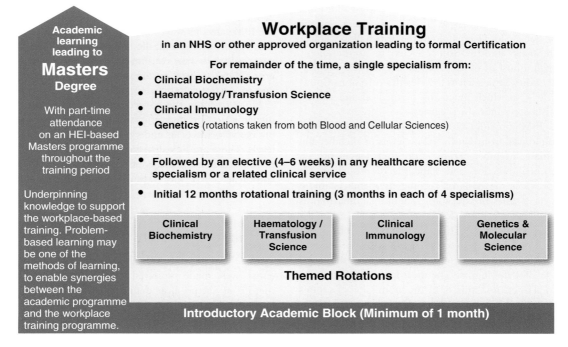

Figure 1.8 Training of blood scientists. (http://www.nhscareers.nhs.uk/explore-by-career/healthcare-science/modernising-scientific-careers/ © Crown copyright).

- Chapter 2 will examine some basic techniques in blood science
- Chapters 3–8 will consider haematology
- Chapters 9 and 10 will look at immunology
- Chapter 11 will discuss blood transfusion
- Chapters 12–21 will focus on biochemistry
- Chapters 22 will have a number of case reports in blood science.

Naturally, although particular chapters will focus on their particular topic, each will also refer to other aspects of the other disciplines of blood science. This is particularly relevant for molecular genetics, which has a role in each of the other disciplines.

Summary

- Historically, biomedical science has arranged a number of well-established disciplines: biochemistry, haematology and blood transfusion, histopathology and microbiology.
- Recently, immunology has been included as a biomedical science, whilst molecular genetics is the most recent addition to the family.
- The biomedical sciences have been reinvented as blood science, cellular science and infection science.
- Biochemistry, haematology, blood transfusion and immunology make up blood science. However, there is also a place for molecular genetics, as this new discipline has a place in each of the established disciplines.
- Blood science is set to become a major force in healthcare-based biomedical science.
- In turn, biomedical science, alongside other diagnostic specialities such as imaging, audiology and cardiology, make up the life sciences.
- The major professional groups in the laboratory are biomedical scientists and clinical scientists, although in time it is likely that these will merge.

References

Iles RK, Butler SA. Clinical cell biology and genetics. In: Iles R, Docherty S (eds). Biomedical Sciences: Essential Laboratory Medicine. Chichester: Wiley-Blackwell, 2012; pp. 89–138.

Further reading

Bench AJ, White HE, Foroni L. *et al.* Molecular diagnosis of the myeloproliferative neoplasms: UK guidelines for the detection of JAK2 V617F and other relevant mutations. Br J Haematol. 2013;160:25–34.

Cree I. Managing implementation of molecular pathology. Bull R Coll Pathol. 2012;159:154–155.

Nielsen C, Birgens HS, Nordestgaard BG, Bojesen SE. Diagnostic value of *JAK2 V617F* somatic mutation for myeloproliferative cancer in 49 488 individuals from the general population. Br J Haematol. 2013;160:70–79.

Reitsma PH. No praise or folly: genomics will never be useful in arterial thrombosis. J Thromb Haemostas. 2007;5:454–457.

Web sites

Modernizing scientific careers and healthcare scientist training:

- http://www.nhscareers.nhs.uk/details/Default.aspx?Id=2105
- http://www.dh.gov.uk/en/Publicationsandstatistics/Publications/PublicationsPolicyAndGuidance/DH_113275
- http://www.nhsemployers.org/PLANNINGYOUR-WORKFORCE/MODERNISING-SCIENTIFIC-CAREERS/MSC/Pages/MSC.aspx.

The Association for Clinical Biochemistry: http://www.acb.org.uk/

The Health and Care Professions Council: www.hpc-uk.org

The Institute of Biomedical Science: www.ibms.org

The National School for Healthcare Science: http://www.nshcs.org.uk/

2 Analytical Techniques in Blood Science

Learning objectives

After studying this chapter, you should be able to:

- appreciate the importance of different anticoagulants and glass or plastic tubes for the various blood tests requested;
- recognize the importance of assay performance, quality control, quality assurance and the audit cycle;
- describe major techniques in blood science, such as spectroscopy, electrophoresis, chromatography, flow cytometry and microscopy;
- list the major techniques used in molecular genetics;
- discuss the value of point of care testing;
- be aware of health and safety issues in the laboratory.

A major feature of blood science, indeed, one of its founding principles, is the commonality of certain aspects of various tests between the different sub-sciences. Blood must first be obtained, and the steps involved in this are described in Sections 2.1 and 2.2. Generic technical and analytical aspects relevant to all scientific testing are explained in Sections 2.3 and 2.4. The major blood science methods are explained in Section 2.5. Measuring molecules with antibodies (immunoassay) is to be found in haematology, biochemistry, immunology and blood transfusion, whilst microscopy is important in haematology and immunology (and in the other life sciences of infection science (microbiology, virology) and cellular science (histology, cytology)). Conversely, certain other tests and procedures are specific for their particular specialism, molecular genetics being an excellent example. Indeed, these techniques are so specialized that they warrant their own section (Section 2.6).

The last 10 years has seen the growth of blood science testing outside the laboratory; this is called point of care testing and is explained in Section 2.7. Finally, the laboratory can be a dangerous place, and health and safety aspects of blood science are summarized in Section 2.8.

2.1 Venepuncture

Although blood sciences, by definition, start with a sample of blood, blood scientists find at least other body fluids of occasional interest, such as cerebrospinal fluid and urine. Clearly, the latter is easy to obtain, but unfortunately it does not provide a great deal of information. Nevertheless, urine can be used as a quick and simple screening tool for conditions such as hyperglycaemia and renal damage. Other body fluids that blood scientists may analyse include bone marrow and synovial fluid.

Venepuncture is the process of obtaining this sample of blood, generally from a vein on the inside of the elbow joint. An alternative is blood obtained by puncturing the skin, perhaps of the thumb – often called a capillary sample – which in some cases is markedly different from venous blood, and this must be recognized. Rarely, arterial blood may be required, such as for determination of blood gases. One of the most popular places for sampling arterial blood is the radial arteries of the wrist.

Until recently, most blood samples were obtained using a needle and syringe. However, this method has been superceded by vacutainers. These glass tubes are sealed and have a partial vacuum, so when the vein is penetrated, blood is sucked into the tube. The amount of vacuum is such that the correct volume is collected. Venepuncture itself is a learned clinical skill that requires a defined training period and assessment of competency. Obtaining a sample of bone marrow is clearly more difficult as it requires driving a needle through bone,

Blood Science: Principles and Pathology, First Edition. Andrew Blann and Nessar Ahmed.
© 2014 John Wiley & Sons, Ltd. Published 2014 by John Wiley & Sons, Ltd.

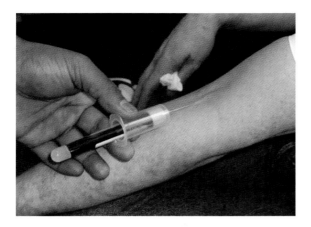

Figure 2.1 Venepuncture. The process of obtaining a sample of blood, generally from a vein near the surface of the skin on the inside of the elbow. Blood is being drawn into a vacutainer with a yellow top (right-hand side) that has no anticoagulant, but a small piece of gel at the bottom (left-hand side) to help the preparation of serum.

often the sternum or the iliac crests of the hips. Figure 2.1 illustrates venepuncture.

2.2   Anticoagulants

Once removed from the body, blood will clot rapidly (within minutes), forming a semi-solid mass of blood cells (the clot) and a fluid. Separation of the clot from the fluid in a centrifuge provides a clear, often yellowish liquid called serum. The serum is often required for the estimation of micronutrients and proteins, and for others tests. A large number of blood cells and coagulation proteins (such as fibrinogen) will make up the clot, so it cannot be used for most tests of red cells, white cells, platelets and coagulation proteins. So in order to be able to analyse these cells and proteins, as do haematologists, we must ensure the blood does not clot, and do so using anticoagulants.

In almost all cases, vacutainers (and non-vacutainer blood tubes) are supplied with an anticoagulant already present. Other vacutainers can be obtained that are free of an anticoagulant and so will therefore ultimately provide serum. These anticoagulants can influence the particular test, so that a variety of these chemicals are available (Figure 2.2). However, some are test specific or analyser-specific, so that at one hospital blood may be taken into a tube with anticoagulant 'X', but at a different hospital it may be taken into anticoagulant 'Y'.

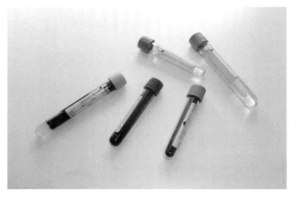

Figure 2.2 Vacutainers. Note the different coloured tops, indicating different anticoagulants (blue: sodium citrate; grey: fluoride oxalate) or no anticoagulant (beige). Note also that vacutainers come in different sizes.

Key anticoagulants

Ethylenediaminetetraacetic acid   Ethylenediaminetetraacetic acid (EDTA) is the required anticoagulant for a full blood count and flow cytometry, and for other tests such as for the integrity of the membrane of the red cell and for investigation of different types of haemoglobin. It is often supplied as a potassium salt (i.e. K-EDTA). Some haematology analysers are designed to be able to provide an erythrocyte sedimentation rate (ESR) result on the same blood as is used for the full blood count. However, one problem with some anticoagulants is that they interfere with the blood itself. The best known artefacts of anticoagulants include a swelling of red cells and platelets by EDTA. It follows that the full blood count needs to be analysed within perhaps 3 h if we are to be sure that the indices of red cell and platelet volume remain accurate.

Sodium citrate   Coagulation tests are invariably performed on plasma that is obtained from whole blood that has been anticoagulated with sodium citrate. The plasma itself is obtained after the vacutainer or other tube containing the blood has been centrifuged. The blood cells, lying below the plasma, are generally discarded. The ESR can be assessed on blood that is held within its own dedicated glass tube; blood clotting in this tube is also prevented by sodium citrate. Clearly, blood anticoagulated with sodium citrate, or indeed K-EDTA, cannot be used for the measurement of these electrolytes in biochemistry.

Fluoride oxalate   This anticoagulant is required for the estimation of glucose, free fatty acids and lactate. The fluoride effectively prevents any white cells from metabolizing any glucose in the blood, which otherwise may lead to incorrect low levels, whilst the oxalate binds calcium.

Lithium heparin   Rarely, blood may need to be anticoagulated with heparin for certain haematology and biochemistry tests, such as for the analysis of different types of white cells or their function. This anticoagulant should not influence levels of calcium.

The problem with anticoagulants is that the nature of the anticoagulant chemicals themselves often interferes with many biochemistry methods. For example, one of the ways that anticoagulants work is by interfering with the coagulation cascade by denying calcium to certain key enzymes. So there is little point in trying to accurately assess the patient's plasma calcium in this setting. Therefore, biochemists seek ways to analyse plasma without chemicals, but the best way to do to this is to simply let it clot and then analyse the serum. In fact, some vacutainers have a small amount of gel to actively promote clot formation, as in Figure 2.1. Assorted vacutainers are shown in Figure 2.2.

2.3   Sample identification and tracking

All patient samples, be they blood or other fluids, tissues excised at surgery or samples for microbiological study, arrive at a central site in the laboratory, generally in an area called 'sample reception' or perhaps 'specimen reception'. Staff in these locations determine whether or not the sample is fit for analysis against a number of criteria, a minimum of which are likely to be as follows:

• the physical integrity of the sample (such as blood not leaking out of the vacutainers)
• information about the patient (hospital number, age, sex, and names etc., often provided on a patient's 'sticky label') is sufficient and is matched between the sample itself and the request form
• details of the requesting practitioner, so that results can be forwarded.
• that the sample is not haemolysed (a pink or red coloration caused by red cell damage and so the cause of incorrectly high potassium, phosphates, and lactate dehydrogenase) and does not have hyperlipidaemia (high levels of lipids influence several biochemical methods).

• a clear and lucid statement of which analyses are requested
• that the sample is appropriate for the request (e.g. full blood count to be taken into a vacutainer with the anticoagulant EDTA, and a sufficient volume of blood taken for coagulation studies into a citrated vacutainer);
• and not crucial, but certainly of help, is an indication of why the analyses are requested (e.g. iron studies are entirely justified if the patient has a microcytic anaemia).

If one of more of these criteria is unacceptable, the sample is likely to be rejected and (if possible) the requester informed. However, not all seemingly correct requests will be processed, as a sample may be vetted and rejected for one of several reasons. For example, because the analyte has a very long half-life, there is likely to be little change in a key index of hyperglycaemia, HbA$_1$c, if a request is repeated within a few weeks, and so the request may be declined. Similarly, in thyroid function testing, a request for thyroid-stimulating hormone (TSH), tri-iodothyronine (T3) and thyroxine (T4) is likely to see TSH being analysed first, with T4 being analysed only if the TSH is abnormal. Many laboratories do not offer T3, arguing it gives little extra information.

If approved, sample reception staff will then affix a unique laboratory identifier, often a bar code that can be read by a computer. The same bar code will be affixed to the request form, which may then be scanned into a computer to allow the practitioner to confirm patient and request details remotely from a computer monitor at the bench. This laser bar code allows the sample to be traced through different parts of different laboratories, and different analyses may be performed on the same sample. Finally, reception staff will direct the sample to the appropriate laboratory.

Once the sample has physically arrived in the laboratory, a blood scientist is likely to draw up a batch of different samples for the same analysis, and so a worksheet, often with the support of the laboratory computer. As each test is performed, and so ticked off the work sheet, the laboratory computer is updated and, once approved by a senior member of staff, the result is communicated to the requestor.

2.4   Technical and analytical confidence

Data in biomedical science arises when blood, tissues or other material taken from an individual is analysed by one or more of the laboratory techniques. These data are likely to be used in diagnosis or in monitoring or changing treatment. Consequently, it is of vital importance that these data are both worthwhile (in terms of identifying disease) and are correct. There are a number of safeguards and checks to ensure that this analysis is not merely adequate but comes up to a defined standard and so is verified. These checks include assay performance, quality control and quality assurance. In seeking and checking methods, the audit cycle is used.

Assay performance

We need to know how well a method or analyser is doing its job, such as in its ability to detect disease and how reliable the result is. This can be achieved in a number of ways.

Confirming or refuting disease Small children are generally very good at determining differences between the adult sexes: daddies are often taller than mummies, and mummies generally have longer hair than daddies. But how reliable is height and hair length in defining sex? Not very, since there are many instances of tall women, short men, hairless (or very short haired) women and long-haired men. These factors bring us to a number of keywords: sensitivity and specificity.

Consider a new blood test, measuring substance 'X', that shows promise in a certain disease. You measure levels of this molecule in a group of patients that you are absolutely sure have the disease. The average result is (say) 50 units/mL, but there is variability so that some have higher concentrations (perhaps 70 units/mL) whilst others have lower concentrations (maybe 30 units/mL). You then measure it in about the same number of people, matched for the same number of men and women, and of roughly the same age, as the patients, and you are absolutely sure that these normal people do not have the disease. This is called the control group; the patients are called the cases. You find an average result of 25 units/mL in the control, but the highest levels, around 41 units/mL, overlap with the cases with the disease (Figure 2.3). This seems to be quite a good outcome – on average, the people with the disease have much more substance 'X' as those free of the disease.

However, note that several points overlap. Which level of the new blood test is best at detecting the disease? If we set the level too low (perhaps 29 units/mL) it will detect all the 24 patients with the disease (=true positive), but it will also incorrectly define 9 normal people as having the disease (=false positive). But setting the level high

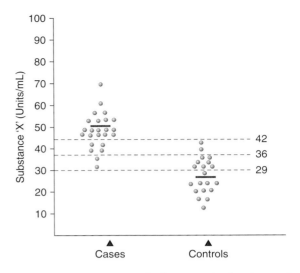

Figure 2.3 Concentrations of substance X in the plasma of patients and controls. The thick bar represents the average (mean) value.

(perhaps 42 units/mL) will exclude all those normal people who do not have the disease (=true negative), but will also exclude six people with the disease (=false negative). These indices, true positive (TP), false positive (FP), true negative (TN) and false negative (FN) are used to derive other indices of clinical sensitivity and specificity:

Sensitivity is the number of TPs divided by the sum of the TPs and FNs.

Specificity is the number of TNs divided by the sum of the TNs and TPs.

Setting the cut-off point at 42 units/mL correctly identifies 18 of the 24 patients with the disease but has failed to identify 6 patients, so that the sensitivity is 75%. It has not identified any of the control subjects, so the specificity is 100%.

We can improve the sensitivity by lowering the cut-off point. At 36 units/mL, we include 21/24 patients, so sensitivity is better at 87.5%, but we incorrectly included one control subject, so the specificity falls to 95.2%. At 32 units/mL, we have missed only one case, so the sensitivity is 95.6% but specificity has fallen to 85.7%. By including all the disease cases, at a result of 27 units/mL, specificity is 100% but it has incorrectly included nine controls, so that sensitivity falls to 57.1%.

So improvements in sensitivity can only be achieved at the cost of a deterioration in specificity. Therefore, in choosing the cut-off point, the key question is the extent to which one index can be traded off against the other. Ideally, there would be no overlap in levels of the new test, so that both sensitivity and specificity would be 100%. In practice, this is an extremely rare phenomenon.

Value in predicting disease An extension of sensitivity and specificity gives the ability of the test to predict disease, as the positive predictive value (PPV). This is defined as the number of TPs divided by the total number of positive results; that is, TPs plus FPs. But we can also derive an estimate of the reverse; that is, the negative predictive value (NPV), which is obtained from TNs divided by the total number of negatives; that is, TNs plus FNs. These points are illustrated in Example 2.1.

Confidence in the result How reliable is a method? Suppose the value of substance 'X' in the blood is 100 units/mL, but the existing analyser, and the method that was dedicated to it and that produced this figure, needs to be replaced. What scientific criteria are considered when assessing a new analyser and method? Consider three new methods (A, B and C) that may be on different analysers that seek to replace this existing method. The same six samples are tested by each method with the following result:

Method A: 75, 100, 124, 106, 93 and 112

Method B: 90, 88, 92, 87, 90 and 91

Method C: 101, 100, 99, 101, 101 and 99

Although the average result from method A is 102, and so only 2% away from an ideal result of 100 units/mL, it has poor reproducibility, with individual results being in error by up to 25%. The results from method B are very tightly clustered around 90, which in itself is excellent, showing high reproducibility, but as a result of 90 is far from 100, it has good precision but poor accuracy.

Example 2.1: Value of a method

Suppose a new method is being developed that the researchers hope will be useful in defining a common condition, such as diabetes or prostate cancer. This needs to be determined in a large group of people in whom it is known with 100% certainty that the disease is present. The new test is applied to the patients and a parallel group (probably healthy) in whom we are 100% confident do not have disease.

	70 patients with the condition	82 subjects free of the condition	
New test: positive result	65 true positives (TP)	9 false positives (FP)	PPV = TP/(TP + FP)
New test: negative result	5 false negatives (FN)	73 true negatives (TN)	NPV = TN/(FN + TN)
	Sensitivity = TP/(TP + FN)	Specificity = TN/(FP + TN)	

Interpretation

PPV = TP/(TP + FP) = 65/(65 + 9) = 65/74 = 0.88 or 88%

NPV = TN/(FN + TN) = 73/(73 + 5) = 73/78 = 0.94 or 94%

Sensitivity = TP/(TP + FN) = 65/(65 + 5) = 65/70 = 0.93 or 93%

Specificity = TN/(TP + TN) = 73/(73 + 9) = 73/82 = 0.89 or 89%

For an ideal method, all these would be 100%. The actual value of the test therefore depends on how close (or distant) to 100% are all these indices. But there are two more mathematically derived indices of interest: the positive likelihood ratio LR = sensitivity/(1 − specificity) and the negative likelihood ratio LR = (1 − sensitivity)/specificity.

Thus, the positive LR is 0.93/1 − 0.89 = 8.45. A result greater than 1 implies the test is associated with the disease, so the test scores well. Similarly, the negative LR is (1 − 0.83)/0.89 = 0.191. A result less than 1 indicates that the test is associated with the absence of the disease, so the test again scores well. These figures are scrutinized closely by authorities that decide whether or not a new test is adopted into clinical practice.

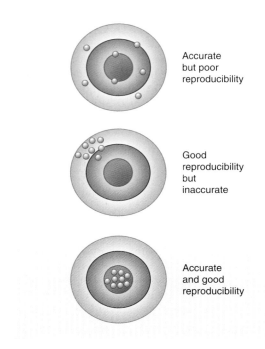

Accurate
but poor
reproducibility

Good
reproducibility
but
inaccurate

Accurate
and good
reproducibility

Figure 2.4 The accuracy and reproducibility of a test illustrated by an archery target. In the top figure, the results are spread out (poor reproducibility) but centre on the bull's eye (accurate). In the middle figure, the results are tightly clustered together (good reproducibility) but are far from the bull's eye (inaccurate). The bottom figure shows results that are both tightly clustered and accurate.

An ideal technique, method C, gives results that are both accurate (very close to 100) and highly reproducible (with a small degree of variation), and so is likely to be adopted (Figure 2.4).

Other nonscientific issues in the choice of a new method include cost of the equipment, cost of reagents, physical size, infrastructure (provision of electricity, water, waste removal), health and safety issues, and so on.

Quality control

Few, if any, analysers are sufficiently reliable that they can be safely left for days or weeks without checking. In practice, many laboratories will routinely pass the same blood or plasma sample through their analyser and then compare results. Naturally, there will be some variation in the results obtained, but all should be within a predefined set of results. Deviation outside these criteria implies a malfunction in the analysers that therefore demands action to correct the fault. For example, consider the following set of consecutive results on the same sample of tissue:

25, 23, 24, 22, 24, 23, 25, 22, 24, 23, 27, 29, 22, 24, 29, 28, 25, 27, 29, 26

Close observation of these data indicate a change in the pattern of results between the first 10 and the second 10. The first 10 range between 22 and 25 (i.e of four different values) with an average of 23.4. The second 10 range from 22 to 29 (i.e. of eight different values) with a higher average of 26.6. It therefore appears that there has been a fundamental change in the performance of the analyser sometime after the tenth sample that seems to have reduced its reliability. This, in turn, should generate activities by senior staff to discover the fault and then rectify it.

We use an index called the intra-assay coefficient of variation (CV) to check the variability of an assay. The CV, obtained by dividing the standard deviation by the mean of the first 10 samples, is 4.6%, which for some methods is acceptable. However, the last 10 samples have a CV of 8.9%. The latter is not only larger than the former, but is also greater than the limits of acceptability, which for this method is <5%. The intra-assay CVs of methods A, B and C outlined above are 16.5%, 2.1% and 1.2% respectively, so that method A is clearly unacceptable.

Quality assurance

An extension of this principle is to provide quality assurance. This is effectively where the host laboratory measures levels of the molecule, antibody or metabolite in question not in a sample of its own, but in a sample provided by another (independent) laboratory. There are several quality assurance schemes, the principal one being the National External Quality Assurance Scheme (NEQAS). This commercial laboratory provides samples of blood and tissues to subscribing laboratories which perform particular tests and then return their results. NEQAS can then provide an idea of how a particular laboratory performs when compared with all the other subscribing laboratories. Example 2.2 provides an illustration of this point.

Subscription to an independent quality assurance organization such as NEQAS is mandatory for all National Health Service (NHS) and other laboratories that provide data upon which clinical decisions are made.

Example 2.2: Quality assurance

Let us suppose you subscribe to a quality assurance provider that supplies a blood sample for you to test. Ten other laboratories using the same analyser also subscribe to this process. The return from the provider may be a list of results on three different tests from the 11 laboratories:

Laboratory	Test 1	Test 2	Test 3
1	100	5.6	46
2	105	6.2	48
3	126	5.7	46
4	104	5.8	48
5	99	5.8	50
6	102	6.0	46
7	104	5.9	46
8	102	5.7	48
9	101	6.1	40
10	103	5.8	54
11	102	6.0	48
Average	104	5.9	47

Several features are noteworthy. In test 1, the difference between the highest (126) and the lowest (99) is 27%. However, 10 laboratories give a result between 99 and 105 (a difference of only 6%), but one value (126, from laboratory 3) is far removed, being 21% higher than the mean value. This outlier is obviously suspicious. In test 2, all 11 laboratories return a result between 5.6 and 6.1 (a difference of 9% between highest and lowest) with no obvious outlier, so the data seem sound.

The difference between the highest and the lowest for the third test is a huge 35% (i.e. 54/40) and this also implies an outlier – obviously the result of 40, from laboratory 9. Removal of this point brings down the highest/lowest difference by a half, to 17%. Perhaps laboratory 9 needs to look closely at the performance of its analyser. A single result is difficult to interpret, but over a series of several quality assurance assessments, if the same error keeps being returned then action may be necessary.

Indeed, paperwork provides proof of adherence to the system and good quality assurance, and may be demanded by inspectors.

In many cases, quality assurance and quality control are easily represented graphically. Indeed, this means that trends can rapidly be noted and acted upon. Commonly used graphics are those of Levey–Jennings and Youden. The best laboratories will place these plots in a prominent position so that the quality assurance/quality control performance can be seen by all their staff. Figure 2.5 shows a Levey–Jennings plot of the data outlined above. The line denoted by 23.5 units is the mean (average) of the first 10 points, and it can be seen that the maximum deviation from this line is 1.5 units. Clearly, the eleventh point is out of the pattern, and this loss of pattern continues for the points that follow, demanding action.

Audit

This is a process that seeks to improve a system by identifying and then clarifying problems. Once a possible improvement has been identified, or a particular problem has been resolved in a satisfactory manner, the audit is repeated, and ideally an enhanced performance will be apparent.

Audit is such a renowned and important process that it has an international standard (ISO 9001) which defines key processes. Indeed, formal audit according to ISO 9001 is now an established part of a quality management system and is required to be followed by Clinical Pathology Accreditation (UK) Ltd. Audit is also part of a defined quality management system, such as that of ISO 15189, which has been tailored for medical laboratories, and requires standards in competency of staff, turnaround times, equipment calibration and methods of reporting.

There are various types of audit. Perhaps the best known follows a sample as it passes through the laboratory, and so identifies opportunities to improve this journey. This is called vertical audit. A horizontal audit looks at several samples at the same stage in the journey. There is also audit of staff as they move around the laboratory and work in specific areas, and also self-audit, the latter probably being part of an annual review and often linked to continued professional development. Audit can generally be viewed as a series of steps:

Step 1: *Plan.* Consider what steps in the process are to be audited, and discuss the approach to be taken with colleagues.

Step 2: *Do.* Undertake the audit; assess the results, identity areas for improvement. Discuss the latter. Consider changes to be made.

Step 3: *Act.* Put the planned changes into operation.

Step 4: *Check.* After a suitable predetermined period, after perhaps 250 analyses, determine the effectiveness of the changes.

This is not the end of the process. audit is continuous, and upon completing the last step, return to step 1.

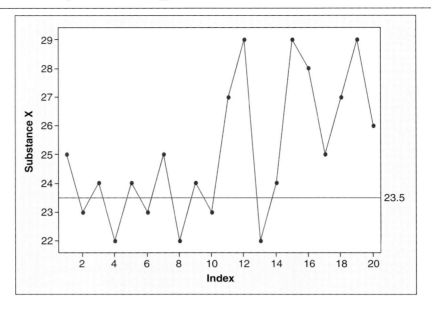

Figure 2.5 This Levey–Jennings plot shows sequential results for substance X, which are acceptably stable up to index point 10, then become highly variable, indicating a problem with the method.

Therefore, audit may correctly be called the audit cycle. Indeed, in many establishments every single method and procedure is audited in sequence in a controlled manner.

2.5   Major techniques

Once a blood sample has been delivered to the laboratory, it must be analysed. Almost all routine blood science results are obtained from tests performed on highly sophisticated machines, or from films of blood that have been stained with special dyes allowing them to be seen with a microscope or an imaging system. The particular method has a chemistry or biochemistry that is specific for the particular technique, although there are similarities, as in different types of electrophoresis.

Although there are marked technical variations, the methods used are becoming standardized in the UK, Europe, and around the world. These methods include spectroscopy, immunoassay, chromatography, cytometry and microscopy. As discussed in Chapter 1, genetics is such a new and focused discipline that its techniques can be summarized in a separate section (Section 2.6).

The standard operating procedure

A common and fundamental aspect of each laboratory method is the 'standard operating procedure', or SOP. This is effectively a highly formalized and sequential set of instructions that any informed and intelligent scientist will be able to follow, such as how to operate a centrifuge. Key points include not simply how to perform the

An erratic analyser

A new analyser is placed in the laboratory. After a period of bedding-in, an audit of data points indicates that the reproducibility is markedly worse after a change in the shift of the staff at 4 P.M., with several unusually high results. There seems to be no clear pattern, and on re-retesting, the rogue results often return to levels that are acceptable. The analyser is checked out by senior staff and by a representative of the manufacturer but the problem remains. The late afternoon staff are subjected to enhanced training and scrutiny, which they naturally resent.

One afternoon a member of the incoming staff appears in the laboratory wearing sunglasses. The penny drops. A member of staff forms the hypothesis that bright sunlight falling on the analyser is the cause of the problem. The hypothesis is tested by drawing curtains around the analyser. An audit of the results shows the problem has been resolved. The intermittent nature of the abnormality is ascribed to occasions of bright sunlight, present only in the afternoon as the laboratory is shaded until then, naturally modulated by clouds.

procedure, but also why it is needed, what it will achieve, and what to do in case of an unusual situation or unexpected result. It also lists reagents and health and safety aspects.

A SOP is different from a log book, which will be present for each analyser. It records details of use, when the service is due, and the quality control/quality assurance record. However, the SOP and log book are commonly merged into a single file. The SOP can also be drawn up for other aspects of laboratory work, such as in sample reception, and in giving information on the telephone (for instance, confirming in writing the time, the hospital number of the patient, the result, and the recipient).

The standard curve

Almost all methods need to include several samples with a known concentration of the analyte in question to which the unknown value of the analyte in the patient sample can be compared. These known samples make up a standard curve, although in many cases the standard curve is a straight line (Figure 2.6). The results obtained from a group of samples with known levels of the analyte are plotted on graph paper, and in this example give a

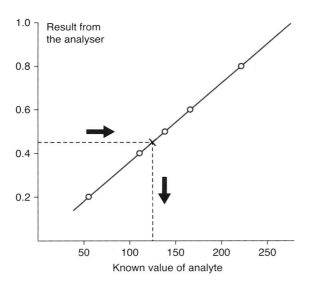

Figure 2.6 A standard curve. Analyser results from a series of five samples of known concentrations of the analyte (open circles) are plotted on a graph. The analyser result from the patient (perhaps 0.47 units on the vertical axis) drawn across the plot meets the standard curve at point 'X'. Drawing down from this point to the horizontal axis gives a result of about 120 units.

straight line. The analyser result from the patient's sample is then compared with those from the known samples, and so the concentration of the analyte in the patient's sample reported.

Upper and lower limits of sensitivity

The standard curve gives two other important indices that apply to all methods. Note that the straight line in Figure 2.6 stops at an analyser result of perhaps 0.17, translating to a known value of perhaps 40 units of the analyte – the line does not go all the way to zero. This is because the assay becomes inaccurate at this point, so that a machine result of perhaps 0.1 cannot be relied upon. This point is called the lower limit of sensitivity. If the patient's result from the analyser is indeed 0.1, then all the scientist can report is a result of less than 40 units.

At the other end of the standard curve, the same principle applies. An analyser result of 1.2 is above the top of the range of the standard curve, so is unable to provide a reliable result. Thus, the upper limit of the analyser's range is a machine result of 0.85 giving a result of 260 units in the patient's sample. If the analyser result is 1.2 units, the correct process is to dilute the sample, probably 50/50 with an appropriate buffer to bring it down into the measurable range of the method. If a 50/50 diluted gives a method result of 0.6, this would give a plasma result of about 200 units, so multiplying back will give a result of 400 units to be reported to the ward or clinic.

Centrifugation

Even before analysis is started, the sample may need additional pre-analytical processing before the amount of the substance of interest (the analyte) can be determined. Serum and plasma need to be prepared by centrifugation. This fundamental machine spins the sample in its vacutainer (Figure 2.2) in a circular manner many thousands of times over perhaps 10–20 mins; as a result, the blood cells and the clot are pushed to the bottom of the sample tube. The serum or plasma can be recovered from the top part of the tubes, and may need to be collected by a member of staff and transferred to a second vessel or test tube. However, some robots are now capable of doing this transfer themselves.

Analysis of metals and nonmetals

Sodium, potassium, calcium, lithium and magnesium are all common metals found in the body and the blood.

When ionized they become positively charged (i.e. cations, denoted as Na^+ and K^+) and can be detected and quantified by a number of methods, of which ion-selective electrodes are the most common. But for every positively charged ion there must be a negatively charged ion (i.e. an anion), the most common one being the chloride ion (Cl^-). Hydrogen is also a nonmetal, and ionizes to form a cation (H^+), the number of which in relation to the number of hydroxide ions (OH^-) confers the pH of a solution.

Ion-selective electrode analysis relies on the electrochemistry of the particular ion being analysed, generally in tandem with a reference chemical reaction, such as that of silver chloride ($AgCl$, which ionizes to Ag^+ and Cl^-). These chemistries are held together in glass structures referred to as electrodes. Specific electrodes are now available for the measurement of hydrogen ions (i.e. as the pH meter), carbon dioxide, oxygen, sodium, potassium, lithium, chloride and other analytes. From the practical point of view, electrodes require a high degree of technical care and need to be subject to regular quality control. Additional chemistries include those of ion-selective field-effect transistors and voltammetric techniques.

Spectroscopy

These techniques determine the interactions between electromagnetic radiation (of many different wavelengths) and material in the sample to be analysed. In many cases this involves differences in the absorption, scatter or emission of radiation. Colorimetry generally uses radiation with wavelengths within the visible range (400–800 nm, useful for assessing haemoglobin), whilst others use other wavelengths that call for more sophisticated detectors (i.e. spectrophotometers). Ultraviolet wavelengths (<400 nm) can be used, for example, to detect vitamins, whilst infrared wavelengths (>800 nm) may be used to assess the make-up of a molecule, such as the presence of carbon–carbon double or triple bonds, or the presence of ketone groups. Infrared and Raman spectroscopy can be used to determine the presence of drugs such as penicillin and amphetamines, whilst nuclear magnetic resonance spectroscopy can differentiate chemicals such as methyl bromide from methyl ethanoate. Spectroscopy includes fluorescence, to be developed in a later section.

The Beer–Lambert law The driving basis of many types of spectroscopy is the Beer–Lambert law. Broadly speaking, this states that the absorbance of light of a given wavelength is proportional to the concentration of the analyte and also to the length of the path that the light takes to pass through the solution. As the path length is fixed, then the absorption of light is directly proportional to the concentration of the material being analysed. Variants of the Beer–Lambert law are applicable to many methods.

Colorimetry This is a very common and popular method that obviously provides a coloured product that can be detected by a spectrophotometer. There are many variants of this method. In one, a chemical reaction is set up so that an enzyme acts on the analyte and another chemical or chemicals (together, the substrates) to produce at least one product that is coloured. Examples of this type of assay are the determination of glucose using the enzyme glucose oxidase. In another, the analyte itself participates directly in a chemical reaction. Examples of analytes measured in this way are calcium, phosphates and urea. Creatinine reacts with alkaline picrate to form a red colour, whilst albumin reacts with bromocresol green at low pH to form a green colour.

Plasma enzymes (such as alkaline phosphatase) themselves can be determined by providing a sample of serum or plasma with a specific substrate (often a clear solution such as paranitrophenol phosphate) that is then cleaved by the enzyme to give a coloured product (in this example, free paranitrophenol, which is yellow). Whilst most colorimetry is based in biochemistry, in haematology the levels of the haemoglobin molecule can be measured by lysing the red cells and mixing the liberated protein with Drabkin's solution. However, this reagent contains potassium cyanide and has been deemed to be a hazard, so that alternatives are available.

Mass spectrometry This technique determines the ratio of the mass of an analyte to its charge. As the precise physics and chemistry of this technique are complex, but freely available elsewhere, we will focus on clinical applications. One is in the determination of inborn errors of metabolism (such as phenylketonuria) in neonates (Chapter 20), another is in determining the presence of drugs of abuse and a third is the presence of the acute-phase reactant amyloid-A.

Variants of mass spectrometry (MS) include electrospray ionization (ESI), and matrix-assisted laser desorption ionization (MALDI), both of which can identify and quantify proteins. ESI-MS can be used to probe for small differences in the structure of beta-globins in thalassaemia, whilst MALDI can be linked to a time-of-flight (TOF) analyser (hence MALDI-TOF) to provide a high-throughput platform that can analyse one sample per minute. MS linked to gas chromatography (GC–MS) is probably the most accurate method for the analysis of steroids, whilst tandem MS is effectively two separate MS set-ups in sequence, and produces spectra called MSMS. Both GC–MS and tandem MS are amenable to routine hospital laboratories and are being widely adopted. For example, tandem MS may displace high-performance liquid chromatography (HPLC) as the method of choice for the analysis of haemoglobin A_2.

Immunoassay

This powerful and common group of assays relies on the identification of a particular molecule by an antibody, the latter often a polyclonal antiserum raised in a rabbit, sheep or other mammal, or a mouse monoclonal antibody. There are many different types of immunoassay that can be used for the estimation of a host of analytes.

Agglutination Microscopic particles of latex can be coated with an antibody to the blood protein of interest. The analyte will bind to antibodies, and so will cross-link the latex particles. In the absence of the analyte, no latex particles are cross-linked and the solution of particles remains milk-like. However, cross-linked particles will aggregate together into clumps which can be seen by eye. Examples of analyses of this type include for the presence of d-dimers in coagulation and c-reactive protein in biochemistry.

An alternative assay is to coat the latex particles with components of the membrane of a pathogen, such as a bacterium. If present, antibodies to the bacteria in the patient's sample will cross-link the particles and cause them to agglutinate. Hence, this can be used as a quick and simple screening test to determine the likelihood that the patient has antibodies to the pathogen. This may imply a present infection.

Red cells can also be agglutinated by cross-linking using antibodies. This process is the basis of the blood transfusion laboratory, where such agglutination is an indication of an incompatibility between the donor blood and the patient recipient. People suffering from glandular fever are characterized by an unusual set of antibodies. Curiously, these antibodies will agglutinate horse red cells. This will be further developed in Chapter 9, on immunopathology.

The advantage with agglutination is that the result can be determined by eye, calls for no expensive hardware and is rapid. However, it has a poor lower limit of sensitivity (i.e. cannot detect very low levels of the analyte) and the precise numeration of the result is often subjective.

Enzyme-linked immunosorbent assay This type of immunoassay relies on presence of a physical link between an enzyme and an antibody. The enzyme is then offered a substrate, generally as a clear solution, that it can then cleave to give a coloured product. Thus, enzyme-linked immunosorbent assay (ELISA) is effectively a form of colorimetry.

The process of ELISA is standard and involves a defined series of steps (Figure 2.7). First, a primary antibody has to be attached to a solid body, such as the side of a test tube or the well of a microtitre plate. Second, after binding, and the removal of unbound antibodies by washing with a buffer, the sample of test fluid is added and the analyte of interest binds to the antibodies, which themselves are bound to the solid phase. Thus, the first antibody is often called the 'capture' antibody. The third step proceeds after a set period of time: the unbound test fluid is washed out of the tube or well, and the addition of a second antibody, conjugated to an enzyme, binds to the analyte held by the primary antibody. This second antibody is often called the 'detection' antibody.

Finally, the unbound second antibody is washed out and a clear substrate (such as orthophenylene diamine) is added, which the enzyme (such as horse radish peroxidase) on the detection antibody then cleaves to give a product that is coloured yellow. The reaction is often terminated by adding acid, which may also enhance the intensity of the colour of the product. The colour is then translated into numbers by a plate reader (effectively a modified spectrophotometer set up to detect the colour of the product). In common with many methods, the concentrations of the antibodies and other reagents are carefully worked out to be in excess, so that the rate-limiting reactions are those between the antibodies and the analyte, so that the strength of the colour is directly proportional only to the concentration of the analyte in the sample.

As the sample is fixed between two antibodies, it is commonly described as a sandwich assay. Most ELISAs are performed in 96-well microtitre plates, and they must include a standard curve as well as quality assurance/quality control samples to ensure that the method has

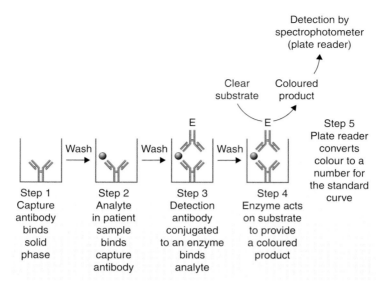

Figure 2.7 Five common steps in an ELISA, where the amount of an analyte in a sample is converted to colour for analysis. The amounts of the reagents are in excess so that the sole rate-limiting factor in determining the amount of colour being developed is the amount of the analyte.

worked and has delivered results within the expected range (Figure 2.8). The entire 96-well microtitre is loaded into a dedicated plate reader with eight detectors, so that the colour in all 96 wells can be quantified rapidly, and so is very efficient. However, time is required to ensure the sample binds the antibodies, so the whole assay may take several hours, and this generally places it outside the acute setting, where a rapid result is demanded.

The use of a 96-well microtitre plate for ELISA is common, especially for unusual analytes, but has a high unit cost. However, technical developments (such as robotics) and the demands of high throughput have led to ELISAs that can be performed not in microtitre plates but in autoanalysers, which markedly reduce unit costs and turn-around time, as in conventional biochemistry and haematology.

A variant of ELISA is enzyme-linked immunospot, which is used, for example, to detect cell secretions such as cytokines and antibodies from different lymphocyte populations.

Chemiluminesence immunoassay   Like ELISA, this technique also relies on an antibody linked to an enzyme such as horse radish peroxidase. However, in contrast to ELISA, where the substrate produces a coloured complex, in chemiluminesence immunoassay (CLIA) a substrate (such as isoluminol, lucigenin or an acridium ester, with

enhancers), when oxidized by hydrogen peroxide, produces a photon of light. A photomultiplier can detect the light and software converts it to a digital signal. Many CLIAs have superior sensitivity than the conventional colorimetric method(s), and do not require long incubations or the addition of stopping reagents, as is the case in some ELISAs.

These technical improvements have led to the introduction of CLIA as a viable laboratory method for analyses such as growth hormone, thyroid and sex hormones, tumour markers such as prostate specific antigen, and cardiac markers such as troponin. However, the entire process demands its own analyser, a chemiluminometer.

Fluorescence immunoassay   This method once more relies on an antibody bound to a detection system. In ELISA the system is an enzyme–substrate combination that generates a coloured product, whilst in CLIA the product of the antibody–enzyme–substrate complex is a photon of light. However, in fluorescence immunoassay (FIA) the antibody is conjugated to a fluorochrome that, when stimulated by light of a particular wavelength, emits light of a different wavelength. This emitted light is detected by a photomultiplier tube, the apparatus being described as a fluorimeter.

The detection of certain populations of cells in immunology and haematology by fluorescence-activated cell

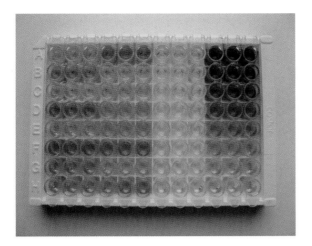

Figure 2.8 A microtitre plate showing results of an ELISA procedure. Samples are loaded in groups of three (triplicates). The three columns (counting from the left: numbers 10–12) of the far right of the plate are the standard curve. The topmost triplet of wells have the highest concentration of a known amount of the analyte (perhaps 100 units/L), and triplets of wells below this have proportionately lower amounts of the known standard (80 units/L, 60 units/L and so on down to maybe 5 units/L). The rest of the plate, to the left of the standard curve, contains triplicates of samples of plasma from different patients. Note that some are highly coloured (such as the three wells in horizontal row F, vertical columns 1–3), and other less so (such as row B, columns 4–6). The standard curve (Figure 2.6) will translate these unknown colours to the concentration (units/L) of the analytes in the plasmas. Columns 7–9 are blank, and so have no colour.

scanning (FACS) is a type of FIA: antibodies to CD molecules are conjugated to fluorochromes such as fluorescein, phycoerythrin and peridinin-chlorophyll proteins. FIA can also be used to detect and quantify antigens at the surface of, and in the cytoplasm of, cells localized in tissues, and also autoantibodies to these antigens. Both these methods are expanded upon in sections that follow.

Immunoturbidimetry and nephelometry These two techniques are based on the absorption or deflection of light by soluble immune complexes in a fluid. Immunoturbidimetry measures the light that is absorbed by the complexes, whilst nepholometry detects the degree of light scatter. The technique is relatively simple: the sample of plasma, serum, urine or cerebro-spinal fluid is mixed with the antibody of the antigen of choice and

placed in the path of beam of light (often provided by a laser) whilst photomultipliers detect changes in that light.

The process can be enhanced by reagents such as polyethylene glycol, but attention must be paid to the dilutions of the antibody and sample to be analysed, because in great antigen excess the expected linear relationship between antibody and antigen is often lost and results are unreliable – the so-called prozone effect. A variant of this process is to measure the speed at which the immune complexes form – described as rate or kinetic nephelometry. The twin techniques of turbidimetry and nephelometry are often used to measure plasma proteins such as total immunoglobulins and the subclasses IgG, IgA and IgM, complement components, c-reactive protein, rheumatoid factor and d-dimers.

Immunoprecipitation This method relies on the complex of an antigen and antibody coming out of solution and forming a precipitate. It is most commonly performed in an agarose gel. In one method, two wells are cut in the gel, antibody is placed in one and plasma/serum sample on the other. The two then diffuse into the gel towards each other and form a precipitate when they meet. This is called double immunodiffusion. In another, single radial immunodiffusion, the gel contains the antibody and a well is cut in the gel into which a small amount of plasma or serum is placed. As the antigen in the fluid sample diffuses out in all directions into the gel, it forms a ring of precipitate with the antibody, which can be measured. The diameter of the ring of precipitate is proportional to the concentration of the analyte in the plasma/serum sample. In both methods, protein dyes such as Coomassie blue can help determine the presence of the precipitate. A further variant of this, immunoelectrophoresis, is discussed later.

Radioisotopes Health and safety issues have largely restricted the use of radioisotopes, as are used in radioimmunoassay (RIA) and the related immunoradiometric assay (IRMA) to a small number of analyses. In RIA, a known concentration of radiolabelled version of the analyte competes for the antibody with unknown levels of the analyte in the serum or plasma. The principle in IRMA is the same as for ELISA, but a radioisotope is linked to the detection antibody in place of the enzyme, the amount of analyte being proportional to the degree of bound radioisotope. In both RIA and IRMA, the amount of the radioisotope is determined by a radio-detector such as a gamma-counter or beta-counter, dependent on the isotope concerned.

In haematology, radioisotopes are occasionally used to determine the total red cell mass of an individual, or an estimation of the lifetime of the cell.

Chromatography

Although one of the oldest techniques, the primary scientific basis of chromatograpy, despite its name, is not to do with colour. When developed in the early part of the 20th century, samples were first separated and the component parts identified by staining with dyes of different colours (hence chroma-). Both the separation techniques and the methods for identifying the samples have become more sophisticated, and the two major types of chromatographic techniques are HPLC and gas–liquid chromatography (GLC).

High-performance liquid chromatography   This powerful technique evolved from a physical separation technique, column chromatography. HPLC is rapid (results available within minutes), accurate, uses a small amount of sample (a few microliters) and is automatable. The sample is driven into a series of columns where the separation takes place, and the products monitored by a detector as they emerge. The method has great flexibility, and with specific columns it can, for example, provide information on different types of steroids and other metabolites, and different haemoglobin species. Accordingly, it finds use in both haematology and biochemistry units. As regards the former, HPLC can determine different types of haemoglobin as in sickle cell disease (Chapter 4). However, the equipment itself can also be placed within biochemistry units as HPLC can also be used to measure glycated haemoglobin, an important measure of the diagnosis and management of diabetes (Chapter 14). Figure 2.9 shows a compact HPLC workstation.

Gas–liquid chromatography   This technique utilizes differences in volatility of molecules as a basis for their separation. In GLC, the molecules to be separated are partitioned between a mobile gas phase and stationary liquid phase. The stationary phase is usually a high boiling point organic liquid such as silicone grease. The inert gas is forced through a coiled capillary column in which the liquid stationary phase is coated on the inner wall and which may be as long as 100 m. The capillary column is enclosed within an oven whose temperature can be varied to increase the volatility of the sample molecules. The separated components leave the capillary

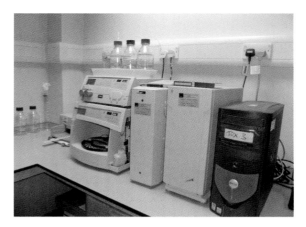

Figure 2.9   HPLC. This compact workstation consists of a series of enclosed units containing the columns, reservoirs for buffers, pumps and (on the far left) the sample area and controls. The Dionex HPLC-DAD offers the separating power of HPLC linked to a photodiode array detector (DAD) for clinical and forensic applications. The DAD records the absorbance spectra of compounds over a range of wavelengths (e.g. 200–595 nm) as they pass through the detector flow cell. This data can then be used to provide definitive identification of a compound, or to select the optimum wavelength for quantitation.

column in the mobile gas and are detected by either flame ionization or electron capture detectors. GLC is a versatile technique and can be used to investigate a wide range of substances, including alcohols, lipids, phenols and some drugs. It has excellent resolution, reproducibility and sensitivity for detection of these substances. GLC is used to introduce samples into a mass spectrometer.

Electrophoresis   As with immunoassay, there are many variants of electrophoresis, although all rely on the electrical properties of the analyte, which most frequently are proteins or nucleic acids. There is no routine need for the electrophoresis of lipids. However, the size and shape of the molecule is sometimes important. The process is not passive; the sample needs to be driven through a solid phase (such as paper, agarose or polyacrylamide gel) by an electrical charge, itself delivered by a separate transformer. The sample to be analysed is placed near one of the electrodes and then 'migrates' to a certain position on the solid phase where it can be detected, generally by staining with a particular dye or fluorescent probe.

One of the best known examples of this is the separation of plasma proteins so that the different types of proteins (such as alpha-, beta- and gamma-globulins) can be quantified. This process is most often found in biochemistry departments. There is also variation of the technique according to the size of the molecules of interest and the pH of the buffer. For example, different types of haemoglobin migrate to different positions on the gel in the presence of different pH values. This is explained in Chapter 4 where, for example, it is used to diagnose sickle cell disease.

In isoelectric focusing, the solid phase, inevitably a gel, is itself a spectrum of different pH values, so once more the analyte migrates to that particular pH most in tune with its own electrical profile. Using a polyacrylamide gel enables molecules to be separated principally by size, and can be used for both proteins and nucleic acids.

Capillary electrophoresis (CE) separates the components of the sample in a narrow-bore tube with an internal diameter of 25–100 nm, but of 40–100 cm length. The method relies on ion migration and electro-osmotic flow, but should not be confused with the electrophoretic mobility as is present in cellulose acetate electrophoresis. These conditions allow a rapid analysis (10–20 min), and with the requirement for a very small sample (5–10 μL) the process may be called high-performance capillary electrophoresis. CE may be applied to the determination of metal ions, amino acids, peptides, proteins, deoxyribonucleic acid (DNA) fragments and drugs. Like 'standard' electrophoresis, CE can be used to analyse paraproteins in myeloma and its related diseases, and different haemoglobin species in sickle cell disease and thalassaemia. However, tandem MS may prove to be suitable for the newborn screening of haemoglobinopathies.

Immunoelectrophoresis, which merges 'standard' electrophoresis with immunoassay, allows defined proteins to be detected. This is important in the diagnosis of myeloma and allied diseases where a more exact biochemical nature of the abnormal protein can be determined, as is explained in Chapter 6.

Counting cells and particles

One of the basic necessities in cell biology is the ability to count particles. In haematology and immunology these are red cells, white cells and platelets, whilst in microbiology these are bacteria.

Impedance With sophisticated software, the passage of electricity through a small aperture can be monitored. Should a small particle also pass through the aperture, the passage of the electrical current will be temporarily disrupted and can be detected by the software. The degree of interruption of the electrical charge is proportional to the size of the particle. In this way, the number and size of blood cells in a given sample can be estimated, and so provides the red cell count, the white cell count and the platelet count. However, this technology, pioneered by Coulter and dominant in the last half of the 20th century, has, in many cases, been superceded by flow cytometry.

Flow cytometry This method of counting cells is fundamentally different from impedance. In flow cytometry, a laser shines a beam of light at a stream of blood cells, and the interruption of the passage of light is noted by sensitive detectors on the far side. The frequency at which the beam of light is interrupted by the passage of cells gives the cell count. The degree to which the beam of light is obstructed gives the size of the cell. The beam of light can be deflected or scattered by the cell, and this gives an index of the nature of the nucleus and of the granularity of the cell (the latter defined by intracellular organelles). Thus, flow cytometry provides information on the number of cells, their size and their granularity. This method is at the heart of modern haematology analysers.

An additional aspect of modern flow cytometers is the inclusion of cytochemistry to allow differentiation of white cell subgroups. White cells can be further classified by the shape of their nucleus and the number of intracellular granules: the nucleus occupies 90–95% of the lymphocyte, whilst neutrophils, eosinophils and basophils contain many different granules. However, this alone cannot differentiate certain subgroups. One way of getting round this problem is to take advantage of the different cytoplasmic granules of certain white cells, which contain different enzymes that would normally participate in a physiological and/or pathological function. Providing cells with a chemical substrate for their particular enzyme profile will lead to different colour and so the ability to identify certain cells (Figure 2.10).

Fluorescence-activated cell scanning An extension of the 'standard' flow cytometry process is the ability to measure different molecules on the surface or in the cytoplasm of cells. Different wavelengths of light emitted by different lasers can activate and detect changes in the wavelength of that light when it falls on certain fluorochromes. By attaching fluorochromes (such as fluorescein isothiocyanate, FITC) to antibodies, then cells bearing the molecules can be identified. Thus, combining the size and granularity of cells with molecules expressed at the cell surface, or within the cell, gives a very powerful technique.

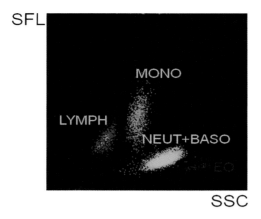

Figure 2.10 Flow cytometry. This technique for counting white cell subpopulations makes use of the size of the nucleus and presence of intracellular granules. The colour coding defined by the analyser's software in this figure makes recognition of each type of cell easier; the cells themselves have no colour. Mono: monocytes; Lymph: lymphocytes; Neut: neutrophils; Baso: basophils; Eo: eosinophils. These cells are described in detail in Chapter 7. (Image courtesy of Sysmex UK).

The FACS method first found a home in immunology laboratories, where the different proportions of T and B lymphocytes, and the subgroups of T lymphocytes are measured (Figure 2.11). However, flow cytometry has also been put to use in identifying particular types of white cell malignancies (leukaemia and lymphoma, as in Chapter 6), in abnormalities in the red cell membrane (Chapter 4) and in investigating platelet disorders (Chapters 7 and 8).

FACS can also be used for multiplex technology. This system allows the multiple determination of several analytes using microscopic polystyrene beads, each of which has bound to it a different antigen or antibody. A second antibody conjugated to a fluorochrome is added, recognizes its particular antigen and is detected by the flow cytometer. The technique is limited only by the number of lasers used in FACS, each of which can only recognize a single fluorochrome.

Cytochemistry

Colour, cells and chemistry come together in cytochemisty. Certain cells contain molecules that react with chemical and dyes. This will become apparent in Chapters 5 and 6 on white cells, as the nucleus of the white cell is easily stained purple. Furthermore, the cytoplasm of different white cells contains specific

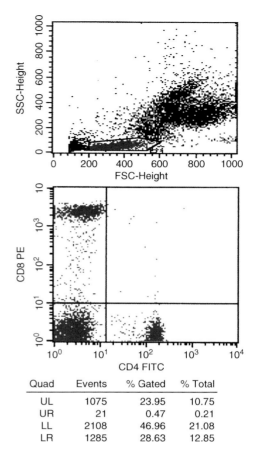

Quad	Events	% Gated	% Total
UL	1075	23.95	10.75
UR	21	0.47	0.21
LL	2108	46.96	21.08
LR	1285	28.63	12.85

Figure 2.11 FACS analysis of T lymphocyte subsets. In the upper figure, the scientist has put a 'gate' around that region of the plot where they expect lymphocytes to be, and the FACS machine software has coloured these in red. The lower figure shows analysis of those cells which have bound an antibody to CD4 (itself bound to the fluorochrome FITC, lower right (LR) quadrant), an antibody to CD8 (itself bound to fluorochrome phycoerythrin, upper left (UL) quadrant), both antibodies (lower left (LL) quadrant) and neither antibody (upper right (UR) panel).

The panel of numbers is the actual number of cells (events) in each quadrant. The key mathematics is the proportion of cells bearing only CD4 (LR: 1285 events) versus those cells bearing only CD8 (UL: 1075). The ratio between the two, the CD4/8 ratio is therefore 1.2.

As is discussed in the case study in Chapter 9, this ratio is reversed in infections with the human immunodeficiency virus, which is associated with a specific loss of CD4-bearing cells and so a reverse of the CD4/8 ratio.

Incidentally, the CD4/8 ratio in this patient is below the reference range for the local laboratory of 1.5/1 to 4/1 because they have chronic lymphocytic leukaemia. Notably, almost half of the cells fail to express either CD4 or CD8 (the LL quadrant). This will include normal B lymphocytes, but is unusual because of the leukaemic cells. (Image courtesy of H. Bibawi, NHS Tayside).

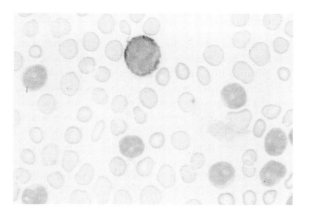

Figure 2.12 Immunocytochemistry detecting an abnormal white cell (stained red). Unstained cells are counterstained light blue/grey. The enzyme is alkaline phosphatase.

granules that take up different chemicals and so can be identified by colour. Red cell precursors and some white cells can be stained blue by Perls' stain for the presence of iron, as demonstrated in Chapter 4.

By adding an antibody we have immuno-cytochemistry. The key is the ability to link an enzyme to an antibody that itself recognizes a molecule of interest, such as may be expressed by a leukaemic cell. This is illustrated by Figure 2.12, which shows a leukaemic cell highlighted by a red colour.

The colour is only present on the particular cell because only that cell has the membrane component picked up by the antibody–enzyme conjugate. As in an ELISA, the reaction is completed by exposing the enzyme to a substrate, which in this example turns red when cleaved by the enzyme. The figure also shows several smaller cells which have failed to pick up the antibody, so that their identity is separate from the one 'red' cell. Their blue colour comes from the use of a counterstain.

Microscopy

By far the most common use of this technique in blood science is light microscopy; use of fluorescence micros-copy is restricted to a small number of indications. Our colleagues in cellular sciences (histology, cytopathology) and infection sciences (principally microbiology) have great skills in microscopy.

Light microscopy The objective of this method is to enable the scientist to view and comment on morpho-logical aspects of cells and tissues. In order to do this, cells are almost always fixed and stained with a dye designed to enhance a particular feature. This is perhaps best illus-trated by the examination of a glass slide smeared with a film of blood or bone marrow where different cells can be identified and enumerated.

Key features of all microscopy include magnification, contrast and resolution. The former is defined by the magnifying power of the lens in the eyepiece (often 10 ×) and of the lens (Figure 2.13). Most microscopes have a number of different objective lens (such as 10 × , 40× and 100 × magnification) that can be rotated to give a range of magnification. Contrast and resolution refer to the ability to differentiate very small but important features, and this depends on factors such as the light provided by the condenser and the use of the condenser: too little light, like too much light, will reduce the discriminating power of the microscope.

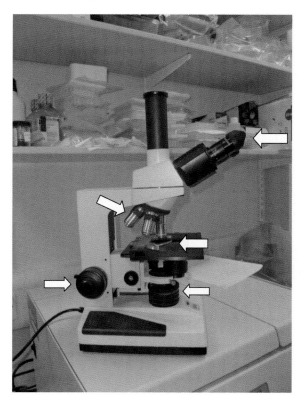

Figure 2.13 A simple bench light microscope. The top arrow indicates the eyepieces. The three objective lenses are high-lighted by the arrow on the left; the stage, where the glass slide is placed, is indicated by the lower right arrow. The lowest arrow on the right points out the light source; the lowest on the left indicates the focusing apparatus.

The microscope is a sensitive item of equipment that requires care, and it must be treated with respect otherwise it can easily lose its ability to detect the fine differences in the morphology of cells. Accordingly, it is likely that senior staff alone will have the expertise to 'set up' the microscope with respect to contrast, resolution and other technical aspects. The large vertical black tube at the top of the microscope in Figure 2.13 allows the attachment of a camera. Accordingly, many microscopes can be attached to video equipment and so to a television screen, enabling many people to look at the same image at the same time, and so providing an excellent teaching tool.

Fluorescence microscopy   This technique, a variant of FIA described earlier, relies on the principle that a specialized chemical (a fluorochrome) can absorb light at a certain wavelength, but emit light at a different wavelength. If this emitted light is at a wavelength outside the visible spectrum, it must be detected by specialized equipment. This method is popular because it cuts out background light from the visual spectrum. Many fluorochromes (such as FITC, which we have already come across in the section on FACS earlier) can, like enzymes in ELISA, be conjugated to antibodies. This enables the detection of molecules on the surface of cells (as in FACS analysis), but also on tissues.

Indirect immunofluorescence describes a technique particularly useful in the immunology laboratory in detecting the presence of autoantibodies. Patient's serum is diluted and a drop placed on a section of tissue, such as epithelial cells, liver, oesophagus, kidney or neutrophils, all located on a glass slide. If present, autoantibodies bind to their antigen on the tissues. Unbound serum is washed off and replaced with an antiserum conjugated to a fluorochrome such as FITC. When unbound antiserum is washed off, the fluorescence microscope is used to determine the presence of the autoantibody by its binding to the tissue and the antiserum. Additional details are presented in Chapter 9.

Other microscopy   Blood scientists also use visible light microscopy to analyse synovial fluid from, for example, patients with rheumatoid arthritis. The same sample may also be analysed for crystals, such as may be formed from uric acid in gout. Very high power microscopy may require a small amount of transparent oil between the sample on the glass slide and the objective lens – this is called 'oil immersion'. Other types of microscopy include dark-field, confocal, phase-contrast and electron microscopy.

Automation

With the advantage of technological developments, machines have become increasingly sophisticated and can perform hundreds of analyses in an hour with the need for a scientist being cut to a minimum. Once the samples are placed on the machine it can be programmed to run a series of different analyses on the same sample, be it serum, plasma or urine. A key aspect of the use of robotic arms is to move aliquots of the sample, alongside reagents, to reaction vessels where analysis proceeds. The reaction vessel will then be washed free and so will be ready to receive the next sample. Figure 2.14 shows a typical biochemistry analyser.

The next step in the development is the physical linking of different analysers that perform different functions. These analysers are linked by a track, which carries a vehicle containing the vacutainer. As the vacutainer passes such analysers, it may or may not be sampled, as the program (as is written by the scientist) instructs. In haematology, different analysers linked by a common track include a machine that produces a full blood count, another that can make and stain a blood film, and another that will perform the ESR (Figure 2.15).

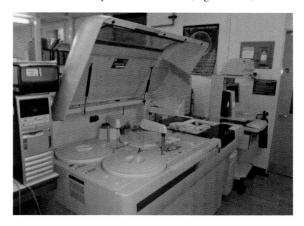

Figure 2.14 A biochemistry analyser. This substantial item of capital equipment is capable of the simultaneous analysis of several different analytes. However, it still needs to be programmed by scientists, who also need to ensure reagents are kept topped up and that waste is being safely disposed of.

The Olympus AU640 is a fully automated and open-system clinical chemistry analyser capable of generating between 800 and 1200 tests per hour. It can be used for routine biochemistry, therapeutic drug monitoring and drugs of abuse screening. It can deliver up to 48 homogeneous immunoassays, which can be monitored by 13 wavelengths between 340 and 800 nm, and has a capacity of 150 samples with continuous loading.

Figure 2.15 A haematology tracking system. A vacutainer is conducted along a track, and different machines are programmed to perform their own specific analyses as required. (Image courtesy of Sysmex UK).

2.6 Molecular genetics

This component of blood science is concerned with detecting abnormalities in genes and chromosomes. The expression 'molecular biology' refers to studies of molecules in biology, and is often and erroneously used to specifically describe gene work. A preferable expression is 'molecular genetics', being the study of the molecules that make up genes, and thus chromosomes.

The building blocks of genes are the four nucleotide bases of adenine, guanine, thymidine and cytosine, and most aspects of laboratory genetics look for direct or indirect signs of alterations in the sequence of the bases that provide evidence of the altered genes that themselves cause disease. Almost all of these genes are present in chromosomes located in the nucleus, and although there is DNA in mitochondria, molecular genetics of the pathology laboratory is concerned almost completely with nuclear DNA. Similarly, the vast majority of diagnostic molecular genetics looks at DNA; studies on ribonucleic acid, although entirely possible, are rare.

Tools in molecular genetics

Purification and extraction of DNA We start with the isolation of DNA, which of course requires some cells. In some cases these may be obtained by washing out the mouth with some saline, so collecting some cheek and tongue cells. However, this also collects bacteria, which have their own DNA and so may cause problems. Perhaps the most 'clean' source of DNA is from the white cells in a blood sample.

Extraction of DNA is reasonably straightforward. The cells themselves are lysed, and the cell pellet washed and digested with various enzymes (proteases to break down proteins), detergents (to solubilize lipids) and buffers to ultimately produce a solution rich in DNA. Use of an alcohol such as ethanol or propanol will precipitate the DNA, allowing it to be purified, followed by solubilization in an alkaline solution. The amount of DNA in a solution, the degree of protein contamination, can be estimated in a spectrophotometer at wavelengths of 260 and 280 nm. If additional DNA is needed, the polymerase chain reaction (PCR) can rapidly provide more.

Analysis of DNA The solution of DNA is now ready for analysis. The next major tools are enzymes that digest DNA at certain specific positions; these enzymes are called restriction endonucleases. They can be used to probe for the presence of a particular sequence of DNA which would be altered (and so absent) if the gene is mutated. Restriction endonucleases will only cut a specific section of DNA, so that if it does not find its specific DNA sequence, no cut will be made. This technique gives sections of DNA of different lengths that can be separated and identified by polyacrylamide gel electrophoresis (Figure 2.16).

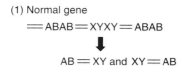

(1) Normal gene

$$===ABAB===XYXY===ABAB$$

⬇

$$AB===XY \text{ and } XY===AB$$

(2) Abnormal gene

$$===ABAB===XYOY===ABAB$$

⬇

$$AB===XYOY===AB$$

(3) Incubate with probe specific
for AB—XY

(4) Electrophoresis and radiography

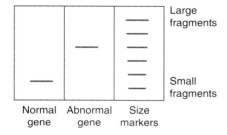

			Large fragments
	—	—	
		—	
—		—	
		—	Small fragments

Normal   Abnormal   Size
gene   gene   markers

Figure 2.16 Identification of mutated genes. Two different restriction endonucleases are mixed with the DNA (step 1). One enzyme cuts the DNA strand at nucleotides AB/AB, the other at XY/XY. In a normal situation, this generates two identical small fragments: AB=====XY. The same enzymes applied to the abnormal DNA, with abnormal nucleotides XY/OY but normal AB/AB sequences, which generate only one large fragment because the enzyme specific for XY/XY cannot cut the mutated sequence (step 2). Mixing the fragments with radio-labelled probe AB-----XY will see the probe binding both fragments as both the short normal and longer mutated fragments contain the matching DNA sequence (step 3). When this mixture is run through an electrophoresis gel, the smaller normal fragments will migrate faster than the larger abnormal fragments. The size of the patient's fragments can be assessed by running a series of fragments of known size.

However, the application of restriction endonucleases to some DNA may well generate millions of fragments of different lengths. The identification of the fragment of interest can be made by using a probe. This is an artificial segment of DNA generated in the laboratory that is designed to bind to a specific section of DNA from the patient. By tagging the probe with a radioisotope and mixing it with the sample of DNA being analysed, the probe should highlight the presence or absence of the

section of DNA in question when the electrophoresis gel is analysed by radiography.

Figure 2.17 shows a practical example of this technique in the investigation of a mutation in the DNA of genes coding for haemoglobin that causes a particular type of haemoglobinopathy.

In other analyses, a single base pair may be different from that expected, and may be significant. This difference, a single nucleotide polymorphism (SNP), is widely exploited in the search for pathological differences. Perhaps the best example of an SNP is that in the gene for the beta globin of haemoglobin that causes sickle cell disease. The use of SNPs is especially useful in searching for polymorphisms in genes in large numbers of people. This genetic 'epidemiology' is called genome-wide association, and has been used to probe for links between SNPs and disease such as chronic renal disease (Chapter 12).

The application of molecular genetics to human disease

Naturally, no analysis proceeds without a firm potential diagnosis that has been made on other (clinical) grounds. Thus, the purpose of molecular genetics is to confirm or refute this diagnosis. There are many examples of the use of molecular genetics in all branches of blood science. In many cases the problem can be detected at the level of the whole chromosome (see Table 1.5 Section 1.4), but in others the examination of the gene is necessary and/or informative.

Molecular genetics in biochemistry   One of the most common defects in steroid metabolism is in the *CYP21A2* gene that codes for the 21-hydroxylase enzyme. Mutations in region of chromosome 6 where this and similar genes are located can lead to problems in steroid and other metabolic pathways. Deficiency in 21-hydroxylase leads to congenital adrenal hyperplasia, as will be discussed in Chapter 18. Familial hypercholesterolaemia is caused by a mutation in the gene found on chromosome 19 that codes for a receptor on the cell surface that binds low-density lipoprotein (LDL) cholesterol. The consequence of this mutation is that the receptor fails to recognize LDL and remove it from the plasma, so that levels remain high. This mutation has a frequency of 1/500 of the indigenous population of western Europe; and there are many other mutations in lipid metabolism, as we shall note in Chapter 14. Several genes are implicated in chronic renal disease (Chapter 12).

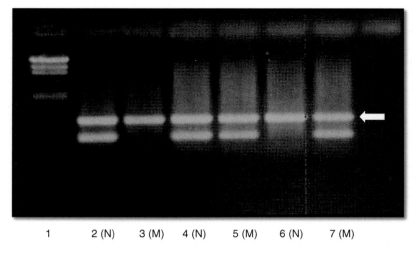

1   2 (N)   3 (M)   4 (N)   5 (M)   6 (N)   7 (M)

Figure 2.17 Gene analysis. This is an amplification refractory mutation system PCR used for diagnosis of a β-thalassaemia mutation. Lane 1 (left-hand side) is molecular markers, lanes 2 and 3 patient 1 (normal and mutant primers respectively), lanes 4 and 5 patient 2 (normal and mutant) and lanes 6 and 7 patient 3 (normal and mutant primers). The arrowed bands represent internal control bands that identify a standard region of the DNA regardless of the thalassaemia mutation. This is very useful in confirming that the PCR reaction is optimized as the absence of the specific primer band (normal or mutant), and is diagnostic.

Patient 1 (lanes 2 and 3) has a normal homozygous genotype; patient 2 (lanes 4 and 5) is heterozygous, whilst the pattern of patient 3 (lanes 6 and 7) indicates a homozygosity for the particular mutation. Further details of gene typing in haemoglobinopathy (including thalassaemia and sickle cell disease) are presented in Chapter 4. (Image courtesy of H. Bibawi, NHS Tayside).

Molecular genetics in haematology

Probably the most dominant aspect of genetics in haematology is in haemoglobinopathy. This condition is caused by mutations in the genes that code for different globin molecules. Qualitative defects, due to a single change in the DNA sequence, lead to abnormal globin species such as haemoglobins S, C and D, and thus sickle cell disease. The most common quantitative gene defect causes the absence of a large part of the globin molecule, and so the disease thalassaemia (Figure 2.16). Multiplex PCR methods can diagnose different deletions in the thalassaemias: a single tube can detect heterozygosity, homozygosity and compound heterozygosity of seven alpha-globin gene deletions that account for lesions of this type.

Defects in several different genes lead to failure in iron metabolism, such as in the formation of haem (leading to porphyria) and in the excessive build-up of iron in various organs (haemochromatosis). These are explained in Chapter 4.

Perhaps the most well-known gene mutation in haemostasis is that which results in the absence of coagulation factor VIII and causes haemophilia. However, this mutation is very rare; the most common genetic haemorrhagic condition is caused by a mutation in the gene for von Willebrand factor, which causes von Willebrand disease. Factor V

Leiden is the consequence of a mutation in the gene for coagulation factor V that leads to an increased risk of venous thrombosis. Indeed, it is the most common cause of unexplained deep vein thrombosis and pulmonary embolism. Additional details are to be found in Chapter 8.

The Philadelphia chromosome was noted in cases of chronic myeloid leukaemia in 1960. The translocation, noted in 1973, brings together parts of two different chromosomes, each carrying different genes (*Abl* on chromosome 9 and *BCR* on chromosome 22), which together form a new gene that has malignant potential. The new fused gene generates an enzyme (tyrosine kinase) that activates genes important in cell cycle control and proliferation, and hence the leukaemia. Further work on this enzyme led to the development of a family of specific drugs (tyrosine kinase inhibitors) that successfully target this particular problem and bring a great reduction in the burden of the disease. We shall revisit this advance, which is a good example of molecular genetics leading directly to treatment, in Chapter 6.

Genetic analysis in acute myeloid leukaemia (AML) can also provide crucial clues in predicting outcome. For example, those cases with a translocation between chromosomes 15 and 17 have a 5-year survival rate of 81%, whereas others with a translocation between chromosomes 9 and 11 have a survival rate of 39%. In patients with cytogenetically normal AML, the presence of a particular mutation in *NPM1* correlates with a better response to intensive induction and with long-term survival, whereas certain mutations in *DNMT3A* are associated with an adverse outcome. However, not all studies concur: in analysis of *c-KIT* mutations in a certain AML phenotype, one group reported an overall survival of some 80% in those with wild-type *KIT*, but approximately 35% in those with a *KIT* mutation. Conversely, another group found overall survival to be around 75% in those with the mutation but in the region of 50% in those with wild-type *KIT*.

Molecular genetics in immunology Geneticists can help immunologists by demonstrating mutations and rearrangements in genes for immunoglobulins and in the T lymphocyte receptor, and in defining X-linked agammaglobulinaemia. Chronic granulomatous disease is characterized by poor bacterial-killing granulocytes, and there are several variants. Historically, it was defined by the nitro-blue tetrazolium test, but genetics can now unequivocally confirm or refute the likelihood that chronic granulomatous disease is X-linked or due to other mutations. All these points are expanded upon in Chapter 9.

The place of molecular genetics in diagnosis and management Although molecular genetics is an exceptionally powerful technique, and often gives unequivocal results, in many cases it merely confirms the presence of a particular disease. However, in many cases, genetic confirmation of a particular mutation, and thus assurance of the cause of the disease, may not necessarily change the clinical management of that disease. For example, haemophilia is reasonably easy to diagnose on clinical grounds (haemorrhage in a male infant), especially if there is a family history, and the absence of coagulation factor VIII activity as defined by standard haematology tests. Confirmation at the gene level may make the healthcare professional feel better, but it will do nothing for the well-being of the child, or its management, which is prophylaxis and treatment of a haemorrhage by the infusion of factor VIII concentrates. A similar argument can be made for the precise genetic nature of a haemoglobinopathy: in almost all cases management is driven by the clinical condition of the patient, not by his or her particular genotype.

However, there are many instances where molecular genetics are of undoubted help in confirming or refuting

a diagnosis. A good example of this is the coincidental finding of a hypercholesterolaemia, perhaps a total cholesterol of 8 mmol/L, in a blood sample: Is this due to a diet rich in cholesterol, or does it have a genetic basis, such as familial hypercholesterolaemia? If the latter is correct, then the gene mutation may be present in other family members, and this information may be important. In Section 1.4 we noted that the *BRCA-1* gene has a powerful penetrance into the risk of breast and ovarian cancer, so that its presence must be acted upon. Less serious is the presence of the gene mutation that leads to factor V Leiden (also caused by an SNP), as this can explain an unexpected deep vein thrombosis in the absence of other risk factors such as obesity and use of female sex hormones (the oral contraceptive pill, hormone replacement therapy). An alternative cause of the unexpected thrombosis is cancer, considerably less desirable a diagnosis than factor V Leiden, as discussed in Chapter 8.

Of course, molecular genetics has many applications outside what we would consider biomedical and blood science. For example, newborn screening for fragile X syndrome is possible using a mutation in the *FMR1* gene. Using a large-scale technique in 14 000 neonates, one child with the full mutation was identified, although the rate of the permutation was present in one female in 209 and one male in 430. This prevalence rate is higher than was previously thought.

Pharmacogenomics

This important development recognizes the fact that different people respond very differently to the same situation, and that this can be clinically very important. For example, in some patients a very small dose of warfarin (such as 0.5 mg daily) is sufficient to give them protection for thrombosis, whereas others patients may need 15 mg daily. This difference, effectively hypersensitivity and resistance respectively, may be due to the activity of certain enzymes on different isoforms of warfarin: the R isomers and the S isomer. The activities of these enzymes are, in turn, governed by isoforms of genes of the cytochrome P450 system. Isoforms of a second gene, *VKORC1*, also influence the patient's response to warfarin. Thus, knowledge of the patient's genome will inform the practitioner on an appropriate dose of a drug. This is developed in more detail in Chapter 8 and Chapter 21.

A second example of the importance of patient-centred pharmacogenomics is in the metabolism of the anti-cancer drug 5-fluorouracil. Up to 12% of patients have a variant of the enzyme dihydropyrimidine dehydrogenase (*DHPD*) resulting in loss of function, which leads to increased levels of this dangerous cytotoxic agent. Knowledge of the isotype of *DHPD* helps the oncologist to ensure the patient receives an appropriate dose of 5-fluorouracil.

2.7 Point of care testing

Advances in technology have brought benefits in the physical size of analysers, and in several cases this includes the ability to be portable, and often powered by a battery, thus freeing the operator from a permanent electrical supply. These mini-analysers can then be moved out of the laboratory and closer to the patient, be they on the ward, in the outpatient clinic or even outside the hospital. This is called 'point of care testing' (POCT), but is also known as 'near patient testing' (NPT).

POCT/NPT has been available for decades in outpatient side rooms and in general practice. Urine 'dipsticks' can be used to crudely identify gross abnormalities in glucose, proteins, red cells, white cells, bilirubin, ketones, specific gravity and pH. However, it is acknowledged that these are simple screening tests, and poor sensitivity means that only the worst cases are identified. A positive dipstick result will often result in a formal blood test. Of course, pregnancy testing is freely available as an over-the-counter test in any high-street pharmacy and many supermarkets. The police have long been using chemistry in suspected cases of driving under the influence of alcohol, where the latter can be detected in exhaled breath.

Mini-analysers

By paring down the number of analyses being performed on a particular blood sample, a physically smaller machine can be developed to cater for a more focused need. The standard laboratory haematology analyser delivers (at least) six red cell indices, six white cell indices and (often) two platelet indices. Whilst all of these 14 analyses are important, it could be argued that they are not all important in all settings all of the time, such as in life-threatening situations. The same principle applies to commonly requested biochemistry tests. Small analysers offering a more focused profile are available and are proving popular in certain settings where the full haematology and biochemistry profiles are not needed.

Mini-analysers are often found in accident and emergency (A&E) units, and in intensive care units (ICUs). The former are likely to measure electrolytes, blood gases, amylase, drugs of abuse (to confirm/refute overdose), cardiac enzymes and troponin (to help diagnose angina and myocardial infarction). Many of these tests are also often required urgently in critical and intensive care, where practitioners are also likely to need urgent glucose, calcium, magnesium, lactate and osmolality.

Micro-analysers

These offer only a single analysis on a single sample, and may also be found in both A&E and ICUs. A good example from A&E is the use of a rapid test for d-dimers, a breakdown product of a clot, which can help diagnose a deep vein thrombosis or a pulmonary embolism. In some cases, micro-analysers can be used by the patients themselves to monitor their particular abnormality and even adjust their therapy. These are becoming popular with both the patient and the hospital, as self-testing results in fewer outpatient appointments. Some of the examples in Table 2.1 are suitable for patient self-testing.

The advantages of POCT

It could be argued that POCT is quicker and cheaper than the alternative; that is, the main laboratory service. POCT is certainly likely to be quicker, but the economics of POCT are complex. The unit cost of a main laboratory analysis is low because of the economy of scale; that is, a high throughput of samples. However, POCT is by definition close to the patient, and this brings possible efficiency as regards the number of patient visits, such as a separate visit to the phlebotomist for venepuncture, and also in general hospital infrastructure, such as transport of the sample to the laboratory, and so on.

Table 2.1 Examples of POCT applications.

Test	Indication
International normalized ratio[a]	Atrial fibrillation
Glucose[a]	Diabetes
Total cholesterol	Hypercholesterolaemia
B-type natriuretic peptide	Heart failure
Platelet function	Thrombosis
Neutrophil gelatinase-associated lipocalin	Acute kidney injury
Blood gases	Acidosis and alkalosis

[a]Potential for use by patients.

A rapid result is also popular with patient and practitioner as treatment may be immediately modified. A next step is to place the POCT device outside the hospital, such as in a pharmacy or general practice, with savings for the NHS trust. Taken to its ultimate, the patient may even have their own POCT device, monitor their own blood, and then self-medicate. However, it seems likely this option will be unusual. Many of us are aware of small devices to measure blood pressure. As this widespread initiative is common and accepted, then personal POCT based on a blood test may follow.

The disadvantages of POCT

The major problem with POCT is rarely the machine itself. Errors in POCT can come from a number of areas, the principal being incorrect use by untrained staff. Other problems arise from physical misuse of the device (inappropriate handling and storage), poor quality control of the analyte, poor calibration against a known standard and use of date-expired reagents. All of these are unlikely to occur in the main laboratory, and all are correctable if the POCT device (and, ideally, its operators) is viewed as part of the laboratory. This is clearly possible (and desirable) for POCT devices in hospitals.

Recognizing the value of a laboratory-accredited POCT device, many NHS trusts have a POCT committee staffed by scientists to ensure good laboratory practice. The committee will advise on the applicability of the use of a POCT device, purchasing, service contracts, training of staff and ensuring quality control, and so on. Naturally, this will include the drafting of POCT-specific SOPs. In recognition of the value of POCT, the Clinical Pathology Accreditation (UK) Ltd now offers the accreditation of the analysis of blood gases, total haemoglobin and blood glucose.

Regulations and guidelines on POCT

In recognizing the potential pitfalls in POCT, several bodies, such as the Medicines and Healthcare Products Regulatory Agency, the Association for Clinical Biochemistry and Laboratory Medicine and the Royal College of Pathologists all offer support. These include technical matters and buyer's guides.

2.8   Health and safety in the laboratory

The laboratory is undoubtedly a workplace full of potential dangers and hazards (Table 2.2). The practitioner is

Table 2.2 Laboratory hazards.

Biological hazards	Non-biological hazards
Bacteria	Physical and mechanical
Viruses	Chemical (including
Fungi and moulds	liquids and gases)
Parasites	Radioactivity
Prions	Electrical
(These may enter the body by	Environmental (e.g.
needlestick injury, through	excessive noise,
broken skin, or via the	extremes of
lungs)	temperature)
	Fire
	Lasers

protected from these by a series of rules and regulations, the principal ones being the Health and Safety at Work Act (HASAWA), and the regulation on the Control of Substances Hazardous to Health (COSHH). The former is often shortened to simply 'health and safety', abbreviated to H&S. Each department must have its safety advisor, whose responsibilities include training and ensuring that local guidelines are widely known and adhered to. The safety advisor will also conduct regular inspections to ensure the workplace is as safe as is practicable.

In parallel, a number of bodies are relevant, the principal being the Health and Safety Executive. Other bodies include the Institute of Occupational Health and Safety.

The Health and Safety at Work Act

This major item of legislation, formalized in 1974, sets the standards of safety throughout the UK in all places of work where the number of employees exceeds a certain minimum. It was, and continues to be, followed by a series of semi-specific regulations and codes of practice for defined situations. For example, one such is issued by the Advisory Committee on Dangerous Pathogens. The HASAWA has two major parts, covering the duties of employers and of employees.

Duties of employers   These are to ensure, as far as is reasonably practicable, the health, safety and welfare of its employees, subcontractors and visitors. These include setting site-specific documents, the appointment of safety advisors and the availability of protective clothing, gloves, eye goggles, and so on. There is also the requirement that manufacturers ensure that their equipment is as safe as is reasonably practicable, and that hazards (such as lasers) are clearly identified. This extends to the providers of reagents, who are required to provide safety information on all chemicals, drawing attention to possible hazards such as risk of poisoning (e.g. sodium azide) and if a powerful oxidizing agent (e.g. concentrated sulphuric acid).

Duties of employees The HASAWA demands that (a) the employee works safely and ensures the safety of others and (b) that they cooperate with employers on health and safety issues. The definition of 'works safely' is enshrined in local and generic documents, regulations, and so on. As regards the latter, the UK subscribes to European law, and the European Commission provides a set of six regulations, covering management of the regulation, the workplace, display screen equipment (such as computer screens), personal protective equipment (such as white coats and disposable gloves), manual lifting and handling, and provision and use of equipment.

The Control of Substances Hazardous to Health (COSHH)

The definition of a hazard is reasonably straightforward, such as a factor that leads to ill health. However, not all hazards are equally dangerous, and not all are present at the same frequency. The likelihood that a potential hazard actually causes harm is called its risk. A further aspect is whether or not a risk or a hazard is well known and, therefore, can be protected against. These factors lead to a scoring system, where a hazard may be given a score of one to five for likelihood of causing danger, and one to five for the consequences of the danger. Multiplying the two gives an overall score of the practical level of hazard. Thus, a dangerous, potentially life-threatening hazard likely to present rarely may be as important in a practical sense as a modestly dangerous hazard that can be expected to occur frequently.

The same principle can be developed for methods or techniques that have varying degree of danger, and so inherently hazardous aspects can be eliminated. An example of this is the use of plastic containers instead of glass, as the former will not shatter if dropped. This is also why radioisotopes in immunoassay are less favoured.

It is the responsibility of the employer, generally passed to a designated safety advisor, to draw up local guidelines applicable to each workplace. All substances, processes and equipment likely to be hazardous to health must be risk assessed, especially during development, and all reasonable steps taken to ensure the safety of staff. Fortunately, and quite rightly, there is a large health

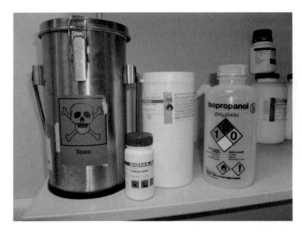

Figure 2.18 Identification of hazards. All laboratory hazards must be identified. As far as chemicals are concerned, this can be marked on the labels. These include skull and crossbones with 'toxic' label on the left, the smaller orange marks on the label of the containers in the middle, and the diamonds on the propylene squeezy bottle on the right.

and safety and COSHH industry that will provide advice, training and documents. Many of us will be aware of brightly coloured stickers which identify hazards, but these are slowly being replaced by a world-wide hazard system (Figure 2.18).

COSHH regulations extend from simply identifying hazards, also demanding that the exposure of employees to hazards is monitored, that health surveillance and occupational health services are available, and that employees are informed, trained and supervised. There are numerous other aspects of health and safety/COSHH, which include:

- awareness of the hazards of fire (e.g. a safe escape route and the identity of the fire warden);
- when and how to use personal protective equipment (gloves, masks, clothing);
- the transport of potentially hazardous material, such as glass and vacutainers (filled with blood or not);
- personal hygiene when moving from the laboratory to a non-laboratory area (offices, canteen).

Summary

- Blood for analysis is most commonly obtained by venepuncture of a peripheral view, such as on the inside of the elbow.

- If necessary, blood must be anticoagulated to prevent it from clotting. Major anticoagulants include EDTA and sodium citrate.
- The specimen reception section of the laboratory will receive samples, log them, affix a bar code and pass the sample on to an appropriate section, such as biochemistry or haematology.
- In assessing the value of a method, we consider factors such as sensitivity and specificity, and also take account of its accuracy and reproducibility.
- The laboratory must invest in quality control and assurance, and also perform regular audit to ensure good laboratory practice.
- Major techniques in blood science include ion-selective electrodes, colorimetry, immunoassay, chromatography, cell and particle counting, cytochemistry and microscopy. Many of these can be automated.
- Molecular genetics provides unequivocal confirmation or refutation of preliminary diagnosis. Its techniques are generally discipline specific.
- Advances in machine technology see the analyser move from the laboratory to the outpatient clinic, the emergency room, the bedside, or even the patient's home. This POCT has both advantages and disadvantages.
- The laboratory can be a place of danger. Accordingly, health and safety and the assessment of hazards are important.

Further reading

Briggs C, Kimber S, Green L. Where are we with point of care testing in haematology? Br J Haematol. 2012;158:679–690.

Daniel YA, Turner C, Haynes RM. *et al*. Quantification of hemoglobin A2 by tandem mass spectrometry. Clin Chem. 2007;53:1448–1454.

De Mare A, Groeneger AH, Schuurman S. *et al*. A rapid single-tube multiplex polymerase chain reaction assay for the seven most prevalent alpha-thalassemia deletions and alphaalphaalpha(anti 3.7) alpha-globin gene triplication. Hemoglobin. 2010;34:184–190.

Eriksson N, Wadelius M. Prediction of warfarin dose: why, when and how? Pharmacogenomics. 2012;13:429–440.

Fan A, Cao Z, Li H. *et al*. Chemiluminescence platforms in immunoassay and DNA analyses. Anal Sci. 2009;25:587–597.

Kalyuzhny AE. Chemistry and biology of the ELISPOT assay. Methods Mol Biol. 2005;302:15–31.

Lewis SN, Nsoesie E, Weeks C. *et al*. Prediction of disease and phenotype associations from genome-wide association studies. PLoS One. 2011;6:e27175.

Ofran Y, Rowe JM. Genetic profiling in acute myeloid leukaemia – where are we and what is its role in patient management. Br J Haematol. 2013;160:303–320.

Strathmann FG, Hoofnagle AN. Current and future applications of mass spectrometry to the clinical laboratory. Am J Clin Pathol. 2011;136:609–616.

Van Kuilenburg AB. Screening for dihydropyrimidine dehydrogenase deficiency: to do or not to do, that's the question. Cancer Invest. 2006;24:215–217.

Web sites

The Health and Safety Executive: www.hse.gov.uk.

See also the dedicated laboratory site: www.hsl.gov.uk. NHS sickle cell and thalassaemia screening programme: http://sct.screening.nhs.uk.

The Institution of Occupational Safety and Health: www.iosh.co.uk.

ISO 9001. Several commercial sites offer this facility, such as http://www.british-assessment.co.uk and http://www.9001superstore.com

Clinical pathology accreditation: http://www.cpa-uk.co.uk/

http://www.clinbiochem.info/poct.html

http://www.mhra.gov.uk

3 The Physiology of the Red Blood Cell

Learning objectives

After studying this chapter, you should be able to:

- explain the importance of effective haemopoiesis;
- list the major components of bone marrow;
- explain the value of the analysis of bone marrow;
- comment on the uses of cytochemistry and flow cytometry;
- explain how the red blood cell is adapted for its purpose;
- list the different types of haemoglobin and understand the relationship between their structure and function;
- describe the structure of the red blood cell membrane;
- discuss the importance of the red blood cell enzymes and metabolic intermediates;
- identify major variations in the morphology of the red blood cell.

3.1 Introduction

The ultimate goal of the pathology laboratory is to deliver to the practitioner an accurate and true result of their particular analyses. In Chapter 2 we learned that the principal haematology blood test is the full blood count. The red blood cell section of the full blood count has a number of analyses, as indicated in Table 3.1. These focus on the number, size and other characteristics of the cells, but other useful information can be obtained from the morphology of the cell once stained by certain chemicals and then viewed by a light microscope or by image-identifying software. However, we cannot fully understand the clinical aspects of these tests until we gain an appreciation of the physiology of the red blood cell. Since this entire chapter is about the red blood cell, we will dispense with the 'blood' part, naming it a red cell. In parallel, white blood cells will be referred to as white cells.

The red cell is one of the most highly specialized cells in the body. There are many manifestations of this, such as that of its membrane, which is modified for its function: that of the free and easy passage of oxygen both in and out of the cell. An additional consequence of specialization is that it has a relatively 'uncluttered' cytoplasm, with only those organelles and molecules directly required for its unique function; that is, the carriage of oxygen by haemoglobin. The degree of specialization is so extreme that it has no need for organelles – the nucleus, mitochondria, ribosomes and endoplasmic reticulum are missing.

A further aspect of the lack of organelles is that the cell is unable to generate fresh protein, and so suffers metabolically when existing proteins are consumed or are exhausted. A consequence of these factors is a relatively short lifespan of about 120 days. However, freedom from intracellular organelles gives the red cell a physical flexibility or deformability that allows it to pass along the smallest capillaries and so deliver oxygen to the tissues – a normal 7 or 8 μm red cell can deform to pass along a capillary half this diameter.

Before examining these specializations in detail, we will review, in Section 3.2, how blood cells develop – haemopoiesis. The process of the maturation of red cells, white cells and platelets develop by a progression from precursor stem cells in the bone marrow to the fully functioning mature cells found in the blood. Much of this knowledge has been obtained from analysis of bone marrow itself.

Having established the importance of haemopoiesis, we will then be in a position to examine the red cell in detail. This will be accomplished by a detailed look at how red cells develop (erythropoiesis, in Section 3.3), and then the structure of the cell and how it is related to function. First, this will consider the membrane of the red cells (Section 3.4), and then the cytoplasm (Section 3.5),

Blood Science: Principles and Pathology, First Edition. Andrew Blann and Nessar Ahmed.
© 2014 John Wiley & Sons, Ltd. Published 2014 by John Wiley & Sons, Ltd.

Table 3.1 Red cell aspects of the full blood count and allied tests.[a]

Index	Typical result	Unit	Reference range
Haemoglobin (Hb)	139	g/L	125–180
Mean cell volume (MCV)	87	fL	79–99
Red cell count (RCC)	4.8	10^{12}/L	4.3–5.7
Mean cell haemoglobin (MCH)	30.3	pg	27.0–34.5
Mean cell haemoglobin concentration (MCHC)	347	g/L	316–365
Haematocrit (Hct)	40.1	L/L	38.0–54.0
Reticulocytes[a]	30	10^9/L	25–125
Red cell distribution width[b] (RDW)[c]	12.5	%[c]	10.3–15.3
Erythrocyte sedimentation rate[b] (ESR)	4	mm/h	<10
Plasma viscosity[b] (room temperature)	1.63	mPa/s	1.5–1.72

[a]The full blood count may also contain comments regarding the shape of the red cells. Relevant expressions include anisocytosis, polychromasia, schistocytes and Howell–Jolly bodies. These will be explained in Section 3.9.
[b]May not be part of a full blood count.
[c]Unit for RDW is coefficient of variation (i.e. the standard deviation divided by the mean).

and then how these come together in the transport of oxygen, in Section 3.6. The fate of the red cell and the recycling of its components are explained in Section 3.7.

We will then be in a position to get to grips with the clinical aspects of the red cell part of the full blood count. Section 3.8 will look again at the tests outlined in Table 3.1, while Section 3.9 will consider what we can learn from the morphology of the cell as viewed under a light microscope. Thus, once we have fully understood the function of the red cell, we will move to a discussion of the diseases of this cell in Chapter 4, of which the principal one is anaemia.

3.2 The development of blood cells

Effective haemopoiesis is crucial to the health of the individual: it generates mature, functional blood cells that transport oxygen, defend us from infection, and participate in haemostasis (the balance between too much clotting (thrombosis) with too little clotting (haemorrhage)). In adult life, haemopoiesis predominantly occurs in the bone marrow, although in exceptional circumstances (such as in certain pathological conditions) it may also happen in other tissues, such as in the liver, lymph node and spleen. However, in the foetus and the neonate it is normal for haemopoiesis to occur in the bone marrow, lymph node, liver and spleen as well as in the yolk sac of the embryo.

In health, some 5–10×10^{11} blood cells are produced by the bone marrow each day by a process that is highly balanced and regulated by cytokines, growth factors and environmental factors such as the amount of oxygen in the body. Ideally, this number of cells produced each day exactly matches the number of cells that have come to the end of their life cycle. However, irregularities in the production of blood cells will lead to disease. A thorough understanding of the structure and function of the bone marrow and haemopoiesis is necessary in order to grasp the concepts of these diseases, some of which are life-threatening, such as aplastic anaemia (where there is a reduction in cell production) and in the haemoproliferative disorders (the most common being leukaemia) where an excess of blood cells is produced. Leukaemia and the haemoproliferative disorders are discussed in more detail in Chapter 6.

The mature cells that are present in the blood all have their origins in the bone marrow, which is the location of the stem cells that ultimately give rise to the mature cells found in blood. As red and white cells mature under the influence of lineage-specific growth factors, they pass through reasonably well-defined stages of maturity, including several blast stages, such as the myeloblast. Mature platelets are fragments of the cytoplasm of their own dedicated precursor, the megakaryocyte.

Bone marrow architecture and cellularity

Haemopoiesis begins in the embryonic yolk sac, and then transfers to the liver and spleen and ultimately to the bone marrow, so that at birth it is normally the only place where haemopoiesis occurs. In the adult the major sites of haemopoiesis are the sternum and iliac crests: other sites include the skull, vertebrae and ribs. This bone marrow haemopoiesis is called intramedullary. However, in certain diseases, such as haemoglobinopathy, myelofibrosis and severe haemolytic anaemia, haemopoiesis can occur outside the bone marrow. This extramedullary haemopoiesis, as can happen in the liver, lymph node and spleen, is therefore a sign of severe stress on the bone marrow. Myelofibrosis and haemoglobinopathy, and their consequences (such as anaemia), will be discussed in Chapters 6 and 4 respectively.

The major structural function of bone is to provide support and pivot points for muscles, ligaments and tendons, protection for delicate tissues (such as the brain) and physical support for body organs. Much of these physical demands follow from the strength of bone as a hard and supportive connective tissue. However, many bones are in fact 'hollow' and are extremely dynamic organs. These spaces within the bones are host to haemopoietic tissues and the blood vessels that serve them. The complex microenvironment that ultimately gives rise to the blood cells can be seen as having three components. These are haemopoietic tissues consisting of the stem cells and their progeny (the immature but developing blood cells), sinuses (vascular spaces or pools of blood that are lined with endothelial cells) and cells that support the bone marrow and often produce growth factors. These include:

• stromal cells, producing the scaffolding that supports other cells;
• macrophages, producing growth factors to promote erythropoiesis, store iron and perform routine debris removal; and
• adipocytes that store energy in the form of fat.

A close physical association of various cells is necessary to ensure the correct development of particular mature blood cells via cell–cell contact with secretion and binding of growth factors. Additional details of the importance of bone marrow and its analysis are presented in Chapters 5 and 6.

Models of differentiation

Blood cells pass thorough a series of well-defined stages of development and maturation. Haemopoiesis begins with a common pluripotent stem cell that can independently replicate, proliferate and differentiate into the various lineage-specific cells that in turn produce mature bloods cells (Figure 3.1). Most of our understanding of haemopoiesis comes from animal models and the laboratory

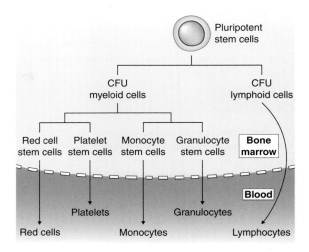

Figure 3.1 Haemopoiesis. The pluripotent stem cell gives rise to two colony forming units (CFUs: myeloid and lymphoid) that in turn produce lineage-specific stems cells. These ultimately produce mature cells that leave the bone marrow and enter the blood.

where stem cells, also known as CFUs, can be grown from samples of bone marrow.

We currently recognize stem cell CFUs dedicated to particular cell lineages. One of these CFUs, often referred to as the common myeloid precursor, ultimately gives rise to granulocytes, monocytes, megakaryocytes and red cells. A second CFU, often described as the common lymphoid precursor, is dedicated to the production of lymphocytes. The myeloid CFU then gives rise to two additional stem cells, one specific for both the erythrocyte and one for the megakaryocyte, the cell that ultimately produces platelets. In parallel, another stem cell gives rise to granulocyte stem cells and monocyte stem cells that produce the mature cells found in the blood.

Many of the cells developing within the bone marrow are called blasts. These cells are almost exclusively found in the bone marrow and are part of the normal and healthy process of haemopoiesis. Blasts often have additional names, denoting membership of a particular lineage of the cell into which it will develop. For example, a myeloblast is a blast of the myeloid pathway, and a megakaryoblast is the precursor cell to the megakaryocyte that in turn produces platelets.

However, disease such as leukaemia is characterized by abnormalities in haemopoiesis so that there may well be increased numbers of these blast cells in the bone marrow and in the blood. The latter is especially undesirable as it implies the tumour is progressing and could be transforming itself into a more aggressive and therefore more dangerous disease. Additional details are presented in Chapter 6.

Growth factors

The bone marrow microenvironment is crucial for successful haemopoiesis – cells must be able to communicate directly with each other. However, there are also a host of crucial hormone-like signaling molecules – cytokine growth factors – glycoproteins released by one type of cell (such as a macrophage or a stromal cell) to act on different stem cells and blast cells to promote the process of cell development. As the target of these growth factors are CFU cells, they are called colony stimulating factors (CSFs). Perhaps the best-known example of a growth factor is erythropoietin, which acts on red cell precursors and blasts. Broadly speaking, levels of erythropoietin correlate inversely with haemoglobin. Other growth factors include cytokines such as interleukins.

Advances in biotechnology now allow the large-scale production of many of these growth factors, and thus their availability as therapeutics in a variety of conditions. For example, after bone marrow transplantation or severe chemotherapy, use of growth factor, such as granulocyte–macrophage colony stimulating factor, can help the donor stem cells to repopulate the bone marrow and so accelerate the production of mature blood cells.

Bone marrow sampling and analysis

Knowledge of the interrelationships between different cells of the bone marrow is important in numerous diseases (such as aplastic and sideroblastic anaemia, and in myeloma, myelofibrosis and leukaemia). One of these is the ratio between myeloid precursors and erythroid precursors (the M/E ratio), which gives us information on the relative growth patterns of different cell growth in health and disease. However, a sample of bone marrow may also be used to look for infiltration by metastatic tumours from a distant primary site such as lung, prostate or breast carcinoma. In these cases the M/E ratio will generally be unaltered. Additional details of the importance of bone marrow and its analysis are presented in Chapter 6.

In practice, the only method of assessing haemopoiesis is to examine the contents of the bone marrow. A sample of bone marrow is obtained via the process of bone marrow aspiration, and can be achieved in one of two ways.

• The first is an extension of the method for obtaining a peripheral blood sample by venepuncture, but in this case a heavy gauge needle is driven part of the way into the bone (such as the sternum or the iliac crest) to aspirate some marrow.
• The second method is to deliberately harvest bone tissue itself in the form of a trephine biopsy, which also provides a view of the internal architecture of the bone marrow. This process demands a considerably more robust needle and an anaesthetic.

Aspirated bone marrow can be smeared on a glass slide and stained as if a sample of peripheral blood, for example by a Romanowsky stain such as one of those pioneered by Jenner, by Giemsa or by Leishman. Once stained, bone marrow examination can provide useful information as to the relative proportions of the different types of cells in different stages of maturation, such as the myeloid/erythroid ratio. Particular white cell precursors may be recognized by certain characteristics, such as the presence of a prominent nucleolus in a myeloblast. However, infiltrating (cancer) cells and even infections (by bacteria or fungi) can be detected.

As a trephine biopsy provides actual bone, this tissue must be processed as if a piece of normal tissue. This will require decalcification before examination, and the trephine biopsy is often processed in the histology laboratory (Figure 3.2).

Much of our knowledge of growth factors such as erythropoietin has been obtained from tissue culture experiments. Bone marrow harvested from volunteers can be grown and characterized in tissue culture, and so be used as a model for the effect of various CSFs on stem cells. For example, incubation of bone marrow with a certain highly purified cocktail of growth factors may well result in the growth or maturation of granulocytes, but not of monocytes. However, in this highly artificial system, we cannot be sure if the events we witness are truly representative of the situation within our own bone marrow in either health or in disease. Nonetheless, use of exogenous CSFs such as thrombopoietin as a specialized therapy in those with a failing or stressed bone marrow (as in cases of chemotherapy, aplastic anaemia or following bone marrow transplantation) is of undoubted benefit.

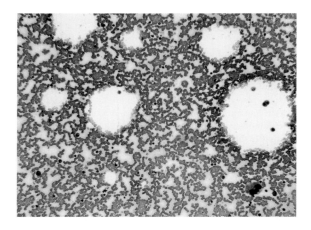

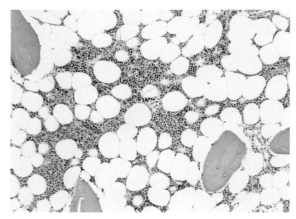

Figure 3.2 Top: A bone marrow aspirate spread on to a glass slide, dried, and stained as if a sample of peripheral blood. Middle: A trephine sample that retains the architecture of the bone marrow. Lower: For comparison, a sample of peripheral blood at the same low power magnification. Note the markedly fewer number of white cells in the latter. (Courtesy of D. Bloxham, Cambridge University Hospitals NHS Foundation Trust).

Special investigations

Two different types of special investigations can provide crucial information regarding the pathology of different cells in the peripheral blood and also in the bone marrow. The first, cytochemistry, utilizes the components of the cytoplasm (generally granules) in chemical reactions in a system akin to conventional haematological staining, such as with the Romanowsky stains. The second, flow cytometry, assesses cell populations according to the presence of different molecules on the surface of the cell, although it can be modified to look at molecules within the cytoplasm.

Cytochemistry This technique can be used to study both red cells and white cells, although the focus is on the latter. Cells are spread onto a glass slide, dried and then subjected to various dyes, chemical and buffers. Developed and refined over decades, this technique is generally used to probe peripheral blood and bone marrow for the presence of different cytoplasmic enzymes and other chemicals. Different patterns of staining can be used to classify particular leukaemias. Without doubt, the greatest use of special stains is in investigating malignancies such as leukaemia, full details of which are present in Chapters 5 and 6. The most common staining techniques focus on enzymes within different cells, such as:

- myeloperoxidase, an enzyme that is present most strongly in granulocytes and their precursors, but it may also be present in red cells and their precursors;
- neutrophil alkaline phosphatase, an enzyme found predominantly in mature neutrophils;
- acid phosphatase, generally used to identify certain lymphocytic leukaemias in peripheral blood, and macrophages and megakaryocytes in bone marrow; and
- the esterases, a family of enzymes that are generally used to investigate different leukaemias.

There are also stains named after particular dyes or the scientists primarily responsible for their development. These include:

- Sudan black, which stains a component of the granules of granulocytes and monocytes;
- periodic-acid-Schiff stains glycogen and related polysaccharides; granulocyte precursors stain weakly, if at all, and mature neutrophils show intense staining.

Perls' stain detects iron-containing molecules inside the cell, and if found in a red cell it is called a siderocyte. If applied to a bone marrow aspirate the stain can

Table 3.2 Commonly used CD markers.[a]

CD molecule	Cell identified
CD3	T lymphocyte
CD14	Monocyte
CD15	Granulocyte
CD19 or CD20	B lymphocyte
CD34	Stem cell
CD61	Resting platelet
CD69P	Activated platelet
CD235	Red cell

[a]Other CD molecules on red cells are discussed in Section 3.4.

identify iron within red cell precursors (erythroblasts) and so are named sideroblasts. The importance of this stain will be explained in Chapter 4.

Flow cytometry   This technique relies on a complex analytical machine (the flow cytometer) and sophisticated reagents (monoclonal antibodies conjugated to a fluorochrome). Together, they define different populations of cells in the peripheral blood and bone marrow according to the presence of certain molecules on the surface of the cell. However, flow cytometry can also investigate molecules in the cytoplasm.

A common use of flow cytometry is to investigate different leukaemias. Most cell surface molecules (and some intracellular molecules) are classified according to an internationally agreed system: the 'cluster of differentiation', abbreviated CD. Those predominantly used in white cell investigations are CDs 1–25 (and possibly elsewhere), whereas many red cell molecules are classified between numbers 230 and 250. However, some CD molecules are tightly specific for a particular cell, while others are found on many different cells.

In practice, many laboratories will focus on a panel of only a few selected monoclonal antibodies that can be used to screen for the most common pathological conditions. However, care is required as the expression of a particular CD molecule may well be altered in diseases such as leukaemia. For example, the expression of markers can be different in cancer cells than on normal cells. The most well-known molecules of this system, and the cells on which they are present, are shown in Table 3.2.

The importance of CD molecules for the investigation of leukaemia and other haematological neoplasia are presented in Chapters 5 and 6, whilst flow cytometry for lymphocyte subsets in immunology is discussed in Chapter 9.

Blood science angle: Flow cytometry

This technique is a classic example of a new method finding more and more applications in routine pathology. Developed in the 1970s as a research tool, it was first used in clinical pathology to explore the different expression of CD molecules on lymphocyte subsets, such as are important in the transition from human immunodeficiency virus infection to acquired immunodeficiency syndrome (Chapter 9). As such, it was firmly in the realm of the immunologist. However, it was soon adapted for use by haematologists in leukaemia analysis (Chapter 6), and then in investigating molecules on the surface of red cell diseases (Section 3.4 and Chapter 4) and platelets (Chapters 7 and 8). Microbiologists can also use it to define various microbes, such as bacteria and certain parasites. Flow cytometry can also be used outside 'routine' pathology by many other healthcare professionals, such as oncologists looking at the cell biology of cells derived from tumours.

3.3 Erythropoiesis

In this section we will look at how the red cell develops, erythropoiesis being the red cell part of haemopoiesis. The earliest stem cells (bearing CD34) in this process ultimately give rise to red cells, but also some white cells and precursors to the platelet lineage. However, the earliest red-cell-specific stem cell is the CFU-E, from which arises the first and largest (14–19 μm diameter) morphologically-recognizable erythrocyte precursor: the proerythroblast. Further steps are marked by a steady reduction in the size of the cell through to the normoblast (or erythroblast, 10–15 μm) stage, where synthesis of haemoglobin begins.

The next step is the nucleated red cell, which then loses its nucleus to become a reticulocyte, the final immature step. Indeed, one may describe the reticulocyte as the 'teenager' to the 'adult' of the mature red cells, the latter having a diameter of 7–9 μm. The reticulocyte can also be recognized because it often has remnants of haemoglobin messenger ribonucleic acid (RNA), and as such can be identified using the dye brilliant cresyl blue.

Reticulocytes can also be identified in the adult by the presence of foetal haemoglobin (as may be present after a woman has had her baby, and some of the baby's blood has got into the maternal circulation). Furthermore, the reticulocyte often stains more bluish with conventional stains than mature red cells do, a feature described as polychromasia. The final step is the transformation of

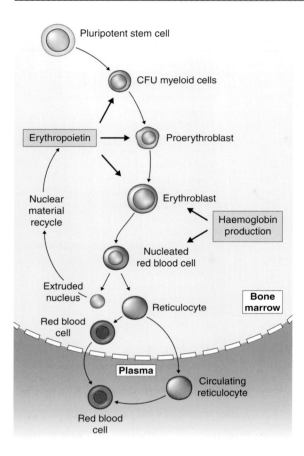

Figure 3.3 Erythropoiesis. Stages in the development of the red cell in the bone marrow. Early stages involve the derivation of the lineage-specific CFU for myeloid cells, which develop into proerythroblasts and erythroblasts under the influence of erythropoietin. The nucleated red blood cell loses its nucleus to become a reticulocyte, and then the mature red cell.

the reticulocyte into the mature red cell, a process taking 3–4 days (Figure 3.3).

Thus, the key steps in the development of the red cell from the blast cell to the fully functioning erythrocyte involve a slow and steady reduction in size, with loss of the nucleus. This process occurs in parallel with the development of haemoglobin, which becomes more and more prominent as the red cell approaches adult life as a mature cell. Several crucial steps in erythropoiesis depend on a red-cell-specific growth factor – erythropoietin – whose levels are often regulated by the amount of oxygen in the tissues. Low oxygen (hypoxia) will promote the generation of this hormone, and thus an improvement in the red cell count.

Insufficient erythropoietin will result in failure to stimulate red cell precursors in the bone marrow, and so a reduction in the production of red cells, leading to the major disease of red cells: anaemia. Conversely, too much erythropoietin can lead to the overproduction of red cells (polycythaemia). However, polycythaemia can result from problems with the process by which erythropoietin acts on the nucleus, via the intracellular second messenger JaK2, a process we will discuss in Chapter 4.

However, erythropoietin is not the only growth factor influencing red cell production. The male sex hormone testosterone also acts as a red cell growth factor, and so partially explains why men have more red cells and haemoglobin, and a higher haematocrit, than do women. Fully functioning erythropoiesis also requires a host of micronutrients and hormones. Furthermore, erythropoiesis (actually, all haemopoiesis) is suppressed by an inflammatory response, so that anaemia may result from a deficiency of any of these hormones and micronutrients, or by chronic inflammation, as will be explored in Chapter 4.

The process of erythropoiesis is generally very tightly regulated so it generates approximately 5×10^{10} erythrocytes each day, enough to replace those cells destroyed by a particular disease process or simply by old age. However, this number may change – many diseases of the bone marrow (such as leukaemia, myeloma and myelofibrosis) lead to reduced production of red cells, and so to the principal disease of these cells, that is anaemia. Increased numbers of red cells can also be produced by the bone marrow. When an increased red cell count is the response of the bone marrow to excessive red cell destruction, the increased count is said to be erythrocytosis. On the other hand, an increased red cell count may be the result of a leukaemia-like disease called polycythaemia. This particular disease is characterized by abnormalities in the stem cells within the bone marrow. Additional details of both these conditions will be presented in Chapter 4.

Blood science angle: Erythropoietin

Erythropoietin is produced almost completely by the kidney. It follows that if there is damage to this organ, there may also be reduced erythropoietin produced, and so anaemia may develop. Consequently, blood scientists will need to be aware of methods of assessing renal function. This can be done with the standard biochemistry test of urea and electrolytes, high levels of which are associated with renal failure (Chapter 12). To complete the link between anaemia and renal failure, erythropoietin itself can be measured in a blood sample, a test which may be performed in the haematology, biochemistry or immunology sections.

3.4   The red cell membrane

A 'normal' cell communicates with other cells and receives messages via its cell membrane. However, the red cell is so highly specialized for oxygen transport that it does not need a complex membrane. Indeed, the only major requirement of the red cell membrane is that it presents no barrier to the movement of oxygen, so from this perspective it needs to be as simple as possible. The red cell membrane has three components:

• a double layer consisting of phospholipids and cholesterol;
• various proteins and glycoproteins – some are expressed only on the external or internal surfaces of the cell, whilst other larger molecules traverse the membrane and have sections on both surfaces;
• an internal cytoskeleton that gives the cell its characteristic round but flattened shape – a biconcave disc.

The membrane is a bilayer (50% protein, 40% lipids, 10% carbohydrates), but the two layers are different. The external layer is rich in phosphatidyl choline and sphingomyelin, whilst the internal layer is mostly phosphatidylserine and phosphatidylethanolamine, although they can 'flip-flop' between the two faces. Cholesterol is believed to be equally distributed. The fluidity of the lipid component of the membrane is crucial in facilitating the lateral movements of floating 'rafts' made up of complexes of glycoproteins that include those defining the blood group.

On the external surface are dozens, perhaps hundreds, of different molecules whose functions include ion and gas transport and adhesion, whilst others are enzymes. Collectively, the molecules on the outside of the cell are called the glycocalyx. However, the most well-known red cell structures are of course those that define the blood groups. Principal among the transmembrane glycoproteins are Band 3, glycophorin A (GPA, expressing blood group structures A and B), glycophorin C (GPC) and Rhesus-associated glycoprotein, the function of many of which have been discovered; see Table 3.3, which also lists other membrane components.

Other molecules that have yet to be included in the CD system can be classified broadly by function, such as a group of ion transporters. These include aquaporin 1 (transporter of water, oxygen and carbon dioxide, linked to blood group Colton), aquaporin 3 (transporter of water and glycerol, linked to blood group Gill), Glut-1 (glucose and L-dehydroascorbic acid transport), Kidd

Table 3.3 Glycoprotein components of the red cell membrane.

CD molecule	Other name, and function
CD35	Complement component 3b/4b receptor
CD55	Decay accelerating factor – a complement regulatory protein
CD59	Membrane inhibitor of reactive lysis – a complement regulatory protein
CD47	Part of the Band3–Rh complex
CD151	Might associate with integrins and laminins
CD233	Anion exchanger (bicarbonate/chloride), part of Diego blood group
CD234	Duffy blood group molecule; receptor for chemokines
CD235 A–D	Glycophorins A, B, C and D, link the membrane to the internal cytoskeleton
CD238	Kell glycoprotein, a peptidase that cleaves endothelin-1
CD239	Lutheran blood group; binds to laminin
CD240D	The D component of the Rh complex
CD240CE	The CcEe component of the Rh complex
CD241	RhAg: required for rhesus molecule expression
CD242	Intercellular adhesion molecule-4, an integrin-binding protein, associated with blood group LW

antigen (urea transport) and several $Na^+/K^+/Cl^-$ transporters. These molecules are concerned with maintaining the balance between the levels of ions and the molecules in the cells compared with levels in the plasma. Other red cell membrane components include enzymes such as Na^+/K^+-ATPase, Ca^{2+}-ATPase, carbonic anhydrase, acetylcholinesterase, 5-nucleotidase and alkaline phosphatase. An important component is glycosyl phosphatidylinositol, which anchors other molecules within the membrane.

The organization of the membrane

The cell surface and cross-membrane molecules associate into two different groupings which interact with the major structural molecules inside the cell (alpha and beta spectrins) with, respectively, the Band 3–Rh complex and the GPC–4.1R complex (Figure 3.4). Notably, a further example of the unusual nature of the red cell membrane is that it lacks human leukocyte antigen (HLA) molecules present on all nucleated cells. The importance of HLA molecules is explained in Chapter 10.

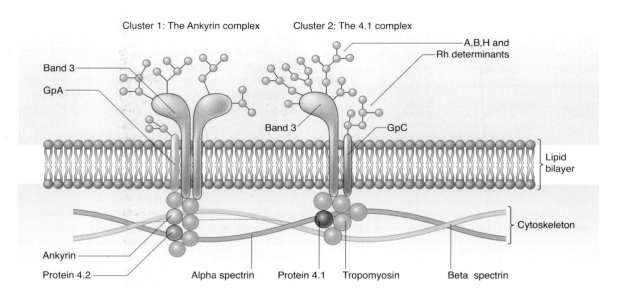

Figure 3.4 The red blood cell membrane. Our current view of the red cell membrane can be explained in this cartoon. The ankyrin complex (left) includes transmembrane molecules Band 3 and GpA that span the lipid bilayer. The intracytoplasmic tails link to ankyrin and protein 4.2, and thus the internal cytoskeletom on alpha and beta spectrins. The 4.2 complex (right) also include Band 3, but also GpC. These molecules link to protein 4.1 and tropomyosin and also the spectrins. Not shown for clarity are molecules such as Duffy and Kell.

Cluster 1   The Band 3–GPA complex links with the intracellular cytoskeleton of alpha and beta spectrins via ankyrin and protein 4.2. The complex may also contain other molecules, such as RhAg, CD47 and GpB. The cytoplasmic tails of these molecules interact with a complex formed from protein 4.2, ankyrin and the enzymes aldolase and carbonic anhydrase, the latter involved in bicarbonate transport. These molecules link with the major proteins alpha and beta spectrin which make up the 'skeleton' of the red cell, suggesting a role in the shape of the red cell.

Cluster 2   A collection of other glycoproteins, Band 3, GPC, Kell, Duffy and Glut 1, form a second complex. The intracellular parts of these molecules interact in turn with other molecules that include adducin, actin, tropomyosin and 4.1R, which themselves are linked to the cytoskeleton.

The cytoskeleton and cell volume   The shape of the red cell is defined by its cytoskeleton of structural proteins: the alpha and beta spectrins. These are crucial in giving the cell its flexibility and its unique double concave shape, enabling flexibility and strength. The cytoskeleton also needs to ensure that the MCV of the cell is maintained between perhaps 77 and 99 fL. This in turn keeps

the haemoglobin concentration of the cell within its own range, generally 315–365 pg/L.

A reverse gradient exists between sodium in the plasma (high) and in the cell (low), and between potassium in the plasma (low) and in the cell (high). The laws of osmosis are constantly trying to equalize these concentrations, so that the cell has to work hard to maintain this gradient, which it does with ion pumps and transporters, as described above. A further complication is the presence of chloride and bicarbonate ions, and the need to balance not only the ionic status, but also the pH of the cell. Some of these pumps require energy in the form of ATP, and other enzymes demand calcium as a cofactor. Therefore, the cell has to perform many complex reactions in parallel if its integrity is to be maintained. It follows that problems with these molecules lead to defects in the membrane, which inevitably mark it out for removal. We will revisit the consequences of these defects in Chapter 4.

Consequences of membrane specialization   The high degree of specialization of the red cell membrane for the ease of oxygen movement means that it is far more fragile than the membranes of nucleated cells. This leads to susceptibility to extremes of homeostasis, such as pH and temperature changes, which the more complex membrane of a nucleated cell could resist. In the laboratory, this extra fragility can be quantified by the response of the cell to different concentrations of sodium chloride in the osmotic fragility test.

Blood science angle: Blood groups

In the past, blood groups were simply annoying structures on red cells that frustrated free and simple blood transfusion. Frustrating because, in many cases, the body responds to an alien red cell as it would do to a pathogenic microbe, by producing antibodies designed to destroy the invader. Any such foreign molecule prompting an antibody response is called an antigen. We now know that blood group structures, of which there are over 30 major types, have crucial roles in maintaining the physiology of the cell, and this knowledge has brought a new focus into the blood transfusion laboratory (Chapter 11).

3.5 The cytoplasm of the red cell

The function of the red cell is to transport oxygen, which it does by the highly specialized protein haemoglobin. However, high concentrations of oxygen may oxidize, and so damage, proteins, fats and carbohydrate. Consequently, the red cell has developed antioxidants, such as glutathione (GSH), to protect itself. The red cell cannot use oxygen to generate energy as it lacks mitochondria, but it has metabolic pathways that enable it to generate a limited amount of energy anaerobically.

Haemoglobin

Haemoglobin is an iron-containing protein of relative molecular mass 68 kDa synthesized by the erythroblasts and other bone marrow precursors. Over 600 million molecules are present in each mature red cell. It absorbs oxygen from areas of high oxygen content (i.e. at the lungs, where it becomes oxyhaemoglobin) and then releases it in areas were oxygen levels are low (e.g. in the tissues), at which point it may be described as deoxyhaemoglobin. Haemoglobin has two parts: a protein part (globin) and a complex non-protein part (haem) that contains the iron.

Haem

Haem is a complex molecule synthesized partly in the cytoplasm and partly in the mitochondria of erythroblasts and nucleated red blood cells.

The initial step in the formation of haem is in the mitochondria, where glycine and succinyl-CoA are converted into aminolevulinic acid (ALA) by the enzyme ALA synthase. ALA then passes to the cytoplasm where it is converted into the pyrrole structure porphobilinogen by the enzyme porphobilinogen synthase. Four porphobilinogen molecules are amalgamated and subsequently converted, via a series of complex metabolic steps, into coproporphyrinogen III, protoporphyrinogen IX and, in the mitochondria, into protoporphyrin IX. The final step is the insertion of an atom of ferrous (Fe^{2+}) iron into the protoporphyrin ring by the enzyme ferrochelatase to create haem. This pathway is summarized in Figure 3.5.

Ferrochelatase is often unable to distinguish zinc from iron, and so sometimes inserts zinc into the protoporphyrin ring. Thus, zinc protoporphyrin may be taken to be an indirect marker of the levels of iron within the erythroblast.

Several of the enzymatic reactions leading to the synthesis of haem require micronutrients that must be provided in the diet, such as vitamin B_{12}, vitamin B_6 and

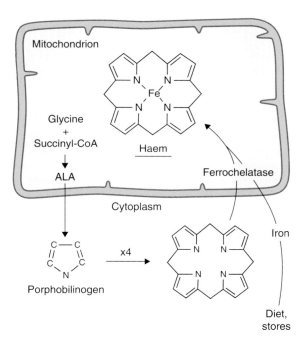

ALA = Aminolevulinic acid

Figure 3.5 Major steps involved in the synthesis of haem in cells such as the erythroblast and nucleated red cell. On the left, within the mitochondrion, glycine and succinyl-coA form aminolevulinic acid, which leaves the mitochondrion and is converted into porphobilinogen, four of which form uroporphyrinogen. The latter enters the mitochondrion, and the enzyme ferrochelatase inserts iron, forming haem.

folate. It follows that lack of these micronutrients leads to poor haem production, and so anaemia. Different anaemic conditions arising from lack of these micronutrients and problems with enzymes are described in Chapter 4.

Blood science angle: Micronutrients

Fully effective erythropoiesis demands trace amounts of essential micronutrients, which include iron, cobalt, vitamin C, copper, vitamin E, vitamins B_6 and B_{12}, thiamine and riboflavin, in addition to the thyroid hormone thyroxine. These can be measured in the blood, but the physical site of the analysis often varies from hospital to hospital: in some the analyses are done by haematologists, and in others by biochemists. However, thyroid function tests are usually performed in the biochemistry laboratory, whilst vitamin analyses are sited within biochemistry, as vitamin D is an important micronutrient rarely called for by haematologists.

Iron

This essential micronutrient is absorbed in the duodenum and jejunum. However, in a diet rich in meat and liver, iron may be locked within haem and other complexes, from which it may be harvested, and the acid environment of the stomach promotes this release. Most well-balanced diets deliver 15–25 mg of iron daily, of which perhaps 10% is required, and so provide sufficient iron to replace that lost in dead cells and by excretion, although this varies with age and between the sexes (Table 3.4).

Notably, healthy women need to replace the iron in the blood lost by menstruation, and provide iron to their child if pregnant or breastfeeding, and to replenish their own stores. Other foods with the highest proportion of iron include eggs, grains (such as lentils) and fruit. Iron in vegetables (high levels being found in broccoli and kale) is mostly ferric (Fe^{3+}), whereas ferrous iron (Fe^{2+}) is more common in animal material.

Table 3.4 Iron requirements.

Physiology	Daily requirement (mg)
Infant, adult male, post-menopausal female	1
Adolescent, menstruating adult	2–3
Menstruating adolescent, pregnancy	3–4
Lactation	1.5–2.5

Free iron is absorbed actively by intestinal epithelial cells (enterocytes) via a cell membrane molecule, divalent metal transporter-1, which preferentially absorbs ferrous iron. Ferric iron is converted to ferrous iron by ferro-reductase enzymes at the surface of the enterocyte, whilst haem has its own transporter, called HCP-1.

Once inside the enterocyte, iron within haem is liberated by the enzyme haemoxygenase. Some iron may be stored in ferritin, but most passes promptly into the plasma through an exporting protein – ferroportin. However, this export of iron is regulated by a small hormone-like peptide called hepcidin. This 25-amino acid molecule is secreted by the liver in response to the amount of iron in the circulation and that which is stored in intracellular deposits. It also acts to regulate the release of iron from macrophages, Kupffer cells and hepatocytes. Figure 3.6 summarizes the regulation of iron movement from cells into the plasma.

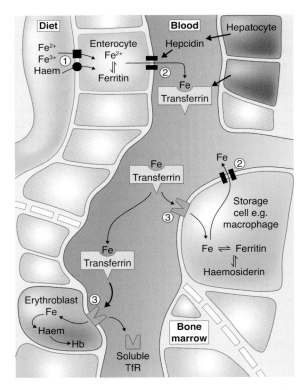

Figure 3.6 Regulation of iron uptake, and its fate. Iron in the diet (1) passes through the enterocyte and is carried in the blood by transferrin (2). In the erythroblast in the bone marrow it is incorporated into haemoglobin (3); in other cells such as the hepatocyte or macrophage, it is stored in ferritin and haemosiderin (3). Tfr: transferrin receptor.

Once in the circulation, free iron is collected by hepatic carrier proteins albumin, lactoferrin and (preferentially) transferrin. Ferric iron is the preferred ion and is converted into this form from ferrous iron by enzymes such as hephaestin and the copper-containing caeruloplasmin. When free of iron, transferrin, a 76–80 kDa protein, is more correctly termed apotransferrin, and many use the expression holotransferrin to denote transferrin actually carrying iron. Each transferrin molecule can transport two atoms of iron, and 1 g can carry 1.25 mg of iron. The amount of iron being carried by transferrin can be a useful indication of the general iron status of the body and can also regulate iron uptake. This is because the saturation of transferrin with high levels of iron will stimulate hepcidin release, which in turn reduces levels of iron passing from the intestines to the blood.

Transferrin delivers iron to the red cell precursors such as erythroblasts in the bone marrow, the interaction being via the transferrin receptor (TfR), of which there can be up to 50000 on each cell. The TfR itself can be shed from the erythroblast surface and be detected in the plasma as the soluble transferrin receptor (hence sTfR). Since levels of sTfR reflect levels of the membrane-bound form, measurement of the sTfR provides a surrogate for increased activity of the erythroblasts and possibly of erythropoiesis itself. However, the TfR is found on other iron-storing cells, such as macrophages and hepatocytes, so caution in interpretation may be required. Once inside the cell, iron is released from the transferrin molecule in an endosome, and the resulting apotransferrin is returned to the plasma from where it can travel back to the intestines to collect more iron.

Iron not destined immediately for erythropoiesis may be stored in organs such as the liver, pancreas and spleen, complexed with proteins ferritin and haemosiderin. Ferritin is a very large spherical molecule (relative molecular mass 450 000–465 000 Da) that can store up to 5000 atoms of ferric iron. Ferritin free of iron is called apoferritin. Haemosiderin, formed by the aggregation of partially digested ferritin, is mostly found in macrophages and hepatocytes. Iron stored in haemosiderin cannot be released quickly and, therefore, represents long-term deposits that cannot readily be accessed. In this form it can be detected using light microscopy by the special stain Perls' Prussian blue. This stain is particularly valuable in assessing iron studies in the bone marrow in both iron deficiency and iron excess. Most (65%) body iron (generally 3–4 g) is in haemoglobin, transferrin, myoglobin and enzymes, but perhaps a third is stored in ferritin and haemosiderin.

> ### Blood science angle: Iron genetics
>
> Several genes have an impact on iron metabolism. Hepcidin is coded for by the *HAMP* gene found on chromosome 19, and can be regulated not only by levels of iron, but also by inflammatory cytokines such as interleukin-6. Several mutations in the *HAMP* are now known to be linked to juvenile onset haemochromatosis (essentially, high iron levels), whilst mutations in *HFE* lead to hereditary haemochromatosis and other problems with iron metabolism. Curiously, *HFE* is located very close to the genes for HLA molecules, as discussed in Chapter 10, and so is of interest to immunologists. Molecular geneticists have also discovered a mutation in a gene coding for ferroportin that confers resistance to hepcidin, so that the passage of iron into the plasma is unregulated. Consequently, many regional genetic services are now able to offer a service to confirm or deny these and other mutations, and so provide assurance regarding pathophysiology. The further implications of these 'iron' gene mutations will be discussed in Chapter 4.

Vitamins

The essential vitamins B_6, B_{12} and folate are needed in steady state throughout life in both sexes (unlike iron), although more is required in pregnancy. The daily requirement of this vitamin B_{12} is in the region of 2 μg (more in pregnancy), which should be easily satisfied by a mixed diet (meat, milk, eggs, fish) providing about 10–30 μg, of which perhaps 2–3 μg are absorbed. The liver stores perhaps 1.5 mg, a quantity which equates to the daily requirement for up to 5 years.

Metabolically active vitamin B_{12} is almost absent from the plant kingdom, being synthesized by bacteria. Consequently, strict vegans and those on a macrobiotic diet may not achieve their daily requirement unless they take supplements. In order to be absorbed, dietary vitamin B_{12} must be dissociated from the ingested foodstuffs, which requires the enzyme pepsin (and possible trypsin) and hydrochloric acid in the stomach. Specialist gastric parietal cells produce intrinsic factor, a 45 kDa glycoprotein, which forms a complex with vitamin B_{12} in the upper part of the small intestine. The complex passes to the distal or terminal section of the ileum where it interacts with a complex and specific receptor on the enterocyte, and so is absorbed.

Once absorbed into the bloodstream, vitamin B_{12} is carried to the bone marrow by plasma proteins transcobalomines I and II. Once delivered to the erythroblast, vitamin B_{12} uncouples from transcobalamine to link with key enzymes in major metabolic pathways in the synthesis of haem.

Vitamin B_{12} can be found as a number of isomers: adenosylcobalamin is the form required by ALA synthase, whilst cyanocobalamin is needed by the enzyme that converts homocysteine and methyl-tetrahydrofolate (for which folate is also required) into tetrahydrofolate and methionine, an essential component of the synthesis of DNA. All vitamin B_{12} isomers are based on a common pattern of a prophyrin ring (like haem), but with an atom of cobalt at the centre instead of iron.

Folate is found widely in plant and animal foodstuffs; the daily requirement is approximately $100–200\,\mu g$, which should be provided by a mixed diet which is able to provide $200–300\,\mu g$. Vitamin B_6 is also found widely in a mixed diet and, like folate, is relatively easily absorbed.

The B vitamins are required by two mitochondrial enzymes. Vitamin B_6 (also known as pyridoxine) is a cofactor for the enzyme that generates succinyl-Coenzyme A, whilst vitamin B_{12} is required by ALA synthase, the enzyme that effectively fuses glycine with succinyl-Coenzyme A, so generating the substrate for porphobilinogen (Figure 3.5). The consequences of the deficiency of these micronutrients will be discussed in Chapter 4.

Globin

Globin is a globular protein with a mass of $16–17\,kDa$, existing in slightly different forms coded by different genes. Two genes, for alpha and zeta globin, are on chromosomes 11, whilst the genes for epsilon, beta, delta and gamma globin are on chromosome 16. These genes each produce specific messenger RNA that in turn generates different types of globin protein molecules. For example, the alpha globin chain contains 141 amino acids, whereas the beta globin chain has 166 residues. The gene for gamma globin has two sub-types, each coding for a globin molecule with a slightly different amino acid component.

Synthesis of globin molecules occurs in the cytoplasm for red cell precursors such as the erythroblast and the nucleated red blood cell in parallel with the synthesis of haem; the two are then combined to form haemoglobin. The mature haemoglobin molecule is a tetramer of four individual monomer globin molecules and a haem ring, so that the entire molecule has a relative molecular mass of approximately $64–68\,kDa$. The different globin

monomer molecules that come together in different types of haemoglobin confer different oxygen-carrying properties on the tetramer. This is important in early life in the uterus and as a neonate.

Embryonic and foetal haemoglobin   The precise make up of the globin tetramer evolves from the embryo and foetus, via the neonate, to the adult. Embryonic haemoglobin is dominant for the first 12 weeks of life, and consists mostly of alpha, zeta, epsilon and gamma globins – making up haemoglobins Portland, Gower I and Gower II. Embryonic haemoglobin gives way to foetal haemoglobin, comprising alpha and gamma globin chains. These different haemoglobin species have evolved because the embryo and foetus is faced with the challenge of obtaining its oxygen not from its lungs but from the placenta, and thus ultimately from the maternal circulation.

Hence, embryonic, and subsequently foetal, haemoglobin must effectively take oxygen from the adult haemoglobin of the mother. It does this by having a considerably greater affinity for oxygen than does the maternal haemoglobin. The foetus and neonate face a similar challenge as it develops, which it partly solves by introducing delta globin in the haemoglobin molecule in the place of gamma globin. The major final switch from fetal to adult haemoglobin in the neonate occurs gradually some 3–6 months after birth as beta globin slowly replaces delta and gamma globin (Table 3.5 and Figure 3.7).

Adult haemoglobin   The dominant haemoglobin species in the adult, making up perhaps 96–98% of this protein, is HbA. This consists of two alpha globin molecules and two beta globin molecules. A minor form of HbA is HbA_2, which has two alpha chains (as does HbA) but has two delta globin chains in place of the beta globin chains of the HbA. This second species makes up about 2% of all haemoglobin in the healthy adult. The remainder, HbF, comprises two alpha and two gamma

Table 3.5 Globin chains in haemoglobin variants.

Growth stage	Globin chains	Haemoglobin
Embryo	2 × epsilon, 2 × eta	Gower 1
	2 × gamma, 2 × epsilon	Portland
	2 × alpha, 2 × eta	Gower 2
Foetus	2 × alpha, 2 × gamma	Hb F
Adult	2 × alpha, 2 × delta	HbA_2
	2 × alpha, 2 × beta	HbA

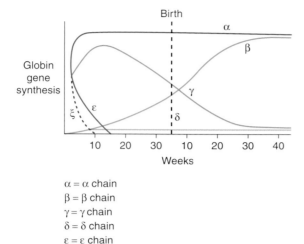

α = α chain
β = β chain
γ = γ chain
δ = δ chain
ε = ε chain
ξ = ξ chain

Figure 3.7 Haemoglobin (Hb) development. Changes in species of Hb in different stages of development in the embryo (weeks 0–10) where zeta, epsilon and alpha globin molecules are synthesized. As the embryo grows into the foetus the zeta and epsilon chains give way to gamma globin, whilst close to birth the beta globin molecules take over from gamma globin. From perhaps 30 weeks, delta globin genes are active. The gamma globin genes slowly shut down so that after 30 or 40 weeks of age only a trace remains, the dominant molecules alpha and beta chains that form HbA. (From Iles & Doherty (eds) 2012, Fig. 8.26, p. 346. Reproduced with permission of John Wiley & Sons, Ltd.).

globin chains, and makes up the remaining small fraction (less than 1%). Thus, the healthy adult is still expressing genes that were most active in foetal and neonatal life. Notably, therefore, all adult haemoglobin consists of two alpha globin molecules paired up with a second type of molecule, which can be either beta, delta or gamma globin (Figure 3.7).

Other haemoglobin species Several other types of haemoglobin can be detected which have arisen not from different genes but by changes to the molecules once in the plasma. These are called post-translational modifications. Four major types are recognized: carboxyhaemoglobin, carbaminohaemoglobin, methaemoglobin and sulphaemoglobin.

Carboxyhaemoglobin is formed by the uptake of carbon monoxide. This occurs at the same site of the haemoglobin molecules that bind to oxygen, so that carbon monoxide effectively prevents the carriage of oxygen. This happens because the affinity (which in this setting means its ability to bind particular molecules)

of haemoglobin for carbon monoxide is over 200 times that of the affinity for oxygen, and results in decreased release of oxygen to the tissues. Normally making up less than 2% of haemoglobin, increased levels are present in habitual tobacco smokers (where levels may rise to 5% of total haemoglobin). High levels of carboxyhaemoglobin can be reversed by hyperventilation with air, although this is quicker if oxygen is breathed. However, death occurs when levels of carboxyhaemoglobin exceed 80% of total haemoglobin, as may follow the inhalation of petrol engine exhaust.

Carbaminohaemoglobin is a form of haemoglobin complexed with carbon dioxide. A small amount of this gas is carried in plasma, but the greater proportion is carried in combination with water as bicarbonic acid (chemical formula H_2CO_3). However, perhaps 10% is carried as carbaminohaemoglobin.

Methaemoglobin is a variant of haemoglobin characterized by the iron being in the oxidized ferric state (i.e. as Fe^{3+}) instead of its usual reduced ferrous state (as Fe^{2+}). Methaemoglobin is continuously being formed by metabolites of oxygen (oxidants), but it can be converted back to 'normal' haemoglobin or deoxyhaemoglobin by the enzyme MetHb-reductase (centre left of Figure 3.8). Under normal conditions, levels of methaemoglobin (which is essentially inert and does not carry oxygen) do not exceed 2% of the total haemoglobin pool. This is because oxygen does not bind to Fe^{3+}. However, if the cell lacks the ability to return the ferric iron of methaemoglobin to the ferrous state of haemoglobin, then methaemoglobin levels will rise, and when this reaches perhaps 10% of total haemoglobin, clinical signs of headaches, breathing problems and cyanosis can develop.

Sulphaemoglobin, an inert nontoxic form of haemoglobin, can be produced by the action of sulphur-containing drugs such as sulphonamides. In this form, oxygen can neither be carried nor delivered, and cyanosis results when levels are 3–5% of that of total haemoglobin. Unlike methaemoglobin, sulphaemoglobin cannot be converted back to haemoglobin.

Red cell enzymes and metabolism

Red cell enzymes are synthesized whilst the developing cell still retains a nucleus, during the erythroblast and nucleated red blood cell stages in the bone marrow. However, these enzymes eventually become degraded and so lose function, which contributes to the eventual demise of the cells.

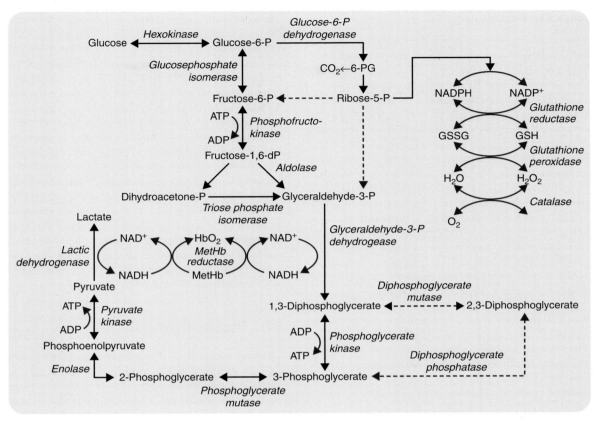

Figure 3.8 Red cell metabolic pathways. The major pathway for the anaerobic generation of energy (in the form of adenosine triphosphate (ATP), nicotinamide adenine dinucleotide phosphate (NADH) and nicotinamide adenine dinucleotide phosphate hydrogen (NADPH)) is the Embden–Meyerhof glycolytic pathway (central spine). See text for details. (From Iles & Doherty (eds) 2012, Fig. 8.19, p. 341. Reproduced with permission of John Wiley & Sons, Ltd.).

Generating energy

Mature red blood cells lack mitochondria, so are unable to obtain energy by normal aerobic respiration (i.e. using oxygen). Energy is needed to enable the cell to maintain its shape and deformability, and to resist the toxicity of oxygen. However, the cell can obtain energy by anaerobic respiration (obtaining energy in the absence of oxygen) by using the Embden–Meyerhoff glycolytic pathway (Figure 3.8, top left and centre). The substrate for this pathway is glucose. An early step in this pathway is the generation of glucose-6-phosphate. Most (90%) of the glucose-6-phosphate proceeds along the glycolytic pathway to form two molecules of glyceraldehyde-3-phosphate. However, perhaps 10% is diverted to a sub-pathway (the pentose phosphate pathway, of which ribose-5-phosphate is a member) that, with NADP, can then generate the hydrogen carrier NADPH and 6-phosphogluconate (6-PG). A further step uses glyceraldehyde-3-phosphate to generate another hydrogen carrier (nicotinamide adenine dinucleotide hydrogen (NADH)). The importance of these hydrogen carriers we shall come to shortly.

The next key step is the potential generation of the metabolic intermediate 2,3-diphosphoglycerate (2,3-DPG) in the Luebering–Rapoport shunt, the synthesis of which sacrifices the production of one molecule of ATP (Figure 3.8, bottom right). A more detailed discussion of the role of this intermediate will follow in a subsequent section on the carriage of oxygen by haemoglobin. Further along the glycolytic pathway, ATP can be generated from phosphoenolpyruvate by the action of the enzyme pyruvate kinase. The final step in this pathway is conversion of pyruvate to lactate (Figure 3.8 centre left).

Protection from oxygen

Apart from the enzymes involved in respiration, the red cell possesses a separate group of enzymes and other molecules designed to help protect it from the cytotoxic effect of oxygen and its metabolites such as the super-oxide radical. One of the principal protective systems is GSH, a tripeptide of glutamine, cysteine and glycine. In its reduced state, GSH is a buffer that can limit the effects of these reactive oxygen species and oxygen itself within the cell by effectively donating hydrogen to the oxygen, so forming water. In this process, GSH becomes oxidized (represented by GS–), and with another molecule forms the dimer GS–SG. Alternatively, GS– can combine with a sulphur group on a bystanding protein (forming, for example, GS–S-protein). These oxidized GSH species can then be converted back to the reduced GSH by the enzyme GSH reductase and the metabolic intermediate, NADPH, itself generated from glucose-6-phosphate and ribose-5-phosphate (Figure 3.8, top right).

An important component in the generation of this NADPH from glucose-6-phosphate and NADP is the enzyme glucose-6-phosphate dehydrogenase (G6PD). The substrate glucose-6-phosphate is converted into 6-PG, and then ribose-5-phosphate, which can be fed back into the metabolic pathway, or can move into another metabolic pathway. Therefore, lack of G6PD leads to the impaired generation of NADPH, which in turns lead to a lack of GSH. Low levels of GSH leave the red cell open to the toxic effects of oxygen.

The interrelationships between these molecules and their metabolic pathways are summarized in Figure 3.8. Several diseases are caused by aberrations in these pathways, as will be explained in Chapter 4.

Clinical aspects of metabolism

There are numerous clinical consequences of errors in these pathways. The importance of G6PD is demonstrated by mutations in the gene that gives rise to the intact enzyme. The consequences of these mutations are that the mutated G6PD is not only less efficient than the normal G6PD, but also that is has a shorter lifespan within the cell, giving rise to what is, in effect, G6PD deficiency. Other antioxidant enzymes include superoxide dismutase and catalase. The former helps protect the red cell by converting oxygen radicals such as superoxide to hydrogen peroxide. However, hydrogen peroxide is still dangerous, but can be neutralized by catalase.

Individuals unable to synthesize pyruvate kinase have decreased levels of intracellular ATP, which leads to early destruction of the red cell and so to a different type of anaemia. Because pyruvate kinase effectively removes the metabolites of 2,3-DPG, then the concentration of this molecule may rise and could adversely influence the way in which the haemoglobin molecule carries oxygen.

An additional example of the importance of these metabolic enzymes is of the conversion of methaemoglobin back to haemoglobin by NADH. Increased levels of methaemoglobin can therefore be a consequence of a deficiency in an enzyme that can use NADH to convert ferric iron to the ferrous iron of haemoglobin. Lack of this enzyme, NADH-linked methaemoglobin reductase, can be congenital and can lead to 10–20% of the total haemoglobin pool being methaemoglobin, a condition associated with mental handicap and cyanosis. The consequences of the deficiencies of pyruvate kinase and G6PD are anaemia (Chapter 4).

3.6   Oxygen transport

The physical and chemical properties of the haemoglobin molecule exhibit a highly specialized property for the uptake and release of oxygen. Haemoglobin (or, more accurately, deoxyhaemoglobin) needs to be able to pick up large amounts of oxygen at the lungs (where it is abundant, and so becomes oxyhaemoglobin) and subsequently to be able to deliver and give it up in those places where oxygen is scarce (e.g. in the tissues), where it reverts to deoxyhaemoglobin. The key to understanding this process is the oxygen dissociation curve.

The amount of oxygen in the blood or tissues can be quantified in terms of partial pressure (denoted pO_2), which has the units of mmHg. At a high pO_2 (such as 90 mmHg) which we expect at the lungs, all the haemoglobin should be saturated with oxygen (so oxygen saturation is as close to 100% as is practicable). Each haemoglobin molecule can carry up to four molecules of oxygen, but the process of uptake is staggered. The uptake and binding of the first molecule of oxygen by deoxyhaemoglobin increases the affinity for the binding of the second molecule. This is called 'facilitation'. Similarly, the uptakes of the third and fourth molecules are increasingly easy. Indeed, the affinity of haemoglobin for the fourth molecule of oxygen is approximately 400 times that of the first.

As the oxyhaemoglobin circulates from the lungs to the tissues where oxygen levels are low (i.e. a low pO_2 is

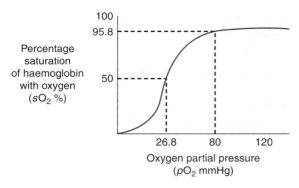

Figure 3.9 The oxygen saturation curve. The normal relationship between the partial pressure of oxygen in the blood and the degree to which haemoglobin is saturated with oxygen is given by the solid line. It is convenient to refer to the degree of oxygen saturation where 50% of the haemoglobin is saturated (i.e. the P50, where in theory each molecule of haemoglobin carries two molecules of oxygen from a maximum of four molecules). This equates to a partial pressure of oxygen pO_2 of approximately 26.8 mmHg. (From Iles & Doherty (eds) 2012, Fig. 8.23, p. 345. Reproduced with permission of John Wiley & Sons, Ltd.).

present, such as 15 mmHg), the haemoglobin is forced by laws of chemistry to give up its oxygen, and so revert to deoxyhaemoglobin. This relationship between the uptake or release of oxygen and the amount of oxygen in the blood or tissues gives the oxygen dissociation curve its typical sigmoid shape, which we can see in Figure 3.9.

Blood science angle: Blood gases

Knowledge of the relative proportions of oxygen and carbon dioxide carried by the blood is often demanded in an acute or emergency setting, but also if there is severe chronic acidosis or alkalosis. This is because these factors influence the ability of haemoglobin to carry oxygen. The blood gases test itself may be performed in the biochemistry laboratory, but also in accident and emergency and in intensive care units. A commonly used biochemical test is for the amount of bicarbonate, which may be low in cases of acidosis, and this may prompt the infusion of an intravenous drip of a bicarbonate buffer to treat the low pH. This whole area is so important that it demands its own chapter (Chapter 13).

A convenient measure of the ability to carry oxygen is the P50. This is the partial pressure of oxygen at which 50% of the haemoglobin is oxygenated, so that there are equal proportions of oxyhaemoglobin and deoxyhaemoglobin. This pO_2 is generally a little under 27 mmHg. However, this figure can vary with disease or other conditions, and is associated with a shift in the dissociation curve to the left or right.

- An increased P50 indicates a *shift to the right* and so a decreased affinity of the deoxyhaemoglobin for oxygen. In practice, this means that it is more difficult for haemoglobin to bind oxygen, so that a larger pO_2 is required to maintain the 50% oxygen saturation. However, it also means that it is easier for oxyhaemoglobin to give up its oxygen to the tissues where it is needed, such as in cases of high metabolic activity.
- Conversely, if the P50 is low, we expect a *shift to the left* in the curve, which indicates increased affinity of the haemoglobin for oxygen and so a reduction in the amount of oxygen that is required to maintain a 50% oxygen saturation. The practical consequences of this are that deoxyhaemoglobin can take up oxygen more easily, but that the oxyhaemoglobin is less willing to release it.

Factors influencing oxygen metabolism

Several factors influence the ability of haemoglobin to absorb and/or release oxygen. A decrease in pH (as in acidosis) shifts the curve to the right, whilst an increase in pH (as in alkalosis) causes a left shift. However, this effect may be influenced by carbon dioxide and the subsequent formation of the bicarbonate anion and a proton. This is important in, for example, exercising muscles, as the metabolic effects of this exercise (such as high levels of carbon dioxide and hydrogen ions – hence low pH) act to shift the oxygen dissociation curve to the right so that, for a given pO_2, more oxygen is released to the tissues.

Conversely, in the lungs, levels of carbon dioxide and hydrogen ions (hence high pH) are both low. This shifts the curve to the left so that more oxygen is absorbed by the deoxyhaemoglobin. An exercising muscle is also generally associated with increased local temperature; this acts to move the curve to the right to enable the release of oxygen to that muscle. At the lungs, where ambient air temperature is generally lower than body temperature, a decrease in temperature shifts the curve to the left, enabling oxygen uptake.

In Section 3.5 (bottom right of Figure 3.8) we briefly examined 2,3-DPG, a metabolic product of anaerobic respiration in the glycolytic pathway. This 2,3-DPG is generally present in the red cell at a similar molar concentration to haemoglobin and can influence the ability of the latter to carry oxygen, because both 2,3-DPG and oxygen compete for the same binding site in the beta globin molecule.

We have already noted that a small amount of carbon dioxide may combine with haemoglobin to form carbaminohaemoglobin, which is associated with a leftward shift in the curve. However, carbon monoxide has a far more powerful effect, as it is bound by haemoglobin far more avidly than is oxygen, and in high concentrations also causes a left shift in the curve. The same leftwards shift is also seen in the presence of high levels of methaemoglobin.

Foetal haemoglobin is a complex of two alpha globin molecules and two gamma globin molecules, whereas adult haemoglobin is composed of two alpha globin molecules and two beta globin molecules. The foetus survives in the uterus because the make-up of the foetal haemoglobin molecule provides an oxygen dissociation curve to the left relative to the maternal adult haemoglobin. In practice, this means that it is easier for foetal haemoglobin to take up oxygen than it is for adult haemoglobin. This difference enhances the placental uptake of oxygen from the maternal circulation.

In addition, the placental microenvironment is associated with a high concentration of 2,3-DPG, assisting the release of oxygen from adult oxyhaemoglobin. The gamma globin chains of foetal haemoglobin do not bind 2,3-DPG, so this metabolite does not promote oxygen release by the foetal oxyhaemoglobin.

Therefore, the biochemistry of oxygen uptake by deoxyhaemoglobin and release by oxyhaemoglobin is complex and is influenced by several factors, some of which may be competing. However, these factors are not present in isolation and they act together. In the tissues, the local pH, CO_2, 2,3-DPG and temperature all act in concert to enable the release of oxygen. At the lungs the

reverse is true, as CO_2 is expelled and oxygen taken up. These effects are summarized in Table 3.6.

3.7   Recycling the red cell

At the end of its 4-month lifespan, the red cell is destroyed and much of it is recycled. As the red cell has no DNA or ribosomes, it is unable to generate new enzyme molecules (such as G6PD and pyruvate kinase), and an important reason for the death of the cell is that these enzymes are eventually depleted. A consequence of this is loss of shape and flexibility, changes that are noted by phagocytic cells of the reticuloendothelial system, such as macrophages in the liver and spleen, which then destroy these senescent red cells. As we shall see in Chapter 4, early and/or inappropriate loss of enzymes such as G6PD can also lead to premature red cell destruction and, therefore, to anaemia.

The destruction of red cells is called haemolysis. When this process occurs in the spleen and liver it is described as extravascular haemolysis (i.e. outside the blood vessels). When it happens within the circulation it is called intravascular haemolysis (inside the blood vessels). Most of this red cell recycling (90%) is extravascular; the remaining 10% is intravascular.

If red cell destruction is intravascular, the majority of free haemoglobin is likely to form a complex with the plasma protein haptoglobin. This complex is removed by the macrophages of the reticuloendothelial system and so prevents iron being filtered by the kidney. Reduced plasma concentrations of haptoglobin may indicate intravascular haemolysis. Excess circulating haem can form a complex with haemopexin, in which form it can also be cleared by the liver. Any free haem not bound to haemopexin may be broken down by plasma haemoxygenase into iron (which may be picked up by apotransferrin), carbon monoxide and biliverdin, which is rapidly converted to bilirubin.

Table 3.6 Factors influencing the oxygen dissociation curve.

	Left shift	Right shift
Physiological effect	Increases O_2 affinity, easier for Hb to bind O_2, harder for Hb to release O_2	Decreases O_2 affinity, harder for Hb to bind O_2, easier for Hb to release O_2
pH	High pH (alkalosis)	Low pH (acidosis)
Carbon dioxide	Low	High
Carbon monoxide	High	Low
2,3-DPG	Low	High

Most of the bilirubin binds to albumin and gives our plasma and serum its mildly lemon yellow colour. In this form bilirubin is described as unconjugated, or indirect bilirubin. In the liver, bilirubin may be combined with glucuronic acid (and as such is described as conjugated, or direct bilirubin) and is then moved into the gall bladder. It is then excreted via the bile duct and duodenum and in this form gives faeces its brown colour. Intestinal bacteria can convert some of this bilirubin to urobilin and urobilinogen. However, some intestinal conjugated bilirubin may be absorbed from the large intestines and so may be found in the circulation. In this form it may be excreted via the kidney, giving the urine its characteristic light yellow colour.

Figure 3.10 summarizes the fate of the products of the red cell.

When healthy, the body is perfectly capable of clearing and recycling millions of senescent red cells each day. However, if there is excess red cell destruction (i.e. pathological haemolysis), then the recycling processes of the liver and other organs can be overwhelmed. One consequence of

this is a build-up of bilirubin which, if deposited in tissues, may lead to the yellowing coloration of the skin – jaundice (from the French for yellow). This may also manifest itself as more heavily (yellow) coloured urine and plasma. Hence, clinically evident jaundice may be present in severe haemolytic anaemia and is a clear pathological sign.

> ### Blood science angle: Jaundice
>
> One of the consequences of excess red cell destruction is a build-up of bilirubin in the plasma. However, the competent blood scientist will be aware that many other factors may cause a raised bilirubin level, and one of these causes is inflammation of the liver (hepatitis), as may be caused by a hepatitis virus. Therefore, knowledge of liver function is crucial in the investigation of jaundice, and calls for liver function tests such as alkaline phosphatase and gamma glutamyl-transferase (Chapter 17). If these two tests are normal, then one is more confident in excluding liver disease as the cause of the jaundice. But if raised, the cause of liver disease must be fully determined, and one place to start may be in sending some serum to the immunology laboratory to test for infection with a virus that causes liver disease.

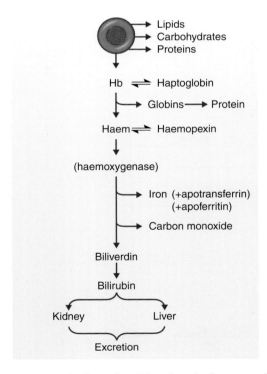

Figure 3.10 Red cell recycling. When the red cell comes to the end of its life, it is broken up and many of its components are recycled. An exception to this is bilirubin, which is excreted.

3.8 Red cell indices in the full blood count

Now that we have adequately examined the workings of the red cell, we are in a position to understand how and why particular tests of these cells have evolved and why they are important. We will take in order those test outlined in Table 3.1.

- *Haemoglobin (Hb)*. The molecule that carries oxygen. Possibly the index that is most referred to in haematology. Levels differ between the sexes, with lower levels in women.
- *MCV*. The size of the average red cell. Small red cells (i.e. those with an MCV below the bottom of the reference range) are called microcytes; large cells (above the top of the reference range) are macrocytes. The MCV is of great importance in the investigation of anaemia. However, the MCV is larger in the neonate, and falls to adult levels after perhaps 3 months. MCV does not vary with sex.
- *RCC*, or perhaps the red blood cell count, is the number of red cells in a standard volume of blood. Like haemoglobin, the RCC is also lower in women than in men, which is curiously unrelated to levels of erythropoietin, as this molecule does not vary with sex.

These three tests are often described as 'primary', because the result is obtained directly from the blood. However, by mathematical manipulation of the three primary indices we can obtain a further three indices, which may be described as 'secondary'.

- *MCH.* The amount of haemoglobin inside the average red cell; it takes no account of the size of the cell. Consequently, a small cell (microcyte) may have the same MCH as a large cell (a macrocyte). Its value is obtained by dividing the haemoglobin result by the RCC. Although both the haemoglobin and RCC results are both lower in women, the mathematics cancels this out so that MCH does not vary with sex.
- *MCHC.* The concentration of haemoglobin inside the average red cell. It is obtained by dividing the haemoglobin value by the product of the MCV and the RCC.
- *Haematocrit (Hct).* Roughly, the proportion of blood that is made up of red cells. It is defined as the product of the MCV and the RCC, divided by 10 or multiples of 10, and can be expressed as a percentage or as a decimal,

usually to two significant figures. Hct is also lower in women than in men.

These six indices are certainly the most frequently referred to, but there can be problems with the mathematics. For example, multiplying a normal RCC of 5.0 with a normal MCV of 80 and then dividing by 10 gives a normal Hct of 40%, or 0.4. The danger is that the same 'normal' Hct would be obtained by multiplying an abnormal RCC of 8.0 with an abnormal MCV of 50. Therefore, caution is advised, and the last three indices should never be used without checking the first three. Other red cell tests that can be useful in certain circumstances include the following:

- *Reticulocytes.* These are juvenile or perhaps immature red cells, increased numbers of which generally imply a pathological process. The MCV of a reticulocyte is greater than that of a mature red cell.
- *RDW.* This index offers information on the 'spread' of the sizes of the red cell pool. This is illustrated in Figure 3.11.

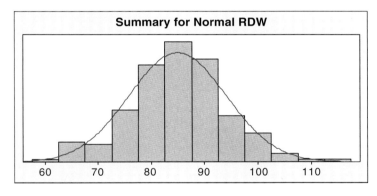

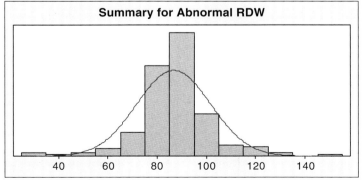

Figure 3.11 The RDW. Although both histograms show a roughly symmetrical distribution, values of the MCV in the normal (upper plot) vary from about 60 to 115, giving a mean of 84.9 and a standard deviation of 9.15, so that the RDW is $(9.15/84.9) \times 100 = 10.8\%$. Note that in the lower (abnormal) plot, although the mean MCV is a little higher, 86.7, the spread of results is much greater, ranging from 30 to 150, giving a larger standard deviation of 14.99. Hence, the RDW in the lower plot is 17.3%, outside the reference range of 10.3–15.3%.

Normally, the spread of MCV values is narrow, but if wide it implies an increase in the proportion of microcytes and/or macrocytes, which in turn implies a particular pathology.

- *ESR*. This is not a cell or a molecule, but a property of red cells. Filling a thin tube with blood and allowing it to settle over an hour will result in a small band of plasma at the top, ideally less than 10 mm. But if the column of clear plasma on top of the column of red cells is, shall we say, 25 mm, this is a sign of a pathological process. Unfortunately, many conditions (inflammation, cancer, anaemia and after surgery, to name but a few) can give rise to an abnormal ESR, so that many find the ESR result unhelpful in trying to determine an exact diagnosis (Figure 3.12).
- *Plasma viscosity*. This test simply quantifies how thick (like treacle) or thin (like water) the plasma is. The major

factors that contribute to the viscosity are plasma proteins, but the test is also strongly dependent on temperature, with lower figures if performed at 37 °C.

There are several tests of iron metabolism, the most common being total iron, transferrin, total iron binding capacity, soluble transferrin receptors and ferritin. At present, hepcidin is not a routine blood test, but may become so in the future. Apart from those related to iron, the other frequently requested micronutrient is vitamin B_{12}, as this is implicated in a common anaemia. Other tests, such as vitamin B_6, folate, haptoglobin, haemopexin, and zinc protoporphyrin may be offered in-house, especially in large teaching hospitals, or will need to be sent off to a reference or specialist laboratory.

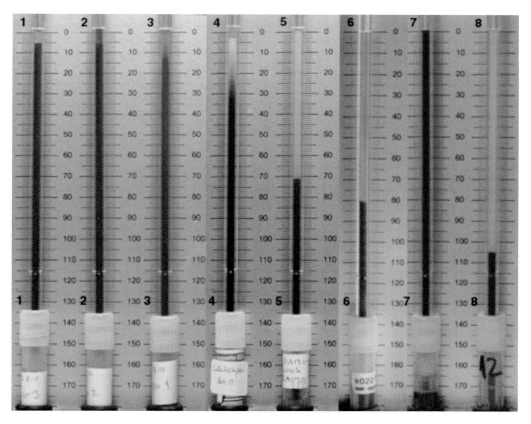

Figure 3.12 The ESR. A series of eight ESR tubes, showing a column of cloudy plasma above the column of red cells. Clearly, there is variation in the level of the red cells. In numbers 1–3 and 7 the plasma is less than 10 mm, so the result is normal. In columns 5, 6 and 8 the column of plasma is much greater, being in the region of 70 mm, 80 mm and 105 mm respectively. All these are abnormal. However, in sample 4, the cutoff is not as clear cut, but many would report a result of 30 mm. (Courtesy of Dr U. Woermann, University of Bern, Switzerland).

3.9 Morphology of the red cell

The basic morphology of a red cell, as with white cells, when viewed under a light microscope is of a disc. However, this very simple descriptor masks a great variety of different sizes, shapes and colours of different populations of these cells. These are illustrated in Figure 3.13, Figure 3.14, Figure 3.15, Figure 3.16 and Figure 3.17.

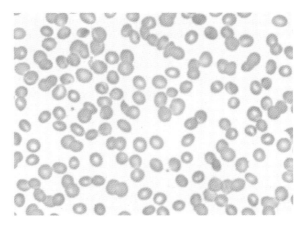

Figure 3.13 Normal red cells with a roughly uniform size and shape. Almost all of them have a small area of pallor right in the centre. (Image courtesy of H. Bibawi, NHS Tayside).

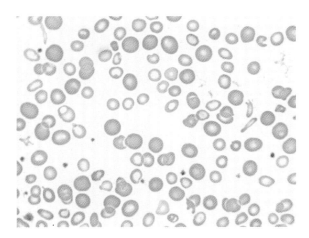

Figure 3.14 Anisocytosis. This photograph shows variety in the size of the cell; unlike Figure 3.15, there is a great variety in the size of the cells – some are clearly much larger than others; there are both macrocytes and microcytes. Many of the large cells are fully coloured, whereas many of the small cells are coloured only around the outside, with lack of staining in the middle. These small cells may therefore be hypochromic as well as microcytic. (Image courtesy of Sysmex UK).

Anisocytosis

The average red cell blood cell is generally taken to have a diameter of around 7–8 μm and to have a shape that is (ideally) perfectly round in two dimensions (Figure 3.13). However, this can vary somewhat according to various pathophysiological conditions. This variation in size is called anisocytosis (Figure 3.14). Just as a human population will have some people who are tall and others who are short, then a red cell population will have some cells which are larger than others and some which are

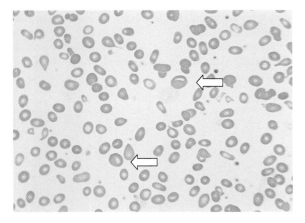

Figure 3.15 Reticulocytes. This high-magnification photograph is dominated by two reticulocytes (arrowed). They have a more 'blue' colour than the other red cells. (Image courtesy of Sysmex UK).

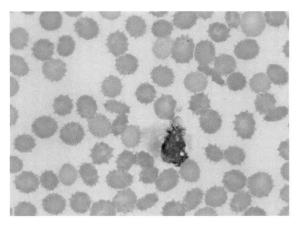

Figure 3.16 The effect of storage. Samples taken into ethylenediaminetetraacetic acid and stored incorrectly or too long will undergo changes resulting in crenation (with lots of 'spikes') of the red cells and deterioration of the white cells. (Image courtesy of Sysmex UK).

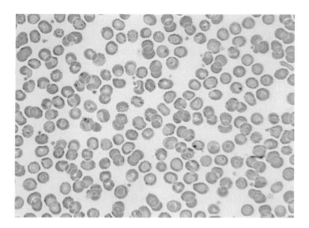

Figure 3.17 The effect of poor fixation. Poor drying and fixation results in trapped water within the cells and poor morphology. (Image courtesy of Sysmex UK).

smaller than others. Thus, anisocytosis is simply another way of looking at the RWD.

Macrocytes These cells can be defined as those over a certain size, such as 97, 98 or 100 fL, but taken by many to be >100 fL). The exact size that defines a macrocyte is generally determined by a senior haematologist, and can therefore be different at different hospitals. As we shall see in Chapter 4, a primary cause of a high MCV is to have a high intake of alcohol; another is to be pregnant, and a third is to be deficient in vitamin B_{12} or folate. However, it should be recalled that reticulocytes are slightly larger than mature erythrocytes, so that, overall, a raised MCV may be due to a large number of these immature red cells.

Microcytes Similarly, these cells may be those cells smaller than 80, 78, or 75 fL. Once again, the exact definition of a microcyte depends on local conditions. Thus, the local reference range is of relevance. Once more, this feature is not merely of academic value, as the most common cause of microcytes is iron deficiency; another is a problem with the structure of the haemoglobin molecule (haemoglobinopathy), such as sickle cell disease and thalassaemia. Thus, the size of the cell can become important, as certain types of anaemia can be identified and classified in this way.

Variation in colour

As with variation in cell size, there will also be variation in their colour. Red cells readily take up dyes and so appear as if coloured. As the stains are taken up by haemoglobin,

the density or strength of the stain can give us an idea of the amount of haemoglobin inside the cell and so of the ability of that cell to carry oxygen. Cells which fail to take up the dye (because they are deficient in haemoglobin), and so are not particularly well coloured, may be described as hypochromic, and if there are many such cells present there will also be hypochromia.

Polychromasia refers to a bluish tinge associated with an immature red cell (early reticulocyte) population. This is illustrated in Figure 3.15: there are two large reticulocytes, both clearly stained more blue than the other, smaller cells. Several cells are both microcytic (small, with a low MCV) and hypochromic (with little colour). In some cells there is colour only in a small band around the outside of the cell – the middle of the cell seems to be 'empty'. This is called 'the area of central pallor'.

Other morphological changes

Close examination of the shape of the red cell can provide a wealth of information regarding the likelihood of different disease, as will be discussed in Chapter 4. For example, sickle cell disease is so named for the curved shape that the red cell tends to form when deprived of oxygen, whilst a schistocyte is simple a damaged cell. Haematologists also recognize various other irregular shapes that red cells may adopt. These include burr cells, target cells and poikilocytes.

There are also changes not due to a particular pathology but due instead to processing and laboratory factors. Figure 3.16 shows some changes due to prolonged storage, and this has also resulted in changes to the morphology of white cells (right-hand panel). Figure 3.17 shows changes that result from poor drying and fixation. In both these examples the morphology of the red cells is irregular – there are clear differences with the red cells compared with those in Figure 3.13, Figure 3.14 and Figure 3.15.

Inclusion bodies

Normally, the healthy red cell, when seen under the light microscope, should be homogeneous with no unusual internal features. However, in various disease and conditions, 'inclusion' bodies can be seen that can be of considerable diagnostic value. Inclusion bodies may be parasites, deposits of iron, remnants of DNA, or denatured haemoglobin. Some of these inclusion bodies demand special techniques as they may not be detectable by conventional staining. More details of particular inclusion bodies will follow in Chapter 4 in diseases of the red cell.

Summary

• The production of blood cells is haemopoiesis. In the healthy adult, it occurs only in the bone marrow; in the neonate it may also occur in the liver and spleen.
• Haemopoietic tissue consists of stem cells and supportive tissue such as fibroblasts. Each major group of blood cell has its own specific haemopoietic pathway; that for red cells is erythropoiesis.
• Bone marrow aspiration is required to confirm diseases suspected of having an origin in this tissue, and to monitor the effect of treatment. Special analyses include cytochemistry and flow cytometry.
• The red cell is highly specialized for the carriage of oxygen, a consequence being the lack of a nucleus and cell organelles. The cell membrane is also highly specialized with a series of structural proteins providing shape and flexibility.
• The cytoplasm contains haemoglobin (consisting of protein (globin) and non-protein (haem) components), metabolites and enzymes.
• Haem is a complex molecule with an atom of iron at the centre. Intracellular enzymes provide energy and resist the toxic effects of oxygen. At the death of the cell, globin proteins, lipids and iron are recycled, but a component of haem is degraded to bilirubin and excreted.
• Anisocytosis describes the variation in size of red cells; cells larger than the top of the reference range are macrocytes, cells below the bottom of the reference range are microcytes and those whose size is within the reference range are normocytes.
• Polychromasia and hypochromia describe variation in the 'colour' of different red cells. Red cells may contain one or more different types of inclusion bodies, which often indicate a particular metabolic problem.

Further reading

An X, Mohandas N. Disorders of red cell membrane. Br J Haematol. 2008;141:367–375.

Mohandas N, Gallagher PG. Red cell membranes: past present and future. Blood. 2008;112:3939–3948.

Metcalf D. Haemopoietic cytokines. Blood. 2008;111:485–491.

Orkin SH, Zon LI. Haematopoiesis: an evolving paradigm for stem cell biology. Cell. 2008;132:631–644.

Wang J, Pantopoulos K. Regulation of iron metabolism. Biochem J. 2011;434:365–381.

4   The Pathology of the Red Blood Cell

After studying this chapter, you should be able to:

- appreciate that the principal diseases of red blood cells can involve both low levels/numbers (leading to anaemia) and high levels/numbers (erythrocytosis and polycythaemia);
- explain how anaemia can arise from changes within the bone marrow;
- describe the consequences of abnormal iron metabolism;
- describe the consequences of poor supply of vitamins B_6, B_{12} and folate;
- outline how disease in different body organs can lead to anaemia;
- explain the causes of haemolysis;
- explain the contribution of molecular genetics to red cell disease;
- describe major feature of the haemoglobinopathies;
- outline how mutation in genes for membrane components and enzymes can lead to anaemia;
- suggest how the laboratory can diagnose these conditions.

In Chapter 3 we discussed how the red blood cell works – how it is marvellously adapted for its specialized role of oxygen transport. We will now look at how and why problems with this cell come about, and so how function is compromised, and so, in turn, how this leads to the problems that the patient experiences. The laboratory is at the forefront in the investigation of disease of the red blood cell, not merely in establishing pathophysiology, but also in that this information often informs treatment. As in Chapter 3, we will drop the word 'blood' and refer simply to 'red cells'. Others call these cells erythrocytes.

In Section 4.1 we will look at the different red cell diseases, the principal being anaemia, which can have many forms. Some arise from damage to the bone marrow (Section 4.2), deficiency of micronutrients (Section 4.3), intrinsic defects in the red cells (such as of the membrane and haemoglobin (Section 4.4)) and external factors acting on an 'innocent' red cell (Section 4.5). Section 4.6 will consider the causes and consequences of increased number of red cells, whilst the contribution of molecular genetics to the study of the red cell diseases will be summarized in Section 4.7. We will conclude the chapter with some case studies.

4.1   Introduction: diseases of red cells

Broadly speaking, the pathology of red cells considers the two polar extremes of the reference range, although, as we have discussed, having a result within the reference range is no guarantee of health. These aspects can be summarized as follows:

- *Anaemia.* This is present when there is a problem with too few red cells, or a problem with their function, that leads to symptoms in the patient. Anaemia is by far the major pathology of red cells.
- *Polycythaemia and erythrocytosis.* These are the reverse of the low red cell count and haemoglobin of anaemia, and can be viewed as a situation where too many red cells are present and, therefore, that the haematocrit and haemoglobin levels are high.

Scientists are interested not only in the disease, but how it came about. This is called aetiology. Understanding the aetiology of a disease is fundamental in understanding the abnormalities we observe, and what can be done to treat it. Different conditions give rise to different

Blood Science: Principles and Pathology, First Edition. Andrew Blann and Nessar Ahmed.
© 2014 John Wiley & Sons, Ltd. Published 2014 by John Wiley & Sons, Ltd.

Table 4.1 The diverse aetiology of red cell pathology.

Aetiology	Consequences
Lack of building blocks (such as iron and vitamin B_{12})	Red cells are, respectively, smaller or larger than normal
Attack by an outside agent (such as malaria or an antibody) on otherwise mature, healthy cells	Cells considered to be foreign, so are attacked and destroyed
Physical loss of cells due to an external cause	Falling red cell count and haemoglobin
Intrinsic defect in haemoglobin (such as sickle cell disease), in the membrane, or in key enzymes	Red cells are recognized as being abnormal, and are eliminated
Stress, or attack, on the bone marrow, perhaps by cancer or drugs	Red cell production reduced
Problem with the regulation of red cell production	High numbers of red cells; blood is too thick and sludgy: risk of malignancy (polycythaemia)
Not enough oxygen in the tissues (for lots of reasons)	High numbers of red cells, blood is too thick and sludgy (erythrocytosis)

red cell abnormalities and so demand different treatments. Examples of different aetiologies in red cell disease are illustrated in Table 4.1.

Polycythaemia and erythrocytosis are considerably less common and are fundamentally different from anaemia. For example, there are many different types of anaemia, but relatively few causes of polycythaemia and erythrocytosis. However, a more fundamental difference between erythrocytosis and polycythaemia is that the latter carries a risk of malignancy. We need to consider the bone marrow, as both too few cells and too many cells may arise from disorders in haemopoiesis.

This chapter will first address the most common diseases involving dysfunction of, or too few numbers of, red cells and will then describe polycythaemia and erythrocytosis, where the reverse is present – too many red cells. Red cell function and haemopoiesis are described in Chapter 3.

Anaemia

This disease is present when there is a problem with too few red cells, or a problem with their function. Some of the different causes for this disease are given in Table 4.1. However, before we examine these aetiologies in detail, we need to understand exactly what anaemia is. Unfortunately, despite the fact that anaemia is the primary pathological condition of red cells, there is a surprising lack of consensus as to how it should be defined.

Recognition and management of anaemia are also complicated by the variety and severity of signs and symptoms in different individuals (Table 4.2), few of which are fully sensitive enough (i.e. are present in a large number of people with the condition) or are specific enough (i.e. are found only in those people with the condition) to be reliable in practice. However, many of them are indicative of insufficient delivery of oxygen to the tissues (hypoxia or ischaemia), of which the red cell is only one component. For example, poor lung function, as in chronic obstructive pulmonary disease, may lead to hypoxia as the lungs cannot provide the blood with oxygen. A weakened heart, as in heart failure, may not be able to pump enough oxygenated blood around the body. In neither of these cases is the blood at 'fault'.

Definitions of anaemia

Many chose a 'numerical' definition of this disease. One textbook defines anaemia (in the adult) as a level of haemoglobin in the blood of less than 140 g/L in adult

Table 4.2 Signs and symptoms of anaemia[a].

Physical signs	Pallor, tachycardia (pulse rate over 100 beats per minute), glossitis (painful and swollen tongue), spoon nails
Symptoms	Tiredness, lethargy, reduced work and exercise capacity, shortness of breath (especially on exertion), palpitations
Severe indications	Jaundice, swollen spleen (splenomegaly), chest pain (possibly angina), swollen liver (hepatomegaly), dark urine

[a]Not intended to be exhaustive; can be caused by many different conditions.

males, or less than 120 g/L in adult females, whilst in another a result of less than 135 g/L or 115 g/L respectively defines the condition. The World Health Organization, taking a global view, defines anaemia when haemoglobin is less than 130 g/L in men and 120 g/L in women (110 g/L if pregnant). Some authorities have proposed that age (with cut-offs at age 20–49 and >50 years for women, and at 20–59 and >60 years for men), and race (white European, black Afro-Caribbean) should also be considered when formulating a reference range. Others suggest that degrees of anaemia exist in three stages: mild (haemoglobin >100 g/L), moderate (85–100 g/L) or severe (<84 g/L). Notably, these definitions do not consider the patient, whose symptoms (or lack of) are not considered.

Others characterize the disease as abnormalities in red cells that, alongside appropriate symptoms, call for treatment, a definition that does not specify the level of haemoglobin or the sex of the subject. It follows from this point of view that someone whose haemoglobin is 90 g/L may not be anaemic. This is amply illustrated by a letter to *The Lancet*, discussed next.

A young woman with sickle cell disease   The patient was diagnosed with homozygous sickle cell disease and was being treated, from the age of 18 months, at a university hospital in the West Indies. The disease ran a fairly benign course, and she was never transfused, despite a haemoglobin of around 50–70 g/L. She qualified as a pharmacist, worked full time and had an uneventful pregnancy. When last seen her haemoglobin was 39 g/L. Was she anaemic? According to the 'numerical' definition, yes.

Three years later she immigrated to the USA, and at a routine visit to her family doctor was found to have a haemoglobin of 38 g/L. She was given a transfusion of six units of blood within 24 h, which saw her haematocrit rise from 11% to 31%. Her blood pressure also increased, she developed headaches, had a cerebral haemorrhage, and died. The jury awarded US$11.5 million.

The report by Serjeant (2003) does not give full details (such as was she actually symptomatic at the family doctor's?), but the case underlines the dangers in the rapid treatment of a low haemoglobin. It follows that someone who is asymptomatic may not necessarily benefit from treatment. A corollary of this is that a low haemoglobin is not necessarily anaemia, and perhaps we should only treat those with an abnormal result and who are symptomatic and are therefore most likely to benefit from the presumed alleviation of their symptoms. Therefore, a new paradigm for medicine may be

$$\text{Disease} = \text{abnormal result} \times \text{symptoms}$$

According to this equation, someone who is tired, lethargic and has pallor (i.e. the symptoms of anaemia) but whose haemoglobin is 150 g/L does not have anaemia. Perhaps the young woman with sickle cell disease, if asymptomatic, was not anaemic despite the obviously low haemoglobin.

Classifications of anaemia

A widespread understanding of anaemia is also frustrated by the different methods of classification, such as those based on the size of the red blood cell, on symptoms or on aetiology.

The size of the red cell   It is convenient to classify anaemia based on the mean cell volume (MCV). This is popular because the analyser objectively provides the definition. For example:

- If there is anaemia, and the red cells are small (with a low MCV, perhaps 70 fL), then the anaemia can be described as microcytic.
- Conversely, the anaemia associated with large red cells (where, for example, the MCV may be 110 fL), is termed macrocytic.
- When the red cells are of normal size *and* the subject is anaemic then the term normocytic anaemia is used.

Symptoms   This classification system is completely patient centred, and responds to how they feel about their disease. It does not refer directly to haemoglobin.

- Mild anaemia may well be asymptomatic, or hardly influence the lifestyle of the individual.
- In moderate anaemia, the patient's lifestyle suffers, in that they are unable to do those everyday tasks many of us take for granted, such as a long walk around the park, going shopping or doing the laundry.
- Those with severe anaemia are likely to have their lives considerably influenced by the disease. For example, they may not be able to walk far without becoming very tired and out of breath. In a worst case scenario, the patient may become bed-bound and need transporting in a wheelchair.

Haemolysis   Certain types of anaemia lead to severe abnormalities in the red cell which can be recognized by cells of the reticulo-endothelial system, such as macrophages. An important function of macrophages is to remove abnormal material, which includes dead and dying cells, and waste material. Scavenging macrophages in the liver, spleen and in the blood itself recognize and attack abnormal red cells, and this leads to haemolytic anaemia.

Aetiology   This system is based on how the disease developed in the first place, and is outlined in Table 4.1. As pathologists, we will use this system as it provides an understanding of the pathophysiology of a particular anaemia (or any disease) and also gives an indication of a management plan (i.e. treatment). This system will consider:

- anaemia resulting from attack on, or stress to, the bone marrow (Section 4.2);
- anaemia resulting from deficiency of key micronutrients (Section 4.3);
- anaemia due to an intrinsic defect in the red cell (Section 4.4);
- anaemia due to an external agent acting on an otherwise healthy red cell (Section 4.5).

No classification system is perfect, and there will inevitably be inconsistencies and overlap. Accordingly, our discussion of anaemia will also consider, where relevant, the severity of the disease as experienced by the patient, haemolysis and the MCV.

4.2 Anaemia resulting from attack on, or stress to, the bone marrow

The bone marrow is the site of red cell production. This process, erythropoiesis, was described in Chapter 3. In this section we will look at:

- reduction in red cells alone;
- reduction in red cells and platelets (thrombocytopenia);
- reduction in red cells, white cells and platelets.

Reduction in red cells alone

The most common anaemia in this area, pure red cell aplasia (PRCA), is due to the severe reduction, if not the absence, of the red cell progenitors, the erythroblasts. The most common congenital form of PRCA (generally apparent in the first year of life) is Diamond–Blackfan anaemia, but even this has an incidence of only four to seven per million live births. In over 50% of cases it is caused by mutations in one of at least nine genes coding for ribosomal proteins (hence, ribosomopathy). A common finding is elevated red cell adenosine deaminase (an enzyme involved in purine metabolism), and accordingly this enzyme is an excellent confirmatory test for the disease, although in 16% of cases levels of the enzyme are normal. An alternative diagnosis to Diamond–Blackfan anaemia is transient erythroblastopenia (i.e. lack of erythroblasts) of childhood, which usually develops in the second year of life and with no family history of anaemia.

PRCA can also be acquired, and classified as primary (idiopathic – where no clear cause can be identified) or secondary (acquired as the result of exposure to a clear pathogenic agent). The list of known causes of acquired PRCA is considerable. The very rare condition congenital dyserythropoietic anaemia is also associated with abnormalities of erythroid precursors. However, in contrast to reduced erythroblasts in PRCA, there is erythroid hyperplasia with increased erythroblasts, which leads to ineffective erythropoiesis and so a reduced red cell count.

Reduction in red cells and platelets

The bone marrow is simply a place of production, perhaps a factory, whose products are blood cells. Any factory whose work space is taken over by external forces or objects, and is thus at a reduced work capacity, will clearly not be able to produce goods, and in this respect the bone marrow is no exception. The major invader is cancer, which can spread from its original site to invade other tissues. However, there may also be cancer arising within the bone marrow itself. In both cases, the cancer tissues invariably grow and slowly take over normal bone marrow tissue. If this is the case then erythropoiesis will suffer and so anaemia may result.

Cancer of the bone marrow includes the haemoproliferative diseases of leukaemia and myeloma, diseases characterized by a failure of the bone marrow to correctly regulate the number of white cells it produces. Excess numbers of malignant white cells build up in the marrow and suppress red cell and platelet production. A third white cell tumour, lymphoma, develops in a lymph node but may invade the bone marrow, and so may also displace those stem cells responsible for producing red cells and so cause anaemia. These diseases are described fully in Chapter 6.

Reduction in all blood cells

Pancytopenia is characterized by low levels of all three types of blood cell: red cells, white cells (leukopenia) and platelets (thrombocytopenia). The accompanying anaemia is called aplastic anaemia. The causes of this reduction in cell numbers can be classified as being congenital, acquired or idiopathic.

Disease arising from the bone marrow itself   The principal congenital cause of pancytopenia (accounting for two-thirds of cases) is Fanconi's anaemia, an inherited autosomal recessive condition with abnormalities in several genes. There are many causes of acquired pancytopenia, such as viral hepatitis. Many of the remaining causes of pancytopenia and aplastic anaemia are idiopathic but often have an immunological aetiology. This theory is supported by the observation that this can be reversed with immunosuppression with drugs such as ciclosporin. Fanconi's anaemia may evolve into myelodysplasia.

Myelodysplasia is a collection of syndromes characterized by the clonal proliferation of multipotential haemopoietic stem cells (i.e. stem cells that give rise to more than one type of blood cell), but these are disordered and inefficient. The ineffective haemopoiesis leads to pancytopenia. With regard to red cell biology, there is often a refractory anaemia, perhaps with ringed sideroblasts (as defined by Perls' stain) with a variable number of myeloblasts and abnormalities in erythroid precursors. As the number of blasts increases, the disease transforms into a myeloid leukaemia (Chapter 6). Chemotherapy and/or radiotherapy for neoplasia are risk factors for myelodysplasia and, therefore, is an acquired cause.

Myelofibrosis is a condition where, like leukaemia and myeloma, the initial problem is with the cells of the bone marrow itself, but the proliferating cell is not haemopoietic. The bone marrow becomes overgrown with fibroblasts – cells not directly involved in haemopoiesis but that normally have a supporting role. The fibroblasts are often driven to proliferate and produce collagen by inappropriate responses to growth factors produced by other cells. Like cancer, this is progressive and eventually leads to poor haemopoiesis, and thus to anaemia.

Disease caused by factors originating outside the bone marrow   Lymphoma is an example of a disease where tumour cells invade the bone marrow and so cause a reduction in red cells, white cells and platelets.

However, any other metastatic cancer (often from the breast and prostate) can send secondary growths to the bone marrow, which can then proliferate and so suppress all haemopoiesis. A treatment of cancer is cytotoxic chemotherapy, but these agents also suppress the bone marrow, causing low cell numbers. Fortunately, suppression is almost always reversible upon cessation of use of the particular drug. Together, medication drugs cause 15–25% of bone marrow suppression. Indeed, a frequent cause of both PRCA and pancytopenia (and therefore also aplastic anaemia) is chemotherapy.

Various nonmedication drugs that attack the bone marrow include industrial hydrocarbons such as benzene and agricultural drugs such as pesticides. Other agents leading to pancytopenia include very high doses of ionizing radiation and certain viral infections.

Treatment of an anaemia resulting from disease of the bone marrow

Treatment depends on the aetiology and on clinical severity. Where the cause of the anaemia is evident, such as being caused by a particular drug, then the disease should be reversible upon withdrawal of the drug. Congenital diseases such as Fanconi's anaemia are curable only by bone marrow transplantation, although androgens may help stimulate erythropoiesis. Some cases of acquired or idiopathic dyserythropoiesis may be successfully treated with immunosuppression by ciclosporin or corticosteroids. Unfortunately, the treatments of most haematological malignancies (such as leukaemia) are with agents (cytotoxic drugs, radiotherapy) that will also further suppress erythropoiesis. However, growth factors such as erythropoietin and granulocyte–macrophage colony stimulating factor can be given to help the recovery of the red cell count.

The role of the laboratory in anaemia following bone marrow changes

The initial signs of PRCA are likely to be those of general anaemia: reduced red cell count and haemoglobin. Generally, in most conditions where the bone marrow is being attacked, the MCV is within the reference range so that anaemia will be normocytic, although there are exceptions. The haematocrit is unlikely to be below the lower end of the reference range unless the red cell count is considerably reduced. Indeed, the red cell count may be on the high side of normal in certain anaemias (as physiological response to the tissue hypoxia) so that the haematocrit may be at the bottom of the reference range. By definition, the white cell and platelet counts in PRCA are within the reference range. However, key investigations involve examination of the bone marrow, which is normocellular with respect to leukocyte precursors and megakaryocytes, but will reveal grossly reduced or absent erythroblasts.

Diamond–Blackfan anaemia is generally associated with a moderate to severe anaemia (haemoglobin 20–100 g/L) but the MCV is often above the top of the reference range, so the anaemia is macrocytic, and there is reticulocytopenia. The bone marrow is normal, except for a paucity of erythroid precursors. The macrocytic anaemia in Diamond–Blackfan disease is in contrast to the transient erythroblastopenia of childhood where the MCV is within the normal range. In the congenital dyserythropoietic anaemias, haemoglobin is often in the range 80–110 g/L, producing in the patient a moderately symptomatic anaemia.

The anaemia that follows from leukaemia is perhaps the easiest to detect, as a high white cell count (and possibly a thrombocytopenia) is present. The anaemia is most likely to be normocytic and normochromic. A bone marrow aspiration will inevitably be performed and should find all the erythroid precursors (including reticulocytes) to be present but in reduced numbers. The bone marrow in myelodysplasia is usually hypercellular with enlarged and abnormal erythroid precursors, although there are abnormalities in the precursor of all cell lineages. However, a hallmark of the bone marrow in myelodysplasia is the presence of ring sideroblasts. These will be described more fully in the section on iron that follows.

The definition of pancytopenia is reduced numbers of red cells, white cells and platelets and so should be relatively simple to diagnose from a full blood count. Generally, the (aplastic) anaemia should be normochromic and normocytic. The bone marrow will be hypoplastic, the normal haemopoietic tissues being grossly reduced and replaced by fat cells and other non-haemopoietic tissues. However, deviations from this picture occur in Fanconi's anaemia and in acquired aplastic anaemia, where the red cells may be macrocytic. The bone marrow in myelofibrosis will show increased collagen and fibroblasts alongside reduced haemopoietic precursors.

Generally speaking, the reticulocyte count, normally 0.5–2.5% of the red cell count (absolute count 25–125 × 10^9 cells/L), should be raised (to perhaps 5%, or even 10% in extreme cases) in most cases of anaemia as a

Table 4.3 The bone marrow and anaemia.

Reduction in red cells alone
- Pure red cell aplasia, Diamond–Blackfan anaemia

Reduction in red cells and platelets
- Leukaemia and myeloma

Reduction in all blood cells
- Invasion by cancer
- Myelofibrosis and myelodysplasia
- Chemotherapy and radiotherapy

Laboratory
- The anaemia is generally normocytic and normochromic (except Diamond–Blackfan anaemia, which is macrocytic)

physiological response to the inability of the depleted red cell mass to supply the tissues with sufficient oxygen for its physiological needs. However, in those anaemias where erythropoiesis is compromised (as in PRCA and aplastic anaemia) then reticulocyte numbers are likely to be low in the peripheral blood and grossly reduced in the bone marrow.

Further details of the use of bone marrow analysis in leukaemia and other neoplasia are presented in Chapter 6. The role of the bone marrow in different forms of anaemia is summarized in Table 4.3.

4.3 Anaemia due to deficiency

The bone marrow is a factory, but even the most modern and efficient factory must have raw materials to fashion into products. In the case of red cells, these include the essential micronutrients iron, vitamins B_{12} and B_6, and folate. Chapter 3 indicated steps in the development of red cells that require the vitamins and folate as cofactors for key enzymes in the metabolism of haem, and the need for iron.

These micronutrients are obtained from the diet, so that inadequate nutrition (i.e. malnutrition) is likely to cause different types of anaemia. However, an apparently adequate diet may lead to disease as the micronutrients may fail to be absorbed by abnormalities of the intestines (i.e. malabsorption). Subsequently, levels of deficiency arise when the micronutrients are unable to pass into the bloodstream, or perhaps cannot be transported to the bone marrow. Once in the erythroblast, iron may not be able to pass into the cytoplasm and then into the mitochondrion, whilst the (lack of) vitamins and folate may not be able to assist as cofactors for the enzymes in haem synthesis.

Iron

Each molecule of haemoglobin must have an atom of iron to which oxygen can (transiently) bind and so be carried from the lungs to the tissues. Thus, lack of iron leads directly to impaired carriage of oxygen. Indeed, iron deficiency, when due to malnutrition and/or malabsorption, is the most common cause of anaemia in the developed and developing world. The World Health Organization estimates that 30% of the world's population (some 2 billion people) is anaemic, and within this group the most common cause is iron deficiency, and the most common cause of this deficiency is nutritional. However, a minor aspect of iron-related disease is sideroblastic anaemia, a topic we will address shortly.

Section 3.5 outlined key steps in the absorption of iron from the intestines (summarised in Figure 3.6). These steps are passage from the lumen of the intestines into the cells of the intestine (enterocytes), crossing the enterocyte, movement from the enterocyte into the blood and carriage to the bone marrow. If any step in this process is impaired, anaemia can develop.

Ferrous iron (Fe^{2+}) is readily absorbed, and although ferric iron (Fe^{3+}) is poorly absorbed it can be converted to ferrous iron by the luminal enzyme iron-reductase. This functions best at low pH, an environment promoted by ascorbic acid and by acid secretions from the stomach. Therefore, disease or surgery to the stomach and intestines can also lead to its malabsorption.

Once imported by the enterocyte, iron is exported into the circulation via ferroportin. Problems with the molecule, perhaps brought on by high levels of hepcidin, may lead to iron-deficiency anaemia as the iron will not enter the plasma. Once in the circulation, iron must be picked up by its transport protein, apotransferrin, whereupon the two become transferrin.

Blood science angle: The liver

This organ is of interest to haematologists as it produces transferrin and hepcidin and stores iron in the form of ferritin and haemosiderin, and problems with these may cause abnormal red cell morphology, such as macrocytosis and target cells. Thus, iron-based anaemia may be the consequence of liver disease, such as may be caused by alcoholic liver disease, cirrhosis, primary biliary cirrhosis and hepatoma. Liver function tests (LFTs) may be ordered in the investigation of these anaemias, as are discussed in Chapter 17.

At the bone marrow, iron-loaded transferrin passes into red cell precursors such as erythroblasts by binding to transferrin receptors on the cell surface. Once inside the cytoplasm, the iron is uncoupled from the transferrin and moves to the mitochondria for incorporation into haem. The iron-free apotransferrin leaves the erythroblasts for the plasma for further duties in iron transport. Lack of iron in a bone marrow aspirate is reasonably easy to demonstrate with Perls' stain, which may also be known as Perls' Prussian blue (Figure 4.1).

(a) Negative: no blue = no iron!

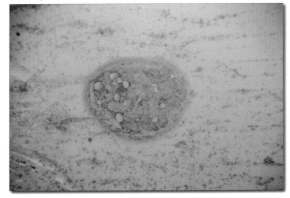

(b) Plenty of blue = lots of iron

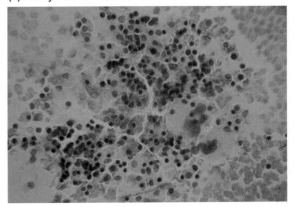

Figure 4.1 Perls' stain. This process stains iron a blue colour, as illustrated in these two samples of bone marrow: (a) complete lack of blue colour, is from a patient with profound iron deficiency; (b) from a subject with normal iron stores. (Image courtesy of H. Bibawi, NHS Tayside).

Vitamins B₁₂, B₆ and folate

Vitamin B_{12} is an essential cofactor for enzymes involved in the synthesis of haem (Section 3.5). Whilst dietary deficiency of vitamin B_{12} is not unheard of (as in strict vegans), the dominant cause of deficiency is failure to be absorbed. In turn, the most common cause of failure to absorb this vitamin is lack of secretion of intrinsic factor (IF) by the stomach, the most frequent reason for this being auto-antibodies. Other diseases of the stomach will also cause failure to produce sufficient IF, such as gastric atrophy, alcoholism and carcinoma, and the effects of surgery for the latter where IF-generating tissue is simply removed.

In addition to the general signs and symptoms of anaemia (Table 4.2), a potential clinical diagnosis of vitamin B_{12} deficiency would be supported by certain well-described observations. These are a painful and swollen tongue (glossitis), occasional hypopigmentation (vitiligo), unsteady gait, and peripheral neuropathy (tingling, loss of sensation), the latter two being a consequence of inadequate myelination of neurones, for which the vitamin is an essential requirement. In long-term and/or severe deficiency, there may also be psychiatric changes (such as irritability, depression and psychosis) and jaundice (resulting from haemolysis).

Vitamin B_6 is an essential cofactor of the enzyme aminolevulinic acid (ALA) synthase, required for the formation of porphobilinogen and ultimately haem, so that deficiency leads to anaemia (Figure 3.5). Deficiency may be a consequence of malnutrition, but is more likely to be due to certain drugs, notably isoniazid (used to treat tuberculosis) and pencillamine (used to treat high levels of copper).

The 'raw' folate molecule from the diet is processed in intestinal cells and enters the blood as methyltetrahydrofolate, from which vitamin B_{12} collects a methyl group, leaving tetrahydrofolate. The enzyme methyltetrahydrofolate reductase (MTHFR) can then regenerate methyltetrahydrofolate. If this metabolic pathway is interrupted, tetrahydrofolate is eventually consumed and stocks are depleted. This eventually leads to folate deficiency, a consequence of which is that it is unable to participate in other pathways such as deoxyribonucleic acid (DNA) synthesis. Lack of the enzyme MTHFR also causes increased levels of homocysteine, reputed to be a risk factor for atherosclerosis.

The additional importance of folate (and vitamin B_{12}) is demonstrated by the fact that most cases of spina bifida and neural tube defects result from maternal deficiency in the peri-conceptual period. If deficiencies persist, cognitive impairment may develop in the adolescent. Therefore, use of supplements in pregnancy may be advised.

The role of the laboratory in anaemia following lack of micronutrients

As with all potential anaemias, laboratory investigation includes a full blood count and a blood film for red cell and white cell morphology. The key index in micronutrient deficiency is the MCV, of which the polar extremes are important.

Microcytic anaemia   This is defined as symptoms of anaemia, haemoglobin below the bottom of the reference and an MCV below the bottom of the reference range. The principal cause of this is iron deficiency, but another cause is haemoglobinopathy, and vitamin B_6 deficiency may also cause microcytosis. In many cases the microcytes also have a reduced amount of haemoglobin, leading to a low mean cell haemoglobin (MCH) and, despite a reduced size, a reduced mean cell haemoglobin concentration (MCHC). These findings lead to the further qualification of the disease as microcytic hypochromic anaemia.

In many cases, more investigations will follow, such as serum iron, transferrin, ferritin, total iron binding capacity, transferrin saturation and the soluble transferrin receptor. The exact panel of tests varies according to the particular laboratory. However, several of these molecules are influenced by the acute-phase response, as is likely to be present in several chronic inflammatory diseases. This is a general perturbation of a large number of physiological systems in response to an actual or presumed infection. In haematology it is characterized by an abnormal erythrocyte sedimentation rate (ESR) and increased levels of certain plasma proteins. Therefore, a genuinely low ferritin may be artefactually raised, and so appear normal, in an acute-phase response.

Figure 4.2 shows a typical blood film in microcytic anaemia; the acute-phase response is explained in Chapter 9.

Should the microcytic anaemia (or any anaemia) be so severe that treatment is called for, the effects of treatment (which may be a blood transfusion or iron) can be noted on a blood film as a dimorphic picture, with two distinctly different populations of cells, the 'old' microcytes and the 'new' normocytes, but there may also be some reticulocytes. This should also be noted by the red cell distribution width (RDW) result, because as reticulocytes appear in the blood they will drive up both the MCV and

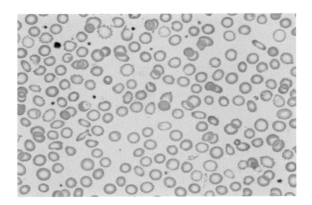

Figure 4.2 Marked microcytic anaemia, in this case due to gross iron deficiency. Almost all of the cells are 'empty' of colour, and so of haemoglobin, and almost all are microcytic. Compared with Figure 3.2 and Figure 3.13. (Image courtesy of H. Bibawi, NHS Tayside).

the RDW, indicating a heterogeneous population. Ultimately, we hope for an MCV within the reference range and a small RDW, indicating a homogeneous population of healthy cells.

Some autoanalysers report the proportion of red cells that it considers to be hypochromic – greater then 2% is generally considered suggestive of iron deficiency. An additional feature of iron-deficient anaemia is the reduction or absence of stored iron (as ferritin and haemosiderin) in macrophages of the bone marrow, liver and spleen. Indeed, it is possible iron stores in the bone marrow become depleted before there is evidence of iron deficiency from the red cell indices.

Macrocytic anaemia This is defined as the triad of symptoms of anaemia (Table 4.2), haemoglobin below the bottom of the reference and an MCV above the top of the reference range. The principal cause of this is vitamin B_{12} deficiency. A raised MCV alone is macrocytosis, which may be present in alcoholism, liver disease, hypothyroidism, myeloma, myelodysplasia, aplastic anaemia, acute leukaemia and pregnancy – some of which may not necessarily be associated with anaemia. Drugs causing macrocytosis include chemotherapy (such as cyclophosphamide) and antimicrobials (such as trimethoprim). Furthermore, neonates may also have a macrocytosis, with an MCV of perhaps 110 fL, becoming 'normal' (e.g. 95 fL) by adult standards by 2 months of age.

However, a bone marrow examination is usually performed to fully confirm the diagnosis, to exclude the other possible causes of macrocytosis. This will be hyperplastic with increased numbers of myeloid precursors. However, caution is required, as increased numbers of these cells (such as myelocytes) are present in numerous conditions, such as leukaemia or myelodysplasia. Erythroblasts and other progenitors become enlarged, and are referred to as megaloblasts. This finding gives the disease its common name: megaloblastic anaemia.

Since reticulocytes are also larger than mature erythrocytes, then a marked reticulocytosis may also cause an overall raised MCV. However, reticulocytes are likely to exhibit a degree of polychromasia, and if present at frequency greater than 10% are more likely to reflect increased erythropoiesis, perhaps in response to haemolysis or an acute bleed. A further consequence of the lack of vitamin B_{12} is hypersegmented neutrophils, which can be detected by blood film microscopy (Figure 4.3). In severe disease there may also be pancytopenia (leucopenia and thrombocytopenia).

Other investigations include the Schilling test, where the (suspected) patient takes oral radiolabelled vitamin B_{12}, alone or with IF, which, if the diagnosis is correct, will fail to be absorbed. Those whose absorption is unimpaired will excrete radiolabelled vitamin in their urine. Some workers find a single cutoff point for plasma vitamin B_{12} to be unreliable, especially with borderline results, and suggest testing for levels of methylmalonic acid, which, if raised, indicates vitamin B_{12} deficiency.

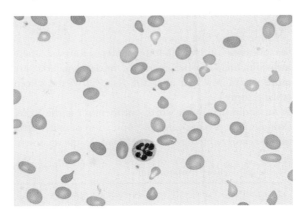

Figure 4.3 Macrocytic anaemia. These red cells are much larger than those of Figure 4.2, and are also much more heavily stained, and so have more haemoglobin. This macrocytosis is due to deficiency in vitamin B_{12}; an additional factor is the increased number of segments of the nucleus of the neutrophil, the so-called hypersegmented neutrophil, an example of which is shown in this figure. There is also a moderate degree of anisocytosis (variation in the sizes of the red cells). (Image courtesy of H. Bibawi, NHS Tayside).

The consequences of vitamin B_6 deficiency include an anaemia that can be normocytic, microcytic or sideroblastic, whilst those of folate deficiency can be similar to, or even identical to, those of vitamin B_{12}, and this includes a megaloblastic anaemia and macrocytosis. Serum folate is a standard laboratory measure; but as there is a high concentration of folate in red cells, then haemolysis may produce a falsely high level. Accordingly, red cell folate is a better indicator, as levels are constant throughout the life of the cell and are not influenced by short-term dietary changes. However, factors causing low serum folate may also cause low red cell folate.

Blood science angle: Immunology

The most common form of vitamin B_{12} deficiency is auto-antibodies to IF (present in 30–50% of patients) and/or to the gastric parietal cells where IF is synthesized (present in 60–85% of patients), thus causing pernicious anaemia. Thus, perhaps 30% of patients will have both types of auto-antibodies. The target of the autoantibody in the gastric parietal cells is the proton pump, resulting in reduced acid secretion. Indeed, perhaps 25% of patients with iron-deficiency anaemia with no evidence of gastrointestinal blood loss may have autoimmune gastritis, with autoantibodies to acid-producing cells. These autoantibodies are relatively easy to detect in the immunology laboratory, and are discussed in greater detail in Chapter 9.

A rare blood film characteristic of folate deficiency is the presence of Howell–Jolly bodies. However, these inclusion bodies of fragments of nuclear material are more commonly seen after the removal of the spleen (splenectomy).

An additional index worthy of mention is the RDW (Section 3.8). A homogeneous population of macrocytes may be expected to have a small RDW. As the treatment proceeds, reticulocytes will enter the blood, and as these large reticulocytes mature into smaller erythrocytes, then the MCV will fall but the RWD will remain high. As with the treatment of microcytic anaemia, we eventually hope for an MCV and RDW both within the reference range.

Whatever the aetiology, the standard treatment of severe vitamin B_{12} deficiency is intramuscular or subcutaneous injection, likely to produce a rise in haemoglobin and fall in MCV within 10–14 days, although the full response (i.e. disappearance of megaloblasts from the bone marrow and hypersegmented neutrophils from the peripheral blood) should occur within 8 weeks. An alternative form of treatment of pernicious anaemia is with steroids to suppress the autoantibodies to IF and/or parietal cells. Successful treatment will, of course, place demands on iron stores, which, if low, may be unable to keep up with demand.

Iron and vitamin B_{12} compared and contrasted

Considerable disease is due to the lack of either of these micronutrients. There is overlap in terms of daily requirements, intestinal absorption, carriage in the plasma, body stores, and pathophysiological consequences. However, deficiency of each micronutrient produces small (i.e. microcytic, in the case of iron deficiency) or large (i.e. macrocytic, in the case of vitamin B_{12} deficiency) red cells (Table 4.4).

Endocrine disorders

Deficiency of hormones is also associated with anaemia. Patients with chronic renal failure, and possibly on dialysis, are at risk of becoming anaemic as this organ produces the growth factor erythropoietin. If so, the anaemia is generally normocytic. The normal inverse relationship between erythropoietin and haemoglobin is lost when serum creatinine rises above 135 μmol/L,

Table 4.4 Deficiency of iron and vitamin B_{12}.

	Iron	Vitamin B_{12}
Descriptor of the anaemia	Microcytic	Macrocytic
MCV	Reduced	Increased
Primary cause of malabsorption	Intestinal disease, genetic (e.g. increased hepcidin)	Autoimmunity
Allied blood tests	Transferrin, ferritin Total iron binding capacity Soluble transferrin receptor	Antibodies to IF, antibodies to gastric parietal cells
Plasma transporter	Transferrin	Transcobalamin

and the degree of hypoxia required to stimulate release of the hormone rises considerably.

The anaemia of hypothyroidism, as thyroxine is a red cell growth factor, may be masked by a reduced blood volume but macrocytosis may be present. Conversely, in hyperthyroidism, microcytosis is common, even if there is no anaemia. Higher red cell indices in men are due in part to testosterone acting on the bone marrow as a growth factor. Those men who have lost their testes to disease or accidents have a fall in red cell indices, and this may contribute to a normocytic anaemia.

Whatever the deficiency, the laboratory is important in assessing the success of treatment, which should be monitored with regular full blood counts whilst on treatment and until deemed successful, and annually thereafter.

4.4   Intrinsic defects in the red cell

This large section includes many conditions; all ultimately caused by gene mutations which themselves are often hereditary, of which changes to the haemoglobin gene (and so the molecule) are perhaps the most well known and best characterized.

Defects in iron metabolism

Defective synthesis of haem   This complex molecule, whose synthesis we examined in Chapter 3, and summarized in Figure 3.5, requires not simply the correct constituents (including iron), but also the correct functioning of a series of enzymes that synthesize the porphyrin ring and thus a functioning haem molecule. Failure of any of these enzymes, which can be hereditary (primary) or acquired (secondary), leads to a collection of conditions known as the porphyrias. One such gene, for ALA synthase, is found on the X-chromosome, so mutations lead to a sex-linked porphyria that affects males. Another mutation causes lack of ferrochelatase activity, responsible for the insertion of iron into the protoporphyrin ring.

A consequence of the failure to incorporate iron into haem is the accumulation of iron, which becomes deposited in ferritin. Iron-rich mitochondria form a ring around the nucleus of the erythroblast, and as such are described as ring sideroblasts which give the most common disease, sideroblastic anaemia, its name (Figure 4.4). However, in some cases red cells can still be produced, but if this

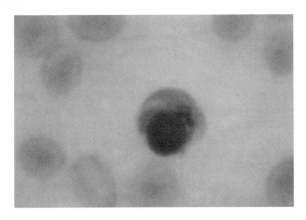

Figure 4.4   Ring sideroblast. Perls' stain is also used to define a sideroblast, which in this figure has a large (blue) deposit in the cytoplasm, but there are also some deposits in the nucleus. Ring sideroblasts can be found in sideroblastic anaemia and lead poisoning. (Image courtesy of H. Bibawi, NHS Tayside).

happens these red cells may contain iron granules; such cells are called siderocytes, and the iron inclusion granules (haemosiderin) are called Pappenheimer bodies.

Sideroblastic anaemia may also be acquired, such as in myelodysplasia, myelofibrosis, myeloma, rheumatoid arthritis and haemolytic anaemia, and may also be caused by drugs such as alcohol, isoniazid and chloramphenicol, by lack of copper (as it is a cofactor for other enzymes involved in haem synthesis) and by excess ingestion of zinc and lead.

In most cases, the treatment of inherited sideroblastic anaemia is oral vitamin B_6, which should be effective in one-third of patients. Severe cases may need to be transfused, but this may lead to iron overload, although chelation therapy may be effective in reducing inappropriately high iron stores. Treatment of acquired cases may be removal or replacement of the causative agent, such as a different antibiotic in place of chloramphenicol.

Iron overload   Just as insufficient iron can lead to problems, so also can too much iron. The body has no effective mechanism for actively eliminating iron, so absorption must be carefully regulated. When this mechanism fails, iron stores rise, causing problems for the tissues where it is stored or deposited. When iron levels are high, but there is no tissue damage, we use the term haemosiderosis. However, if iron levels are so high that there is indeed damage to organs, it becomes haemochromatosis.

A key regulator of iron absorption is the HFE protein, coded for by the *HFE* gene, present on chromosome 6. The precise function of the protein product is unclear, but several possible roles have been proposed. However, mutations in *HFE* cause hereditary haemochromatosis (HH), the most common genetic cause of iron overload, with perhaps 83% of cases caused by a single mutation (C282Y). This mutation is present in Europeans in its heterozygous form in perhaps 9.2% of people, although this figure varies greatly with region – ranging from approximately 1% in southern Europe, to 25% in Ireland. Heterozygotes do not have an increase risk of clinically evident HH. African, Middle Eastern and Australian populations have a prevalence of less than 0.5%. In its homozygous form HH is present in 0.4% of Europeans and inevitably has clinical consequences.

Several other types of genetic variants of HH have been described. These are haemojuvelin-associated HH (due to a mutation in the gene, named *HFE2*, for a poorly understood molecule called haemojuvelin that leads to juvenile-onset haemochromatosis), a mutation in the hepcidin gene resulting in no or inactive hepcidin and a mutation in the gene for the transferrin-receptor 2, which also leads to low levels of hepcidin.

The major nongenetic cause of iron overload is hypertransfusion. One unit of transfused blood contains approximately 200–250 mg of iron, and although blood transfusion may ameliorate anaemia a consequence may well be iron overload. Generally, after the transfusion of perhaps 15 units, patients become overloaded and iron chelation therapy is recommended as overload can have serious clinical sequelae. Iron deposit in the heart muscles leads to myocarditis and cardiac fibrosis, which eventually causes severe arrhythmias or heart failure. Other complications include deposition of iron in the liver, leading to fibrosis, cirrhosis or cancer, and diabetes mellitus as a result of islet cell destruction in the pancreas.

Treatment of iron overload depends on aetiology. The introduction of iron chelating therapy has seen deaths from iron overload significantly decrease, and treatment can be monitored by the ferritin level. Chelation is usually administered by subcutaneous infusion and the treatment should be started after 10–15 units of blood transfusion or when ferritin is over 2000 μg/mL. An alternative to iron chelation therapy is to simply venesect the patient, and is the preferred method in HH. Venesection should proceed regularly (perhaps monthly) until the ferritin levels are at the lower end of the reference range. Dietary advice is to minimize foodstuffs with high iron content.

The role of the laboratory in iron-related pathology

Porphryria is failure to synthesize the porphyrin ring that houses the iron, and various porphyrin metabolites can be detected in urine. A useful alternative test of porphyria is for the presence of zinc protoporphyrin. Porphyria is also found in lead poisoning, where a useful sign is basophilic stippling (the presence of partially degraded ribosomal and messenger RNA) in red cells (Figure 4.5). However, this is not specific for lead poisoning as it may be seen in megaloblastic anaemia.

Sideroblastic anaemia is diagnosed from an examination of bone marrow where greater than 15% of erythroblasts are ring sideroblasts. However, a small number of ring sideroblasts may be present in other conditions, such as myelodysplasia and copper deficiency, also diagnosed by examination of the bone marrow. The blood film of X-linked sideroblastic anaemia is characterized by microcytic anaemia, in the bone marrow by ineffective erythropoiesis and in the tissues by iron overload. The key laboratory test is Perls' stain (Figure 4.1).

HH is characterized by low serum hepcidin, which fails to prevent the export of iron by ferroportin into the blood, leading to the hyperferraemia that is characteristic of the disease. Liver biopsy is the best method for the assessment of iron as it holds 90% of body stores as ferritin and haemosiderin, and is quantitative, specific and sensitive. Increased LFTs (such as raised transaminase enzymes) imply the involvement of this organ.

Although serum ferritin levels give a good indication of the status of iron stores, this test is subject to a major confounding influence: inflammation. Consequently, as

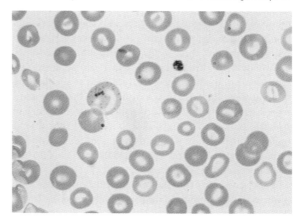

Figure 4.5 Basophilic stippling. These inclusion bodies, consisting of over a dozen very small blue or purple dots, can be very hard to detect. Notably, the red cell in which they occur is often larger than other cells. (Image courtesy of H. Bibawi, NHS Tayside).

in the investigation of microcytic anaemia, interpretation of a ferritin result must be cautious in infections and in inflammatory diseases such as rheumatoid arthritis. If transferrin saturation exceeds 45% and there is raised ferritin (>380 μg/L in men and >365 μg/L in post-menopausal women), then HH may be suspected.

Blood science angle: Micronutrients

With established methods for metal ions such as sodium, potassium and calcium, the biochemistry laboratory seems the natural home for the analysis of iron, and is also likely to be able to measure vitamins, although using a different technology. However, defined proteins such as ferritin, transferrin and transcobalamin can be measured as easily in an immunology lab as in a haematology lab – it depends on where the particular analyser is sited.

Membrane defects

The principal causes of anaemia relating to abnormalities in the red cell membrane are hereditary spherocytosis (HS), hereditary elliptocytosis (HE) and paroxysmal nocturnal haemoglobinuria (PNH). The principal enzyme defects are of pyruvate kinase (PK) and glucose-6-phosphate dehydrogenase (G6PD). All of these defects cause physical deformities that therefore lead to haemolysis and so haemolytic anaemia.

Hereditary spherocytosis   HS is the most common hereditary haemolytic anaemia in northern Europeans (frequency estimated at 1/2000 to 1/5000), 75% of cases displaying an autosomal dominant manner of inheritance with variable clinical presentation ranging from a severe neonatal haemolytic anaemia to an asymptomatic state. Several mutations can cause a defect in proteins such as Band 3, protein 4.2, ankyrin, and alpha and beta spectrins that are involved in the maintenance of the shape of the cell. A consequence is that cells become spherical.

These abnormal cells (Figure 4.6) have a considerably reduced life span (between 6 and 20 days, compared with 120 days in health), are unable to pass through the splenic microcirculation and so are eliminated. Accordingly, patients with the most severe disease (haemoglobin 60–80 g/L, reticulocyte >10%), or even moderate disease (haemoglobin 80–120 g/L, reticulocytes <6%), benefit from splenectomy and enjoy an increase in haemoglobin.

Hereditary elliptocytosis   The principle of HE is the same as HS; that is, loss of structural integrity. One underlining defect is deficiency of protein 4.1, another

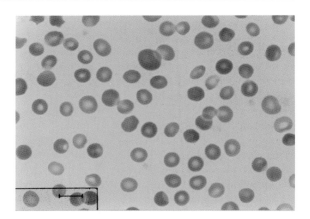

Figure 4.6 Hereditary spherocytosis. This is manifest as cells which have lost their central pallor. There is also a modest degree of anisocytosis. The bar represents 10 μm. (Image courtesy of Drssa Wilma Barcellini, Fondazione IRCCS Ca' Granda, Ospedale Maggiore Policlinico, Milano, Italy).

involves alpha spectrin, beta spectrin and glycophorin C. On a blood film the phenotype shows a broad spectrum of red cell morphology ranging from slightly oval cells to extreme elliptocytosis. Most patients do not demonstrate haemolytic anaemia and there is no correlation between haemolysis and the degree of elliptocytosis (Figure 4.7).

A variant of this disease is South-East Asian ovalocytosis, commonly found in Indonesia, the Phillipines and Malaysia, where the prevalence ranges from 5% to 25%. The abnormal red cell shape is caused by mutations that

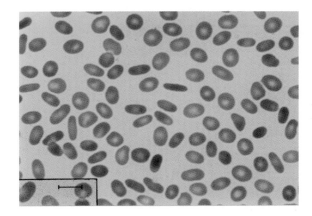

Figure 4.7 Hereditary elliptocytosis. Few cells have retained their circular shape, the remainder varying from being slightly oval to grossly elliptical. The bar represents 10 μm. (Image courtesy of Drssa Wilma Barcellini, Fondazione IRCCS Ca' Granda, Ospedale Maggiore Policlinico, Milano, Italy).

lead to defective membrane component Band 3. The consequence of this red cell rigidity, which (as for HS cells) brings about their detection and elimination, often in the spleen. Like many hereditary membrane, enzyme and haemoglobin diseases, the high frequency may have arisen as a protection against malaria.

Hereditary stomatocytosis The abnormality in this condition, like HS, is due to defects in the internal cytoskeleton component Band 3 (see Figure 3.4). Non-hereditary causes include dehydration, and problems with maintaining the correct balance of sodium and potassium within the cell. Red cells are large and osmotically fragile with a low MCHC (Figure 4.8).

Paroxysmal nocturnal haemoglobinuria The molecular basis of PNH is the defective synthesis of glycosyl phosphatidylinositol on the cell surface. This molecule is an anchor which supports the integrity of several proteins, including CD55 and CD59, which are important in resisting the activation of complement. As a consequence, the red cells become sensitive to complement-mediated intravascular haemolysis. Since CD55 and CD59 are also present on platelets and white cells, other clinical feature includes recurrent thrombosis of large veins and intermittent abdominal pain. Thus, laboratory findings include anaemia (ranging from mild to severe), leucopenia and thrombophilia (for which anticoagulants, such as warfarin, may be required).

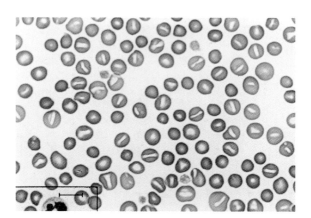

Figure 4.8 Hereditary stomatocytosis. This condition is characterized by a slot-shaped area of central pallor, although smaller cells have retained their normal round area of central pallor. The bar represents 10 μm. (Image courtesy of Drssa Wilma Barcellini, Fondazione IRCCS Ca' Granda, Ospedale Maggiore Policlinico, Milano, Italy).

The laboratory in membrane defects The full blood count in HS shows an increased reticulocyte count (5–20%), low haemoglobin (e.g. 70 g/L), reduced MCV, raised MCH and raised MCHC (such as 360 g/L). The blood film shows dense micro-spherocytes, with no central pallor. Most cases of HE are heterozygous with no haemolysis. The full blood count may show normal to slightly low haemoglobin, whilst the blood film may show typically 80% of cells are oval. However, and in cases where numbers of abnormal cells are not large, an alternative diagnosis may be acquired elliptocytosis, as in iron deficiency. Homozygous HE presents with severe haemolytic anaemia and a haemoglobin result less than 100 g/L. The full blood count from a patient with PNH demonstrates mildly reduced haemoglobin, with an increase in reticulocytes, mild leucopenia ($<2.5 \times 10^9$/L) and thrombocytopenia (platelets $<50 \times 10^9$/L). Consequently, PNH should be excluded in patients with pancytopenia.

In all conditions, assessment of the membrane is essential. In both HS and HE, polyacrylamide gel electrophoresis of a preparation of red cell membranes will display absence of particular spectrins. Functional methods test the ability of the membrane to withstand stress.

• HS and HE may be investigated with the osmotic fragility test. Blood is added to buffered hypotonic sodium chloride solutions ranging from 0.1% to 0.8%. After 30 min at room temperature, during which time some cells (the weakest) lyse, the degree of haemolysis is recorded and is expressed as a percentage of haemolysis at each sodium chloride concentration. Interpretation relies on marked increased fragility of the red cell population.
• Ham's test involves suspending red cells in normal serum which has been acidified to pH 6.5–7.0, followed by incubation at 37 °C and examination for haemolysis.
• In the sucrose haemolysis test, cells are incubated in isotonic solutions of low ionic strength with a small amount of serum present in the mixture. Haemolysis of <5% is negative, 5–10% haemolysis is borderline, but >10% is consistent with PNH.
• In the acidified glycerol lysis test, red cells are suspended in slightly acidified phosphate-buffered sodium chloride–glycerol reagent and the time taken for 50% haemolysis to occur is measured. Interpretation relies on a shortened haemolysis time in HS compared with normal red cells.
• The cryohaemolysis test employs a similar principle. The stress is to warm the cells to 37 °C and then to transfer them promptly to an ice bath, at which point weakened cells, such as those for a patient with HS, will lyse.

Flow cytometry is useful in two settings. It can be used to probe the absence of membrane components CD55 and CD59 on the surface of the cell in PNH (Figure 4.9). The fluorescent probe eosin-5-maleimide binds to normal red cells, but does so less avidly if there are abnormalities in membrane components spectrin, protein 4.2 or Band 3. Consequently, the test may be useful in the diagnosis of HS and South-East Asian ovalocytosis. In all tests, good positive and negative controls are essential.

Other membrane defects Hereditary pyropoikilocytosis is a rare autosomal recessive disorder characterized

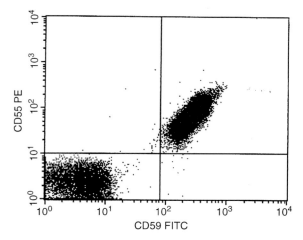

Quad	Events	% Gated	% Total
UL	34	0.14	0.14
UR	17479	73.46	69.92
LL	6268	26.34	25.07
LR	12	0.05	0.05

Figure 4.9 Paroxysmal nocturnal haemoglobinuria. This flow cytometry plot is the result of mixing red cells with one antibody to CD55 and another to CD59, both linked to different fluorescent probes (phycoerythrin and fluorescein isothiocyanate respectively). The machine counts the number of cells binding both antibodies (in the upper right (UR) quadrant), either antibody alone (upper left (UL) and lower right (LR)), or neither antibody (lower left (LL)). In health, all cells would express both CD55 and CD59, and so bind the antibodies, so that the UR quadrant should have almost 100% of the events. However, only 26.34% of the cells express neither CD55 nor CD59, making the diagnosis of PNH. (Image courtesy of H. Bibawi, NHS Tayside).

on the blood film by microcytes and bizarre poikilocytes with marked red cell fragments. These cells, with a defect in alpha spectrin, show exceptional heat sensitivity, and their spectrin cytoskeleton denatures at a lower temperature $(45–46\,°C)$ instead of the expected $49\,°C$.

The RDW and membrane defects Three of these membrane-defect diseases provide a tutorial on the RDW. In HS (Figure 4.6), the RDW is high at 20.1% (reference range 11.5–14.5%), indicating increased variability in the size of the cells. However, in the HE red cells (Figure 4.7) the RDW is considerably greater, being 28.1%. This seems obvious in comparing the two figures, as there is likely to be more variability in the elliptocytes than in the spherocytes. However, in hereditary stomatocytosis (Figure 4.8) the RDW is normal at 13.7%, which seems acceptable if the figure is compared with HE and HS.

Metabolic defects

Chapter 3 has full details of red cell metabolic pathways (Figure 3.8).

G6PD deficiency G6PD is a key enzyme in the formation of glutathione, a crucial antioxidant which counters the damaging and toxic effects of oxygen by effectively neutralizing or quenching it. Patients with G6PD deficiency therefore suffer reduced antioxidant capacity, the consequences of which are that the red cells are more susceptible to oxidant stress, leading to increased levels of methaemoglobin, damage to their membrane and ultimately to haemolysis.

The gene for G6PD is found on the X chromosome, and so the inheritance is sex linked. The condition affects up to 1% of the world population, but is considerably higher (13%) in those of West African descent. Deficiency shows marked clinical heterogeneity and produces different manifestations, including sensitivity to different drugs. The most common consequence of G6PD deficiency is drug-induced haemolysis. The oxidant drugs that can cause haemolysis include anti-malarials, antibiotics and analgesics. The ingestion of fava beans also precipitates acute crises of haemolytic anaemia with haematuria and pain. However, the 'steady-state' condition is associated with chronic haemolysis and therefore jaundice.

Treatment for G6PD deficiency is avoidance of the precipitating factor(s), such as fava beans and the resolution of an infection. In severe cases a blood transfusion would be considered. There is no replacement therapy. In the absence of haemolytic crisis, the full blood count is normal, although during acute intravascular exacerbations the blood film demonstrates features of haemolysis, such as red cell ghosts (without haemoglobin – 'bite' and 'blister' cells) and polychromasia. Heinz bodies (haemoglobin denatured by the high levels of oxidants) may be seen in reticulocytes, although they may also be seen in chemical poisoning and in unstable haemoglobins, generally after splenectomy.

Commercial screening and assay kits are available for G6PD deficiency assessment, and these tests are based on the measurement of nicotinamide adenine dinucleotide phosphate hydrogen level by fluorescence. In health, specimens fluoresce brightly indicating normal G6PD screen, but in G6PD deficiency there is reduced or no fluorescence. Intermediate degrees of fluorescence indicate a specimen from heterozygotes or patients with a mild G6PD variant. Reticulocyte cells demonstrate higher G6PD activity.

Pyruvate kinase deficiency This enzyme helps generate adenosine triphosphate (ATP), so that failure of this enzyme leads to reduced energy within the cell. Reduced ATP leads to a mild to moderate chronic non-spherocytic haemolytic anaemia (haemoglobin typically 40–100 g/L) with splenomegaly and increased reticulocyte count. The blood film shows poikilocytes and a reticulocytosis is common. Reduced red cell survival and chronic haemolysis result in increased iron turnover – increased ferritin is found in 60% of untransfused PK patients. A further consequence of PK deficiency is a rise in levels of another metabolite in the metabolic pathway, 2,3-diphosphoglycerate (2,3-DPG).

In Chapter 3 we discussed the importance of 2,3-DPG in the movement of oxygen in and out of the red cell in the lungs and in the tissues. Thus, in PK deficiency, high levels of 2,3-DPG shift the oxygen dissociation curve to the right, which effectively reduces the clinical symptoms in comparison with the degree of anaemia. The laboratory confirmation of PK deficiency relies on the ability of PK to convert phosphoenolpyruvate and ADP to pyruvate and ATP. Lactate dehydrogenase catalyses the reduction of pyruvate to lactate with the oxidation of nicotinamide adenine dinucleotide hydrogen to nicotinamide adenine dinucleotide and the change in fluorescence is measured at 340 nm. Specimens from normal healthy individuals demonstrate reduced or no fluorescence. In contrast, samples from PK deficiency show bright fluorescence.

Table 4.5 Defects in the membrane and in enzymes.

HS and HE	Defects in structural component of the cytoskleleton leading to loss of shape
PNH	Absence of anchoring glycoprotein leads to susceptibility to attack by complement components
G6PD deficiency	Loss of defence from oxidation
PK deficiency	Failure to generate ATP, increased levels of 2,3-DPG

Table 4.5 summarizes pathophysiological aspects of membrane and enzyme defects.

Haemoglobinopathy

Section 3.5 highlighted the genetics and structure of haemoglobin. In the adult, globin genes code for different variants of globin proteins alpha, beta, delta and gamma, various combinations of which lead to different types of haemoglobin, such as HbA, HbA$_2$ and HbF. However, gene mutations give rise to different types and amounts of haemoglobin that cause certain types of anaemia – that is, haemoglobinopathy – the most common form being sickle cell disease and thalassaemia. Although the vast majority of conditions have a hereditary component, they may also arise in the newborn of parents who themselves are subsequently shown to be healthy. Nevertheless, we can classify all haemoglobinopathies by the number and types of genes that are abnormal.

The two genes for alpha globin and two genes for beta globin produce HbA ($\alpha_2\beta_2$). However, if these genes carry a mutation (such as may cause sickle cell disease; that is, $\alpha_2\beta^s\beta^s$) then the person is said to be homozygous, and they have HbS. However, someone with only one mutated gene, but a normal second gene, is said to be heterozygous (such as with genes $\alpha_2\beta\beta^s$, generating HbAS). These heterozygotes are 'carriers', and the condition being called a 'trait', and, as a whole, subjects have less severe clinical disease. It is also possible to have a combination of mutated genes, such as one in an alpha globin gene and one in a beta globin gene, or perhaps a different mutation in each of the two beta globin genes. These points will be revisited in the sections that follow.

Sickle cell anaemia Sickle cell disease, the most common haemoglobinopathy, is so named because of the characteristic sickle shape of the red cells as viewed by

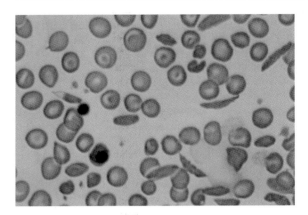

Figure 4.10 The sickle cells are evident. Also present are a nucleated red blood cell and what is likely to be an extruded nucleus. There are also some target cells, so this may be from a patient with a mixed haemoglobinopathy. (Image from Blann AD. *et al.* (2009) Haematology Morphology. The IBMS Training CD-ROM. © IBMS).

light microscopy on a blood film (Figure 4.10), or in a wet (unfixed and unstained) preparation of blood. Disease follows as sickle cells do not flow normally through capillaries, they carry oxygen poorly and they are detected and removed from the circulation, leading to anaemia. The characteristic change in shape occurs only at low oxygen tension (i.e. hypoxia), and re-oxygenating blood can lead to the reversal of a sickle cell back to normal shape.

In normal DNA, the nucleotide bases generate a globin molecule with the amino acid glutamine at position six (HbA). However, in sickle cell disease the beta globin gene is mutated (β^s) so that valine is present at position 6 in the globin chain instead of glutamine (HbS). This gives the beta globin molecule, and so the whole haemoglobin molecule, a different shape which translates to a reduced ability to carry oxygen. This mutation is called qualitative because the change does not alter the size of the molecule. Mutations in the alpha gene leading to abnormal alpha globins are rare.

If permanently damaged, the elongated rigid sickle red cells cause vaso-occlusion and obstruct small blood vessels so that other red cells cannot deliver oxygen to the tissues, leading to infarction. Sickle cells also develop increased 'stickiness', which promotes adhesion to the vessel wall and increases the risk of thrombosis. These changes lead to a chronic haemolytic anaemia, because the sickled cells are detected as abnormal and so are eliminated by macrophages in the liver and spleen.

In heterozygotes (i.e. with HbAS), the presence of normal globin molecules (alpha, delta and gamma, giving HbF and HbA$_2$) makes the polymerization less severe and easier to reverse. Thus, sickle cell carrier status leads to a clinical picture that is generally asymptomatic, although there are reports of occasional renal disturbances, and painful crises may occur at times of physiological stress. HbAS red cells do not sickle unless the oxygen saturation is <40%, a level rarely achieved in the circulation. The symptoms of anaemia are mild in relation to the low haemoglobin levels, and this is due to a right shift in the oxygen dissociation curve; that is, when HbS releases oxygen to tissues more readily when compared with normal HbA.

The frequency of the haemoglobin S gene (and therefore including both heterozygotes and homozygotes) varies in different racial, geographical and ethnic groups: West African 1:5, Afro-Caribbean 1:10, Asian 1:50, Mediterranean and Middle East 1:100. Areas where haemoglobin S is common run parallel with endemic falciparum malaria, leading to the hypothesis that haemoglobin S has risen as a protective mechanism against the parasite. In the UK more than 12 500 people have a clear sickle cell disorder. Over 240 000 are seemingly 'healthy' carriers, and each year 1 in 200 neonates are born with sickle cell disease. In the USA, perhaps 8% of African Americans are heterozygous for this mutation.

Other qualitative beta globin gene disorders Haemoglobin C is also caused by a mutation in the beta gene, leading to lysine being substituted for glutamic acid, but at position 6. Perhaps the highest frequency of this haemoglobin is in parts of West Africa, where the incidence may reach 20%. Approximately 2–3% of African Americans are heterozygous carriers, such that 1/5000 live births in this population are homozygotes. Haemoglobin D disease is a collection of different mutations in the beta globin gene. The variant with the highest frequency is in the western (Punjab) region of India and Pakistan (hence HbD$_{Punjab}$) and is caused by substitution of glutamine for glutamic acid at position 121. In the homozygous state, there is a mild haemolytic anaemia with target cells, but the heterozygote form (i.e. haemoglobin AD) has minor abnormalities.

Haemoglobin E disease is common in South-East Asia, with a frequency varying from 8% to 50%, and is caused by replacement of glutamic acid by lysine at position 26 on the beta globin chain. Haemoglobin E has a higher incidence in malarial areas, leading to the suggestion that heterozygotes have some protection. Homozygous haemoglobin E disease results in a mild haemolytic anaemia with target cells, and a compensatory erythrocytosis. The heterozygous state is asymptomatic, but the film shows microcytosis. There are several hundred other mutations in alpha and beta genes, giving rise to abnormal alpha and beta globin.

Thalassaemia This group of diseases is characterized by the reduced synthesis of either or both globin chains (hence 'quantitative' defect), this results in serious abnormalities in the intact haemoglobin molecule and the poor oxygen transport that leads to the clinical signs and symptoms. Those red cells carrying the abnormal haemoglobin do not, as in sickle cell disease, go markedly out of shape, but are still detected as defective and so are removed from the blood, leading to haemolytic anaemia.

Thalassaemia is predominantly found in the eastern Mediterranean region, the Middle East, the Indian subcontinent and South East Asia, all regions where sickle cell disease is also highly prevalent. Therefore, in common with sickle cell disease, this leads to the theory that thalassaemia carriers are protected from malaria. Medical care focuses on disease severity, regardless of genotype, of which there are two major types:

- Alpha-thalassaemia, following alpha gene mutations, is frequently present in Chinese, whereas in Thailand the gene frequency for the various forms of alpha-thalassaemia reaches 25%.
- Beta-thalassaemia, the consequence of beta gene mutations, is found in Cypriots, Asians, Chinese and Afro-Caribbeans. In southern Italy and Greece, 5–10% of the population is heterozygous for beta-thalassaemia.

Chromosome 16 carries two alpha globin genes, so that with two chromosomes there are four genes overall, which can be represented as $\alpha/\alpha/\alpha/\alpha$. Various combinations of defects are known. Mutation in only one gene is represented by $-/\alpha/\alpha/\alpha$, and so alpha globin output is 75% of normal, is generally clinically silent and without an anaemia. Mutations on two genes may be on the same $(-/-/\alpha/\alpha)$ or different $(-/\alpha/-/\alpha)$ chromosomes. This is described as alpha-thalassaemia minor, or as alpha-thalassaemia carrier. In both cases, alpha globin production is 50% of normal.

Haemoglobin H disease results from problems in three of the four alpha-globin genes, that is, $-/-/-\alpha$. Clinically, it is a type of thalassaemia intermedia and is characterized by levels of alpha globin 25% of normal, leading to jaundice, hepatosplenomegaly, leg ulcers, gall stones and folate deficiency. To compensate for the anaemia, extramedullary haemopoiesis is common, and intensive marrow hyperplasia leads to thinning of the bones and so to easy fractures. Complete loss of all four genes $(-/-/-/-,$ thus no alpha globin) results in haemoglobin Bart's, also called hydrops foetalis syndrome, and is incompatible with life. Impaired alpha globin synthesis leads to excess gamma and beta globin chains that form unstable and physiologically afunctional tetramers. In haemoglobin Barts this becomes gamma (γ_4), whilst in haemoglobin H the tetramer is β_4.

There are more than 200 different mutations of the beta globin genes (present on chromosome 11) leading to beta-thalassaemia, with reduced or absent synthesis of beta globin chains. There is also increased synthesis of gamma globin chains leading to increased HbF. As with alpha-thalassaemia, the severity of the disease, at both clinical and laboratory levels, is dependent on the number and character of the abnormal genes.

Several levels of disease activity are recognized:

- Thalassaemia major, the consequence of inheritance of two different mutations, leads to interference with beta-globin chain production, and includes both β^0-thalassaemia (where the globin chains are absent) and β^+-thalassaemia (where globin chains are partially present). Hypertrophy of the ineffective bone marrow is associated with skeletal changes and hepatosplenomegaly.
- Thalassaemia intermedia may be caused by a variety of genetic defects, and so is often classified clinically, as opposed to genetically or haematologically. In one manifestation there is homozygous beta-thalassaemia, but effects of the anaemia are countered by high HbF. In another form there are mild defects in the synthesis of beta chains.
- Heterozygotes for beta-thalassaemia, including β^0-thalassaemia and β^+-thalassaemia carrier status are often asymptomatic. These conditions are not associated with anaemia as the remaining non-mutated beta-globin gene is able to synthesize sufficient beta globin to permit oxygen carriage. Consequently, clinical presentations are mild.

Compound and other haemoglobinopathy There are many combinations of different mutations that effectively create hybrids of quantitative and qualitative defects, or indeed two different quantitative defects such as in alpha- and beta-thalassaemia. Given the overlapping geographical areas where these mutations are endemic, these associations are to be expected, and arise because of the asymptomatic nature of many types of heterozygous disease.

Examples of combined qualitative defects include combinations of haemoglobin S with both C and D, forming HbSC and HbSD. Beta-thalassaemia and sickle cell disease can combine to form a mixed defect – both HbS/β^0 and HbS/β^+ are known. Because of the dominance of the sickle haemoglobin, all of these are clinical sickling conditions, and the clinical picture resembles homozygous sickle cell disease (e.g. with splenomegaly). Unequal crossing over of globin genes leads to hybrid molecules, such as Hb Lepore, where delta globin and beta globin fuse.

Another form of mutations in both delta and beta genes lead to impaired production of two species of globin molecules, hence $\delta\beta$-thalassaemia. However, foetal haemoglobin (haemoglobin F, $\alpha_2\gamma_2$) is increased, giving at least some (limited) capacity to carry oxygen and the production of alpha and gamma globin is generally unaffected. This syndrome, hereditary persistence of foetal haemoglobin (HPFH), presents clinically as thalassaemia intermedia. However, there are other forms of HPFH caused by deletions, point mutations or cross-over of beta and gamma genes. Indeed, measurement of the level of HbF can be very informative. Table 4.6 summarizes the molecular genetics of the haemoglobinopathies.

Laboratory definition of haemoglobinopathy As in the investigation of any anaemia, a full blood count (including reticulocytes) and blood film are essential. However, a range of other tests are specific for the haemoglobinopathies, and many are based on properties of the abnormal globin molecules and of the genes

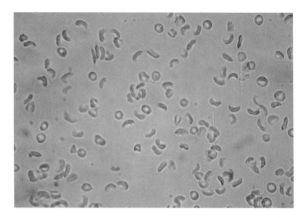

Figure 4.11 A wet preparation of whole blood from a patient with sickle cell disease that has been incubated with a buffer to induce hypoxia, and this has resulted in many red cells adopting the sickle shape. (Image courtesy of H. Bibawi, NHS Tayside).

themselves. These are essential in confirming a preliminary diagnosis and in monitoring treatment, and also have a place in screening at-risk populations.

Wet preparations are performed directly on whole blood. The simplest relies on the definition of the sickling process – that phenotypically normal-looking red cells adopt the sickle phenotype when hypoxic (as they would in the circulation) (Figure 4.11). A reducing agent (such as sodium metabisulphite) lowers the oxygen content of a blood sample, whereupon HbS cells adopt the characteristic sickle shape. The test must be performed with positive and negative controls.

A similar test is of the solubility of sickle cells. Under deoxygenated conditions HbS becomes insoluble in buffer supplemented with saponin and sodium dithionite. Normal red cells lyse and the solution becomes clear. However, the blood solution being opaque or turbid (due to the red cells being intact and in suspension) implies sickle haemoglobin. This test can be extended by centrifugation, where, in the presence of sickle haemoglobin, the buffer solution gives a clear dark red or purple coloration at the top of the tube, with a thin 'scum' of precipitated protein and red cell stroma, whilst the solution below will be pink or colourless (Figure 4.12).

A false negative may be due to deterioration of the dithionite, unstable saponin, anaemia and a post-transfusion sample. Another source of error is high levels (10–20%) of HbF, which counter the polymerization of the sickle haemoglobin and so may give a false negative. Accordingly, the test is unreliable in neonates during the first few months.

Table 4.6 Molecular genetics of the haemoglobinopathies.

Type of defect	Nature of mutation	Examples
Qualitative	Beta gene	Sickle cell disease (HbS)
		Haemoglobins C, D, E
Quantitative	Alpha gene	Alpha-thalassaemia
	Beta gene	Beta-thalassaemia
Mixed qualitative	Beta genes	HbSC, HbSD
Mixed quantitative	Delta and beta genes	$\delta\beta$-thalassaemia
Mixed qualitative and quantitative	Beta genes	HbS/β^0thal, HbS/β^+thal

(a)

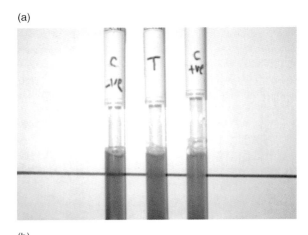

(b)

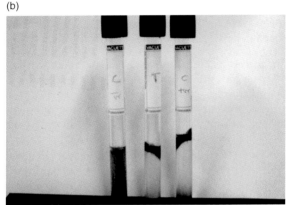

Figure 4.12 The solubility test. (a) Incubation of red cells in a lysing buffer will indicate those samples likely to come from a patient with sickle cell disease. Of the three tubes, that on the left is the negative control – the black line can easily be seen through the lysed cells. On the right is the positive control – the red cells have not been lysed and so the black line is almost completely obscured. In the middle is the test patient's sample, which clearly gives the same picture as the positive control, thus supporting the diagnosis of sickle cell disease. (b) After centrifugation, a positive result gives a dark red band at the top of clear plasma (middle and right tubes), whereas a negative result is a red column (left tube). (Images courtesy of H. Bibawi, NHS Tayside).

Different types of electrophoresis can be employed to detect different haemoglobin species as different globin molecules have their own physicochemical characteristics. Electrophoresis methods use various buffers (such as urea), different ranges of pH (pH 6.0–6.3 or pH 8.4–8.7) and different physical supports (such as cellulose acetate or agarose) (Figure 4.13).

(a)

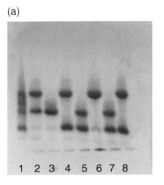

(b)

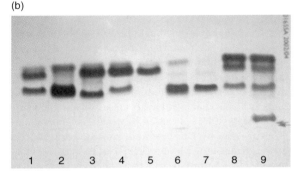

Figure 4.13 Electrophoresis. (a) At alkaline pH (8.5) together with a commercial control (AFSC) at lane 1. Haemoglobin is a negatively charged protein at alkaline pH and will migrate towards the anode (+) in an electrical field. According to their electrical charges, different Hb variants will separate into different bands. Hb variants can then be identified and compared with known control bands. The results from different patients were as follows: lane 1, AFSC control; lane 2, AS; lane 3, SS + F; lane 4, AC (or A/O-Arab, or A/E); lane 5, SC (or S/O-ARAB, or S/E); lane 6, A; lane 7, SC (or S/O-ARAB or S/E); lane 8, AC (or A/O-Arab or A/E). (b) Haemoglobin electrophoresis results on citrate agar at acidic pH (6.0) together with a commercial control (FASC). The corresponding mobility for the same patients at alkaline pH (8.5) were (from left to right) as follows: lane 1 AC; lane 2, SC + F; lane 3, AS; lane 4, AC; lane 5, A; lane 6, SC + F; lane 7, SS; lane 8, AS + F; lane 9, FASC control. (Images courtesy of H. Bibawi, NHS Tayside).

Foetal haemoglobin (HbF, $\alpha_2\gamma_2$) is dominant in the foetus, but levels fall in the first few months of life, and is present as perhaps 0.5–0.8% of total adult haemoglobin. However, increased levels are present in various haemoglobinopathies and, as such, are an aid to diagnosis and the monitoring of treatment. HbF level below 12% can be accurately quantified by high-performance liquid chromatography (HPLC) or the alkali denaturation method (also called the Betke method), which have replaced the traditional acid elution method of Kleihauer. HbF-bearing cells can also be detected by flow cytometry, immunodiffusion and by enzyme-linked immunosorbent assay.

HPLC is quite possibly the 'gold standard' technique for determination of haemoglobin variants. Although the capital investment and running costs are considerable when compared with electrophoresis, but the method can be automated (giving high throughput), uses as little as 5 μL of blood, and is accurate, rapid (perhaps 5 min), reproducible and reliable. The method is particularly suited to a high-throughput laboratory that may be required to perform antenatal screening for haemoglobinopathy (Figure 4.14).

The principle relies on different ionic properties of different haemoglobins. Positively charged haemoglobin is adsorbed onto a negatively charged stationary phase in the column. A buffer detaches the bound haemoglobin proteins from the column and the optical density of the eluted solution (carrying the haemoglobins) is measured spectrophotometrically. This result is converted into a chromatogram that shows all the haemoglobin peaks in a particular sample. The net charge on a particular haemoglobin molecule gives a unique retention time on the column that enables haemoglobin identification. As with electrophoresis, positive controls of known haemoglobin variants must be present in each batch. An additional aspect of this technique is that it can also provide levels of glycated haemoglobin (HbA_1c).

Isoelectric focusing (IEF), related to electrophoresis, also relies on the different electrical properties of haemoglobin variants. IEF plates are made from either polyacrylamide or agarose and both contain amphoteric molecules with various isolectric points, thus establishing a pH gradient, generally between pH 6 and pH 8. Different haemoglobins have different net charges, thus different haemoglobin molecules will have unique isoelectric positions on the gel (their isoelectric point). Once completed, the gel can be stained and the position of the various haemoglobin species can be quantified by densitometry. The densitometry peaks can be superimposed on traces of known variants to assist in the diagnosis.

Like HPLC, IEF demands capital investment, but advantages include the ability to separate many different variants, clear demarcation due to sharp bands, ability to separate haemoglobins D and G from S, and the requirement of a small blood sample, so that it is suitable for neonatal haemoglobinopathy investigation and screening.

Estimation of HbA_2 ($\alpha_2\delta_2$), comprising around 2% of total adult haemoglobin, may be called for in the investigation of a suspected thalassaemia. The most common method is anion-exchange chromatography, where negatively charged haemoglobin is adsorbed to positively charged resin. The definition of the reference range for this test is crucial and must be defined locally. In beta-thalassaemia carrier status, the HbA_2 level is elevated (perhaps 5%) compared with reference values (which may be up to 3.0%). HbA_2 may also be estimated by cellulose acetate electrophoresis and by HPLC. Haemoglobin variants can be detected by immunoassay. Commercially prepared kits are available for the detection of common haemoglobins S, C, D, E and A_2 with sensitivity down to 5–10%.

DNA analysis, inevitably the preserve of a reference laboratory, is used for the confirmation of haemoglobin variants, and prenatal diagnosis of serious disorders of haemoglobin synthesis. It can also be used for foetal diagnosis of haemoglobinopathy. The general area of molecular genetics is intensely technical and so very demanding of skilled scientists, very likely to be in a regional referral laboratory. The basic technique may be modified according to the mutations expected in the particular ethnicity of the patient. These are used to define the exact nature of a particular gene mutation in sickle cell disease, beta-thalassaemia and alpha-thalassaemia (Figure 4.15).

Laboratory findings in haemoglobinopathy The consistent finding from a full blood count and blood film is a reduced haemoglobin and MCV, and so a microcytosis. If the patient is symptomatic, this makes the diagnosis of microcytic anaemia. However, the sensitivity and specificity of a reduced MCV is poor as a common alternative diagnosis is iron deficiency, and this should be excluded, either by iron studies (such as serum ferritin, iron, transferrin, transferrin saturation) or additional haemoglobinopathy testing.

Bone marrow changes include a reactive erythrocytosis as the bone marrow attempts to maintain the ability of the blood to deliver oxygen to the tissues. Further evidence of this stress erythropoiesis is a reticulocytosis, and as these cells are larger than mature erythrocytes, and are certainly larger than microcytes, then the MCV may not be far below the bottom of the reference range. Thus, an increased RDW may be present, and this should be detected and reported by the haematology autoanalyser.

The blood film can also be informative with evidence of red cell destruction with schistocytes, the hallmark of a

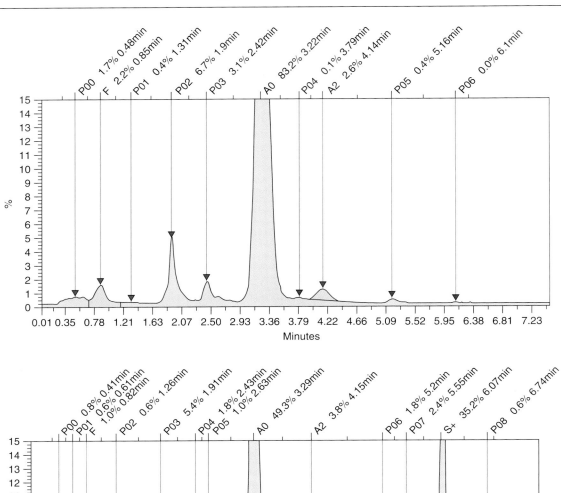

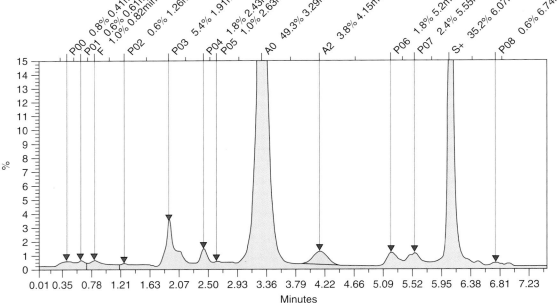

Figure 4.14 HPLC for different haemoglobin species. (a) Normal HPLC scan showing the major peaks of HbA (83.2%) and minor peaks (in red) of HbA_2 (2.6%) and HbF (2.2%). (b) The major peak in the middle of the plot, at the HbA position, makes up only 49.3% of total haemoglobin. Note the new large peak to the right, making up 35.2% of all the haemoglobin, and this is HbS. Minor peaks are HbA_2 (3.8%) and HbF (1.0%). (Courtesy of Dr S. Marwah, City Hospital, Birmingham).

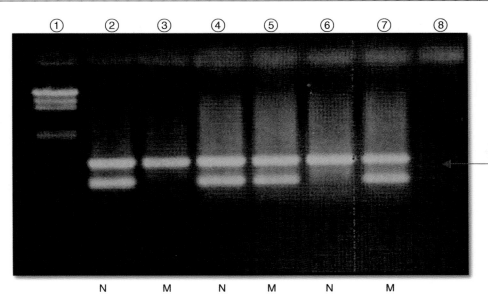

Figure 4.15 Gene analysis by polymerase chain reaction (PCR). BSu 36 I is a restriction enzyme with known restriction sites on the beta globin gene. In sickle cell disease some of these restriction sites disappear and therefore the enzyme cannot digest DNA from PCR products. Accordingly, larger PCR products can be viewed by electrophoresis in sickle cell disease (SS) compared with smaller (digested) bands in normal conditions. In heterozygous conditions (AS) both bands can be seen. Lane 1 is molecular marker, lanes 2 and 3 patient 1 (normal (N) and mutant (M) primers respectively), lanes 4 and 5 patient 2 (normal and mutant), lanes 6 and 7 patient 3 (normal and mutant primers), lane 8 is blank, the arrow points out the DNA variants. (Image courtesy of H. Bibawi, NHS Tayside).

haemolytic anaemia (Figure 4.16). These fragments are interpreted by the imaging software of haematology autoanalysers as being true cells, so that a high RDW is often found. Indeed, in Figure 4.16 the RDW is 45 (reference range 11.5–14.5). The presence of sickle cells is also a clear sign, but high numbers of target cells could be due to several alternative pathologies, including those of haemoglobin (Figure 4.17). Nucleated red cells may also be found in the blood, indicating a hyperactive or stressed erythropoiesis, although these cells are found

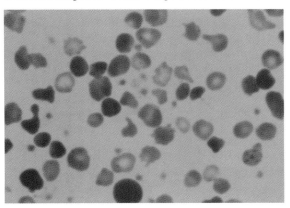

Figure 4.16 Schistocytes. This figure is from a patient with hereditary pyropoikilocytosis, which leads to a haemolytic anaemia. There are numerous examples of damaged cells and also fragments. However, there are also spherocytes, the cells that are very darkly stained. (Image courtesy of H. Bibawi, NHS Tayside).

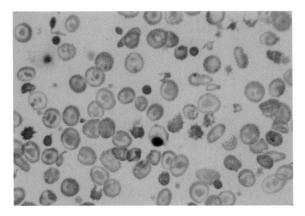

Figure 4.17 Target cells. There are many target cells present, but also microspherocytes, schistocytes, elliptocytes and a nucleated red blood cell. Consequently, the patient will have a complex and severe haemolytic anaemia. (Image courtesy of H. Bibawi, NHS Tayside).

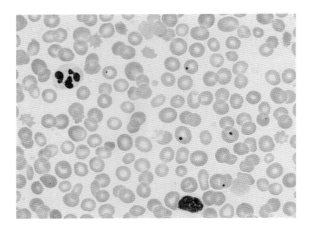

Figure 4.18 Howell–Jolly bodies. These are small but distinct purple bodies inside six red cells composed of fragments of nuclear material, often found after splenectomy or splenic disease. There are also two white cells (a neutrophil and a lymphocyte). (Image courtesy of H. Bibawi, NHS Tayside).

in numerous conditions, such as a myeloid malignancy. A consequence of splenectomy (often offered to prolong the life of the abnormal red cell) is the appearance on the blood film of Howell–Jolly bodies within red cells, although these bodies (being nuclear remnants) may also be present in any splenic pathology, and so are common in the haemoglobinopathies (Figure 4.18).

Blood science angle: Thalassaemia

A study of 236 beta thalassaemia major patients (with an average age of 24 years) found 9% to have delayed puberty, 51% to be hypogonadal, 12% to have hypothyroidism, 13% to have growth hormone deficiency and 14% to have diabetes mellitus. This 'extra' disease is not confined to the endocrine system as 8% were taking medications for heart disease, and 32% were seropositive for hepatitis C. With the exception of the latter, this pathology may be due to infiltration of endocrine and cardiac tissues with iron, although 88% were on iron-chelating therapy. This illustrates the diverse nature of many diseases and the need for Blood Scientists to have a broad knowledge of pathology.

Patients with sickle cell anaemia have a haemolytic anaemia, with a broad range of haemoglobin concentration in the range 60–90 g/L and elevated reticulocyte count of perhaps 10–20%. The haematocrit ranges between 0.18 and 0.30. Haemoglobin electrophoresis of homozygous sickle cell disease (HbSS) will report >80% HbS, with variable levels of HbF (5–15%) but a normal (trace) amount of HbA_2. In HPLC, the chromatogram shows a small peak at HbF retention time and a large peak in the HbS window. In heterozygous disease (HbAS) there is generally an excess of HbA (50–60%) compared with HbS (perhaps 35–45%). This picture can be contrasted with the relatively common combined defect sickle/beta-thalassaemia, where electrophoresis reveals that 60–90% of the haemoglobin is S whereas 10–30% is haemoglobin F. However, if the patient retains some alpha gene activity then haemoglobin A may be present. Haemoglobin A_2 is moderately elevated in sickle/beta-thalassaemia disease.

In alpha-thalassaemia carrier status (when one or two alpha genes are deleted), haemoglobin may be normal but is generally slightly reduced, providing an asymptomatic phenotype. However, the MCV (<77 fL) and MCH (<25 pg/cell) are both reduced but the red cell count may be raised ($>5.5 \times 10^9$/L). The blood film looks very similar to iron depletion. Measurement of the relative rates of alpha- and beta-synthesis may support alpha-thalassaemia confirmation.

When three alpha genes are deleted, HbH disease is present. Haemoglobin is typically 60–110 g/L, and the blood film demonstrates marked microcytosis with anisocytosis, poikilocytosis, target cells and basophilic stippling. In the neonate, electrophoresis or HPLC shows Hb Barts (γ_4) up to 25%, the remainder being HbA ($\alpha_2\beta_2$) and HbF ($\alpha_2\gamma_2$) with a small amount of HbH (β_4). As the gamma-chain production switches to beta-chain synthesis, HbH (β_4) gradually replaces Hb Barts together with disappearance of HbF. In adults the haemoglobin pattern is HbA and HbH which may range from 5–25%. HbA_2 levels are slightly decreased and the HbF level is normal or slightly increased. The high levels of beta globin can form deposits or precipitates within the cell, which can be detected on a blood film by a supra-vital stain such as brilliant cresyl blue.

Overall, the beta-thalassaemias demonstrate a broad range of clinical and haematological variability due to the heterogeneous nature of the molecular defects affecting beta-chain production. Heterozygotes for beta-thalassaemia (carriers) are generally asymptomatic, but have microcytic, hypochromic red cells with reduced MCV (typically 65–75fL), reduced haemoglobin (90–110 g/L) and a reduced MCH (20–22 pg/cell). There is also likely to be raised HbA_2 (>3.5%). In thalassaemia intermedia a very broad phenotypic picture may be found, ranging from symptomless to the requirement of blood transfusions, and the consequent iron chelation therapy. Those with a haemoglobin range from 60 to 100 g/L may not always require blood transfusion. The red cells are very microcytic and hypochromic, and there is increased erythropoiesis. These patients may demonstrate some bone deformity, enlarged spleen and liver, and features of iron overload.

Severe beta-thalassaemia major is associated with a profound anaemia (haemoglobin 20–60 g/L), reduced MCV (<65fL) and MCH, and raised reticulocyte count. Increased levels of erythropoietin are a likely response to hypoxia. The blood film shows marked changes with anisocytosis, target cells, hypochromic microcytic cells, basophilic stippling and nucleated red cells. Investigation of the bone marrow is generally unnecessary (unless a differential diagnosis needs to be excluded), but if performed should show hypercellularity with erythroid hyperplasia. Electrophoresis or HPLC demonstrates absence (in β^0-thalassaemia) or almost complete absence (β^+-thalassaemia) of normal adult haemoglobin (HbA) with almost all the circulating haemoglobin being foetal haemoglobin (HbF) and varying amounts of HbA_2 ranging from low to slightly raised.

Compound and rare haemoglobinopathies offer a challenge as there is often clustering of various laboratory findings (such as a low MCV and target cells). For example, both HbS/beta-thalassaemia and HbSC are associated with prominent target cells, basophilic stippling and a microcytosis. Furthermore, the electrophoresis and HPLC patterns in compound HbS/thalassaemia resemble HbSS as both will lack HbA. Target cells and spherocytes are commonly found in both homozygous and heterozygous haemoglobins C and E diseases with microcytosis and hypochromasia. These two variants co-migrate on cellulose acetate electrophoresis but fortunately separate on acid citrate agar electrophoresis.

Clinical aspects

A curious feature of the haemoglobinopathies is the varying relationship between a particular mutation, blood results, symptoms and clinical picture. A common finding is that two people with seemingly identical genetic and laboratory features have widely different symptoms, implying other factors yet to be discovered.

The major clinical aspects are vaso-occlusive crises due to the blockage of blood vessels by irreversible sickle cells, leading to ischaemia, bone pain and acute chest syndrome caused by sickling of red cells in the lungs. Other clinical complication features includes leg ulcers, chronic liver damage, gall stones and kidney damage. The spleen is often enlarged in infancy and early childhood due to trapping of sickle cells. Regardless of the underlyng genetic basis, severely affected patients are sustained with blood transfusions, but after repeated transfusions iron overload becomes important and may be treated with chelation (Section 4.4).

Oral hydroxycarbamide (hydroxyurea) is commonly used to reduce both the frequency and duration of sickle cell crisis. This drug increases HbF synthesis, decreases intracellular HbS level by increasing the MCV, lowers the white cell count and reduces the adhesiveness between sickle cell and endothelium. Increased platelet activation brings a risk of venous thromboembolism and thus attendance at an oral anticoagulation clinic (Chapter 8). Regardless of aetiology, and in common with many haemolytic anaemias, splenectomy is considered at the age of 5 to reduce the blood requirement, and is followed by oral penicillin therapy for life as subjects are susceptible to infections.

Bone marrow transplantation, perhaps from an human leukocyte antigen-matching sibling, may offer the prospect of permanent cure of haemoglobinopathy if carried out in early life. The success rate is over 80% in well-chelated young children without liver complications.

Prenatal diagnosis and prevention of haemoglobinopathy

The severity of many haemoglobinopathies has prompted initiatives aimed at prevention. Almost all sickle cell disease and thalassaemias are genetically inherited conditions, and the affected patients (carriers) are at risk of transmitting the disease to their descendants and those who are homozygous or double heterozygote for a disease. Hospitals with a high prevalence perform universal haemoglobinopathy screening for all antenatal attenders (when a first-time booker for pregnancy), regardless of ethnic or racial origin. Mothers who are tested positive for significant haemoglobinopathy are invited to call the father for screening.

If a couple is diagnosed with haemoglobinopathy then there is 25% chance that the foetus is homozygous, or doubly heterozygous, and 50% chance that the foetus is a carrier. In at-risk couples, prenatal diagnosis is offered using either DNA (chorionic villous or amniotic fluid) or foetal blood. The foetal DNA is amplified by using PCR and the DNA mutation(s) are detected. In homozygous disease of the foetus, the couple should be counselled and if appropriate termination may be offered.

4.5 External factors acting on healthy cells

We have already seen how attack on an 'innocent' blood marrow causes anaemia. Another is that perfectly good red cells in the blood are attacked or destroyed by outside agent or agents: the cells themselves are not intrinsically defective. This haemolysis is therefore fundamentally different from that in the haemoglobinopathies and other conditions where red cells are destroyed because they are intrinsically defective. These external agents include antibodies, physical damage, drugs and infections. Anaemia may also follow disease in certain organs, in which case the red cell is again an innocent bystander.

Antibodies

Ideally, antibodies should be directed only towards pathogenic microbes. However, for various reasons, some recognize, and thus target for destruction, the body's own tissues. This is called autoimmunity, which is fully explained in Chapter 9.

Autoimmune haemolytic anaemias These anaemias are caused by antibodies directed towards molecules on the surface of the red cell, causing their destruction, hence autoimmune haemolytic anaemia (AIHA). However, autoantibodies are not always pathogenic. Perhaps 1 in 10 000 apparently healthy blood donors tests positive for autoantibodies, a frequency which is age dependent with an increased likelihood in the elderly.

The attachment of autoantibodies to the surface of red cells can be detected in the laboratory with the characteristically positive direct antiglobulin test (DAT). A variant of this test is the indirect antiglobulin test (IAT). Dependent on the temperature at which the antibody reacts with red cells, the anaemias may be 'cold' or 'warm'.

Warm AIHA is the more common form, causing 80–90% of cases. Red cells become abnormally coated with immunoglobulin (generally of the IgG class), possibly in association with complement, which may lead to the lysis of the cell. However, complement alone can bind to the cell, and with an appropriate antibody can cross-link and thus agglutinate the red cells. The coating of red cells by antibody and/or complement leads to their targeting by phagocytes (macrophages and neutrophils) in the spleen (hence the frequent sign of splenomegaly) and the destruction of the cell. Warm AIHA may occur alone or in combination with other conditions such as idiopathic thrombocytopenic purpura and systemic lupus erythematosus (SLE),

although chronic lymphocyte leukaemia and lymphoma are associated with perhaps half of all cases.

In cold AIHA, reactive antibodies are predominantly IgM cold agglutinin antibodies that bind to red cells optimally at 4 °C and readily fix complement leading to haemolysis both within (intravascular) and outside (extravascular) blood vessels, the latter in the spleen and liver. Patients may have chronic haemolytic anaemia compounded by cold and often associated with intravascular haemolysis, mild jaundice and splenomegaly. Laboratory findings include spherocytes, red cell agglutinates in the cold, and a positive DAT.

Paroxysmal cold haemoglobinuria (N.B. not to be confused with PNH, explained in Section 4.4) is caused by the Donath–Landsteiner antibody. This IgG autoantibody specifically binds to P blood group antigens on the surface of the red cells in the cold (such as in the skin, which is cooler than the rest of the body). The problem arises when these antibody-coated cells return to the warmth of circulation and attract the attention of the complement system, which becomes activated and so destroys the cell.

Treatment of all types of AIHA includes suppression of autoantibody production with steroids (such as prednisolone). In severe disease, with intractable and life-threatening anaemia, blood transfusion may need to be considered. However, a further option is splenectomy, which is effective because the spleen is the major site of the removal of damaged cells, so that, in effect, these damaged cells remain in the circulation and so may still be able to contribute to oxygen transport. Treatment of a cold AIHA includes keeping the patient warm.

Alloimmune haemolytic anaemia This condition is caused when an antibody produced by one individual reacts with the red cells of another. It follows that there must be the physical introduction of someone else's blood (that person generally being the 'donor') into the body of the patient (i.e. the 'recipient'). The only condition where this occurs is mismatched blood transfusion, haemolytic disease of the newborn (Chapter 11) and in the aftermath of bone marrow transplantation (Chapter 10).

The role of the laboratory in antibody-mediated haemolysis A full blood count, reticulocyte count and blood film are essential initial investigations in all haemolytic anaemias, and we expect a reduced haemoglobin and, if the reticulocyte count is grossly raised, there may also be a raised MCV. Nucleated red cells may be present, reflecting (like reticulocytosis) a reactive increase in erythropoiesis. Bone marrow examination is rarely needed, but if so is likely to demonstrate erythroid hyperplasia.

The blood film is likely to reveal polychromasia (reflecting reticulocytosis), schistocytes and spherocytes (Figure 4.16 and Figure 4.17). There may also be indications of the cause of the haemolytic anaemia. For example, raised white cell count in leukaemia or low platelets in idiopathic thrombocytopenia purpura are diseases where AIHA may occur.

In Chapter 3 we learned of the fate of the red cell and how one sign of this is the bilirubin that (broadly speaking) makes our plasma and urine yellow and makes our faeces brown. Thus, in haemolysis there should be increased bilirubin, haemoglobin and haptoglobin in the plasma, and increased urobilinogen in the urine. In severe cases there may also be free haemoglobin and haemosiderin in the urine.

The DAT and IAT can be performed at different temperatures to search for warm or cold antibodies. However, in investigating a warm antibody, all reagents (such as saline) and equipment (including pipette and plastic test tubes) must be pre-warmed to 37 °C. Similarly, the ambient temperature of the laboratory (unless air conditioned and controlled) may vary by several degrees during the day, and so reactions positive at 9 A.M. may not be at 5 P.M. and vice versa.

Both DAT and IAT are susceptible to artefact, such as due to high levels of immunoglobulin, to heterophile antibodies (as are found in infectious mononucleosis), syphilis, and the presence of rheumatoid factor and other autoantibodies. The specialist laboratory will also be able to offer specialist variants of the DAT and IAT which can be used to probe for rare or unusual antibody reactions. These include techniques such as the use of polybrene to overcome the natural electrostatic repulsion between red cells, of albumin to enhance a weak agglutination reaction, or treatment of serum with 2-mercaptoethanol or dithiothreitol to denature IgM antibodies and so reduce their ability to agglutinate red cells. These and other complex tests are explained fully in Chapter 11 on blood transfusion.

Blood science angle: Antibodies

Faced with the possibility of AIHA (where the exact specificity of the antibody is unknown) and PNH (where the antibody is likely to be directed against blood group P), the haematologist will need the help of colleagues in the blood transfusion laboratory who can offer DAT and IAT. These will help confirm or refute an initial diagnosis.

Physical damage

Mechanical causes of haemolysis include prolonged and/or unaccustomed exercise or physical activity, or endurance sports such a running a marathon. Micro-angiopathic haemolytic anaemia is associated with physical damage to red cells either on abnormal surfaces (such as an artificial heart valve) or by damage caused by red cells travelling through fibrin strands deposited in capillary microvessels (as may happen in disseminated intravascular coagulation).

Drug-induced haemolytic anaemia

Haemolytic anaemia can occur as a result of a particular drug being potentially toxic, possibly by accidental or deliberate overdose. Drug effects are moderately common precipitants, accounting for 10–20% of haemolytic anaemias. Several mechanisms exist:

• Direct chemical destruction of the membrane, or perhaps generating an increased oxidative stress leading, for example, to oxidation of the membrane and of haemoglobin. The DAT is negative.
• Development of an antibody that is directed against a drug, such as penicillin or tetracycline. In this case the drug binds or otherwise sticks to the cell membrane, which is recognized and bound by an antibody, thus rendering the cell more susceptible to attack. This mechanism may also be called 'neoantigen', or 'immune complex', and the DAT is positive.
• The drug itself (such as alpha methyl dopa) stimulates the production of an autoantibody to a component of the normal red cell membrane, but the drug itself does not directly take part in the reaction. The DAT is positive.
• The drug itself (such as hydrochlorothiazide) directly binds to the red blood cell, and this combination induces the production of an antibody towards the combination of the drug and the red cell. The DAT is generally positive.

Infections

Infection can lead to haemolytic anaemia via multifactorial routes, a good example being *Clostridium perfringens* septicaemia, which is associated with acute haemolysis and spherocytes. The blood film can be informative as meningococcal infection may be associated with fragmented red cells (schistocytes). A high temperature, such

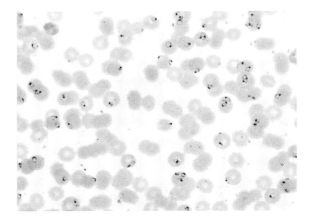

Figure 4.19 Malaria. A marked infection of the ring-form trophozoites of *Plasmodium falciparum*. Perhaps 20–25% of cells are carrying a parasite. (Image courtesy of H. Bibawi, NHS Tayside).

as may be present in response to influenza, may also cause red cell destruction. This is likely to be semi-physiological, as a small number of red cells, perhaps those which are aged, are likely to be more sensitive to brief adverse changes such as a fever. This explains the frequent reports of more yellow-coloured urine during and shortly after an episode of influenza. Snake, bee and spider venoms may produce acute intravascular haemolysis.

Infections with the intracellular parasite malaria (of which there are several types) (Figure 4.19) are also likely to precipitate a haemolytic anaemia, in addition to other laboratory and clinical changes, such as a raised ESR and fevers respectively. These occur because the parasite-loaded cell is detected as being abnormal, and so is eliminated. Different species of the parasite that causes malaria enter the red cell via particular cell-surface molecules: *Plasmodium falciparum* binds to glycophorins A, B and C, whilst *Plasmodium vivax* and *Plasmodium knowlesi* gain entry via the Duffy blood group molecule (Chapter 11).

Haemorrhage

Blood loss, and thus anaemia, may be the result of various conditions. Normal healthy women lose blood approximately each month as part of their menstrual cycle, and this also contributes to lower haematology indices in their sex. However, excessive blood loss (heavy periods – menorrhagia) or other uterine disease can lead to a normocytic anaemia.

Anaemia may arise from the physical loss of blood from bleeding intestinal lesions. In the upper section this may be caused by oesophagitis and oesophageal varices. In the lower section of the intestines, blood may be lost from haemorrhagic ulcers, colorectal cancer, inflammatory bowel disease (colitis, Chron's disease) and haemorrhoids. Indeed, occult bleeding is an early sign of gastrointestinal cancer and may lead to normocytic anaemia. However, if chronic blood loss persists, and iron stores fall, the anaemia may become microcytic.

Other 'external' causes of anaemia

Red cells may also be destroyed as a consequence of many other types of disease, examples of which include thrombotic thrombocytopenia purpura and haemolytic uraemic syndrome. Red cells may also be adversely affected by high levels of urea (i.e. uraemia) and of bilirubin (i.e. hyperbilirubinaemia). The peripheral blood film generally shows fragmented and distorted red cells (schistocytes), often of triangular shape, indicating the 'tearing' or 'shearing' mechanism. By definition, the DAT and IAT are negative and are therefore tests of exclusion.

The anaemia of chronic diseases

Many inflammatory diseases are associated with a normocytic anaemia. These include SLE, rheumatoid arthritis, thyroiditis, glomerulonephritis and hepatitis, key drivers of which are pro-inflammatory cytokines such as interleukin-6. There is experimental evidence of bone marrow suppression, so that the targets are not mature red cells but their stem cells and precursors such as erythroblasts. This has also been described as the anaemia of inflammation, and is normocytic.

However, the anaemia of inflammation may also be associated with low serum iron, low serum iron-binding capacity and normal to elevated ferritin concentrations. Other experimental data show that inflammatory cytokines contribute to the low iron status by inducing the synthesis of hepcidin (high levels of which may inhibit the absorption of iron), so that the liver would seem to be the target. If present, the anaemia is likely to be microcytic.

Several aspects of anaemia in cancer must be considered, and we have already looked at some of these in the section on bone marrow (Section 4.1). One aspect is a direct effect of the tumour on erythropoiesis by suppressing renal erythropoietin production or by a direct effect on the bone marrow that is likely to produce a normocytic anaemia. An alternative is the secondary effect of the tumour. Examples of this include the chronic loss of blood from a bleeding gastrointestinal cancer (which can be detected by faecal occult blood) and the effects of liver cancer. Finally, many malignancies have an inflammatory component, independently causing an abnormal ESR, and this too may lead to a normocytic anaemia.

4.6 Erythrocytosis and polycythaemia

These conditions are characterized by an increase in the total red cell mass of the body, and also a high haematocrit. The increased red cell mass brings hyperviscosity, and so a stress on the heart and cardiovascular system, which may precipitate a stroke. However, polycythaemia and erythrocytosis, and their consequences, are the product of very dissimilar pathologies.

Erythrocytosis

This is defined by a red cell count being above the top of the reference range, but haemoglobin and haematocrit are also likely to be increased. Precise definitions vary: one suggests a haematocrit >0.52 in males and >0.48 in females, another of a haemoglobin >185 g/L in men and >165 g/L in women.

Pathology The simplest explanation for erythrocytosis is a reduction in the volume of plasma, as may be present in dehydration, and is termed 'relative erythrocytosis'. Alternatively, 'absolute erythrocytosis' is applicable in those where red cell mass is within the reference range. Most cases of absolute erythrocytosis can be classified as being congenital or acquired. A principal cause of the former is an abnormal haemoglobin, but most causes are acquired and are generally reactive to a state of hypoxia. One of the functions of the kidney is to sense levels of oxygen, and upon a preset low level (i.e. hypoxia) will increase the release of erythropoietin to stimulate the bone marrow to produce more red cells. Another cause of hypoxia may be lung disease, such as congestive obstructive pulmonary disease. Erythrocytosis may also arise from cyanotic congenital heart disease.

The erythrocytosis of high altitude (as may be found in the Andes and Himalayas) is a required physiological response to the atmospheric hypoxia. When the subject descends to sea level, the erythrocytosis (as it is at 'normal' atmospheric oxygen levels) is an unnecessary burden and slowly resolves. It follows that those unaccustomed to high altitude need weeks of acclimatization to allow their bone marrow to increase the red cell mass, and is the basis of altitude training by athletes.

The laboratory in erythrocytosis In addition to changes in red cell indices (raised haemoglobin, red cell count, haematocrit), there is also likely to be a neutrophilia and a thrombocytosis for reasons which are unclear. However, smokers may have a neutrophil leukocytosis, possibly the result of chronic pulmonary inflammation. Bone marrow examination is rarely required but may be called on to exclude polycythaemia. Estimation of red cell mass requires the use of radioisotopes and is therefore reserved for specialist centres. However, estimation of erythropoietin is probably the most useful diagnostic tool, and levels are typically raised in erythrocytosis secondary to hypoxia, but also in polycythaemias of various aetiologies.

Management of erythrocytosis The first step is to identify and treat contributing factors (where possible) such as smoking, alcohol and treatments of hypertension that are based on fluid elimination by diuretics. The standard treatment is venepuncture to reduce the red cell mass, and generally when the haematocrit exceeds 0.54. Patients having had a thrombosis, or who are deemed at being at risk, perhaps by virtue of thrombocytosis and/or hyperviscosity, may require anticoagulation (Chapter 8). In this high-risk group, venesection may be called for at a lower haematocrit, such as 0.45, especially in the presence of diabetes or hypertension.

Chuvash polycythaemia This disease results from a mutation in the von Hippel–Lindau tumour suppressor gene leading to an abnormal protein product, which effectively fails to switch off a hypoxia-sensing mechanism (HIF-1α and HIF-2α). Consequently, the body continues to behave as if hypoxia is present, and so results in high levels of erythropoietin (up to 10 times normal) and so raised haemoglobin of perhaps 180 g/L. There is also increased frequency of varicose veins (that would seem to be the product of venous congestion) and thrombosis.

Polycythaemia

The aetiology of this disease is completely different from that of erythrocytosis, in that it is primarily the consequence of molecular changes in genes involved in responses to hypoxia. The disease extends from red cells to other blood cells and can transform into life-threatening malignancy such as myelofibrosis or chronic myeloid leukaemia.

Pathology The key to diagnosis is the relationship between erythropoietin and its receptor (the EpoR), which activates a cytoplasmic tyrosine kinase protein called Janus kinase 2 (Jak2) that ultimately results in transcription of various genes within the nucleus of the erythroid precursor (such as the erythroblast). Jak2 is also allied to the thrombopoietin receptor pathway. Therefore, failure of this system, whether by lack of erythropoietin or loss of function mutation in the *EpoR* gene, will result in failure to switch on the erythroblast and so falling red cell numbers.

However, a mutation in the *Jak2* gene confers on erythroblasts an increased sensitivity to erythropoietin, and so an erythrocytosis. In addition, this condition can progress to full-blown multicell lineage pancytosis with increased platelets (thrombocythaemia) and leukocytes, and also to myeloid leukaemia. Accordingly, this disease is described as polycythaemia vera (where 'vera' derives from 'true').

The laboratory in polycythaemia Proposed criteria for the diagnosis of polycythaemia vera include raised red cell mass or haematocrit greater than 0.60 in males and 0.56 in females, absence of a firm alternative cause of secondary erythrocytosis, thrombocytosis (platelets greater than 450×10^9/L) and a neutrophil leukocytosis (neutrophils greater than 10×10^9/L in nonsmokers, greater than 12.5×10^9/L in smokers). Other definitions include an increase in the red cell mass (>36 cm^3/kg in men and >32 cm^3/kg in women). As with erythrocytosis, the primary specific laboratory test is for levels of erythropoietin.

Bone marrow examination is not crucial in simple unequivocal polycythaemia, but may be necessary to determine transformation to more severe multi-lineage disease. It may also serve as a baseline for subsequent investigations, and an aspirate is expected to demonstrate marked erythroid hyperplasia with mild to moderate hyperplasia of granulocyte precursors and megakaryocytes. Cytogenetic abnormalities (such as trisomy of chromosomes 8 and 9) are found in 10–20% of patients and are a strong risk factor for progression to acute leukaemia and the myelodysplastic syndrome.

Management As presenting signs include thrombosis, haemorrhage and splenomegaly, aims of treatment include reducing the risk of these events, such as the use of aspirin. Polycythaemia rubra vera (PRV) refers to a clinical sign – a ruddy complexion resulting from a high red cell mass. Venesection will reduce the red cell mass (as monitored by the haematocrit, aiming at less than 0.45) and therefore the risk of cardiovascular complications, but will not alleviate the root cause of the disease. Accordingly, bone marrow suppression with radiotherapy, radioactive phosphorus and chemotherapy (such as anagrelide) is advocated.

Risk factors for survival and the development of leukaemia include age >70 years, a white cell count $>13 \times 10^9$/L and thromboembolism at diagnosis. Table 4.7 summarizes erythrocytosis and polycythaemia.

Blood science angle: Polycythaemia and erythrocytosis

The leading cause of polycythaemia rubra vera (PRV) is mutation Val617Phe in *Jak2*. A regional molecular laboratory is likely to be able to offer advice on identification of this mutation, and presence of this mutation considerably influences the diagnosis. Indeed, it is rapidly becoming the case that the mutation defines the disease, so that if Val617Phe is not present the disease is not PRV – thus initiating the search for a real cause.

Similarly, if the regional genetics laboratory finds a mutation in the von Hippel–Lindau tumour suppressor gene (leading to Arg200Trp or to Val130Leu), this makes the diagnosis of Chuvash polycythaemia. However, other mutations are known which lead to the diagnosis of non-Chuvash polycythaemia. Other genetic analysis has led to further relevant findings, such as the involvement of transferrin and the transferrin receptor and low levels of hepcidin.

Table 4.7 Erythrocytosis and polycythaemia.

	Erythrocytosis	Polycythaemia
Primary pathophysiology	Dehydration, abnormal haemoglobin, Response to hypoxia, abnormal erythropoietin	Mutation in gene(s) (primarily JAK V617F) regulating cell numbers
Laboratory	Raised red cells alone	Initially raised red cells, then raised platelet and white cells
Treatment	Address hypoxia, venesection	Bone marrow suppression, venesection

4.7 Molecular genetics and red cell disease

Whole sections of this chapter considered the consequences of defects in various genes for iron metabolism, membrane components, enzymes and haemoglobin that cause haematological disease. It could be argued that molecular genetics has revolutionized our understanding of many of these diseases. Indeed, the case of sickle cell disease is widely taken as the best example of a single gene defect causing a great deal of human morbidity and mortality. In many cases, the fine points of diagnosis are not entirely relevant to the therapist, as the link between the precise genetic basis of a disease and how it affects the lifestyle of the particular patients is often imperfect.

Nevertheless, geneticists have discovered many different mutations in assorted genes coding for different components on the cytoskeleton that cause shape deformities that shorten the life of the cell, thereby leading to anaemia. Complex electrophoresis can demonstrate abnormalities in the proteins coded for by these genes. However, knowledge of the exact molecular cause of the disease is often of minimal value in management, which is concerned with patient well-being. Examples of some of these are presented in Table 4.8. However, in

other cases, such as certain leukaemias, the exact genetic basis of a disease is of crucial importance both in diagnosis and in management. This we will revisit in Chapter 6.

4.8 Inclusion bodies

The normal red cell cytoplasm contains only haemoglobin and metabolic enzymes, and with conventional staining, there should be no detectable intracellular particles or other material. Therefore the presence of 'inclusion bodies' implies a pathological process, and the finding of nucleated red cells (Figures 4.17 and 4.20) and malarial parasites (Figure 4.19) are prime examples. The most common molecular inclusion bodies, and their composition, are as follows:

- Howell-Jolly bodies: remnants of DNA (Figure 4.18)
- Pappenheimer bodies: particles of iron-rich protein (haemosiderin), therefore defining a siderocyte in the peripheral blood and a sideroblast in the bone marrow (Figure 4.4).
- Basophilic stippling: denatured RNA and ribosomes (Figure 4.5)
- Heinz bodies: oxidised denatured haemoglobin, generally the result of failure of anti-oxidants, and so may be present in G6PD deficiency. However, they may be found in haemoglobinopathy and disease of the liver and spleen.
- Dohle bodies: small bodies found in the cytoplasm of neutrophils. Their exact composition and pathological significance are unclear.
- Cabot rings: These are fine, purple filamentous loops or figures of eight that are believed to be microtubule remnants of mitotic spindles. They have been associated with leukaemia, lead poisoning, and anaemia.

4.9 Case studies

Case study 1

An 8-year-old boy presents with an increasing tiredness and lethargy, which his parents say is similar to cousins on both sides of the family. On examination he appears well. A blood sample was obtained and was sent for a full blood count, ESR and blood film. Results are as follows:

Table 4.8 Examples of molecular genetics in red cell pathology.

Condition	Genetic lesion
Diamond–Blackfan anaemia	Genes coding for ribosomal proteins
Certain megaloblastic anaemias	Genes involved in vitamin B_{12} and folate metabolism, e.g. methylenetetrahydrofolate reductase
Porphyria	Genes coding for enzymes involved in haem synthesis, e.g. ALA synthase
Hereditary haemochromatosis	HFE
Hereditary spherocytosis	Genes for membrane and cytoskeleton components such as Band 3 and ankyrin
Certain types of haemolytic anaemia	Loss of function in genes coding for glucose-6-phosphase dehydrogenase and pyruvate kinase
Sickle cell disease and thalassaemia	Genes for alpha and beta globin

Analyte	Result	Reference range
Haemoglobin	123 g/L	133–167
Red blood cell count	5.5×10^{12}/L	4.3–5.7
Haematocrit	0.42	0.35–0.53
MCV	76 fL	77–98
MCH	22.4 pg	26–33
MCHC	293 pg/L	330–370
ESR	8 mm/h	<10
RDW	14.5%	10.3–15.3
Reticulocytes	135×10^9/L	25–125

Blood film: see Figure 4.20.

Discussion   There is low haemoglobin, MCV, MCH and MCHC. The reticulocyte count is above the top of its reference range. The red cell count, haematocrit, RDW and ESR are within their reference ranges. The case is symptomatic and has several red cell abnormalities, so is anaemic. As the MCV is below the bottom of the reference range, it is a microcytic anaemia. The major causes of microcytic anaemia are insufficient iron reaching the bone marrow and haemoglobinopathy.

The film is dominated by a cell with a purple nucleus; this is not a lymphocyte but is a nucleated red cell. This is likely because the cytoplasm is very blue, and the nucleus does not take up >90% of the cell. There is a reticulocyte above to the left of the nucleated cell. There is also a minor degree of anisocytosis (variation of the size of the cells).

The presence of a nucleated red cell implies stress on the bone marrow and overactive erythropoiesis. This is supported by the high reticulocyte count. Although the RDW is not raised, it is very close to the top of the reference range, and this is probably supportive of a red cell problem. The haemoglobin is not particularly low and the film does not look hypochromic. This rather counts against iron deficiency (where nucleated red cells are very rare), so is suggestive of a haemoglobinopathy. There are clearly no sickle (Figure 4.10) or target cells (Figure 4.17), so this may be a thalassaemia or a non-sickling beta globin abnormality.

Iron studies will obviously help, but diagnosis of haemoglobinopathy calls for sickling tests (Figure 4.11 and Figure 4.12) and haemoglobin electrophoresis (Figure 4.13) and/or HPLC (Figure 4.14). If in doubt, and if necessary, DNA studies will be helpful (Figure 4.15). The final diagnosis is (relatively mild) alpha thalassaemia trait. There is no evidence of haemolysis.

Case study 2

A 67-year-old man from northern Europe presents with increasing tiredness. On examination, there was a tender liver. A blood sample was obtained and was sent for a full blood count, ESR and blood film. Results are as follows:

Analyte	Result	Reference range
Haemoglobin	129 g/L	133–167
Red blood cell count	4.9×10^{12}/L	4.3–5.7
Haematocrit	0.50	0.35–0.53
MCV	102 fL	77–98
MCH	26.3 pg	26–33
MCHC	258 pg/L	330–370
ESR	12 mm/h	<10
RDW	12.5%	10.3–15.3
Reticulocytes	55×10^9/L	25–125

Blood film: see Figure 4.21.

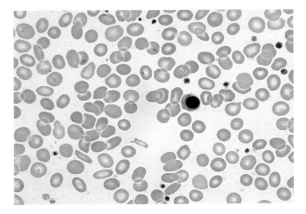

Figure 4.20 Blood film for case study 1. (Image courtesy of H. Bibawi, NHS Tayside).

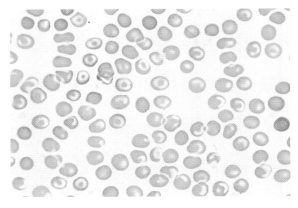

Figure 4.21 Blood film for case study 2. (Image courtesy of H. Bibawi, NHS Tayside).

Discussion   There is low haemoglobin, high MCV, low MCHC and raised ESR. The patient is symptomatic with abnormal red cell indices, so is anaemic. The MCV is above the top of the reference range, defining a macrocytic anaemia. The major cause of macrocytic anaemia is insufficient vitamin B_{12}, so that this is the next likely test needed to confirm the diagnosis. The blood film shows many target cells (Figure 4.17) and some macrocytes (Figure 4.3, which accounts for the raised MCV).

The major cause of target cells is haemoglobinopathy, indicating another possible diagnosis, and this too can be confirmed or refuted by additional tests such as the sickle solubility test, although there are clearly no sickle cells present (Figure 4.10). Tests for thalassaemia are more complicated and require electrophoresis and/or HPLC.

The number of target cells is marked, and this abnormality is rare in the macrocytic anaemia of vitamin B_{12} deficiency, so this seems an unlikely diagnosis. A haemoglobinopathy seems unlikely as the patient's racial and ethnic origin is atypical for this condition, although a non-familial mutation could account for an abnormal haemoglobin. One diagnosis that does fit the picture is alcohol liver disease, which causes the red cell morphological abnormality and the macrocytosis, and may also explain the anaemia as this organ produces the molecule (transferrin) carrying iron from the intestines to the bone marrow. Notably, the patient has a tender liver, but there are numerous explanations for this. LFTs will be in order (Chapter 17).

Summary

- There are several alternatives for the definition and classification of anaemia.
- Anaemia can follow damage to, and/or stress of the bone marrow.
- Deficiency in iron, perhaps due to malabsorption, is the most common cause of microcytic anaemia.
- Deficiency of vitamin B_{12}, perhaps due to autoimmune disease, is the most common cause of macrocytic anaemia.
- Increased iron, caused by HH or hypertransfusion, can cause disease.
- Defects in membrane components and metabolic enzymes lead to haemolytic anaemia.
- The principal haemoglobinopathy conditions are sickle cell disease and thalassaemia.
- External causes of haemolytic anaemia include antibodies, drugs and mechanical processes, whereas haemorrhage is associated with a normocytic anaemia.

- Erythrocytosis is an increase in the number of red cells, the haematocrit and haemoglobin, often a response to hypoxia.
- In polycythaemia, there are also raised red cell indices, but in addition there are raised platelets and/or white cells. The likely cause is genetic.
- Molecular genetics has demonstrated the exact basis of many red cell diseases.

References

Serjeant GR. Blood transfusion in sickle cell disease. A cautionary tale. Lancet. 2003;361:1659–1660.

Further reading

Ataga KI, Cappellini MD, Rachmilewitz EA. Beta thalassaemia and sickle cell anaemia as paradigms of hypercoagulability. Br J Haematol. 2007;139:3–13.

Barcellini W, Bianchi P, Fermo E. *et al.* Hereditary red cell membrane defects: diagnostic and clinical aspects. Blood Transf. 2011;9:274–277.

Beutler E. Glucose-6-phosphatase deficiency: a historical perspective. Blood. 2008;111:16–24.

Bolton-Maggs PH, Langer JC, Iolascon A. *et al.* Guidelines for the diagnosis and management of hereditary spherocytosis. Br J Haematol. 2012;156:37–49.

Cullis JO. Diagnosis and management of anaemia of chronic disease: current status. Br J Haematol. 2011;154:289–300.

Horos R, von Lindern M. Molecular mechanisms of pathology and treatment in Diamond–Blackfan anaemia. Br J Haematol. 2012;159:514–527.

Packman CH. Haemolytic due to warm autoantibodies. Blood Rev. 2008;22:17–31.

Parker C, Omine M, Richards S. *et al.* Diagnosis and management of paroxysmal nocturnal hemoglobinuria. Blood. 2005;106:3699–3709.

Ryan K, Bain BJ, Worthington D. *et al.* Significant haemoglobinopathies: guidelines for screening and diagnosis. Br J Haematol. 2010;149:35–49.

Spivak JL, Silver RT. The revised World Health Organisation diagnostic criteria for polycythaemia vera, essential thrombocytosis and primary myelofibrosis: an alternative proposal. Blood. 2008;112:231–239.

Thein SL, Menzel S. Discovering the genetics underlying foetal haemoglobin production in the adult. Br J Haematol. 2009;145:455–467.

5 White Blood Cells in Health and Disease

Learning objectives

After studying this chapter, you should be able to:

- understand how white blood cells develop in the bone marrow, and the importance of the thymus, lymph node and spleen;
- explain the functions of the different members of the white cell family;
- recognize the different types of mature white blood cells in a Romanowsky-stained blood film;
- describe numerical abnormalities in the white blood cell count, and their causes;
- explain the value of the CD system of classification and importance of the fluorescence-activated cell scanner in enumerating different groups of leukocyte;
- appreciate how the different white blood cells work together to defend us from microorganisms such as viruses and bacteria.

In this chapter we will look at white blood cells, also known as leukocytes (or leucocytes), how their structure and function are interrelated, and how they protect us against attack by microbial pathogens. We will leave the malignancies of the white cell for Chapter 6, and certain other pathological aspects of white cell function (i.e. immunodeficiency and autoimmunity, where the body attacks itself) for Chapter 9. However, as we will see, certain white cells also perform other physiological and pathological duties not directly related to defence. As with red blood cells, there is only one type of white blood cell in the blood, so we will refer to them simply as white cells and the number of the cells as the white cell count (WCC).

Following an introduction, we will look at how white cells develop from the bone marrow (Section 5.2), and then examine each of the five types of white cell in Sections 5.3–5.7. How these cells defend us from infection, by inflammation and immunity, is explained in Section 5.8. The chapter closes in Section 5.9 with a discussion of increased numbers of the different white cells in the blood.

5.1 Introduction

White cells defend us from infection by microbial pathogens. In some cases, this defence is in-built (or innate), but in other cases it is acquired after encountering a foreign body for the first time, and also develops (is adaptive), improving its efficiency. The process of defence is twofold: immunity and inflammation; and we expect an increased white cell count (a leukocytosis) whilst this defence is in process. However, they may also be increased in several conditions, such as in leukaemia and after surgery. The reverse, a low white cell count, is leukopenia. All blood cells develop in the bone marrow, and there are also important roles for the thymus, lymph node and spleen. The meanings of several important descriptive terms, and some examples, are presented in Table 5.1.

The number of white cells in the peripheral blood is usually maintained within fairly tight limits, with a reference range of perhaps $(4.0–10.0) \times 10^9/L$, although these figures can vary between different laboratories. However, recall that this reference range is not a normal range, as many people can mount perfectly adequately immunological and inflammatory responses with a white cell count of $3.9 \times 10^9/L$ or $10.1 \times 10^9/L$. Nevertheless, results consistently outside the reference range are likely to indicate a disease process.

Although it is simple to say that white cells defend us from microbial attack, this statement belies an extremely

Blood Science: Principles and Pathology, First Edition. Andrew Blann and Nessar Ahmed.
© 2014 John Wiley & Sons, Ltd. Published 2014 by John Wiley & Sons, Ltd.

Table 5.1 Meaning of common terms.

Term	Meaning	Example
-cyte	Cell	Leukocyte
Leuco-, or leuko-	Of white blood cells	Leukocyte
-aemia	Of the blood	Leukaemia
-myelo	Of the bone marrow	Myelocyte, myeloma
-penia	Low levels, generally below the bottom of the reference range	Neutropenia
-cytosis, -philia	High levels, generally above the top of the reference range	Leukocytosis, neutrophilia
Lympho-	Of lymph nodes	Lymphocyte, lymphoma
-blast	An immature or unusual cell	Myeloblast

complex process, as there are five types of cell operating in the blood and various organs. These five cells are, in order of frequency in the blood, the neutrophil, lymphocyte, monocyte, eosinophil and basophil. Neutrophils, eosinophils and basophils all have many granules in their cytoplasm, so are collectively granulocytes. The cells are identified by their size, the shape of their nucleus and the colour of granules in the cytoplasm (if present) after staining with suitable dyes.

In some conditions, most of which are pathological, white cell precursors can leave the bone marrow and appear in the blood, so that further classifications may be necessary. Certain white cells are also to be found outside the blood in other organs and tissues, and also in the lymphatics – in many cases this is normal, but it may also be part of their pursuit of pathogens.

Each of these species of white cells has a very particular function, and fortunately each of the five cell types also have a reasonable specific appearance (their morphology) when viewed under a light microscope. Consequently, we can usually suggest a particular diagnosis for a patient based on the different proportions of the cells in the full blood count, and their appearance. Indeed, the different proportion of white cells is called 'the differential', often shortened to 'diff' (Table 5.2).

A haematology analyser recognizes different cell populations by the process of flow cytometry against a set of rigid criteria. However, when faced with a cell of unusual morphology it cannot easily classify as one of the five major families, it often describes the cell as 'atypical', or perhaps 'unclassifiable'. This is not to say that one or two unusual cells implies some kind of pathology, but if the number of these cells exceeds more than a handful, a genuine disease state becomes more and more likely, and may be serious.

CD molecules

Further subclassification of white cells (and, indeed, any cell) can be achieved by identifying particular families of cells by the presence of certain molecules on their cell membranes. Many of these molecules are classified according to a system called 'cluster of differentiation', or CD. This process, calling for antibodies linked to fluorescent chemicals, requires a machine called a fluorescence-activated cell scanner. We have already met the CD system in Table 3.2. All the different cells in the blood have their own set of CD molecules providing clues as to their identity. For example, red cells express CD235 and activated platelets express increased amounts of CD62P. Before the development of the CD system, a molecule called the leukocyte common antigen defined all white cells, but it has been renamed as CD45.

In considering the value of CD molecules, factors such as sensitivity and specificity must be addressed (as in the value of a cancer marker, explained in Chapter 6). In practice this condenses to the extent to which a CD

Table 5.2 A white cell differential.

Cell population	Typical result	Reference range
Total white cell count[a]	6.9	4.0–10
Neutrophils	3.8	2.0–7.0
Lymphocytes	1.2	1.0–3.0
Monocytes	0.5	0.2–1.0
Eosinophils	0.3	0.02–0.5
Basophils	0.1	0.02–0.2

[a]The unit for white cell counts is 10^9/L, although others may quote 10^6/mL or 10^3/μL. Fortunately, this is often academic and overlooked, as whatever the reason, a result of 5 is within the reference range, whereas a result of 15 is definitely raised.

molecule is found on all cells of a particular type, and on what other cells the marker is present. For example, although CD4 is taken to be a very good marker of a subset of lymphocytes, it can also be found on a small number of monocytes and macrophages. Fortunately, the latter are very much larger than lymphocytes, and the fluorescence-activated cell scanner can also detect this. Although some contend that CD14 is a specific marker of monocyte/macrophages, others suggest it can be found at a low level on granulocytes. Some markers have a distribution that defies a simple explanation, such as CD178 being found on a subset of lymphocytes and also on cells of the testis.

Two other problems with CD markers arise. The first is that some markers are present only when the cell is physiologically activated, and are not present when the cell is resting. For example, CD70 is found on certain activated lymphocytes that are about to transform into the cells that make antibodies and on those preparing to attack cells loaded with viruses, but not on cells in their resting state. Fortunately, these different types of lymphocyte cells can be differentiated by co-staining with, for example, CD20 and CD3 respectively. Secondly, many cancers are characterized by alterations of the expression of surface molecules, and neoplastic leukocytes are particularly prone to this phenomenon. This therefore demands careful consideration in diseases such as leukaemia, as will be discussed in Chapter 6.

Table 5.3 shows a list of selected CD molecules, many of which are valuable markers of a variety of pathological situations. We will revisit this list in Chapters 6 and 10, in considering a panel of antibodies that are useful in diagnosis and management of malignancies.

The CD system, although powerful, does not define all cell groups or activities. We also need to be able to recognize antibodies on the surface of certain cells, and different types of human leukocyte antigen (HLA) molecules – neither of these are CD markers. All these will be revisited in Chapters 6, 9 and 10. Before we examine the different families of white cells, we must review the process of how these cells develop in the bone marrow.

5.2   Leukopoiesis

Haemopoiesis is the process of the generation of all blood cells, which in the healthy adult happens only in the bone marrow, although in the infant it may also happen in the liver and spleen. In certain diseases in the adult, to be discussed in Chapter 6, haemopoiesis can also occur in

Table 5.3 Selected leukocyte CD molecules

Marker	Distribution
CD2	T lymphocytes, some NK cells
CD3	T lymphocytes
CD4	T helper cells
CD8	T cytotoxic cells
CD11	Monocytes, macrophages, some B lymphocytes
CD14	Monocytes
CD15	Granulocytes
CD16	NK cells, monocytes, neutrophils
CD19	B lymphocytes
CD20	B lymphocytes
CD24	Granulocytes, B lymphocytes
CD25	Activated T lymphocytes
CD30	NK cells
CD34	Stem cells, endothelial cells
CD45	All white cells
CD56	NK cells
CD91	Monocytes
CD114	Granulocytes, monocytes
CDw125	Eosinophils, basophils
CD133	Stem cells, endothelial cells
CD159c	NK cells
CD163	Monocytes, macrophages

the organs. The production of the different types of cells has its own descriptor:

- erythropoiesis, production of red cells (Chapter 3);
- thrombopoiesis, the production of platelets (Chapter 7);
- leukopoiesis, the production of white cells, with three lines of cell development:
  - lymphopoiesis, the production of lymphocytes;
  - monopoiesis, the production of monocytes;
  - granulocytopoiesis, the production of granulocytes.

In common with red cells and platelets, white cells ultimately arise from a self-perpetuating (pluripotent) stem cell in the bone marrow (usually in the sternum). Under the direction of growth factors, this cell then gives rise to a series of unipotent stem cells that transform with increasingly more focus on defined cell lineages. Stem cells are stimulated to divide and differentiate by growth factors, some of which are cytokines produced by other white cells and non-leukocytes in the bone marrow and possibly elsewhere. Stem cells can also be identified by the presence of the molecule CD34 on the cell surface. Little is known of the exact function of this molecule, which is also present on endothelial cells.

When examining the morphology of immature white cells, we must understand that each characteristic stage of development (such as myeloblast, promyelocyte, myelocyte, metamyelocyte, band form and mature cell) is formed through a dynamic process. The precise definition of each cell is an approximation, and each stage merges into another, so cells will often share common features between different stages of maturation.

We recognize two 'daughter' cells of the pluripotent stem cell: a common lymphoid precursor, from which lymphocytes develop, and a common myeloid precursor, from which red cells, platelets and all other white cells eventually arise.

Lymphopoiesis

The first, major committed stem cell in lymphopoiesis is the common lymphoid precursor, also known as the colony-forming unit-lymphoid (CFU-L). As is explained in detail in Section 3.2, much of our knowledge of the development of cells in the bone marrow has come from animal models, and tissue culture experiments, bringing the concept of the colony of cells of a particular lineage (hence colony-forming unit) (Figure 5.1).

Compared with the development of other leukocytes, lymphopoiesis is surprisingly simple. The unipotent stem cell, the CFU-L, is likely to give rise to lymphoblasts. These cells, with a diameter 14–20 μm, are characterized by a high ratio of nucleus to cytoplasm, transform into prolymphocytes that then develop into one of three types of mature lymphocytes, each with a diameter of 9–11 μm, and so are only slightly larger than red cells (generally 7–8 μm). It is unclear whether or not the transformation from lymphoblast to prolymphocyte is specific for each of the three types of lymphocytes: T cells, B cells and natural killer (NK) cells.

T lymphocytes (or T cells) These cells are so named as they must pass through the thymus, at which point they may be described as thymocytes, in order to develop their full potential. The thymus itself is located between the heart and the sternum, and most active in infants, becoming smaller as the child grows to adolescence, and then shrinks in the middle aged to become rudimentary in the elderly.

One of the first molecules to appear on the surface of the embryo T cell is CD7 (at the pre-T cell stage), then CD2 and CD5 as the cells develops into a thymocyte. As it

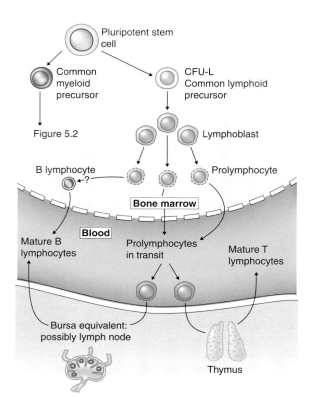

Figure 5.2

Figure 5.1 Lymphopoiesis. The process of the development of mature lymphocytes begins in the bone marrow. T lymphocytes must pass through the thymus to become fully functional, but the site of the transformation of pre-B lymphocytes into mature B lymphocytes is unknown.

acquires maturity, CD3, CD4 and CD8 appear on the cell surface.

B lymphocytes (or B cells) Just as T cell development demands a thymus, in birds the B cells require passage through the bursa of Fabricius (hence 'B'). However, there does not seem to be a human equivalent of the bursa, and it has been suggested that, in our species, immature B lymphocytes become functional by passing through any combination of the liver, spleen and lymph nodes. Alternatively, B lymphocytes may mature within the bone marrow.

CD10 marks pro-B and pre-B lymphocytes; CD20 and CD79 appear from the pre-B cell stage. As the B cell becomes functional, immunoglobulins appear in the cytoplasm and then on the surface of the cell.

Natural killer cells (NK cells) These rare and mysterious cells are difficult to classify, but fulfil a key role in defence against unusual pathogens. There are no morphological differences between T, B and NK lymphocytes when using conventional light microscopy (i.e. they all look the same). However, each sub-type has its own CD signature, and so can be recognized and measured by a fluorescence-activated cell scanner.

Myelopoiesis

The second unipotent stem cell that the pluripotent stem cell produces is the common myeloid progenitor, so called because it ultimately gives rise to granulocytes, erythrocytes, myeloid cells and megakaryocytes (hence CFU-GEMM) (Figure 5.2). This stem cell then generates two other progenitors, one for red cells and megakaryocytes (the CFU-EMk) and another for both granulocytes and monocytes (CFU-GMo).

The CFU-GMo then produces four separate and specific CFUs, for monocytes and for each of the three types of granulocytes – the neutrophil, eosinophil and basophil.

Monopoiesis The CFU-Mo (for monocytes) first produces monoblasts (which seem identical to myeloblasts), then promonocytes and then the mature monocyte. As these cells mature they become increasingly smaller, and with a progressive reduction in cytoplasmic granules (Figure 5.2, right). Many believe that monocytes in the blood are in fact immature forms of the ultimate cell of the lineage, the macrophage, which is found in the tissues and various organs. Further differentiation of macrophages into more complex cells occurs in the liver, spleen, lymph nodes, bone and elsewhere.

Because the nucleus is generally round and unsegmented in monocytes and lymphocytes, these are often both referred to as mononuclear leukocytes. This is in contrast to the irregular nucleus of granulocytes.

Granulocytopoiesis The maturation of the granulocytes (neutrophils, eosinophils and basophils) follows a common path (Figure 5.2, left and centre). Each particular CFU first produces a myeloblast, which in a normal bone marrow make up to 4% of nucleated cells. These cells have a high ratio of nucleus to cytoplasm – that is, the nucleus of the cell is almost as large as the cell itself – and a fine open lacy chromatin pattern is apparent, accompanied by prominent nucleoli. The cytoplasm of the cell stains a light shade of blue and has no granules. With standard staining, it is not possible to differentiate

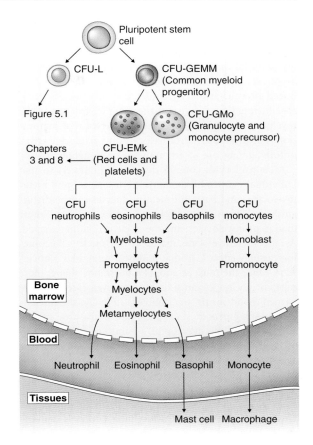

Figure 5.1

Chapters 3 and 8

Figure 5.2 Myelopoiesis. This process gives rise to red blood cells and platelets, as well as monocytes and granulocytes. As with lymphopoiesis, the key cell is the CFU, which generates the three granulocyte lineages, and the monocyte/macrophage lineage.

myeloblasts of a particular granulocyte lineage. The size of a myeloblast is 12–20 μm in diameter.

Promyelocytes are significantly larger than myeloblasts, measuring 15–25 μm, although they demonstrate a lower nucleocytoplasmic ratio and occasional nucleoli. The cytoplasm has a deep blue coloration in comparison with the paler myeloblast, and contains certain primary granules, the function of which will be explained in a following section. However, it is still not possible to differentiate a neutrophil promyelocyte from an eosinophil promyelocyte or a basophil promyelocyte. The nucleus of the promyelocyte is also slightly indented.

Myelocytes are smaller than promyelocytes, approximately 10–20 μm in diameter, and show evidence of chromatin clumping. Nucleoli are no longer visible. Secondary (or specific) granules are now present throughout the cytoplasm, enabling us to identify the granulocytic lineage to which these myelocytes belong. Secondary granules are so called because they are produced after the primary granules, and they possess a cell-specific function. Gradual changes in the structure of the myelocyte result in the metamyelocyte stage of maturation. The myeloblast, promyelocyte and myelocyte are all mitotic, giving rise to any number of 'daughter' cells.

Metamyelocytes contain distinctive granules, so that individual neutrophil metamyelocytes, eosinophil metamyelocytes and basophil metamyelocytes can be distinguished. The nucleocytoplasmic ratio is much lower than in promyelocytes, and the nucleus is kidney shaped. As the metamyelocyte matures, the nucleus becomes increasingly curved, so that it almost resembles a horseshoe. Once the nucleus adopts a horseshoe shape, these cells are no longer considered metamyelocytes, but are termed 'band forms' or 'stab cells'. Finally, the nucleus becomes segmented and the mature features of the effector cell – the neutrophil, eosinophil or basophil – become clear. The process of granulopoiesis takes approximately 5–6 days in the bone marrow.

Granules

The importance of these intracellular bodies is demonstrated by the pathological consequences of their absence or malfunction. Dyes and stain have always been important to pathologists, with that developed by Romanowsky being the most popular. These stains, such as those of Jenner, Giemsa and others, colour the nucleus various shades of purple. An early step in the classification of granulocytes was the realization that different stains could be used to identify subgroups of cells.

• Neutrophils are so-called because their granules preferentially take up dyes at a neutral pH. These dyes are various shades of grey.
• Similarly, eosinophils are so named because their granules take up dyes based on the dye eosin at an acid pH. These dyes are various shades of red.
• Basophils take up basic (alkali) dyes that are various shades of blue/black/purple.

Most granules contain enzymes designed to attack and destroy microbes such as bacteria and parasites, but they can also interact with, and possibly even digest, the body's own tissues. However, they may also contain other chemicals that have other actions, such as in anticoagulation and the storage of energy-rich glycogen. Whilst most granulocyte biology focuses on neutrophils, and their primary role in defence against bacteria, the other members of the family, the eosinophils and basophils, have their own specific granules, but they also share some with neutrophils.

Granules can be classified as to their contents, and as primary, secondary or tertiary. As the cell matures, the characteristics of the granules change. Primary granules are formed early in the life of the granulocyte, at the promyelocyte stage, whilst secondary granules and then tertiary granules appear in myelocytes and metamyelocytes. These different patterns allow the particular cells to be recognized. The contents of the granules (the first five of which are primary) are as follows:

• Myeloperoxidase is the enzyme that catalyses the conversion of hydrogen peroxidase to hypochlorous acid and the oxidation of the amino acid tyrosine to tyrosyl radicals. These toxic chemicals are important in the destruction of bacteria. The eosinophil has its own form of peroxidase, which is also bactericidal and can form reactive singlet oxygen and hypobromous acid in the presence of hydrogen peroxide and the bromide anion.
• Bacterial permeability-inducing factor contributes to the destruction of bacteria by weakening the cell wall.
• Lysozyme digests N-acetylglucosamine and N-acetylmuramic acid residues, components of the bacterial cell wall, especially those staining positive for Gram's stain.
• Elastase is particularly effective at degrading fungi and Gram-negative bacteria. However, this enzyme is also active against the body's own tissues, so can cause significant injury if released inappropriately.
• Acid hydrolases include a wide range of different enzymes, including acid phosphatase, glucuronidase and enzyme-digesting carbohydrates.
• The antibacterial activity of lactoferrin, in a secondary granule, is accounted for by its ability to sequester iron, thus denying it to those bacteria whose growth relies on this metal.
• Gelatinase (a tertiary granule), like elastase, digests proteins such as collagens. This property enables neutrophils to digest the extracellular matrix, such as the subendothelial basement membrane, and so in theory enable their migration into the tissues.
• Major basic protein, also known as proteoglycan-2, is an unusual molecule rich in arginine that is particularly adept at disrupting the cell membranes of helminth worms, but it is also toxic towards bacteria. It is present particularly in eosinophils.

Anti-microbial defence extends to secretory vesicles. These bodies contain factors such as alkaline phosphatase and components of the complement system. The latter, to be discussed in Chapter 9, are an important defence against bacteria and have several roles in inflammation.

Granules may fuse with each other, and they move to the surface to expel their contents into the local environment. This process is called degranulation. Alternatively, the granulocyte may ingest pathogens and other material, exposing them to the contents of the granules, so intracellular digestion can proceed (phagocytosis).

The role of growth factors

Red cell differentiation is driven, in part, by erythropoietin, and likewise (as is described in Chapter 7) thrombopoietin stimulates megakaryocytes to produce platelets. However, there is no obvious single 'leukopoietin', probably because the pathways are too complex. Nevertheless, we recognize granulocyte colony-stimulating factor (G-CSF) and granulocyte-macrophage colony-stimulating factor (GM-CSF), both of which are useful in attempting to stimulate a damaged or suppressed bone marrow. There are also roles for cytokines, such as interleukins (ILs) and transforming growth factor, which can be classified in several ways:

- those acting specifically, such as IL-7, acting only on pre-B cells via the IL-7 receptor, and IL-6 on plasma cells;
- those of restricted activity, such as IL-4, acting on B cells, T cells and basophils;
- those acting nonspecifically, such as IL-3, which acts on all non-lymphoid cells.

Some cells respond to a 'panel' of growth factors. For example, the development of eosinophils is under the control of IL-3, IL-5 and GM-CSF. The proof of concept of these growth factors is their use in clinical practice. Several drug regimes suppress different aspects of haemopoiesis, so that GM-CSF and G-CSF are used as therapy to help the bone marrow to recover, and so aid the return of leukocytes in the blood.

As growth factors are necessary for cell growth and differentiation, it follows that lack of growth factors leads to cell death, and this is often by apoptosis, or 'suicide'.

Blasts and malignancy

Precursor cells such as myeloblasts, promyelocytes and myelocytes should not be present in the peripheral blood.

As is explained in Chapter 6, an increase in the number of 'early' blasts and precursor cells in the bone marrow in the absence of other intermediate stages of maturation is associated with certain leukaemias.

Whilst it is possible for a small number of metamyelocytes to be present in the blood in particular 'normal' conditions, such as an acute infection, in health these and other precursor cells should be absent. It follows that any such cell found in the blood is likely to be defined as abnormal by the imaging software of the analyser, and so labelled 'atypical', 'unrecognized' or 'blast'. Therefore, a differential with >5% (or less, in some circumstances) of atypical cells or blasts must be checked by an experienced staff member as it may be an early sign of a malignancy. However, successive generations of imaging software are becoming increasingly sophisticated and able to define precursor cells.

In some cases 'blasts' may in fact simply be normal cells responding to an abnormal and temporary external situation, such as activated (virus-seeking) T lymphocytes with an unusual morphology. In this case there is no malignancy. Indeed, there may be a place for morphologically defined 'large granular lymphocytes'.

Now we have gained an understanding of the development of white cells, we are in a position to get to grips with their structure and function. This we will undertake by working through the white cell family in order of frequency.

5.3   Neutrophils

These are the most common type of white cell, making up perhaps 75% of the total. With a diameter of 9–15 μm, their half-life may be as short as 6–8 h. Whilst the greater proportion of neutrophils circulate in the blood, a considerable proportion are stuck to the insides of blood vessels – these are called marginating cells. There may also be reservoirs of resting neutrophils in the bone marrow and spleen that can be called upon rapidly at times of high demands (such as an acute infection).

Identification

Neutrophils are characterized morphologically by a nucleus that is broken up into segments (hence 'segmented'), also called lobes, and a cytoplasm that appears granular (Figure 5.3). As indicated above, many granules are loaded with enzymes; others are storage granules, such as of glycogen. The 'normal' number of lobes is three to five,

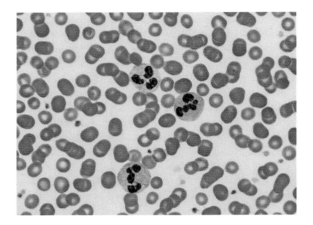

Figure 5.3 Three neutrophils. Note the different number and layout of the three or four purple-stained lobes. The fine structure of the cytoplasm is not smooth but irregular, due to the presence of granules. (Courtesy of D. Bloxham, Cambridge University Hospitals NHS Foundation Trust).

a cell with two being described as hyposegmented, and is effectively one stage on from a metamyelocyte. Hyposegmentation may be part of the congenital condition Pelger–Huet anomaly, many cells having a bi-lobed nucleus, but a single-lobed nucleus may also be present, described as a stab cell.

Similarly, some neutrophils have five or more lobes. It follows that the number of lobes is a marker of the maturity of the cell. However, we have already noted in Chapter 4 that, in the case of pathological vitamin B_{12} deficiency, the number of segments can increase, perhaps to seven or eight, when the cell is called hypersegmented (Figure 4.3). Many cells contain twice as much DNA as normal because the cell cycle is arrested in the S phase. Hypersegmentation may also be present in high serum urea, iron deficiency and infections.

The process of lionization is of the inactivation of one of the X chromosomes in females. In the neutrophil, this can be demonstrated as a small dumb-bell protrusion referred to as a Barr body.

These features form the basis of other names for the neutrophil, such as polymorphonuclear leukocyte (or simply 'polymorph') and granulocyte. However, the nuclei of eosinophils and basophils are also segmented (but generally with only two lobes), and their cytoplasm is also rich in granules, so these cells may also be described as polymorphs and as granulocytes.

Neutrophils can also be identified as they have receptors for a certain part of the immunoglobulin G (IgG) molecule – the Fc (fraction crystallizable) section, to be discussed in Section 5.8. Fluorescence activated cell scanning (FACS) can be used to confirm the presence of the Fc receptor (FcR), one type of which is CD16. However, careful interpretation is required as CD16 is also found on NK cells, and monocytes/macrophages. Fortunately, flow cytometry can distinguish these cells on the basis of size (monocyte/macrophages are large, NK cells are small) and granularity (neutrophils have many granules, monocytes few, NK cells none).

Function

Neutrophils are key mediators of inflammation, attacking and destroying pathogenic microbes such as bacteria, often in the process of phagocytosis (detailed in Section 5.8). Increased numbers of neutrophils (a neutrophilia, or neutrophil leukocytosis) are therefore expected (and desired) during infection with bacteria. However, numbers of neutrophils may also be increased in certain diseases (examples of which are cancer and autoimmune disease such as rheumatoid arthritis – Chapter 9) regardless of the presence of micropathogens. Neutrophils also seem likely to be important in tissue repair after injury.

5.4 Lymphocytes

Lymphocytes are the second most frequent circulating white cell, accounting for 20–40% of circulating leukocytes. However, many more lymphocytes are resident within lymph nodes, and there are also considerable numbers in the spleen, liver and bone marrow. Many lymphocytes have a half-life of perhaps 70 h, although many (probably resident in lymph nodes) can live for decades.

Identification

Lymphocytes are the smallest white cell, with a diameter of 9–12 μm, so being only slightly bigger than a red cell, which generally has a diameter of 7–8 μm (Figure 5.4). Nevertheless, some lymphocytes may have a diameter of up to 15 μm, and because they occasionally have granules, especially when activated, may be described as large granular lymphocytes. The regular nucleus occupies 90–98% of the cell, the cytoplasm being reduced to a small ring surrounding the nucleus.

Three subtypes of lymphocytes are T cells (making up 65–80% of circulating lymphocytes), B cells (10–30%) and NK cells (<5%, often called non-T, non-B lymphocytes).

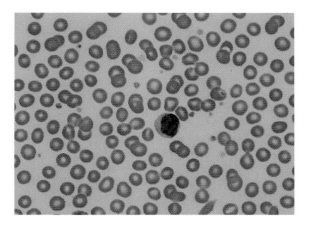

Figure 5.4 The single lymphocyte is characterized by a single, roughly circular (purple) nucleus. Compare the fine details of the cytoplasm (which lacks granules) with the grainy nature of the cytoplasm on neutrophils. (Courtesy of D. Bloxham, Cambridge University Hospitals NHS Foundation Trust).

These different types of lymphocytes are indistinguishable by conventional Romanowsky staining, but may be separated by the presence of specific molecules on the cell membrane using immunofluorescence microscopy and FACS analysis (Figure 5.5).

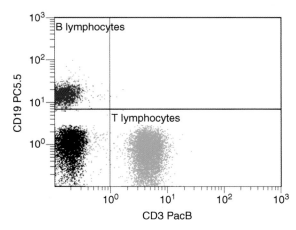

Figure 5.5 Different lymphocyte groups can be enumerated by the presence of particular CD molecules. This example uses monoclonal antibodies to CD3 and CD19, both linked to different fluorochromes (Pac B and PC 5.5 respectively). Cells binding CD3, but not CD19 are defined as T cells (lower right quadrant), those binding CD19 but not CD3 (upper left quadrant) and B cells, whilst those binding neither antibodies are NK cells (lower left quadrant). There are almost no cells binding both antibodies (upper right quadrant). (Image courtesy of M. Hill, Heart of England NHS Foundation Trust).

Function

Lymphocytes are the primary cell involved in immunological processes, but (unlike granulocytes and monocytes) different subtypes have precise and nonoverlapping functions:

- T lymphocytes are defined by the presence of CD3, and there are several subtypes. Those carrying CD3 and CD8 seek and destroy cells they regard as foreign, such as those infected with viruses, or cells from another individual, such as a transplanted kidney, and so are called cytotoxic. The most efficient production of certain antibodies by B lymphocytes calls for the cooperation of T helper cells, characterized by CD3 and CD4. However, it is also clear that, in some circumstances, cytotoxic CD4-bearing cells can be demonstrated in response to many diverse pathogens.
- B lymphocytes make antibodies. This is most unlikely to occur in the plasma, but in the lymph node, spleen, liver and bone marrow, and generally requires dendritic cells and T helper cells. However, in some circumstances, B cells can make antibodies to certain bacteria without these helper cells. The key recognition markers on mature B lymphocytes are CD19 and CD20. Those B cells already committed to the production of antibodies may be identified by immunoglobulins on their cell surface. The process of antibody generation is explained in Section 5.8.
- NK cells do not need to be instructed about which cells to kill: their cytotoxic ability is innate. Even less is known about how the NK cells mature than how the other lymphocytes develop. CD30 and CD91 are the most commonly used markers of NK cells. Although clearly important in certain circumstances, the enumeration and functional assessment of NK cells is rarely called for, but if so, then the presence of certain CD molecules will be sought.

Some authorities refer to perhaps 10% of lymphocytes as large granular lymphocytes, and this population includes many NK cells. Additional details of the function of all lymphocytes follow in Section 5.8 and in Chapters 9 and 10.

5.5   Monocytes

These are the third most frequent circulating white cell. They circulate in the blood for 2–3 days, and then move

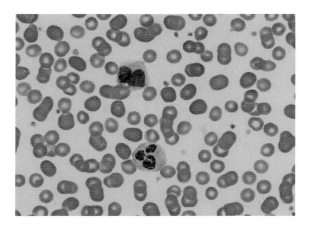

Figure 5.6 A monocyte. There are two cells in this figure; the monocyte is the upper cell. Note that about two-thirds to three-quarters of the cell is taken up by the purple nucleus, which has an indent. Contrast this with the lower cell, which is a three-lobed neutrophil. (Courtesy of D. Bloxham, Cambridge University Hospitals NHS Foundation Trust).

into the tissues where they become macrophages, and may live there in an immunosurveillance role for several months.

Identification

Monocytes are the largest leukocyte in the blood with a diameter of 15–25 μm (Figure 5.6). The nucleus is a single round shape, but may be indented, with a C or U shape, and takes up 60–80% of the cells. Granules are often present. The total body monocyte pool is split equally between the blood and the spleen. However, the monocyte is not the last cell of its lineage – in the tissues monocytes are transformed into macrophages. Extending this process, some macrophages undergo further changes in certain anatomical locations.

Functions

Monocytes can be shown to have several different functions *in vitro*, some of which, such as phagocytosis, are likely to occur in the body. Key markers are CD14 (the receptor for bacterial lipopolysaccharide) and CD16 (the receptor for FcR).

Among the specialized cells that seem to have started life as monocytes, and then becoming macrophages, are dendritic cells (often found in lymph nodes), Kupffer cells (in the liver), Langerhans cells (various tissues,

frequently the skin), histiocytes and osteoclasts (in the bone). These specialized cells often lose classical monocyte/macrophage morphology, but some bear particular CD molecules (such as CD19, CD23 and CD83 on dendritic cells, CD1a, CD14 and CD83 on Langerhans cells, and CD53, which is shared between leukocytes, dendritic cells and osteoclasts). Use of CD markers demands care because of poor sensitivity and specificity – CD19 is also found on B lymphocytes and C16 is also found on neutrophils.

Subtypes of monocytes have important immunological and inflammatory functions. Dendritic cells, alongside T helper lymphocytes, help to present antigens to B lymphocytes in the process of antibody generation. Such monocytes are called antigen-presenting cells. It is also likely that hepatic Kupffer cells and dermal Langerhans cells function as antigen-presenting cells. Osteoclasts are important in bone turnover (perhaps to do with their ability to release bone-dissolving enzymes) but seem to have minimal or no immunological or inflammatory activity.

Monocytes/macrophages also generate and release inflammatory cytokines such as tumour necrosis factor, interferons and ILs which promote the activation of a number of cells including leukocytes (other monocytes, lymphocytes), and non-leukocytes (such as endothelial cells). Monocytes/macrophages are also participants in haemostasis by generating tissue factor and coagulation factor V (Chapter 7).

5.6   Eosinophils

These granulocytes have a diameter of perhaps 12–17 μm. Almost all reside in the blood, and have a circulating half-life of some 6 h, but are more long-lived in the tissues, where they may remain for 10 days or so. However, the bone marrow contains more eosinophils, and so can be seen as a reservoir.

Identification

With standard Romanowsky stains, the typical eosinophil has a bi-lobed nucleus and a cytoplasm dominated by reddish granules (Figure 5.7), which first appear in eosinophil metamyelocytes of the bone marrow. Mature eosinophils express CD9, CD15, CD23, CD35 and CDw125, although many other cells also bear these molecules. Occasionally viewed by light microscopy, Charcot–Leyden crystals are composed of lysophospholipase and are specific for eosinophils.

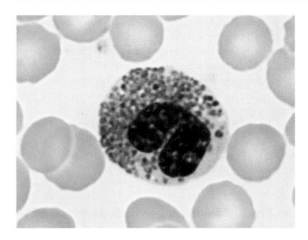

Figure 5.7 An eosinophil. The key features are a bilobed nucleus, and a cytoplasm dominated by reddish granules. (From Hoffbrand & Moss, Essential Haematology, Sixth Edition, 2011, Fig 8.1 (b) p. 109. Reproduced with permission of John Wiley & Sons, Ltd.).

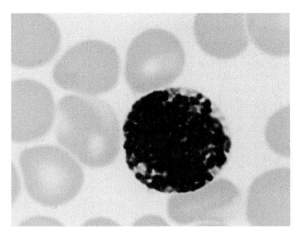

Figure 5.8 A basophil. The key features are a bilobed nucleus and a cytoplasm dominated by blue–black–purple granules. However, the granules may be so large and numerous that the nucleus becomes obscured. (From Hoffbrand & Moss, Essential Haematology, Sixth Edition, 2011, Fig 8.1 (c) p. 109 . Reproduced with permission of John Wiley & Sons, Ltd.).

Function

Key to eosinophil function are its granules, dominated by those that contain major basic protein, and peroxidase. However, there are also granules specific for the eosinophil, such as that containing a cationic protein rich in arginine, another which has enzymes to degrade ribonucleic acid, a third containing a neurotoxin and a fourth rich in histamine.

Although they have general 'housekeeping' antibacterial duties, eosinophils have two major roles. They are important mediators of allergic (hypersensitivity) reactions and are important in our defence primarily against helminth parasite infections (worms), such as filariasis, hookworm, schistosomiasis and trichinosis. However, an eosinophilia may be present in protozoal parasite infections.

5.7 Basophils

Basophils are the least frequent easily recognized leukocyte, and are slightly smaller than eosinophils at 10–14 μm. In common with the monocyte, the basophil is not the terminal cell of its lineage as some transform into mast cells when they migrate into the tissues.

Identification

The nucleus of the basophil is bi-lobed, and often obscured by the large number of purple or black granules (Figure 5.8). Basophils are relatively poor in CD molecules, although they share CD9, CD68 and CDw123 with several other cells and CDw125 with eosinophils. However, a key feature is the presence of the FcR not for IgG, as in other granulocytes, but for immunoglobulin E (IgE), of which there are two species: a low-affinity receptor and a high-affinity receptor. In migrating into the tissues, basophils become mast cells, which bear CD49b.

Function

Like the eosinophil, the basophil shares several of the types of granules of its co-granulocytes (such as those that include elastase and other proteases). However, again in common with eosinophils, it has its own specific granules, which include the anticoagulant heparin and the smooth muscle relaxant histamine. Other granule constituents include leukotrienes and cytokines.

Basophils have much in common with eosinophils, such as being active in defence against parasites and in hypersensitivity responses, but it is the latter that basophils focus upon. These include allergic and anaphylactic responses to factors such as certain pollens, and both basophils (in the circulation) and mast cells (in the tissues) degranulate in response to a variety of stimulants. A key activation pathway is the docking of an IgE molecule into its FcR. When basophil and/or mast cells activation and degranulation are excessive, consequences include several clinical conditions, as will be discussed in Chapter 9.

5.8 Leukocytes in action

White cells defend us from pathogenic organisms by the processes of inflammation and immunity. These organisms include bacteria, viruses and parasites, which may all be present within the blood but also in the tissues, whilst some may be on the skin alone. If the infection is marked and widespread, it is likely that there will be profound changes to the numbers of different white cells in the blood. However, some infections that are modest and localized, especially on the skin alone, may not be associated with such changes to groups of circulating leukocytes.

In almost all cases of an infection, the leukocytes are activated, and an active immune response is almost always accompanied by a raised white cell count (a leukocytosis). However, as is explained in Chapters 6 and 9, a leukocytosis is present in many pathological situations. Conversely, a low white cell count, a leukopenia, is never found in physiology (to be reviewed in Chapter 6). However, certain healthy ethnic and racial groups (such as those of African origin and from the Indian subcontinent) have a lower reference range for neutrophils than white Europeans.

Additional changes during an infection include an abnormal erythrocyte sedimentation rate (ESR) and an acute-phase response. The latter, almost certainly driven by cytokines, is characterized by increased levels of certain proteins in the blood, such as the inflammatory marker c-reactive protein (CRP). Broadly speaking, we recognize two different mechanisms for defence against an infection: inflammation and immunity.

Inflammation

There are a large number and variety of bacteria living on and within us, and they are kept in check by white cells and other mechanisms, such as the mechanical barrier of the skin. When the number of bacteria rise, or new bacteria are introduced, we expect the neutrophil count to rise in response. Once the neutrophil count has passed above the top of the reference range, we describe a neutrophilia, or perhaps a neutrophil leukocytosis. There may also be an increased monocyte count: a monocytosis.

Inflammation is a very 'primitive' process, by which we mean not only that it is present in relatively unsophisticated organisms such as starfish, but it does not require a complex set of developmental steps. Indeed, inflammation is a key part of 'innate' defence, which includes the barrier function of skin, mucus and digestive enzymes of the intestines (including saliva), all of which are not specific for a particular pathogen. An inflammatory response also involves soluble mediators such as cytokines, histamine, serotonin and leukotrienes, which are secreted principally by granulocytes, monocytes and macrophages. These inflammatory mediators have both local and distant effects, the former including inducing pain, itching and vasodilation in the region of the infection.

Recruitment of cells In many cases, a bacterial infection will have its focus in a particular tissue, such as a surgical skin wound or the lung. In which case, the neutrophil and monocyte must pass out of the blood and move towards the infection, the first step of which is mediated by adhesion molecules on the endothelial cells that line the blood vessel. Once in the tissues, the neutrophils and monocytes are drawn towards the highest concentration of bacteria by the process of chemotaxis. The cells sense increasing levels of a number of soluble mediators, including leukotrienes, bacterial lipopolysaccharide, complement components, ILs and chemokines. These small molecules are called chemotaxins. Complement, a collection of perhaps 20 molecules that come together to help defend us from pathogens, is described in detail in Chapter 9.

Contact of cells with pathogens Once neutrophils and monocytes encounter bacteria, the process of attack begins. However, recognition and attack are enhanced if the particular organism is coated with molecules called opsonins. These include immunoglobulins (which can interact with the cell via its FcRs) and certain components of the complement system. The principal process by which leukocytes destroy bacteria is phagocytosis.

Phagocytosis This complex process requires the bacteria to be drawn into the cell by endocytosis. Once inside the cell, the vesicle in which the organism is absorbed fuses with other vesicles and granules to produce a phagolysosome, allowing bacteriocidal enzymes to start digesting the bacterium. A second method for destroying bacteria is to expose them to a variety of toxic reactive oxygen species that include the superoxide anion, hydrogen peroxide, the hydroxyl radical and hypochloride. These are generated by enzymes such as myeloperoxidase and are associated with the uptake of oxygen by the cell in a so-called 'respiratory burst'. These pathways are as follows:

$$2O_2 + NADPH \rightarrow 2O_2^- + NADP^+ + H^+ \quad (5.1)$$

$$O_2^- + O_2^- + 2H+ \rightarrow H_2O_2 + O_2 \quad (5.2)$$

$$O_2^- + H_2O_2 \rightarrow OH^\bullet + OH^- + O_2 \quad (5.3)$$

$$Cl^- + H_2O_2 \rightarrow HOCl + H_2O \quad (5.4)$$

$$2H^+ + 2O_2^- \rightarrow O_2 + H_2O_2 \quad (5.5)$$

Reaction (5.1) generates the superoxide anion (O_2^-) by the action of nicotinamide adenine dinucleotide phosphate hydrogen (NADPH) oxidase on NADPH and oxygen. Reaction (5.2) converts two of these anions to hydrogen peroxide and oxygen, which can then be substrates for additional reactions. In reaction (5.3), the Haber–Weiss reaction, the products are a hydroxyl radical, the hydroxide anion and oxygen. In reaction (5.4), myeloperoxidase catalyses the formation of hypochlorous acid from a chloride ion and hydrogen peroxide. All these reactions therefore generate (and to some extent, recycle) toxic oxygen groups that can damage cell membranes.

However, these reactive oxygen species can also damage the host cell, which would be a problem were it not for defence systems such as reaction (5.5), where the enzyme superoxide dismutase (of which there are several isoforms) neutralizes the superoxide anion. This reaction also produces hydrogen peroxide, but this molecule can be degraded by the common enzyme catalase. Notably, some bacteria produce their own protective superoxide dismutase. The generation of the oxygen species is often referred to as an oxidative or respiratory burst, and can be detected in the laboratory as a proof of the biochemical potential of the cell.

Degranulation will result in the export of these potentially damaging molecules to the local extracellular environment. These include the proteolytic enzymes and the reactive oxygen species outlined above, but also physiologically active molecules such as histamine, also present in basophils, mast cells and platelets. Histamine (alongside kinins, prostaglandins and leukotrienes) is a powerful vascular dilator that also increases vessel permeability. This promotes the movement of various leukocytes and inflammatory mediators such as complement components in and out of the blood towards the site of the infection.

As is presented in Chapter 9, deficiency of these pathways is associated with failure to combat certain organisms, and therefore repeated infections. Figure 5.9 illustrates the process of phagocytosis.

(a)

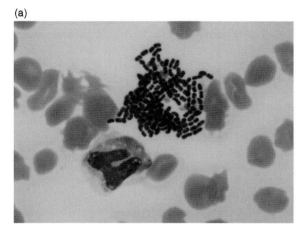

(b)

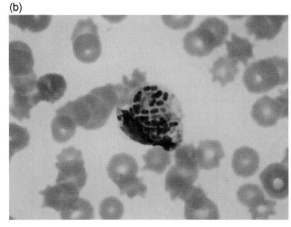

Figure 5.9 Phagocytosis. (a) A monocyte (with its characteristic horseshoe-shaped nucleus) close by a colony of bacteria. The latter have strongly taken up a blue dye. (b) A monocyte that has ingested over a dozen bacteria it has absorbed by the process of phagocytosis. It is tempting to speculate that the bacteria are coated with antibodies and complement opsonins to increase the efficiency of this process. (Images courtesy of Dr Colm Keane).

Acute inflammation If the infection is heavy and acute, the neutrophil population in the blood can be reinforced by cells in the spleen and elsewhere, and also by those adhering to the inside of the blood vessels (marginating cells). The bone marrow can also rapidly increase its generation of neutrophils, and in these cases the neutrophil count can be so great that it resembles leukaemia, and accordingly is called a leukaemoid reaction. Therefore, consideration of other factors, especially a typical history, is essential. The expression 'left shift' is used to describe a picture of an increased number of immature neutrophils, generally as part of a bacterial infection.

A further aspect of an inflammatory response is that it does not confer long-lasting defence against the stimulating pathogens. But if the acute inflammation is not resolved, it may progress to a chronic inflammation.

Chronic inflammation If an acute inflammation fails to eradicate the pathogen, the infection may persist (possibly because the immune system is compromised and/or overwhelmed) and chronic inflammation may result. This pathological process is characterized by increased and persistent damage to the tissues (possibly due to uncontrolled degranulation) and so the release of proteases and other cytotoxic factors, and results in a great deal of morbidity.

There are numerous examples of chronic inflammation. For example, a persistent lung infection that is unresponsive to antibiotics may lead to, or contribute to, pulmonary disease such as emphysema and pneumonia. Alternatively, chronic inflammation may be the consequence of an autoimmune disease such as rheumatoid arthritis, thyroiditis and systemic lupus erythematosus, as is discussed in detail in Chapter 9. There are other undesired laboratory consequences of chronic inflammation, such as in the anaemia of chronic disease (Chapter 4).

Immunity

Whilst most inflammation can be seen as a rapid, rough and ready defence, immunity is generally slow, steady and subtle. In many cases the immune system is 'learning' and improving its defensive capacity, and so may be described as 'adaptive'. The most active cell in the latter is the lymphocyte. The purpose of these cells is

- to make antibodies;
- to provide assistance in the generation of antibodies; and
- to detect and destroy cells infected with a virus.

The production of antibodies by B lymphocytes (with or without input from T helper lymphocytes) happens within the lymph nodes, liver, spleen and bone marrow. It follows that if there are increased lymphocytes in the blood (a lymphocytosis), then the increase in cell count is inevitably due to antiviral lymphocytes seeking to attack these pathogens.

Lymph nodes These small organs (of which there are hundreds scattered around the body) are packed with lymphocytes and are a major site of immune responses. They are linked by lymphatic vessels in chains, and eventually the lymphatic circulation empties in the venous circulation near the heart. The location of lymph nodes is not random: they are sited at key points through which the body may become infected (Table 5.4).

Some specialist lymph nodes are found in defined anatomical sites, and have focused functions. These

Table 5.4 Location and function of selected lymph nodes.

Location	Area defending, and function
Tonsilar	The throat: defending us from pathogens in food and drink
Pulmonary, bronchial and tracheal	The airways: inhaled pathogens (including fungal spores)
Abdominal	Intestines: pathogens in food and drink
Axillary (armpit)	The arms: pathogens moving up the arms, perhaps from lesions in the hands
Inguinal (groin)	The legs: pathogens moving up the legs, perhaps from lesions in the legs
Para-uterine and cervical	Female reproductive tract: sexually transmitted diseases (N.B. these lymph nodes are enlarged in pregnancy, definitely not a disease!)

tissues, described collectively as mucosal associated lymphoid tissues, include:

- Bronchus-associated lymphoid tissues, consisting of lymph nodes attached to the bronchus, although some commentators include lymph nodes in the lungs and lining the trachea. These tissues are likely to protect against air-borne pathogens.
- Gut-associated lymphoid tissues, such as Peyer's patches, mesenteric and gastric lymph nodes, and the appendix. These seems primed to protect against intestinal pathogens that have been ingested alogside food.
- Nasopharynx-associated lymphoid tissues (NALTs), which includes the tonsils and lymph nodes of the nose and pharynx, and also protect against air-borne pathogens. It has also been suggested that the cervical and clavicular lymph nodes of the front and sides of the neck should be part of the NALT family.

There are also clusters of lymph nodes in the armpit (axilla) and groin (inguina) which protect against infections moving up the lymphatic vessels of the arm and legs respectively. Also worthy of note are Peyer's patches, aggregates of lymphoid tissues in the lowest portion of the small intestines. It could be argued that they are not classical lymph nodes, as they are unconnected to the lymphatic circulation, but they are certainly rich in lymphocytes, which are organized into follicles that resemble those in lymph nodes and are therefore important in immunological defence.

The process of antibody production happens in specific areas within the lymph node called germinal centres (Figure 5.10). However, these areas can also be host to the development of cytotoxic T cell activity. If the lymphatic vessels bring in a great deal of pathogens and soluble mediators then the activity and the size of the lymph node increase and there is lymphadenopathy. An excellent model of physiological lymphadenopathy is infectious mononucleosis (IM); another is tonsillitis, often the consequence of a bacterial infection in the throat. Once the immune response has passed, and the pathogen eliminated, then the lymphadenopathy resolves and the particular node returns to its normal size.

It follows that persistent lymphadenopathy, especially in the absence of a clear and active immune response to an identified organism, is pathological. This we will consider in Chapters 6 and 9. The process of antibody production also happens in bone marrow, the liver and the spleen, so that plasma cells may be found in these organs. However, plasma cells are very rarely seen in the

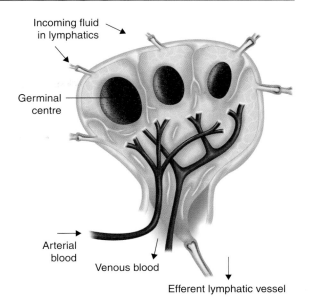

Figure 5.10 Anatomy of the activated lymph node. Lymphatic vessels carry lymph fluid (possibly loaded with pathogens and cytokines) from the tissues to the lymph node. A germinal centre is primarily the focus of the activity of T helper cells and antigen-presenting cells in promoting antibody production. Each node is fed by an arterial blood vessel, whilst vein carries effluent blood back to the circulation. A lymphatic vessel may carry cells and lymph fluid to the next node in the chain.

circulation, and if so are likely to accompany an acute infection. The major disease of plasma cells, myeloma, will be discussed in Chapter 6.

B lymphocytes

The purpose of the adaptive immune response is to produce focused and effective tools to attack and destroy specific pathogens. These tools include antibodies, produced by B lymphocytes, cells characterized variously by the presence of CD10, CD19, CD20 and CD22 on the cell surface, but also by the presence of antibodies at the surface (hence surface membrane immunoglobulins, SmIgs) and in the cytoplasm (in those cells actively making antibodies). CD molecules and SmIgs are detected by immunofluorescence with a modified microscope or a fluorescence-activated cell scanner.

Almost all functional antibody production by B lymphocytes in the lymph node and elsewhere requires cooperation with antigen-presenting cells and T helper cells lymphocytes. Perhaps the best example of a T-independent response is that of bacterial lipopolysaccharide, which at low concentration drives the production of specific antibodies. A popular model of B/T cell interaction suggests that fragments of a particular pathogenic organism that have been digested by a phagocyte are physically presented to the B lymphocyte by a combination of a T helper cell and an antigen-presenting cell, in conjunction with HLAs (described in Chapter 10). Some antigenic material may have been delivered to the lymph node by the drainage system that is the lymphatics.

The process of antigen presentation to B lymphocytes is tightly regulated and demands that cells recognize each other by a combination of highly specialized surface molecules. This recognition structure, often referred to as the B cell receptor, is formed from an immunoglobulin molecule and two molecules of CD79. The parallel molecules on the surface of the T helper lymphocyte include CD4, the T cell receptor (to be detailed below and in Chapter 10) and HLA molecules. Cells must also be stimulated by certain cytokines, most of which are ILs. Once the B cell receptor has received the appropriate cell

signals, intracellular signal transduction is mediated by CD79 and antibody production begins on a large scale. Once in progress, the cytoplasm fills with immunoglobulin protein and the B lymphocyte enters its final stage of differentiation, losing expression of CD19 and CD20, but upregulating expression of CD27. At this point it is described as a plasma cell.

Antibodies Antibodies are formed to a standard pattern of two 'heavy' chains and two 'light' chains, together forming a Y-shaped molecule (Figure 5.11). These four protein chains form a molecule with a mass of some 150–200 kDa.

The antibody molecule binds other molecules (antigens) it regards as non-self (foreign). This happens at the two 'tips' of the Y part of the molecule. The part of the antibody molecule that binds antigens is called the fraction antigen-binding (or Fab). The other end, or 'base', of the Y-shaped immunoglobulin is called the fraction crystallizable region (i.e. Fc). The Fab region is thus formed from a region of the heavy chain and a section of the light chain.

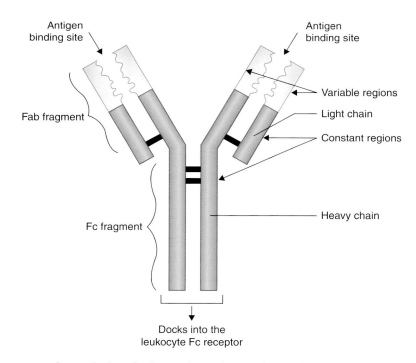

Figure 5.11 Basic structure of an antibody molecule. Two heavy chains and two light chains come together to form a Y shaped molecule that binds antigens with its F_{ab} sections. The other end of the molecule (the F_c section) has the capacity to dock into special receptors on certain leukocytes. The section of the heavy and light chains that find the antigen are called the variable region – the remainder is the constant region.

Antibody classes   The immune response to a potentially infinite variety of micropathogens demands an extraordinary degree of flexibility of its antibody response. This it achieves with different types of antibodies, and is partially achieved with variation in both heavy and light chains that confer on the molecule different properties. There are five different classes of heavy chain, the product of the manipulation of sections of particular genes, being alpha, delta, epsilon, gamma and mu, and so define the immunoglobulins IgA, IgD, IgE, IgG and IgM. Each of the five antibody classes has its own characteristics:

• IgA is the second most abundant immunoglobulin with a plasma concentration of 0.8–4 g/L, making some 12–14% of the plasma immunoglobulin pool. It can exist as a monomer or as a dimer, the latter formed by fusion of two monomers and stabilized by a 'J' chain. In the plasma, dimeric IgA is further stabilized by a 60 kDa secretory component generated by epithelial cells. IgA dominates in tissues of the intestines, lungs and the urogenital systems.
• IgD is found in only trace amounts in the plasma (<1% of the total immunoglobulins), but can be found on the surface of B lymphocytes that have yet to be activated by contact with antigens. It may also have a role in basophil and mast cell function.
• IgE is found in the plasma in low levels, but its major function is to become bound to basophils and mast cells, and so provide the trigger for these cells to degranulate, releasing heparin and histamine. Plasma levels are increased in parasitic infections and, as is outlined in Section 9.5, in hypersensitivity.
• IgG is the most abundant of the immunoglobulins, present at perhaps 6–16 g/L, so forming 80% of this class of protein. There are four subtypes (IgG1, IgG2, IgG3 and IgG4), with different concentrations, and properties such as complement activation and defence against parasites. Each to a different extent provides the primary defence against bacteria (and their toxic products) and viruses, particularly in extravascular spaces.
• IgM is unusual as it exists as a pentamer of five individual molecules, and as such is one of the largest molecules in the plasma with a relative molecular mass cited at 1×10^6 Da. Like the IgA dimer, the IgM pentamer is held together by J chains. However, it is present at a relatively low concentration of 0.5–2 g/L, so comprising about 6% of all immunoglobulins. With such a large size, it is a very effective agglutinator, as it has 10 different antigen binding (Fab) sites, and its large size ensures it remains in the plasma.

Similarly, there are two light chain classes, kappa (κ) and lambda (λ). These can combine with any of the five types of heavy chain so that, for example, there are IgGκ and IgGλ molecules. The different light chains also provide variability in the other antibody classes. Figure 5.12 illustrates the structure of the major antibody classes.

Antibody class switch   As the precursor lymphocytes develop in the bone marrow, they express the mu heavy chain as pre-lymphocytes, and on leaving for the blood, become naive B cells, expressing IgM and IgD. Entering the lymph node, the cell may then undergo a switch in its heavy chain, which can change to IgG, IgA or IgE, whilst retaining the same variable regions that recognize the antigen. This switch of heavy chain relies on an interaction between CD40 and its ligand, CD154, the latter expressed by the CD4 positive T helper lymphocyte. The same class switch changes the IgG classes from IgG_1 sequentially to IgG_4. These changes broaden the range of different antibody responses to the same antigen.

Receptors for antibodies   Several white cell classes have molecules on their cell surface that act as receptors for the Fc sections of antibodies (hence FcR). Since IgG is the dominant antibody, it is this which locates primarily into the cell via the FcR, and in this way the cell is primed and ready to be activated once the antibody has bound its antigen. It follows that the pentameric IgM and dimeric IgA molecules do not expose their Fc regions, and so do not interact with white cells in this way (Figure 5.12). Furthermore, there are subtypes of the FcR for different antibody classes. These are FcRγ for IgG (with subtypes CD16, CD32 and CD64), predominately expressed by neutrophils and some NK cells, and FcRε for IgE (CD23), the latter being found on basophils and mast cells.

Thus, the FcR can form a bridge between a cell and an antigen. Should this antigen be a structure on a bacterium, then immunoglobulin binding to the bacterium at one end and to the granulocyte via the FcR at the other end gives the cell an enhanced opportunity to kill the pathogen.

T lymphocytes   These cells are so called because of the requirement to pass through the thymus as part of their development (Figure 5.1); they can be classified into several groups:

• Cellular aspects of immunity call for the production of cytotoxic T lymphocytes (defined by the presence of CD3 and CD8) that can recognize and so remove cells that are

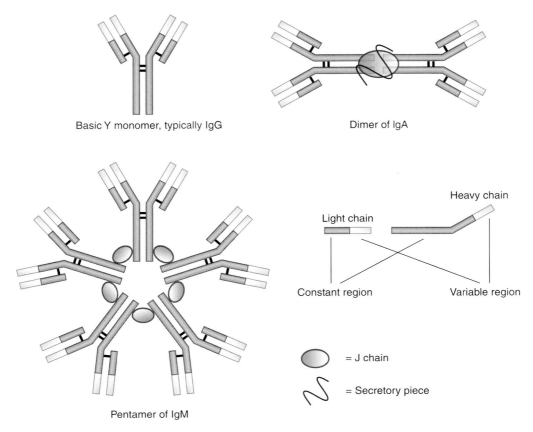

Basic Y monomer, typically IgG

Dimer of IgA

Pentamer of IgM

Heavy chain

Light chain

Constant region

Variable region

= J chain

= Secretory piece

Figure 5.12 Structure of major antibody classes. The standard 'Y'-shaped molecule is the prototype, and is typified by IgG, consisting of two heavy chains and two light chains, each with a variable region and a constant region. Two such monomers form an IgA dimer, whilst IgM is composed of five monomers, both being stabilized by J chains.

infected with a particular virus, such as measles or the Epstein–Barr virus. The ultimate mode of action of cytotoxic T cells is the contents of their granules, consisting of molecules perforin and granulysin; both form pores in the cell membrane of the target, whilst the latter also induces apoptosis.

• As described above, a second class of T lymphocytes are helper cells (defined by CD3 and CD4), so called because they cooperate with B lymphocytes and modified macrophages (which serve as antigen-presenting cells) in the production of antibodies.

• Upon first encounter with a pathogen, helper cells can be further subgrouped into Th1 (mostly providing help in bacterial, fungal and viral defence, and producing the cytokine gamma-interferon) and Th2 cells (involved in several responses, including defence against parasites, and producing IL-4). A third type of helper cell, Th17,

is characterized by their secretion of pro-inflammatory IL-17, and may be involved in help against bacteria and fungi. An additional requirement in some cases is that CD40 on naive B cells must link to CD40 L on the CD4 T helper cell.

• There is also evidence for a population of CD4-bearing regulatory T cells (formerly known as suppressor cells), some of which bear CD25, which modulate the functions of other T cells.

• However, one of the more recent developments in leukocyte biology is the description of myeloid-derived suppressor cells, a heterogenous population of immature myeloid cells. Their numbers dramatically increase in chronic and acute inflammatory diseases, and in this role may be attempting to limit the immune response. They seem able to suppress antigen-specific CD8$^+$ and CD4$^+$ T-cell activation, and the presence of CD11 implies their origin lies with monocytes/macrophages. There is also evidence the B cells can also be suppressive, and they may, as with the myeloid suppressors, do so via the inhibitory effects of IL-10.

Natural killer cells in action   The precise day-to-day role of NK cells in immune responses is unclear when compared with our knowledge of B and T cells. It seems that NK cells, like cytotoxic T cells, also kill with cytotoxic molecules, such as granulysin, stored in granules. T lymphocytes use a complex set of cell surface molecules to recognize their target (the T cell receptor) in the context of HLA molecules. However, NK cells lack the T cell receptor, though they have their own unique and poorly understood receptors that recognize certain virally infected or damaged cells and some tumour cells. There is evidence that this recognition system involves CD94 and HLA-E (the latter a part of Chapter 10).

Some NK cells express other surface molecules which allow them to participate in the defence against bacteria. These include CD16 (the receptor for the Fc of IgG), which also endows on some NK cells the ability to participate in antibody-dependent cellular cytotoxicity. They also produce inflammatory cytokines such as tumour necrosis factor and ILs, and so are likely to promote the action of T cells and macrophages.

Antigen recognition and genetics

The processes by which T cells, B cells and macrophage subtypes come together to generate antibodies and learn which virus-infected cells to kill are complex. A key aspect is the recognition of 'self', a concept developed in Chapters 9 and 10, so that anything that is not 'self' must be foreign, and so a potential danger.

The definition of self is found in our HLA molecules, which are a set of cell surface structures unique to the individual. The precise form is defined by variation in genes, and a similar type of gene variation gives rise to different antibody responses. This is called immunogenetics. A practical consequence is transplantation, for which histocompatibility is required. These terms are explained in Chapter 10.

5.9   White cells in clinical medicine

On a day-to-day basis, it is likely that the blood scientist will focus on high or low levels of leukocytes (leukocytosis

and leukopenia respectively), according to the white cell differential.

A high white cell count in response to infections is highly desirable. This is in contrast to malignancy (Chapter 6), where there are increased numbers of neoplastic white cells in the blood and tissues. However, it could be argued that the malignant cells responsible for these diseases are pathologically overactive by their sheer presence of numbers. Nonetheless, in this section we will look at the nonmalignant increased numbers and overactivity of white cells that cause illness and disease.

By far the greatest clinical area of the overactivity of white cells is their role as the cause of autoimmune diseases. Inevitably defined by an autoantibody, itself therefore the product of an abnormal B lymphocyte, these diseases are discussed in detail in Chapter 9, on immunopathology. That chapter also looks at the consequences of white cell underactivity: immunodeficiency.

What remains is the non-autoimmune overactivity of white cells, which is inevitably associated with high levels of individual particular white cells, generally in the blood, but often also in the tissues.

Neutrophilia

We need increased number of neutrophils, which may also be described as a neutrophil leukocytosis, to fight microbial pathogens, such as in acute inflammation. This physiological process may well generate a degree of pathology, but the entire process is of crucial importance, and the discomfort and perhaps limited tissue damage is a price worthy paying. However, if the stimulus for the acute response is great, then so too will be the response, and this may cause more serious disease. Furthermore, if an acute inflammation fails to shut down, it may lead to chronic inflammation, which is always pathological.

A consequence of neutrophilia may be a change in the spectrum of cells of the neutrophil lineage. Normally, there should be very few, if any, metamyelocytes in the blood. However, at times of high demand, the bone marrow may well export some such immature cells, and possibly some myelocytes. The blood film may be described as 'left shifted', indicating some immature cells. Working on the same principle, a leukaemia could be described in the same terms, but the left shift would be far more extreme. In parallel, granules in left-shifted neutrophils may be small and immature, leading to the expression 'toxic granulation', a historical term that does not imply greater or lesser toxicity.

An excessive acute inflammation   An excellent example of this is septicaemia (sepsis, blood poisoning), a potentially life-threatening and overwhelming infection, generally with a bacterium, such as *Escherichia coli*. The body's response to this must reflect the severity of the infection, and can be summarized as the systemic inflammatory response syndrome to a defined pathogen. If uncontrolled, this will lead to multisystem organ failure and death.

The typical response of the body to septicaemia is multifactorial, with pyrexia (temperature $>38\,^{\circ}$C), tachycardia (heart rate >100 beats/min) and hyperventilation (>20 breaths/min). The white cell count is certainly above the top of the reference range (perhaps 10×10^9/L), and may be as high as 20–25×10^9/L, of which over 90% will be neutrophils. The bone marrow may also respond, with increased numbers of metamyelocytes. Increases in other leukocytes are not uncommon, but none are as marked as the neutrophilia.

Parallel abnormalities will be a raised ESR and an acute-phase response. The latter includes raised plasma molecules such as CRP and coagulation factors (produced by the liver, as outlined in Chapter 17) and immunoglobulins (as detailed in Chapter 9). As the disease progresses, and so the clinical picture and prognosis worsen, there may be lactic acidosis (a falling pH), renal failure, oliguria (very little urine) or anuria (no urine), with raised urea and liver failure (raised LFTs). Much of the widespread abnormalities flow from the industrial levels of inflammatory cytokines such as IL-1 and IL-6 probably released from macrophages. It is also possible that the neutrophils are contributing to the picture with hyper- and possibly misdirected-phagocytosis, with inappropriate release of proteases and inflammatory mediators. There may also be disseminated intravascular coagulation, with thrombocytopenia and hypofibrinogenaemia (which are discussed fully in Chapter 8).

Treatment must address the bacteraemia, generally with high doses of intravenous antibiotics. The patient is also likely to need other support, such as prophylaxis with anticoagulants in view of the risk of venous thromboembolism. Hopefully, as the bacteraemia resolves, so will the neutrophilia, and the clinical picture.

Leukaemoid reaction   As the name suggests, this is a leukaemic-like picture, with many neutrophil precursors (mostly metamyelocytes, with a smaller number of myelocytes). However, it is the 'normal' response to very severe infections (perhaps a septicaemia), drugs or a malignancy. The key difference is that a leukaemia is a monoclonal disease that has no apparent cause, and where there is also anaemia and thrombocytopenia, whereas a leukaemoid reaction is a polyclonal response to an external stimulus with normal red cell and platelet counts. It is therefore a matter of finding a likely provoker of the reaction (such as a bacterium, if present) to confirm the diagnosis.

Chronic inflammation   This may be present in autoimmune disease, where there is no clear microbial pathogen (Chapter 9). However, in this setting the inflammation is initiated by a pathogen, such as in the lungs, and leads to physical incapacity (in this case, perhaps chronic obstructive pulmonary disease). Physiology requires that the immune system addresses a local infection and eliminates it, and so the inflammatory response abates. However, if the pathogen is resistant and/or persistent, the inflammatory response will continue to the extent that many continue after the pathogen has been destroyed.

A manifestation of chronic inflammation is a neutrophilia, which is likely to be mature neutrophils and a leukocyte count of perhaps 12–16×10^9 cells/L and a neutrophil count that exceeds 7.5×10^9/L. There are also likely to be chronically high levels of acute-phase reactants.

There are a number of potential reasons for the persistence of the inflammatory response. One is a cycle of mutual- and self-activation by neutrophils and macrophages (with cross-talk by inflammatory mediators) that, between them, misdirect their normal function towards healthy tissues. Another is that the collateral damage linked to the initial inflammatory response has altered normal tissues so that the leukocytes consider it to be foreign, and so a bona fide subject of their attentions.

But whatever the aetiology, the treatment is immunosuppression with high-dose nonsteroidal anti-inflammatory drugs, although steroids may be necessary, and these should reduce the neutrophilia. Theoretically, antibiotics will be ineffective.

Blood science angle: The acute-phase response

There are many responses of the body to an infection or inflammatory stimulus. These include increased numbers of white cells, but many plasma proteins, such as CRP and caeruloplasmin, both made by the liver and therefore of importance to biochemists. The response also leads to raised coagulation factors, hepcidin, ferritin and haptoglobin (thus engaging haematologists), complement and immunoglobulins (so of interest to immunologists, as in Chapter 9).

Lymphocytosis

A high white cell count (leukocytosis) is not seen only in viral infections, it may be present in bacterial infections, possibly as a 'back-up' mechanism. However, as indicated, a principal reason for a lymphocytosis is as a response to a viral infection, one of the best examples of which is IM, also known (incorrectly) as glandular fever (lymph nodes are not glands, and the swelling in the neck is not swollen salivary glands). This common infectious condition is transmitted by saliva and so has a high prevalence in teenagers and young adults. Symptoms include fever, malaise and chills with generalized aches and pains, much like those of influenza. However, a major sign is a sore throat and lymphadenopathy of the lymph nodes in the neck. These nonspecific aspects are also present (as mentioned) in tonsillitis and other infections, so more focused investigations are demanded. Treatment is symptomatic (analgesia, rest), but over a quarter of those with IM may also have a concurrent bacterial infection (such as a streptococcal infection of the throat) that may respond to antibiotics.

The primary causative factor in the relatively benign lymphoproliferative disorder that is IM is the Epstein–Barr virus, a member of the herpes family of viruses. The target of this very common virus is CD21, preferentially expressed by B lymphocytes, although it may also infect T cells and epithelial cells. B cells infected with the virus are sought, detected, attacked and destroyed by cytotoxic T lymphocytes, and it is these cells, also described as atypical or reactive, that dominate the circulating lymphocytosis. These activated lymphocytes are illustrated in Figure 5.13.

However, a second aspect of antibody production is of note: there is a surge in antibodies that are not directed towards a specific antigen but nonspecifically and coincidentally recognize antigens present on the cells of other animals. These heterophile antibodies include those that bind red blood cells of the sheep, ox and horse, and are relatively easy to detect with commercial kits, marketed under names such as the Paul Bunnell test and the Monospot test.

But there is a problem with the diagnostic value of these tests, in terms of less than perfect sensitivity and specificity (i.e. a positive result in the absence of IM (a false positive) and vice versa). Furthermore, the exact timing of the test may be crucial, and the levels of the antibodies may be rising or falling. A general nonspecific heterophile response by the body's B lymphocyte population is also found in other viral and bacterial infections,

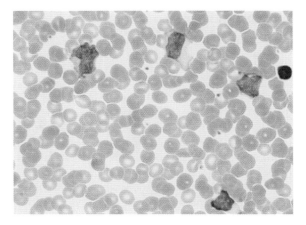

Figure 5.13 Activated lymphocytes. The increased white cell count often found in IM is due largely to activated lymphocytes, which are larger than resting lymphocytes and have more cytoplasm. This figure shows four activated and one normal lymphocyte. (Image courtesy of H. Bibawi, NHS Tayside).

not only in IM, and also in leukocyte malignancy such as lymphoma and leukaemia. For example, another heterophile antibody, the Forssman antibody, which binds to glycosphingolipid antigens on guinea pig red cells, is present in many infections but is rarely present in IM.

Other viral pathogens include cytomegalovirus (CMV), herpes simplex and zoster, hepatitis A, B and C, measles, influenza and the human immunodeficiency virus (HIV). All of these can cause specific and possibly lethal disease (as will be developed in Chapter 9), but in many cases the body is successful in preventing this. For example, up to 70% of the population in industrialized nations is seropositive for cytomegalovirus, proving a successful defence.

Monocytosis

Increased numbers of monocytes alone are very rarely encountered in physiology. Conversely, high levels are most often found in both inflammatory and immune responses, alongside a neutrophilia and/or a lymphocytosis, suggesting that monocytosis is simply a nonspecific response. This is particularly relevant in the septicaemia case described above. It has been suggested that the CD14 and CD16 can differentiate certain monocyte phenotypes, such as those committed to phagocytosis versus those involved in antigen presentation, and in the production of inflammatory cytokines.

Eosinophilia

The physiology of these cells is in defence against parasites (such as *Schistosoma*), low level participation in inflammation, and tissue scavenging (as do neutrophils, such as the removal of fibrin). However, pathological raised numbers are also found in auto-immune and other diseases such as polyarteritis nodosa, lymphoma and vasculitis, and in the response to certain drugs. In these instances it is unclear whether or not the cells are active in contributing to the pathology, or are merely bystanders, but levels seem to be secondary to the original disease. This leaves a number of situations where a high eosinophil count contributes directly to illness.

Allergic disease   We cannot escape infections, but ideally these would be countered with a minimum of collateral disease. When the response of the immune system is inappropriately high, we can describe it as hypersensitivity. This is principally the province of basophils and mast cells, but eosinophils may have a role in allergy. This may become manifest in a number of situations, one of which features the release of eosinophil cationic protein from granules. It has been suggested that this contributes to airways inflammation, particularly in those with asthma. More details are present in Chapter 9.

Chronic eosinophilic leukaemia   The aetiology of this disease is as for the other leukaemias, as explained in Chapter 6. It is caused by a genetic lesion in an eosinophil stem cell/precursor that leads to increased numbers of phenotypically mature eosinophils and their myelocytes and metamyelocytes. The most common cytogenetic abnormality is a fusion gene in chromosome 4. The leukaemic cells are likely to invade the tissues (such as the heart, skin, intestines and lungs) where they may degranulate and so cause tissues damage. However, the leukaemic cell is immature, so that degranulation may simply release ill-formed and inactive metabolites.

Hypereosinophilic syndrome   This can closely resemble chronic eosinophilic leukaemia, but there is no clonality and the blast count is <5%. It is effectively a diagnosis of exclusion, present when other potential diagnoses have been eliminated. The high mortality rate demands treatment, generally with marrow suppressants such as steroids, although use of a monoclonal antibody against IL-5, a major eosinophil stimulant, may be helpful.

Other conditions   There may be an eosinophilia in Churg–Strauss syndrome, a type of autoimmune vasculitis, but the significance of this is unknown. Loeffler's syndrome is characterized by eosinophil accumulation in the lung (hence eosinophilic pneumonia) due to a parasitic infection, but is now used for any such accumulation. Similarly, eosinophils in the myocardium causes Loeffler's endocarditis. However, it could be argued that these two conditions are simply consequences of the hypereosinophilic syndrome.

A variety of drugs may precipitate an eosinophilia, which can become manifest as a syndrome that includes rash, fever, lymphadenopathy and other abnormalities, collectively named the DRESS syndrome (drug reactions, eosinophilia, systemic symptoms).

Basophilia

The primary reason for a healthy raised basophil count is in the attack against external parasites (that is, ectoparasites), such as may be invading the skin. Under these conditions tissue mast cells are an effective defence. However, inappropriate activation of basophils and mast cells are most frequently found in allergic conditions, such as to pollen. This will be re-visited in Chapter 9. A basophilia may be seen in certain leukaemias, as is discussed in Chapter 6.

The most common cause of increased circulating basophils is in myeloproliferative disorders, typically CML and polycythaemia rubra vera. Other increases are found in infections and ulcerative colitis. However, recall that basophils are not the end cell of the lineage, but transform into mast cells in the tissues. Furthermore, the key feature of basophils and mast cells is that they bear receptors for the Fc section of IgE antibodies. The docking of the antibody into the receptor, and subsequent antigen bonding, initiates a sequence of metabolic changes resulting in degranulation, which is desirable in the normal response to a potentially pathogenic situation.

Although limited basophil and mast cell activation are beneficial, inappropriate (over)activation of these cells leads to pathological degranulation. The contents of the granules include heparin and large amounts of histamine, which promotes various inflammatory processes and also increases vascular permeability. This cell activation can be detected in the laboratory by FACS, determining changes in the expression of CD203c (a cell surface marker, upregulated within 5 min of activation) and CD63 (a component of granules, upregulated within 10 min). A clinical consequence of the activation of mast cells in the tissues, and allergy in general, is discussed in more detail in Chapter 9.

Leukopenia

This is defined as a white cell count below the lower end of the reference range, of which there are several well-established causes, the two most common being:

- Malignant invasion of the bone marrow, which will eventually cause erythropenia and thrombocytopenia as well as leukopenia (that is, pancytopenia). This may be invasion by a distant cancer or a tumour of the bone marrow itself (such as myeloma and leukaemia).
- Chemotherapy (not necessarily cytotoxic, but including immunosuppressants) and radiotherapy, as in the treatment of cancer and bone marrow transplantation.

There may also be isolated reduction in particular cell families.

Neutropenia   A reduced neutrophil count may be due to drugs, autoimmune disease (such as Felty's syndrome, a complication of rheumatoid arthritis) or viral infections such as the hepatitis B virus. The very rare Kostmann's syndrome is a severe autosomal–recessive congenital neutropenia, and in most cases is linked with genes encoding an elastase, whilst other cases are linked to abnormalities in apoptosis. The neutrophil count can vary over a 21-day cycle, so that at a trough there may be a neutropenia (hence the diagnosis cyclic neutropenia). Just as rare is autoimmune neutrophilia, caused by autoantibodies.

Lymphopenia   The principal causes of an isolated low lymphocyte count include an infection with the HIV, which preferentially attacks and destroys cells bearing CD4 (some T lymphocytes and monocytes). Lymphopenia may also be caused by other viruses that simply nonspecifically suppress their function.

Myelodysplasia and myelofibrosis   These two conditions, often classified as malignancies, are also relevant in diseases of the red cell, as they both cause anaemia. Myelodysplasia is a collection of syndromes characterized by the disordered and inefficient clonal proliferation of multipotential haemopoietic stem cells: the marrow is

described as hypercellular. The ineffective haemopoiesis leads to a reduction in the numbers of all circulating blood cells (pancytopenia), but there are several notable signs, such as ringed sideroblasts in the bone marrow, and hypolobulated neutrophils on the blood film (pseudo Pelger–Huet syndrome). The condition may follow chemotherapy and/or radiotherapy for neoplasia, or exposure to environmental toxins such as benzene. However, this disease may transform to acute myeloid leukaemia.

The pathophysiology of myelofibrosis relates to the cells of the bone marrow itself, but the proliferating cell is not haemopoietic. The bone marrow becomes overgrown with fibroblasts, and these cells are driven to proliferate and produce collagen by inappropriate responses to growth factors and cytokines produced by other cells. Like cancer, this is progressive and eventually leads to poor haemopoiesis, and thus to leukopenia (as well as anaemia and thrombocytopenia, and hence, once more, a pancytopenia). In an attempt to maintain levels of all blood cells, haemopoiesis may develop in extramedullary sites such as the liver and spleen, leading to enlargement of these organs.

Table 5.5 summarizes key functions of leukocytes.

Table 5.5 Principal functions of white blood cells.

Cell	Functions
Neutrophils	Phagocytosis, participation in inflammation
	Scavenging and removal of debris
Lymphocytes	Generation of antibodies (B cells)
	Assisting antibody generation (T helper cells)
	Killing virus-infected cells (T cytotoxic cells)
Monocytes	Phagocytosis, participation in inflammation
	Scavenging and removal of debris
	Assisting antibody generation (antigen-presenting cells)
	Release of cytokines
	Participation in haemostasis
Eosinophils	Protection against parasitic infection (such as helminths)
	Participation in allergic responses
	Release of histamine
Basophils	Participation in hypersensitivity reactions
	Release of histamine and heparin

5.10 Case studies

Case study 3

A 20-year-old student presents with a weeks' onset of influenza-like symptoms, with general malaise, sweating, pyrexia, an intermittent skin rash and a sore throat. As the present symptoms seemed to be more severe and persistent than general 'flu', a blood test was performed:

	Result (unit)	Reference range
Haemoglobin	131 g/L	118–167
Platelets	285×10^9/L	140–400
WCC	12.1×10^{12}/L	4.0–9.5
RCC	4.9×10^9/L	3.9–5.7
MCV	88 fL	79–98
Haematocrit	0.43	0.33–0.53
ESR	23 mm/h	<10
Differential:		
Neutrophils	4.4×10^9/L	1.7–6.1
Lymphocytes	6.6×10^9/L	1.0–3.2
Monocytes	0.35×10^9/L	0.2–0.6
Eosinophils	0.25×10^9/L	0.03–0.46
Basophils	0.05×10^9/L	0.02–0.09
Blasts	0.45×10^9/L	<0.02

Interpretation There are two abnormalities: a raised white cell count and a raised ESR. The former justifies the differential, which reports a lymphocytosis but also increased blasts. The primary cause of a lymphocytosis is viral, and this is supported by the raised ESR, reflecting an acute-phase response.

Notably, the number of blasts exceeds that of the monocytes, which cannot be physiological. However, these are not true malignant blasts, but are simply atypical lymphocytes, perhaps activated by the acute-phase response and seeking cells infected with viruses. A likely primary diagnosis is IM, confirmed by the Paul Bunnell/Monospot test. Note that the sex of the student is not specified, so the reference range for red cells and haemoglobin has to include that for males and females.

Case study 4

A blood sample is received from a general practitioner, whose patient is a 72-year-old man who has reported a third infection within 3 months, each responding to a course of antibiotics. The full blood count is as follows:

	Result (unit)	Reference range
Haemoglobin	141 g/L	133–167
Platelets	302×10^9/L	140–400
WCC	3.7×10^{12}/L	4.0–9.5
RCC	5.2×10^9/L	3.9–5.7
MCV	84 fL	79–98
Haematocrit	0.44	0.33–0.53
ESR	12 mm/h	<10
Differential		
Neutrophils	1.0×10^9/L	1.7–6.1
Lymphocytes	1.5×10^9/L	1.0–3.2
Monocytes	0.7×10^9/L	0.2–0.6
Eosinophils	0.4×10^9/L	0.03–0.46
Basophils	0.06×10^9/L	0.02–0.09
Blasts	0.00×10^9/L	<0.02

Interpretation There is a raised ESR and a leukopenia, the latter justifying a look at the differential, which shows a neutropenia (probably explaining the bacterial infections). However, there is also a slight monocytosis, which could simply be a response to the lack of neutrophils.

The general practitioner rings back in a couple of days to say her patient was diagnosed with a form of atrial fibrillation, and so was started on the drug flecainide. This drug is known to cause neutropenia, so the patient was moved to another antiarrhythmic drug (amiodarone) and the neutropenia slowly resolved.

Summary

- White blood cells develop in the bone marrow by the process of leukopoiesis.
- Each of the five major types of white cell has its own development pathway, although several are shared between different cells (such as the granulocytes).
- White cells defend us from attack by pathogenic microbes such as bacteria and viruses. At these times we expect increased number of cells in the blood – a leukocytosis.
- A Romanowsky-stained blood film is used to identify white cells under light microscopy. Key features are the size of the cell, the morphology of the nucleus (mononuclear or polymorphonuclear) and granules.
- The CD system is inevitably used in flow cytometry to define different white cell populations; it requires a fluorescence-activated cell scanner.
- An inflammatory response is characterized typically by phagocytes (principally neutrophils) attacking bacteria.

- Immune responses against viruses are principally mediated by T lymphocytes, which also support the generation of antibacterial and antiviral antibodies by B lymphocytes. Most antibody generation occurs in lymph nodes.
- Monocytes are also phagocytes, but can also promote immune responses by secreting pro-inflammatory cytokines and presenting antigens to lymphocytes.
- Eosinophils and basophils are involved in defence against parasites, but may also mediate hypersensitivity responses.
- A leukocytosis is generally found in response to an infection. If bacteria, then a neutrophilia is most likely, whereas a viraemia should produce a lymphocytosis.
- Leukopenia is most likely to result from bone marrow suppression by drugs or malignancy. There may also be reduced numbers of neutrophils or lymphocytes (neutropenia and lymphopenia).

Further reading

Akuthota P, Xenakis JJ, Weller PF. Eosinophils: offenders or general bystanders in allergic airway disease and pulmonary immunity? J Innate Immun. 2011;3:113–119.

Allen CD, Cyster JG. Follicular dendritic cell networks of primary follicles and germinal centers: phenotype and function. Semin Immunol. 2008;20:14–25.

Bouma G, Ancliff PJ, Thrasher AJ, Burns SO. Recent advances in the understanding of genetic defects of neutrophil number and function. Br J Haematol. 2010; 151:312–326.

Chiche L, Forel JM, Thomas G. et al. The role of natural killer cells in sepsis. J Biomed Biotechnol. 2011; 2011:986491.

Geissmann F, Manz MG, Jung S. et al. Development of monocytes, macrophages, and dendritic cells. Science. 2010;327:656–661.

Harvima IT, Nilsson G. Mast cells as regulators of skin inflammation and immunity. Acta Derm Venereol. 2011;91:644–650.

LeBien TW, Tedder TF. B lymphocytes: how they develop and function. Blood. 2008;112:1570–1580.

Sica A, Mantovani A. Macrophage plasticity and polarization: in vivo veritas. J Clin Invest. 2012;122:787–795.

Stone KD, Prussin C, Metcalfe DD. IgE, mast cells, basophils, and eosinophils. J Allergy Clin Immunol. 2010;125(2 Suppl 2):S73–S80.

Summers C, Rankin SM, Condliffe AM. et al. Neutrophil kinetics in health and disease. Trends Immunol. 2010; 31:318–324.

Topham NJ, Hewitt EW. Natural killer cell cytotoxicity: how do they pull the trigger? Immunology. 2009; 128:7–15.

White Blood Cell Malignancy

After studying this chapter, you should be able to:

- explain the molecular and cellular basis of white blood cell malignancies;
- recognize a malignant phenotype on a blood film;
- appreciate the value of flow cytometry in the diagnosis and management of white cell malignancies;
- recognize the different types of leukaemia and their consequences;
- describe the different types of lymphoma;
- outline the major features of myeloma.

Having examined the physiology and some aspects of the pathology of white cells (leukocytes) in Chapter 5, we now move to the causes and consequences of the most serious diseases of these cells: cancer, the collective term for all blood cell cancers being haemato-oncology. The most well-known malignant disease, leukaemia, and its allied diseases of myeloma and lymphoma, are all characterized by the failure of the particular white cell to do its job. The initial focus of lymphoma is the lymph node, whereas leukaemia, myeloma and other diseases start in the bone marrow. However, there are other malignant diseases of the bone marrow and other tissues. These, such as polycythaemia (of red cells) and essential thrombocythaemia (platelets) are discussed in their respective chapters. These diseases also have an impact on white blood cells. Myelofibrosis, a disease of the proliferation of non-leukocytes in the bone marrow, also influences white cells. The closely named myelodysplasia is a further bone marrow disease that generally presents first with an anaemia but that often develops into an acute leukaemia.

We begin our exploration of white-cell cancer in Section 6.1 with a review of the genetic basis of the disease, whilst Section 6.2 will look at laboratory methods in haemato-oncology. Having set the scene, Sections 6.3, 6.4 and 6.5 look at each of the three major diseases of leukaemia, lymphoma and myeloma. The chapter concludes with a brief revision of myelodysplasia and myelofibrosis, and then two case studies.

6.1 The genetic basis of leukocyte malignancy

There are several common themes in all types of cancer, such as the fact of uncontrolled growth that is eventually fatal and that all are based on changes to the deoxyribonucleic acid (DNA) of the cell that gives rise to this new growth, or neoplasia.

An agent that causes cancer is a carcinogen. Some may be hydrocarbons, such as benzene and naphthylamine, some are natural products (aflatoxin causing hepatocellular carcinoma), and others may be viruses (herpes, hepatitis, papilloma viruses). But in many cases an exact physical carcinogen cannot easily be identified, and this is especially true in haemato-oncology. However, observations made in Japan in 1945–46 clearly indicate that ionizing radiation (caused, in this case, by atomic bombs) is a major cause of white cell malignancies (and many other cancers as well). This has been supported over the decades that followed by an increased incidence of cancers following accidents at nuclear power plants. In the current era, uses of murine knock outs and zebrafish are providing new aspects on the genetics of these diseases.

Whilst Chapter 1 has some details of the role of genetics and mutation in cancer, many chromosome abnormalities are present in leukocyte cancer, and Chapter 2 has looked at key methods in molecular genetics. Furthermore, some abnormalities are taken as examples of genetics and cancer that apply to many aspects of

Blood Science: Principles and Pathology, First Edition. Andrew Blann and Nessar Ahmed.
© 2014 John Wiley & Sons, Ltd. Published 2014 by John Wiley & Sons, Ltd.

human oncology. One of the most well-known genetic diseases is Down syndrome, caused by an entire extra copy of chromosome 21 (the karyotype therefore being 47XX or 47XY), and is linked with an increased risk of an acute leukaemia. However, there are also abnormalities within a particular chromosome, and these can be described in three groups: deletions, inversions and translocations (Figure 6.1).

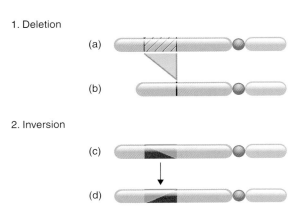

1. Deletion

(a)

(b)

2. Inversion

(c)

(d)

3. Translocation

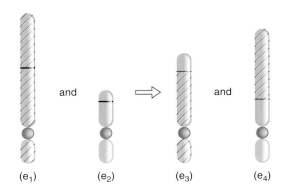

(e₁) and (e₂) ⇒ (e₃) and (e₄)

Figure 6.1 Deletions, inversions and translocations. Compared with a normal chromosome (a), a deletion is characterized by a missing nucleotide, gene or larger section of DNA, leading to a shorter chromosome (b). In an inversion (c) the chromosome is of the same length but a section of DNA or part of the chromosome is reversed (d). Most translocations see sections of DNA being reciprocally transferred between different chromosomes (e₁ and e₂), often resulting in new hybrid chromosomes of different lengths and containing different genes (e₃ and e₄).

Deletions

These lesions are characterized by sections of missing DNA, which can vary between a single nucleotide (a micro-deletion) to whole sections of chromosomes. In some cases deletions are within a chromosome, whereas others are at one of the terminal sections; both result in a shorter chromosome. So if we describe a series of genes on a particular chromosome as *ABCDEFGHIJ*, then the deletion of genes *DEF* leaves the sequence *ABCGHIJ*.

Inversions

In this abnormality, sections of DNA are switched around, and there is a system of naming the particular genetic lesion. For example, an inversion between different parts (sections 21 and 26) of the long arm (named q; the short arm is named p) of chromosome 3 is described as *inv(3)(q21:q26.2)*. Note that this nomenclature is in italics. Using the same system of letters for genes, an example of this might be a change from *ABCDEFGHIJ* to *ABCGFEDHIJ* (inversion underlined).

Translocations

These are the most common form of abnormalities. A section of DNA from one chromosome is grafted onto another chromosome. This often happens in pairs, where part of chromosome A is pasted onto part of chromosome B, and part of chromosome B is in turn pasted on to part of chromosome A. This is called a reciprocal translocation. In the same way as an inversion is named, an example of a translocation is *t(11:17)(q23:q21)*. This means that sections 23 and 21 of the long (q) arms of chromosomes 11 and 17 have swopped over.

An example of this might by the normal sequences of *ABCDEFGHIJ* on one chromosome and *STUVWXYZ* on another. A translocation would see, for example, new sequences of *ABCDEWXYZ* and *STUVFGHIJ*.

What this means

It is certain that the precise sequence of a DNA nucleotide base (adenine, thymine, guanine and cytosine) is crucial for a correctly functioning gene. It is also likely that the order of genes along a section is important. There can be abnormalities in a single nucleotide of a single gene (as in haemoglobin in sickle cell disease) or in a whole section of DNA (as in thalassaemia) (Chapter 4).

But the whole point about many of these malignancies is that the mutation brings together sections of DNA that by themselves have normal function, but when fused have an altogether more sinister function. Using our model of letters of the alphabet for the sequence of nucleotide bases, or genes, then the sequence *DANGEXYZXYZ* on one chromosome and *XYZXYZOUS* on another may be brought together by a translocation to create *DANGEROUS* and *XYZXYZXYZ*. The latter may not have any sinister implications, but the former certainly does, and may initiate a cancer.

Blood science angle: Cancer genetics

There are hundreds, if not thousands, of examples of cancer and other diseases caused by gene mutations. Chapter 5 mentioned haemoglobinopathy, but haematologists are also aware that a mutation in the gene *Jak2* leads to a common form of polycythaemia, whilst a mutation in the von Hippel–Lindau tumour suppressor gene leads to another form of polycythaemia (Chapter 11). Genetic disease also causes nonmalignant (in the cancer sense) abnormalities such as haemophilia (Chapter 8) and a whole host of inherited metabolic diseases, including cystic fibrosis and phenylketonuria (Chapter 20).

In this setting, *DANGEROUS* and *XYZYZXYZ* are neogenes (neo = new), and if causing cancer they are called oncogenes. The process by which a cell (or, indeed, a gene) becomes neoplastic is called malignant transformation.

Although mutations in chromosomal DNA are the cause of almost all haemato-oncology, there can be subtle alterations to normal DNA or ribonucleic acid (RNA) which changes its properties. These epigenetic changes include adding or removing key chemical groups such as methy- and acetyl-, leading perhaps to hypermethylation or hypoacetylation. These can in turn lead to changes in gene and cell function that may result in a malignancy.

Genes and chromosomes

An important technique is fluorescence *in-situ* hybridization (FISH), where a gene probe linked to a fluorescent probe is exposed to a sample from the patient. The probe will link with its particular gene, which can be visualized with a fluorescence microscope. By using probes with different fluorochromes, different genes can be located, and of course the absence of a signal implies the normal gene sequence is missing (perhaps mutated). Conversely, a probe for an abnormal gene can

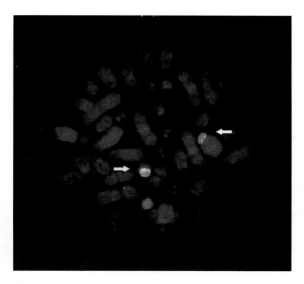

Figure 6.2 Fluorescence *in-situ* hybridization. Red and green fluorochromes are linked to probes for different genes. Normally, the genes binding the two probes would be on different chromosomes. However, the arrows highlight two chromosomes with both colours, thus defining a reciprocal translocation. Some of the 'red' gene has moved to the 'green' chromosomes and vice versa. (From Hoffbrand & Moss, *Essential Haematology*, Sixth Edition, 2011, Fig. 11.12(c), p. 161. Reproduced with permission of John Wiley & Sons, Ltd.).

be used to help investigate the mutated genes in a particular malignancy (Figure 6.2).

An equally valuable method is transcriptional profiling of RNA. Messenger RNA (mRNA) is extracted from the patient's cells, is linked to a fluorescent probe and then applied to a microarray to which are fixed perhaps dozens of pre-prepared genes, each to a different mutation. If present, mRNA in the patient's sample will hybridize to the gene, which can then be visualized by the colour of the fluorochrome. By repeating this scores of times on the same microarray, a picture can be built up, which may resemble that of an established leukaemia, so supporting a particular diagnosis (Figure 6.3).

Consequences

Changes to genes and chromosomes such as these disrupt the normal life cycle of the cell, and they can lead to unregulated growth and reproduction or perhaps to failure of cells to develop into fully mature cells. This can result instead in many immature cells (maturation arrest), whilst in other cells there is a failure of the cell to die

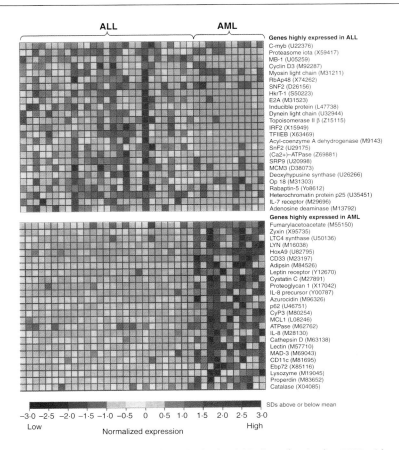

Figure 6.3 RNA microarray analysis. The upper panel shows (red colour) binding of patient's mRNA with genes consistent with a diagnosis of acute lymphocytic leukaemia, collected on the left of the array. In the lower panel, the patient's mRNA binds to a different pattern of genes, where the red colour is dominant on the right, suggesting a diagnosis of acute myeloid leukaemia. (From Hoffbrand & Moss, *Essential Haematology*, Sixth Edition, 2011, Fig. 11.13(b), p. 162 Reproduced with permission of John Wiley & Sons, Ltd.).

(the process of apoptosis). However, whatever the mechanism, there are simply too many cells in the bone marrow, which is said to be hypercellular.

These changes generally occur in one single precursor cell, which then generates its own colony of abnormal 'daughter' cells. This process is called malignant transformation. If this happens in the bone marrow, it is usually of a precursor cell, such as a blast, and generally leads to leukaemia or myeloma within the bone marrow. Conversely, lymphomas are generally caused by the malignant transformation of a more mature cell within a lymph node.

An abnormal growth that has started from a single malignant cell is a clone, hence the term clonal proliferation. Confusingly, some malignancies are characterized by cells of more than one lineage being abnormal, an

example being myelodysplasia, initially characterized by an anaemia, but which may progress to leukaemia. Finally, myelofibrosis is the uncontrolled proliferation of fibroblasts within the bone marrow. Although difficult to classify as a cancer, it nonetheless kills because of the slow and steady manner in which blood cell production is suppressed by the excessive growth of the fibroblasts.

Clinical presentation of white cell malignancies

Of course, the patient is unaware of these changes to the constituents of their blood. Unless discovered fortuitously, in most cases the patient presents themselves to a healthcare practitioner with a range of symptoms caused by their disease. These can vary with the exact type of disease, but there are common features.

As with other cancers, a consequence of the growth of the leukaemic and myeloma tumour within the bone marrow is that other functions of this organ start to suffer. As these are the production of white cells, red cells and platelets, then numbers of these cells fall as the disease progresses, and these also cause symptoms. These we can classify in three broad areas:

• Infections due to the reduced ability of the marrow to maintain the production of normal healthy white cells. In addition, the leukaemic cell itself is immature, and although there will be many of them, they are unable to attack and destroy microbial pathogens like the mature leukocyte can. This is likely to lead to infection, perhaps of the lungs, or maybe the throat or urinary system. It may also be responsible for night sweats and a 'flu-like' illness, and these may in turn be driven by inappropriate release of inflammatory cytokines such as interleukins 1 and 6. These cytokines are the drivers of the acute-phase response (Chapter 9), and there may be several other changes in the components of the blood.
• As in all diseases of the invasion of (or damage to) the bone marrow, a fall in the red cell count and haemoglobin will lead to the symptoms of anaemia (Chapter 4), such as tiredness, lethargy, pallor, and so on. If an anaemia is present, it is likely to be normocytic, and may possible be hypochromic.
• As the platelet count falls, the patient will experience bruising (such as ecchymoses or petechiae) and bleeding (the latter may be nosebleeds, excessive bleeding when brushing teeth or shaving), or (more seriously) blood in the urine (haematuria). This is more likely to happen when the platelet counts fall below 100×10^9/L (reference range generally 150–400 $\times 10^9$/L), a condition known as thrombocytopenia (Chapter 8).

These latter two bullet points are rarely present in the early stages of a lymphoma, though they may well be present in later stages if the disease spreads to the bone marrow. Nevertheless, the patient with a lymphoma may well complain of a generalized malaise, and also recurrent infections caused by the failure of their lymph nodes to participate in defence against microbial pathogens. If the malignant lymph node is near the surface of the body, it may be detected as a lymphadenopathy.

A combination of these symptoms is likely to prompt the practitioner to request a full blood count, often with an erythrocyte sedimentation rate (ESR), which is likely to be abnormal, generally >10 mm/h (reference range <10 mm/h). However, a grossly abnormal ESR, tending towards 100 mm/h, is a major feature of myeloma.

6.2   Tissue techniques in haemato-oncology

We have already looked at some general techniques in molecular genetics (which are also described in Chapter 1), but there are several specialist tests we must address that are based on peripheral blood, the bone marrow and the lymph node. White cell malignancies may well involve the liver, spleen and other organs, but these are very rarely part of the laboratory investigative process. Non-laboratory methods often of value in haemato-oncology (and in other aspects of pathology) include imaging techniques such as X-ray, ultrasound and magnetic resonance imaging (MRI)/computerized axial tomography (CAT) scanning.

Peripheral blood

All leukaemias are ultimately characterized by a high white cell count, and inevitably by the presence of unusual cells (often blasts). An abnormal peripheral blood leukocyte picture is rare in the early stages of lymphoma, myeloma and other malignancies, but in advanced and terminal stages it is not unusual to find atypical cells in the blood. Indeed, distinctive plasma cells are a marker of myeloma disease progression. In order to further investigate and characterize the malignancy we use the following:

• *Morphology* – to determine what the cells look like under a light microscope using conventional Romanowsky staining. To some extent, this is needed in order to provide information about which types of precursors are present, the value of which is shown in Figure 5.3 and Figure 5.4.
• *Flow cytometry* – for the presence of certain CD molecules, not only those in Table 5.3 and Figure 5.3 that are found on mature cells, but also those expressed only (or preferentially) by normal precursors or malignant cells (Figure 6.4).
• *Cytochemistry* – the staining of cells according to the make-up of the enzymes in their cytoplasm and in their granules. The most commonly used stains/enzymes detected are acid phosphatase, alpha-naphthyl acetate esterase, chloro-acetate esterase, myeloperoxidase, periodic acid Schiff and Sudan black B. Most of these are found only on cells of the myeloid lineage.

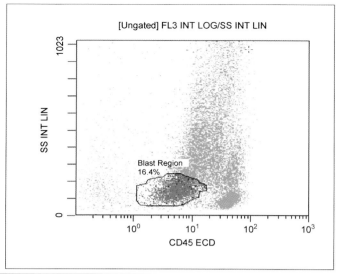

Figure 6.4 Use of flow cytometry to quantify the proportion of CD34-bearing blast cells (16.4%; reference range <0.5%) in acute myeloid leukaemia. (Image courtesy of M. Hill, Heart of England NHS Foundation Trust).

- *DNA analysis and cytogenetics* – which can be looking for abnormalities and gene mutation in the whole cell (perhaps by the process of FISH), or by exacting and analysing DNA from the cell.

Bone marrow

In leukaemia and myeloma, this organ is where the disease starts. There are two commonly types of bone marrow analysis used: an aspirate and a trephine. The former involves driving a heavy gauge needle in the bone (generally the sternum or iliac crests) and sucking out the marrow, which is then treated as if a blood sample (Figure 3.2). The problem is that, by pure coincidence, the sampling site may be relatively free of malignant changes, or it may be an active focus of cancerous cells. In either case, the result may bring false reassurance or concern. However, this is unavoidable and grudgingly accepted.

A trephine sample attempts to retain the specialized architecture of the bone marrow, with its separate sections of haemopoiesis. It also gives information about how particular cells interact physically whilst in their 'natural' state. The trephine sample includes bone, and so is generally treated as if it was a histological section, and so may require decalcification and other processing.

Both these analyses tell us of the different proportions of precursor cells such as myelocytes and erythroblasts, the developments of which are described in Figure 5.2 and Figure 3.3 respectively. It can also tell us of other cells such as fibroblasts and adipocytes, and any cells that should not be in the bone marrow, perhaps those that have metastasized from distant tumours such as the breast or prostate. We can also assess the impact of the disease on red blood cell and platelet production (erythropoiesis and thrombopoiesis respectively), such as the ratio between erythrocyte precursors and myeloid precursors.

In many cases, bone marrow analysis is necessary to confirm an otherwise unclear diagnosis, and to quantify the extent of the malignant growth. Table 6.1 illustrates differences in the constituent haemopoietic cells in the bone marrow in health, in an acute leukaemia and in a myeloma. Flow cytometry can also be used to search for malignant cells in a bone marrow aspirate, but this

Table 6.1 Bone marrow cells in health and malignancy.

Cell type	Normal bone marrow (range)	Bone marrow in an acute leukaemia	Bone marrow in a myeloma
Myeloblasts (%)	0–3	31.9	4.2
Promyelocytes (%)	3.2–12.4	18.1	8.9
Metamyelocytes (%)	2.3–5.9	9.5	4.7
Neutrophils (%)	23.4–45	17.5	15.2
Eosinophils (%)	0.3–4.2	1.0	0.7
Basophils (%)	0–0.4	1.5	0.7
Monocytes (%)	0–2.6	2.0	1.5
Erythroblasts (%)	13.6–38.2	10.9	12.2
Lymphocytes (%)	6–20	7.6	22.5
Plasma cells (%)	0–1.2	0	29.4
Myeloid:erythroid ratio	1.3–4.6	8.8	7.2

Note the increased proportions of myeloid precursors in the acute leukaemia and increased lymphocytes and plasma cells in the myeloma. These are increased at the expense of the erythrocyte precursors, and is reflected by grossly abnormal myeloid:erythroid ratios.
Source: Table adapted from Moore G, Knight G, Blann A, 'Haematology', Oxford University Press, 2011, Table 10.2.

request is far less frequent than the analysis of peripheral blood.

The lymph node

Chapter 5 has details of the physiology of the lymph node and its role in antibody production. Lymphomas are almost always based in lymph nodes and are very likely to start in one single node, inevitably spreading to others. These small pea- or bean-sized bodies are found extensively all over the body, such as in the axilla (armpit), inguina (groin), lungs and in the walls of the intestines (where they may be described as mucosa-associated lymphoid tissue (MALT)), and so too can a lymphoma. Rarely, a malignant focus of extra-nodal lymphoma-like tissue may develop in the liver or spleen.

The major clinical feature of lymph node activity is lymphadenopathy (swelling of the lymph node), although there are many cases of this in a healthy immune response (such as tonsillitis and infectious mononucleosis – the latter is described in Chapter 5). However, both leukaemia and myeloma may spread from the bone marrow (i.e. metastasize) to lymph nodes, and this clearly indicates a worsening prognosis.

The key lymph node investigations are biopsy and aspirate, the latter being essentially the same as for bone marrow. A needle is driven in the target node, and some of the contents simply sucked out. Like aspirating the bone marrow, this method has a degree of 'hit and miss',

as it may be that by coincidence the needle aspirates some normal tissues, so giving false reassurance.

A biopsy, like the marrow trephine, retains the architecture of the lymph node, and must also be processed by histological methods, often with standard haematoxylin and eosin staining. Very rarely will an entire lymph node be surgically removed, as it is still needed to allow the lymphatic fluids to circulate.

Having introduced general aspects of major white cell malignancies, we are now able to look at each of them in more detail.

6.3 Leukaemia

This is the most common leukocyte malignancy. It was first described in the first half of the 19th century, with Virchow's work being the most frequently referred to (he coined 'leukaemia', meaning white blood), whilst Erlich's development of stains allowed the abnormal white cells to be characterized. Subsequent work in the 19th and 20th centuries demonstrated that there are several different types of leukaemia, depending on which cell lineage (lymphoid, granulocyte, monocyte, eosinophil, basophil) is the object of the malignant transformation. A second aspect of leukaemia is the rate at which it develops, which is partly determined by the rate at which the patient presents with different sets of symptoms. If the patient reports a rapid deterioration in health, then we may define the presentation of the leukaemia as acute.

Conversely, if the rate of appearance of symptoms is slow, it may be a chronic leukaemia.

However, symptoms are often misleading, and because of this the laboratory plays an important part in defining the stage of the disease. Indeed, some formal classification systems have dispensed with the clinical aspect of the disease as reported by the patient and define a leukaemia as acute if greater than 30% of cells in the bone marrow are blasts.

Classification

It is convenient to classify leukaemia by the lineage of the particular cell, and by the (presumed) rate of progression of the disease. This gives us:

- chronic myeloid leukaemia (CML);
- acute myeloid leukaemia (AML);
- chronic lymphocytic leukaemia (CLL);
- acute lymphoblastic leukaemia (ALL).

Chronic granulocytic leukaemia (CGL), once considered a separate disease and present as such in many old textbooks, has now been subsumed into CML. Similarly, note that ALL is lymphoblastic, not lymphocytic leukaemia, reflecting the classification system that relies primarily on the number of blasts, not on clinical aspects. The leukaemias described above are the most common; there are also several other types less often encountered. We will address all in order.

Chronic myeloid leukaemia

The myeloid lineage includes neutrophils, eosinophils and basophils, but the neutrophils dominate. However, we do not really talk about chronic neutrophil leukaemia, as the neutrophil is the mature end-stage cell of the lineage, and leukaemia is all about problems with precursor cells. CML constitutes perhaps 15–20% of all leukaemias and occurs with a frequency of around 10–15 per million people per year, with a median age at diagnosis of 67 years.

The key point about CML is that it is perhaps one of the best (if not the best) examples of how research into a particular disease has provided an explanation not only of the pathophysiology but has also directed the search for a treatment. This is based on the 'Philadelphia chromosome'.

The Philadelphia chromosome This is actually a two-chromosome disease, because it is a reciprocal translocation, defined as t(9:22)(q34:q11). The mutation is of the transfer of a portion of chromosome 9 to chromosome 22, and vice versa. This brings together two sections of DNA that then create a new gene. A section called 'breakpoint cluster region' (*BCR*) on chromosome 22 is moved next to a gene on chromosome 9 that has strong homology with DNA from a virus (the Abelson virus) known to cause a type of leukaemia in mice (hence *ABL1*). This gene codes for an enzyme, tyrosine kinase (of molecular weight 210 kDa, hence p210), that is involved in cell signalling and proliferation, and is regulated in part by a certain sequence called SH3.

The movement of *ABL1* next to *BCR* creates a new fused gene: *BCR–ABL*. The crucial point is that the new gene codes for a new type of tyrosine kinase which lacks the SH3 section, which means that it is unregulated and (it seems) constantly active. The consequence of this active tyrosine kinase is the inappropriate activation of several other genes involved in cell proliferation, and this is the basis of the increased number of immature cells in the circulation. This is illustrated in Figure 6.5. The reverse translocation, *ABL–BCR*, seems to have no major consequences.

The importance of the measurement of *BCR–ABL1* is demonstrated by the publication of a specific guideline based purely on methodology.

Cellular basis of chronic myeloid leukaemia The initial diagnosis of CML is made in the laboratory by a full blood count and an examination of the blood film. However, the blood itself has only been obtained from a patient who has presented to his or her practitioner with a series of symptoms that have prompted the venepuncture. These symptoms vary from case to case, and the cases themselves may present when symptoms are mild and early or more marked and established.

A typical blood film consists of a varied collection of myeloid precursors, possibly all of them. Although Table 6.1 refers to an acute leukaemia, these precursors are present in the bone marrow of a patient with CML, but in different proportions, and are those which 'spill over' into the blood. It is not possible to accurately define very early precursors, such as colony-forming units, but a differential in a case of CML is likely to quantify myeloblasts, promyelocytes, myelocytes and metamyelocytes. Figure 6.6 shows a typical CML blood film.

Treatment of chronic myeloid leukaemia Historically, CML was treated with a variety of sophisticated poisons whose purpose was to destroy the tumour –

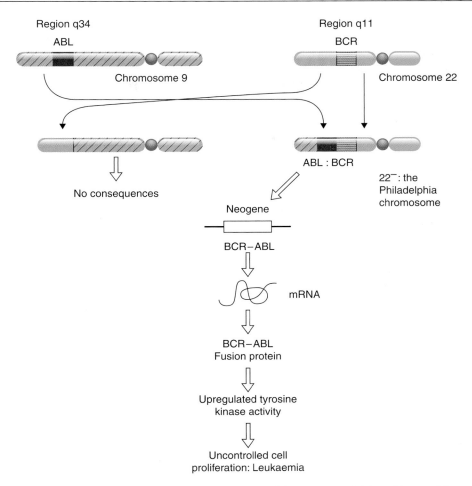

Figure 6.5 Formation of *BCR–ABL*. This fused gene is generated by part of chromosome 22 joining with part of chromosome 9. The fused product is a variant of a tyrosine kinase that is effectively continuously active. The consequences of this are the activation of several genes involved in cell proliferation and signalling, so that the cell continues to generate abnormal 'daughter' progeny – that is, leukaemic cells.

cytotoxic chemotherapy. However, knowledge of the genetic lesion in CML is not merely academic: it not only can help with diagnosis, but it also directly drives treatment. The abnormal fusion product of the *BCR–ABL* oncogene – a form of tyrosine kinase – has been targeted by a group of drugs – the tyrosine kinase inhibitors (TKIs). These drugs are very successful as they target only those cells which carry the particular defective species of tyrosine kinase, and they have revolutionized the management of this disease.

The objective of this treatment is to reduce the tumour burden, and ultimately eradicate the tumour altogether. This can be monitored by the disappearance of myeloid

precursors from the peripheral blood, and then from the bone marrow. However, these methods, searching for any so-called 'minimally residual disease' based on microscopy, are relatively insensitive. Better methods for checking the success of treatment and the defeat of the tumour include those of molecular genetics, such as searching for evidence of *BCR–ABL*.

Unfortunately, cancers (and many other diseases) have the irritating habit of biting back, and it is now clear that, in some cases, the leukaemia becomes resistant to TKIs, so that other strategies are required. Despite the emphasis on *BCR–ABL*, there is a variant of CML that does not carry this mutation, and is described as atypical CML.

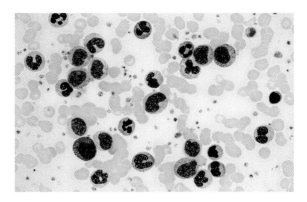

Figure 6.6 Chronic myeloid leukaemia. This film shows granulocytes at several stages of differentiation. The larger cells are promyelocytes and myelocytes; the smaller cells are metamyelocytes. Note also the degree of granulation in the cytoplasm of the different cells. On the far right is a lymphocyte, with a darker nucleus that occupies almost all of the cell. (From Hoffbrand & Moss, *Essential Haematology*, Sixth Edition, 2011, Fig. 14.3, p. 195. Reproduced with permission of John Wiley & Sons, Ltd.).

Consequently, TKIs are ineffective, and conventional chemotherapy is required.

Chronic neutrophil leukaemia This is a very rare condition, and is unusual amongst the leukaemias as >90% of the cells are mature neutrophils, with only a small number of myeloblasts, myelocytes and metamyelocytes. This picture would have been typical of an infection, but in chronic neutrophil leukaemia there is no such infection, inflammation or other obvious cause of a neutrophilia or CML (such as *BCR–ABL*). However, as with other leukaemias, the later stages of the disease see the development of anaemia and thrombocytopenia, neither of which are present in an inflammatory neutrophilia.

Acute myeloid leukaemia

This disease is not simply a more aggressive form of CML, although it can appear as such. Whilst almost all cases of CML are Philadelphia chromosome positive, there are dozens of different genetic lesions linked to AML, so that it is certainly not one disease. In addition, the phenotype of the cells, both in the bone marrow and the peripheral blood, can also vary markedly (Table 6.2).

In almost all cases, AML is a primary disease arising from a genetic mutation in a stem or precursor cell, although it may arise from a background of myelodysplasia, which generally presents as a disease of red blood cells. The variety of cellular abnormalities has led to the development of formal classification systems for AML, such those of a group of French, American and British haematologists (hence FAB), and of the World Health Organization (WHO).

Classification The FAB system, using standard Romanowsky staining and cytochemistry, recognizes several different conditions, starting with the less well

Table 6.2 Partial FAB classification of AML.

Sub-group (frequency)	Name	Romanowsky morphology	Cytochemistry
M0 (5%)	Undifferentiated AML	Unclear: other analyses required. No Auer rods	<3% of cells positive for SBB, MPO or CAE
M1 (15%)	AML without maturation	≥90% blasts, ≤10% granulocytes/monocytes. Occasional Auer rods	<3% of cells positive for SBB, MPO or CAE
M2 (25%)	AML with maturation	30–89% blasts, granulocytes >10%, monocytes <20%. Auer rods may be present	MPO/SSB/CAE positive
M3 (10%)	Promyelocytic	Promyelocytes dominate, <30% myeloblasts. Auer rods frequent.	MPO/SSB/CAE and AP positive
M4 (20%)	Myelomonocytic	>30% blasts, >20% granulocytes and monoblasts. Auer rods may be present	CAE/ANAE
M5a (5%)	Monoblastic	>80% monocytic, of which >80% monoblasts. Auer rods may be present	ANAE positive, MPO and SBB negative
M5b (5%)	Monocytic	>80% monocytic, of which <80% monoblasts. Auer rods may be present	ANAE positive, MPO and SBB negative

SSB: Sudan black B; MPO: myeloperoxidase; CAE: chloroacetate esterase; AP: acid phosphatase; ANAE: alpha naphthyl acetate esterase.

differentiated, and others that are increasingly 'mature' (Table 6.2). A further aspect of the increasing maturity of the cell is the appearance of Auer rods, modified eosinophilic granules containing peroxidase and lysosomal enzymes. These intracellular features are absent from mature granulocytes and monocytes. The FAB and WHO systems also classify other haematology malignancies (of red cells, and of megakaryocytes).

Guidelines recommend immunophenotyping for CD3, CD7, CD13, CD14, CD33, CD34, CD64, CD117 and HLA-DR (Chapter 10). Cytochemical analyses include myeloperoxidase, combined esterase and Sudan black.

Molecular genetics of acute myeloid leukaemia - Different chromosomal abnormalities may also be used in staging. For example, FAB class M3 acute promyelocytic leukaemia by t(15;17)(q22;q21), M4 myelomonocytic leukaemia by inv(16)/t(16;16) and M5 acute monoblastic leukaemia by del(11q) and t(9,11). M2 AML may be characterized by t(8;21)(q22;q22), which brings together *RUNX1* (a transcription factor) and *RUNX1T1* (a co-repressor). The new fused gene generates a product that blocks differentiation, leading directly to the build-up of blasts that characterizes the disease. Many of these are relevant in childhood AML, although this disease also has its own genetic lesions.

Guidelines recommend reverse transcription polymerase chain reaction for genes such as *AML1–ETO*, *CBFB–MYH11* and *PML–RARA*, with FISH in selected cases, although the practitioner will refer to the regional molecular genetics service. As with CML and *BCR–ABL*, the genetic lesion in AML can also guide not only diagnosis but also treatment. Ninety-eight per cent of cases of acute promyelocytic leukaemia (PML) are associated with the translocation t(15;17)(q22;q12). This astonishingly high proportion has led some to suggest that the remaining 2% actually have some other disease distinct from PML. The result of this mutation is the fusion of part of an unknown gene with the gene for the retinoic acid receptor (RAR).

The fused *PML–RARα* is responsible for the maturation arrest in the myelocyte. Unlike conventional cytotoxic chemotherapy, the treatment for this particular leukaemia, *all-trans*-retinoic acid, actually promotes cell maturation, and provides the missing signal allowing the cell to continue its development. Unfortunately, the dose required to achieve this often leads to side effects such as shortness of breath. Certain cytogenetic abnormalities are associated with prognosis (Table 6.3).

Table 6.3 Prognosis and cytogenetic abnormality.

Prognosis	Abnormality
Good	t(8;21)(q22;q22)
	inv(16)(q13;q22)
Intermediate	Trisomy 8, trisomy 21
	Most breakpoint 11q23 lesions
Poor	t(9;22)(q34;q11)
	t(4;11)(q21;q23)

Wilms tumour gene (*WT1*) encodes a transcription factor highly expressed in many haemopoietic tumours, including AML. Expression of *WT1* predicts disease progression in AML patients treated with conventional chemotherapy, and after stem cell transplantation is a strong predictor of relapse, independent of the phase of the disease. It follows that *WT1* expression may be used to guide early interventional therapy.

A wider view of acute myeloid leukaemia This disease is certainly complex, and can involve other cells. One of these is the 'evolution' of the disease into one which also results in a megakaryocyte abnormality. Myeloblasts may leave the bone marrow to 'infect' muscle cells, leading to myeloid sarcoma. The trisomy of Down syndrome brings an increased risk of AML and acute megakaryoblastic leukaemia. Figure 6.7 shows a blood film from a patient with AML.

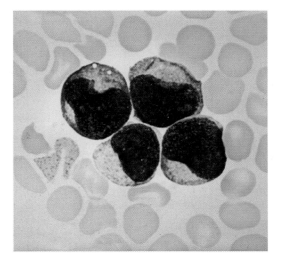

Figure 6.7 A blood film in AML. This film shows four large blast cells. That of the top right has an Auer rod. (From Hoffbrand & Moss, *Essential Haematology*, Sixth Edition, 2011, Fig. 13.4, p. 184. Reproduced with permission of John Wiley & Sons, Ltd.).

Chronic lymphocytic leukaemia

This disease is the most common leukaemia in the Western world, and was probably the first to be described by Victorian pathologists, who noted massive splenomegaly in certain patients. It presents in the UK and USA at a rate of 42 cases per million per year, with a median age of 72 years. Epidemiology indicates a sevenfold increased risk of CLL and 2.5-fold increase in other lymphoid malignancies in the relatives of those with CLL, clearly pointing to a susceptibility gene(s).

The FAB and WHO classification systems recognize numerous lymphoid neoplasias. The former has been updated into the Revised European American Lymphoma classification system, focusing on B-cell neoplasia and T-cell/natural killer (NK)-cell neoplasia. Because of the close association between leukaemia and lymphoma, these diseases are often classified together because, for example, although a particular CLL may start in the bone marrow, it may well 'metastasize' to a lymph node, and so resemble a lymphoma. A recent guideline is available from the British Committee for Standards in Haematology.

Conversely, in the early stage of many lymphomas the disease remains with the lymph node, but in later stages the lymphoma cells may move to the bone marrow and appear in the peripheral blood. Indeed, one system focuses on the maturity of the malignant cell, regardless of its anatomical origin, suggesting that CLL and small cell lymphocytic lymphoma are essentially the same disease. Monoclonal B-cell lymphocytosis, present in 3% of the general population aged over 50 years, may be a precursor of CLL.

The blood film   The classical laboratory feature of CLL is an otherwise unexplained persistent (>3 months) lymphocytosis. The blood film shows an increased number of lymphocytes, as the morphology of the mature CLL cell is indistinguishable from a normal lymphocyte. However, what differentiates CLL from a reactive lymphocytosis (perhaps in response to a virus as in infectious mononucleosis) is the presence of smear cells in the former.

A feature of all leukaemias is that the leukaemic cell itself is not simply immature but is also often physically fragile. This means that it is less able to resist the damaging effects of being spread out on a glass slide and stained than are normal cells. Hence, smear cells are likely to be disrupted leukaemic cells.

Although the typical CLL cell closely resembles a mature lymphocyte, it may actually be a prolymphocyte,

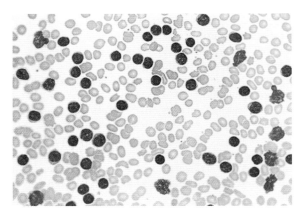

Figure 6.8 A blood film in CLL. Note that the leukaemic cells in this film are markedly smaller than those of the AML or ALL blasts in Figure 6.7 and Figure 6.9, being only a little larger than the red cells. (From Hoffbrand & Moss, *Essential Haematology*, Sixth Edition, 2011, Fig. 18.3, p. 236. Reproduced with permission of John Wiley & Sons, Ltd.).

the cell that lies between a lymphoblast and mature cells. However, in some CLLs the malignant cell may be identifiable as being slightly larger and perhaps with more cytoplasm than the healthy cell. Figure 6.8 shows a blood film in CLL.

Cytogenetics, prognosis and treatment   A number of mutations are pertinent.

- Del13q14.3 is present in over a third of CLLs, and is associated with an improved prognosis.
- Del11q23 is present in 20%, and carries a poor prognosis.
- Perhaps 7% of CLLs carry Del17p13, a mutation associated with a particularly poor prognosis. The exact problem may be complete loss, or loss of function, of *TP53*.

The expression of high levels of CD38 at the cell membrane also indicates a poor outcome, although its genetic basis is unclear.

In as much as molecular genetics has shown us the link between *BCR–ABL* and malignant transformation in CML, and mutations in *RUNX1* and *WT1* in AML, in CLL the *NOTCH* system may be important. This gene codes for a receptor that contributes to lymphocyte differentiation, and gain of function mutations are common in a variety of cancers, including CLL, where one product, NOTCH2, is linked to the expression of the activation marker CD23. Gliotoxin is a fungal metabolite that targets NOTCH2, and efficiently induces apoptosis, and so may be a new potential tool for therapy of CLL.

Other laboratory indices Beta-2-microglobulin is a surface membrane component associated with certain human leukocyte antigen (HLA, explained in Chapter 10) molecules, measurable by immunoassay, often in the immunology laboratory. Lactate dehydrogenase (LDH) is a general metabolic enzyme, released from many cells and easily measured in the biochemistry laboratory. Increased levels of both of these molecules in the serum carry a poor prognosis.

B-cell chronic lymphocytic leukaemia Immunophenotyping by flow cytometry is essential in identifying the true basis of the disease. A B lymphocyte CLL is characterized by the presence of CD19, CD20, CD23, CD79a and CD79b, and the absence (or very weak expression) of CD22. However, CD20 may also be present on NK cells, and CD23 is also found on monocytes, macrophages and eosinophils, although these are, of course, easily distinguished from lymphocytes.

B cells make antibodies. Therefore, a clear sign of a B-CLL is the presence of immunoglobulins within the cytoplasm or at the surface of the cell (hence surface membrane immunoglobulin: SmIg). There may also be changes to the genes coding for the variable regions of the heavy and light chains of the antibody molecule. If these genes are in 'germ line' configuration, they are unmutated and therefore the cell is immature. However, if the genes have undergone somatic (hyper)mutation, the cell is clearly more mature, and this carries a better prognosis than the unmutated genotype. This is explained in more detail in Chapter 10.

The antibody molecule is constructed from heavy and light chains (see Figure 9.1), the latter having two variants, kappa and lambda, generally present in equivalence. Gross deviations in the kappa/lambda ratio are associated with a poor prognosis.

B-cell prolymphocytic leukaemia This is a rare (1–2% of CLLs) and aggressive disease. Commonly presenting with a very high white cell count ($>100 \times 10^9$/L) and splenomegaly, the prolymphocyte is larger and with more cytoplasm than the mature lymphocyte. The immunophenotype resembles mature B-CLL, but there may also be CD5.

As in mature CLL, genetic lesions del17p13, del13q14.3 and del11q23 are noted, but the translocation t(11;14) (q13;q32) may also be present. Regrettably, none of this

information is as yet used in treatment or predicting outcome.

Hairy cell leukaemia Hairy cell leukaemia (HCL) is so called because of the irregular pattern of the cytoplasm, this disease (accounting for perhaps 2% of all leukaemias and 8% lymphoid leukaemias), has a median age at diagnosis of 50 years and a male/female excess of four- to five-fold. Splenomegaly is present in 65% of cases, hepatomegaly in 45% and abdominal lymphadenopathy in 10%. At the cell level it is characterized by CD11c, CD25, CD103 and CD123, all of which may be found on cells of varying lineages. However, the presence of CD20 and SmIg firmly defines this disease as a B lymphocyte malignancy. Furthermore, in over three-quarters of cases there is rearrangement of genes for variable regions of antibody heavy and light chains, and in some 40% there is the co-expression of immunoglobulin (Ig)G, IgM or IgA.

Unlike the other leukaemias and their leuocytoses, in HCL there is often pancytopenia (in 65% of cases), with thrombocytopenia in 80% and, in particular, low monocytes (present in 90% of cases). However, the white cell count may be raised because of the high number of abnormal lymphocytes and infections. Fortunately, the disease is treatable with a cocktail of interferon, purine analogues and a monoclonal antibody to CD20, which can produce a remission in 90% of cases.

Diagnosis follows flow cytometry with a B cell panel of CD19, CD20, CD22 and SmIg, and with a HCL panel of CD11c, CD25, CD103 and CD123. However, in one study, 100% of HCL patients and 0% of non-HLC cases were positive for a mutation in *BRAFV600E*. This is likely to revolutionize the diagnosis of this disease.

Abnormal antibodies An established characteristic of B-cell leukaemias (and often of lymphomas) is the aberrant production of antibodies. Indeed, in Section 6.5 we see how this is taken to its extreme by myeloma and its related diseases. A feature of this generalized B-cell perturbation is the presence of autoantibodies, most often directed towards red cells, although antibodies to platelets and neutrophils have been described. These may lead, respectively, to an autoimmune haemolytic anaemia, immune thrombocytopenia and neutropenia. The key investigation is the direct antiglobulin test, as is commonly performed in the blood transfusion laboratory and explained in Chapter 11.

T-cell and NK-cell leukaemias Like its B-cell equivalent, T-cell prolymphocytic leukaemia is also characterized by a very high white cell count and spleno-megaly, but (unlike the B-cell form) many patients also have hepatomegaly or lymphadenopathy. The key immu-nophenotype finds increased expression of CD2, CD3 and CD7. The most common cytogenetic abnormalities are inversion or translocations in chromosome 14, but this is of little help as the prognosis is poor with a median survival of less than a year.

The WHO recognizes three disorders of large granular lymphocytes: T-cell large granular lymphocytic leukae-mia, chronic lymphoproliferative disorders of NK cells, and aggressive NK-cell leukaemia. The former is rare (<3% of all small lymphocytic leukaemias) and, curi-ously, is associated with autoimmune disease. Leukaemic T cells often express CD3, CD8, CD16 and CD57, the disease being relatively benign, often being found during an unrelated investigation, and many patients survive for 10 years after diagnosis. In contrast, NK-cell leukaemias (some linked to Epstein–Barr virus infections), express-ing CD16 and CD56 but not CD3, are generally more severe with a worse prognosis.

Acute lymphoblastic leukaemia

This disease is characterized by the increased number of lymphocyte-like blasts, hence lymphoblastic, not lym-phocytic, leukaemia. Once more, it is worthwhile to emphasize, as with CLL, the close relationship between ALL and lymphoma, and that the same phenotypes can be found in both peripheral blood and in lymph nodes.

Most malignancies are of B cell, with 15% of paediatric cases and 25% of adult cases being of T cell lineage. As with myeloid leukaemias, there is an FAB classification system that relies primarily on morphology, and recog-nizes three types:

• Type L1 is characterized by a small homogeneous lymphoblast, and accounts for 80% of ALLs.
• Type L2 lymphoblasts are larger and have a greater proportion of cytoplasm; 18% of ALLs are of this type.
• Type L3 disease is the remaining 2% of ALLs and is the only type based on B cells; L1 and L2 can be B or T malignancies. The blasts are large with much basophilic cytoplasm. The correct identification of L3 is important as there is a specific treatment.

These conditions may also be characterized immuno-logically. B-cell ALLs are generally positive for the enzyme terminal deoxynucleotidyl transferase (TdT), HLA-DR and CD10, although some may also express CD79a. T-cell ALLs also express TdT, but are defined by the T cell marker CD3. They may also variably express CD2, CD4, CD5, CD7, CD8 and CD10, and there are likely to be rearrangements in the genes for the T cell receptor (Chapters 9 and 10). However, perhaps more pertinent are the particular genetic lesions that cause the particular sub-disease, not merely because they provide insights into molecular genetics. These include:

• t(9;22)(q34:q11.2). This is the same lesion as causes Philadelphia positive CML, and is characterized by *BCR–ABL1*. It is found in 30% of adult and 3% of childhood ALLs, and many express CD25. As with the CML, TKIs are effective.
• There are a number of ALLs associated with the variable translocation of another section of a chromo-some on to 11q23, hence denoted t(v;11q23). The com-mon pattern is of a high white cell count, central nervous system involvement, and a poor prognosis.
• t(12;21)(p13;q22). The most common mutation, accounting for 25% of childhood ALLs, this disease has a relatively good prognosis, with a cure rate in children of 90%. At the cell level, the disease is mostly likely caused by the fusion of the *RUNX1* gene with a second gene (such as *ETV6*) in a similar manner to that of FAB M2 class AML (an 8:21 translocation), and of course *BCR–ABL* in CML, which we have already discussed.
• t(5;14)(q31:q32). Although very rare (<1% of all ALLs), it provides an interesting model, as the induced inflammation (excessive release of interleukin (IL)-3) precipitates an eosinophilia.

There are, of course, many others. There are also increased and reduced numbers of entire chromosomes (hyperploidy – particularly of 51–65 chromosomes – and hypoploidy respectively), as is present in AML and Down syndrome. Figure 6.9 shows a representative blood film in ALL.

Other types of leukaemia

The four diseases we have looked at so far (AML, ALL, CML, CLL) comprise 95% of all leukaemias. The remain-ing 5% include monocytic (Figure 6.10), eosinophilic and basophilic leukaemia, which present with decreasing frequency, reflecting their proportions in health. How-ever, there are instances where the phenotype of the disease is not clear cut, and can involve what appears to be two different lineages, in contravention of the clonal nature of the disease. Other diseases include acute

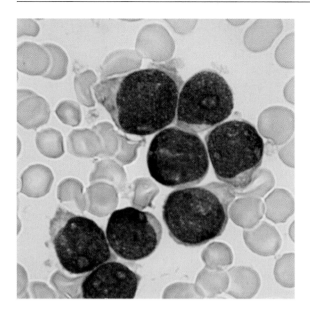

Figure 6.9 A blood film in ALL. These blasts are clearly much larger than nearby red cells, and are also agranular, unlike the AML blasts in Figure 6.7. The presence of CD10 on these cells (as defined by flow cytometry) marks them as malignant B cells. (From Hoffbrand & Moss, *Essential Haematology*, Sixth Edition, 2011, Fig. 17.3 (a), p. 227. Reproduced with permission of John Wiley & Sons, Ltd.).

promyelocytic leukaemia and both acute and chronic myelomonocytic leukaemia (CMML) (Table 6.2).

CMML, with a median survival of only 12–18 months, is often associated with abnormalities in *JAK2*, *TET2*,

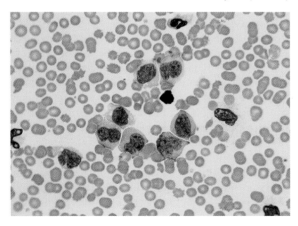

Figure 6.10 A blood film in chronic monocytic leukaemia. These blasts are relatively mature, with a low cytoplasm-to-nucleus ratio. (Image from Blann AD, *et al*. (2009) Haematology Morphology. The IBMS Training CD-ROM. © IBMS).

CBL, *IDH* or *RUNX1* and *RAS* genes. Flow cytometry is used to probe for myelomonocytic antigen CD33, but there is often a depressed expression of 'classical' monocyte marker CD14 with overexpression of CD56. An increase in CD34 may be associated with an increased risk of transformation to AML.

NK-cell malignancies are heterogeneous: they may present as an aggressive leukaemia but also as a type of lymphoma. Although the presence of CD56 and absence of CD3 are common features, some are associated with Epstein–Barr virus, and loss of chromosomes 6q, 11q, 13q and 17p are recurrent findings. Further confusion is provided by the description of some as a large granular lymphocyte leukaemia, which is often a descriptor given to T-cell (CD3-bearing) leukaemias.

6.4 Lymphoma

The causes of lymphadenopathy include a normal response to an infectious agent (such as in tonsillitis and infectious mononucleosis) and should resolve when the infection has passed. However, a more sinister cause of persistent lymphadenopathy is tuberculosis, another is human immunodeficiency virus infection, and more adverse is lymphoma. The scientific basis of this disease pre-dates that of leukaemia, and it is possible that some early descriptions of leukaemia were actually lymphoma.

The natural history of lymphoma

In the absence of a clear and persistent lymphadenopathy, the practitioner will mostly likely be faced with a patient complaining of perhaps several nonspecific symptoms, such as fever, night sweats, weight loss and itching. At this stage a full blood count is unlikely to be helpful, as at the early stages the disease is only active in lymph nodes. However, an abnormal ESR (if present) will clearly indicate some kind of abnormality, and a raised c-reactive peptide is also evidence of a nonspecific pathology.

Symptoms will increase in number and severity, perhaps with infections, cough, itching, tiredness and lethargy, until eventually the search for the lymphoma begins, most likely with computerized imaging (MRI, CAT, etc.), possibly supported by tissue- or organ-specific signs and symptoms. As the disease progresses, more and more lymph nodes become involved, and there will be extra-nodal disease with bone marrow infiltration (with consequent variable anaemia, thrombocytopenia and leukopenia), hepatomegaly and splenomegaly. Some rare lymphomas have their basis within the central nervous system.

Aetiology and classification

As with all haemato-oncology, radiation undoubtedly causes lymphoma, and there are also established chemical carcinogens. Viruses may be important in this particular disease as many patients have high titre of antibodies to these pathogens. However, this does not imply causation, and the antibodies may be the consequence of disease. Nevertheless, the disease may be the result of a response to several viruses at the same time, as there are numerous examples of oncogenic viruses. Indeed, the Epstein–Barr virus is a cause of Burkitt's lymphoma, although it also causes infectious mononucleosis. Any of these processes, perhaps in sequence, may cause the genetic abnormality that often defines a particular variant of the disease.

These diseases can be classified in a number of ways, such as low grade and high grade, which is based on the proliferation rate, and in this respect cells may be described as centrocytes or centroblasts. A second system considers the extent to which the disease is focused on well-defined regions within the lymph node (nodular), or where this is widespread (diffuse). Cytochemistry is rarely called upon to characterize the disease, but the use of CD analysis is common, and this helps define the lymphoma as of T cell or B cells. However, perhaps the simplest classification is historical, named after the undoubted 'father' of lymphoma, Thomas Hodgkin, who first described the disease in detail in 1832. However, it is likely that only three of the seven patients he described actually had this disease that bears his name. Present data classification relies on the identification of the 'Reed–Sternberg' (RS) cell, a large cell characterized by a multi- or bilobed nucleus, and if found it defines the disease as a Hodgkin lymphoma (HL).

Hodgkin lymphoma

The RS cell is the key to this disease, and almost all are neoplastic B lymphocytes. Molecular genetics show that the genes coding the variable regions of immunoglobulin have rearranged, indicating a degree of maturity. However, these mutated genes fail to generate functioning antibodies, a consequence of which is that these cells fail to self-destruct (the process of apoptosis), and so gain a degree of immortality.

Other abnormalities of the RS cell include lack of B cell markers CD19 and CD20, as well as leukocyte marker CD45, although HLA class II molecules, CD15, CD30 and CD40 are often present. The Epstein–Barr virus can be found in 45–50% of HLs, but this does not imply causation. The key genetic lesion is constitutive activation of NF-kappaB, possibly linked to other genes such as *TRAF3* and *MAP3K14*.

Further classification   Using the internal morphology of the lymph node, the WHO classification system recognizes two major groups. The lesser (5% of lymphomas) are described as nodular and, confusingly, are characterized by the lack of RS cells but the predominance of lymphocytes. The second (95%) group are classical HLs with RS cells, and can be subdivided according to morphology:

- In nodular sclerosis HL (the most common form of HL), nodules are often surrounded by fibrotic material such as collagen. Eosinophils may be present.
- In mixed cellularity HL (the second most common form), there are many RS cells but also lymphocytes, histiocytes, neutrophils and plasma cells.
- Lymphocyte-depleted HL is characterized by many RS cells and few lymphocytes, but there is often diffuse connective tissue involvement and an irregular architecture.
- In lymphocyte-rich HL there are few RS cells, eosinophils and plasma cells, but there are many small lymphocytes

It is clear that a major weakness is the subjective nature of defining the relative proportions of constituent cells.

Advanced disease   Once the diagnosis has been secured, treatment with radiotherapy and/or cytotoxic chemotherapy should begin promptly, depending on the stage of the disease (generally, the number and anatomical sites of nodes involved). Although the radiotherapy may be targeted with some accuracy to the particular malignant node(s), chemotherapy will be systemic and (as with all such agents) brings the standard side effects of nausea, vomiting and diarrhoea, jaundice, alopecia and rash. Suppression of the bone marrow will result in pancytopenia, and thus infections. Bone marrow transplantation is a late option in those resistant to standard chemotherapy.

Even without the effect of treatment, the developing disease brings factors such as a normocytic anaemia, leukopenia and thrombocytopenia (and thus infections, bruising and bleeding). A lingering problem with HL is the increased risk of a second malignancy, potentially a consequence of the radiotherapy and chemotherapy.

Non-Hodgkin lymphoma

As the name implies, non-Hodgkin lymphoma (NHL) is effectively a diagnosis of exclusion, so that NHL is any lymphoma that is not an HL, and so is a large and diverse group of conditions, often with few features in common. However, one feature commonly found is the upregulation of the *BCL-2* gene, which codes an apoptosis regulator protein. Although first described in B-cell lymphomas (hence BCL) it has been found in many diverse malignancies and decreases the likelihood of the cell going into apoptosis, so it remains (inappropriately) alive. The presence of high levels of CD20 on the malignant cells provides the opportunity for therapy with a monoclonal antibody, and is often very successful.

Classification   NHL is considerably more complex than HL. It may also be classified as being of B cell (85%), or of T or NK cell origin (15%), by the degree of maturity of malignant cells (as pre-B bone-marrow-like cells, or cells of the germinal centre; as thrombocytes or mature T cells), or by the grade of the disease (highly proliferative, and therefore immediately dangerous, or slowly smouldering but with the capacity to become malignant). The most common forms of NHL are as follows:

- Lymphocytic lymphoma. The invasive cell in this low-grade variant closely resembles that in CLL. There is rarely extra-nodal disease (e.g. metastases to bone marrow), and often little or no treatment is required. SmIg is rarely present, but cells often bear CD5, CD20 and CD23, and typical cytogenetic abnormalities include trisomy of chromosome 12, and deletions of 13q14, 17p and 11q, with the expression of *BCL-2*.
- Follicular lymphoma (25–30% of NHLs). This is often associated with t(14:18)(q32;q21), and the expression of SmIg, CD10, CD19, CD20 and sometimes CD23 (but not CD5). This low-grade and common disease is the consequence of the fusion of *BCL-2* with an immunoglobulin heavy chain gene (*IgH*), but there may also be abnormal activation of *BCL-6*, whose product is a transcription factor. This results in reduced apoptosis and so increased survival. The disease is called follicular because the malignant cells are localized within well-demarked follicles, resembling germinal centres, and immunohistology is useful in confirming the diagnosis.

- Mantle cell lymphoma (5% of NHLs). This is characterized by the expression of SmIg (often IgM and IgD), CD5, CD19, CD20 (thus being similar to CLL), and t(11;14)(q13;q32) with *BCL-2* expression, the lymphoma cells being found in a ring (the mantle) around the outside of follicles. The cytogenetic abnormality brings together *cyclin D1* with *IgH*. Lymphoma cells are often found in the blood, and bone marrow, liver and spleen involvement is common.
- Diffuse large B-cell lymphoma (30–40% of NHLs) is a mixed group of high-grade diseases, and there is variable expression of SmIg and CD10, although CD20 expression is common. Translocations t(3;14) and t(14;18) are often found – these are responsible for expression of *BCL-2*, as is also found in follicular lymphoma, but also *BCL-6*. Many of these lymphomas exhibit rearrangement of their immunoglobulin variable genes, implying exposure to antigens, with the malignant cell often being twice as big as nearly normal lymphocytes.
- Splenic marginal zone lymphoma (~2% of NHLs) focuses on the spleen (hence a splenomegaly), not the lymph node, so there may not actually be a lymphadenopathy. Bone marrow infiltration is common, so there is often anaemia and thrombocytopenia but also neutropenia. The dominant malignant cell is CD20 positive, but there is also likely to be increased expression of CD19, CD22 and CD79a. These cells may also appear in the blood, giving the appearance of a lymphocytosis, and increased gammaglobulins are often present, reflecting the aberrant B-cell nature of the lymphomas.
- Marginal zone lymphomas (MZLs) develop from mucosa-associated lymphoid tissues and are a heterologous group of diseases accounting for 5–8% of NHLs. A typical site for this is the Payer's patches of the intestines, and the dominant aetiology seems to be chronic antigenic stimulation by defined infections or inflammation, perhaps due to, for example, *Helicobacter pylori*. Most (60%) of MZLs are associated with three translocations:
  - t(11;18)(q21;q21) results in an aberrant apoptosis gene (*BCL-10/MALT1*) that leads to inhibition of this process and so increased cell survival;
  - t(1;14)(p22;q32) places *BCL-10* alongside an *IgH* gene, which result in increased cell survival via the B-cell receptor;
  - t(14;18)(q32;q31) can be seen as a hybrid of the other translocations as *MALT1* is placed alongside *IgH*.

Burkitt's lymphoma   This NHL deserves special mention, as the dominant aetiology is infection by the

Epstein–Barr virus. The disease was first characterized as a tumour in children in Africa in areas where malaria is endemic, leading to the hypothesis that some degree of malaria-induced immunosuppression allows the virus to escape from the normal viral defence systems. This hypothesis is supported by the increased incidence in cases of human immunodeficiency virus infection.

Malignant cells bear SmIg, CD10, CD19 and CD20, but they do not express *BCL-2*. Although there may be cytogenetic abnormalities of t(2:8) and t(8:22), the most common is t(8;14)(q24;q32), present in 85% of cases, places *c-myc* next to an *IgH* gene. *Myc* is an oncogene, named because of its homology to the myelocytomatosis viral oncogene, and when active generates a molecule that promotes cell transcription. This produces a phenotype with a very high rate of proliferation, but fortunately the prognosis is excellent and the tumours respond to chemotherapy such as methotrexate and cyclophosphamide.

Lymphoplasmacytoid lymphoma Lymphoplasmacytoid lymphoma (LPL) also deserves recognition as it is an interesting and precise entity. Until reclassified by the WHO as an NHL, this disease, formerly known as Waldenstrom's macroglobulinaemia, was part of the myeloma family of conditions. As the name implies, the disease centres on plasma cells, but in LPL an aberrant plasma cell produces abnormal amounts of IgM (hence macro-, as this antibody is large), which makes the blood very viscous. Cytogenetic analysis points to chromosome 6(p21.3), but there may also be deletions and translocation at various places in the genome (such as 6q23 and 13q14) which may be responsible for the overexpression of growth factor receptors.

In the laboratory this disease is characterized by the high levels of IgM, detectable by electrophoresis (fully explained in a section that follows) and a raised ESR (a consequence of hyperviscosity). In the bone marrow, the malignant cells express B cell markers CD19, CD20 and CD22. CD40 is often present, but CD10, CD23, CD103 and CD138 are often negative. There is surface and intracellular IgM. As with other NHLs, the patient is likely to complain of a range of nonspecific symptoms such as fatigue (a likely consequence of anaemia and so possible sign of bone marrow involvement), weight loss and visual disturbances. The upregulation of IL-6 may contribute to the anaemia, which is similar to anaemia of chronic disease. Bleeding may be linked to interference of the abnormal protein with the coagulation system. The biochemistry laboratory may find

raised calcium (reflecting bone invasion), LDH and uric acid.

T-cell lymphomas These are comparatively rare, and a small number of variants are recognized. The most common is mycosis fungoides, and predominantly develops in the skin. A closely related but much less common and more malignant variant is Sezary syndrome. Malignant cells of both species express classic T cell markers of CD2, CD3 and CD5, but CD8 is usually not present, although Sezary cells may be CD4 positive and CD7 negative.

Other rare T-cell lymphomas include angioimmunoblastic T-cell lymphoma, anaplastic large cell lymphoma (that may bear CD30), and a CD8-expressing variant. An adult T-cell leukaemia/lymphoma is linked with a specific virus (human T-cell lymphotropic virus type 1). Cytogenetic analyses can be helpful (such as certain translocations), and rearrangement of T cell receptor genes (Chapter 10) gives a clue as to staging.

6.5 Myeloma and related conditions

These are diseases of malignant B lymphocytes, whose normal function is to produce antibodies. A normal B cell that produces antibodies is a plasma cell, generally found in the lymph nodes, although they may also be present in the liver, spleen and bone marrow. However, this family of conditions is characterized by a failure of the normal antibody-generating process. In a minor proportion of cases, antibody production simply stops, but in the majority of cases abnormal antibody molecules are produced, often in copious amounts, which are called paraproteins.

These abnormal proteins pass into the plasma, in which case they form a paraproteinaemia. Since these are gamma globulins, and are inevitably associated with a disease, they are also described as monoclonal gammopathies, to distinguish them from polyclonal gammopathies, which is a generalized increase in gamma globulins, as may result from an infection. Paraproteinaemia, which is generally of large molecules, is easily detected by protein electrophoresis. However, if the abnormal protein is small (less than 67 kDa) it will pass through the glomerulus, and can be detected in urine as Bence–Jones protein.

Myeloma (or perhaps malignant myeloma, or myelomatosis) is certainly the most well-known disease of this type. In 8% of cases, the disease is relatively benign, and is referred to as 'smouldering', although it may transform to a more aggressive variant. Allied diseases include Waldenstrom's macroglobulinaemia, LPL and monoclonal gammopathy of undetermined significance (MGUS). In many of these cases there is also a paraproteinaemia. Rarely, a malignancy may focus outside the bone marrow, such as in bone or skin; these are called solitary plasmacytomas.

Myeloma

This disease causes about 2% of all cancer deaths, 10% of haematology malignancies, and presents at a rate (in Caucasian populations) of some 35 persons per million per year, with a higher rate in black populations. This means there are 3000–3500 new cases in the UK per year. The prognosis is relatively poor, with a 5 year survival of about 50%.

The laboratory   Key laboratory characteristics include the accumulation of abnormal antibody-producing B cells (plasma cells) within the bone marrow, only demonstrable with a bone marrow aspirate or trephine (Table 6.1). A normal aspirate should contain only about 1% of plasma cells, but in myeloma the content typically exceeds 30%.

Plasma cells are characterized by heavy SmIg staining and the standard B cell marker, CD19, but also by the presence of CD38 (an ectoenzyme), CD126 (the IL-6 receptor), CD56 and CD138. In the early stages of the disease the profile of the abnormal protein often cannot be distinguished from that in MGUS, leading to the hypothesis that MGUS is pre-myeloma.

If present, the paraproteinaemia is responsible for a grossly elevated ESR. However, this may also be present in inflammatory disease such as rheumatoid arthritis, so additional tests and investigations are required to exclude the latter.

Genetics and pathophysiology   The aetiology is unknown, but genetic analysis shows that immunoglobin genes are rearranged, implying a signalling event, possibly an aberrant antigen response. Common gene abnormalities include mutation in *RAS* (coding for a molecule involved in cell growth and survival), *TP53* (a tumour suppressor) and *MYC* (a transcription factor, also active in Burkitt lymphoma). However, a major clinical factor is the metastatic movement of the disease from the bone marrow to other bone, such as the skull and ribs. This transition is linked to overexpression of a series of genes

that include the receptor activator of nuclear factor κB (*RANK*) and its ligand (*RANKL*), osteoprotegerin and ILs, and results in the overactivity of osteoclasts. This in turns leads to weakness in the bone (osteolytic lesions, demonstrable on X-ray), with pain and spontaneous fractures. Increased plasma levels of cytokines such as tumour necrosis factor, IL-1 and IL-6 are present, and may be involved in bone disease as some activate osteoclasts.

Perhaps half of all myeloma tumours have extra chromosomes, whereas many of those that have a normal complement of chromosomes demonstrate translocations that place *IgH* adjacent to D-cyclin genes. Genetic variation at the 8q24 locus confers risk of this disease.

Blood science angle: Cytogenetics of white cell malignancy

It is abundantly clear that our knowledge of these diseases has benefitted enormously from genetics. The principal example of this is the discovery and exploitation of the *BCR–ABL* mutation that defines and also points to the treatment of CML. However, considerable advances in understanding the pathophysiology of many other leukaemias, lymphomas and myelomas have followed from cytogenetic studies.

Myeloma as a systemic disease   Myeloma provides a useful example of the integration of the different subtypes of blood science with the clinic. Although the electrophoresis could in practice be performed in any laboratory, it is generally in the biochemistry or immunology sections.

- *Haematology aspects.* As the tumour develops within the bone marrow, red cell, white cell and platelet numbers will fall, leading ultimately to a normocytic anaemia, leukopenia and thrombocytopenia. These will be noted in the clinic with tiredness and lethargy and so on, recurrent infections, and with bruising and bleeding (as with leukaemia). The high plasma proteins of the paraprotein may also suppress platelet function, contributing to haemorrhage. However, in potential contradiction, the high plasma and blood viscosity may precipitate purpura and (by 'sticking' platelets together) venous thrombosis.

  However, the most marked blood test is a grossly abnormal ESR, being the consequence of the increased plasma viscosity caused by the paraprotein. The blood film may have a gentle blue background because of the effects of the paraprotein in causing high plasma proteins, and this high protein causes many red cells to be arranged in 'columns', called rouleaux. As the disease progresses, plasma cells will leave the bone marrow and can be identified with relative ease in the peripheral blood. This stage may be described as plasma cell leukaemia. This is especially so in the terminal stages, and there may also be infiltration of lymph nodes leading to lymphadenopathy.

- *Biochemistry aspects.* Many patients, predominantly in advanced disease, complain of bone pain. Spread of the tumour requires dissolution of bone, which will lead to hypercalcaemia, and also a likely change in serum phosphate. The high calcium should produce, via negative feedback, low levels of parathyroid hormone. There may also be increased alkaline phosphatase, as this enzyme is also involved in bone metabolism.

As the paraprotein develops there is an increasing risk of chronic renal failure with rising urea and creatinine, and so a reduced glomerular filtration rate. There will be a high plasma protein, whilst serum albumin should be within the reference range (as per Table 6.4).

- *Immunology aspects.* These are generally in respect of the protein immunoelectrophoresis and the quantification of light chains, and this laboratory will also quantify levels of the 'normal' immunoglobulin classes and so define a hypogammaglobulinaemia. This laboratory may also measure beta-microglobulin, often raised in myeloma. Myeloma makes up about 20% of all monoclonal gammopathies.

Table 6.4 Protein subsets in health and myeloma.

	Normal profile		Myeloma profile	
	(g/L)	(%)	(g/L)	(%)
Total proteins	75	100	95	100
Albumin	46.2	61.6	46.2	48.6
Alpha globulins	3.8	5.1	3.8	4.0
Beta globulins	7.6	10.2	7.6	8.0
Gamma globulins	17.4	23.1	37.4	39.4

In myeloma, the total protein result is higher, and although the absolute albumin, alpha globulin and beta globulin results are the same, the proportion of each as a percentage is reduced. This is accounted for by the gamma globulin result, higher in absolute terms and percentage.

Conditions allied to myeloma

The paraproteinaemia often brings a number of sub-conditions, which may be found in any myeloma-related disease, but also in lymphoma or certain leukaemias. Indeed, a paraprotein may be present in up to 50% of patients with B-cell CLL. Other conditions related to myeloma include MGUS, amyloidosis, hyperviscosity syndrome, cryoglobulins and LPL.

Monoclonal gammopathy of undetermined significance   MGUS is present in 1–3% of the over-50s, 3–5% of the over-70s and in 10% of those over 80 years of age, and is probably underdiagnosed. As the name implies, it is a paraproteinaemia (predominantly of IgG) whose origin is obscure, but which makes up about half of all monoclonal gammopathies. Almost half express the translocation t(11;14)(q13:q32), which activates the *cyclin D1* gene.

The disease itself is often benign, with little evidence of bone marrow or peripheral white cell abnormalities, and is often found coincidentally. However, perhaps 1–2% transform to a myeloma or lymphoma each year, a rate high enough to require outpatient surveillance. This leads to the hypothesis that MGUS is effectively pre-myeloma. In support of this is the observation that progression is linked to the size and type of the paraprotein (IgM and IgA carrying a worse prognosis than IgG) and an abnormal plasma kappa/lambda ratio (<0.26 or >1.65). Bone-related problems in those at high risk involve treatment with oral calcium and vitamin D supplements.

Amyloidosis   Amyloid is an umbrella term for the inappropriate deposition of insoluble proteins, and can take many forms, such as in Alzheimer's disease. As regards paraproteins, high levels of the unusual protein, inevitably a fraction of light chain material, form complexes in tissues such as the heart, kidneys and tongue, causing dysfunction (such as heart and renal failure). Amyloid comprises 11% of all monoclonal gammopathies, has an incidence one-fifth that of myeloma and is present at a rate of perhaps nine people per million per year, which equates to 600 new patients in the UK each year.

Hyperviscosity syndrome   This is the consequence of thick plasma caused by the high levels of paraproteins, and is present when plasma viscosity exceeds the top of the reference range (>1.72 mPa). It is separate from whole blood hyperviscosity, as is found in polycythaemia, and is due to the high red cell count. At high levels

(>2.5 mPa) it contributes to renal failure (as determined by raised urea and electrolytes (U&Es)) and occlusive venous thrombotic disease, which may call for plasmapheresis.

Cryoglobulins Literally, cold globulins, which may be present in any condition where there are high levels of an abnormal protein, which includes rheumatoid factors and immune complexes. At body temperature (i.e. $37\,^{\circ}$C) these proteins are soluble, but when the temperature falls, as in the skin, fingers and toes, the paraproteins may come out of solution, become insoluble and potentially obstruct blood flow. This can lead to Raynaud's syndrome, purpura, vasculitis and other clinical signs.

Lymphoplasmacytoid lymphoma We have already met this disease, previously known as Waldenstrom's macroglobulinaemia, in the section on lymphoma. It accounts for 1–2% of haematological malignancy with an incidence of three people per million per year. Until relatively recently it was part of the myeloma family of conditions, as it was seen as a B-cell neoplasm characterized by a lymphoplasmacytic infiltrate in the bone marrow with an associated paraprotein.

Blood science angle: B-cell malignancy

A broad hypothesis, based on generalized and specific genetic and antigenic aetiologies, links all B-cell malignancies. A nonspecific B-cell lymphocytosis often precedes B-cell CLL (which is often accompanied by increased gammaglobulins), whilst MGUS may be an early stage of myeloma. Furthermore, the same genetic mutation is often present in a leukaemia and a lymphoma, suggesting a common background aetiology, but other, as yet unknown, factors that lead to disease in the bone marrow or the lymph node. Indeed, a non-secreting myeloma may be described as a lymphoma of the bone marrow, whilst a paraprotein-secreting lymphoma may be described as a myeloma of the lymph node.

The paraprotein The major serum aspect of these diseases is the nature of the unusual protein. An abnormal serum monoclonal gammopathy is present in almost all cases of myeloma, and in 80% there is Bence–Jones protein. A paraprotein is less likely to be found in other related diseases such as MGUS, but it cannot be excluded. The paraprotein is detected by serum electrophoresis, a process described in Chapter 17 on the liver and plasma proteins. To briefly recapitulate, groups of proteins are separated according to their electrical nature into albumin, alpha-globulin, beta-globulins and gamma-globulins. Immunoglobulins are found in the latter fraction.

A second aspect of protein electrophoresis is the ability of the laboratory to convert each of the different plasma protein bands into a defined plasma concentration. This is performed on a densitometer, as again described in Chapter 17, where Figure 17.8 applies. This is illustrated in Table 6.4, where an increased absolute gamma-globulin result (20 g/L higher than the normal profile) contributes to an increased percentage, and reduced percentages of other constituents. However, the phenomenon of immune paresis may be present, where there is an active suppression of the other antibody classes by the malignancy, causing a hypogammaglobulinaemia that contributes to the increased frequency of infections.

Sub-typing the paraprotein An additional aspect of the analysis of the paraprotein is its nature in terms of the components of the normal antibody molecule, which is composed of a mixture of heavy chains and light chains. Heavy chains can have one of five different identities: IgA, IgD, IgE, IgG and IgM. A second aspect is the nature of the light chain, which can be either kappa or lambda.

The paraprotein can be any fraction of a heavy chain or a light chain. The exact identity is defined by immuno-electrophoresis with fixation, hence immunofixation electrophoresis (IFE). Other methods include haemagglutination (now rarely used), nephelometry, and Western blotting. In IFE, probably the most widely used technique, first there is standard electrophoresis, followed by the probing of the paraprotein band with an anti-human antiserum to each of the heavy chains and light chains (Figure 6.11). These antibodies are linked to a detection marker, often an enzyme, to allow the paraprotein to be visualized, and the method can be used to probe serum and urine. This method has been used to demonstrate the most frequently encountered paraproteins, which are IgG (53%, of which the major species is IgG1), IgA (22%), light chain (15–20%), IgD or IgE (1.5%) and IgM (0.5%). However, perhaps 3% of myelomas are non-secretory.

Light chain analysis In health, the two light chains (kappa and lambda) are produced in slight excess by the B lymphocytes, and these are exported into the plasma as free light chains (FLCs). The proportion of each of the two FLCs each slightly favours lambda chains, leading to an average kappa/lambda ratio of perhaps 0.75, although this masks considerable variation. Nevertheless, it follows

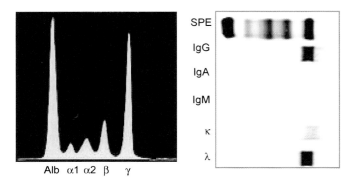

Figure 6.11 Immunofixation electrophoresis. Serum protein electrophoresis (SPE). On the left is the densitometer profile showing a large paraprotein peak in the gamma (γ) region. This will have been derived from the SPE trace at the top of the panel on the right. The second aspect of this analysis is the typing of the gamma-globulin band by IFE, which has produced a heavy band in the IgG and the lambda (λ) regions, thereby identifying the paraprotein as an IgGλ. (Images courtesy of Dr JA Katzmann).

that a kappa/lambda ratio that differs markedly from a range of maybe 0.2–2.0 must be due to an excess of one of the two types of FLC. The major reason for this is a paraprotein, and, in turn, the major reason for a paraprotein is a B-cell malignancy such as a myeloma.

The ratio can easily exceed 1:1500 in either direction, and the FLC ratio can be a good marker of the progression of the disease, or the effects of treatment with drugs such as melphalan, thalidomide, vincristine and adriamycin. An abnormal FLC ratio is also a risk factor for the progression of MGUS to myeloma. However, measurement of FLCs is pertinent as they are likely to contribute to renal failure (present in up to 50% of patients with a myeloma); accordingly, it may be worthwhile removing these proteins from the blood by the process of haemodialysis.

Urine analysis   Electrophoresis and immunoelectrophoresis may also be performed on urine to help define the nature of the paraprotein that is likely to be a Bence–Jones protein. However, the protein in urine may need to be made more concentrated before analysis. In most cases it is strongly advised to run the urine electrophoresis alongside a sample of serum from the patient, enabling a comparison of any abnormal bands (Figure 6.12).

In Figure 6.12, samples 1 and 2 have dense bands in the alpha2 and beta-globin regions, the latter in sample 2 being very strong, and also present in the urine. Sample 3 is unremarkable, but in sample 4 there is very strong staining of the urine gamma-globin region, and even a small albumin band, implying a massive proteinuria. This would be confirmed by standard biochemistry.

Sample 5 has a strong paraprotein in the gamma-globin region, which may arise from a myeloma, but notably there is no parallel band in the urine, although there is some albumin in the urine. In sample 6 the gamma-globin band is diffuse, which could represent the normal polyclonal response to an infection.

The urine sample 7 has a weak band in the albumin and beta-globin regions, but no gamma-globin band, which may reflect renal disease. In sample 8, the urine sample seems protein free, and in sample 9 there is a trace of protein, but no serum sample. Sample 10 is also

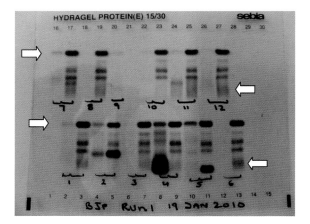

Figure 6.12 Serum and urine electrophoresis. There are 12 sets of analyses. In each case the urine sample in on the left, and clearly has great deal less 'blue' than its paired serum sample to the right. The arrows on the left highlight the albumin band; the arrows on the right point to the gamma globulin region.

Table 6.5 Major aspects of white malignancies.

	Leukaemia	Lymphoma	Myeloma
Cellular basis	Can be any white cell	90% B lymphocytes, 10% T lymphocytes	Always B lymphocytes
Basis of classification	Cell lineage, proportion of blasts defines acute nature	Presence of RS cells defines Hodgkins disease, otherwise non-Hodgkin	Nature of the paraprotein (if present), Bence–Jones protein (if present)
Primary organ base	Bone marrow	Lymph nodes	Bone marrow
Leukocytosis	Always present (except HCL)	Uncommon, but often present in advanced disease	Only present in advanced disease
Bone involvement	Not uncommon	Rare	Common (leading to raised serum calcium)
Common features	Anaemia, thrombocytopenia, infections		

unremarkable, but sample 11 has a broad gamma-globulin band in the urine, which may reflect moderate proteinuria. Sample 12 has a weak gamma-globin band in the urine sample.

In each case it is possible to quantify the intensity of the blue staining by densitometry, and so estimate the amounts of protein in each band in both the serum and urine samples.

Blood science angle: Urine and serum electrophoresis

Despite our emphasis on malignancy and Bence–Jones protein, this analysis is also important in investigating renal aspects of proteinuria, and so in excluding myeloma-based disease. Therefore, U&Es are a required analysis in the interpretation of the electrophoresis of urine. In cases of haemolytic anaemia, there may be haemoglobin in the urine, which will give its own pattern, and so demand a full blood count.

Table 6.5 summarizes major aspects of the three white cell malignancies.

6.6 Myelofibrosis and myelodysplasia

These two conditions are described in Chapter 4 on red cells as causes of anaemia, but they also have roles in white cell disease. The aetiology of myelofibrosis is of the proliferation of non-haemopoietic tissues in the bone marrow, but it can follow polycythaemia rubra vera (PRV) or essential thrombocytosis (ET) – see Chapters 4 and 8. However, in the white cell setting, the blood film may resemble CML, with a leukoerythroblastic picture.

The genetic lesions include *JAK2 V617F* and *ERK/MAPK*, and AML or CML may develop from PRV or ET, and (as in standard CML) JAK may be targeted with TKIs.

The aetiology of myelodysplasia is of stem cell abnormality, and the production of blood cells is unregulated. This generally manifests itself as reduced red cells and platelets with increased numbers of white cells, although there may be pancytopenia. Nevertheless, perhaps a third of myelodysplasia patients progress to AML. The most common genetic lesion is a deletion in the q arm of chromosome 5, although trisomy of chromosome 8 has been described.

However, studies of *TP53*, linked to chromosome 5q, seem to be providing new insights. Mutations in this gene are present in slightly under 10% of myelodysplasia patients, and in one study were associated with increased expression of TP53 protein, a higher blast count and progression to leukaemia, and a shorter survival time (9 months versus 66 months). These finding suggest that genetic typing for 'wild-type' or mutated *TP53* may become an important clinical tool.

6.7 Case studies

Case study 5

A 50-year-old man presents to his health care practitioner with an 8-month history of progressive tiredness. The practitioner notes he has had two courses of antibiotics in the past 4 months. On examination the man's left upper abdomen is firm and painful upon gentle pressure. A full blood count is ordered. The white cell component is as follows:

	Result (10^9/L)	Reference range
White cell count	4.6	4.0–10.0
Neutrophils	1.5	2.0–7.0
Lymphocytes	4.0	1.0–3.0
Monocytes	0.04	0.2–1.0
Eosinophils	0.01	0.02–0.5
Basophils	0.01	0.02–0.1
Blasts/atypical cells	0.04	~0.01

The haematology autoanalyser flags up an abnormality calling for a manual check of the blood film. The scientist reports the lymphocytes have an unusual morphology (Figure 6.13) and calls for CD marker analysis by fluorescence activated cell scanning. This reveals that most lymphocytes express CD19, CD20, CD22 (strongly), CD25 (weakly) and CD103 (strongly).

Interpretation   Although the white cell count is within the reference range, the lymphocyte count is raised and levels of all other normal white cells are low. The number of blasts equals that of the monocytes. This prompted the call for CD markers: CD19 and CD20 are B lymphocyte markers, CD25 is found on activated T and B cells and their progenitors, CD103 on activated lymphocytes and other cells. These results, alongside the morphology, points to HCL, which is supported by the pain on the left abdomen – the spleen, hence splenomegaly.

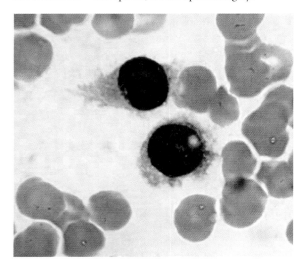

Figure 6.13 Blood film for case study 5. Two abnormal lymphocytes. Note the irregular border of the cytoplasm, best determined with high-magnification lenses. (Image from Blann AD, *et al.* (2009) Haematology Morphology. The IBMS Training CD-ROM. © IBMS).

Case study 6

A 62-year-old woman reported a year's history of increasing tiredness and weight gain. More recently, she complained of drenching night sweats. Venous blood samples were obtained, which found a raised ESR and LDH, and the following white cell differential:

	Result (10^9/L)	Reference range
White cell count	9.8	4.0–10.0
Neutrophils	3.7	2.0–7.0
Lymphocytes	4.2	1.0–3.0
Monocytes	1.4	0.2–1.0
Eosinophils	0.2	0.02–0.5
Basophils	0.2	0.02–0.1
Blasts/atypical cells	0.1	~0.01

The blood film confirmed the lymphocytosis but failed to report any abnormalities in the morphology of the cells.

Interpretation   The raised ESR and LDH indicate some abnormality, but provide no specific direction. Although the white cell count is just within the reference range, the pattern of the differential is unusual, with raised lymphocytes, monocytes, basophils and blasts cells.

There is clearly some leukocyte abnormality, but no clues. The next step would be a CD marker screen to investigate if there are any abnormalities in the white cells. But with no clear directions the only option is a disappointing 'watchful waiting', which, as implies, waits for further signs. Not all diagnoses are either quick or simple, and this is such an example.

In the case of this woman, her symptoms worsened and eventually a whole-body MRI/CAT scan type imaging would be ordered. This could perhaps identify several small masses in the abdomen, believed to be lymphoma. This would need to be confirmed with ultrasound-directed biopsy of the masses, and then a multidisciplinary team meeting to discuss management options based on Hodgkin and other classifications, such as stage.

Summary

- The greater proportion of diseases of white blood cells are malignancy. These are leukaemia, lymphoma and myeloma.
- In the vast majority of cases, malignancy is due to abnormalities in chromosomes. The expression of CD markers is a useful diagnostic aid.

- The dominant form of CML is caused by a translocation that generated a fused oncogene: *BCR–ABL*. This genetic lesion can be targeted by TKIs
- AML is a diverse collection of diseases characterized by high numbers of blasts in the peripheral blood and a variety of genetic abnormalities.
- Cytogenetics are also important in many cases of CLL, and most of these are B lymphocyte diseases.
- ALL (like AML) has many different presentations, and translocations are common.
- The cellular basis of many lymphomas is similar to those of leukaemia. HL is defined by the RS cell; if not present, the disease is an NHL.
- Myeloma is a bone marrow tumour of B lymphocytes, almost all of which secrete abnormal parts of antibody molecules and can be detected by electrophoresis. In the late stages the tumour appears in the blood as plasma cells and may metastasize to lymph nodes, spleen and liver.

Further reading

Bladé J, Cibeira MT, Fernández de Larrea C, Rosiñol L. Multiple myeloma. Ann Oncol. 2010;21(Suppl 7): vii313–vii319.

Chan AO, Lau JS, Chan CH, Shek CC. Cryoglobulinaemia: clinical and laboratory perspectives. Hong Kong Med J. 2008;14:55–59.

Costa R, Abdulhaq H, Haq B. *et al.* Activity of azacitidine in chronic myelomonocytic leukaemia. Cancer. 2011; 117:2690–2696.

Dearden C. Large granular lymphocytic leukaemia pathogenesis and management. Br J Haematol. 2010; 152:273–283.

Fonseca R, Hayman S. Waldenstrom macroglobulinaemia. Br J Haematol. 2007;138:700–720.

Grever MR. How I treat hairy cell leukemia. Blood. 2010;115:21–28.

Hubmann R, Hilgarth M, Schnabl S. *et al.* Gliotoxin is a potent NOTCH2 transactivation inhibitor and efficiently induces apoptosis in chronic lymphocytic leukaemia (CLL) cells. Br J Haematol. 2013;160:618–629.

Hochhaus A, La Rosée P, Müller MC. *et al.* Impact of *BCR–ABL* mutations on patients with chronic myeloid leukemia. Cell Cycle. 2011;10:250–260.

Kennedy-Nasser AA, Hanley P, Bollard CM. Hodgkin disease and the role of the immune system. Pediatr Hematol Oncol. 2011;28:176–186.

Korde N, Kristinsson SY, Landgren O. Monoclonal gammopathy of undetermined significance (MGUS) and smouldering multiple myeloma (SMM): novel biological insights and development of early treatment strategies. Blood. 2011;117:5573–5581.

Kraszewska MD, Dawidowska M, Szczepański T, Witt M. T-cell acute lymphoblastic leukaemia: recent molecular biology findings. Br J Haematol. 2012;156: 303–315.

Kulasekararaj AG, Smith AE, Mian S.A. *et al. TP53* mutations in myelodysplastic syndrome are strongly correlated with aberrations of chromosome 5, and correlate with adverse prognosis. Br J Haematol. 2013;160:660–672.

Landgren O, Kyle RA. Multiple myeloma, chronic lymphocytic leukaemia and associated precursor diseases. Br J Haematol. 2007;139:717–723.

Liang X, Graham DK. Natural killer cell neoplasms. Cancer. 2008;112:1425–1436.

Magrath I. Epidemiology: clues to the pathogenesis of Burkitt lymphoma. Br J Haematol. 2012;156: 744–756.

O'Connell TX, Horita TJ, Kasravi B. Understanding and interpreting serum protein electrophoresis. Am Fam Physician. 2005;71:105–112.

Payne E, Look T. Zebrafish modelling of leukaemias. Br J Haematol. 2009;146:247–256.

Rosenwald A, Ott G. Burkitt lymphoma versus diffuse large B-cell lymphoma. Ann Oncol. 2008;19(Suppl 4): iv67–iv69.

Guidelines

Foroni L, Wilson G, Gerrard G. *et al.* Guidelines for the measurement of *BCR–ABL1* transcripts in chronic myeloid leukaemia. Br J Haematol. 2011;153: 179–190.

Jones G, Parry-Jones N, Wilkins B. *et al.* Revised guidelines for the diagnosis and management of hairy cell leukaemia and hairy cell leukaemia variant. Br J Haematol. 2012;156:186–195.

Milligan DW, Grimwade D, Cullis JO. *et al.* Guidelines on the management of acute myeloid leukaemia in adults. Br J Haematol. 2006;135:450–474.

Oscier D, Dearden C, Erem E. *et al.* Guidelines on the diagnosis, investigation and management of chronic lymphocytic leukaemia. Br J Haematol. 2012;159: 541–564.

Reilly JT, McMullin MF, Beer PA. *et al.* Guideline for the diagnosis and management of myelofibrosis. Br J Haematol. 2012;158:453–471.

7

The Physiology and Pathology of Haemostasis

The common final factor in most causes of death, leaving aside trauma and some other unusual situations, is inevitably cardiac arrest. In turn, the most likely reason for this is because this organ has failed to receive sufficient oxygen and glucose to keep it beating. This modified muscle receives its nutrients from arteries on the outside (hence coronary arteries) that deliver the nutrients to the cells (cardiomyocytes). The principal reason for the failure of this delivery is atherothrombosis – a mixture of atherosclerosis and thrombosis. However, clots in veins (venous thrombosis) are also responsible for morbidity and mortality (the latter via clots in the pulmonary circulation).

Consequently, clots generally get a bad press. This is undeserved, as without them we would haemorrhage, perhaps fatally. Indeed, conditions such as haemophilia clearly demonstrate the importance of the need for clotting. However, it is also clear that too much clotting, as in cardiovascular disease, is also undesirable and can be just as deadly.

The balance between too little clotting (haemorrhage) and too much clotting (thrombosis) is therefore crucial. This balance (i.e. the homeostasis of thrombosis) is called haemostasis, and loss of this balance can lead to fatality. However, before addressing coagulation diseases, we must first understand the mechanics of the process. Having done so, in Chapter 8 we will address the management of these diseases. The major indices in haemostasis, and their reference ranges relevant to this chapter and the next, are presented in Table 7.1.

Virchow's triad

As long as 150 years ago, Rudolph Virchow recognized three requirements for thrombogenesis, these being: (i) abnormal blood flow, (ii) abnormalities in the vessel wall and (iii) abnormalities in the constituents of the blood. This basic concept has been extended and modified by modern knowledge.

Blood flow has a place, as sluggish and/or turbulent blood flow promotes thrombosis. We also know a great deal more of the blood vessel cell, and of the endothelium and its function. As regards blood constituents, the importance of blood cells and molecules are established; that is, the platelet and coagulation factors (Figure 7.1).

Although Virchow's original hypothesis referred to venous thrombosis, we now know it is equally applicable to arterial thrombosis and can be revised as the blood vessel wall (Section 7.1), platelets (Section 7.2) and coagulation factors (Section 7.3). In Section 7.4 we will examine haemostasis as the balance between thrombus formation and removal, whilst Section 7.5 will look at the haemostasis laboratory. The chapter will conclude in Section 7.6 with the pathology of thrombosis.

Blood Science: Principles and Pathology, First Edition. Andrew Blann and Nessar Ahmed.
© 2014 John Wiley & Sons, Ltd. Published 2014 by John Wiley & Sons, Ltd.

Table 7.1 Major indices in haemostasis.

Index	Typical result	Unit	Reference range
Platelets	300	10^9/L	143–400
Prothrombin time	12	s	11–14
Activated partial thromboplastin time (APTT)	29	s	24–34
Thrombin time	17	s	15–19
Fibrinogen	2.7	g/L	1.5–4
Von Willebrand factor (vWf)	89	IU/dL	50–150
D-dimers[a]	125	Units/mL	<500
Antithrombin	1.12	Units/mL	0.86–1.32

[a]Result and reference range are strongly dependent of the particular manufacturer.

7.1   The blood vessel wall

The anatomy of the blood vessels consists of a series of concentric layers, the inner lining of which, that interfaces with the blood, is the endothelium. Outside this is the media, a layer of smooth muscles cells, and the outer later is the adventitia, rich in connective tissue cells and fibres such as collagen. In veins, the media is often thin (partially because it needs not support high blood pressure, as do arteries), whilst in capillaries this layer is absent.

Forming a continuous layer, endothelial cells are flattened, orthogonal cells that line all blood and lymphatic vessels and the inside of the heart. In the adult the endothelium consists of $1–6 \times 10^{13}$ cells, weighs approximately 1 kg and covers a surface area of approximately 4000–7000 m^2. Each cell is anchored to an underlying basic elastic lamina, rich in connective tissue components such as (again) collagen, but also actin, elastin and fibronectin. Should the endothelial layer be disturbed, these fibres rapidly attract platelets. Individual endothelial cells are attached to their neighbours by specialized junctions in the gaps between the cells. These gaps regulate the passage of various cells and substances moving between the blood and the tissues.

Once considered simply an inert lining, the endothelium is in fact a highly dynamic organ, with several roles:

- In inflammation, endothelial cells mediate the passage of leukocytes out of the blood and into the tissues, partially by the expression of adhesion molecules such as E selectin. They also release and respond to inflammatory cytokines such as interferon, interleukins 1 and 6, and tumour necrosis factor.
- As participants in blood pressure control, endothelial cells release nitric oxide (a vasodilator) and endothelin (a vasoconstrictor), which together act on the smooth muscle cells of the middle section of the artery. The loss of balance between these molecules contributes to hypertension.
- In haemostasis, endothelial cells express and release molecules that counter thrombosis (i.e. are antithrombotic) and others that promote thrombosis (i.e. are prothrombotic) (Table 7.2).

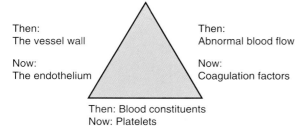

Then:
The vessel wall

Now:
The endothelium

Then:
Abnormal blood flow

Now:
Coagulation factors

Then: Blood constituents
Now: Platelets

Figure 7.1 Virchow's triad considers different roles for the blood vessel wall, blood flow and the constituents of the blood in the pathogenesis of thrombosis.

Table 7.2 Involvement of the endothelium in haemostasis.

Antithrombotic	Prothrombotic
Heparin	Coagulation factor V
Protein C and protein S	Tissue factor
Tissue plasminogen activator	Plasminogen activator inhibitor-1 (PAI-1)
Prostacyclin, nitric oxide	Thromboxane
Thrombomodulin	Platelet activating factor
Protein C receptor	vWf
Tissue factor pathway inhibitor (TFPI)	

Antithrombotics

These consist of components of the cell membrane such as thrombomodulin and the endothelial protein C receptor, and molecules that are secreted or otherwise released into the circulation.

Heparin is a complex mucopolysaccharide, a glycosaminoglycan rich in sulphate groups that can be both attached to the cell membrane and released to circulate in the blood. It has no anticoagulant activity in itself: it acts as a cofactor for the inhibitor molecule, antithrombin. As we shall see in Chapter 8, heparin can be given as a therapy to reduce the risk of thrombosis.

Prostacyclin and nitric oxide are both vasodilators, and so have a role in blood pressure regulation, and at the level of the capillary may promote blood flow. However, they can also suppress platelet function.

Prothrombotics

These are also components of the cell membrane or molecules that are secreted or released (Table 7.2). The leading procoagulant produced by the endothelium is vWf.

The endothelial cell can be defined by the presence of specific Weibel–Palade bodies in the cytoplasm. These organelles can be observed only by electron microscopy, and are storage granules for vWf. Upon stimulation of the endothelium (perhaps by desmopressin, cytokines, changes in blood flow, or thrombin), or damage (perhaps by toxins), Weibel–Palade bodies move to the surface of the endothelial cell and release large amounts of vWf, some of which passes directly to the plasma whilst some remains at the cell surface.

vWf is a polymer of many individual monomers, each of about 2000 amino acids, and may combine to give a large molecule of a relative molecular mass of perhaps a million Daltons. This large molecule has a number of sites (domains) that recognize a variety of other molecules, including heparin, collagen, elastin, vitronectin, and platelet membrane components GpIb and GpIIb/IIIa. This means that vWf is effectively a platelet glue, sticking these cells to one another to form microthrombi, but also is mediating the adhesion of the platelet to the subendothelium. However, large multimers of vWf (which are the most effective prothrombotics) may be digested down to smaller units (which are less prothrombotic) by the action of a complex enzyme abbreviated to ADAMTS-13.

Another crucial function of vWf is an essential cofactor for coagulation factor VIII. The importance of this will become clear in Chapter 8, when the consequences of the lack of vWf, leading to von Willebrand's disease, are discussed. Other prothrombotics include coagulation factors and platelet activators.

Pathophysiology of the endothelium

Despite the potential of the endothelium in both aspects of haemostasis, it has proved very difficult to influence directly (such as by pharmacotherapy) and is also very difficult to assess functionally in the individual patient. There are several ways to increase the procoagulant capacity of the endothelium (such as with desmopressin) but none that increase its anticoagulant nature.

Nevertheless, it is likely that a policy of the avoidance of cardiovascular risk factors (smoking, diabetes, etc.) promotes an anticoagulant nature, whilst the adoption of risk factors promotes a prothrombotic nature. Indeed, damage to the endothelium is the first step in the development of atherosclerosis. In support of this is the observation that smokers and diabetics have an increased level of vWf, implying vascular damage.

7.2 Platelets

The second part of the reinterpretation of Virchow's triad considers cells, and the dominant cell is clearly the platelet. This tiny and enucleate particle of the cytoplasm of the megakaryocyte contains a host of substances that all promote thrombus formation and integrity. Although it is unwise to deny a direct role for white cells in thrombosis (an example being the expression of tissue factor by monocytes and granulocytes), there is evidence that red cells can promote thrombus formation, probably by assisting in platelet activation by providing adenosine diphosphate (ADP). Indeed, red cells are more prevalent in clots that occur in veins.

Thrombopoiesis and megakaryocytes

Platelets are the end-cell of a long process of cell development that begins with a pluripotent stem cell, and progresses through other colony-forming units. The latter produces a joint precursor with erythrocytes, then the megakaryoblast. This transforms into a promegakaryocyte and then the megakaryocyte itself. A key hormone in the process is thrombopoietin, synthesized principally by the liver, the platelet analogue of the red cell hormone erythropoietin (Figure 7.2). However,

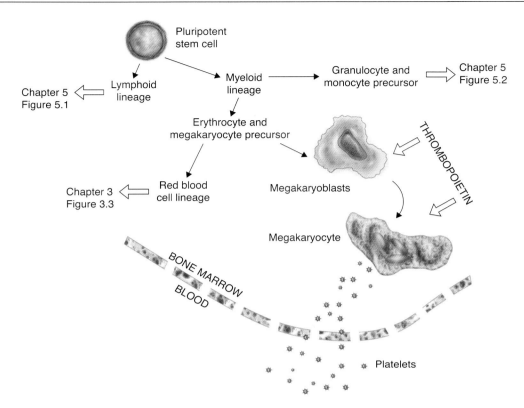

Figure 7.2 Thrombopoiesis. Platelets are the end cell of this process, intermediates being stem cells, and precursors the megakaryoblast and the megakaryocyte. The leading hormone promoting this process is thrombopoietin.

platelet precursors may also respond to cytokines such as interleukins.

The megakaryocyte is a large cell (50–100 μm in diameter), with up to 32 sets of chromosomes, present in the bone marrow with a frequency of perhaps 1 in 10 000 bone marrow cells. Some 2000–5000 platelets are produced by each cell by the budding-off of pseudopodia, and the 'exhausted' remains of the megakaryocyte is consumed by phagocytes.

Platelets

Also called thrombocytes, these appear on a Romanowsky blood film as a small purple body, with a diameter of perhaps 2–3 μm (Figure 7.3). Produced by megakaryocytes at a rate of perhaps 10^{11} daily, they circulate in a resting stage for perhaps 6–9 days before being cleared by scavengers of the reticulo-endothelial system. In order to participate in haemostasis, they must be activated, most often by the binding of agonists to cell membrane receptors.

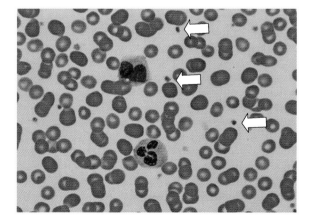

Figure 7.3 Platelets in a blood film. This film shows two different white cells (each with a heavy purple nucleus), red cells (coloured pink) and platelets (several small light purple bodies, arrowed). (Courtesy of D. Bloxham, Cambridge University Hospitals NHS Foundation Trust).

Table 7.3 The platelet membrane.

Component	Function
GpIb–IX–V	Binds vWf
GpIIb/IIIa	Binds fibrinogen
GpIc–IIa	Binds fibronectin
CD62P (P selectin)	Adhesion molecule
Thrombin receptor	Initiates platelet activation when bound by its ligand (thrombin)
ADP receptor	Initiates platelet activation when bound by its ligand (ADP)
GpVI	Binds collagen

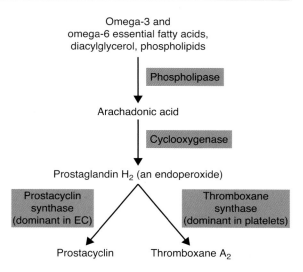

Figure 7.4 Platelet metabolism. The primary metabolic pathway in the platelet is the generation of thromboxane via arachadonic acid and prostaglandin H_2. They key enzyme in this pathway is cyclooxygenase. EC = endothelial cell.

The platelet membrane As with other cells, the membrane is a fluid mosaic of glycoproteins embedded in a phospholipid bilayer. The principal components of the membrane are listed in Table 7.3. The majority are named by glycoprotein bands determined by electrophoresis – hence Gp – followed by a Roman numeral. Many are involved in adhesion of the platelet to another platelet, or to the subendothelium (via vWf). Others have specific names which denote their function as receptors, and many span the membrane and so communicate with intracellular second messenger systems within the cytoplasm. Each receptor can be bound by a specific ligand, and upon doing so a chain of actions is initiated which results in the activation of the cell.

The platelet cytoplasm Several of the transmembrane glycoproteins are anchored by a connection with the internal cytoskeleton of the cell, such as fibres of actin. The cytoplasm also contains a network of canalicular and dense tubular systems to help the movements of small molecules and ions, potentially from granules. This internal cytoskeleton allows the shape change when platelets are activated and participate in haemostasis. There are two major organelles participating directly in haemostasis:

- The contents of dense bodies, of which there are likely to be less than a dozen per cell, include adenosine triphosphate, ADP, calcium and serotonin, the latter a powerful vasoconstrictor.
- Alpha granules are far more numerous, up to 75 or 80 per cell. They contain platelet factor 4, beta-thromboglobulin, TFPI, protein S, vWf, plasminogen activator inhibitor-1 (PAI-1), fibrinogen and coagulation factors V, VII and XIII.

The membrane of this granule contains P-selectin (CD62P), an adhesion molecule whose ligand is PSGL-1, present on several white cells. It follows that increased membrane P-selectin reflects alpha degranulation.

The platelet also contains mitochondria, granules of glycogen (a source of energy) and lysosomes; the latter are presumed to release proteolytic enzymes upon cell activation. Within the cytoplasm are a host of intracellular metabolic pathways. The first step in the principal pathway in haemostasis is the conversion of omega-3 and omega-6 fatty acids, diacylglycerol and phospholipids to arachadonic acid by enzymes such as phospholipase. Arachadonic acid is the substrate for cyclooxygenase, converting it to prostaglandin H_2, an endoperoxide (Figure 7.4). Two pathways then follow: the synthesis of prostacyclin (via prostacyclin synthase) and of thromboxane A_2 (via thromboxane synthase). In the platelet, the latter pathway dominates, whereas in the endothelium most prostaglandin is converted to prostacyclin.

7.3 The coagulation pathway

The final part of the reinvention of Virchow's triad concerns the coagulation factors.

Coagulation factors

These molecules are a collection of mostly liver-produced proteins, many with the curious property of being activated by another factor, which then in turn activate another molecule (i.e. are zymogens – inactive precursors of an enzyme). Several factors have names (such as prothrombin); others are denoted by Roman numerals, such as factor V. The 'resting' or perhaps 'inactive' form is noted by the simple factor number, but if activated carries the notation 'a', an example being factor Va. Table 7.4 lists the coagulation factors, a key factor of which is their half-life.

Other coagulation molecules include prekallikrein (which is converted to kallikrein), kininogen (a cofactor for factor XI binding to phospholipids) and tissue factor, the latter being present on endothelial cells and monocytes. The liver requires vitamin K for the development of several of these molecules, a fact exploited by the anticoagulant warfarin, discussed in detail in Chapter 8.

The coagulation pathway

The function of the coagulation pathway is to generate fibrin. This happens by the interactions of several highly regulated enzyme–substrate reactions, where the product of one reaction acts on another. The cascade itself can be seen as proceeding in a number of ordered steps (Figure 7.5).

Initiation   This short pathway (also known as the extrinsic pathway) begins with factor VII, which needs to be converted to factor VIIa in the presence of tissue factor. In the presence of calcium, phospholipids and tissue factor (a membrane component exposed by tissue damage and/or cell activation, perhaps the endothelium), this complex has the ability to convert factor X to factor Xa, and so is often referred to as the extrinsic tenase complex. It also generates factor IXa, and most (if not all) of these processes occur at, or close to, the endothelial surface. This pathway is part-regulated by TFPI, which suppresses factor Xa.

Amplification   Factor Xa generated in the initiation phase in turn generates a small amount of thrombin from prothrombin. This thrombin then acts on factors V and VIII (generating Va and VIIIa), and also on platelets, which upon activation expose a large amount of phospholipids. The process has therefore moved from the endothelium to the platelet, which provides the platform for an increase in the coagulation process. Therefore, activation of this cell and the procoagulants of its surface are crucial requirements for the progression of the pathway.

The importance of factor Xa is demonstrated by the fact that is has an enzyme commission number: EC 3.4.21.6.

Propagation   Although the extrinsic tenase complex was important starting the process, it is physically restricted to, or close to, the surface of the endothelium and subject to suppression by TFPI. An alternative, platelet-based intrinsic tenase complex can be assembled on or near to the platelet surface, which is composed of factor IXa, factor VIIIa, phospholipids and phosphoserine, and is also calcium dependent. This complex is 50 times more efficient than that of the extrinsic complex, generating over 90% of factor Xa.

Table 7.4 Coagulation factors.

Factor	Function	Half-life (hours)
Fibrinogen (factor I)	Source of fibrin (Ia)	72–120
Prothrombin (factor II)*	Source of thrombin (IIa)	60–70
Factor V	Source of factor Va; cofactor for factor Xa	12–15
Factor VII*	Source of factor VIIa; activates factors VII, IX and X	3–6
Factor VIII	Source of factor VIIIa; cofactor for factor IXa	8–12
Factor IX*	Source of factor IXa; activates factor X	18–24
Factor X*	Source of factor Xa; activates prothrombin	30–40
Factor XI	Source of factor XIa; activates factor IX	45–52
Factor XII	Source of factor XIIa; activates factor XI and prekallikrein	24–30
Factor XIII	Source of factor XIIIa; stabilizes the fibrin clot	200

a: activated. Missing factors (III, IV, VI) were subsequently found to be calcium, errors of purification and phospholipids.
*Vitamin K dependent.

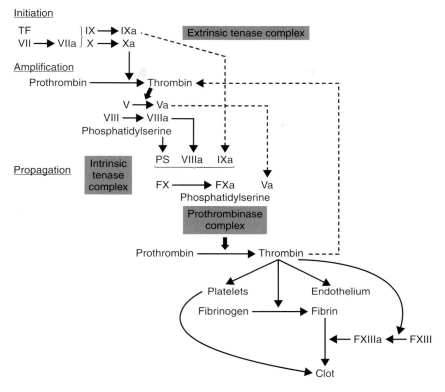

Initiation

Amplification

Propagation

Figure 7.5 The coagulation pathway consists of three overlapping stages, initiation, amplification and propagation, and involves two complex super-enzymes, the tenase complex and the prothrombinase complex. These happen in close proximity to the endothelium and the platelet.

The final stage is the construction of the prothrombinase complex from factor Xa, factor Va, phospholipids and calcium. This super-enzyme generates industrial quantities of thrombin, which has several functions, including fibrin formation from fibrinogen (Table 7.5). Indeed, the importance of thrombin is demonstrated by the fact that it has its own dedicated inhibitor (i.e. antithrombin). Thrombin can also contribute to a positive feedback in the amplification stage.

The importance of fibrinogen, a large molecule with mass of 340 kDa, is demonstrated by its high plasma concentration (around 3 g/L), and the fact that more can

Table 7.5 Selected functions of thrombin.

Function	Implications
Proteolysis of fibrinogen	Generation of fibrin, and so formation of the fibrin clot
Activation of factors V, VII, VIII, XI and XIII	Promotion of additional coagulation pathway activity (Va, VIIa, XIa) Stability of the fibrin clot (factor XIIIa)
Activation of platelets (thrombin receptor)	Increased expression of phospholipids and phosphatidylserine, thus promoting tenase and prothrombinase Mobilization of intracellular granules
Activation of protein C	Suppression of the coagulation pathway
Activation of the thrombin-activatable fibrinolysis inhibitor (TAFI)	Inhibits fibrinolysis, thereby effectively promoting thrombosis

be produced rapidly as part of the acute-phase response. Cleavage by thrombin generates fibrin, but also small remainder molecules, fibrinopeptides A and B. Single molecules of fibrin polymerize to large, strand-like molecules.

The final part of thrombogenesis is activation of factor XIII by thrombin, which, as factor XIIIa, stabilizes the clot, making it mechanically stronger by cross-linking stands of fibrin. As the clot becomes more mature, internal forces bring about retraction, which helps platelet-rich thrombi to withstand the shear forces of the circulation, and may also help with wound repair.

Blood science angle: Anticoagulants

Many coagulation factors are produced by the liver, and so levels may be low in hepatic disease. However, this organ also produces anticoagulants such as antithrombin, proteins C and S, TFPI and plasminogen. Thus, haematologists may need to call on their biochemist colleagues to provide liver function tests, especially in the face of signs such as jaundice (Chapter 17), because, if present, the doses of therapeutic anticoagulants may need to be changed.

Inhibitors

These important molecules ensure the coagulation pathway does not develop too rapidly or too extensively: they can be explained in terms of the brakes on a motor vehicle without which it would career out of control and crash. The key inhibitors are antithrombin, TFPI, and the protein C system.

Antithrombin The primary regulator of coagulation is antithrombin, which in addition to inhibiting thrombin also suppresses factors IXa, Xa and XIa. It also inhibits the factor VIIa that is part of the factor VIIa/tissue factor complex of the initiation phase. However, in itself, antithrombin is a relatively weak inhibitor: 90% of its bioactivity accounted for by its binding to heparan sulphate on the surface of the endothelium. The binding of a key pentasaccharide in the heparan molecule to one part of the antithrombin molecules acts to effect a conformational change on the entire molecule that results in more avid binding of thrombin at a second site. This fact is exploited in the provision of extra purified heparin as a therapeutic anticoagulant, as discussed in Chapter 8.

Tissue factor pathway inhibitor We have already discussed the role of this membrane-bound single-chain molecule in regulating the initiation phase of the coagulation cascade via an action on factor VIIa/tissue factor. It can also act on factor Xa and thrombin. TFPI can be detected in the plasma, but it is not known whether or not this is biologically active.

Protein C system This is so called because several separate factors are required:

• The endothelial membrane component thrombomodulin binds thrombin, and in this form converts inactive protein C to an active molecule (hence activated protein C). It follows that the failure of the endothelium to produce thrombomodulin, or the removal of this molecule by proteases, is essentially prothrombotic.
• This activation is enhanced by the binding of protein C to the endothelial protein C receptor, and is stabilized by the presence of protein S. Once again, loss of this receptor from the endothelial surface is procoagulant.
• Activated protein C inhibits factors Va and VIIa, and, to a much lesser extent, thrombin.

Interestingly, therefore, thrombin is required to activate a molecule that counters its whole purpose. Both protein C and protein S are vitamin K dependent.

Other inhibitors These include heparin cofactor II (which also inhibits thrombin) and protein-Z-dependent protease inhibitor (which inhibits factors Xa and XIa). Alpha-2-macroglobulin and alpha-1-antitrypsin (which has structural similarities to TFPI) are both nonspecific enzyme inhibitors, and have had 'backup' roles should the other mechanisms be insufficient.

7.4 Haemostasis as the balance between thrombus formation and removal

The previous sections have explained the three aspects of Virchow's triad and their interrelationships. The coagulation system by itself generates a fibrin clot, but this is weakly effective in preventing blood loss. For a fully effective thrombus, platelets must be present, so that the factors that activate this cell are crucial.

Activation of platelets

The platelet can be activated in a number of ways, such as the docking of agonists (ADP, fibrinogen, thromboxane, thrombin, etc.) with their receptors (Table 7.3). These agonists may be provided by other platelets, the coagulation cascade or damage to the blood vessel wall. These result in a number of changes to the physiology of the platelet. These are:

• Shape change, from discoid to spherical, and then the extrusion of pseudopodia, thus increasing surface area. This also enables spreading onto the subendothelium, and the formation of lamellipodia (sheets of cytoplasm between the pseudopodia).
• Changes to the nature of the membrane, with increased expression of phospholipids and phosphatidylserine, thus promoting the activation of the coagulation pathway.
• The mobilization of intracellular granules, resulting in the extrusion of a large number of biologically active mediators (calcium, ADP, factor V, etc.).
• Degranulation of alpha granules also results in the appearance of P-selectin on the cell surface, which, it is assumed, promotes adhesion.
• The appearance of other adhesion molecules, promoting platelet to platelet adhesion/aggregation, and platelet to sub-endothelial adhesion. Native GpIIb/IIIa has a low affinity for its ligand fibrinogen. However, activated GpIIb/IIIa has a far higher affinity, and strongly promotes platelet–platelet interactions. Indeed, this is a target for therapy, as will be discussed in Chapter 8.

Several of these actions promote not only the progress of the coagulation cascade (such as factor V), but can also activate other platelets (such as ADP), so creating an accelerative cycle (Figure 7.6).

Thrombus formation

With the generation of fibrin from the coagulation cascade and the activation of platelets, the two come together to form a thrombus. Once formed, additional activation of the cascade and platelets is promoted to build and consolidate the thrombus. It may pass into the circulation or remain close to the original site of blood vessel damage, which is likely to be nearby the endothelium, which may itself be damaged or activated. Note also that the platelet itself provides a platform for the coagulation cascade with expression of phosphatidylserine and other phospholipids.

An activated endothelium (perhaps stimulated by thrombin) can promote thrombus formation in a number of ways by providing factor V, tissue factor, thromboxane and platelet activating factor, but also by downregulating thrombomodulin.

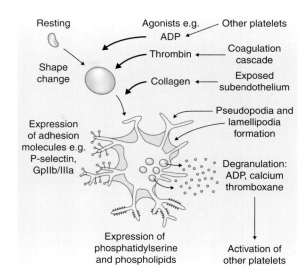

Figure 7.6 Platelet activation. A resting platelet undergoes shape change upon stimulation by agonists such as ADP. Further changes include the degranulation, the development of pseudopodia, and the expression of adhesion molecules and phospholipids.

A further aspect is the adhesion of platelets to the subendothelium, which may be exposed by the loss of damaged endothelial cells. Indeed, many of these features are synergistic: the increase in platelet surface area, the pseudopodia and increased expression of adhesion molecules all promote adhesion. This is mediated by adhesion molecules such as vWf, which can link to both connective tissue proteins (such as collagen) and the platelet with the GpIb–IX–V complex, and is often described as tethering.

Fibrinolysis

Once a thrombus has performed its function and reduced blood loss, and the damage repaired, it must be removed. This is performed largely by the enzyme plasmin in the process of fibrinolysis, although many also use the expression thrombolysis. Plasmin itself is derived from the zymogen plasminogen by the action of tissue plasminogen activator (tPA), the latter an endothelial product. Plasmin has several inhibitors: TAFI, alpha-2-antiplasmin and the nonspecific alpha-2-macroglobulin.

tPA also has a regulator, plasminogen activator inhibitor, of which there are several types (hence PAI-1), and which can be released from platelets, endothelial cells and

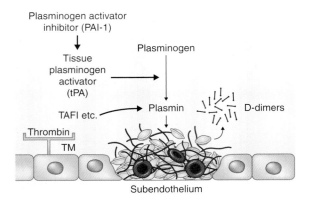

Figure 7.7 Fibrinolysis. Plasminogen is converted into plasmin by tissue plasminogen activator, itself regulated by an inhibitor. Plasmin, which can be regulated by TAFI, acts on fibrin, generating fragments called d-dimers.

other cells. Since tPA and PAI-1 are believed to react in a stoichiometry of 1:1, the balance between the two is crucial for the process of fibrinolysis. Other plasminogen activators include those produced by certain bacteria, snakes and vampire bats.

The action of plasmin on fibrin (and fibrinogen) is to generate quite specific protein fragments that are easily identified in the plasma. These fragments are named 'X', 'Y', 'D' and E', but two form a dimer, hence d-dimers. High levels of d-dimers are considered proof of active fibrinolysis, and therefore of a high general burden of thrombus within the body (Figure 7.7).

The dynamics of haemostasis

An old view of haemostasis considered it to be a 'stop–start' start model, where various factors would initiate the process, which would proceed along a defined pathway and eventually stop with the formation of a clot. This view has been superseded by the dynamic hypothesis.

In this model, the coagulation system is permanently active, but at a low level, and is held in check by inhibitors. The platelet pool is at rest and very few are activated. Upon stimulation, coagulation activity increases, escapes from inhibitor regulation and thrombus formation follows. However, the inhibitors soon catch up and eventually slow down and prevent the process from expanding too rapidly. Quite possibly, in parallel, increased numbers of platelets become

activated (perhaps by thrombin), degranulate and so promote the coagulation pathway, and thus thrombosis. The platelet shape change favours adhesion and aggregation, and binding to fibrin results in thrombus formation.

The same model predicts that there are always background levels of active factors generating a small amount of clot, but that this is degraded by low levels of plasmin generated from the fibrinolysis pathway. Indeed, the small but measurable amounts of plasma d-dimers, reflecting active fibrinolysis, that are present in health support this hypothesis. Therefore, it may well be that it is physiologically normal to have a small degree of thrombus at certain parts of the circulation, and that clinically evident thrombosis is the result of the failure of the body to control this clot or clots.

Thus, the balance between a semi-active coagulation system, regulation by inhibitors, and fibrinolysis is crucial in thrombosis (Figure 7.8). The attraction and strength of this model is that it explains many of the causes of clinical thrombosis, predicts outcomes and provides opportunities to intervene. Some of these we will examine in the next section, the management of which we will leave for Chapter 8.

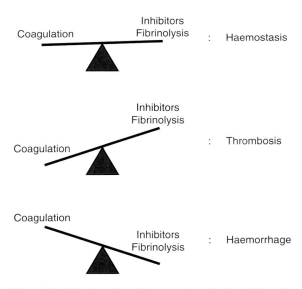

Figure 7.8 The dynamics of haemostasis activation. When the forces of coagulation are balanced by those of inhibitors such as antithrombin, and of fibrinolysis, then haemostasis is in balance. However, if the coagulation is dominant then thrombosis is likely. Conversely, if inhibitors and fibrinolysis are stronger, then there is haemorrhage.

7.5   The haemostasis laboratory

Platelets

The platelet count is provided as part of the full blood count, inevitably a vacutainer anticoagulated with ethylenediaminetetraacetic acid. Reference ranges are in the region of $140–400 \times 10^9$ cells/mL. Levels above this range are thrombocytosis and below the range are thrombocytopenia, with the latter being a risk factor for haemorrhage.

Also potentially useful is the size of the platelet, provided as the mean platelet volume (MPV). In the red cell maturation process, functioning red cells are smaller than reticulocytes, as defined by the mean cell volume. The same is true of platelets, where immature platelet reticulocytes are larger than mature cells. The clinical aspect of this is that large platelets are a risk factor for cardiovascular disease such as stroke. A problem is that, as for red cells, platelets become larger in the presence of anticoagulants, so that blood must be analysed as soon as possible.

However, simply counting these cells or knowing their size does not tell us of their function. For this, other tests are required.

Multiplying the platelet count by the MPV gives the 'plateletcrit', this being the platelet version of the haematocrit, although the clinical significance of this is unknown. A further index is platelet size deviation width (PDW), this being the platelet version of the red cell distribution width. An increased PDW implies an increased number of large platelets; although at present the clinical significance is unclear, it may become useful in the future.

Platelet aggregation   This is probably the gold standard for assessing platelet function in their ability to form self-aggregates. It can be performed in solution of purified platelets (platelet-rich plasma, PRP) or in whole blood. The production of PRP requires a gentle centrifugation of venous blood anticoagulated with sodium citrate, which is then placed in a small glass tube and placed in an aggregometer. The cloudy PRP will occlude the passage of a beam of light, but when an agonist such as ADP or arachadonic acid is added, the platelets are activated and form micro- and then macro-thrombi. As this proceeds, the cloud of PRP becomes clear and the passage of light increases. Thus, the rate of change in light transmission gives us a functional view of the ability of platelets to respond (or not) to the agonists.

The method is very powerful and intrinsically attractive: the platelets clearly form thrombi in real time in the aggregometer, and it is assumed that the same process occurs in the body. However, a problem is that it requires platelets to be purified out of their natural state (in blood) and so may produce a functional artefact. It also denies any role for red cells or white cells, which could be important.

An alternative is the aggregation of whole blood. This cannot be done using changes in light transmittance, so uses another method, such as the inhibition of movement of small 'paddles' by the formation of the clot in a small plastic cuvette. Both methods have good performance characteristics (with low coefficients of variation) and are linked to small computers.

Fluorescence flow cytometry   This powerful technique is generally used to investigate leukocyte subsets (Chapters 5, 6 and 9), but fluorescence activated cell scanning (FACS) can also analyse platelets with monoclonal antibodies linked to fluorochromes. This can be performed in whole blood or in PRP, but the former is preferred. The method can quantify the number of cells bearing different surface markers, and this is useful clinically as certain clinical conditions are linked with the appearance or the nonappearance of certain cell membrane markers.

Platelet microparticles   Fine tuning of FACS allows the detection of small bodies, much smaller than even the smallest platelets. Use of platelet marker antibodies can define which of these are of platelet origin, and hence platelet microparticles. Although their measurement has yet to enter the world of the routine laboratory, they may become so as there is evidence that they are procoagulant.

Coagulation factors

Many of the tests for assessing haemostasis were developed during the last century, when our antecedent scientists were trying to understand the coagulation cascade; to do so needed appropriate tools. The most common, the prothrombin time and the partial thromboplastin time (the latter sometimes preceded by 'activated') are the current version of these early tests. Blood for these tests needs to be taken into a vacutainer with sodium citrate. In some cases, the method can be performed on whole blood; in others platelet-free plasma must be prepared by centrifugation.

Prothrombin time The key to this test, developed by Quick and Owren, is thromboplastin, an artificial platelet substitute that provides many of the essentials for the generation of a fibrin clot, such as phospholipid. However, it does not provide fibrinogen, prothrombin or factors V, VII and X, and these must be present in the patient's sample. In the laboratory, the patient's sample is mixed with thromboplastin and then calcium, and the time for a fibrin clot to form is recorded (hence prothrombin time); this is generally about 11–14 s.

However, if any of the required factors in the patient's sample are deficient, the mixture will take longer to form a clot, hence a prolonged prothrombin time. Thus, a prothrombin time (perhaps 18 s) may be due to deficiency of, for example, factor V. Despite its value in dissecting errors in the coagulation pathway, the test is most frequently used to assess the efficacy of warfarin in reducing vitamin-K-dependent coagulation proteins, and so a reduction in the risk of thrombosis. This is explained in more detail in Chapter 8.

Activated partial thromboplastin time Sometimes simply described as the partial thromboplastin time, this test is similar in practice to the prothrombin time: plasma is added to a platelet substitute. Another similarity is that today's APTT is the current version of a test that can be traced back in history to the kaolin cephalin clotting time, which, as the name implies, used a suspension of brain tissue mixed with a semi-solid clay/chalk-like material (kaolin). The APTT requires the presence of prothrombin, thrombin and factors V, VIII, IX, X, XI and XII, but not the presence of factors VII or XIII.

It follows that the normal APTT (perhaps 24–34 s) will be prolonged if any of the required factors are deficient or missing. Perhaps the best examples of this are haemophilia, with its absence of factor VIII, and severe von Willebrand's disease, where lack of vWf leads to reduced factor VIII activity. But a major problem with this test is that antiphospholipid antibodies, perhaps as part of the lupus anticoagulant may interfere with the APTT, although mixing tests can be used to determine the presence of these factors. If present, inhibitors should be nullified by the partial presence of factors from some healthy plasma; 50:50 mixtures are common.

However, like the prothrombin time and warfarin, the greatest use of the APTT is to monitor the effect of unfractionated heparin in its use as an anticoagulant. This will also be explained in more detail in Chapter 8.

Thrombin time This third clot formation time relies on adding an excess of thrombin to a sample of patient's plasma. The thrombin will act on its substrate, fibrinogen, to generate fibrin, and thus a clot. Therefore, as the only requirement is fibrinogen, the thrombin time is effectively a test of the presence of functional fibrinogen, and is prolonged if hypofibrinogenaemia is present.

Individual factor assays There are several circumstances where knowledge of individual factors are needed. Examples of this include factor V, factor VII, factor VIII, factor IX and factor X. In some cases these are justified to help pin down a diagnosis (factor VIII for haemophilia), and in others to ensure adequate anticoagulation (factor X for low molecular weight heparin). Others include:

- *Fibrinogen.* This molecule can be measured by a functional assay (generally, the Clauss technique) or by immunoassay. A disadvantage of the latter is that, as a method that looks at the structure of the molecule, it provides information of how much is present, not whether or not it has any functional activity.
- *vWf.* This large molecule can be measured globally by immunoassay (generally enzyme-linked immunosorbent assay or latex immunoassay), but the fine structure of the proportion of monomers and polymers can be determined by electrophoresis. A functional assay is to determine the ability of vWf to bind to collagen, another is the ristocetin-cofactor assay (ristocetin is an antibiotic, no longer used). Another curious property is the ability of ristocetin to promote the aggregation of platelets in the presence of vWf and GpIb.

Functional plasma haemostasis A clear disadvantage of partial or single factor assays is that they fail to provide a full picture of the entire coagulation pathway. One way around this is to determine the ability of an aliquot of plasma to generate thrombin; another is the rate of formation of a physical clot (much like the prothrombin time and APTT), which can be measured by a device called a thrombelastograph. This machine can also quantify the rate of fibrinolysis.

7.6 The pathology of thrombosis

The dominant cause of thrombosis (and its reverse, haemorrhage) is the loss of haemostasis.

Loss of haemostasis

Figure 7.8 illustrates the importance of the balance between coagulation on the one side and the dual processes of inhibition and fibrinolysis on the other side. Factors that favour haemorrhage include failure to form a clot, perhaps by lack of platelets or coagulation molecules, and/or excessive fibrinolysis. Conversely, factors promoting thrombosis include excessive stimulation of the coagulation system, the excessive activation of platelets, failure of the inhibitors and a reduction in fibrinolysis. From a clinical perspective, these changes become evident with thrombosis in arteries and in veins.

Arterial thrombosis

The risk factor hypothesis Few of us are unaware of the four major risk factors for atherosclerosis: diabetes, smoking, hypertension and dyslipidaemia. The latter encompasses increased low-density lipoprotein cholesterol (which may become oxidized, and so toxic) and reduced high-density lipoprotein cholesterol. Each of these risk factors, in the absence of clear atherosclerosis, promotes thrombosis. For example, smoking and hypertension increase plasma fibrinogen, and diabetes is linked with increased platelet activation. This can lead to the formation of small and medium-sized microthrombi, which circulate freely in those with risk factors, especially those with multiple factors.

However, in fulfilment of Virchow's triad, a second aspect of cardiovascular disease is endothelial damage, as is promoted by the risk factors. This leads to the deposition of lipids and other material at several key anatomical points, such as the bifurcation of arteries. These deposits accumulate and lead to atherosclerotic plaque, which can 'grow' into the arterial lumen, causing stenosis and disrupting the normal healthy laminar blood flow. There is also evidence that the endothelium overlying plaque is disrupted (exposing the subendothelium), losing its anticoagulant nature, and so is the target for the attachment of platelets, enabling the formation of thrombus.

Pathophysiology The exact reasons why coronary arteries, the carotid arteries, the cerebral circulation, and the aorta, iliac and femoral arteries in particular are singled out by atherosclerosis is unclear. For example, the brachial arteries are of similar structure and carry oxygenated blood at high pressure, but are generally spared.

Stenosis of these arteries restricts blood flow, and so the delivery of oxygen, leading to ischemia and hypoxia of downstream tissues, especially at times of high oxygen demands. In such conditions, the tissues are forced to respire anaerobically, leading to the build-up of lactic acid and the symptoms of pain. This is the likely cause of angina and intermittent claudication. The brain feels no pain, but the consequences of hypoxia/ischaemia are likely to cause transient ischaemic attacks (TIAs). However, occlusion of these arteries by circulating thrombi, or perhaps those resulting from the rupture of the atherosclerotic plaque, causes prolonged hypoxia and, ultimately, infarction.

Two other disease processes are worthy of mention: atrial fibrillation and heart failure. The former is the most common cardiac arrhythmia, the irregular heart beat and atrial function being the likely cause of the increased rate of TIA and stroke. These may be caused by turbulence-driven thrombosis, or the embolization of mural intra-cardiac thrombosis, both of which may run up the carotid arteries to occlude the cerebral circulation.

Heart failure is defined by the inability of the heart to deliver to the body sufficient (oxygenated) blood for its needs, which is generally the result of left ventricular failure and a reduced ejection fraction. The aetiology of heart failure may be hypertension, cardiomyopathy or myocardial infarction, but the consequences include atherothrombosis. Both of these conditions require anticoagulation, to be discussed in Chapter 8.

Venous thrombosis

Risk factors Although the four risk factors for arterial thrombosis are also risk factors for venous thrombosis, their effect is not as strong as in arterial disease. The more powerful risk factors for venous thrombosis are obesity, high levels of female sex hormones (as occur in pregnancy (before and after), the oral contraceptive pill and hormone replacement therapy), prolonged surgery (especially orthopaedic), cancer (especially with cytotoxic chemotherapy) and thrombophilia. However, not all risk factors are of equal effect; some are more likely to cause thrombosis than others.

Table 7.6 Pathophysiology of arterial and venous thrombosis.

	Arteries	Veins
Vessels most often affected	Coronary, aorta, carotid, cerebrovascular, iliac, femoral	Saphenous, femoral and iliac
Organs most often affected	Heart, brain, legs	Lungs, legs
Major risk factors	Diabetes, hypertension, dyslipidaemia, smoking	Cancer, thrombophilia, obesity, male sex, hormone replacement therapy
Dominant pathological cause	Platelets	Coagulation factors
Primary therapeutic option	Anti-platelet (e.g. aspirin, clopidogrel)	Anticoagulant (e.g. warfarin, heparin, dabigatran, rixaroxaban)

Pathophysiology The primary manifestation of venous thromboembolism (VTE) is deep vein thrombosis (DVT), generally of the saphenous and femoral veins. The typical DVT is anchored to the venous intima (perhaps permitted by a damaged endothelium), often close to a valve. The fact that varicose veins are a risk factor for DVT supports the view that venostasis and/or increased blood turbulence (in direct support of Virchow's hypothesis) are precipitating factors.

Many DVTs grow by elongation, spiralling up the vein. However, the base of the clot becomes more well-formed and consolidated, but the tip of the thrombus, with fresh clot (not unlike scrambled eggs) may well break off and is driven up the venous circulation. The process of clots breaking off is embolization; the clots themselves are emboli.

Such an embolus will be driven up the inferior vena cava, around the right side of the heart, and eventually become lodged in the capillary bed of the pulmonary circulation, where it becomes a pulmonary embolus (PE). If large, or if added to by local growth or additional emboli, PEs can cause serious haemodynamic complications, such as pulmonary hypertension and right heart disease, and death.

Contrasting arterial and venous thrombosis

Broadly speaking, there are two major types of thrombus: the white thrombus and the red thrombus. The former is mainly composed of platelets in a mesh of insoluble coagulation proteins such as fibrin, whilst the latter is simply white thrombus plus erythrocytes. However, red clot is the dominant form in venous thrombosis, where white clot is found in arterial thrombosis. Naturally, it is inevitable that there will be some 'pink' clot; that is, white clot with a very small number of red cells. This colour coding is purely for our convenience. As it happens, white clot is particularly amenable to anti-platelet drugs, whilst red clot is best treated with drugs that target the coagulation cascade.

Key factors in the pathophysiology of arterial and venous thrombus are summarized in Table 7.6. These issues of anti-platelet drugs for arterial thrombosis (such as heart attack and stroke) and anticoagulant drugs for venous thrombosis (such as DVT) we will address in detail in Chapter 8.

Blood science angle: Inflammation and haemostasis

Quite often, coagulation and thrombosis are needed at times of trauma (such as an encounter with a large carnivore) in order to stem blood loss, and in which case the immune system may also be called upon. Therefore, it is not surprising that there are close links between the two, and inflammation drives many aspects of haemostasis.

An example of this is the increase in fibrinogen as a part of the acute-phase response, and is a link with the complement system, as is discussed in Chapter 9. Furthermore, bacteria can bind to, and so activate, platelets. This may be enhanced by antibodies, as platelets have receptors for the fraction crystallizable section of the immunoglobulin molecule.

Summary

- Haemostasis can be summarized in terms of Virchow's triad, consisting of the endothelium, platelets and coagulation factors.
- Platelets arise from megakaryocytes by the process of thrombopoiesis.
- The membrane of the platelet comprises a number of receptors and adhesion molecules; the cytoplasm contains granules of procoagulants.
- The coagulation cascade sees factors come together in three stages (initiation, amplification and propagation) to generate fibrin.
- Fibrin and activated platelets together form thrombus, which can be anchored to the subendothelium or can circulate free in the blood.
- Thrombus is digested by the enzyme plasmin in the process of fibrinolysis, a product of which are d-dimers.
- Assessment of the platelet in the laboratory is by the platelet count, aggregation and analysis by FACS.
- The coagulation cascade is assessed by the prothrombin time, the APTT and the thrombin time. Specific factor analyses include fibrinogen and vWf.
- Haemostasis balances coagulation with fibrinolysis and inhibition. When coagulation dominates, the result is thrombosis. When fibrinolysis and inhibitors dominate, the result is haemorrhage.
- Platelets are important in arterial thrombosis, where the coagulation factors are more important in venous thrombosis. This dichotomy drives clinical management.

Further reading

Cesari M, Pahor M, Incalzi RA. Plasminogen activator inhibitor-1 (PAI-1): a key factor linking fibrinolysis and age-related subclinical and clinical conditions. Cardiovasc Ther. 2010;28:e72–e79.

Cox D, Keerigan SW, Watson SP. Platelets and the innate immune system: mechanism of bacterial-induced platelet activation. J Thromb Haemost. 2011;9: 1097–1107.

Kasthuri RS, Glover SL, Boles J, Mackman N. Tissue factor and tissue factor pathway inhibitor as key regulators of global hemostasis. Semin Thromb Hemost. 2010;36:764–767.

Kaushansky K. Historical review: megakaryopoiesis and thrombopoiesis. Blood. 2008;111:981–986.

Mosesson MW. Fibrinogen and fibrin structure and functions. J Thromb Haemost. 2005;3:1894–1904.

Morrissey JH, Davis-Harrison RL, Tavoosi N. et al. Protein–phospholipid interactions in blood clotting. Thromb Res. 2010;125(Suppl 1):S23–S25.

Pendurthi UR, Rao LV. Factor VIIa interaction with endothelial cells and endothelial cell protein C receptor. Thromb Res. 2010;125(Suppl 1):S19–S22.

Rezaie AR. Regulation of the protein C anticoagulant and antiinflammatory pathways. Curr Med Chem. 2010;17:2059–2069.

Roberts LN, Patel RK, Ayra R. Haemostasis and thrombosis in liver disease. Br J Haematol. 2009;148: 507–521.

Siller-Matula JM, Schwameis M, Blann A. et al. Thrombin as a multi-functional enzyme. Thromb Haemost. 2011;106:1020–1033.

Wei AH, Schoenwaelder SM, Andrews RK, Jackson SP. New insights into the haemostatic function of platelets. Br J Haematol. 2009;147:415–430.

Yano Y, Ohmori T, Hoshide S. et al. Determinants of thrombin generation, fibrinolytic activity, and endothelial dysfunction in patients on dual antiplatelet therapy. Eur Heart J. 2008;29:1729–1738.

Zuern CS, Schwab M, Gawaz M, Geisler T. Platelet pharmacogenomics. J Thromb Haemost. 2010;8: 1147–1158.

8

The Diagnosis and Management of Disorders of Haemostasis

Learning objectives

After studying this chapter, you should be able to:

- appreciate that the consequences of the loss of haemostasis are thrombosis and haemorrhage, and that these may be directly life-threatening;
- discuss the overactivity and/or increased numbers of platelets as a cause of thrombosis;
- recognize that thrombosis may also follow the overactivity of the coagulation pathway, and the factors that lead to this;
- explain how and why both qualitative and quantitative defects in platelets lead to haemorrhage;
- describe how and why abnormalities in the coagulation pathway lead to haemorrhage;
- be aware of the role of the laboratory in detecting diseases of haemostasis and in monitoring anticoagulant treatment;
- explain why disseminated intravascular coagulation (DIC) presents such a challenge to the patient and the practitioner;
- appreciate the value of molecular genetics in haemostasis.

Chapter 7 explained the mechanisms and functions of haemostasis, and of the importance of the balance between coagulation and the dual processes of inhibition and fibrinolysis. Factors that favour thrombosis include excessive stimulation of the coagulation system, the excessive activation of platelets, failure of the inhibitors and a reduction in fibrinolysis. These changes become evident in two areas:

- Thrombosis in arteries, generally due to the overactivity of platelets, is the leading cause of death in the developed world, mostly by myocardial infarction and stroke.

- Venous thrombosis (or perhaps venous thromboembolism, VTE), generally expressing itself as deep vein thrombosis (DVT) and pulmonary embolism (PE), caused by overactivity of the coagulation pathway, is responsible for 500 000 deaths in Europe annually.

However, failure of haemostasis can also be viewed from the opposite perspective, the inability to form a thrombus when one is desired. This leads to excessive bleeding (haemorrhage). Factors that favour haemorrhage include failure of the platelet and/or the coagulation pathway to come together to form a clot. It is rarely due to an excess of inhibitors or fibrinolysis. Haemorrhagic disease can be viewed as of two parts: that which is 'natural' and so is likely to be genetic or perhaps acquired as the result of another pathology, and that which is brought on by our profession. Treatments are aimed at replacing the missing parts of the haemostatic process (if possible) and by ensuring that treatment is safe and effective.

The laboratory is involved in all stages of the diagnosis and management of abnormal haemostasis, including the investigation of the causes and consequences of these conditions. This chapter will address these issues, which are summarized in Table 8.1.

Section 8.5 will look at the most severe coagulopathy, DIC, whilst Section 8.6 summarizes the contribution of molecular genetics to these conditions. We conclude in Section 8.7 with some case studies.

8.1 Thrombosis 1: overactive platelets and thrombocytosis

Pathophysiology

Most causes of arterial and venous thrombosis are qualitative: a small yet subtle increase in platelet activity. An alternative is a change in the number of platelets that may

Blood Science: Principles and Pathology, First Edition. Andrew Blann and Nessar Ahmed.
© 2014 John Wiley & Sons, Ltd. Published 2014 by John Wiley & Sons, Ltd.

Table 8.1 An overview of major aspects of thrombosis and haemorrhage.

	Overactive platelets and thrombocytosis (Section 8.1)	Overactive coagulation (Section 8.2)	Underactive platelets and thrombocytopenia (Section 8.3)	Underactive coagulation (Section 8.4)
Causes	i. Risk factors for atherosclerosis ii. High platelet count	i. Risk factors for VTE ii. Failure of inhibitors (often genetic)	i. Intrinsic defect ii. Overtreatment with anti-platelet drugs iii. Low platelet count	i. Intrinsic defect ii. Overtreatment with anticoagulant drugs
Treatments	i. Avoidance of risk factors ii. Anti-platelets	i. Avoidance of risk factors ii. Anticoagulants	i. No specific treatment ii. Correct treatment	i. Replacement therapy ii. Correct treatment
Laboratory assessments	i. Monitoring risk factors (where possible) ii. Full blood count	i. Monitoring risk factors (if possible) ii. Coagulation factor assays	i. Platelet function studies ii. Full blood count	i. Coagulation factor assays ii. Coagulation factor assays

not necessarily be activated; this is called quantitative, and refers to a high platelet count (thrombocytosis).

Qualitative disease There is no major and recognized intrinsic defect in the platelet that leads to its overactivity. When this does happen it is caused by an external factor acting on an 'innocent' platelet. Section 7.6 introduced the major causes of platelet overactivity, and each of the four risk factors for cardiovascular disease (smoking, dyslipidaemia, diabetes and hypertension) are associated with this overactivity, and is powerful evidence of their link with myocardial infarction and stroke. The platelet count itself is not often increased, but if so, it is rarely very much above the top of the reference range.

Quantitative disease: thrombocytosis Modestly increased numbers of platelets are commonly found in acute inflammation as part of the acute-phase response, and in chronic inflammatory disease such as rheumatoid arthritis. However, more sinister is a platelet count considerably above the top of the reference range.

Essential thrombocythaemia (ET) is often defined by a platelet count $>600 \times 10^9$/L, or a platelet count of $>450 \times 10^9$/L plus other factors, such as bone marrow changes or a reactive thrombocytosis. Notably, ET is closely related to polycythaemia rubra vera (PRV), as discussed in Chapter 4, and this must be excluded. A thrombocytosis is often present in certain leukaemias,

especially of the myeloid lineage, and is therefore a secondary aspect of the malignancy (Figure 8.1).

Our understanding of ET (and PRV) has been revolutionized by the discovery of the *JAK2V617F* mutation. Janus-associated kinase 2 (JAK-2) is an enzyme crucial to

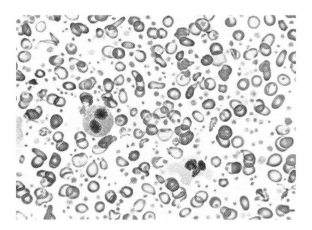

Figure 8.1 The blood film in ET. This blood film shows a remarkably high platelet count, but also an eosinophil (on the left) and a monocyte (to the lower right). There is also a great variation in the size of the platelets (anisocytosis), suggestive of a malignancy, However, before a diagnosis of ET is made, causes of reactive thrombocytosis (e.g. infection, following surgery, post-splenectomy, inflammation, chronic blood loss), chronic myeloid leukaemia, PRV and idiopathic myelofibrosis should be excluded. (Image courtesy of H. Bibawi, NHS Tayside).

intracellular signalling pathways that result in cell proliferation, and may be driven by growth factors such as erythropoietin and thrombopoietin. The *JAK2V617F* mutation effectively leads to unrestricted JAK-2 activity, and so unregulated cell proliferation that translates to increased cell counts in the blood.

JAK2V617F is present in the majority of cases of ET and PRV, and there is a growing consensus that ET or PRV in the absence of *JAK2V617F* mutation is actually a different disease, a situation reminiscent of the relationship between *BCR–ABL* and chronic myeloid leukaemia described in Chapter 6. In many cases, the pathology extends from platelets into white cells and the development of acute myeloid leukaemia, leading to the concept of myeloproliferative leukaemia, or perhaps myeloproliferative neoplasia, to take account of this multilineage aspect.

An alternative cause of thrombocythaemia is mutations in the gene (*Mpl*) for the thrombopoietin receptor (CD110), one of which results in the constitutive activation of the receptor, and so thrombocytosis. Four different mutations in the thrombopoietin gene itself (*THPO*) all cause increased levels of thrombopoietin, which also result in the overproduction of platelets by the megakaryocyte. These conditions, therefore, have a fundamentally different aetiology from the ET and PRV diseases caused by *JAK* mutations, and accordingly are called familial thrombocythaemia. This name also reflects their heritability, which is mostly autosomal dominant.

Assessment

Despite the proven importance of platelet overactivity and its link with arterial thrombosis, there are few simple and effective methods for assessing platelet function. Clearly, the first step is simply the platelet count, but other platelet indices such as the mean platelet volume (MPV) are rarely used in practice, although there are instances of 'giant' platelets, as will be discussed in Section 8.3. Nevertheless, increased MPV (>9.5 fL) brings a 30% increased risk of VTE compared with those with low MPV (<8.5 fL). However, the 'gold standard' platelet function method, aggregometry, is set up to detect deficiencies, not overactivity (Section 8.3).

Thrombocytosis may be secondary to inflammation, in which case a white cell count, erythrocyte sedimentation rate and c-reactive protein may be useful. Molecular genetics are required to confirm or refute the presence of the *JAK2V617F*, *Mpl*, and thrombopoietin gene mutations, and this may contribute to management.

Treatment

In the quantitative defect of ET, bone marrow suppression may be called for, such as with hydroxyurea, anagrelide (which suppresses the release of platelets from megakaryocytes), or interferon-alpha. However, the key to the treatment of the qualitative defect of an overactive platelet is the physiology of the cell itself (see Figure 8.2). This indicates three potential routes by which suppression is possible.

Metabolic inhibitors   There are two major drugs in this class, both are taken orally: aspirin and dipyridamole. Aspirin is actually three drugs for the price of one, with analgesic, antipyrexial and anti-inflammatory activity,

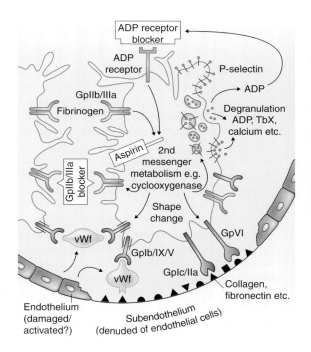

Figure 8.2 Inhibition of platelet function. Platelets may be activated by the occupancy of various receptors (such as the ADP receptor and GpIIb/IIIa) by their particular ligands (ADP and fibrinogen respectively). This leads to the activation of the cyclooxygenase pathway, which in turn initiates platelet shape change, degranulation (releasing ADP, thromboxane and other mediators) and other features that result in the promotion of thrombosis. This includes binding to other platelets (linked by fibrinogen and vWf) and to the subendothelium (via endothelial-derived vWf, collagen, GpVI and GpI//IX/V).

the latter applying to the platelet. Aspirin passively crosses the membrane and irreversibly inhibits cyclo-oxygenase by acetylation of the amino acids adjacent to the active site. This is the rate-limiting step in synthesis of thromboxane A_2 from arachadonic acid and prostaglandin endoperoxides, and occurs in the megakaryocyte, so that all budding platelets are dysfunctional. Figure 7.4 shows the central position of cyclooxygenase in this pathway.

Without a nucleus, platelets are unable to produce more of this enzyme and, therefore, the effect of aspirin is as long as the lifespan of the platelet, generally in the region of 7–10 days. However, leaving aside the established problems of gastrointestinal intolerance, a severe weakness of aspirin is that its specificity for cyclo-oxygenase means it has little effect on the phospholipase-C-dependent or other pathways of platelet activation. Thus, aspirin fails to prevent aggregation induced by thrombin and only partially inhibits that induced by adenosine diphosphate (ADP) and high-dose collagen.

Aspirin 75 mg daily is the primary therapy in the treatment and prevention of secondary cardiovascular disease, although its value in primary prevention is speculative. There is also a place for aspirin in the treatment of ET, although this is simple antithrombotic prophylaxis. Technically, aspirin is a nonsteroidal anti-inflammatory drug, other members of this group include ibuprofen and naproxen. They also have antipyrexial and analgesic activity, and inhibit cyclooxygenase, thus having anti-platelet activity.

Dipyridamole (Persantin) is an inhibitor of phospho-diesterase, which prevents the inactivation of cyclic adenosine monophosphate. Hence, intra-platelet levels of cyclic adenosine monophosphate are increased, resulting in reduced activation of second messengers within the cytoplasm. It may also effect prostacyclin release (which inhibits platelet activation), inhibit thromboxane A_2 formation, and inhibit pro-inflammatory cytokines. These result in reduced platelet aggregability and adhesion *in vitro*. Its effect is relatively short-lasting, and repeated dosing or slow-release preparations are required in order to achieve 24-h inhibition of platelet function.

Adenosine diphosphate receptor blockage ADP is a powerful platelet stimulant, acting via a specific purino-receptor. Three drugs can interfere with this process and so reduce platelet activation. All are used orally.

Clopidogrel (Plavix) is a thienopyridine derivative, metabolized through cytochrome P450 in the liver. It dramatically inhibits platelet aggregation induced by the binding of ADP to its receptor on the platelet surface, an effect independent of cyclo-oxygenase. There is also impairment of the platelet response to thrombin, collagen, fibrinogen and von Willebrand factor (vWf). It is used after a myocardial infarction or the placement of an intra-coronary artery stent for up to a year. Like aspirin, the dose is 75 mg daily. However, clopidogrel is an option as a monotherapy in those intolerant of aspirin.

Prasugrel (Effient), like clopidogrel, is a thienopyridine prodrug, also metabolized via the liver cytochrome P450 system to its active metabolite, which irreversibly inhibits the platelet P2Y$_{12}$ receptor. However, because it is metabolized rapidly, the maintenance dose of 10 mg daily produces an effect on platelets that is more potent and consistent. The government body National Institute for Health and Clinical Excellence (NICE) has released technology appraisal 182 on the value of Prasugrel in acute coronary syndromes.

Ticagrelor (Brilique), a cyclo-pentyl-triazolo-pyrimidine, is a direct and reversible P2Y$_{12}$ antagonist, with a short half-life that requires twice-daily dosing, generally with a 90 mg tablet. Unlike clopidogrel and prasugrel, it is not a prodrug but acts directly and so rapidly, although there may be effects on the lung and heart. It is recommended for up to 12 months, alongside aspirin, after an acute coronary syndrome.

Preventing platelet–platelet interactions In many cases, signal transduction in the platelet occurs when specific receptors on the surface are occupied by ligands such as ADP and serotonin. For example, platelet activation leads to structural modification of the GpIIb/IIIa receptor on the surface of the platelet (as explained in Chapter 7). This is the most common receptor on the platelet surface and represents the final common pathway for platelet aggregation, binding to fibrinogen and vWf and resulting in cross-linking of platelets. Three agents are available, each of which must be given by injection or infusion.

Abciximab (ReoPro) has a long history, being first in its class, not only in GbIIa/IIIb blockage, but also as a therapeutic monoclonal antibody. It is an established agent in the prevention of aggregation in acute coronary settings (alongside heparin and aspirin) and inhibits aggregation by 90% within 2 h of its infusion.

Integrellin (eptifibatide) and tirofiban (Aggrastat) also interact with GpIIb/IIIa, so interfering with platelet–platelet interactions. They are licensed for the prevention of early myocardial infarction in patients with unstable angina, but contra-indications include bleeding, and hepatic and renal impairment.

Table 8.2 Mechanism for suppressing platelet function.

Mechanism	Route of administration	Examples of agents
Inhibition of metabolic pathways	Oral	Aspirin, dipyridamole
GpIIb/IIIa blockade	Injection/ infusion	Abciximab, tirofiban, eptifibatide
ADP-receptor blockade	Oral	Clopidogrel, prasugrel, ticagrelor

Established antithrombotics are summarized in Table 8.2, whilst Figure 8.2 illustrates how our knowledge of platelet physiology has enabled us to inhibit its activity.

Other agents Iloprost is a prostacyclin analogue that exerts its effects by promoting vasodilatation and inhibiting ADP-induced platelet aggregation, thereby opposing the effects of thromboxane A_2, but must be continuously infused. Cilostazol is a phosphodiesterase inhibitor, and so reduces platelet aggregation. Its use is restricted to intermittent claudication, a symptom of atherosclerosis of the arteries of the legs.

A number of agents derived from snake venoms are in development. Bitistatin, an 83 amino acid protein derived from the venom of the viper *Bitis arietans*, inhibits platelet aggregation via GpIIb/IIIa. Another peptide snake venom, trigramon from *Trimeresurus gramineus*, can block interactions between vWf, collagen, laminin and fibrinogen with GpIIb/IIIa.

8.2 Thrombosis 2: overactive coagulation

Virchow's triad applies equally to arterial and venous thrombosis, but certain aspects differ. For example, stasis is far more likely to bring venous disease, whilst antithrombotic treatment is far more effective in arterial disease. Many of the risk factors for each variant are also different, although there is overlap.

Pathophysiology Recall that the efficacy of the coagulation pathway is a balance between increased activity of

coagulation factors and the ability of the inhibitors to suppress this activity. Therefore, thrombosis may follow (a) increased levels of coagulation factors that the normal levels of inhibitors are unable to contain, or (b) reduced levels of inhibitors that allow normal levels of coagulation factor to generate thrombus. There are no specific and major causes of thrombosis that arise purely from failure of thrombolysis, although in principle this may occur. Notably, there is evidence that impaired fibrinolysis may be risk factor for venous thromboembolism. There may, of course, be a degree of all three aspects acting concurrently to promote thrombosis.

Increased coagulation factors The production of coagulation factors by the liver is generally stable, but increased levels follow an acute-phase response. However, certain inhibitors are also increased in this manner, so to some extent they cancel each other out, although the platelet count will also increase in the acute-phase response, so that overall acute and chronic inflammation are risk factors for thrombosis. Coagulation factor levels are sensitive to liver cancer, where fibrinogen levels >10 g/L are not uncommon.

The clinical problem with loss of haemostasis and overactive coagulation is VTE. This is such a well-established phenomenon that a large number of risk factors have been identified, such as surgery, cancer, obesity and immobility. However, not all factors bring the same risk of VTE (Table 8.3).

The greater part of these factors are brought about by existing illness, but more so by interventions undertaken in hospital (surgery, chemotherapy). Accordingly, all patients entering hospital must be assessed for their risk of developing VTE, and prophylactic action taken if warranted (which it almost always is).

Lack of inhibition The major inhibitors of the coagulation pathway are antithrombin and the protein C system.

Antithrombin is a liver-produced serine-protease inhibitor (serpin) with a half-life of approximately 3 days, a mass of 58 kDa and a plasma concentration in the region of 0.12 mg/mL. It also inhibits factors IXa, Xa, XIa, and the factor VIIa/tissue factor complex, and its inhibitory effects are markedly enhanced by heparin. Consequently, antithrombin deficiency is a major risk factor for VTE, and can be acquired or inherited. Examples of the former include nephrotic syndrome (where the molecule is lost through a damaged glomerulus and is not reabsorbed), sepsis and liver disease.

Table 8.3 Stratification of the risk factor for VTE.

Strong risk factors (increased risk >10-fold)	Hip, pelvis or leg fracture, hip or knee replacement, major general surgery, major trauma, spinal cord injury, hospital or nursing home confinement, homozygous factor V Leiden, deficiencies of inhibitors
Moderate risk factors (increased risk 2- to 9-fold)	Arthroscopic knee surgery, central venous lines, malignancy (alone, 2- to 4-fold, but with chemotherapy 4- to 6-fold), heart failure, hormone replacement therapy, use of oral contraceptives, paralytic stroke, post-partum pregnancy, previous VTE, heterozygous factor V Leiden (and most other thrombophilias), varicose veins at age 45
Weak risk factors (increased risk <2-fold)	Bed rest >3 days, immobility due to sitting (e.g. prolonged car or air travel, wheelchair), increasing age, laparoscopic surgery (e.g. cholecystectomy), obesity (body mass index >30), ante-partum pregnancy, varicose veins at age 60

Inherited deficiency in antithrombin (which is auto-somal dominant) is present in between 1/20 000 to 1/50 000 of the population. It can be quantitative (type I), a reduction of the amount of the molecule in the blood, generally to 50% of normal, or qualitative (type II), where the molecule is present but dysfunctional, such as with an abnormal heparin binding site.

The protein C system includes protein C (a 62 kDa two-chain molecule), protein S (a 70 kDa single-chain glycoprotein), thrombomodulin and the endothelial protein C receptor (ePCr). Protein C (like protein S, a vitamin-K-dependent product of the liver) must be activated by thrombin in the presence of thrombomo-dulin and the ePCr. An essential cofactor for this process is protein S. Consequently, loss of any part of this process (including endothelial damage) leads to failure to acti-vate protein C. This leads directly to failure to inhibit its targets, principally factor Va and factor VIIIa, which therefore continue to act and so precipitate a VTE.

The principal cause is deficiency in protein C itself, present in a mild (heterozygous) form in perhaps 1/200 to 1/500 of the population, although it is unclear whether or not the inherited variant is autosomal dominant or recessive. Type I deficiency (where levels are generally

50% of normal) is more frequent than type II disease, and the most common cause is mis-sense mutations levels. Congenital homozygous protein C deficiency is associ-ated with massive thrombosis.

Heterozygous protein S deficiency (levels 50% of normal) also leads to a 50% increase in the risk of VTE. Most cases are caused by mutations in the gene, although perhaps 50% of protein S circulates bound to a complement component and does not participate in haemostasis. Deficiencies of protein C, protein S type 1 and antithrombin each confer a relative risk of VTE of 15–20 times that of normal levels of each particular molecule.

Thrombophilia   This is a collective term for a number of causes of an increased risk of VTE. These include antithrombin, protein S and protein C deficiency, but also several other seemingly unrelated conditions and situations:

• A mutation in the noncoding region of the gene for prothrombin (named *G20210A*) that causes increased levels of this protein and a two- to three-fold increase in the risk of VTE. Both heterozygous and homozygous cases are known.
• Factor V Leiden (FVL, also known as F5 R506Q). Factor Va is an essential cofactor for factor Xa in the prothrombinase complex that generates thrombin. Fac-tor Va is inhibited by activated protein C, which cleaves a key sequence of the factor V molecule, thus inactivating it. However, a point mutation in the gene for factor V presents a different amino acid sequence to activated protein C which the latter fails to recognize, and so does not inactivate, leading to persistence of factor Va (i.e. FVL), and so an increased risk of VTE.

The gene for FVL is present in 3–5% of the indigenous population of north-west Europe (and its colonies around the world: the USA, Canada, South Africa, Australia, etc.), and in its heterozygous form confers a five- to seven-fold risk of VTE. In its homozygous form the risk is an astonishing 80-fold. Other mutations in the factor V gene that cause activated protein C resistance are also named after their place of discovery, such as FV Cambridge.
• The antiphospholipid syndrome (APS) is an auto-immune disorder characterized by antiphospholipid antibodies (APAs) that target molecules such as beta-2 glycoprotein-1, prothrombin or protein–phospholipid complexes. The consequences of this include arterial and venous thrombosis, and are often found in disorders

of pregnancy (pre-eclampsia and placental insufficiency) and systemic lupus erythematosus (SLE). Indeed, the latter lends its name to the term 'lupus anticoagulants' (LAs – as it was first described in two patients with SLE) to be described in more detail below.

• Although APAs interact with platelets, the antibodies may also operate via the protein C system, by the activation of coagulation factors, and by adversely influencing the endothelium. However, in defining APAs, the antibodies must persist over a period of months, as short-lived antibodies, perhaps to a microbial pathogen (such as human immunodeficiency virus or syphilis), can mimic a true APA.

• Factor VIII. The inclusion of increased levels of this molecule as a risk factor is justified as there is a familial link with VTE. A possible mechanism is that the normal protein C system is unable to inhibit all the factor VIIIa, the remainder of which escapes to promote VTE. High levels confer a risk of venous thrombus of three to five times that of normal levels. However, for those in the highest 10% the risk of VTE increases almost sevenfold.

Other laboratory markers that may contribute to thrombophilia include raised levels of the amino acid homocysteine (which carries a 1.5- to 2.5-fold increased risk), non-O blood group (1.5- to 1.8-fold) and high levels of other clotting factors, notably factor IX (two- to three-fold risk) and factor XI (1.5- to 2.5-fold risk). The increased risk due to activated protein C resistance (three- to five-fold) may in part be due to FVL.

In common with the risk of arterial thrombosis, the risk of VTE increases in the presence of multiple risk factors, as is illustrated by FVL and obesity (Table 8.4). The risk of DVT after pregnancy rises with the age of the mother, and is also linked to polymorphisms in genes for factor V, factor VII and P-selectin. The combination of obesity and use of the oral contraceptive pill brings a risk

almost 24 times that of women free of both factors. However, quite possibly the greatest is the combination of air travel over 4 h with high-risk surgery in the previous 4 weeks, which carries an increased risk of VTE of 141 times that of no surgery and no air travel.

Blood science angle: Risk factors for VTE

Whilst it is abundantly clear that thrombosis sits fully within haematology, inputs from biochemistry and immunology are often required. Additional risk factors for VTE include chronic kidney disease (1.3- to 1.7-fold), nephrotic syndrome (3- to 10-fold), renal transplantation (3- to 8-fold) and hyperthyroidism (1.5- to 3-fold), calling on the need for urea and electrolytes (U&Es), thyroid stimulating hormone and thyroxine measurements. Indeed, an unexplained VTE may be the first sign of renal failure. Similarly, thrombosis may be an early sign of SLE, which carries a three- to eight-fold increased risk, whilst inflammatory bowel disease and high c-reactive protein carry an increased risk of three- to four-fold and 1.2- to 1.8-fold respectively.

It follows, therefore, that an individual's risk of VTE is the sum of all their individual single risk factors, as illustrated in Figure 8.3.

Race Genetics seems likely to account for up to 60% of the risk of VTE. One meta-analysis reported that FVL, the prothrombin G20210A and obesity may account for an increased risk in European populations, whereas high factor VIII, high vWf, low protein C and the increased prevalence of obesity may explain some of the increased risk in African-Americans. The rate of VTE in African-Americans is some fivefold that of Asian populations in the Far East (China, Japan, Hong Kong), with Europeans carrying an intermediate risk.

Assessment

The initial approach to the assessment of the risk of thrombosis is to simply measure levels of the coagulation factors. The prothrombin time (PT) and activated partial thromboplastin time (APTT) are first-line investigations, but are of limited value. It is established that a shortened APTT ratio is a risk factor for recurrent VTE, but as the test is a compound of several individual coagulation factors, the information is of empirical use only as, for example, it may be due to increased factor VIII and/or factor IX.

Table 8.4 Increase in the risk of VTE with multiple risk factors.

	Normal factor V	FVL
Normal body mass index	1	2.6
Overweight	1.3	3.6
Obese	2.3	5.3

Data are relative increase in risk of VTE compared with normal body mass index and wild-type factor V. At each step, risk increases (Adapted from Severinsen MT. *et al.*, Br J Haematol 2010:149;273–9).

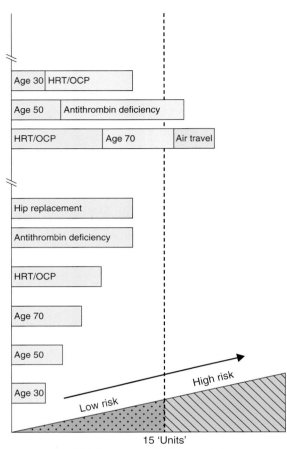

Figure 8.3 The risk of a VTE is the sum of individual factors. Suppose that each factor has a particular numerical number: some factors are acquired, others natural. In this model, a score exceeding 15 arbitrary units puts the individual at high risk of a DVT or PE. This may be achieved by a combination of any factors: age 50 plus antithrombin deficiency (12 points plus 5 points equals 17 points) does, whereas age 30 plus hormone replacement therapy (HRT) or oral contraceptive pill (OCP; 3 points plus 9 points equals 12 points) does not. However, this simple model predicts that adding another 40 years (and so 4 points) at age 70 may well precipitate a thrombosis.

The major considerations for the risk of VTE (Table 8.3) are almost all clinical. The laboratory contributes to this assessment by testing for inhibitor deficiency and the other aspects of thrombophilia.

Inhibitors   Assays of inhibitors focus on the immunological and the functional. The former consists mostly of enzyme-linked immunosorbent assays (ELISAs), but there is also latex-bead-based assay, radial immunodiffusion and immunoelectrophoresis. However, these assays merely provide information about the structural presence of the inhibitor.

Functional assays provide evidence that the particular inhibitor is effective in the laboratory, and so is presumed to be effective *in vivo*. The biological activity of antithrombin is defined by its ability to inhibit the ability of thrombin or factor Xa (in the presence of heparin) to cleave a clear engineered chromogenic substrate into a coloured substrate. The enzyme–substrate reaction is set up and patient's plasma is added, which should result in less colour formation than a parallel reaction vial, to which buffer has been added instead of the test plasma.

Determination of protein C is more complicated because of the requirement for several components (thrombin, thrombomodulin, protein S) to be present in the correct sequence and in the correct concentration. However, one way around this is to use the venom of the southern copperhead snake (*Agkistrodon contortrix*) which directly activates protein C in a similar manner to that of thrombin. The functional activity of the so-generated activated protein C can then be measured in one of two ways (a) using a chromogenic assay as in the thrombin assay described above, or (b) an assay based on ability of the activated protein C to prolong the clotting of plasma, which it should do by virtue of its inhibition of factor Va and factor VIIIa.

As protein S is not an enzyme, or has any enzymatic activity that can be manipulated, the chromogenic approach will not work. A somewhat unwieldy functional assay measures the ability of an engineered protein C reaction, where all the components are present except protein S. Therefore, the only source of protein S to allow the reaction to proceed will be that in the test sample of plasma from the patient. However, this assay is also subject to interference from, for example, high levels of factor VIII in the patient's plasma, and these may give false positives and false negatives.

Factor V Leiden and the prothrombin G20210A mutation   Whilst in theory it is entirely possible for the biological effects of these mutations to be sought in complex functional assays, they are fraught with technical difficulties. Accordingly, it is now perhaps easier and certainly specific to go directly to the particular deoxyribonucleic acid (DNA) sequence in the genome. These molecular genetic techniques can also advise on homozygosity/heterozygosity.

Antiphospholipid syndrome The large number of tests in this section underlines the heterogeneity of the syndrome; therefore, considerable thought is required not only in commissioning but also in interpretation. For example, antibodies may be coincidental, may be the consequence of the disease or may have a direct role in pathophysiology (such as in activating or damaging platelets or endothelial cells).

ELISAs are commonly used to detect antibodies to prothrombin (possibly complexed to phosphatidylserine), beta-2-glycoprotein-1 and to biphosphatidyl glycerol (formerly cardiolipin, hence anti-cardiolipin antibodies). One study found that a very high level of anti-cardiolipin antibodies carried a 5-year risk of VTE of over five times that of very low levels.

LAs (also present in many diseases other than SLE) are a heterogeneous group of antibodies that become manifest in a number of ways, such as binding to platelets or coagulation factors. However, the classical findings for an LA are:

• prolongation of a phospholipid-based clotting test;
• demonstration of the phospholipid dependence of the inhibitor;
• demonstration of the presence of an inhibitor by mixing tests.

Regarding the latter point, some forms of inhibition can be overcome by mixing the patient's plasma with some normal plasma and then re-performing the assay. If the abnormality is due to a factor deficiency, the normal plasma will correct it and generate a normal result. However, if there is an inhibitor in the patient's plasma, it should continue to interfere with the assay. This mixing may be 50/50, but titrations can also be helpful.

Although no one single test can define all LAs, almost all are based on their ability to interfere with, and so prolong, a calcium-dependent coagulation assay, such as:

• APTT – the coagulation pathway is initiated by the activation of factor XII and the phospholipid is diluted;
• the kaolin or silica clotting time – either of these two chemicals will activate factor XII, and so the pathway; no phospholipid is required;
• the dilute PT, where thromboplastin activates factor VII in the presence of phospholipid;
• the activated seven LA – pre-activated recombinant factor VIIa acts directly on factor X;
• dilute Russell's viper venom time (DRVVT), which activates factor X.

The British Committee for Standards in Haematology (BCSH) recommends that the DRVVT and one other test be employed for LA detection. All of these are subject to numerous interfering factors, and false positives may be due to vitamin K antagonists (VKAs) and heparins (which should therefore be stopped for testing), and deficiencies of coagulation factors. However, in warfarin therapy (to be discussed later in this chapter), the venoms of the coastal taipan snake (*Oxyuranus scutellatus*) and Australian eastern brown snake (*Pseudonaja textilis*) both contain prothrombin activators that generate a functional thrombin. A third snake venom, from the saw-scaled viper (*Echis carinatus*), is the basis of the ecarin clotting time, which also directly activates prothrombin. This it does independently of phospholipid and calcium, so is not influenced by an LA, and is also effective in warfarinized plasma. We will meet this test again in sections to follow.

Problems and pitfalls in antiphospholipid syndrome and lupus anticoagulant testing It is clear from the above that testing for these syndromes is highly complex and demands a high level of expertise that may exceed that of most routine laboratories. For example, the plasma sample must be absolutely free from platelets as the phospholipid in these cells will interfere with the assay. Freezing and thawing should be kept to an absolute minimum as this can influence proteins. The reagents themselves can vary markedly between suppliers.

For additional details, the reader is referred to the chapter on thrombophilia by Moore and Jennings (2010), which inspired much of this section.

Testing for thrombophilia A potential drawback to the ease of access to information is the high frequency of requests for thrombophilia screening. In many cases the penetrance of these abnormalities into actual clinical disease (i.e. a clot in the leg) is weak, and therefore of little statistical likelihood of provoking an event.

Subjects with thrombophilia are not usually treated until they have had an episode of thrombosis, or are placed in a high-risk situation (such as needing surgery) because anticoagulants such as heparin and warfarin are associated with a relatively high risk of bleeding. Hence, long-term oral anticoagulant therapy with a VKA is only recommended for subjects with recurrent spontaneous thrombosis, or the presence of additional genetic defects such as antithrombin deficiency or LA. However, these recommendations are likely to change as new anticoagulants become licensed.

The clinical issues in deciding in whom prophylaxis is warranted do not directly involve the laboratory if the diagnosis has been made, but blood scientists must be aware of the relevant issues. The BCSH regularly publishes guidelines on these and others issues, such as one for testing for heritable thrombophilia. However, NICE, in its Clinical Guideline 144, made the following recommendations:

1. Do not offer thrombophilia testing to patients who are continuing anticoagulation treatment.
2. Consider testing for antiphospholipid antibodies in patients who have had unprovoked VTE if it is planned to stop anticoagulation treatment.
3. Consider testing for hereditary thrombophilia in patients who have had unprovoked VTE and who have a first-degree relative who has had VTE if it is planned to stop anticoagulation treatment.
4. Do not offer thrombophilia testing to patients who have had provoked VTE.
5. Do not routinely offer thrombophilia testing to first-degree relatives of people with a history of DVT or PE and thrombophilia.

The guideline recognizes the importance of cancer as a precipitant of thrombosis. It recommends that all patients diagnosed with unprovoked VTE who are not already known to have cancer are investigated for this condition.

Treatment

For decades the treatment and prevention (prophylaxis) of VTE, regardless of cause or precipitant, was dominated by warfarin and heparin, the latter being refined to a low molecular weight form (hence low molecular weight heparin, LMWH). However, a new group of oral anticoagulants was developed over the period 2008–2012 and are set to replace the older agents in the most common conditions.

Warfarin

The drug The long cold winters of the late 1920s forced farmers of the USA–Canada border to feed their cattle on 'rotten' silage that contained sweet clover. Many of the cattle subsequently died of haemorrhage; from this silage, workers at the University of Wisconsin, led by Karl Link, isolated the causative agent, bis-hydroxycoumarin. As the research was funded by the Wisconsin Alumni Research Foundation, the substance was named

warfarin. Successfully used as a rat poison (where the unfortunate animals bled to death), warfarin entered the clinic in the 1950s.

By the late 1970s, the exact biological activity of warfarin had been defined, and was accounted for by its inhibition of the ability of the liver to synthesize prothrombin and coagulation factors VII, IX and X. A crucial part of these molecules is a carboxyglutamic acid residue that binds calcium and facilitates the insertion of the molecule into the platelet membrane. Warfarin exerts its effects by interfering with the particular metabolic enzyme, a microsomal carboxylase, which requires vitamin K as a cofactor, so the enzyme fails to act on its substrate. Thus, warfarin is often described as a VKA. For those unable to tolerate warfarin, other VKAs are available: phenindone, acenocoumarin and phenoprocoumarin.

A direct consequence of the action of warfarin is that functional levels of these coagulation factors fall (although 'structurally' defined levels may be normal), which in turn translates to a slower rate of thrombin generation, then fibrin formation, and so protection from inappropriate thrombosis.

The effect of warfarin is very variable, and must be monitored in each patient using the PT. This is based on a test developed by Armand Quick, and mixes patient's plasma with a platelet substitute – thromboplastin – giving the PT. There are several types of thromboplastin, which themselves have evolved from preparations of human and rabbit brain, although bovine brain is commonly used at present. The sensitivity of the particular thromboplastin must be assessed against a reference sample perhaps provided by a reference laboratory, so generating an 'international sensitivity index', or ISI, which varies between 1 and 2. Calcium must also be added to the plasma–thromboplastin mixture, as this has been sequestered by the sodium citrate anticoagulant.

Good laboratory practice demands high standards to ensure accuracy and reproducibility. Ideally, a laboratory will settle on one particular preparation of thromboplastin, obtain the ISI, and so be able to use the product over a long period of time. The laboratory will tailor its PT to the ISI of the batch of thromboplastin.

Laboratory management of warfarin The problem is that the effect of this subtle poisoning may be mild or marked, so that some patients require a small amount of warfarin (perhaps 1 mg/day), whilst others need a great deal more (a dose of 15 mg a day is not unusual – most patients take 3–4 mg daily). This may be due to genetic factors (see blood science box on p. 189).

In practice this means that warfarin use must be monitored by its effect on the PT when compared with the normal PT (when free of warfarin), known as the international normalized ratio (INR). However, the quality control of the thromboplastin reagent and INR are related by an equation that includes the ISI:

$$INR = \left(\frac{PT \text{ on warfarin}}{PT \text{ of standard}}\right)^{ISI}$$

Ideally, the ISI will have a value of 1. If so, and a normal (standard) PT is perhaps 12 s, and the PT on warfarin is 30 s, then the INR is 2.5. This number is significant because it is the target value for the great majority of patients taking this drug.

Warfarin dosing is managed in a number of settings. For most patients this is in an oral anticoagulant clinic; that is, patients also taking VKAs other than warfarin. Patients have their INR checked on a sample of blood from a finger prick, although a venous sample may be taken.

The 'dosing officer' will then assess the result and may recommend a change in the daily oral dose of warfarin by up or down titration to give an INR generally in the range 2–3. Information and records are generally retained on a laptop computer, which also recommends the next appointment, and details are recorded in a 'yellow book' which the patient retains. However, some patients will be at a higher risk of thrombosis, and so need additional protection, as is offered by an INR in the range of 3–4, although some patients are best protected by a target INR of 2.75 or 3. A second aspect is the duration of treatment, which varies between 3 months or for as long as the risk remains, which may be for life (Table 8.5).

Pros and cons of warfarin No drug is perfect; overall, the advantages must outweigh the disadvantages. Warfarin has undoubted efficacy in reducing the risk of thrombosis in many conditions (Table 8.5), such as the risk of stroke in atrial fibrillation (AF). A further advantage is that it is taken orally, is easily stored and has a long shelf-life. Although the drug itself is very cheap, the costs of management (the staff) are not. However, there are several notable disadvantages; one is the variability between subjects we have already noted, but there are others (Table 8.6).

Table 8.5 Target INRs and recommended duration of anticoagulation.

Indication	INR target	Duration
Pulmonary embolus	2.5	6 months
Distal DVT due to temporary risk factors (such as pregnancy)	2.5	3 months
Proximal DVT or DVT of unknown cause or those associated with ongoing risk factors	2.5	6 months
VTE associated with malignancy	2.5	6 months then review
Recurrence of VTE – whilst *not* on warfarin	2.5	Long term
Recurrence of VTE – whilst *on* warfarin	3.5	Long term
AF	2.5	Long term
AF for cardioversion	2.75	4 weeks pre CV minimum 4 weeks post
Cardiomyopathy	2.5	Long term
Mural thrombus	2.5	3 months
Rheumatic mitral valve disease	2.5	Long term
Mechanical prosthetic heart valves (aortic)	3.0	Long term
Mechanical prosthetic heart valves (mitral)	3.5	Long term
APS (venous)	2.5	Long term
APS (arterial)	3.5	Long term
Thrombophilia	Discuss with haematologist	
Most cases of surgery	2.5	6 months

AF: atrial fibrillation; APS: Anti-phospholipid syndrome; DVT: deep vein thrombosis; PE: pulmonary embolism; VTE: venous thromboembolism.
According to BCSH guidelines the target INRs may vary depending on valve type; if unsure, use generic target for valve location. Patients referred to the anticoagulant clinic will be assigned a target INR as per BCSH guidelines *unless* reasonable evidence is given to the contrary.

Furthermore, a large number of drugs are known to influence the effect of warfarin (and the other VKAs), most by enhancing or suppressing the effect. These medications include allopurinol, many antibiotics, anti-arrhythmics, clopidogrel, several cytotoxics, thyroid hormones, anti-epileptics and St John's wort.

The problems with this list are (a) that it is difficult to predict whether or not a particular drug will influence

Table 8.6 Patient factors that influence the efficacy of warfarin.

Enhanced effect	Reduced effect
Excess alcohol ingestion	Weight gain
Increased age (e.g. >80 years)	Diarrhoea and vomiting
Heart and renal failure[a]	Relative youth (e.g. age <40 years)
Impaired liver function[b]	Non-white European background

[a]Present because warfarin is cleared by the kidney, so that failure to do so leads to higher levels.
[b]A damaged liver may be more sensitive to warfarin and/or produce lower levels of coagulation factors.

the INR in a particular patient, and (b) if this is so, how the INR should be managed. Regarding the latter, a problem with warfarin is its long half-life (35–45 h), so that changing the daily dose will take several days to work its way through to a change in the INR. A practical example of this is the difficulty in managing a short course of a week of antibiotics, which may be over by the time the INR has been adjusted. For a longer course of a new medication, weekly INR monitoring is advised, with adjustment to the daily dose, should it be necessary. This long half-life and the inflexibility it brings mean that it is not favoured in surgery, where a rapid change in anti-coagulant status is called for. This flexibility is provided by heparin.

In addition, there are a number of cautions and contraindications to warfarin, which include recent surgery, dementia, breast-feeding, use of cranberry juice, peptic ulcer, severe hypertension, thrombocytopenia and bacterial endocarditis.

Inhibitors protein C, protein S and protein Z are also vitamin K dependent, and so production of these molecules is itself inhibited by VKAs. This may have an interesting and unfortunate consequence for those starting on this drug. We saw in Chapter 7 (Table 7.4) the differences in the half-lives of the different coagulation factors, one of the shortest of which is that of factor VII, with a half-life of perhaps 3–6 h, whereas prothrombin levels remain high for days after starting therapy. However, the half-life of protein C is also very short, so that levels of this inhibitor may well fall before those of factor VII and other factors, leading to a short period of increased risk of thrombosis. This may become manifest

as warfarin-induced skin necrosis, and is an unfortunate example of a treatment that ironically causes the problem it seeks to prevent.

A greater disadvantage of warfarin is haemorrhage, which is discussed in Section 8.4.

Patient power – warfarin   Advances in the technology of generating an INR have facilitated the development of small machines which allow an INR to be determined away from the hospital, such as in a GP clinic or even in the patient's home. This is called 'near patient testing' (NPT), or 'point of care testing' (POCT). This has proved popular with many patients who find a visit to their hospital outpatient department to have their INR checked inconvenient. Thus, a growing number of patients are now monitoring and treating themselves with their own NPT machine, which, at face value, seems a good idea as patients are no longer required to visit their hospital. However, by definition, any move away from a centre of excellence (i.e. the haemostasis laboratory) must be detrimental, as there may be problems in training, interpretation and monitoring the quality of the INRs produced by NPT. Additional comments on general concepts in NPT/POCT are to be found in Section 2.7.

Heparin
The drug   The modern era of heparin can be traced to 1880 with the description of anticoagulant preparations that evolved into heparin that ultimately became clinically useful. Subsequent research demonstrated the requirement of heparin and another substance in order for anticoagulation to be effective, a factor that we now recognize as antithrombin. Commercial heparin was thus ready for industrial/pharmacological-scale production, and by the 1970s its value in prophylaxis for VTE was becoming established.

Heparin is a natural product, present on the cell membrane; but that which is provided for therapy is a mixture of glycosaminoglycans, polysaccharides that are composed of long chains of repeating disaccharide units (hexosamine and glucuronic or iduronic acid), although the composition of different macromolecules varies markedly. However, heparin must be given intravenously or subcutaneously, with the former providing a more rapid effect. In its whole, unfractionated form (hence unfractionated heparin (UFH)) molecular weights range from 3 to 30 kDa (although most is in the range 12–15 kDa), but perhaps only one-third of a standard heparin preparation has anticoagulant activity.

Blood science angle: Molecular genetics of antithrombotics and anticoagulants

One possible explanation for the wide variation in responses to warfarin is in the difference in the metabolism of this molecule, much of which relies on one of the cytochrome P450 enzymes. Mutations in the genes for certain of these and other enzymes, such as *CYP2C9* and *VKORC1* (the latter being more promising), may be responsible for defining the sensitivity of an individual's response to warfarin. It follows that personal genotyping may be valuable in helping to find the right dose of this drug. Similarly, alleles in *CYP2C19* may influence the metabolism of the inactive prodrug of clopidogrel, and so the effective active level in the blood that actually binds to the platelet ADP receptor and so the degree of protection from thrombosis.

However, the penetrance of these genes (the genotype) into the actual reduction in the risk of thrombosis (the phenotype) is generally poor, meaning that other non-genetic causes of the variation in response are important (typically, the effects of other medications such as antibiotics).

Strictly speaking, heparin, by itself, is not an anticoagulant; it is a cofactor in the activity of antithrombin, a 58 kDa single-chain polypeptide synthesized in the liver. Heparin binding to specific sites on antithrombin induces a conformational change, exposing a site that binds serine proteases factor Xa, thrombin (factor IIa) (in approximately equal proportions) and, with less affinity, factors IXa, XIa, XIIa, kallikrein, plasmin and C_1-esterase, although almost all of its effects are against thrombin and factor Xa.

The major inhibitor of coagulation, the effects of antithrombin are accelerated some 1000-fold in the presence of heparin. It follows, therefore, that special measures are required in subjects deficient in antithrombin (whether by genetics, loss in renal disease or excess consumption). In high doses, heparin also prolongs the bleeding time by inhibiting platelet aggregation *in vitro* and may exert some effect on the endothelium, although it is unclear (and possibly academic) whether or not these influence clinical efficacy.

Laboratory management of heparin As with warfarin, there is marked variability in the effect of heparin in different individuals, mostly due to complex pharmacokinetics (such as degree of binding to numerous proteins including fibronectin and vWf). Again, as with warfarin, laboratory monitoring of the effects of heparin is necessary and can be followed with the relatively straightforward APTT. This test derives from the ability of citrated plasma to clot an artificial 'platelet' substitute of phospholipids and other substances to which calcium has to be added; quality control is crucial, as the quality of reagents can vary markedly. In practice, most clinicians aim for a degree of anticoagulation that prolongs the APTT by anything from 1.5 times to three times that of normal, uncoagulated blood; that is, an APTT patient/control ratio of 1.5–3.0.

Pros and cons of heparin Heparin is undoubtedly an extremely effective anticoagulant. However, it has many drawbacks, such as the variability between patients, which demands monitoring with the APTT, and the requirement for the use of the needle. As with aspirin, the major clinical problem with heparin is dose-dependent excess bleeding (haemorrhage), which we will discuss in detail in Section 8.4. A second problem is the irritating propensity of heparin to destroy platelets – as discussed in Section 8.3 – and is called heparin-induced thrombocytopenia (HIT).

Other uncommon side effects of heparin (especially in long-term and high-dose use) include fractures and osteopenia, hyperkalaemia, elevations in liver enzymes, skin necrosis at the site of administration, alopecia, hypersensitivity, priapism and hypoaldosteronism. Thus, despite the proven safety and effectiveness of continuous infusions of heparin, its limitations and adverse side-effect profile, unpredictable pharmacokinetic response, daily laboratory monitoring with dose adjustments and requirement for hospitalization leaves room for alternatives. One such alternative is a low molecular weight variant of the standard (unfractionated) heparin preparation.

Low molecular weight heparin

The drug The laboratory demonstration of a more efficacious fraction of whole UFH (i.e. LMWH) has ushered in a new era of anticoagulation. LMWH can be manufactured from UFH by several methods, such as benzylation with alkaline hydrolysis, nitrous acid depolymerization or digestion, heparinase digestion, and isoamyl nitrate digestion. Each provides a different LMWH with molecular weights varying from 4.4 to 5.9 kDa, so that each has different pharmacokinetics and activities. However, there are class-common advantages over UFH, most of which address mode of action and side effects. Broadly speaking, compared with UFH, LMWH has a longer subcutaneous half-life and, therefore, has potential for outpatient use, causes much less HIT, has a more predictable anticoagulant response requiring less monitoring and has better anti-factor Xa effect.

Laboratory management of low molecular weight heparin The very predictable pharmacokinetics and pharmacodynamics and the safety profile of LMWH mean it can be used without routine laboratory testing for efficacy. However, should it be necessary, that assay of choice is not the APTT, as LMWH is a far more effective direct inhibitor of factor Xa than it is a cofactor for the inhibition of thrombin by antithrombin.

Should the effect of LMWH need to be determined, its anti-factor Xa activity can be assessed in a coagulation-based assay or in an amidolytic (chromogenic) assay. Commercial kits for the latter are widely available. The patient's plasma is compared with healthy plasma and plasma known to be deficient in factor X, and factor Xa can be sourced commercially.

Building on early trials, several large studies have unequivocally demonstrated the value of LMWH in the initial treatment of DVT. The value of LMWH compared with UFH now extends to prevention of thrombosis in general surgery, orthopaedic surgery, hip fracture, multiple trauma and neurosurgery, whilst efficacy and safety issues in medical patients have been addressed.

Table 8.7 summarizes the differences between UFH and LMWH.

Until recently, hospital treatment of, and prophylaxis against, VTE generally began with the use of heparin on the ward, moving to oral anticoagulants (such as warfarin) following discharge, for varying periods up to 6 months. However, this practice is evolving and the patient may no longer need to be treated with a VKA (to be discussed below).

The major advantage of heparin and LMWH is that it almost immediately reduces the risk of thrombosis when compared with the slow-acting VKAs, so that prophylaxis for those at risk of VTE can begin promptly. However, not all risk factors are equal, and no drug is free of side effects, so that patients must be assessed as to their total risk of VTE and treated accordingly. These 'patient' risks can be stratified, many of which are precise medical conditions (Table 8.8).

A second perspective is exactly why the patient is coming into hospital, and this is pertinent in surgery. The various types of surgery carry their own risk, the greatest being orthopaedic; and there is also a time aspect, as very quick and superficial surgery carries a minimal risk of thrombosis (Table 8.9).

There is focus on orthopaedic surgery as this is particularly thrombogenic, and in 2010 there were nearly 120 000 hip and knee replacements in England and Wales, representing a huge overall risk of VTE. So by adding the scores from Table 8.8 and Table 8.9, a crude risk factor estimate can be constructed, and this can directly guide anticoagulant management.

Table 8.7 Differences between low molecular weight heparin (LMWH) and UFH.

	LMWH	UFH
Mean molecular weight (kDa)	5	15
Saccharide units	13–22	40–50
Anti-Xa to antithrombin activity	2: 1 to 4: 1	1: 1
Platelet inhibition	+ +	+ + + +
Inhibited by PF4	No	Yes
Bioavailability (%)	92–100	30–50
Half-life (h)	IV: 2; SC: 4	IV: 1; SC: 2
Mode of clearance	Kidney	Liver and kidney
Frequency of HIT (%)	<1	~2.5
Monitoring	Anti-Xa assay	APTT

APTT: activated partial thromboplastin time; HIT: heparin-induced thrombocytopenia;. IV: intravenous; PF4: platelet factor 4; SC: subcutaneous.

Table 8.8 Risk factors for VTE for patients about to go on LMWH.

Risk factors (Score 1)	Risk factors (Score 2)	Risk factors (Score 3)
Age >60	Oestrogen-containing pill (OCP, HRT)	Immobile (>72 h)
Obesity (BMI >30)	Pregnancy and post-partum	History of DVT/PE
IHD, CHF or previous stroke	Known thrombophilic conditions	Active malignancy
Myeloproliferative disorders	Malignancy	
Nephrotic syndrome	Sepsis	
Inflammatory bowel disease	Known family history in two first-degree relatives	

BMI: body mass index; CHF: congestive heart failure; DVT: deep vein thrombosis; HRT: hormone replacement therapy; IHD: ischaemic heart disease; OCP: oral contraceptive pill; PE: pulmonary embolus.

Table 8.9 Additional risk factors for surgical in-patients.

Score	Surgical procedure
4	Major trauma, e.g. lower limb fractures
4	Major joint replacement, surgery for fractured neck of femur
3	Total abdominal hysterectomy, including laparoscopic assisted
3	Thoracotomy or abdominal surgery involving mid-line laparotomy
2	Vascular surgery (not intra-abdominal)
2	Intraperitoneal laparoscopic surgery lasting >30 min
1	Surgery lasting >30 min
0	Surgery lasting <30 min

Table 8.10 shows an example of how the sum of different risk factors can guide treatment with LMWH and graduated elastic compression stockings. However, as with warfarin (and, in fact, any drug), there are a number of cautions and contra-indications for the use of LMWHs. The former include severe hepatic or renal impairment, or major trauma or surgery to the brain, eye or spinal cord. Contra-indications include known uncorrected bleeding disorders such as haemophilia, heparin allergy, HIT, heparin-induced thrombosis, patients on existing anticoagulation therapy, bleeding or potentially bleeding lesions, active peptic ulcer, recent intracranial haemorrhage, intracranial aneurysm or vascular malformation.

Table 8.10 Application of risk assessment for the use of LMWH.

Low risk Score 0–1	Moderate risk Score 2–3	High risk Score ≥4
Early ambulation Consider graduated elastic compression stockings (GECS)	GECS plus low-dose LMWH for surgical patients, high dose of LMWH for medical patients	GECS plus high-dose LMWH (not exceeding 14 days for medical inpatients). Surgical patients – consider intermittent pneumatic compression in theatre

Important point

As advances are always being made, guidelines and recommendation are always being revised. The practitioner **must** refer to their own most recent up-to-date protocols. This applies to all such material in this book.

Pros and cons of low molecular weight heparin Like UFH, LMWH is also an excellent anticoagulant, but with a much improved side-effect profile. An example of this is better inter-patient variability so that APTT monitoring is not required. However, there is still a risk of haemorrhage, although there is far less HIT (Table 8.8). The improvements have led to the development of patient self-medication (as with warfarin).

Patient power – low molecular weight heparin Apart from its clinical effectiveness and better side-effect profile, a further advantage of LMWH is in home use by appropriate patients with DVT who are not severely ill. In one clinical trial of LMWH, patients remained in hospital for a mean of 1.1 days, suffering 13 embolic events, compared with 6.5 days for those on UFH, with 17 events, of which two were fatal. Data demonstrating the superiority of LMWH compared with UFH (e.g. shorter time to effective anticoagulation and more days of effective anticoagulation) continues to be published. At the practical level, manufacturers are now packaging their LMWH into single-use syringes loaded with the particular anticoagulant.

Whilst the weight of literature focuses on (unprovoked) DVT and PE, LMWH is also effective in those with conditions predisposing to VTE, such as cancer, although more data on other groups (such as the obese and thrombophilia) are needed, and these are slowly becoming available. However, despite the weight of literature on LMWH, it is still far from perfect, leaving space for other agents.

Use of LMWHs In the UK, practice in National Health Service (NHS) hospitals is dominated by the British National Formulary. This book, also available online, is updated each 6 months and lists all government-approved pharmaceutical agents and their licensed indications. It follows that in other countries the doses and guidelines may differ, and that doses in pre-filled syringes may also differ.

All LMWHs are given subcutaneously, and regimes are markedly different for prophylaxis of VTE versus treatment of VTE. If monitoring in pregnancy is called for, blood should be taken for monitoring 3–4 h after a dose, and ideally the result will be in the range of 0.5 to 1 unit/ mL of anti-factor Xa activity. Monitoring is generally not required for once-daily treatments. Notably, this desired range is higher than that for prophylaxis of a VTE, which is 0.1–0.3 units/mL anti-factor Xa.

There are numerous guidelines for the use of LMWHs for prophylaxis in various settings, such as in obstetrics and gynaecology, and after surgery, and many are available free and online. Practice in surgery in the UK is dominated by guidelines from NICE, such as Clinical Guidelines 92 and 144, although with regular updates these may have a short half-life. NICE recommends that all patients about to undergo surgery should be assessed to identify their risk factors for developing VTE, which can be assessed by a scoring system, and this is now mandatory in NHS hospitals. The section above applies to the prevention of a VTE. However, once a patient actually has a DVT or PE, then a different dosing regime is called for, which is effectively more anticoagulant.

Fondaparinux (Arixtra) This agent may not be a 'heparin' or even an 'LMWH', but it certainly does inhibit the coagulation pathway in the same manner. It is a novel, selective and reversible Xa-inhibitor, although based on the structure of heparin, is different from both heparin and LMWH. Like LMWHs, it does not affect the PT and has very weak effects on APTT, but its activity can be determined by specific anti-Xa assays, if necessary. Thrombocytopenia occurs even less commonly than with LMWH. Use of fondaparinux has been approved by the UK's NICE as an alternative to LMWH in certain types of surgery, such as orthopaedic.

The closely related idraparinux is a synthetic penta-saccharide inhibitor of factor Xa with a very long half-life that requires once-weekly dosing. Although the once-weekly regime seems attractive, fears of overdose and the lack of an antidote were among the factors that led to the development of a biotinylated version (idrabiotaparinux), whose effect can be reversed by avidin.

Heparinoids, hirudins and other agents There are a small number of other agents with precise indications, generally used when heparin and LMWHs are inappropriate. Danaparoid can be injected subcutaneously to reduce the risk of VTE following orthopaedic surgery and in cases of HIT. There are two hirudins, drugs derived from an anticoagulant found in the mouthparts of the leech:

- Lepirudin is licensed for anticoagulation in patients with type II (immune) HIT who require parenteral antithrombotic treatment. The dose of lepirudin is adjusted according to the APTT, as is heparin, generally to a ratio of 1.5–2.5.
- Bivalirudin, a hirudin analogue, is a thrombin inhibitor which is licensed for unstable angina or myocardial infarction in patients planned for urgent or early intervention, and as an anticoagulant for patients undergoing percutaneous coronary intervention. Accordingly, it is used in conjunction with anti-platelets.

Hirudins can be monitored in the laboratory with the ecarin clotting time. This test was described in the section on APS: the snake venom acts on prothrombin to produce meizothrombin, which has thrombin-like activity resistant to inhibition by heparin or prior treatment with VKAs. This activity is inhibited by hirudin and other direct thrombin inhibitors.

Argatroban is the drug of choice for patients with HIT, and is generally given as a continuous infusion at a dose that depends on the weight of the patient. An oral anticoagulant can be given alongside argatroban, but it should only be started once thrombocytopenia has substantially resolved.

Prostacyclin (epoprostenol) is an alternative to heparin during renal dialysis. It is effective in inhibiting platelet aggregation and is also a potent vasodilator. With a short half-life of 3 min it must be given by continuous intravenous infusion, and there is no routine laboratory monitoring.

New oral anticoagulants

Although LMWHs are very effective agents, they still carry some problems such as a residual risk of HIT and reliance on parenteral introduction. These, and other, problems are being overcome with the introduction of new agents. Some operate directly against thrombin, whereas others target coagulation factor Xa. All of their licensed uses stem from comparisons with an LMWH or with warfarin.

Direct thrombin inhibition Dabigatran etexilate (Pradaxa, molecular weight 628 Da) is the first in a new class of anticoagulants that works by directly inhibiting thrombin. It possesses various qualities which make it potentially an attractive and promising novel oral anticoagulant with its predictable pharmacokinetics and pharmacodynamics. The drug has rapid absorption (within 2 h) and distribution with estimated half-lives of 8–14 h and 14–17 h for single- and multiple-dose administrations respectively. Nearly 80% of dabigatran etexilate is excreted unchanged by the kidneys with average bioavailability of 6.5%, hence high doses are required to maintain adequate plasma concentrations.

Dabigatran has been available in the UK since April 2008 and was subsequently included in the DVT prevention guidance issued by NICE (TA157) and is licensed for the prevention of stroke in AF (NICE TA249). As is common practice with almost all anticoagulants, patients with moderate renal impairment (GFR 30–50 mL/min per 1.73 m^2) require reduced dosages.

Laboratory monitoring of dabigatran is not normally required, but should it be sought then the ecarin clotting time (which assesses thrombin activity) may be appropriate. The PT (hence the INR) is of poor sensitivity and is not recommended. Both the APTT and thrombin time may be useful as qualitative measures to detect an excess of anticoagulant activity. At the high dose of 150 mg twice daily, expected values of APTT and the ecarin clotting time at peak concentration (perhaps 2 h after dosing) are likely to be two to three times the baseline value.

Factor Xa inhibition   These agents reversibly block free factor Xa and that which is bound to platelets and within the prothrombinase complex. The reliability of these drugs, like dabigatran, is such that their effect should not normally need to be monitored. However, should it be required, plasma levels of Xa inhibitors show a straight line with the PT, although some advocate use of a factor Xa assay.

Rivaroxaban (Xarelto, molecular weight 436 Da) is the first selective oral direct factor Xa inhibitor. It has favourable pharmacokinetic characteristics with a bioavailability of 60–80%, achieves peak plasma levels in 3 h and has a half-life of 5–9 h in healthy, young subjects and about 12 h in elderly subjects. It is metabolized by the liver via CYP3A4 with up to two-thirds of the drug being eliminated by the kidneys. The use of rivaroxaban in patients with renal impairment has to be cautious because of its renal clearance. It does not significantly interact with platelet function, where it demonstrated an excellent correlation between its plasma levels and achieved clotting times, while the bleeding risk was comparable to that of an LMWH.

Rivaroxaban is licensed for the prophylaxis of VTE in orthopaedic surgery (NICE TA170) and for the prevention of stroke in AF (NICE TA256). As regards laboratory monitoring, rivaroxaban prolongs PT and APTT, but these are reagent and device dependent; it does not prolong thrombin time, reptilase time and does not affect fibrinogen or d-dimer measurement.

Apixaban (Eliquis, molecular weight 460 Da) is a highly selective and potent inhibitor of factor Xa. Its bioavailability is >50% with a half-life of 9–14 h. Like rivaroxaban, it inhibits both free and bound factor Xa. Apixaban has fixed twice-daily dosing and is metabolized in the liver via CYP3A4, with about 25% excreted by the kidneys and the remainder by intestinal excretion. Apixaban is influenced by potent CYP3A inhibitors such as macrolides, protease inhibitors and ketoconazole. NICE TA245 refers to the licence of apixaban for the prevention of VTE after orthopaedic surgery.

Other novel oral anticoagulants in various stages of development include edoxaban, betrixaban and eribaxaban, and a novel VKA tecarfarin, may all be expected to be licensed for various conditions. Other potential drugs include a factor IXa inhibitor.

An ideal antithrombotic drug

What factors are important? Clearly there must be efficacy over several conditions such as prophylaxis and treatment in the common situations of orthopaedic surgery, cancer, AF and other medical conditions. There should be few side effects, notably lack of haemorrhage and actions towards platelets. Other factors include convenient mode of delivery (for the patient and hospital), lack of drug and diet interactions, a short half-life and an effective antidote. These are compared in Table 8.11. Notably, the newer drugs have more of these features than the established agents.

8.3   Haemorrhage 1: platelet underactivity and thrombocytopenia

As with platelet overactivity, the focus is on quantitative and qualitative problems. In the former we look at reduced numbers of (generally) normally functioning platelets (thrombocytopenia). In the latter the platelet count is generally normal but the cells themselves have a defect which means their participation in thrombosis is reduced. Both these processes lead to haemorrhage regardless of the efficacy of the coagulation pathway.

Table 8.11 Features of common anticoagulants.

	Monitoring	Half-life	Target	Delivery
UFH	APTT	Short	Thrombin > factor Xa	Parenteral
LMWH	Generally not required	Short	Factor Xa > Thrombin	Parenteral
Fondaparinux	Not required	Short	Factor Xa	Parenteral
Warfarin	INR	Long	Several factors	Oral
Dabigatran	Not required	Short	Thrombin	Oral
Rivaroxaban	Not required	Short	Factor Xa	Oral
Apixaban	Not required	Short	Factor Xa	Oral

APTT: activated partial thromboplastin time; INR: international normalized ratio.

Pathophysiology

The bone marrow   A primary reason for thrombocytopenia is a fall in production due to bone marrow problems. The process of thrombopoiesis is described in Chapter 7, but isolated problems with this are exceptionally rare, such as congenital amegakaryocytic thrombocytopenia and megakaryocytic aplasia, which has an autoimmune aetiology. Lack of production of thrombopoietin (by the liver) is most unusual, whereas the lack of erythropoietin as a cause of a low red cell count is established. However, there may be mutations in the gene for the thrombopoietin receptor (*Mpl*) which lead to failure to bind its ligand and/or failure to pass this message to intracellular metabolic pathways. This is effectively the reverse of essential and familial thrombocythaemia.

There are numerous causes of secondary 'bone marrow' thrombocytopenia. Chapter 6 explains why white cell malignancies (leukaemia, myeloma and lymphoma) cause a low platelet count, and Chapter 4 describes Fanconi's anaemia, aplastic anaemia and other factors that are linked to thrombocytopenia. The treatments of many cancers, including those of red and white cells, are cytotoxic, and so a low platelet count may also be present. Some drugs, such as immunosuppressants like prednisolone, are not directly cytotoxic but indiscriminately suppress all haemopoietic activity, leading to pancytopenia.

Intrinsic platelet defects   There are a small number of well-characterized qualitative defects in platelet function that lead to haemorrhage. The most common are:

• Glanzmann's thrombasthenia, characterized by qualitative or quantitative defects in GpIIb/IIIa, resulting in

failure to adequately bind to fibrinogen, and so lack of platelet activation and reluctance to participate in haemostasis. The disease can be caused by mutations in the genes for GpIIb or for GpIIIa. The population frequency is likely to be less than one per million.
• Bernard–Soulier syndrome, due to the lack of the GbIb/V/IX complex, the receptor for vWf. Failure of the complex to bind vWf leads to insufficient platelet–platelet and platelet–subendothelium adhesion. These platelets are also larger than normal platelets, and the disease has a frequency of one in a million.
• Storage pool diseases include those of a deficiency in dense granules (such as in Chediak–Higashi syndrome) and grey platelet syndrome (caused by a lack of alpha granules). In the latter, there is often a degree of thrombocytopenia, and the cells also tend to be larger than usual.
• Platelets from patients with Scott syndrome have a defect in the expression of phosphatidylserine, lipoproteins required for the stable formation of tenase and prothrombinase on the cell surface. The precise cause of this is dysfunction of the scramblase enzyme which, alongside flippase and floppase, regulates the differential expression of phosphatidylserine, phospholipids and phosphatidylethanolamine on the inner and outer surfaces of the platelet membrane.

Each of these conditions is inherited in an autosomal recessive manner. There are, of course, numerous rare conditions of platelet function, such as defects in the ADP, thromboxane, epinephrine and collagen receptors. Abnormalities in platelets are also found in other conditions, such as the X-linked disease Wiskott–Aldrich syndrome, where there is also immunodeficiency and eczema. MYH9 is the abbreviation for myosin heavy chain 9, coded for by a gene on chromosome

22q12.3–q13.2. A manifestation of this is the May–Hegglin anomaly, an autosomal dominant condition, in which there are Dohle bodies in white cells, with thrombocytopenia and large platelets.

Destruction of platelets Several diverse disease processes cause secondary thrombocytopenia.

• Immune thrombocytopenia purpura (ITP, purpura meaning bruising) is caused by antibodies to platelets. There is a neonatal variant where maternal antibodies cross the placenta. It is rare, but not exceptionally so, with 50–100 per million cases diagnosed annually. The most common cause is an immunoglobulin G (IgG) auto-antibody that targets GpIIb/IIIa or the GpIb/V/IX complex. Notably, ITP may also be taken to be *idiopathic* thrombocytopenia purpura, where the cause of the low platelet count has yet to be identified.

• Pregnancy-associated thrombocytopenia is part of the HELLP syndrome, composed of *h*aemolysis (with increased lactate dehydrogenase), *e*levated *l*iver enzymes (such as raised asparate and alanine aminotransferase) and *l*ow *p*latelets. It is present in <1% of pregnancies, but in 10–20% of those suffering pre-eclampsia. There is little evidence of a single clear aetiology, but the red cell and platelet damage may be caused by fibrin deposits and endothelial cell damage, hence the term microangiopathic haemolytic anaemia.

• Thrombotic thrombocytopenia purpura (TPP), with a reported incidence of six per million per year, has an interesting aetiology. A function of vWf is to promote platelet aggregation, which it does so by cross-linking platelets, which is most efficiently performed by large multimers. The frequency of these is part-controlled by the enzyme ADAMTS13, which cuts the multimers down to smaller fragments that fail to cross-link the platelets as efficiently. Therefore, lack of ADAMTS13 leads to large vWf molecules that promote excess numbers of micro-thrombi, so that the remaining depleted platelet pool is unable to provide support for haemostasis, resulting in bruising. An incidental pathology is that passing red cells are damaged by the microthrombi, leading once more to a microangiopathic haemolytic anaemia.

• Haemolytic uraemic syndrome is a consequence of renal failure, itself the result of a bacterial infection, such as by *Escherichia coli* or a *Shigella* species. Both these organisms produce a toxin that causes glomerulo-nephritis. Increased levels of urea are toxic to platelets and red cells, the latter being haemolysed. Diagnosis can be difficult as there is overlap with TTP.

Heparin-induced thrombocytopaenia Technically, this unpredictable and life-threatening condition should be part of the 'Destruction of platelets' section, but it is such an established and serious condition it warrants special attention. Apart from dosing issues, the greatest risk factors for HIT are autoimmune disease, gout and haemodialysis.

Approximately 2.5% of patients receiving UFH therapy and perhaps 0.5% of patients on LMWH suffer this problem. Surprisingly, HIT (generally a platelet count <100 000 cells/μL, and reversible) has been described for over 50 years, with two forms (non-immune and immune – often an IgG recognizing a heparin/platelet factor 4 complex) being recognized, resulting in platelet activation and their removal by the reticulo-endothelial system, generally in the spleen. When present, cessation of heparin/LMWH treatment is mandatory and alternate anticoagulant cover (such as a hirudin) is necessary.

If present, HIT is likely to be apparent 3–10 days after initiation (such that a platelet count is advised on days 4–6), but may also occur very rarely within hours, with the platelet count returning to normal within 4 days of discontinuation. Curiously, 0.4% of patients with HIT suffer arterial thrombosis (that may follow platelet aggregation *in vivo*) or VTE (that may result from heparin resistance caused by the neutralizing effect of the heparin-induced release of platelet factor 4).

There are many rare causes of thrombocytopenia, such as viral infections, nutritional disorders (vitamin B_{12} and folate deficiency, alcoholism) and splenomegaly. A guideline on HIT is available from the BCSH.

Assessment

The full blood count provides a platelet count (for thrombocytopenia) and mean platelet volume, the latter being helpful in a variety of conditions such as MYH9-related disease and Bernard–Soulier syndrome, where platelets are large. However, 'young' platelets, as in young red cells (reticulocytes), are larger than their mature counterparts and are a possible source of mis-interpretation. Examination of the blood film may be instructive as thrombocytopenia may be due to artefact (Figure 8.4).

All methods must take account of the artefact that platelets may become activated by the process of vene-puncture. Accordingly, platelet function testing must be performed on blood from a second vacutainer with 1/10 volume of tri-sodium citrate, the first perhaps being used for a full blood count or other measures. The patient

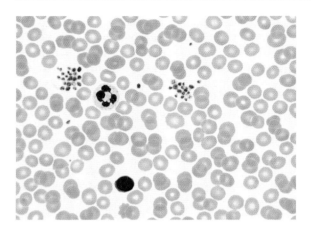

Figure 8.4 A blood film showing microthrombi. To the left and right of the upper leukocyte (a neutrophil) are two micro-thrombi. The blood scientist needs to ask if these are genuine clots formed in the body by some pathogenic process (such as an autoantibody), or if they have formed in the vacutainer after the blood has been drawn (and so are an artefact of failed anti-coagulation). (Image courtesy of H. Bibawi, NHS Tayside).

Table 8.12 Platelet CD molecules commonly used in flow cytometry.

CD number	Platelet molecule name
CD41	GpIIb
CD42a	GpIX
CD42b	GpIb-alpha
CD42c	GpIb-beta
CD42d	GpV
CD61	GpIIIa
CD62P	P-selectin

must also be fasted and have refrained from smoking and caffeine as these all influence platelet function. Analysis must be complete in 30 min to 2 h after venepuncture. Platelet function testing involves detection of cell surface molecules by flow cytometry and functional studies, of which there are several methods. As we have noted, numerous medications can influence platelet function.

Flow cytometry   This powerful technique tells us of the presence of various molecules on the surface of the cell. Each of the major platelet glycoproteins has its own CD number, but some are not specific for platelets, being found on other cells. Examples of this include CD31 (PECAM), also found on endothelial cells, and CD63, also found on monocytes and macrophages. Whilst CD110 (Mpl, the thrombopoietin receptor) is found almost exclusively on platelets and megakaryocytes, there are several other choices (Table 8.12).

However, analysis of this process is not merely academic, as the value of flow cytometry is that it can rapidly help in the diagnosis of several qualitative platelet disorders:

• Bernard–Soulier syndrome is characterized by a significantly reduced number of GpIb molecules.
• In Glanzmann's thrombasthenia there is a reduction in the number GpIIb/IIIa molecules.

• Scott syndrome platelets have reduced expression of phosphatidylseine, which can normally be detected by annexin-V.
• The activation of platelets by an agonist such as ADP should result in alpha degranulation and the appearance of increased levels of P-selectin at the cell surface. Therefore, failure of this process implies abnormality in the particular signalling pathway.

Another flow cytometry technique assesses the efficacy of the ADP-receptor blocker clopidogrel. A consequence of ADP docking into its receptor is the initiation of a train of metabolic processes that result in the phosphorylation of vasodilator-stimulated phosphoprotein. This can also be used in investigations of purinoreceptor $P2Y_{12}$ and $P2Y_1$, often in conjunction with other tests. An important technical note is that the standard flow cytometer is set up to look at white cells; it must be recalibrated to look at platelets.

Platelet aggregation   This is certainly the most well-known and best-established method, of which there are two variants. In light transmission aggregometry (LTA), a beam of light passes through a suspension of platelet-rich plasma. Addition of an agonist causes the formation of micro- and then macro-thrombi, which clarifies the suspension, resulting in increased light passage. Agonists of choice include ADP, collagen, arachadonic acid, ristocetin and epinephrine, although thrombin and thrombin receptor activation peptide (which activates the protease-activated receptor-1) may be used. Non-physiological activators include endoperoxides analogue U46619 (which targets the thromboxane receptor), phorbol myristate acetate (protein kinase C metabolism) and calcium ionophore A23187, which have value in more complex investigations.

LTA is a popular and powerful method that clearly identifies particular platelet problems, such as failure to aggregate when presented with a particular agonist (Figure 8.5). However, despite its popularity, it is often poorly standardized with wide variation in laboratory practice, such as the duration and speed of centrifugation in the preparation of the platelet suspension.

A second aggregation method uses whole blood, to which the agonist is added in a similar manner to LTA, which then proceeds to clot. The time to clot is generally assessed by the clot interfering with a mechanical motion, such as a vibrating 'paddle', or by platelet adhesion to very fine wire electrodes, which influences electrical impedance. The whole-blood method is arguably more

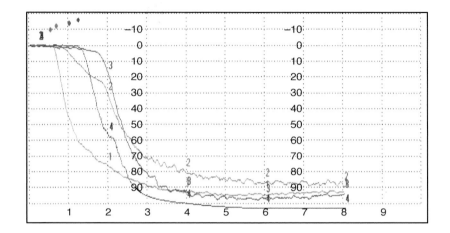

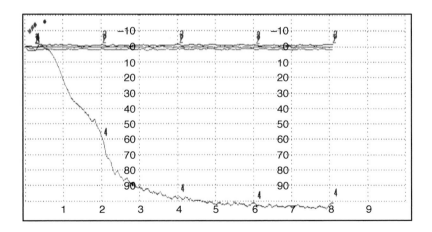

Figure 8.5 Light transmission aggregometry. (upper panel) Changes in light transmission (vertical axis) over 8 min (horizontal axis) when ADP (trace 1), epinephrine (2), collagen (3) and ristocetin (4) are added to platelet-rich plasma. After 4 min, the light transmission to all agonists is greater than 80% of a sample of the subject's platelet-free plasma, indicating the formation of thrombi. (lower panel) Changes in the response of platelet-rich plasma from a 15-year old female whose only symptom was prolonged bleeding from the gums. There were no abnormalities in her coagulation pathway. The aggregation plot shows a normal response to ristocetin (trace 4) of over 90% at 4 min. However, ADP, epinephrine and collagen (traces 1–3, the continuous lines at the top of the printout) have failed to induce any platelet aggregation. This profile supports the diagnosis of Glanzmann's thrombasthenia: the molecular lesion is lack of functioning GpIIb/IIIa. (Image courtesy of H. Bibawi, NHS Tayside).

physiological as it allows the participation of red cells, a potential source of the agonist ADP. The whole-blood methods and LTA show broadly similar results to common abnormalities, although there are differences in sensitivity to certain agonists.

Thromboelastography This may be seen as a variant of whole-blood aggregometry. In thromboelastography, whole blood is placed in a rotating cuvette, but aggregation proceeds without a standard platelet agonist; instead, thrombosis may be initiated by an activator of the coagulation pathway, such as kaolin. The clot thus formed is the sum of all natural components of the process of haemostasis. A further aspect of this technique is that the rate of formation of the clot, its physical strength and its elasticity can all be quantified. By allowing the process to develop over 30 min, the consolidation of the clot can be determined, and also its degradation by fibrinolysis. In a variant of the method, a rotating sensor is placed into a static pool of blood, so that as the blood clots the inhibition of the motion of the sensor is detected and interpreted by software.

The platelet function analyser (PFA-100) This machine provides assessment of the ability of the platelets in a sample of citrated whole blood to occlude an aperture once stimulated by collagen and epinephrine, or by collagen and ADP, and so is (with the exception of the citrate anticoagulant) a good representation of physiology. Accordingly, the technique is widely used as a first-line screening test in otherwise unexplained haemorrhage. However, like any technique, there are caveats: thrombocytopenia will result in an abnormality in the end-point result (the closure time) that resembles a genuine platelet defect, and the method is sensitive to levels of vWf, which itself varies with ABO blood group, infections, inflammation, cancer, and atherosclerosis and its risk factors. The technique is most useful in the investigation of Bernard–Soulier syndrome and Glanzmann's thrombasthenia.

Other tests of platelet function As platelet activation results in degranulation of the dense and alpha granules, increased levels of their contents can be used as surrogates. ELISAs can determine plasma platelet factor 4, soluble P-selectin and beta-thromboglobulin, whilst nucleotides can be detected by a test that relies on luminosity. The crucial intermediate metabolite thromboxane A_2 has such a short half-life that it is difficult to measure in a routine clinical setting. However, its inactive metabolite thromboxane B_2 is stable, long-lived and can be measured by ELISA in plasma or urine; levels are reduced by aspirin. A lumi-aggregometer can simultaneously measure aggregation to agonists and the release of ATP from dense granules.

Platelet function and near-patient testing In certain circumstances, the ability to rapidly assess the platelet status of a patient in an acute setting (such as the coronary care unit) is important. In response to this need, a number of near-patient devices have been developed, the dominant machine providing an assessment of the response of the cell to aspirin and clopidogrel.

Treatment

The causes of thrombocytopenia are primary and secondary. Its consequences, haemorrhage, are discussed in Section 8.4.

Primary platelet function, following from a genetic lesion, whether quantitative or qualitative, is effectively incurable outside of bone marrow transplantation, although its consequences may be treatable. The pathway to the treatment of secondary thrombocytopenia is in addressing its various causes. This would be, where relevant, conservation and protection of the bone marrow, the use of recombinant thrombopoietin to stimulate new platelet production, and the cessation of heparin and LMWH in HIT.

Treatment of ITP focuses on immunosuppression to reduce levels of the pathogenic autoantibody, but this crude measure has wider consequences for the patient's immune system. There are two commercial agonists of the thrombopoietin receptor that can be used to stimulate the differentiation of stems cells to megakaryocytes. Both can be used to promote platelet numbers, and as a consequence there is a reduction in haemorrhage.

Immunosuppression may also have a role in HUS, but the primary objective is the removal of the pathogenic microorganism responsible for the disease. The acute treatment of acute TTP focuses on plasma exchange to remove autoantibodies and replete ADAMTS13. Other treatments include immunosuppression with agents such as methylprednisolone, ciclosporin and rituximab; but in the future, infusions of purified or recombinant ADAMTS enzyme may become available.

As HELLP is effectively a maternal response to her infant, treatment is prompt delivery, natural or assisted. However, with low platelets there may be a risk of haemorrhage, so that prophylactic cover with a transfusion of platelets may be required. If the pregnancy is in too early a stage, treatment is symptomatic, such as red cell transfusion for anaemia and anti-hypertensive drugs for high blood pressure.

Therapeutic platelet transfusion is a widely accepted option in life-threatening thrombocytopenia and/or platelet dysfunction. Details are provided in Chapter 11.

8.4 Haemorrhage 2: coagulation underactivity

In this section we will look at the consequences of failure of the molecules of the coagulation pathway to form a clot. It is convenient to classify coagulation-based haemorrhage as primary or secondary.

Pathophysiology of primary defects

Although defects in almost all coagulation factors are known, in practice only three command sufficient frequency and severity to warrant detailed attention: insufficient factor VIII, insufficient vWf, insufficient factor IX.

Insufficient factor VIII Haemophilia, sometimes haemophilia A, is certainly the haemorrhagic condition with the greatest public awareness, but for scientists it also provides a clear example of a hereditary single gene defect that leads directly to a major clinical problem. It is present with a frequency of about 30–100 per million (1 in 10 000 males), but in perhaps 30% of presenting cases the genetic lesion is spontaneous, appearing without a family history.

The disease manifests itself according to the severity of the bleeding and levels of factor VIII in the blood.

- In severe disease (factor VIII <1.0 IU/dL, reference range 50–150 IU/dL) there is frequent and spontaneous epistaxis (nosebleeds) and bleeding into joints, muscles and internal organs. There is severe bleeding after trauma. It presents in infancy and is responsible for 50% of all cases of haemophilia.
- Moderate disease generally presents before the second year, with far fewer spontaneous bleeds, and less loss of blood after trauma. Accounting for 30% of cases of haemophilia, factor VIII levels are generally 1.0–5.0 IU/dL.
- Mild disease is present if factor VIII is >5 IU/dL. Present in 20% of cases, usually after the age of 2 years. There are no spontaneous haemorrhages, but these do occur after trauma and surgery.

The very large gene for factor VIII (180 kb, with 26 exons) is located towards the tip of the long arm of the X chromosome, at Xq2.8, and encodes a molecule of 2351 amino acids which breaks down to a 'heavy' chain of 2332 residues and a 'light' chain of 19 residues. Post-translational changes bring the molecule down to a 73 kDa, although smaller 50 and 43 kDa fractions are able to contribute to haemostasis.

This large size of the gene provides many opportunities for loss-of-function mutations, and deletions, duplications and insertions have all been described, although point mutations are the most prevalent, being present in 90–95% of patients. However, in those with the most severe disease, an inversion in intron 22 is the dominant form, and occurs principally as an error of DNA replication during spermatogenesis. Some mutations compromise the activation site of the molecule (so it does not transform to factor VIIIa), resulting in a partially active or inactive molecule, whilst others prevent the interaction between factor VIII and its cofactor vWf, without which it cannot function.

The fact that patients with haemophilia do not bleed continuously underlines the potential of the tissue factor/factor VII pathway (sometimes known as the extrinsic or tissue activation pathway), which can activate factor X (i.e. act as a tenase) in the absence of factor VIII.

Insufficient von Willebrand factor Many aspects of this molecule, abnormalities of which cause von Willebrand disease (vWd), contrast markedly with haemophilia and abnormalities in factor VIII. For example, haemophilia is rare, whereas vWd is common; haemophilia has 100% penetrance, but the effect of vWd is highly variable. The exact frequency of vWd is difficult to define exactly as in many the condition is virtually asymptomatic and may never be proven, but has been estimated to be as high as 1% of the population. However, like haemophilia, there are grades of severity of vWd, and the most common forms are the most mild, with the most severe forms being the least frequent.

The gene for vWf, like the molecule itself, is large. Found on chromosome 12 (at 12p13.2), it has 52 exons spanning 178 kb and codes for a molecule of 2813 amino acids which breaks down to a mature molecule of 2050 amino acids, regarded as a monomer of perhaps 250 kDa. The molecule has a number of function-specific domains that include binding regions for collagen, heparin, platelet GpIb (as part of the GpIb/IX/V complex) and factor VIII. The monomers come together to form a polymer that may exceed 20 000 kDa.

Lack of vWf, causing vWd, therefore has several consequences, which translate to failure to promote platelet aggregation and adhesion to the subendothelium, and also failure to support the function of factor VIII in haemostasis. Both these factors lead to haemorrhage. vWd may be classified as follows:

- Type I vWd is the most common (60–80% of cases) and the most mild. Indeed, in many it is asymptomatic, with only the occasional nosebleed. Levels of vWf are perhaps 20–50% of normal.
- In type II vWd (20–30% of cases) the plasma levels may be normal or slightly reduced. The defect relates to qualitative defects, of which there are four variants.
    - Type 2A is characterized by a lack of high molecular weight polymers, required for platelet cross-linking, due either to defective multimer assembly or to accelerated proteolysis. Factor VIII cofactor activity is normal.
    - In type 2B, high molecular weight forms are again reduced, but this is because they are bound with abnormally high affinity to the platelet surface GpIb complex.
    - Type 2 M disease patients have normal levels of vWf, but a qualitative defect is present.
    - In type 2N disease, vWf fails to bind to factor VIII.
- Type III vWd is the most severe form with very low or absent levels of vWf. The phenotype of this form therefore resembles haemophilia, but is also found in females, and is probably the variant of vWd identified by Erik von Willebrand in 1926.

Insufficient factor IX Sometime called haemophilia B, or Christmas disease, this condition is also X-chromosome linked, and occurs in 1 in 50 000 males. The gene is much smaller than that for factor VIII, with only eight exons spanning 34 kb, located on the long arm of the X chromosome at Xq27. Like factor VIII, mutations of the activation site render the molecule biologically inactive and unable to transform to factor IXa. Another mutation results in the inability to bind calcium, with the same consequence.

The view that haemophilia and factor IX deficiency is an exclusively male disease is incorrect as it may occur in females in some unusual situations, such as lyonization of the healthy X-chromosome in a carrier, a mutant gene inherited from both parents, and some chromosome abnormalities.

Deficiencies in other factors These present with increasingly rare frequency, such as factor VII being 1 in 500 000, and those of fibrinogen, factor V, factor X and factor XI (which may also be called haemophilia C) at

one in a million. The degree of deficiency of these molecules is variable, and therefore so is their haemorrhagic potential. The frequency of factor XIII deficiency has been reported as being between one and five people per million, and haemorrhage results because soluble fibrin is not cross-linked in the absence of this molecule.

Pathophysiology of secondary defects

Acquired haemorrhagic disease can have a number of alternative aetiologies. An obvious reason for low levels of coagulation factor is the use of VKAs. Others include liver disease (therefore lack of coagulation factor production), dietary deficiency of vitamin K and vitamin C, prolonged use of steroids, and antibodies to the coagulation factors that render them nonfunctional. The latter are a problem for the treatment of haemophilia, to be discussed below.

Acquired vWd has been reported as being caused by a broad series of situations. These include lymphoproliferative and myeloproliferative syndromes (including monoclonal gammopathy), autoimmunity (typically SLE), solid tumours, hypothyroidism and some cardiovascular disease.

Similarly, there are many causes of acquired haemophilia (apart from antibodies), such as myelodysplasia, treatment with interferon or rituximab, ET, hepatitis, SLE and antiviral agents. There are also reports of acquired factor V, factor IX and factor XIII deficiency, the latter two possibly caused by inhibitors.

Assessment

General screening tests The first measurements will be the general screening tests PT (prothrombin, fibrinogen, factors V, VII and X), APTT (as PT but also factors IX, XI, XII and high molecular weight kininogen) and thrombin time. It follows that a prolonged APTT may be due to the absence of a number of factors, whereas an isolated PT is inevitably due to factor VII deficiency. The most common cause of a prolonged APTT is mild factor XII deficiency, found in 3% of the population.

Factor-specific assays More sophisticated clotting assays are available for each factor, using specific factor-deficient plasma. A good example of this is the two-stage coagulation assay, such as in the detection of functional factor VIII. In the first stage, a cocktail of factors is assembled, but factor VIII is omitted, which must be provided by the patient's sample. The factors should come together to form a tenase complex, which will generate factor Xa from factor X.

In the second stage, the factor Xa acts on a commercial substrate to produce a coloured product, which can be measured in a colorimeter. By very careful attention to the levels of the constituent factors, the rate-limiting step in the generation of colour should be the amount of factor VIII in the patient's sample. Several positive and negative controls are required.

A second useful strategy is mixing tests, in which a defect in one plasma sample can be repaired by adding increasingly large aliquots of normal plasma. This can also help with the investigation of an inhibitor, as may be present in the APS.

Measurement of von Willebrand factor   This can be affected by a number of methods. Immunoassays (ELISA and latex immunoassay) provide information on the mass of vWf present, regardless of activity. Platelet binding activity can be assessed by the ristocetin cofactor assay, and collagen binding by a modified immunoassay. Factor VIII binding activity can be assessed by sandwich ELISA, or a hybrid assay taking parts from an immunoassay and a chromogenic substrate assay. However, a problem with the latter is that some vWf may already be carrying some factor VIII.

A more sophisticated method is required if knowledge of the degree of multimerization is required. This is important because only the large vWf multimers are active in crosslinking platelets – normal vWf will have factor VIII cofactor activity regardless of the multimer profile. The method involves the electrophoresis of plasma vWf through a gel that separates the multimers by size (Figure 8.6).

Treatment

The definition of haemorrhage is difficult, and the patient's concerns may not be relevant. Accordingly, if symptoms are minor, treatment may be minimal. However, clear loss of blood, perhaps in vomiting, nosebleeds (epistaxis), in urine (haematuria), or by the rectum or vagina all demand attention.

Treatment of primary deficiency   Treatment of coagulation factor deficiency is by replacement. This can be achieved in the case of haemophilia by recombinant or human plasma-purified factor VIII. However, this is often frustrated by the development of autoimmune and alloimmune antibodies that inhibit coagulation. A porcine factor VIII product often stimulates xenoantibodies. If

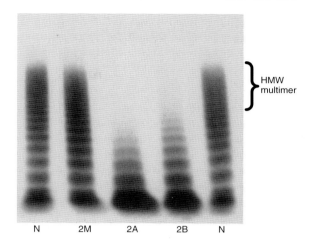

Figure 8.6 vWf multimer analysis. The plasma sample loaded onto the top of an SDS agarose gel, and staining of the resultant electrophoresis plot reveals a series of bands that equate to vWf of different sizes. The topmost are the high molecular weight species that fail to enter the gel, or fail to penetrate very far. Small multimers penetrate further, the smallest being at the bottom. The mass of each band can be estimated by densitometry.

Columns 1 and 5 illustrate normal plasma with a variety of bands of different intensity along the full length of the plot. Column 2 represents a sample from a patient with vWd type 2 M, column 3 is of vWd type 2A, and column 4 is from a patient with type 2B disease. Types 2A and 2B show lack of high molecular weight multimers that would promote thrombus formation, so their absence instead promotes haemorrhage. (From Provan & Gribben (eds), *Molecular Haematology*, 3rd Edition, 2010, Fig. 19.5, p. 239. Reproduced with permission of John Wiley & Sons, Ltd.)

inhibitors to factor VIII are present (as they may be in up to 30% of patients), and cannot be addressed, the patient can be treated with recombinant factor VII (Novoseven), which has its own tenase activity and so generates factor Xa independently of the factor VIII–factor IX tenase complex. The potentcy of inhibitors can be quantified by the Bethesda assay, where patient's plasma is mixed with plasma known to contain 100% factor VIII. One Bethesda unit is the amount of an inhibitor that will neutralize 50% of the coagulation activity of normal plasma.

In the future there may be a place for transplantation of an entire healthy liver, and possibly also of the use of gene therapy, but these are far from entering routine practice. Many of the same issues arise with factor IX deficiency, such as the use of factor IX concentrates and recombinant factor IX. However, gene therapy is more advanced than in haemophilia, with the use of an adenovirus that targets the liver.

The large size of the vWf gene, and its many introns, has proved a problem in attempting to generate a recombinant product. However, cryoprecipitate can be prepared, which is enriched for vWf and other molecules, including factor VIII and fibrinogen. This, therefore, provides the opportunity to repopulate the plasma with functioning vWf, and therefore functioning factor VIII. However, as a product of human plasma, it needs to be clarified of pathogens. Desmopressin, a synthetic analogue of vasopressin, can be used to stimulate the release of vWf from endothelial cells, and is used in both prophylaxis and in the treatment of haemorrhage.

If there is a great deal of blood loss in any condition, it may need to be replaced with a transfusion of packed red cells, and perhaps also plasma expanders (Chapter 11). An additional point about the use of replacement therapy is whether or not it should be used prophylactically to reduce the risk of haemorrhage, or if it should be used only to treat actual haemorrhage.

Treatment of secondary deficiency By far the greater part of clinical haemorrhage is not due to hereditary or pathogenically acquired factor deficiency, but to problems with therapeutics. It is most unlikely that haemorrhage (whether it be simple bruising and bleeding, or more serious blood loss) when a patient is taking aspirin, clopidogrel, UFH, LMWH or warfarin is caused by anything other than high levels of the drug.

Haemorrhage caused by overdosing with antiplatelet drugs aspirin and clopidogrel is initially treated by stopping the drug, as both have a short plasma half-life, although their effect on platelet function will be longer.

Haemorrhage secondary to UFH or LMWH overdose is also initially treated by cessation of the drug. Should the excessive anticoagulation be mild, both the APTT and factor Xa activity will return to an acceptable level relatively rapidly. However, in the case of UFH, the risk of haemorrhage (as is implied by a very prolonged APTT) can be reduced with protamine sulphate (generally 1 mg per 80–100 units of UFH), given by slow intravenous infusion. Protamine sulphate may also be used in LMWH-induced haemorrhage. Should the haemorrhage be life-threatening, then recombinant factor VIIa should be considered.

The situation with warfarin deserves special attention as it has a long biological half-life on the liver, so that immediate cessation off the drug will do nothing for the haemorrhage, or levels of clotting factors that contribute to the INR. Many oral anticoagulant services have a set protocol for acting on a raised INR, which includes reducing or stopping the drug, and giving the effective antidote, vitamin K, orally or by injection. A further problem is that a patient with a high INR may be completely asymptomatic; and vice versa, a patient whose bruising resulted from a trauma, such as a fall, may have a perfect INR. In many cases, treatment may well be according to the INR, as this is where almost all science and epidemiology data have been derived.

Therapeutic decisions are dependent on the INR, whether there is minor or major bleeding, other risk factors for bleeding (e.g. age >70 years, previous bleeding complications) and reason for anticoagulation. Table 8.13 is applicable to a high INR in an outpatient, and Table 8.14 for an inpatient. However, these tables are offered for guidance only – practitioners will refer to their own local guidelines. In the UK, each NHS hospital will have its own documents.

In all cases the patient's current warfarin dose stability, concurrent medications and clinical condition must be considered when advising dose change and date of retest. Significant bleeding (e.g. haematuria) with any INR result requires focussed attention. Any outpatient receiving oral

Table 8.13 Action in response to a high INR in an outpatient.

INR	Action
3.1–3.5 (only if target INR 2.5)	Consider same dose (e.g. if 3.1) or reduce by about 5% (e.g. if 3.5). Retest in 2–5 days
3.6–3.9 (only if target INR 2.5)	Reduce dose by about 5–10%. Retest in 2–5 days
4.0–5.9	Stop warfarin for 1–2 days. Recommence on reduced dose (by 10–15%). Retest in 2–5 days
6.0–6.9	Stop warfarin for 2–3 days. Recommence on reduced dose (by 15–20%). Retest in 2–4 days
7.0–7.9	Stop warfarin for 2–3 days. Recommence on reduced dose (by 20–30%). Retest next day
8.0–8.9	Stop warfarin. Consider vitamin K oral 1.0 mg, retest next day
9.0–11.9	Stop warfarin. Consider vitamin K oral 1.0–2.0 mg, retest next day
12.0–14.9	Inform consultant/head of service. Give vitamin K, oral, 2 mg. Stop warfarin, retest next day
15.0–20.0	Inform consultant/head of service. Give vitamin K oral 2.5–5.0 mg. Stop warfarin, retest the next day

Table 8.14 Action in response to a high INR in an inpatient.

INR	Action
3.1–6.0 (only if target INR 2.5)	Reduce warfarin dose (by 10–20%) or stop. Restart when INR < 5.0
4.0–6.0 (only if target INR 3.5)	Reduce warfarin dose (by 10–20%) or stop. Restart when INR < 5.0
6.0–8.0 with no or minor bleeding	Stop warfarin. Restart when INR < 5.0. Consider vitamin K 0.5–1.0 mg orally
>8.0 with no or minor bleeding	Stop warfarin. Restart when INR < 5.0. If other risk factors for bleeding are present, give 2.5 mg vitamin K oral or intravenously (for INRs 12–20, give 5 mg vitamin K). Repeat dose of vitamin K after 24 h if INR still high

anticoagulants who may be experiencing a major bleeding episode should be advised to seek urgent medical attention from the nearest accident and emergency department.

In managing a high INR in a hospitalized patient, the situation is likely to be more serious, as by definition the patient requires close care. The practitioner will need to know the INR, any current or previous bleeding, reason for anticoagulation, the patient's age and diagnosis (elderly patients are more likely to bleed). Bleeding at a therapeutic INR needs investigation for a potential local cause. Other diagnoses, such as liver dysfunction, are relevant in predicting responses to treatment.

If there is major bleeding in an in- or out-patient, warfarin should be stopped immediately. Four-factor prothrombin complex concentrate (e.g. 25–50 units/kg) should be given, alongside 5 mg vitamin K intravenously, and repeat dose of vitamin K after 24 h if the INR is still high. Recombinant factor VIIa is not recommended, and fresh frozen plasma used only if there is no prothrombin complex concentrate.

Intravenous vitamin K rarely causes allergy and has a good safety profile. The degree and speed of reversal with vitamin K varies on an individual basis. For example, patients with prosthetic valves may require fresh frozen plasma only and a very small dose of vitamin K to avoid oral anticoagulant resistance later. Intravenous vitamin K reverses more rapidly than oral vitamin K and should be used if reversal is urgent.

There are no specific antidotes for the new oral anticoagulants dabigatran, apixaban and rivaroxaban, and the first line is cessation. Bleeding in patient who have taken a high dose of dabigatran may be treated with oral activated charcoal, haemoperfusion and haemodialysis. In situations with ongoing life-threatening bleeding in any drug, prothrombin complex concentrate and recombinant factor VIIa should be considered.

Tranexamic acid   This agent is anti-fibrinolytic, acting by inhibiting the activation of plasminogen to plasmin, and so maintains the integrity of the thrombus. It may be used in mild to moderate primary or secondary haemorrhage, and is available over the counter for the relief of menorrhagia. It may also be used in orthopaedic and cardiac surgery, but care is required for the treatment of haemorrhage after major trauma.

8.5   Disseminated intravascular coagulation

This extremely dangerous and life-threatening condition provides the opportunity to bring together many of the aspects of haemostasis, thrombosis and haemorrhage we have been examining in this chapter and Chapter 7.

Pathophysiology

The basic problem with DIC is that the coagulation system has become permanently activated, perhaps overactivated, by any one of a series of pathological conditions. These include protracted abdominal or thoracic surgery, obstetric problems, massive burns, septicaemia (with the release of endotoxin) and acute myeloid leukaemia. A potential cause is increased levels of tissue factor, leading to the activation of factor VII and so the generation of high levels of extrinsic tenase complex and so thrombin. If the aetiology is inflammatory, high levels of tissue factor may be the result of cytokine-driven activation of the endothelium and monocyte/macrophages. However, some malignancies (lung, pancreas, prostate) may be associated with overexpression of plasminogen activator inhibitor, and so the inhibition of fibrinolysis.

As a consequence, the coagulation system generates not one thrombus (nor a small number) at a small number of anatomical locations, but perhaps millions of microthrombi that become deposited on the endothelium in various capillary beds. Should the particular vascular bed be the skin, a vasculitis rash will develop. Similarly, DIC may also be accompanied by pulmonary problems (leading to respiratory distress, oedema and failure), hepatitis (leading to liver failure and jaundice) and acute renal failure (with glomerulonephritis). This is why DIC is such a clinically demanding syndrome, often requiring a stay in an intensive care unit, and why it carries a high risk of mortality.

Deposition of the microthrombi leads to activation and damage to the endothelium, which, as we saw in Chapter 7, is normally anticoagulant; for example, in expressing heparin. Therefore, vascular insult will lead to the loss of this anticoagulant nature and the development of a procoagulant nature, as in Virchow's triad, which contributes to the coagulopathy (such as loss of thrombomodulin and the release of factor V). A consequence of the hyperactivation is therefore the consumption of platelets and coagulation factors, and so the inability to prevent haemorrhage.

Normally, the inhibitor system should act against an overactive coagulopathy. However, the whole point of DIC is that the response is supra-normal, and consequently antithrombin, tissue factor pathway inhibitor and the protein C system are overwhelmed, so that this regulation is absent.

Assessment

The full blood count will reveal thrombocytopenia, and coagulation studies prolonged PT, thrombin time and APTT with hypofibrinogenaemia. These are all the consequences of consumption for coagulation molecules and platelets in microthrombi. It is also likely that other, nonroutine, tests such as vWf (reflecting vascular perturbation, and further contributing to the coagulopathy) and soluble P-selectin (marking platelet activation) will also be raised. The fibrinolytic system is generally unimpaired in DIC, so that with the dissolution of thrombi the levels of d-dimers will be high.

Other haematological abnormalities depend on the aetiology, but a leukocytosis is inevitable, and there may also be a nonimmune haemolytic anaemia.

Treatment

At first sight it would seem obvious that the treatment of thrombocytopenia and coagulation factor deficiency in the face of haemorrhage would be treated by platelet transfusion and fresh frozen plasma, both to replace missing components. However, because the basic problem is the overactivation of the entire system, any infused material will simply be consumed by the coagulopathy, and indeed may worsen the picture.

A clinical conundrum is provided by the increased risk of thrombosis, which fails to discriminate between arterial and venous thrombosis, the latter possibly leading to PE, which may be fatal. Normally, of course, this risk would be addressed with anticoagulants, probably an LMWH, were it not for the likelihood that the patient is already pathologically anticoagulated by virtue of the disease.

Blood science angle: DIC

This condition carries a number of consequences that call on the blood scientist. An aetiology based on leukaemia clearly calls for knowledge of white cells, whilst an inflammatory aetiology (or a developing inflammatory response) requires skills in immunology. Renal failure is a consideration, requiring U&Es, and liver function must be monitored with liver function tests as this organ generates coagulation proteins. If the disease spreads to the lungs, possibly requiring artificial ventilation, blood gas analysis and knowledge of respiratory acidosis and alkalosis are required.

In many cases, we, as a broad profession, are in our infancy in our ability to treat this disease. A great deal of knowledge of the pathophysiology nevertheless fails to provide a firm direction for an effective treatment. It was hoped that infusion of a commercial preparation of activated protein C would be able to inhibit the coagulopathy, which seemed likely, but regretfully this failed to provide a clear advantage, and has been abandoned. However, there are indications that infusion of recombinant soluble thrombomodulin may be useful in selected cases.

8.6   Molecular genetics in haemostasis

Molecular genetics has been, and continues to be, very important in determining the precise causes of many diseases of haemostasis.

Coagulation factors

Major contributions to our understanding of the coagulation cascade have been:

- Demonstration of the *G20210A* mutation of prothrombin as a risk factor for VTE.
- FVL as a risk factor for VTE.
- The genetic deficiencies in factor VII, factor VIII and factor IX.

- Mutations in the gene for vWf, leading directly to qualitative and quantitative defects which explain the different types of vWd (Figure 8.6).

The efficacy of the metabolism of various drugs, including warfarin, can be part-determined by knowledge of the particular variant of the cytochrome P450 and *VKORC1* genes.

Platelets

Similarly, the identification of the precise molecular lesion in platelet diseases such as Bernard–Soulier syndrome and Glanzmann's thrombasthenia (leading to defects in membrane glycoproteins) has only been possible via the use of molecular genetics.

Broader value of molecular genetics

An additional aspect of the 'academic' approach is the possibility that increasing knowledge of the different types of genetic diseases may one day lead to a disease-specific treatment. The best example of this is the development in tyrosine kinase inhibitors in *BCR–ABL* positive leukaemia that could only have been made with the benefit of knowledge of the particular genetic lesion. Full knowledge of the structure of a receptor, available only from studies of the gene, can lead to the development of precisely engineered inhibitors.

Confirming diagnosis

However, a second aspect of the use of these techniques is in confirming the diagnosis, not only in the patient, but also in their first-degree relatives. Recall that perhaps 30% of cases of haemophilia are spontaneous mutations, and that the parents will need to be aware of their genotype (as in haemoglobinopathy). Probing for the *JAK2V617F* mutation will confirm or refute a preliminary diagnosis of ET. Similarly, investigation of familial thrombocythaemia will benefit from studies of the genes for thrombopoietin and its receptor (*Mpl* and *THPO*).

A further example is a VTE in a middle-aged patient apparently caused by a thrombophilic prompt such as FVL. In almost all cases, they will have acquired this gene from their parents, and in turn may well have passed this gene, and therefore the risk factor, to their children, in whom the effect may or may not become apparent until perhaps decades in the future. This may be important for the patient's daughter when she is pregnant, more so if obese, and so could guide prophylaxis.

8.7   Case studies

Case study 7

A 65-year-old man with no English visits the oral anticoagulation clinic with his wife and an interpreter. The indication for warfarin is DVT and his duration of treatment is 6 months. His yellow book gives the following history, starting with a daily dose of 6 mg, but 7 mg on Wednesdays and Sundays.

Date	INR	Recommended daily dose	Next visit (weeks)
8 October	2.5	Same dose	2
22 October	1.5	Same dose	1
29 October	2.9	Same dose	2
12 November	4.0	Miss one dose, then resume	1
19 November	3.1	Same dose	2
3 December	1.2	6 mg/7 mg alternate days	1
10 December	2.8	Same dose	2
24 December	8.5	Stop warfarin, 1 mg vitamin K given	(4 days)
28 December	3.2	5 mg daily	(3 days)
31 December	2.1	6 mg daily	1
7 January	1.5	6 mg/7 mg alternate days	1
14 January	1.3	7 mg daily	1
21 January	3.6	What to do . . . ?	

Interpretation and plan   Although management started well on 6 mg, but 7 mg Wednesdays and Sundays, the decision to keep the dose the same after the low dose on 22 October was rewarded with an acceptable INR the following week. However, the high INR on 12 November required action, and this too was rewarded. However, on 3 December the low INR demanded an increase in the dose, and the possible additional use of an LMWH, but this prompted an overshoot and a very high INR requiring vitamin K and close monitoring. Although the situation seemed to settle down briefly in early January, it rapidly deteriorated.

This is clearly a difficult case to manage. A major risk factor is that he requires an interpreter, so is likely to fail to fully understand the instructions of what tablets to take and when. The worrying changes in his INR may also be due to changes in other medications. At least half of his dosing record suggests the lack of anticoagulation and so risk of VTE, or a high INR and so risk of haemorrhage. As his duration of anticoagulation is only a few more months, 150 mg aspirin may be better, so long as there is compliance. Aspirin is certainly inferior to warfarin in reducing the risk of a VTE, but it is markedly preferable to no treatment.

Case study 8

A 14-year-old female presents with a history of menorrhagia. On close questioning, the mother recalls her daughter seems to bruise easily, and that there was a lot of bleeding after a dental extraction. A full blood count and coagulation screen are ordered. The full blood count is normal (platelets 243×10^9/L, reference range $(140–400) \times 10^9$/L). The coagulation results are as follows.

	Result	Reference range
PT	12 s	10–14
APTT	50 s	30–40
Thrombin time	10.5 s	9–11
Fibrinogen (Clauss)	2.6 g/L	2.0–4.0

The only abnormality is the modestly prolonged APTT, which may in theory be due to deficiencies of prothrombin, fibrinogen or factors V, VIII, IX, X, XI and XII. However, there is clearly plenty of fibrinogen, and since the PT, which relies on prothrombin, fibrinogen and factors V, VII and X, is normal, we seem able to exclude problems with these factors. This leaves factors VIII, IX, XI and XII, which will be focussed upon.

Deficiencies in factors VIII, IX, XI and XII are either extremely rare and/or very unlikely to present for the first time in a teenage female. From the epidemiological viewpoint, vWd is the most likely diagnosis, which would call for vWf analysis by ELISA, ristocetin cofactor activity and multimer pattern. In view of the only moderately prolonged APTT, the final diagnosis may be type 1 or any of the type 2 variants. Genetic analysis and additional tests are of potential use and would need the collaboration of a reference laboratory. The young woman may be

helped by tranexamic acid and endocrine therapy (synthetic oestrogens).

Summary

• The consequences of the loss of haemostasis are thrombosis and haemorrhage. These may be directly life-threatening in both acute and chronic presentations.
• Overactive and/or increased numbers of platelets are a cause of arterial thrombosis. A high platelet count (thrombocytosis) may be reactive, or result from malignancy (essential and familial thrombocythaemia).
• The overactivity of the coagulation pathway, which may be driven by risk factors or result from gene mutation, leads to VTE.
• Prophylaxis and treatment of arterial thrombosis is with anti-platelet agents such as aspirin and clopidogrel. In VTE, drugs of choice are anticoagulants UFH, LMWH, warfarin, and new oral agents that target thrombin and factor Xa.
• Defects in platelet function that lead to haemorrhage include Bernard–Soulier syndrome and Glanzmann's thrombasthenia. These may be detected with methods such as platelet aggregation and flow cytometry.
• Haemorrhage resulting from low levels of coagulation factors may be classified as primary or secondary. The former include haemophilia, vWd and factor IX deficiency.
• Haemorrhage resulting from secondary factor deficiency is inevitably due to errors in the management of anticoagulants such as UFH, LMWH and warfarin.
• DIC, the consequence of unregulated activation of haemostasis, is the ultimate challenge for the scientist and clinician.

References

Moore GW, Jennings I. Thrombophilia. In: Moore GW, Knight G, Blann AD (eds). Haematology. Oxford: Oxford University Press, 2010; pp. 528–565.
Severinsen MT, Overvad K, Johnsen SP. *et al.* Genetic susceptibility, smoking, obesity and risk of venous thromboembolism. Br J Haematol. 2010;149: 273–279.

Further reading

Alamelu J, Liesner R. Modern management of severe platelet function disorders. Br J Haematol. 2010;149: 813–823.

Angiolillo DJ, Ueno M, Goto S. Basic principles of platelet biology and clinical implications. Circ J. 2010; 74:597–607.

Castellone DD, Van Cott EM. Laboratory monitoring of new anticoagulants. Am J Hematol. 2010;85:185–187.

Chong BH, Ho SJ. Autoimmune thrombocytopenia. J Thromb Haemost. 2005;3:1763–1772.

Davenport R, Khan S. Management of major trauma haemorrhage: treatment priorities and controversies. Br J Haematol. 2011;155:537–548.

De Groot PG, Derksen RHWM. Pathophysiology of the antiphospholipid syndrome. J Thromb Haemost. 2005;3:1854–1860.

Harrison C. Rethinking disease definitions and therapeutic strategies in essential thrombocythaemia and polycythaemia vera. Hematology Am Soc Hematol Educ Program. 2010;2010:129–134.

Huang LJ, Shen YM, Bulut GB. Advances in understanding the pathogenesis of primary familial and congenital polycythaemia. Br J Haematol. 2010;148:844–852.

Kaneko T, Wada H. Diagnostic criteria and laboratory tests for disseminated intravascular coagulation. J Clin Exp Hematop. 2011;51:67–76.

Lhermusier T, Chap H, Payrastre B. Platelet membrane phospholipid asymmetry: from the characterization of a scramblase activity to the identification of an essential protein mutated in Scott syndrome. J Thromb Haemost. 2011;9:1883–1891.

Lijfering WM, Rosendaal FR, Cannegieter SC. Risk factors for venous thrombosis – current understanding from an epidemiological point of view. Br J Haematol. 2010;149:824–833.

Nugent D, McMillan R, Nichol JI, Slichter SJ. Pathogenesis of chronic immune thrombocytopenia: increased platelet destruction and/or decreased platelet production. Br J Haematol. 2009;146:585–596.

Perry DJ, Fitzmaurice DA, Kitchen S. et al. Point-of-care testing in haemostasis. Br J Haematol. 2010;150: 501–514.

Refaai MA, Phipps RP, Spinelli SL, Blumberg N. Platelet transfusions: impact on hemostasis, thrombosis, inflammation and clinical outcomes. Thromb Res. 2011;127:287–291.

Rodeghiero F, Castaman G, Tosetto A. Optimizing treatment of von Willebrand disease by using phenotypic and molecular data. Hematology Am Soc Hematol Educ Program. 2009;2009:113–123.

Wijeyeratne YD, Heptinstall S. Anti-platelet therapy: ADP receptor antagonists. Br J Clin Pharmacol. 2011;72:647–657.

Guidelines

Baglin T, Gray H, Graves M. et al. Clinical guidelines for testing for heritable thrombophilia. Br J Haematol. 2010;149:209–220.

Collins PW, Chalmers E, Hart DP. et al. Diagnosis and treatment of factor VIII and IX inhibitors in congenital haemophilia: (4th edition). Br J Haematol. 2013;160: 153–170.

Harrison P, Mackie I, Mumford A. et al. BCSH guidelines for the laboratory investigation of heritable disorders of platelet function. Br J Haematol. 2011;155:30–44.

Keelings D, Baglin T, Tait C. et al. Guidelines on oral coagulation with warfarin – fourth edition. Br J Haematol. 2011;154:311–324.

Levi M, Toh CH, Thachil J, Watson HG. Guidelines for the diagnosis and management of disseminated intravascular coagulation. Br J Haematol. 2009; 145:24–33.

Makris M, Van Veen JJ, Tait CR. et al. British Committee for Standards in Haematology. Guideline on the management of bleeding in patients on antithrombotic agents. Br J Haematol. 2013;160:35–46.

NICE, CG144. Venous thromboembolic diseases: the management of venous thromboembolic diseases and the role of thrombophilia testing.

Watson H, Davidson S, Keeling D. Guidelines on the diagnosis and management of heparin-induced thrombocytopenia: second edition. Br J Haematol. 2012;159:528–540.

Web sites

www.pathologyoutlines.com/coagulation.html
www.bcshguidelines.com/
www.nice.org.uk/

9 Immunopathology

Cognosce te ipsum: Know thyself

Learning objectives

After studying this chapter, you should be able to

- appreciate that the key to immunology, and so immunopathology, is the ability to distinguish self from non-self;
- be aware of the cellular and humoral components of the immune system;
- describe the functions and diversity of white cells and antibodies;
- outline the role of complement;
- discuss the causes and consequences of immunodeficiency;
- recognize the key features of hypersensitivity;
- explain the basis of major autoimmune diseases.

The central tenet of immunology is the concept of *self*. It follows that anything which is not 'self' must be 'non-self'. Over the millennia, natural selection has taught us that any non-self cell or molecule is likely to be dangerous, and therefore we must have a defence mechanism, which is our immune system. All immunological processes, operating at the level of the cell and molecule can be reduced to this simple concept, and Chapter 5 on white blood cells is crucial to the understanding of immunology. As all white cell functions involve immunology and/or inflammation, it could be argued that this discipline is effectively a subdiscipline of haematology. Similarly, Chapter 11 shows that the entire basis of blood transfusion (principally antibodies reacting with antigens on red cells) is also immunological.

Several other chapters of this book describe immunological processes. Chapter 2 outlines the value of enzyme-linked immunosorbent assay (ELISA) as a method for assessing levels of many different types of molecule in the blood (such as hormones), whilst Chapters 12 and 17 point out that renal and liver disease may be caused by a misdirected immune response. Similarly, Chapters 15 and 18 explain how an inappropriate immune response can cause bone disease and endocrine disease. An important group of red cell diseases (the autoimmune haemolytic anaemias) are the result of the destruction of these cells by the immune system, as is explained in Chapter 4. Before we can discuss the pathology of the immune system, we must first appreciate its physiology. As Chapter 5 has details of the function of the immune system, there is no need to reproduce it fully, although there will be a brief summary.

The immune system exists to defend us against pathogens, and failure to deal with these leads to immunopathology (Section 9.1). To address these issues we will review the immune system in Section 9.2, and soluble factors in the plasma, which are often described as humoral (these being antibodies and complement, Section 9.3).

Once we have an appreciation of how the immune system works, we will address diseases of immunology and inflammation. These can be classified as underactivity (Section 9.4) and overactivity, of which there are two types: hypersensitivity (Section 9.5) and autoimmunity (Section 9.6). The immune system is amenable to being primed to defend us from certain threats; aspects of immunotherapy are discussed in Section 9.7. Section 9.8 summarizes the different tests that the immunology laboratory has at its disposal, and the chapter will conclude with some case studies in Section 9.9.

9.1 Introduction

In as much as you (biologically) are your body (self), the function of the immune system is to protect you from

Blood Science: Principles and Pathology, First Edition. Andrew Blann and Nessar Ahmed.
© 2014 John Wiley & Sons, Ltd. Published 2014 by John Wiley & Sons, Ltd.

microbiological attack, by recognizing what is not you (i.e. non-self), and then destroying it. However, this concept does not apply universally; it generally expresses itself in the recognition, and so destruction, of factors that would cause us harm; that is, pathogens.

Pathogens

These disease-causing objects are best classified by size. Atoms are too small to be recognized by the immune system, although several are toxic, as are certain molecules. One of the most common experimental chemicals that can stimulate the immune system is a modified benzene molecule with a relative molecular mass of about 120 Da; any molecule smaller than this does not generally cause a response. However, in the real biological world, the immune system focuses on four groups of biological pathogens that include thousands of different organisms and that are studied by microbiologists. A further point is that these microbial pathogens are parasites (i.e. an organism that benefits at the expense of another).

• The smallest are viruses, effectively a section of either deoxyribonucleic acid (DNA) or ribonucleic acid (RNA) enclosed within a protein coat. Examples of viral pathogens include polio, measles, influenza, hepatitis and human immunodeficiency virus (HIV). Generally sized 0.1 μm, these can only be viewed directly by electron microscopy.
• Generally with a size of 0.3–5 μm, bacteria can be viewed by light microscopy individually, or as clusters or chains of spherical (cocci), rod-shaped (bacilli) or spiral (spirilla) organisms. The vast majority are so called 'free living', whilst a small proportion must pass a portion of their life cycle as intracellular parasites. Their outer layers are composed of complex polysaccharides cross-linked by peptides, the precise composition of which varies and gives rise to different staining with Gram's stain as positive or negative, and which defines their 'non-self' nature.
• Microbial fungi include *Aspergillus* species (of which there are hundreds, and which characteristically grow in long chains called hyphae), *Candida albicans* (yeast, causing the infection thrush), histoplasmosis species and *Pneumocystis jirovecii*.
• Larger, more complex parasites can be protozoa (single celled organisms) and metazoa. The former include the various *Plasmodium* species that cause malaria (*P. falciparum*, *P. vivax*, *P. ovale*, *P. knowlesi* and *P. malariae*) and infect red cells (Chapter 4), amoeba (*Entamoeba*

histolytica, often causing amoebic dysentery), leishmania and trypanosomes. Metazoan parasites include filaria, schistosomes and various worms (tapeworm, liver flukes). Many of these parasites gain entry to the body, others remain on the outside of the body as ectoparasites (such as leeches and certain arthropods – fleas, ticks and lice).

Not all microbes are pathogenic; indeed, there is evidence of 'good' bacteria that prevent pathogenic bacteria from occupying a particular niche. The key defence systems against all these pathogens are white blood cells and two classes of molecules: antibodies and complement. However, other defences include mucus and the acid and enzymes of the intestines, and the skin itself is an excellent defensive barrier. These are part of the 'innate' system of defences, and are not specific for any particular organism or group of organisms.

Immunopathology

Disease is the consequence of failure of physiology. As regards immunology (and its closely related process, inflammation), disease can result from an insufficient immune response (i.e. failure to defend against the pathogen listed above) or an excessive or inappropriate immune response (where the defence is overactive and/or misdirected):

• immunodeficiency allows pathogens free reign over the body, causing local and systemic disease (Section 9.4);
• overactivity of the immune system includes allergy and hypersensitivity, being manifest in conditions such as asthma and hay fever (Section 9.5);
• A considerable burden of morbidity and mortality is due to the body attacking itself, this being called autoimmunity, where antibodies erroneously target not pathogenic microbes but the body's own tissues (i.e. become autoantibodies) (Section 9.6).

Before we embark on the examination of immunopathology we will briefly review the working of the immune system.

9.2 Basics of the immune system

Chapter 5, featuring the physiology of white cells, has details of the structure and function of five types of white cells, which in order of frequency are neutrophils,

Table 9.1 Principal physiological functions of white blood cells.

Cell	Functions
Neutrophil	Phagocytosis, participation in inflammation
Lymphocyte	Production of antibodies, defence from viruses
Monocyte and macrophage	Phagocytosis, participation in inflammation, release of cytokines, presentation of antigens to lymphocytes in antibody production
Eosinophil	Defence from parasites, release of histamine
Basophil	Defence from parasites, release of histamine and heparin

lymphocytes, monocytes, eosinophils and basophils. Monocytes are found in the blood, but if they move out and into the tissues, such as the skin and lungs, they become macrophages. Similarly, basophils move out of the blood, and in the tissues become mast cells.

The different types of white cells can be distinguished in the laboratory by their morphology, such as the proportions of the nucleus and cytoplasm, and the granularity, and by the expression of CD molecules at the surface of the cell. Each white cell has different functions (Table 9.1).

The anatomy of the immune system

Clearly, white blood cells are found in the blood. However, certain parts of the body are modified to provide specialist sites where the immune system can perform its functions. All white cells arise by the process of leukopoiesis from the bone marrow (mostly the sternum, pelvis, ribs and long bones of the limbs, as is outlined in Chapter 5), although in some circumstances the liver and spleen can also produce these cells.

The second organ of note is the lymph node, of which there are some 500–600 of varying sizes scattered all over the body. These organs are where white cells (primarily lymphocytes and monocytes, the latter described as dendritic cells, or antigen-presenting cells) come together to produce antibodies. However, antibody production can also happen in the spleen and the liver, the latter organ being the site of production of most complement components. Some specialist lymph nodes are found in defined anatomical sites and have defined functions.

A third crucial organ is the thymus. T lymphocytes must pass through this organ to complete their development, and severe disease follows if this transformation is incomplete, typified by the lack of this organ in diGeorge syndrome (Section 9.4).

Inflammation

When a particular organ or tissue is subjected to inflammation (i.e. it is inflamed) it carries the suffix itis. Hence, inflammation of the nephron (the basic unit of the kidney) is nephritis; but more precisely, inflammation of the glomerulus is glomerulonephritis. Acute inflammation is generally a healthy and short-lived response to a pathogen, but if this gets out of hand it can lead to long-lasting disease: chronic inflammation. This we shall return to in Section 9.6.

The acute-phase response

The whole-body effects of inflammation include the acute-phase response, which sees a rise in blood pressure, heart rate, and in temperature, but also the production of a large number of plasma proteins, many produced by the liver (as described in Chapter 17). The likely drivers for this response are inflammatory cytokines such as interleukins (ILs) and tumour necrosis factor (TNF), produced by leukocytes.

Proteins increased by the acute-phase response include serum amyloid A, complement components, coagulation proteins, plasminogen, ferritin, caeruloplasmin, orosomucoid, alpha-1-antitrypsin, alpha-1-antichymotrypsin and immunoglobulins (Igs). Conversely, levels of certain proteins, including albumin, transferrin and antithrombin, are reduced. However, the gold-standard marker is c-reactive peptide (CRP), and is widely used to define and monitor the acute phase.

The acute-phase response also acts on the bone marrow, resulting in an increase in the white cell count, whilst all these act to increase the erythrocyte sedimentation rate (ESR). A further consequence of the acute-phase response is the mobilization of white blood cells that are temporarily located (marginating) on the blood vessel wall, so that the total white cell count and neutrophil count both rise, inducing a leukocytosis and a neutrophilia respectively.

All these factors are desirable, and are (within limits) the response of the body to the threat of, or actual, microbial attack. However, if these responses are too powerful, actual disease may result, as in septicaemia. Furthermore, if acute inflammation persists it may become transformed into chronic inflammation, which is certainly undesirable and pathological. Nevertheless, in chronic inflammation, typified by rheumatoid arthritis (RA; Section 9.6), the markers are still referred to as 'acute'-phase reactants, not as 'chronic'-phase reactants.

Immunity

In contrast to inflammation, immunity is complex and to be fully effective needs to develop over hours and days. Accordingly, it is described as 'adaptive', and in further contrast to inflammation it is specific for the particular infective pathogen, and is present only in vertebrates. Central to the concept of immunity is the lymphocyte, of which there are two types, the T cells and B cells, although many consider the natural killer (NK) cell to be a type of lymphocyte.

An additional aspect is that, unlike inflammation, adaptive immune responses are long-lived, as memory T lymphocytes and B lymphocytes can be reawakened to provide support in defence against defined pathogens decades after an initial exposure. Indeed, this is the basis of vaccination. The 'self' versus 'non-self' aspect of inflammation is crude, but in immunity it is highly developed.

9.3   Humoral immunity

The two major aspects of humoral immunity are antibodies and complement, although there are several other molecules (such as enzymes and mucopolysaccharides) that help defend us from pathogens. In many laboratories, the study of these soluble molecules is called serology.

Antibodies

This group of molecules is briefly mentioned in Chapter 1, but full details are presented in Chapter 5, which refers to their production by B lymphocytes. The previous section has underlined the importance of the lymph node as the site of this production, but antibodies may also be produced in liver, spleen and bone marrow. Chapter 17 explains the classification of these molecules as gamma globulins, and how they are defined by a migration pattern in electrophoresis.

The salient points of antibodies are as follows:

- Antibodies are constructed to a common 'Y' shape, from four polypeptide chains: two 'heavy' chains of approximately 50 kDa each and two 'light' chains of approximately 25 kDa each (Figure 9.1). The polypeptide chains are linked by disulphide bonds.
- From the protein chemistry viewpoint, these molecule belong to the globulin family, and since they are involved in immunological processes are therefore also called Igs.
- Each antibody molecule recognizes antigens by the combined shape of the ends of the heavy and light chains; this is called the 'fraction antigen binding', or Fab region. It follows that each individual antibody molecule has two Fab regions.
- At the other end of the molecule, the two heavy chains form the 'fraction crystallizable', or Fc region. This region can locate into specialized structures on certain white cells called 'Fc receptors', or FcRs.
- There are five different types of heavy chain: alpha (α), delta (δ), gamma (γ), epsilon (ε) and mu (μ). There are two different types of light chains: kappa (κ) and lambda (λ). There are four subclasses of gamma chains and two subclasses of alpha chains.
- The particular heavy chain make-up of the molecule confers a different structure and function, and so are classified as IgG if a gamma heavy chain is present, IgA if alpha, IgM if mu, IgD if delta and IgE if epsilon. Both kappa and lambda light chains can associate with any heavy chain.
- IgG is the dominant antibody in plasma (at a concentration in the region of 12 mg/mL) and in the tissues. There are four sub-classes (IgG1–IgG4), each with different physicochemical and functional properties. IgG molecules are always monomers of the basic structure (shown in Figure 9.1).
- IgA is present at the second highest concentration (approximately 3 mg/mL) and can exist as a monomer or as a dimer. The latter is two monomers connected by an extra polypeptide – a J chain. However, IgA may also be found in secretions (saliva, tears, nasal fluids, sweat, etc.), where it helps defend the mucosal surfaces, such as of the intestines. This secretory type of IgA is protected from digestion by a protective secretory component.
- IgM circulates in the plasma as a pentamer of the basic Ig unit, and so has a molecular weight of around 900 kDa. However, IgM at the surface of the B lymphocyte is a monomer. With a plasma concentration of approximately 1.5 mg/mL, it is very efficient at triggering complement activation, and its large size enables it to cross-link separate antigens, as may be present on two adjacent pathogens such as bacteria.

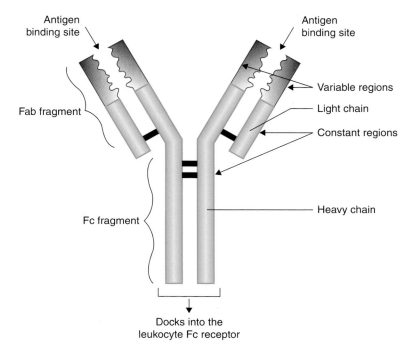

Figure 9.1 Simplified structure of an immunoglobulin molecule.

- IgE is present at very low levels in the plasma (approximately 0.05 μg/mL, although others prefer a reference range of 0–81 kU/L) and is a monomer. Notably, it is able to bind to basophils and induce them to degranulate. This is the basis of many hypersensitivity reactions (Section 9.5).
- IgD is also present in small amounts in plasma (approximately 30 μg/mL, 0–100 kU/L), but is expressed primarily on the surface of B lymphocytes where it may have a role in cell activation and suppression.
- IgG, IgA and IgM, and free kappa and lambda light chains can be easily measured by immunoturbidimetry and nephelometry, IgD and IgE often by immunoassay such as ELISA and chemiluminescence immunoassay (Chapter 2).
- Other techniques for detecting and quantifying the Igs and their subclasses include gel and capillary electrophoresis, and immunofixation (Chapter 6). These are important in investigating diseases such as myeloma.

Complement

This is a series of some 20 proteins (most being designated C and a number, or a name) that, like the coagulation pathway, come together in a set pattern to perform a precise function. The name complement is historical, as it was discovered and developed by Jules Bordet and Paul Erlich in the late 19th century as a property of serum that enhances the destruction of bacteria. Table 9.2 summarizes the key aspects of the complement system.

Table 9.2 Complement components.

Name of component	Pathway	Molecular weight (kDa)	Typical plasma concentration (g/L)
C1	Classical	800	0.18
C2	Classical	102	0.02
C3	Common	185	1.3
C4	Classical	205	0.35
C5	Terminal	190	0.07
C6	Terminal	120	0.065
C7	Terminal	110	0.055
C8	Terminal	152	0.055
C9	Terminal	69	0.06
Factor B	Alternative	93	0.21
Factor D	Alternative	24	0.02
Properdin	Alternative	Multiples of 53	0.05
C1 inhibitor	Classical	105	0.25

Complement pathways This system has several similarities with coagulation, such as the presence of different sub-pathways, the amplification of the process, the enzymatic activation of one molecule by another and the crucial role of a central enzyme. In the coagulation pathway, this is thrombin generated from prothrombin by the prothrombinase complex, whilst in complement, both C3a and C3b are generated from C3 by C3 convertase. There are three pathways for the generation of C3 convertase:

- The classical pathway, often activated by antibody–antigen complexes on the surface of cells and bacteria, requires three isoforms of C1: C1q, C1r and C1s. Together, these bind to the heavy chain of the antibody and then catalyse the formation of C2a and C2b from C2, and of C4a and C4b from C4. C4b can be degraded to C4c and C4d, the latter binding to the endothelium where it can be detected and so used as a marker of complement activity.
- The alternative pathway, where C3 is the substrate for factor B and factor D, which together form C3bBb, a complex stabilized by properdin, which also has a sub-unit structure and can vary in mass. The alternative pathway cascade can be triggered by C3b binding bacteria. This pathway operates at a background low level and ensures some C3 convertase is always present.
- The lectin pathway, which is the most recently discovered, but whose clinical relevance, and so value to immunologists, is as yet unclear.

The relationship between the different pathways is illustrated in Figure 9.2. Although often described as part of the innate aspects of the immune response, the classical pathway requires the presence of antibodies, the most effective being IgG and IgM, and may be seen as contributing to the adaptive arm of the response. However, the lectin and alternative pathways are fully innate, and can participate in the non-specific destruction of bacteria.

In the common pathway, the C3 convertase generates C3a and C3b. The latter then generates C5a and C5b. C3a and C5a have multiple pro-inflammatory functions that include being chemoattractants for phagocytes (neutrophils, monocytes), vasodilators, causing mast cells to release histamine, stimulating an oxidative burst (consumption of O_2) from neutrophils and promoting cytokine release from hepatocytes and leukocytes. If excessive, these effects can be related to a series of pathological conditions described as anaphylaxis, so that C3a and C5a are also described as anaphylatoxins, to be discussed fully in Section 9.4.

C3b and C4b have another role separate from that of propagating the complement pathway. Macrophages have receptors for these components (complement receptor 1, CR1), enabling the cell to home onto and begin the phagocytosis of bacteria that are coated with C3b and C4b. This activity is called opsonization, the facilitating molecule being called an opsonin. CR1 also binds to C1q, whilst a second receptor, CR2 (CD21), binds the breakdown products C3d, C3dg and iC3b. CR3, also known as Mac-1 or CD11b/CD18, binds iC3b and so supports the phagocytosis of microbes coated with this opsonin.

C5b is a key founder of the final part of the pathway, which generates a membrane attack complex with factor C6 to C9. The objective of the latter is to punch a hole in membranes large enough to facilitate the escape of cytoplasm and the influx of tissue fluid, both of which will lead to the death of the particular cell; hence 'complement-mediated lysis'.

The regulation of complement The parallel with coagulation extends to the activity of inhibitors that suppress the complement pathway. Soluble regulators include C1-inhibitor (a serpin that binds C1r and C1s), factor I (a protease that degrades C3b to iC3b and further to C3d, and of C4b to C4d, cofactors for which include factor H), vitronectin and carboxypeptidase N (which inactivates anaphylatoxins).

Membrane-bound complement receptors include CD35 (CR1, present on various cells, and which also degrades C3a and c5a), CD46 (a cofactor for the factor I inhibition of C3b and C4b), CD55 (accelerates the decay of C3 convertase on cells, hence is also known as decay accelerating factor) and CD59 (which regulates the formation of the membrane attack complex by inhibiting the interaction between C8 and C9).

Blood science angle: Complement, coagulation and blood transfusion

The striking parallels between the coagulation and complement pathways imply not only a common evolution but underline their importance in physiology and pathology. Thrombin can convert C5 to C5a, whilst, conversely, C1-inhibitor also blocks parts of the kinin, fibrinolytic and coagulation factors such as factors IX and XII.

Failure of regulation of the complement pathway leads to disease with thrombotic complications. Isoforms of C4a and C4b are the basis of the Chido/Rodgers blood group, whilst CR1 carries the Knops blood group antigens, some of which may be important in resisting malaria.

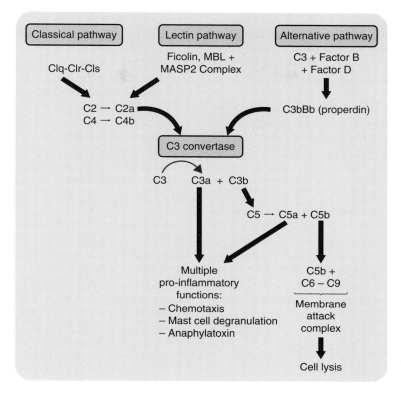

Figure 9.2 The complement pathway consists of three initial pathways that come together to form different types of C3 convertase. The products of this enzyme, C3a and C3b, go on to have other functions – the former as an inflammatory mediator and the latter as the generator of C5a (another inflammatory mediator) and C5b, the latter coming together with C6–C9 to form the membrane attack complex. MBL = mannose binding lectin, MASP2 = MBL-associated serine protease-2.

Cooperation

The sections we have just examined may give the impression that the innate aspect of inflammation and the acquired features of immunity are separate and independent parts of the overall immune response. This is not so, as there are many examples of the cooperation between these different processes. The adaptive response of antibodies combines with the innate complement system to facilitate cell lysis. Another example of cooperation is in phagocytosis, where macrophages and neutrophils ingest and destroy organisms such as bacteria and yeast as part of the innate defence response. The efficiency of this process is markedly enhanced if the target organisms are coated with antibodies, which are the product of the acquired immune response.

Now that we have an understanding of what the immune system is for and how it operates in health,

we can look at the consequences of its dysfunction: first, when it fails to operate correctly (immunodeficiency) and then when it is overactive (as in hypersensitivity) or acts inappropriately (leading to autoimmunity).

9.4 Immunopathology 1: immunodeficiency

In this section we will look at the failure of the immune system – that is, immunodeficiency – which can be viewed in terms of quantitative and qualitative changes in white cells and in the humoral factors: antibodies and complement. Some aspects of the underactivity of white cells are described briefly in Chapter 6, but now warrant additional discussion. Immunodeficiency may be primary (which is effectively genetic, being caused by a mutation) or secondary (where an outside agent causes disease in an otherwise healthy individual). Another key aspect is where the disease is present in only one particular part of the immune system (such as in T lymphocytes) or where several components are abnormal, the most well-described being severe combined immunodeficiency (SCID).

The consequences of an insufficient immune response are, unsurprisingly, recurrent microbial infections, the precise organisms depending on the particular defect. At first sight, a reasonably simple classification is that T lymphocyte deficiency leads to viral and fungal infections, whereas B lymphocyte (and therefore antibody responses) and complement deficiencies lead to bacterial infections. However, there are numerous exceptions to this general scheme.

A further aspect of immunodeficiency is its link with autoimmune disease and cancer, which seems to be causal but in many cases the precise aetiology is unclear. However, the failure of regulatory T lymphocytes to control self-destructive immune responses may lead to autoimmune disease (Section 9.6).

Quantitative deficiency in white cells

The number and concentration of most cells and molecules in the blood depend on the balance between the rate of production and the rate of loss. There are a few cases where white blood cells are lost or destroyed faster than they are produced, and similarly there are a few rare syndromes where white cells fail to develop. However, most cases of low numbers of white cells seen in everyday clinical and laboratory practice are inevitably due to failure of production in the bone marrow.

Bone marrow suppression   This can be due to a number of factors, such as:

- Suppression of leukopoiesis as a result of a viral infection or radiotherapy, although the most common cause is chemotherapy. The latter is any therapy with drugs, includes cytotoxic drugs in the treatment of cancer (methotrexate, vincristine), immunosuppressive drugs in the treatment of autoimmune disease (azathioprine, cyclophosphamide) and in the treatment of transplantation (ciclosporin) (details of which are presented in Section 9.4 and Section 9.5 respectively, and in Chapter 10). However, 'everyday' drugs not normally thought of as being dangerous can also suppress the bone marrow. These include anticonvulsants, antithyroid drugs, antibiotics, nonsteroidal anti-inflammatory drugs (NSAIDs) and some antipsychotic and antihelminth drugs.
- Cancer, where there is invasion of the bone marrow by a tumour whose primary site is elsewhere (lung, breast, prostate) and which effectively takes over those parts of the bone marrow where leukopoiesis is present.
- Myelofibrosis and myelodysplasia. In the former there is the overgrowth of 'supportive' bone marrow tissues, typically fibroblasts, which in many cases is driven by abnormal levels of growth factors and cytokines. Myelodysplasia is characterized by abnormal haemopoiesis, so that numbers of all blood cells often fall.

All of these have the potential to cause low levels of all classes of white cells, which is described as pan-leukopenia. However, some are more specific and cause low levels of only some white cells, such as that of carbimazole, used to treat hyperthyroidism, which causes neutropenia. Fortunately, most drug- or virus-induced leukopenia is reversible upon withdrawal of the causative agent. Leukopenia in the face of malignancy is likely to improve following the resolution of the disease, but this is of course considerably more problematical. All diseases of this type therefore cause secondary immunodeficiency.

Genetic causes of neutropenia   Several forms of severe congenital neutropenia (characterized by a neutrophil count $<0.5 \times 10^9/\text{L}$) are recognized, and most are due to the arrest of the maturation of precursors, mainly at the promyelocyte stage. These sporadic and mostly autosomal dominant disorders are linked to a variety of abnormal genes, such as *ELA2* (encoding neutrophil elstase), *CSF3R* (encoding the granulocyte colony-stimulating factor receptor) and *HAX-1* (coding for a mitochondrial anti-apoptotic protein). Other causes include abnormalities in the Wiskott–Aldrich syndrome protein, which lead to decreased proliferation and increased apoptosis of myeloid progenitors. Children with the closely related autosomal-recessive Kostmann's syndrome have an even lower neutrophil count ($<0.2 \times 10^9/\text{L}$). All diseases of this type are therefore primary immunodeficiency.

Human immunodeficiency virus-1   Few of us can be unaware of the importance of this virus. Its pathogenicity can be accounted for by the binding of a viral protein (gp120, the 'docking' glycoprotein) to the CD4 molecule. This molecule is present mostly on T helper lymphocytes, but is also present on some NK cells and a subgroup of monocyte/macrophages and their relatives (dendritic cells and the microglial cells of the central nervous system). Infection of the cell also requires a chemokine receptor – CXCR4 to gain entry to lymphocytes and CCR5 for macrophages – and this triggers fusion of the viral envelope with the cell membrane. Viral RNA is thus allowed access to the host cell's cytoplasmic processes, which it hijacks so as to reproduce itself.

The consequences of this for the cell are direct killing by the virus, an increased rate of apoptosis and targeting by the remainder of the immune response, all of which lead to falling numbers of T lymphocytes. Shortly after infection (which may resemble influenza or infectious mononucleosis with a transient lymphadenopathy), IgM and IgG antibodies to HIV can be detected, but following this the disease can be latent for years. However, there will be a slow and steady reduction in CD4 positive lymphocytes, which can be plotted in the laboratory by a fall in the absolute number of cells or a reversal of the CD4/CD8 ratio, which normally favours CD4 cells. Monitoring by flow cytometry is the most common method; CD4 T cells are the major lymphocyte group, but a reduction will not show up in the total lymphocyte count from the white cell differential until levels are profoundly low.

Counter to pathophysiology, where low levels of antibodies would be expected to be due to lack of T helper lymphocytes, there is often hypergammaglobulinaemia. This may be the result of dysregulated ('confused') B lymphocytes, which, lacking correct direction, synthesize abnormal antibodies that provide no protection. Falling numbers of CD4-bearing T cells are a direct marker of the progression of the disease to the acquired immunodeficiency syndrome (AIDS). The scientific literature on this virus and the disease it causes is considerable and easily accessible, and so accordingly does not require additional discussion.

DiGeorge syndrome This is caused by a deletion in a section of the long arm of chromosome 22, and is present in approximately 1 in 3000 live births. The consequences of this can include any combination of abnormalities in the thyroid, parathyroids and the palate, cardiac problems, learning disability and thymic aplasia. The latter can vary between a small thymus and no thymus at all, and because this organ is a requirement for T lymphocyte development (Chapter 5), T cells may be reduced in number or absent altogether. The laboratory method of choice for confirming the diagnosis is flow cytometry for lymphocyte subsets, although DiGeorge syndrome inevitably brings other clinical issues that aid diagnosis. The

closely related Nezelof syndrome is also associated with thymic dysplasia, and so T cell abnormalities.

Severe combined immunodeficiency This phenotype, with a frequency of perhaps 1 in 100 000 live births, is characterized predominantly by T lymphocyte abnormalities, and in several cases B cell function is unaffected. SCID has a number of causes, but two dominate.

- The X-chromosome-linked variant is caused by mutations in genes for receptors for cytokines, including IL-2 and IL-7. Accordingly, T cell and NK cell development are impaired. This phenotype is similar to a variant caused by a mutation in genes for the T cell receptor.
- A second major type of SCID is caused by a loss-of-function gene for the enzyme adenosine deaminase. This leads to increased levels of purines and dysregulation of nucleic acid metabolism, and so the inhibition of T and B lymphocyte and NK cell proliferation.

Qualitative defects in white cells

These diseases are characterized not by low levels of white cells, but that the white cells that are present fail to perform their functions. There are few diseases in this group, but one dominates.

Chronic granulomatous disease This condition is caused by a defect in the ability of the phagocyte (a neutrophil or a macrophage) to produce reactive oxygen species, such as the superoxide anion (O_2^-) and hydroxyl radical ($OH^\bullet$), which are of great importance in the destruction of bacteria. Failure to eradicate the bacteria leads to their consolidation in a small and localized nodule called a granuloma; hence the term chronic granulomatous disease (CGD). It has an incidence of approximately 1 in 225 000 live births.

A key enzyme in the generation of these bacteriocidal groups is nicotinamide adenine dinucleotide phosphate hydrogen (NADPH) oxidase, a complex enzyme with at least five subunits and two associated regulatory molecules that is located in the wall of the phagolysosome. CGD results from mutations in the genes for these subunits, which are therefore incorrectly assembled. This leads to low or absent NADPH oxidase activity, which in turn leads to poor (if any) oxygen radical production and so lack of bacteriocidal activity, resulting in infections. The gene for one of these subunits ($gp91^{phox}$) is found on the X-chromosome, abnormalities in which lead to approximately two-thirds of CGD cases.

Molecular genetics are rarely called upon to confirm the diagnosis of CGD as good immunological and biochemical tests are available. In a common laboratory test, the normal granulocyte metabolizes nitro-blue tetrazolium (NBT), converting it to the insoluble blue product formazan. Those cells lacking the appropriate pathway fail to metabolize the NBT, so that no blue product is formed. For those laboratories with a flow cytometer, the dye dihydrorhodamine can be used, as it is oxidized to rhodamine in normal cells. Consequently, failure of cells to generate rhodamine strongly suggests CGD.

> **Blood science angle: Leukopenia**
>
> Many of the causes of panleukopenia may also cause a reduction in numbers of red cells, leading to anaemia (Chapter 4), and of platelets, leading to thrombocytopenia (Chapter 8). If all cell types are decreased then the condition becomes pancytopenia. However, in the case of cancer chemotherapy, other organs may also be damaged. Hence, liver function tests (LFTs) may be needed to check for liver failure, and urea & electrolytes (U&Es) to assess damage to the kidneys.

Other neutrophil defects   A key feature of the Chediak–Higashi syndrome (alongside partial albinism) is the inability to correctly assemble lysosomes because of a loss-of-function mutation in the *LYST* gene. This leads to poor bacterial killing, and so repeated infections, whilst in the laboratory the blood film shows giant intracellular granules of acid hydrolases and myeloperoxidase. These are present in all granulocytes and monocytes, and platelets also have granule abnormalities, but in their dense granules.

Other defects in neutrophils leading to repeated bacterial infections include loss of the ability to adhere to cells, tissues and microbes because of a defect in the adhesion molecule CD18, and defects in IL-12 or the receptor for gamma-interferon.

Other white cell defects   Omenn's syndrome is caused by mutations in genes (*RAG-1* and *RAG-2*) involved in the generation of both the B lymphocyte receptor (i.e. antibodies) and the T cell receptor (details of these receptors are in Chapter 10). The syndrome is also characterized by enlarged liver, spleen and lymph nodes, raised IgE and an eosinophilia. There is often a lymphocytosis, but it is due to the expansion of clones of abnormal T cells that leads to a form of SCID. The

immunodeficiency of leukocyte malignancy will be discussed in a section that follows. Cell-mediated immunity is often compromised in ataxia telangiectasia and the Wiskott–Aldrich syndrome. In the former, this causes lack of IgA and IgE responses, and so upper respiratory tract infections, the latter being associated with low IgM responses.

Clinical aspects of defective cell-mediated immunity

Regardless of the aetiology, low numbers of neutrophils or dysfunctional neutrophils lead to bacterial infections. In the case of CGD this is often organisms such as catalase positive staphylococci. Treatments include antibiotics for the bacterial infection (as is required for all such infections), but ultimately the only cure is bone marrow transplantation.

Primary and secondary functional T lymphocyte deficiency is present in DiGeorge syndrome and HIV infection respectively. In the former, infants are often able to deal with bacterial infections but are overwhelmed by viral infections such as varicella and vaccinia. Those with defective T cell responses are at risk of intracellular pathogens such as certain bacteria (*Listeria, Mycobacterium, Legionella* species) and fungi (*Histoplasma* and *Cryptococcus* species). Furthermore, lacking T cells also leads to poor antibody production, as B cells are not provided with helper cells. Treatment includes thymus engrafting. Fungal infections caused by *Candida albicans* and *Aspergillus* species have their own specific treatments, such as clotrimazole and amphotericin respectively.

SCID is defined at the cell level by incompetent T and B lymphocyte responses, but this can be restored by bone marrow transplantation. As with its pathophysiology, the clinical consequences of the progression of an HIV infection, AIDS, are established and widely available. Briefly, the loss of T lymphocytes and CD4-positive monocyte/macrophages leads to any number of a large series of opportunistic bacterial, viral, fungal and parasitic infections. However, some pathogens dominate, such as the fungus *Pneumocystis jirovecii* (causing a pneumonia), the parasite *Toxoplasma gondii* (often leading to neurological disease), *Mycobacterium tuberculosis* (which also causes lung disease), and hepatitis, cytomegalovirus, herpes and papilloma viruses. The latter two may be responsible for some of the increased risk of lymphoma and cervical cancer that is associated with AIDS.

Blood science angle: HIV/AIDS

Leaving aside the huge worldwide problems with this disease, it represents a condition that is of interest to all aspects of blood science. The major infective route of the CD4 molecule we have outlined, but the consequences of immunodeficiency are also highly pertinent. Increased serum triacylglycerols alongside low total, high-density lipoprotein and low-density lipoprotein cholesterols are present, whilst treatment with the highly active antiretroviral therapy regime leads to combined hyperlipidaemia. Transmission by all forms of blood products is established and justifies they are screened for this virus before being given to a patient.

Although early drug treatments suppressed the bone marrow, leading to anaemia, more recently developed drugs are considerably less toxic. Genome-wide association studies and other techniques in molecular genetics are identifying genes linked to resistance and susceptibility to HIV infection, progression to AIDS, and the effects of therapy. Molecular genetics may also be called upon to determine the plasma viral load through levels of RNA. Although not directly part of blood science, microbiology has (as indicated above) a key role in identifying pathogenic infections.

Leukocyte malignancy

Leukaemia, myeloma and lymphoma deserve their own section, other details being found in Chapter 6. The former two are primary cancers of the bone marrow, and as such, like other cancers invading this organ, slowly suppress the production of healthy white cells, so that there is a functional deficiency, leading to infection as outlined in the previous section. The raised white cell count in leukaemia is due to abnormal white cells that are dysfunctional and so do not provide defence against infections. The myeloma tumour also interferes with normal haemopoiesis, as does leukaemia, but the dysfunctional tumour cells themselves (plasma cells) generally remain in the bone marrow until the final stages of the disease, at which point they may emerge and so contribute to an increased white cell count.

The tumour that is lymphoma grows within a lymph node, and so prevents normal antibody production. As the disease progresses this is compounded by the spread of the tumour to other lymph nodes, and so the eventual wider suppression of the generation of antibodies. Most lymphomas are of B cells, so that, theoretically, T lymphocyte responses are unaffected in the early stages of the disease, but if the tumour spreads more widely, and into the bone marrow, deficient cell-mediated immunity will occur. The malignant and dysfunctional lymphocyte that is the basis of the disease may appear in the blood, leading to an increased total white cell count.

Defects in humoral immunity

Antibodies These proteins are produced by the concerted action of T helper lymphocytes, antigen presenting cells and B lymphocytes, generally within the germinal centres of lymph nodes (Chapter 5). It follows that problems with any of these cells, or the microenvironment of the lymph node (such as lymphoma), will lead to low levels of antibodies (hypogammaglobulinaemia), or in extreme circumstances to their absence (agammaglobulinaemia). The precise definition depends on the total protein count and quantitative protein electrophoresis (Chapter 17).

Perhaps the most well-developed condition is X-linked agammaglobulinaemia, and so most unlikely to be present in females. With a frequency approximately 1 in 100 000, first described by Ogden Bruton in 1952, the disease is caused by a loss-of-function mutation in a gene for tyrosine kinase, of which 400 different mutations have been described. This enzyme is required for the maturation of pre-B lymphocytes into mature B cells, leading to loss of the ability of the latter to generate antibodies. Diagnosis relies on the absence of mature B lymphocytes, generally measured by flow cytometry of peripheral blood for CD19 and/or CD20. A mutation may be sought by standard molecular genetic techniques.

Hyper-IgM syndrome is caused by the failure of Ig class switch, so that an IgM response is fixed and does not develop. This leads to low levels of IgG, IgA and IgE antibodies. Five types have been characterized, the most common form involves defective CD40/CD154 signalling by T lymphocytes to B lymphocytes.

Complement Deficiencies in many complement components are known: they may be due to decreased synthesis by the liver, increased loss (perhaps in renal failure) and increased consumption in inflammatory disease. An autoantibody (C3 nephritis factor) to the C3 convertase complex can provide stability, leading to the consumption of complement components and so a deficiency in C3. Found at the centre of the complement system, C3 deficiency is clinically important. It may arise as above, by excessive consumption, or in a familial manner, inherited as an autosomal recessive mutation.

Deficiencies in several other complement components (such as C1, C2 and C4) are recognized, and may be inherited in an autosomal recessive manner. C2 deficiency is most common, with a heterozygous frequency of 1% of the Caucasian population, and a homozygous frequency of 1 in 10 000. Deficiency in C1 inhibitor is well described, and leads to the unregulated activation of C1, and therefore reduced levels of C3 and C4. These abnormalities are often present in diseases associated with immune complexes, such as systemic lupus erythematosus (SLE), systemic vasculitis and sub-acute bacterial endocarditis.

Genetic deficiencies in alternative pathway components factor D and properdin have been described but are exceedingly rare. Lack of inhibitor factors H and I are known and lead to excessive complement activity with the consumption of C3. The consequences of this include haemolytic-uraemic syndrome and membranoproliferative glomerulonephritis.

Deficiency in all components of the terminal pathway (C5–C9) are known, and may be autosomal recessive. Lack of C5 has additional implications as there will also be reduced levels of C5a and so impaired chemotaxis. Loss-of-function mutations in genes coding for integrins and adhesion molecules lead to failure to express receptors such as CR3.

Lack of C1 inhibitor, an essential regulator of the complement system, leads to the important clinical syndrome of angioedema, of which there are two types:

• With an estimated prevalence of approximately 1 in 50 000, hereditary angioedema (HAE) is caused by an autosomal dominant mutation in the gene *SERPING1* on chromosome 11. However, in perhaps 20% of cases the mutations are spontaneous. In type I HAE, present in 85% of all cases of HAE, C1 inhibitor is absent, or is present at low levels. In the remaining 15% of cases, type II HAE, the molecule is present or even elevated, but the molecule itself fails to provide inhibition of the complement pathway.
• Angioedema may be acquired (hence AAE) in lymphoproliferative (leukaemia, lymphoma) and autoimmune diseases. The most common aetiology is excessive consumption, possibly because an autoantibody renders it susceptible to degradation by serine proteases, and is found in non-organ-specific diseases such as RA and SLE (type I AAE). The presence of the autoantibody in lymphoproliferative disease confirms type II AAE.

Clinical aspects of defective humoral immunity As defective cell-mediated immunity leads to bacterial

infection, direct lack of antibodies leads to the same problem, and so requires treatment with antibiotics. However, another treatment is the infusion of purified gammaglobulins harvested from donated whole blood taken for transfusion (i.e. a blood product, Chapter 11). This is a requirement in agammaglobulinaemia, with its risk of infections with *Staphylococcus*, *Streptococcus*, *Neisseria* and *Haemophilus* species, but needs to be repeated at regular 3–4 week intervals. The laboratory is called upon to monitor plasma levels of total and individual gammaglobulin classes.

In hyper-IgM syndrome, the considerable size of IgM means it cannot leave the blood. Therefore, extravascular tissues (such as of the lung) are undefended, leading to recurrent bacterial infections and opportunistic infections, such as that of *Pneumocystis jirovecii*. Treatment is to replace the IgG, IgA and IgE in intravenous Ig infusions. CD40 ligand deficiency has been successfully treated with bone marrow transplantation.

There is no replacement therapy for complement deficiencies; instead, preventative therapy with antibiotics will be required for the typical *Neisseria gonorrhoea* and *Neisseria meningitidis* infections, which are common in failure to form a functioning membrane attack complex. Chapter 4 carries details of a disease called paroxysmal nocturnal haemoglobinuria, caused by the defective synthesis of glycosyl phosphatidylinositol, a molecule that anchors CD55 and CD59 into the membrane, and which are important in resisting the activation of complement. As a consequence, the red cells become sensitive to complement-mediated intravascular haemolysis, leading to anaemia.

In both cases of HAE and AAE due to C1 inhibitor deficiency, the clinical presentation is similar. Affected individuals suffer from swelling of the soft tissues of the face and neck, but may also have painful gastrointestinal symptoms. At the cellular level, symptoms may be due to increased levels of bradykinin, which are normally part-regulated by C1 inhibitor. In HAE, symptoms may be controlled by infusions of C1 inhibitor, but in AAE this is less successful, where treatment is aimed at the underlying aetiology.

C1 inhibitor is a further example of the close relationship between coagulation and complement: deficiency in inhibitors leads to excessive activation of the particular pathway and so clinical syndromes such as deep vein thrombosis and angioedema respectively.

Table 9.3 summarizes key aspects of immunodeficiency.

Table 9.3 Immunodeficiency.

	Primary immunodeficiency	Secondary immunodeficiency
Cellular responses	Severe congenital neutropenia, DiGeorge syndrome, severe combined immunodeficiency, CGD, Chediak–Higashi syndrome, Omenn's syndrome	Bone marrow suppression, leukocyte malignancy, HIV infection
Humoral responses	X-linked agammaglobulinaemia, individual complement components, C1 inhibitor deficiency, hyper-IgM syndrome	All result from impaired cellular immunity

9.5 Immunopathology 2: hypersensitivity

As with immunodeficiency, we can break down an over-active immune response into a number of different aspects. The two major areas are hypersensitivity and autoimmune disease. The former is generally an excessive and damaging response to a normal stimulation or situation, whilst the latter may be described as effectively the normal response to an abnormal situation.

Hypersensitivity was classified decades ago into four types, a system which remains valid. In three, the particular condition is linked to an excessive antibody response; the remainder is an abnormal white cell response. The clinically dominant form, and therefore that which demands the attention of the immunologist, is type I. The remaining types are certainly worthy of mention, but are less important.

Type I hypersensitivity

This form of hypersensitivity, allergy, is dominant, and in its several forms may be present in 10% of the UK population. The biological basis is the relationship that IgE has with basophils and mast cells, as these cells have receptors for the FcR of IgE, although there can be non-IgE-mediated responses, generally IgG$_4$. However, complement components C3a and C5a can trigger mast cell degranulation independently of IgE (Figure 9.3).

The evolutionary basis of the physiology of most IgE responses is likely to have developed from a response to parasites or other 'external' pathogens. These would include those inhaled, and therefore in the lung (and so would include particulate material such as pollen and fungal spores), as well as ectoparasites on the skin. The molecular mediators of the response are the products of the granules of basophils and mast cells, which include heparin, histamine, elastase, leukotrienes and cytokines.

There may also be mediators that activate platelets, and leukocyte chemotactic factors that recruit neutrophils and eosinophils that are likely to inflate the general inflammatory picture.

Of these inflammatory mediators, histamine is perhaps the most active, and exerts its actions on variety of cells via one of four receptors (H$_1$ to H$_4$) to cause a number of actions that include constriction of the bronchus, vasodilation, increased vessel permeability, pain and itching. The latter may have developed as a method of scratching the skin to remove ectoparasites.

The basis of allergy is of an overresponse to the body to what it presumes is a pathogen, and thus are referred to as allergens (Table 9.4). These allergens bind to IgE antibodies that are themselves linked to mast cells and basophils via the FcR, leading to activation of the cell and degranulation. A characteristic of type I hypersensitivity reactions is that they can develop rapidly, often within minutes of exposure to the allergen.

The physical consequences of this degranulation depend on the location of the particular cell. These can be:

- Anaphylaxis – this is perhaps the most serious reaction, and has been reported as being present in up to 2–5 people per 10 000 per year. It often develops rapidly (minutes after exposure), acts on different physiological systems and may be fatal. One of these effects can be a drop in blood pressure resulting from vasodilation, and subjects are often described as being in anaphylactic 'shock'. The major issue is airway obstruction, resulting in breathing difficulties, and many patients often feel a sense of impending doom.
- Angioedema – swelling of the skin and tissues resulting from increased blood vessel permeability. The major issue is breathing problems if the tissues of the throat become swollen and the passage of air is impaired.
- Asthma – problems with breathing. Degranulation of pulmonary and bronchial mast cells causes constriction

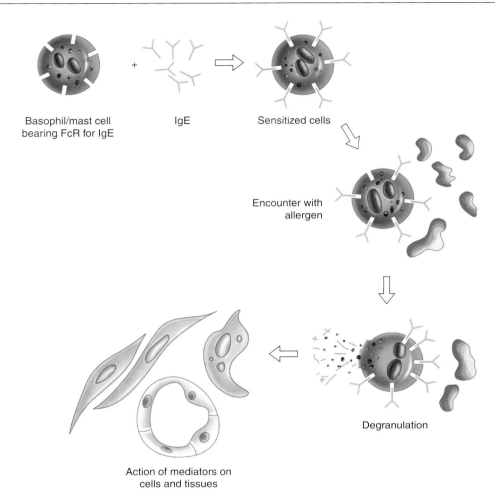

Figure 9.3 Cellular basis of type I hypersensitivity. The basis of this process is the sensitization of resting mast cells and basophils by IgE, that itself then binds to the allergen. Cell degranulation releases inflammatory mediators (such as histamine and tryptase) that act on other cells such as smooth muscle cells and endothelial cells to produce the clinical symptoms such as respiratory distress and itching.

of the airways and so difficulty in breathing, with tightness of the chest, cough and wheezing.

• Allergic conjunctivitis, similar to rhinitis, but being inflammation of the conjunctiva, with the same pain, irritation and itching, and with tears. This is a modified response (like a runny nose) in trying to flush out the stimulant.

• Eczema – also known as atopic or contact dermatitis (inflammation of the skin), although this can be allergic or

Table 9.4 Common allergens.

Foodstuffs	Peanuts, pecans, pine nuts, walnuts, shellfish, egg albumin, wheat, maize, kiwifruit, milk, soy, chestnut
Invertebrate material	House dust mites, storage mites (integument and faeces)
Plant material	Pollen (typically ragweed), poison ivy, certain oaks
Drugs	Antibiotics, muscle relaxants, painkillers, aspirin, NSAIDs
Venoms	Insect bites and stings

nonallergic. This is an umbrella term for several skin conditions that include rash, swelling, dryness and itching. It may be caused by plants such as poison ivy, but also by exposure to 'industrial' allergens such as nickel, latex and even talcum powder. However, there are a host of non-allergic skin diseases, so that differential diagnosis is crucial.

- Intestinal problems, typically pain and cramps, probably due to the action of histamine from mucosal mast cells, themselves triggered by allergens in the diet, such as ovalbumin.
- Rhinitis, meaning inflammation of the nose. However, it is also taken to be the soft tissues of the nasal passage, with itching, blockage and a mucinous discharge. This response, of course, mimics that of the response to the common cold virus.
- Urticaria, also known as hives, is a raised rash (wheal) most often found on the arms, legs and back, whilst facial urticaria is common with food allergies. However, there is a degree of cross-over with eczema and angioedema.

Several of these symptoms can be present concurrently. For example, the most common manifestation of type I hypersensitivity is hay fever, likely to be a mixture of rhinitis, conjunctivitis and often some pulmonary symptoms such as asthma.

The precise allergen itself can be identified by challenging the patient by introducing a variety of potential allergens and then observing which ones evoke a response. This is most often done on the skin, and is referred to as an intradermal (skin) prick test.

Treatment depends on stimulus. Generally, an antihistamine to block the histamine receptor is first line: Piriton is commonly used. In the case of severe allergy, patients will carry an EpiPen, which is a syringe loaded with adrenalin. An additional treatment is with agents that stabilize the mast cell membrane and so suppress degranulation, the most common being sodium cromoglicate. Other targets of treatment include inflammatory mediators such as leukotrienes.

A variety of drugs can be delivered by aerosol directly to the lung in cases of asthma. The agents in these inhalers include salbutamol (a beta-2 adrenoreceptor agonist) and other adrenaline agonists (such as epinephrine). Anticholinergics are also effective, but for severe disease glucocorticosteroids are preferred for long-term control.

Type II hypersensitivity

This is antibody-dependent cell-mediated cytotoxicity (ADCC). It relies on the docking of the Fc section of

(generally) IgG and IgE antibodies into their respective FcRs on the surface of granulocytes, monocytes, macrophages and NK cells. The leukocyte is therefore primed to attack any other cell that is expressing the antigen that the antibodies recognize. It is likely that ADCC is effective in defence against pathogens such as parasites too large to be the object of phagocytosis.

A number of clinical manifestations of ADCC are recognized:

- In incompatible blood transfusions and other forms of transplantation, donor cells are recognized as being non-self and are destroyed (to be revisited in Section 9.7).
- Similarly, a subject may develop antibodies to their own tissues. These autoantibodies may bind to and so target for attack their own cells such as of the kidney, skin, red cells and thyroid (to be developed in Section 9.5).
- Some drugs bind to normal tissues, and so may form an unusual structure that the immune system regards as foreign and so worthy of destruction. Examples of this include chlorpromazine, phenacetin and sedormid.

ADCC is not always a bad thing; in fact, it is exploited in the deliberate destruction of foetal red cells that have passed into the maternal circulation. Anti-D antibodies are injected into the mother to destroy the foetal cells and so minimize the risk of future cases of haemolytic disease of the newborn (Section 9.7 and Chapter 11). There are few other 'real-world' examples of ADCC that have a practical application. Figure 9.4 illustrates the cellular basis of type II hypersensitivity.

Type III hypersensitivity

This is immune-complex-mediated hypersensitivity. The concept of antibodies binding to large bodies such as bacteria and cells is easy to grasp. However, antibodies can also bind to much smaller soluble molecules and other noncellular material in the blood, an excellent example being particles or remnants of the bacterial cell walls (perhaps the product of the action of antibiotics) and viruses. These combinations of soluble antibodies and their antigens are called immune complexes.

Normally, immune complexes should not pose too much of a problem, as they should be safely scavenged by phagocytes. However, if these are present in large amounts then they may become insoluble and fixed or lodged in certain anatomical sites where they are likely to attract the attention of phagocytes. The problem then

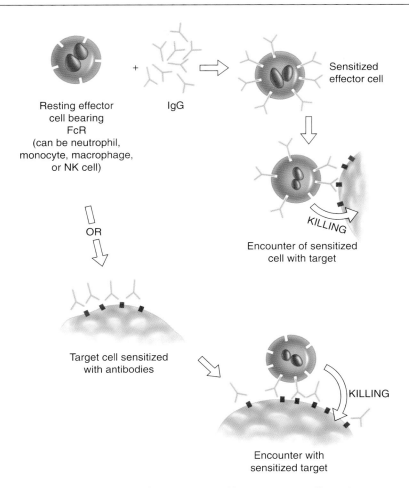

Figure 9.4 Cellular basis of type II hypersensitivity. There are two possible routes: resting cells may become sensitized by the location of IgG into their FcR. The leukocyte may then attack target cells that bear antigens specific for the particular antibody. Alternatively, a leukocyte may directly recognize the FcR of antibodies already bound to a target cell.

arises that the phagocytes are programmed to respond to immune complexes, and if found on normal tissue, such as the endothelium, will inevitably cause a local inflammatory response and damage other healthy cells. White cell responses include release of proteolytic enzymes, mast cell mediators and cytokines, which will recruit other cells that will accelerate the process.

There may also be the involvement of platelets, which can release their own inflammatory mediators and form microthrombi (Chapters 7 and 8), and immune complexes may also promote the activation of the complement system. These processes will in turn cause damage to the nearby cells, the consequences of which depend on the particular site (lungs, skin, kidney) (Figure 9.5).

- In the case of blood vessels, this will lead to vasculitis (inflammation of the blood vessels), with a rash, increased vessel permeability and oedema. A damaged endothelium loses its natural anticoagulation nature and so promotes thrombosis.
- Immune complex deposition in the alveoli will lead to pulmonary problems. This may be precipitated by inhaled substances such as bacteria and fungal spores and other material in house and other dust. This is recognized as 'farmer's lung' and 'pigeon-fancier's lung'.
- Deposition in the kidneys will cause immune-mediated glomerulonephritis, and so renal failure. This condition is often associated with certain bacterial infections, such as streptococci.

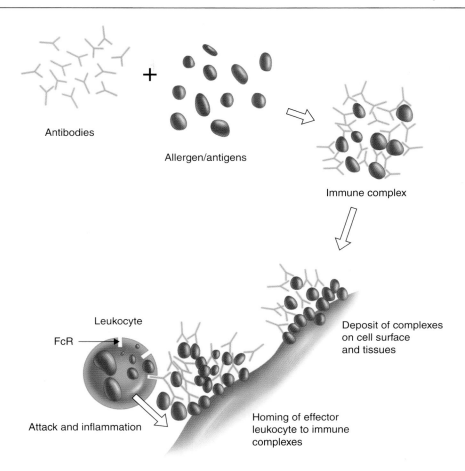

Figure 9.5 Cellular basis of type III hypersensitivity. Antibodies and antigens come together to form soluble or insoluble immune complexes which can bind nonspecifically to innocent cells and tissues. These bound complexes then attract the attention of leukocytes, which proceed to attack the underlying tissues and initiate a local inflammatory response.

As with type I hypersensitivity, several anatomical sites may be influenced at the same time, leading to a more severe overt clinical picture. Clinically, type III disease is of greater importance than is type I hypersensitivity, but the general picture of antibodies causing tissue destruction is dominated by autoimmune disease, as is discussed in Section 9.6.

Type IV hypersensitivity

This final type of hypersensitivity is characterized by a slowly developing cell-mediated response, and can be seen as a two-step process. However, as is evident by the brevity of this section, it rarely occupies the routine immunology laboratory. The pathology is as follows:

in the first phase, an important aspect of the normal immune response to the pathogen is of course its destruction, but a further action is the generation of memory T lymphocytes, which in general is desirable as it allows the immune system to respond more rapidly to a subsequent challenge.

In the second phase, which may be days, weeks or months later, a second presentation of the same pathogen evokes a disproportionately overactive response from lymphocytes, monocytes and macrophages. The mediators of this response include inflammatory cytokines such as IL-2, IL-4 and gamma-interferon from memory T cells. These recruit cytotoxic leukocytes which cause damage to the tissues surrounding the pathogen.

9.6   Immunopathology 3: autoimmune disease

It could be argued that autoimmune disease is effectively the result of a normal immune response to an abnormal situation. Normally, of course, we should not respond to our own tissues, but for a number of reasons 'self' becomes 'non-self' and so, as far as the immune response is concerned, is a legitimate target. The basis of auto-immunity is the autoantibody, which defines an auto-immune disease and is always present. It follows that the real culprit is the abnormal B lymphocyte which pro-duces the autoantibody. There are numerous reasons why this develops, one of which is the failure of T lymphocyte regulation and the overactivity of the Th17 cell which secretes the pro-inflammatory IL-17.

Autoimmune disease is relatively easy to classify by clinical standards; that is, of the cells and tissues that are attacked.

Connective tissue disease

Arthritis is inflammation of the connective tissues (which include bones, joints, ligaments and tendons, and some consider skin to be a connective tissue), and there are many types. The dominant condition is rheumatoid arthritis (RA), which strikes four times as many women as men, although some contend that the closely related non-autoimmune disease of osteoarthritis (OA) causes as much morbidity. These two common diseases are often misdiagnosed cases of other types of arthritis, such as the septic arthritis that is the consequence of the infection of a joint by bacteria, and the lyme arthritis that follows contact with a tick arthropod that carries a bacteria of the *Borrelia* family.

Other arthritides include psoriatic arthritis and juve-nile arthritis, although joint pain and arthritis is often present in other (often inflammatory) conditions such as viral infections, hepatitis and gout.

Rheumatoid arthritis   RA is a complicated, widespread disease whose correct diagnosis requires several factors, such as early morning stiffness, the same swollen joints on each side of the body (such as the left wrist and the right wrist) and the presence of rheumatoid factor (RhF), the defining autoantibody, although it has poor specific-ity for RA. Curiously, RhF is an antibody to the Fc region of another antibody, generally of the IgG class, and so the two together form an immune complex. However, there may also be antibodies to proteins whose arginine

groups have been converted to citrulline – hence cyclic citrullinated protein (CCP), of which the dominant species is mutated citrullinated vimentin (MCV). Like RhF, these anti-CCP anti-MCV antibodies are useful in diagnosis and pathogenesis of this disease.

The complex nature of RA is also demonstrated by the fact that it is possible to have disease but to be negative for RhF – this is called seronegative disease. However, in those 80% of people with RA who are seropositive, increased levels of RhF are generally linked to more severe disease, such as the inflammatory destruction of synovial joints (synovitis, typically of the knee and hip) (Figure 9.6). Further laboratory factors of this chronic inflammatory disease include the persistence of the acute-phase response, with raised ESR, CRP, comple-ment components and white blood cell count.

The disease may spread out of the joints to cause extra-articular disease, such as in the lungs (resulting in fibrosis), skin (where there may be subcutaneous nod-ules), bone marrow, kidney and the heart, where it causes cardiovascular disease. Indeed, so many patients with RA succumb to myocardial infarction and stroke that it may be considered the fifth risk factor for atherosclerosis. Treatment (as with almost all auto-immune disease) is by immunosuppression, the exact nature and strength of which depends on disease sever-ity. This generally inflates from aspirin and other NSAIDs to disease-modifying agents such as hydroxy-chloroquine and methotrexate to steroids, azathioprine and cyclophosphamide. An alternative is more focused treatment with monoclonal antibodies (mAbs) aimed at suppressing the raised levels of inflammatory cyto-kines such as TNF, IL-1 and IL-6. The use of a mAb such as rituximab (which recognizes CD20 present on B lymphocytes) is also successful.

Osteoarthritis   Although not an autoimmune disease, discussion of OA is justified as it is a differential diagnosis to RA. The major difference between RA and OA is that the former is a widespread disease, whereas the latter is restricted only to the joints, and is generally to the large joints of the hips and knees. The primary aetiology of OA is overweight, leading to the 'wear and tear' hypothesis, and so has no sex bias. Accordingly, OA is rarely associ-ated with abnormal laboratory tests such as raised ESR and certainly not RhF. Indeed, if these laboratory mark-ers are present at high titres, the disease may not be OA at all, but actually RA. Alternatively, increased levels of inflammatory markers may indicate that the OA may be evolving into RA. There may be an increase in synovial

Affected joint   **Normal joint**

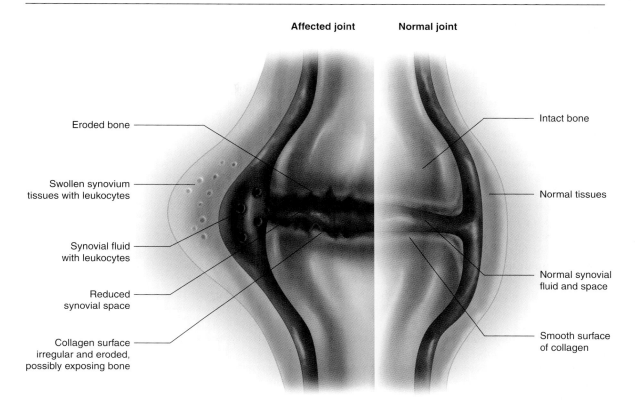

Eroded bone

Swollen synovium
tissues with leukocytes

Synovial fluid
with leukocytes

Reduced
synovial space

Collagen surface
irregular and eroded,
possibly exposing bone

Intact bone

Normal tissues

Normal synovial
fluid and space

Smooth surface
of collagen

Figure 9.6 Synovial joint destruction in RA. The affected joint in RA exhibits a number of abnormalities. These include eroded bone and a reduced synovial space (detected by X-ray) and collagen, the presence of leukocytes and inflammatory mediators in the synovial fluid (detected by aspiration of the turbid fluid, and then biochemistry and microscopy), and swollen synovial tissues which have become infiltrated by leukocytes (as detected by histological examination of a biopsy). (Image courtesy of M. Hill, Heart of England NHS Foundation Trust).

fluid in OA, but (unlike RA) it is clear and lacks white cells and inflammatory mediators.

Systemic lupus erythematosus Like RA, this is also a multisystem disease, and leading presenting symptoms include a rash (often on the face) and arthritis. There may be limited disease with few clinical manifestations, or widespread with many organs involved. A further parallel with RA is the tendency to spread to the lungs, blood vessels, nerves (including the central nervous tissue) and the cardiovascular system, but the disease has a greater sex bias, with nine women being affected for each man. SLE is certainly a more aggressive disease than RA, with many succumbing to renal failure (with lupus nephritis).

The major autoantibody in SLE is towards antigens of the nucleus, hence anti-nuclear antibodies (ANAs), but

there are often antibodies to extractable nuclear antigens (ENAs), ribonuclear proteins, histones, and to single-stranded DNA (ssDNA) and double-stranded DNA (dsDNA). The latter are almost exclusive to SLE, and are present in perhaps two-thirds of patients with active disease. Other autoantibodies in SLE are towards Sm (almost exclusive to SLE), SSA/Ro (present in 40% of patients) and to SSB/La. However, the latter are also present in 50% of patients with another autoimmune disease: Sjögren's syndrome.

RhF may also be raised in SLE, but both RhF and ANAs are often slightly raised in many inflammatory diseases. SLE is also characterized by a continuing acute-phase response (as in RA), but complement components may be low because of excessive consumption. Treatment of SLE, like that of RA, is essentially of the symptoms and with immunosuppression.

Sclerodema Also known as systemic sclerosis, this disease is characterized by the progressive fibrosis of a variety of organs, and is most often evident in the skin. There are two polar extremes: limited (few manifestations) and diffuse (widespread disease); and of course, many patients lie in the mid-ground. Once more, the disease is more frequent in women, and may develop from any of a collection of diverse syndromes that sum to the acronym CREST (calcinosis, Raynaud's disease, (o)esophagitis, sclerodactyly and telangectasia), but other organs such as the kidney may also be affected.

In addition to ANAs, autoantibodies in scleroderma include anti-topoisomerase antibodies (also known as Scl-70, more common in diffuse disease) and anti-centromere antibodies (more common in limited disease). However, broadly speaking, autoantibodies are of limited value because of their poor sensitivity and specificity for a disease (a feature of many other autoantibodies in many other inflammatory diseases).

Other inflammatory connective tissue disease

Other diseases in this broad family include Sjögren's syndrome (attacking salivary and tear glands and part characterized by antibodies to SSA/Ro and SSB/La), polymyositis (the muscle) and dermatomyositis (the skin and muscle). Diseases with a nonspecific target range include Wegener's granulomatosis, polyarteritis nodosum, Takayasu's arteritis, giant cell arteritis and polymyalgia rheumatica (the latter literally meaning rheumatic pain in many muscles).

The astute practitioner is likely to find some evidence of an abnormal immune response in all these conditions, but regrettably the laboratory is generally of limited help in diagnosis, monitoring disease progression or in assessing the effects of treatment, unless the disease process is unusually active. However, the presence of anti-nuclear cytoplasmic antibodies in Wegener's granulomatosis is an example where specificity and sensitivity are good, and so provide assurance that the diagnosis is correct.

> **Blood science angle: Autoimmune connective tissue disease**
>
> Whilst the clinical aspects of RA have been described as a challenge to the physician, the blood scientist will also be challenged by the variety of abnormal blood tests.
>
> The haematological consequences of RA and SLE include anaemia (the so-called anaemia of chronic inflammation) with raised white cells and ESR, although leukopenia is also common. Felty's syndrome may be present in RA, and is linked to a neutropenia. The platelet count is often raised in RA, and antiphospholipid antibodies (leading to the antiphospholipid syndrome) in many diseases bring a risk of thrombosis, manifesting as DVT and PE.
>
> The biochemist is likely to be called upon for CRP, U&Es to check renal failure (most prevalent in SLE), and LFTs for hepatitis. Measurement of uric acid may be useful in differential diagnosis of an arthritis associated with gout, often caused by uric acid crystals.

A further example is temporal arteritis: inflammation of the artery of the temple of the head, just in front of the ear. It is commonly characterized clinically by headache, pain in the muscle of the jaw, and ocular symptoms, and in the laboratory by an abnormal ESR (often >60 mm/h, reference range <10 mm/h). Once a firm diagnosis has been made, treatment must be started promptly and aggressively if blindness is to be avoided because the temporal artery part-supplies the retina. The success of treatment with the oral steroid prednisolone (often 40–60 mg/day) can be followed with a reduction in the ESR, which will in turn guide a reduction in the steroid dose. There may also be a reduction in the high pretreatment level of CRP.

The pathology of gout is not inflammatory but metabolic. High levels of uric acid/urate precipitate as crystals in locations such as skin and joints. In the case of the latter, this can disrupt cells and tissues, causing the pain and inflammatory which can be interpreted as an arthritis. Accordingly, the laboratory will be called upon to confirm of refute a preliminary diagnosis.

Table 9.5 summarizes the major inflammatory connective tissue diseases.

Table 9.5 Major inflammatory connective tissue diseases.

Condition	Dominant autoantibodies	Primary tissues affected
RA	RhF anti-cyclic citrullinated protein	Synovial joints, and then extra-articular disease
OA	None (not autoimmune)	Synovial joints
SLE	ANAs, RhF, ENA (Ro/La), Sm and dsDNA	Skin, kidneys, joints
Scleroderma	ANAs, anti-centromere, anti-topisomerase (=Scl70)	Skin, CREST
Sjögren's syndrome	ANAs, RhF, anti-SSA/Ro, anti-SSB/La	Tear glands, salivary glands

Organ-specific disease

Whilst the connective tissue diseases are characterized by a pathophysiology that may attack a number of organs and tissues, there are several examples of autoimmune disease that strike only one organ, tissue, cell or even a molecule. However, there are some autoantibodies that strike more than one target, and so are called cross-reactive.

The endocrine system   Chapter 18 outlines thyroid physiology and pathology. The pituitary releases thyroid-stimulating hormone (TSH) to induce the production of thyroid hormones. Inflammation of this organ (thyroiditis) may be related to a number of autoantibodies:

- Antibodies to the TSH receptor are the cause of Graves' disease, a form of hyperthyroidism. This autoantibody cross-reacts with muscles of the eyelids, leading to more of the white of the eye (sclera) being exposed, and the appearance of bulging eyes (exophthalmia).
- Antibodies to thyroid peroxidase cause the hyperthyrodism of Hashimoto's thyroiditis. However, there are cases where these antibodies cause destruction of thyroid tissue and so hypothyroidism.
- Antibodies to thyroglobulin, the protein important in the production of the thyroid hormones, may be present in Graves' disease and in Hashimoto's disease.

IgG autoantibodies may cross the placenta and cause thyroid disease in the foetus, often apparent at birth as neonatal Graves' disease, or perhaps post-partum thyroiditis. The disease and symptoms generally resolve as the maternal antibody slowly dissipates.

There is a considerable burden of autoimmune disease that attacks other endocrine tissue. Type 1 diabetes follows the destruction of the beta cells of the pancreatic islets of Langerhans where insulin is produced. There may also be autoantibodies to insulin itself. Autoantibodies to 65 kDa glutamic acid decarboxylase (GAD-65) and to a tyrosine phosphatase named IA-2 (insulinoma-associated) are becoming increasingly relevant to the pathophysiology, treatment and management of this disease. However, perhaps 60% of patients with stiff person syndrome are positive for autoantibodies to GAD-65.

A major cause of Addison's disease is autoantibodies to 21-hydroxylase, a key enzyme in the cells of the adrenal cortex that produces cortisol. These, alongside thyroid disease, and others, may be part of an autoimmune polyendocrine syndrome.

The kidney   Inflammation of the kidney is nephritis, but a more specific aspect is inflammation of the glomerulus, being glomerulonephritis. The dominant disease is Goodpasture's syndrome, characterized by antibodies to the basement membrane (the alpha 3 chain of type IV collagen) that supports the endothelial cells of the glomerulus. This is distinct from the type III immune-complex-mediated hypersensitivity described above, but the clinical consequences (renal failure) are the same. In many cases of autoimmune nephritis, antibodies that react with granules in the cytoplasm of neutrophils are present. These are anti-neutrophil cytoplasmic antibodies (ANCAs), recognizing enzymes such as myeloperoxidase (p-ANCA) and proteinase-3 (c-ANCA).

The liver   Inflammation of the liver is hepatitis. As discussed in Chapter 17, hepatitis may be secondary to infections with a virus (hepatitis A, B or C), bacteria or by parasites. The most well-developed autoimmune liver disease is primary biliary cirrhosis, characterized primarily by anti-mitochondrial antibodies, but there may also be anti-nuclear antibodies and anti-centromere antibodies (as in many inflammatory connective tissue diseases). These autoantibodies recruit infiltrating leukocytes, which damage the hepatocytes, causing fibrosis, raised LFTs, jaundice and eventually chronic liver failure.

The intestines   Autoimmune disease can strike the intestines.

- The pathological basis of many cases of coeliac disease is an abnormal reaction to gluten, a heterologous mixture of proteins found particularly in wheat, but also in barley and rye. Key autoantibodies, often IgA, are directed towards tissue transglutaminase and the endomysium (hence endomysial antibodies), the latter being a component of part of muscles that are rich in reticulin. However, IgA deficiency must be tested for, otherwise there may be a false negative diagnosis. These autoantibodies have a varying degree of sensitivity and specificity for the disease, and some participate in the disease process and stimulate T lymphocytes. This leads to destruction of intestinal tissues and so to the malnutrition, weight loss, abdominal discomfort and diarrhoea that characterize the disease.

National Institute of Health and Clinical Excellence (NICE) clinical guideline 86 offers advice on coeliac disease, and on laboratory testing. It recommends against

the use of IgG or IgA antigliadin antibody tests in the diagnosis of coeliac disease. Instead, it recommends that, when requesting serology, laboratories should:
- use IgA tissue transglutaminase (tTGA) as the first-choice test,
- use IgA endomysial antibodies testing if the result of the tTGA test is unclear,
- check for IgA deficiency if the serology is negative, and
- use IgG tTGA and/or IgG endomysial serological tests for people with confirmed IgA deficiency.

• The stomach can be targeted by autoantibodies to the gastric parietal cells that produce intrinsic factor, the molecule essential for the absorption of vitamin B_{12}. Lack of this vitamin leads to a type of megaloblastic anaemia commonly called pernicious anaemia. The damaged parietal cell is also unable to generate hydrochloric acid, so that the pH of the gastric juices rises, leading to poor digestion and a risk of infections, causing gastroenteritis.

Crohn's disease and ulcerative colitis, two other common chronic inflammatory bowel diseases, are not thought of as having an autoimmune aetiology as no autoantibody has been defined. Instead, the former may be caused by a genetic predisposition and environmental factors, whereas the latter may be due to an abnormal response to certain bacteria.

There are, of course, dozens of autoimmune diseases, presenting with an increasingly reduced frequency, that we are unable to address. Table 9.6 summarizes the major autoimmune diseases.

The spectrum of autoimmune disease

Some autoimmune diseases are tightly organ specific, whilst others (such as Graves' disease) are manifest in more than one organ (in this case, the thyroid and the eyelids). Many of the symptoms of autoimmune connective tissue disease (such as rash, renal disease, arthritis and vasculitis) are present in varying degrees and with varying severity in RA, SLE and scleroderma. Indeed, to underline the broad and overlapping nature of these conditions, there is even a formal diagnosis of 'mixed connective tissue disease'.

Those people with one autoimmune disease are at risk of another. Many patients with RA also have auto-antibodies to the thyroid, and, conversely, thyroid patients often have a raised RhF titre. This leads to the hypothesis of a susceptibility to autoimmune disease, evidence in support of which is that such patients often share certain human leukocyte antigen (HLA) types, especially HLA-DR types, the precise nature of which are discussed in Chapter 10.

> ### Blood science angle: The haematologist and autoimmunity
>
> In addition to the autoimmune cause of pernicious disease, red cells can be the target of autoantibodies, causing auto-immune haemolytic anaemia (Chapter 4). Similarly, auto-antibodies to platelets cause immune thrombocytopenia purpura (Chapter 8). Autoantibodies to erythropoietin and its receptor on erythrocyte precursors and haemopoietic cells are present in a subset of patients with anaemia and may be responsible for their impaired erythropoiesis.

Table 9.6 Other autoimmune diseases.

Condition	Dominant autoantibodies	Primary tissues affected
Goodpasteur's syndrome, glomerulonephritis, chronic renal failure	Anti-glomerular basement membrane	Kidney, but cross-reaction with other tissues
Graves' disease, Hashimoto's thyroditis, hyper- and hypothyroidism	Anti-TSH receptor Anti-thyroperoxidase Anti-thyroglobulin	Thyroid
Diabetes	Anti-islet cell Anti-insulin Anti-GAD, anti-IA2	Pancreas
Addison's disease	Anti-21-hydroxylase	Adrenal cortex
Primary biliary cirrhosis	Anti-mitochondria	Liver
Coeliac disease	Anti-transglutaminase, anti-endomysium	Intestines
Pernicious anaemia	Anti-parietal cell Anti-intrinsic factor	Stomach

The complex nature of autoantibodies

A characteristic of autoimmune disease is the lack of specificity for the target tissue, and that many autoantibodies are nonspecific for a particular clinical syndrome. For example:

- Different antibodies to the glomerular basement membrane may also react with the basement membrane that supports alveoli, whilst others may cross-react with antigens of the basement membrane of skin.
- ANCAs are present in 80% of patients with Wegener's granulomatosis and 20% of patients with microscopic polyarteritis (inflammation of many small blood vessels), and often in patients with glomerulonephritis, SLE, RA, Churg–Strauss syndrome, ankylosing spondylitis and inflammatory bowel disease such as ulcerative colitis. However, the picture is compounded by different ANCAs (p-ANCAs and c-ANCAs recognize myeolperoxidase and proteinase-3 respectively).
- RhF and ANA are present primarily in RA and SLE but are also found (albeit at low levels) in many other autoimmune and inflammatory diseases, including systemic infections and type 1 (chronic active) hepatitis. These autoantibodies may also be induced by drugs such as hydralazine and tetracyclines. Low titres are also found in seemingly healthy elderly people, which may be part of aging and the loss of control of immune regulation.
- Antibodies to dsDNA are most prevalent in patients with SLE but may also be found in type 1 (chronic active) hepatitis.
- Antibodies to mitochondria are found most frequently in primary biliary cirrhosis, but also in the connective tissue diseases of Sjögren's syndrome, scleroderma and RA.

This broad range of antibody specificities (or, more accurately, nonspecificities) reflects the general B lymphocyte perturbation which, lacking precise direction from T lymphocytes, produces nonspecific polyclonal antibodies. The complex nature of autoantibodies can be illustrated by ENAs, of which there are several sub-antigens linked to different conditions (Table 9.7).

Blood science angle: Autoimmune disease and leukaemia

Since autoimmune diseases are all caused by abnormal autoantibodies generated by abnormal B lymphocyte responses, it is perhaps not a great surprise that autoimmune haemolytic anaemia (AIHA) can develop on a background of leukaemia. This is likely to be because the cell biology of the leukaemia, if based on a malignant B lymphocyte, may well give rise to a clone of leukaemic cells that generate anti-red cell autoantibodies.

However, it is now clear that this disease process also works around the other way. Subjects with the diagnosis of AIHA are at a staggeringly increased standardised incidence ratio (SIR) of over 46 (that is, have a 46-fold increased risk) of a diagnosis of leukaemia within a year. This may well be because the AIHA is in fact simply the first manifestation of the leukaemia. None the less, AIHA brings a SIR of over 7 for the development of leukaemia a year after diagnosis of AIHA.

The same principles apply to other autoimmune disease, such as pernicious anaemia, rheumatoid arthritis and systemic lupus erythematosus, where a diagnosis brings SIRs of over 5-fold, 2-fold and 7-fold respectively for the development of leukaemia within a year. These risks fall to 1.86, 1.33 and 1.79 respectively a year after the diagnosis of the autoimmune disease.

Table 9.7 Extractable nuclear antigens.

Type of ENA	Disease association
Ro (SSA)	Sjögren's syndrome, SLE, congenital heart block, neonatal lupus, RA, primary biliary cirrhosis
La (SSB)	Sjögren's syndrome, SLE, RA
Ribonucleoprotein (RNP)	Mixed connective tissue disease, SLE
Sm (RNA-binding proteins)	SLE
Scl-70 (topoisomerase)	Scleroderma
Jo-1	Myositis, Raynaud's disease, interstitial lung disease, arthritis

9.7 Immunotherapy

We have already seen how the immune system can be suppressed with drugs. This ranges from cytotoxic chemotherapy in aggressive autoimmune diseases such as SLE vasculitis, to anti-inflammatory drugs in autoimmune disease and antihistamines in hypersensitivity. It could also be argued the use of granulocyte-macrophage colony-stimulating factor to promote stem cell regeneration is also immunotherapy. Similarly, different forms of interferons are used in the treatment of multiple sclerosis, hepatitis B and C infections, and hairy cell leukaemia. In Section 9.4 we read how the use of intravenous Igs and infusions of granulocytes in autoimmune diseases and immunodeficiency are examples of passive immunotherapy.

However, all of these forms of treatment are relatively nonspecific, but in certain circumstances the immune system can be actively manipulated to provide defence against specific pathogens.

Therapeutic antibodies

Haemolytic disease of the newborn This condition is the result of foetal red blood cells expressing Rh D entering the circulation of a woman who is Rh D negative and invoking an immune response. At a subsequent pregnancy, the mother may make antibodies to these red cell blood groups, which may then cross the placenta and attack her child, causing haemolytic disease of the newborn (Chapter 11). One way of minimizing this is to destroy those foetal red cells in her blood before they can provoke an immune response. This can be achieved by injecting her with preformed antibodies to Rh D (i.e. anti-D), and is one of the few examples of the positive use of antibody-dependent cellular cytotoxicity.

Polyclonal antibodies Plasma from subjects who have suffered from infections such as with hepatitis B virus contains antiviral antibodies. These antibodies may be harvested and used to provide support for those with an active hepatitis B virus infection. The same principle applies in the use of immunotherapy against tetanus toxin, rabies, varicella-zoster and cytomegalovirus. Horse, dog and sheep polyclonal antibodies have been produced to help combat botulism, snake and spider venoms, and stonefish and jellyfish stings. Rabbit and horse anti-thymocyte globulin has been very useful in depleting T lymphocytes in transplant rejection.

Despite their success, these xenobiotic antibodies are likely themselves to provoke an immune responses and may cause 'serum sickness', a form of immune complex type III hypersensitivity. Accordingly, there use is diminishing, especially in the light of improved vaccines and the use of antigen-specific mAbs.

Monoclonal antibodies There are many examples of the use of these exquisitely precise molecules, which can target molecules as part of the cell surface, and soluble molecules in the plasma. Examples include:

- OKT-3, historically important as one of the first mAbs to be developed, recognizes CD3 and so all T lymphocytes. It has proved valuable in reducing acute rejection in organ transplants such as the kidney, but also the heart and liver. Other mAbs to CD3 may be useful in inflammatory bowel disease, and there is also a mAb to CD4 that specifically targets T helper cells.
- Antibodies to CD20 to attack B lymphocytes in lymphoproliferative diseases such as lymphoma, hairy cell leukemia and 'standard' leukaemia. It may also be used to suppress plasma cells secreting pathogenic autoantibodies in autoimmune diseases such as RA and autoimmune haemolytic anaemia. However, these mAbs, of which there are several, rituximab being one of the first, will also recognize CD20 on the surface of healthy B cells, with the possible development of immunodeficiency.
- Similarly, a mAb to CD53, present on T lymphocytes and B lymphocytes, may be useful in the treatment of chronic lymphocytic leukaemia and some lymphomas.
- Antibodies to cytokines such as TNF in autoimmune diseases such as RA, psoriasis, ulcerative colitis and Crohn's disease. Leading agents include infliximab and adalimumab. Mepolizumab targets IL-5, an eosinophil growth factor, and may have a role in asthma.
- Antibodies to the IL-2 receptor (CD25), high levels of which are found on activated T lymphocytes, can be used in the acute rejection of renal transplants.
- Other mAbs target complement component 5 (preventing its cleavage into C5a and C5b), soluble IgE (of possible use in allergy and asthma), IL-6 (of potential anti-inflammatory use), IL-13 (possible use in asthma) and IL-12 and IL-23 (may be valuable in psoriasis).

Although not a full mAb, etanercept is worthy of mention because it is a fusion product of part of the TNF receptor and the Fc portion of an IgG molecule. It is used in the same group of diseases as are the mAbs to

TNF. Similarly, anakinra, a recombinant version of the IL-1 receptor antagonist, blocks the activity of the pro-inflammatory cytokine IL-1. There are two fusion molecules that are both formed from the T lymphocyte molecule CTLA-4 and the Fc part of an IgG molecule. These products block T cell activation.

The problem of xeno-immunization of the patients to the mouse mAbs may be addressed by 'humanizing' the antibodies, by the genetic engineering of human protein genes onto those of the mouse.

The literature has numerous referrals to mAb and fusion products, but, as in all drug development, many failed clinical trials. Those that have succeeded will be listed in the British National Formulary, and be cited by NICE.

Vaccination

Probably the most common manipulation of cell-mediated immunity is in vaccination. The process of deliberately introducing smallpox scabs to those free of the disease with a view to preventing the development of the full-blown disease (variolation, from the name of the virus, variola, that causes smallpox) was known for some time before the work of Jenner. His advance, in the late 18th century, was to inoculate subjects with scabs of those with a related but different virus, the relatively benign cowpox. In one of the first clinical trials, he then introduced smallpox, which failed to produce the disease. The cowpox virus was subsequently replaced by the vaccinia virus; hence vaccination.

Half a century later, Pasteur noted that an 'old' culture of chicken cholera bacillus no longer produced the disease when introduced into chickens, but those same animals failed to succumb to the disease when inoculated with a fresh culture of the bacillus known to be pathogenic. This 'attenuation' of an otherwise pathogenic organism set the scene for the development of vaccines against the virus that causes rabies, and the bacilli that causes anthrax and tuberculosis. It now established that the mode of action of vaccines is to stimulate the development of very long-lived B lymphocytes and T lymphocytes, often called memory cells. However, this memory requires 'reminders' with booster vaccination at intervals.

Vaccination is, of course, now a fully established public health initiative. If sufficiently widespread, it can lead to the virtual disappearance of an infection (herd immunity), as in the case of the eradication of smallpox. However, this requires a high rate of vaccination, and failure to achieve this can lead to continuing infections, as in the cases of measles. Other pathogens that are the subject of vaccines include the viruses influenza, polio, human papilloma virus, rotavirus, Japanese encephalitis virus, yellow fever virus, hepatitis A and B viruses, mumps, rubella and chickenpox.

Among the bacteria that can be immunized against are *Mycobacterium tuberculosis* (with a live attenuated strain of *Mycobacterium bovis*), *Haemophilus influenzae* type B (N.B. this organism does not cause influenza), *Salmonella typhi* (causing typhoid fever), *Neisseria meningitidis*, *Corynebacterium diphtheria*, *Vibrio cholerae*, pertussis (the cause of whooping cough) and *Streptococcus pneumoniae* (or pneumococcus, causing pneumonia).

Technically, the symptoms of infections with *Clostridium tetani* (tetanus) are caused by its neurotoxin: the vaccine is an inactivated version of this toxin. A convenient public health measure is to immunize against several pathogens at the same time, such as the triple vaccine mumps, measles and rubella, and the combination of agents against diphtheria, pertussis and tetanus.

The most successful vaccines are viral protein based, as these provoke a strong cell-mediated response. However, carbohydrates can clearly invoke an immune response, although these are often weak. However, coupling these with proteins improves immunogenicity. These strategies may help develop vaccines for outstanding pathogens such as the fungi *Cryptococcus neoformans* and *Candida* and *Aspergillus* species, parasites *Plasmodium falciparum* (a leading cause of malaria), *Tricinella spiralis*, and leishmania and schistosoma species, and HIV.

Cancer

The fact that many viruses are oncogenic means that vaccination also acts against the development of cancer. This is applicable in hepatitis B virus and liver cancer, and in human papilloma virus and cervical cancer. Direct immunotherapy against solid tumours has proved difficult as there are few good tumour-specific antigens that the immune response can be directed towards. Nevertheless, mAbs to vascular endothelial growth factor and the epidermal growth factor receptor in colorectal and other cancers have been trialled despite the fact that these targets are physiologically important. The use of mAbs in haemoproliferative neoplasia was highlighted earlier.

Indirect anti-cancer methods include attempts to non-specifically stimulate NK and other leukocytes. It is possible that gamma/delta T lymphocytes (which comprise 1–10% peripheral blood T cells (Chapter 10)) can be manipulated *in vitro* into providing anti-tumour activity.

Allergy desensitisation

Allergen-specific immunotherapy has a surprisingly long history, having been described over a hundred years ago, in a study of increasing the tolerance to conjunctival challenge testing with grass pollen extract. The principle of repeated exposure to small doses of allergens has remained essentially unchanged, and can lead to reduced responses to allergens in the treatment of allergic rhinitis and insect venom.

Allergens may be introduced as subcutaneous injection immunotherapy (SCIT) or as sub-lingual immunotherapy (SLIT). A particular successful approach is SCIT for insect venom, with standard (a 12-week protocol of injections), rush (4–7 days) and ultra-rush (1–2 days) protocols. SLIT has been used to desensitize patients with allergies to house dust mite and grass pollen. As briefly mentioned earlier, a mAb to the Fc portion of IgE (omalizumab) is effective in asthma, probably related to its removal of large amounts of IgE from the plasma, and so reduced activation of basophils. This agent is recommended by NICE in technology appraisal TA133.

9.8   The immunology laboratory

The day-to-day workings of the immunology laboratory can be classified as cellular and serology. The latter includes antibodies, complement and other proteins, and is so called because, historically, tests were performed on serum, although in many cases plasma is also acceptable. The purpose of this testing is to assist in the diagnosis and management of disease. These can be grouped together according to the clinical presentations, such as:

• Repeated bacterial infections suggest antibody and/or complement deficiency, so that potentially useful initial investigations include Igs, functional antibodies, C3, C4 and CH50. Hypogammaglobulinaemia (as is determined by serum protein electrophoresis) may be caused by a B cell defect such as a malignancy or cytotoxic chemotherapy.
• Viral and/or fungal infections suggest a T cell deficiency – an initial investigation would include full blood count for total lymphocyte numbers followed by flow cytometry for T cells and their subsets.
• Staphylococcal skin sepsis and pulmonary and dermal fungal infections suggest a neutrophil defect, so that functional studies of these cells would be considered.

• Confirmation of the diagnosis of SLE would be supported with results of ANA, RhF and ESR tests. Those with a positive ANA on screening will be tested for antibodies to dsDNA, ENAs, and for C3 and C4 levels.
• A suspected immunological aspect in renal disease (probably prompted by increased U&E and/or reduced estimated glomerular filtration rate) would be investigated with ANAs, C3, C4, Igs, cryoglobulins, antibodies to the glomerular basement membrane and C3 nephritic factor (only if C3 is absent).

Cellular immunology

The major clinical interest is in neutrophils and lymphocytes, which must of course be taken into an anticoagulant, of which heparin is most often requested, although EDTA may be acceptable. The practitioner must always check with the laboratory. Assessment of cell immunity can be functional or by the expression of CD molecules, and must be done as soon as possible after the blood has been drawn.

Neutrophils   In population terms, of the congenital neutrophil diseases, CGD is the most important, and is where there is no respiratory burst. The clinical consequences are recurrent bacterial infection, generally caused by the inability of the cell to kill these organisms. The molecular lesion can be in any one of several genes coding for enzymes generating the toxic reactive oxygen species, the most common being in those for NADP oxidase. The particular metabolic formulae are outlined in Section 5.8.

The disease is tested for by challenging neutrophils with NBT (actually, a light yellow colour), which in normal cells is acted on by reactive oxygen species to produce the blue product tetrazolium, easily identifiable by light microscopy. In CGD, there is no blue product, thus strongly indicating the diagnosis. Myeloperoxidase deficiency can also be determined with this test.

An alternative test feeds the cells with dihydrorhodamine, which are then stimulated to express the enzymes to generate oxygen radicals. If present, these oxygen groups oxidize the dihydrorhodamine to rhodamine in cells with normal function, which is also easily detected as it is fluorescent. A practical but more complex test of phagocytosis is to feed live neutrophils with yeast, which they should ingest. If successful, a blood film can be made and, using a light microscope, the yeast seen inside the neutrophils, just as bacteria can be seen within a phagocytic monocyte in Figure 5.9.

Lymphocytes Flow cytometry can be used for total T cells (CD3), NK cells (CD56 and CD16) and B lymphocytes (CD19) (Chapter 5) and for helper/cytotoxic subsets (CD4 and CD8). An absolute CD4 count is important in monitoring HIV infection and, ideally, a successful response to treatment (see case study in Section 9.9). Functional activity of lymphocytes can be assessed by the proliferative responses to stimulants such as pokeweed mitogen, phytohaemagglutinin, concanavelin A, purified protein derivative, tetanus toxin and candida antigens.

Basophil function is important in allergy testing, to be discussed below.

Serology

The majority of the work of most routine immunology laboratories is serology, and much of this is in the detection and quantification of autoantibodies. The most common are listed in Table 9.8. Key methods are indirect immunofluorescence and ELISA, although newer techniques such as robotics and multiplex technology are becoming available.

Immunoglobulins A key measurement is of total Ig and the five isotypes. It is likely that total Igs will be measured by electrophoresis and densitometry, perhaps in the biochemistry section of the laboratory. The techniques are explained in Section 17.4 on plasma proteins. However, the five isotypes are unlikely to be measured by standard electrophoresis but by immunoassay (IgG, IgA and IgM perhaps using nephelometry and turbidimetry, IgD and IgE by ELISA), as will be levels of the four subtypes of IgG and two subtypes of IgA. Expected results for the five isotypes are shown in Table 9.9.

Cryoglobulins are Igs that precipitate out of solution when below body temperature. The exact temperature varies, but as far as the patient is concerned, it happens in the cold, generally in the winter months. The consequences are pain and parasthesia in the fingers, toes and any extremity exposed to low temperatures.

The likely aetiology is the deposition of cryoglobulin in arterioles, capillary and venules, causing occlusion and so ischaemia. It is common in Raynaud's syndrome, although the latter may also be caused by arterial spasm and not necessarily by cryoglobulins.

In investigating cyroglobulins, the key aspects are to keep all the venepuncture equipment and vacutainer at 37 °C and to keep it warm during transfer to the laboratory, where a warm centrifuge will be needed to separate the plasma or serum. Following separation, one aliquot is kept at 37 °C and another at 4 °C. The two are checked regularly for the appearance of precipitate or turbidity in the latter.

The exact nature of the cryoglobulin can be determined, and so classified. It is obtained by centrifugation in the cold, and then is resolubilized in the warm. Type 1 forms rapidly and is of a monoclonal gammopathy, perhaps due to a myeloma or related disease. Types 2 and 3 may take up to a week to form: type 2 (the most common form – 60% of cases) being a mixture of Igs but with a single monoclonal species, whilst type 3 (accounting for 30% of cases) is a mixture of polyclonal antibodies and is often seen in bacterial infections and autoimmune diseases.

Many other requests are for levels of defined plasma proteins (Table 9.10), where methods include enzyme immunoassay, ELISA and nephelometry.

Laboratory assessment of complement C3 and C4 can be assessed by standard immunoassays such as nephelometry and ELISA, and so can be measured in most routine laboratories (Table 9.10). However, there is considerably more to complement than these two molecules.

The CH50 (complement haemolysis 50) method is a bioassay, and relies on the ability of the complement pathway to lyse 50% of a suspension of a target such as a sheep red blood cell that has been coated with antibodies. It is best performed in a microtitre plate, which allows a titration of the patient's sample against a standard suspension of red cells, the quantified end point being haemoglobin released from lysed red cells. A poor CH50 result may be due to the lack of one or more of the factors that constitute the pathway, as are summarized in Figure 9.1.

Similarly, the AP50 (alternative pathway 50) method tests the components that make up the alternative pathway. A common method uses rabbit red cells as these are extremely susceptible to non-antibody-dependent haemolysis. The end point, as in CH50, is the release of haemoglobin, which is easily assessed by a colourimeter. CH50 and AP50 are quantified in arbitrary units or per cent; and as with bioassay, extra attention must be paid to assay conditions, such as buffers and temperature, which must be standardized and run with positive and negative controls.

Immunologists can also use changes in the complement cascade to demonstrate the acute-phase response. C3, C4, CH50 and AP50 testing can be used to determine

Table 9.8 Commonly requested autoantibodies.

Test	Indication/comment	Method
ANCAs	Reported as positive or negative with three patterns: C, P and atypical. Further testing for reaction with myeloperoxidase or proteinase-3	IIF on neutrophils
ANAs	Reported as positive or negative, and with titre and pattern	IIF on Hep2 cells
Aspergillus antibodies	Positive in allergic bronchopulmonary aspergillus. Often IgE and IgG antibodies	ELISA
Avian antibodies	Present in extrinsic allergic alveolitis	ELISA
Cardiolipin antibodies	Frequently found in SLE and thrombophilia. Can be IgG and/or IgM. Linked to antiphospholipid antibodies	ELISA
Centromere antibodies	Usually found in the CREST syndrome (positive or negative)	IIF on Hep2 cells
Component-resolved diagnostics	IgE antibodies against peanut, hazelnut, egg, bee venom, wasp venom, latex, wheat (allergen specific reference range)	Enzyme immunoassay
Cyclic citrullinated peptide antibodies	More specific for RA than RhF	Enzyme immunoassay
DNA antibodies	Positive or negative by screen, then numerical result by ELISA	IIF on crithidia, then ELISA
Endomysial antibodies	Performed as a follow up if IgA tissue transaminase antibody is positive	IIF on monkey oesophagus tissue
Epidermal antibodies	Antibodies to desmosomes or basement membrane differentiate pemphigus and pemphigoid (positive or negative)	IIF on monkey oesophagus tissue
Extractable nuclear antibodies (ENAs)	Many different antigens linked with different diseases/syndromes (positive or negative, and titre)	ELISA
Functional antibodies	To pneumococcal antigens, *Haemophilus influenzae* type B, tetanus and pneumococcal serotypes	ELISA
Gastric parietal cell antibodies	Linked to pernicious anaemia (positive or negative)	IIF on rat liver, kidney and stomach
Glomerular basement membrane antibodies	Positive in Goodpasteur's syndrome (positive or negative, and titre)	Enzyme immunoassay
Glutamic acid decarboxylase (GAD-65) antibodies	Can differentiate different types of diabetes; also present in stiff person syndrome	ELISA
Intrinsic factor antibodies	Present in most cases of pernicious anaemia (positive or negative, and titre)	ELISA
Mitochondrial antibodies	Present in most cases of primary biliary cirrhosis (positive or negative)	IIF on rat liver, kidney and stomach
Myeloperoxidase antibodies	Present in certain inflammatory connective tissue disease: microscopic polyangiitis/Churg–Strauss syndrome	ELISA
Proteinase 3 antibodies	Present in Wegener's granulomatosis	ELISA
RhF	Present in 80% of cases of RA: if positive, titre or U/mL	Nephelometry
Smooth muscle antibodies	Present in 75% of cases of autoimmune type 1 (chronic active) hepatitis (positive or negative)	IIF on rat liver, kidney and stomach
Thyroid peroxidase antibodies	Present in autoimmune thyroid disease	ELISA
Tissue transglutaminase (TTG) antibodies	TTG is the major antigen in coeliac disease	ELISA

IIF: indirect immunofluorescence.

Reference ranges have not been shown as they need to be defined by the local reference/specialist laboratory and depend on the methodology and manufacturer of the reagents.

Table 9.9 Levels of immunoglobulins.

Isotype	Adult reference range
IgG	5.9–16.1
IgA	0.7–4.1
IgM	0.4–2.1
IgD	0–110
IgE	0–80

N.B. As in many cases, reference ranges vary according to local practice and methodology, and use of good quality control material such as CRM 470 is essential.
(Adapted from Hall A & Yates C, *Immunology*, Oxford University Press, 2010).

whether the classical or alternative pathways are activated and whether this is reflective of an acute-phase response.

Allergy testing

Clinical testing Obtaining a good history from the patient is essential as this may give clues as to the exposure to particular allergens (eating an unusual meal or visiting a particular location). However, the best method of assessing allergy is to deliberately expose the patient in a controlled setting and observe the result. In some cases this may be immediate, within minutes, whereas other reactions may be sought for up to 48 h.

Table 9.10 Defined protein analyses.

Test	Indication/comment (Reference range)
Beta-2-microglobulin	Levels raised in renal failure and several B cell tumours, including myeloma (1.2–2.4 mg/L)
C1-inhibitor	Low levels in many cases of hereditary and acquired angioedema (0.15–0.35 g/L). Also a functional assay possible (>70% normal)
Complement components C3 and C4	(C3 0.75–1.75 g/L) (C4 0.14–0.54 g/L)) Low levels imply consumption
Mast cell tryptase	Enzyme released from mast cells (2–14 µg/mL). High levels imply an ongoing allergic reaction or mastocytosis

The skin prick test pierces the skin to allow the potential allergen access to the sub-dermal tissues. Patch testing is supra-dermal – the potential allergen remains at the surface, and is useful in investigating contact dermatitis. In testing for asthma, a formal clinical lung function setting is required. Challenge tests expose the patient to increasingly high doses of the allergen, and for all investigation a negative control is essential.

Immunoglobulin E The immunology laboratory will test for overall concentrations of IgE (perhaps by ELISA or chemiluminescence), but also allergen-specific IgE. This testing, also described as component-resolved diagnosis, can identify IgE reactive towards certain grasses, fruit, egg and nuts (Table 9.4), and so which are to be avoided. One of the more comprehensive systems is that of ImmunoCap, which can test for IgE against dozens of potential allergens. Using micro-arrays, simultaneous testing for many allergens using a small sample of plasma is possible.

Although focus is on IgE, there are IgG antibodies that cause hypersensitivity reactions, often manifesting with pulmonary symptoms, and ultimately diagnosis as farmer's lung, allergic alveolitis and pigeon fancier's lung.

Leukocyte testing The clinical consequences of an allergic reaction are caused by the micro-products of basophil and mast cell degranulation, itself caused by cross-linking cell-bound IgE. Consequently, an alternative approach is to test basophil reactivity towards putative allergens. When stimulated with an appropriate allergen (such as house dust mite allergen), the resting basophil upregulates CD203c, and CD63 appears at the cell surface. As the latter is a component of the membrane of intracellular granules, it is proof of degranulation. Whilst this is a very powerful technique, it relies on blood cells and tells us nothing of mast cells in the tissues.

Products of degranulation Basophil and mast cell degranulation results in the release of molecules such as histamine, heparin, cytokines, leukotrienes and prostaglandins. However, for various technical and other reasons, these molecules are difficult to measure. One that is stable and can be measured is mast cell tryptase, levels of which peak an hour after degranulation, returning to baseline after a further 12 h. The two isoforms, alpha and beta, may be measured by immunoassay, and increased levels imply mastocytosis (increased numbers of mast cells in the tissues) and anaphylaxis (if measured within an hour of the attack).

Although allergy testing focuses on basophils, mast cells and IgE, there is also a role for the eosinophil. The granules of this cell contain cytotoxic proteins, enzymes and leukotrienes, but also eosinophil cationic protein. The molecule, which can stimulate the release of histamine from basophils and mast cells, may be measurable by fluorescence immunoassay, and may be of use as a marker of airways inflammation in patients with asthma.

The reference laboratory

No general laboratory has the resources to offer all tests; the more complex and unusual are usually offered by a regional reference or specialist laboratory, often in a university teaching hospital. This is frequently the case in investigating cell-mediated immunity, such as neutrophil and lymphocyte function.

A reference laboratory is also likely to be able to test for uncommon autoantibodies such as to acetylcholine receptor antibodies, adrenal antibodies, beta-2-glycoprotein antibodies (often raised alongside anticardiolipin antibodies), aquaporin 4, basal ganglia, interferon, voltage-gated calcium and potassium channels, C1q, diphtheria, meningococcus, histones, thyroglobulin, and to components of the ovary, parathyroid and testes (N.B. this list is not intended to be exhaustive). However, with improvements in methods and reagents, suppliers can now offer kits to define these antibodies.

Tissue typing to define the particular HLA polymorphism of a subject, often as a prerequisite to transplantation, will be performed by a specialist histocompatibility laboratory, and is expanded upon in Chapter 10.

9.9 Case studies

Case study 9

A 36-year old intravenous drug abuser, of 15 years' duration, reports to the Accident and Emergency Department with increasing shortness of breath, cough and tiredness. She admits to having been to her general practitioner for an infection in her urinary tract, which was treated with antibiotics. Blood is sent for a full blood count, ESR, CRP, d-dimers and pathogenic virus screening, and the patient is sent for a chest X-ray.

Analyte	Result	Reference range
Haemoglobin	129	118–148 g/L
Red blood cell count	4.8	$(3.9–5.0) \times 10^{12}$/L
Haematocrit	0.45	0.35–0.53
Mean cell volume	86	77–98 fL
Platelets	243	$(140–450) \times 10^9$/L
White blood cell count	7.4	$(4.0–10.0) \times 10^9$/L
Neutrophils	5.5	$(1.7–6.0) \times 10^9$/L
Lymphocytes	1.1	$(1.0–3.2) \times 10^9$/L
Monocytes	0.4	$(0.2–0.6) \times 10^9$/L
Eosinophils	0.35	$(0.03–0.46) \times 10^9$/L
Basophils	0.05	$(0.02–0.09) \times 10^9$/L
Blasts	<0.01	$<0.01 \times 10^9$/L
ESR	15	<10 mm/h
CRP	7.5	<5 g/L
D-dimers	625	<500 Units/mL

Interpretation At first sight the full blood count is unremarkable, but the total lymphocyte count is only just within the reference range, and the neutrophil count in on the high side. The high ESR and CRP suggest a low-grade inflammation, whilst the d-dimer was taken to try to exclude a pulmonary embolus, the possible cause of the lung symptoms. As the result is high, a clot in the lung cannot be excluded. Later, the viral screen result comes back negative for hepatitis A, B and C but positive for the HIV.

Flow cytometry The positive HIV result demands flow cytometry for CD4 and CD8, although the lowish lymphocyte count itself may trigger a request for T lymphocyte subsets which, if acted upon rapidly, could be performed on the same blood sample. This reports a reduced CD4/CD8 ratio of 1.2/1 (reference range 1.5/1 to 4/1) (Figure 9.7). Conversion of the flow cytometer data with the lymphocyte count gives normal numbers of CD8 cells but low numbers of CD4 cells (315 cells/mm^3, reference range 500–1200 cells/mm^3). Some T cells are 'missing' as the virus targets these cells, which are then destroyed. This low T cell count is strongly supportive of the diagnosis of HIV, and possibly the early stages of the AIDS. It may also prompt treatment with antiretroviral drugs.

It is tempting to speculate that the patient has a lung infection, so that the case may be passed to the microbiologists to search for an opportunistic infection.

Case study 10

A 75-year-old man, with a body mass index of 33.5 kg/m^2, complains to his general practitioner of 6 months of

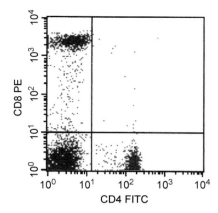

Quad	Events	% Gated	% Total
UL	1075	23.95	10.75
UR	21	0.47	0.21
LL	2108	46.96	21.08
LR	1285	28.63	12.85

Figure 9.7 Flow cytometry of CD4 and CD8 subsets. The typical four-quadrant result showing 1075 events staining above background for CD8 in the upper left quadrant and 1285 events staining for CD4 in the lower right quadrant. These translate to a CD4/CD8 ratio of 1.2.

The upper right quadrant has only 21 events, theoretically cells staining for both CD4 and CD8. The lower left quadrant is events failing to stain for either CD4 or CD8. Since the flow cytometer procedure has used a marker to detect all lymphocytes, therefore those in the lower left quadrant are non-T cells (i.e. B cells) and possibly some NK cells.

Since the number of NK cells will be very small compared with the B cell population, we effectively have a total B cell count. By combining the CD4- and CD8-bearing cells into one (i.e. T cells), we can compare the relative proportions of these cells. In health, a rough guide to T/B cell populations is 70%/30%. However, in this case the numbers of the two groups of cells (events) are 2360 and 2108 respectively, giving a distribution of 52.6%/47.4%, which is again removed from that which is expected in health, but which is expected in the diagnosis.

worsening pains in both knees, the pain in the right being markedly greater than that in the left. The pain is minimal upon rising in the morning, but gets worse during the day. The patient is sent for X-rays, a full blood count, ESR, CRP and RhF. Results were as follows:

Analyte	Result	Reference range
Haemoglobin	139	133–167 g/L
Red blood cell count	4.8	$(4.3–5.7) \times 10^{12}$/L
Haematocrit	0.45	0.35–0.53
Mean cell volume	86	77–98 fL
Platelets	213	$(140–450) \times 10^9$/L
White blood cell count	6.6	$(4.0–10.0) \times 10^9$/L
Neutrophils	4.0	$(1.7–6.0) \times 10^9$/L
Lymphocytes	2.0	$(1.0–3.2) \times 10^9$/L
Monocytes	0.4	$(0.2–0.6) \times 10^9$/L
Eosinophils	0.15	$(0.03–0.46) \times 10^9$/L
Basophils	0.05	$(0.02–0.09) \times 10^9$/L
Blasts	<0.01	$<0.01 \times 10^9$/L
ESR	12	<10 mm/h
Rheumatoid factor	1/20	Titre <1/40
CRP	<5	<5 g/L

Interpretation The X-ray points to moderate erosions in the right knee and mild erosions in the left, consistent with the history of pain.

The only abnormality is the very marginally increased ESR. The RhF titre is well within the reference range. The history points to a chronic connective tissues disease such as OA or RA. The normal RhF does not exclude the latter, but the lack of any gross abnormalities points to the former. Overweight is the major risk factor for OA, and if it were RA the presentation would be different with several sets of swollen and painful joints, and perhaps other factors such as early morning stiffness and more abnormalities in the blood analysis. Erosions in bone generally appear later in the natural history of RA. Normal CRP counts against an overt inflammatory condition.

Summary

- The immune system defends us from pathogenic viruses, bacteria and parasites. When this system fails it leads to immunodeficiency, hypersensitivity and autoimmune disease.
- The key mediators of an immune response are white blood cells, antibodies and a group of molecules called complement.
- Immunodeficiency is present if the body cannot defend itself from infections. The aetiology may be primary or secondary.
- Primary immunodeficiency diseases include DiGeorge syndrome (where there is a lack of T lymphocytes) and X-linked agammaglobulinaemia (where antibodies are not produced).
- Secondary immunodeficiency may result from the suppression of the bone marrow, and so impaired

production of white blood cells. The most common causes are malignancy and cytotoxic chemotherapy.
• There are four types of hypersensitivity: the most prevalent is type I, of which a common symptom is asthma, characterized by an excessive response to pathogens in inspired air.
• Autoimmunity is characterized by an antibody to an antigen on the body's own tissues. The most prevalent conditions in this group are inflammatory connective tissue diseases such as RA and SLE.

References

Hall A, Yates C. Immunology. Oxford University Press, 2010.

Hemminki K, Liu X, Forsti A *et al*. Subsequent leukaemia in autoimmune disease patients. Br J Haematol. 2013;161:677–687.

Further reading

Alimonti JB, Ball TB, Fowke KR. Mechanisms of CD4+ T lymphocyte cell death in human immunodeficiency virus infection and AIDS. J Gen Virol. 2003;84:1649–1661.

Ballow M, Notarangelo L, Grimbacher B *et al*. Immunodeficiencies. Clin Exp Immunol. 2009;158(Suppl 1):14–22.

Barnett D, Walker B, Landay A, Denny TN. CD4 immunophenotyping in HIV infection. Nat Rev Microbiol. 2008;6(11 Suppl):S7–S15.

Braza MS, Klein B. Anti-tumour immunotherapy with Vγ9Vδ2 T lymphocytes: from the bench to the bedside. Br J Haematol. 2013;160:123–132.

Conley ME. Genetics of hypogammaglobulinemia: what do we really know? Curr Opin Immunol. 2009;21:466–471.

Ehrnthaller C, Ignatius A, Gebhard F, Huber-Lang M. New insights of an old defence system: structure, function, and clinical relevance of the complement system. Mol Med. 2011;17:317–329.

Goronzy JJ, Weyand CM. Developments in the scientific understanding of rheumatoid arthritis. Arthritis Res Ther. 2009;11:249–259.

Krishna MT, Huissoon AP. Clinical immunology review series: an approach to desensitization. Clin Exp Immunol. 2011;163:131–146.

Kuijpers T, Lutter R. Inflammation and repeated infections in CGD: two sides of a coin. Cell Mol Life Sci. 2012;69:7–15.

Kuper CF. Histopathology of mucosa-associated lymphoid tissue. Toxicol Pathol. 2006;34:609–615.

Ram S, Lewis LA, Rice PA. Infections of people with complement deficiencies and patients who have undergone splenectomy. Clin Microbiol Rev. 2010;23:740–780.

Rinaudo CD, Telford JL, Rappuoli R, Seib KL. Vaccinology in the genome era. J Clin Invest. 2009;119:2515–2525.

Sarma JV, Ward PA. The complement system. Cell Tissue Res. 2011;343:227–235.

Simpson E. Special regulatory T-cell review: regulation of immune responses – examining the role of T cells. Immunology. 2008;123:13–16.

Souza-Fonseca-Guimaraes F, Adib-Conquy M, Cavaillon JM. Natural killer (NK) cells in antibacterial innate immunity: angels or devils. Mol Med. 2012;18:270–285.

Tsirogianni A, Pipi E, Soufleros K. Specificity of islet cell autoantibodies and coexistence with other organ specific autoantibodies in type 1 diabetes mellitus. Autoimmun Rev. 2009;8:687–691.

Van der Burg M, Gennery AR. Educational paper. The expanding clinical and immunological spectrum of severe combined immunodeficiency. Eur J Pediatr. 2011;170:561–571.

Ward AC, Dale DC. Genetic and molecular diagnosis of severe congenital neutropenia. Curr Opin Hematol. 2009;16:9–13.

Zhu J, Paul WE. CD4 T cells: fates, functions, and faults. Blood. 2008;112:1557–1569.

Guidelines

NICE clinical guideline CG86 (May 2009), from www.nice.org.uk.

Web sites

Allergens: http://www.phadia.com/en/Allergen-information/.

Autoimmune disease: www.labtestsonline.org.uk/; http://www.medterms.com/script/main/art.asp?articlekey=2402.

Immunodeficiency: www.centreforimmunodeficiency.com/.

Hypersensitivity: http://pathmicro.med.sc.edu/ghaffar/hyper00.htm.

10 Immunogenetics and Histocompatibility

Cognosce te ipsum etiam meliores: Know thyself even better

Learning objectives

After studying this chapter, you should be able to:

- discuss the genetics of B and T lymphocyte responses to antigens;
- recognize the key features of HLA molecules;
- explain the role of histocompatibility in transplantation;
- be aware of the relationship between autoimmunity and HLA types.

Chapter 9 introduced the concept of self and non-self. This chapter will expand on this concept and so why it warrants additional discussion. A key feature of this is the requirement to be able to differentiate self from non-self with a series of complex molecules called human leukocyte antigens (HLAs). There are several reasons why we need to know about HLAs.

The first is that HLA underlie the mechanism of the ability of our immune system to defend us against the possible attack by millions of potential pathogens, as will be explained in Section 10.1. The variation in the structure of different HLA molecules is governed by genetics, and this is crucial in the responses of lymphocytes to different antigens, as will be explained in Section 10.2. Variation of other genes also has an impact on the recognition and processing of pathogens, and certain different HLA alleles are linked to autoimmune and other diseases. The study of how these genes impact on immunological health and disease is called immunogenetics.

The second is that HLA plays a crucial role in transplantation, where the immune system regards an organ transplant simply as a giant virus, bacteria or parasite. Accordingly, a transplant will attract the attention of the

both the innate and adaptive arms of the immune system, features of which will be discussed in Section 10.3. We use the expression histocompatibility to describe the differences in the HLA alleles of the donor with those of the recipient.

A third feature is the association between certain HLA types and disease, which implies a breakdown in the immune system. The concept of how self comes to be regarded as non-self, and therefore the object of attack, and so cause autoimmune disease, will be described in Section 10.4.

10.1 The genetics of antigen recognition

The immune system has evolved to its present level of complexity because of the need to be able to respond to a potentially infinite and random series of stimuli that may cause disease; that is, pathogenic microbes. Conversely, the same microbes have been and are consistently evolving systems to evade our defences. Fortunately, herd immunity ensures that we have the upper hand.

Two systems, one in T cells and another in B cells, operate to give the immune system the required flexibility at the point where the molecules of the particular lymphocyte encounter foreign antigens. These processes require a third cell, a dendritic cell, to present antigens to the lymphocytes (hence antigen-presenting cell). Many dendritic cells are modified macrophages, and are found in key anatomical sites such as the lymph node, skin and spleen. However, there are also plasmacytoid dendritic cells that are lymphoid in origin and are found only in lymph nodes.

B lymphocyte responses to antigens

An immune response to a complex pathogen such as a bacteria or a virus results in a broad spectrum of antibodies that recognize different antigens on that pathogen. The process of antibody generation begins with the

Blood Science: Principles and Pathology, First Edition. Andrew Blann and Nessar Ahmed.
© 2014 John Wiley & Sons, Ltd. Published 2014 by John Wiley & Sons, Ltd.

presentation of the antigen to the B cell receptor (BcR) of a naive B lymphocyte by an antigen-presenting cell.

The B cell receptor   The BcR consists of an antibody molecule anchored within the cell membrane, and two copies of a second immunoglobulin-like molecule, CD79. The ability of the BcR to recognize the antigen depends on the conformation of the variable region of the heavy and light chain of the antibody; that is, the Fab. This variability is in turn governed by particular heavy- and light-chain genes. Chapter 5 has details of how the physical shape of each of the two Fab regions of the 'Y'-shaped immunoglobulin molecule is defined by the coming together of a section of the heavy chain and a section of the light chain to construct the variable region (Figure 5.11). The remaining part of the heavy and light chain does not change, and so is called the constant region. The sequence of amino acids that make up the light and heavy chain proteins are formed from instructions in the deoxyribonucleic acid (DNA). Variability in the DNA gives rise to different amino acid sequences, and so different proteins.

Thus, the variability in the physical form of the Fab is defined in the DNA. In the vast majority of the body's genes, the DNA sequence that codes a particular molecule is very strictly controlled so that exactly the same molecule is always generated. Indeed, in numerous cases, unwanted variations in the DNA sequence (mutations) cause disease (as in certain leukaemias and lymphomas, as discussed in Chapter 6). However, exactly the reverse is desirable in our defence against pathogens. We need to be able to generate a range of different forms of Fab structures that are able to interact with the different antigens present on these pathogens. Thus, hyper-mutation in the genes coding for the variable section of the heavy and light chains is an essential component of our immune system as it enables a flexible defence.

The genetics of antibody variability   The five antibody heavy chain genes (alpha, delta, epsilon, gamma and mu), are found on chromosome 14. Genes coding for two different types of light chain, the kappa gene and the lambda gene, are on chromosomes 2 and 22 respectively. Each part of the chromosome that has these genes is called a locus, and consists of a set of genes that ultimately code for the constant region of the particular light or heavy chain. There are three loci coding for genes that give rise to different variable regions:

- up to six genes code for a joining (J) region;
- up to 44 genes code for a variable (V) region;

- in addition, the heavy chain locus has up to 27 genes for diversity (D) regions.

A series of enzymes called VDJ recombinases cuts the DNA and re-splices it to bring together, for example, a constant-region gene, a D gene, a J gene and two V region genes to form a new VDJ-constant supergene. This forms the template for a messenger ribonucleic acid (RNA) molecule that can, in turn, generate the relevant protein chain (Figure 10.1).

This process of somatic recombination (or perhaps somatic mutation) generates the diversity of Fab sections required for a comprehensive antibody response. The nature of this recombination is a marker of the transformation of an immature, naive B lymphocyte into a mature antibody-producing cell. The arrangement of the genes themselves (the sequence) in an immature B cell is the same as in any non-leukocyte; that is, in so-called germ-line configuration. Once a lymphocyte has become activated, and the genes are rearranged into a sequence that can generate antibodies, the germ-line sequence is no longer present.

This has implications for lymphocyte malignancy (leukaemia, myeloma, lymphoma), as the degree of recombination can be assessed by molecular genetic techniques. This is not merely academic, as it may have implications for diagnosis and/or treatment, as in the case of *BCR–ABL* and tyrosine kinase inhibitors (as discussed in Chapter 6). A further example is the malignant recombination of immunoglobulin genes that generate 'nonsense' heavy and light chains (or even their fragments) and so nonfunctioning antibodies. These paraproteins (including urinary Bence–Jones proteins) are most prevalent in myeloma, as is described in Chapter 6.

The BcR is completed by two immunoglobulin-like molecules, CD79a and CD79b, coded for by genes on chromosome 19. There are no variations in these genes, so the molecules are effectively 'constant'. Once the antigen has bound the surface immunoglobulin part of the BcR, the CD79 molecules transmit signals to the interior of the cell which result in the clonal expansion of the activated B-cell.

T lymphocyte responses to antigens

In the same way that B lymphocyte responses are mediated via the BcR (soluble forms of which are antibodies), T lymphocyte responses are mediated via the T cell receptor (TcR), which is part of the cell membrane.

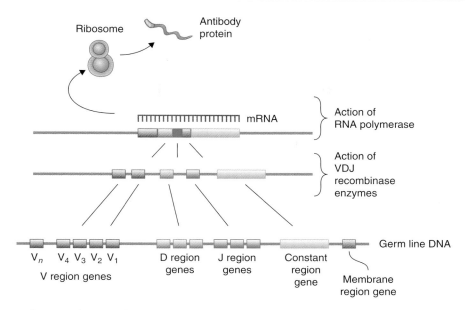

Figure 10.1 Recombination of immunoglobulin genes. The variability in antibody responses resides in the combined variable regions of heavy and light chains that form the Fab. The VDJ recombinase enzyme selects a J region gene, a D region gene and any one of several variable region genes to form a single combined VDJ-constant gene. This is the template for the RNA polymerase, which generates messenger RNA and ultimately the production of the protein chain by the ribosomes. The inclusion of a membrane region gene traffics the molecule to the cell membrane; without this section it will be exported into the plasma as an antibody molecule.

Immature T cells leave the bone marrow unable to respond to antigens; this they learn to do as thymocytes within the thymus, and so develop a mature TcR.

The antigen-recognition aspect of the TcR resides in two molecules, of which there are two isoforms: an alpha and beta dimer, and a gamma and delta dimer. Each of these four molecules has a constant region and a variable region coded for by specific genes, much like antibody genes and molecules. The variable regions are outermost and so can engage with antigens; constant regions are part of the cell membrane. The alpha gene locus is present on chromosome 14, whilst that for the beta gene is on chromosome 7.

As in antibody genetics, variation in the antigen specificity of the alpha–beta dimer or the gamma–delta dimer is governed by recombination of V, D and J genes, a process that occurs in the thymus under the control of recombinase enzymes (Figure 10.2).

The TcR is completed by CD3, which is formed from five molecules (a gamma chain, a delta chain, two epsilon chains and a zeta molecule), coded for by nonvariable genes on chromosome 11. Like CD79 and its relationship to the BcR, these CD3 molecules provide support and stability to the two molecules that recognize antigens.

Despite similarities with the BcR and immunoglobulins, the functioning of the TcR is more complex: the antigen recognition capacity is not designed to be secreted, as are antibodies, but are signalling structures, passing activation messages to the nucleus. The flexibility of the TcR relies on the ability of the genes for the variable regions to be able to recombine, so giving rise to a diverse response to antigens. This is analogous to the ability of the genes for the variable regions of the heavy and light chains of the antibody molecule to generate the BcR. Similarities between the two receptors are illustrated in Figure 10.3.

The diversity of antigen recognition

The combination of different V, D and J genes gives a possible 4.3 million different combinations of the gamma–delta TcR proteins, and 4500 million different alpha–beta TcR protein combinations. Similarly, random assortment of V, D and J immunoglobulin genes has the potential to produce 1300 million combinations of each heavy chain and a kappa light chain, and 1000 million combinations of each heavy chain and a lambda light chain. These figures are likely to be conservative.

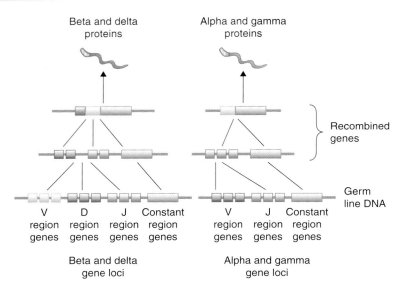

Figure 10.2 Recombination of TcR genes. The antigen recognition site of the TcR is composed of two molecules: an alpha–beta dimer or a gamma–delta dimer, coded for by genes on different chromosomes. Like immunoglobulin genes, the recombination of germ-line V, D and J genes generates a new gene that in turn provides the protein.

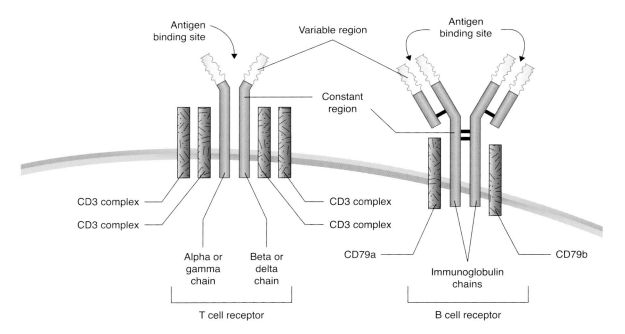

Figure 10.3 The TcR and BcR. The TcR is composed of an alpha–beta dimer or a gamma–delta dimer, and a complex of five molecules that make up CD3. The BcR consists of an immunoglobulin molecule and two CD79 molecules.

Antigen-presenting cells

The development of the immature B and T cells into an effector (i.e., respectively, one which produces soluble antibody molecules and one which is able to kill cells infected with viruses) requires not simply interaction with the particular antigen, but with an antigen-presenting cell, such as a dendritic cell. However, before it can support lymphocyte activation, the immature dendritic cell, which is non-motile and highly phagocytic, must itself be activated by cytokines. The source of these cytokines is often a mystery, but many probably arise from T lymphocytes and macrophages. The latter secretes interleukin (IL)-1, which acts on T cells, but also on the liver, which responds by upregulating acute-phase reactants and IL-12, a cytokine that acts on T lymphocytes. Most aspects of antigen presentation come together best in the dendritic cells, although macrophages, epithelial and endothelial cells, and even B lymphocytes can present antigens.

Once matured, the antigen-presenting (dendritic) cell is motile and poorly phagocytic. The maturation process is also marked by the increased expression of HLA molecules, of which there are several types, each used by different species of T cells depending on the presence of CD4 or CD8:

- the TcR combines with CD4 on T helper cells, forming a complex that recognizes antigen and HLA class II molecules (the HLA D series) on antigen-presenting cells;
- the TcR combines with CD8 on cytotoxic T cells, and so recognizes antigens in the context of HLA class I molecules (HLA A, B and C) on antigen-presenting cells.

This complex association of the TcR, CD3 and HLA molecules ensures that the T cell recognizes antigens strictly in the context of self. Once it has assured itself that whatever it has bound is non-self, the activation signal proceeds, and the cell starts those processes that ensure the destruction of the antigen. In the case of T helper cells, this involves promoting antibody production, but for T cytotoxic cells this results in the lysis of the cell bearing the antigen, which may be a virus. Before we examine the fine details of these processes, we must appreciate the complexity of HLA molecules.

10.2 Human leukocyte antigens

These are a complex series of molecules on the surface of almost all nucleated cells, and which define immunological self, and so non-self. First developed in the field of transplantation, and discovered on lymphocytes, they are present in all vertebrates. Pioneering work in mice pointed to a number of genes that influenced the ability of the mouse to accept or reject skin grafts from mice of other strains. Some genes did not evoke a very powerful effect, and so were designated minor. However, other sets of genes induced a rapid and powerful rejection, and so the region where they are located was designated the major histocompatibility locus (MHC).

In our species the MHC consists of (at least) three major groups of molecules, coded for by genes on the short arm of chromosome 6 at 6p21.31. There are three sets of HLA molecules: class I, class II and class III. The structure of class I and class II molecules resembles those of immunoglobulins, where the protein chains form globular domains. The parallel extends to the presence of constant and variable regions, as are also present in the structure of the alpha and beta chains of the TcR and the Fab of antibodies. Thus, the variation of the HLAs in each person resides in the differences in the sequence of a number of genes that are inherited from that person's parents. The combination of several different HLA genes confers a unique set of HLA molecules, which is often described as an individual's HLA type.

However, there is a major difference between the variability in antibody and T cell responses to antigen and the variability in HLA molecules. Variability in the former is generally short-lived, can change and is different in millions of different soluble molecules (antibodies) in the blood and on a small number of circulating cells (T lymphocytes) respectively. Conversely, the variability in HLA molecules is life-long, is stable, is unique to the individual and is expressed on all nucleated cells of the body. A further aspect is that variability in the TcR and the BcR (antibodies) is generated by somatic gene rearrangements, which are not heritable, whereas HLA variation is heritable, with limited variability within an individual, but a much greater degree of variation within the whole population. This is termed polymorphism.

Class I molecules

The three structures in this group are HLA-A, HLA-B and HLA-C. Their physical structure is of three globular domains, giving a combined relative molecular mass of 45 kDa. A crucial groove in the protein structure of the outer two domains enables the molecule to bind short 8–10 amino acid peptides. A second aspect of the expression of MHC class I molecules is their dependence on the

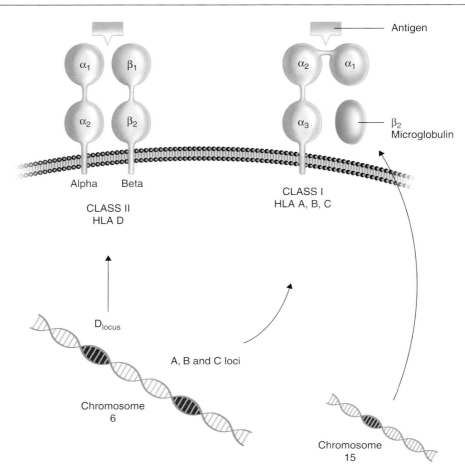

Figure 10.4 The structure of HLA molecules. HLA molecules bear similarities to immunoglobulins, with their globular domain structure. Class I molecules have three domains; each of the two class II molecules (alpha and beta) have two domains. A single-domain-sized molecule of beta-2-microglobulin is associated with the class I molecule. The variation in antigen binding resides in the terminal domains in a manner similar to that of the Fab of antibodies. In class I molecules, the antigen binding site is a groove between domains 1 and 2; in class II molecules, it is formed by the terminal domains of each chain.

presence of a molecule of beta-2-microglobulin (coded for by a gene on chromosome 15). This provides essential stability to the HLA complex, but notably it is not anchored in the membrane (Figure 10.4).

The function of HLA-A, HLA-B and HLA-C is to cooperate in the recognition of self and non-self with T cells bearing CD8. Potentially foreign peptides are presented to the TcR of cytotoxic T lymphocytes bearing CD8, which is then licensed to attack those cells bearing the antigen. Fine details of this process are to follow. Other class I molecules include HLA-E (which may have a role in natural killer (NK) cell function, possibly involving CD94), HLA-F (whose function has yet to be clarified) and HLA-G (with potential roles in pregnancy and NK cell function).

Class II molecules

These molecules make up the HLA-D family, which includes sub-types DM, DP, DO, DQ and DR. As with the components of the TcR, each HLA-D molecule is composed of two globular domains, and two molecules form a dimer on the cell surface. Thus, for example, HLA-DR is composed of an alpha chain (with a mass of perhaps 31–34 kDa) of HLA-DRA and a beta chain (with a mass of approximately 26–29 kDa) of HLA-DRB. The outer domains of the alpha and beta chains come together to form a groove that can, like the groove between the two outer domains of the class I molecules, bind a small peptide (Figure 10.4).

The function of HLA-D molecules is to cooperate in the recognition of self and non-self with T cells bearing CD4. These helper T lymphocytes are therefore primed to initiate the production of antibodies by B lymphocytes.

Class II molecules are more complex than those of class I, which consists of only one polymorphic molecule alongside beta-2-microglobulin. For example, considering HLA-DR, there is only one copy of *HLA-DRA* that gives rise to the DR alpha chain (which is invariable), whereas there are (at least) nine copies of *HLA-DRB* genes, not all of which may give rise to a viable beta chain, as several of these genes are nonfunctioning pseudogenes. However, each haplotype can express up to two *DRB* genes, but some haplotypes express only a single *DRB* gene. Both alpha and beta genes that make up HLA-DQ dimers and HLA-DP dimers are polymorphic.

Class III molecules

Class III genes do not code for HLA molecules, but for certain complement components (C2, C4, factor B), cytokines (tumour necrosis factor), heat shock protein and the receptor for advanced glycation endproducts (possibly important in hyperglycaemia and diabetes). There is no similarity in the structure of the molecules with the class I and II molecules. Although there may be variations in the structure of class III molecules, they are rarely a major issue in immunopathology. However, if deleted, loss of these genes can give rise to significant immunological disease, such as immune complex disorders.

The protein structure of class I and class II molecules is summarized in Figure 10.4, whilst Figure 10.5 shows the arrangement of their genes on chromosome 6.

The generation of antibodies

To recap: an antigen-presenting cell must first process the pathogen (probably by a form of phagocytosis), breaking it down into bite-sized chunks that are small enough to be recognized by the appropriate site on the BcR (although the BcR usually recognizes intact antigen) and TcR. This is often described as a small peptide, although it is known that there are antibodies to saccharides and lipids, so it must be possible to present these molecules to lymphocytes. Within the antigen-presenting cell, the antigen is digested in proteasomes, and fragments channelled to the endoplasmic reticulum where they form a complex with HLA molecules. This complex is then trafficked via the Golgi apparatus to the cell surface where it appears as altered-self (Figure 10.6: top section).

Once processed, the antigen-presenting cell, perhaps a dendritic cell, presents the antigen peptide (perhaps 8–10 amino acids long), located in the groove between the two outer domains of its HLA-D dimer, to the parallel groove in the TcR of a T helper lymphocyte. The CD4 molecule of the latter must interact with the HLA-D molecules of the antigen-presenting cell, that being the defining interaction of self-recognition. This interaction also demands the soluble signals (cytokines such as ILs and interferons) and additional molecular recognition signals between the cells (Table 10.1). These signals, the occupancy of the TcR by a peptide and the recognition of an HLA class II molecule by CD4 together result in the activation of the zeta (ζ) molecule of CD3 and so of the entire cell. This process is marked by the activation of a key intracellular enzyme, ZAP-70 (zeta-chain-associated protein of 70 kDa), the absence of which results in a form of immunodeficiency (Figure 10.6: middle section).

The activated T helper cell is then free to present the antigenic peptide to any B lymphocyte whose BcR recognizes that peptide. This interaction also requires the recognition by the CD40 molecule on the B cell with its

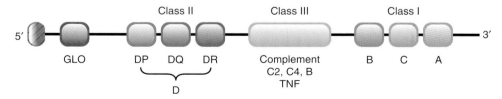

Figure 10.5 The layout of HLA genes. The HLA genes are arranged in a sequence on chromosome 6 and span some 3600 kilobases. From the 3′ end these are the three class I genes, the class III genes and the three class II gene loci. Between these genes and the centromere is the gene for glyoxylase (GLO).

Table 10.1 Molecular recognition systems between antigen presenting cells and T helper lymphocytes.

Molecule on the antigen-presenting cell	Molecule on the T lymphocyte
HLA-D plus antigen	TcR and CD4
CD58 (LFA-3)	CD2 (LFA-2)
CD106 (VCAM-1)	CD49d/CD29 (VLA-4)
CD54 (ICAM-1)	CD18/CD11a (LFA-1)
CD80/CD86	CD28[a]
CD80/CD86	CTLA4[a]

[a]CD28 and CTLA4 compete for CD80/CD86. If CD28 is dominant, stimulation proceeds. If CTLA4 is dominant, the signal is suppressed.

ligand, CD40L, on the T cell, and very possibly other cytokine signalling, such as IL-2, IL-4 and IL-5. Once the Fab of the BcR has recognized its antigen, CD79a and CD79b are activated, and this is transmitted to its own intracellular second messengers that will in turn activate the B cell, which then undergoes clonal expansion.

Having been activated, the B cell then proceeds with the mass generation of antibodies recognizing the same antigen as that presented to it by the T helper cell, resulting in the export of antibodies into the plasma. During the process of antibody shedding, it is inevitable that these proteins appear in large numbers in the rough endoplasmic reticulum of the cytoplasm, and on the cell surface, marking the lymphocyte as a plasmablast, which ultimately matures into a plasma cell (Figure 10.6: lower section).

This maturation process is accompanied by changes in CD marker expression. Naive B cells are CD19 and CD20 positive, but CD27 negative. However, once started on the pathway to antibody production, CD27 appears and is present in high levels on plasma cells, as is CD38, CD78 and the IL-6 receptor, whilst expression of CD19, CD20 and CD45 all fall as the cell matures. The additional importance of these molecules we shall return to shortly.

These processes are likely to occur in the germinal centres of lymph nodes and the spleen. Some macrophages bear CD4, leading to the hypothesis that they can provide T-independent help to B lymphocytes, perhaps in the production of antibodies to polysaccharides and other molecules.

Other forms of antigen recognition

There are a number of situations where effector cells can respond to antigens in the absence of HLA and/or T helper cells.

The B-cell co-receptor   It has been known for decades that antibody responses can occur without the need for T lymphocytes. There is now evidence that antibodies to certain antigens present on the surface of bacteria can be recognized via the B-cell co-receptor. This is a complex of CD19, CD21, CD81 and CD225. The key molecule is CD21, which functions as the complement receptor 2 that recognizes complement component C3d, the product of the activation of C3. Data from CD81 double knockout mice indicate that this molecule may be a regulator, in that its absence leads to enhanced responses to T-cell-independent antigens.

The model suggests that recognition of a bacterial antigen by the BcR and of C3d on the surface of that bacteria are sufficient to induce the activation of the lymphocyte via the CD79 dimer and the intracellular sections of the co-receptor (Figure 10.7). Notably, therefore, antigen recognition and antibody generation are independent of the HLA system and T cells. However, like the T-cell-dependent recognition system, other evidence suggests that this model may also need to be revised, as it has been suggested that there may also be a role for complement receptor 1 (CD35) in the B-cell co-receptor.

Natural killer cells   These large granular leukocytes (described in Chapter 5) are interested not so much in altered-self as missing-self; that is, cells with reduced or absent expression of HLA. The method of killing may be by the mobilization of cytotoxic granules and the release of proteases and a pore-forming molecule called perforin, and/or by the delightfully named death receptors. The latter are a subset of tumour necrosis factor receptors, including the Fas receptor (CD95), whose ligand (i.e. FasL, CD95L) binding results in the activation of caspase enzymes and ultimately the death of the cell by apoptosis. However, to add to the confusion, we have already noted

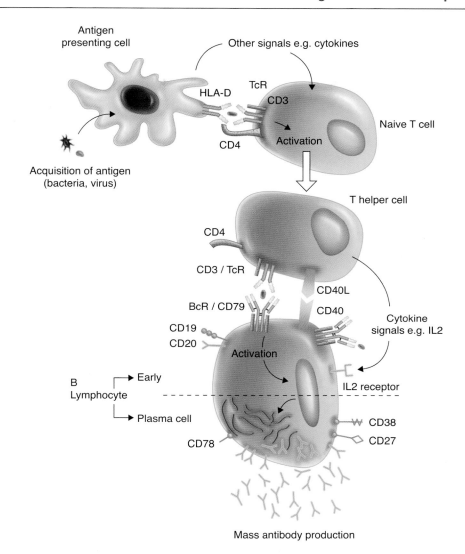

Figure 10.6 T–B interactions generating antibodies. A summary of the interactions between an antigen-presenting cell and a T helper lymphocyte, which receives the antigenic peptide via its TcR. The T cell in turn presents the antigen to the BcR of the 'early' B lymphocyte. This is effectively the 'go' signal for the B cell to transform into a plasma cell, and so generate antibodies specific for the presented antigen.

that HLA-E is recognized by the CD94/CD159a complex on NK cells (and on cytotoxic T lymphocytes).

CD1 This molecule, coded for by genes on chromosome 1, and therefore outside the HLA locus, nevertheless has strong homology for HLA class I molecules, and requires beta-2-microglobulin. CD1 seems to be involved in presentation of lipid or glycolipid antigens

(as may be part of organisms such as *Mycobacterium tuberculosis*) to T cells.

The generation of cytotoxic T lymphocytes

The initial stages of the generation and activation of cytotoxic T lymphocytes have parallels with those of B cell activation. An antigen-presenting cell processes

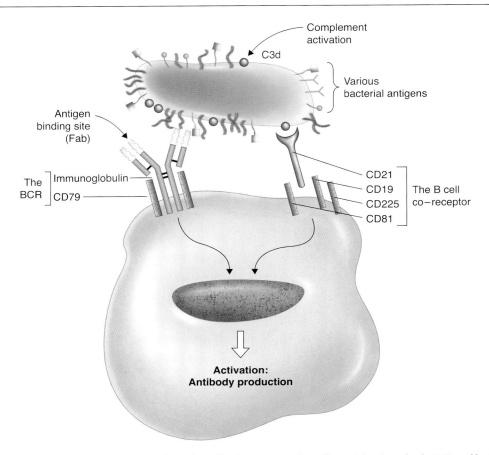

Figure 10.7 The B-cell co-receptor. B-cells may be activated by the co-recognition of bacterial antigens by the BcR, and by C3d on the surface of the bacteria by CD21.

potentially damaging molecules, perhaps from a virus, and places an antigenic peptide in the groove formed from parts of the outer two domains of an HLA class I molecule. This complex is recognized by the TcR, but a molecule of CD8 must recognize the self-aspect of the HLA molecule (Figure 10.8: top section). As with the activation of T helper cells, cytokine signalling (such as IL-2) is important in generating cytotoxic T cells, and the activation of the CD3 zeta molecule initiates the process of cell activation and final maturation.

This process activates the T cell and transforms it into an effector cell capable of killing a cell expressing the particular HLA molecule and the antigenic peptide. This latter complex is described as altered-self (Figure 10.8: lower section). However, this model may be too simple, as there

is evidence of the need for other signalling, such as between CD28 on T cells and CD80 on antigen presenting cells, and between CD95L (also known as FasL) on the T cell and CD95 (Fas) on the target cell, as indicated above. Several of these interactions also stabilize antigen presentation to T helper cells (Table 10.1).

The activated cytotoxic T cell kills its targets by the release of cytotoxins such as perforin to punch holes in the cell membrane of the target. This leads, via various routes, to cell death, as we have already noted in the section on NK cells. Interestingly, perforin has structural similarities to C9, part of the final stage of the complement pathway that forms the membrane attack complex.

Having grasped the concepts of self and non- or altered-self and the role of HLA, we can now look at

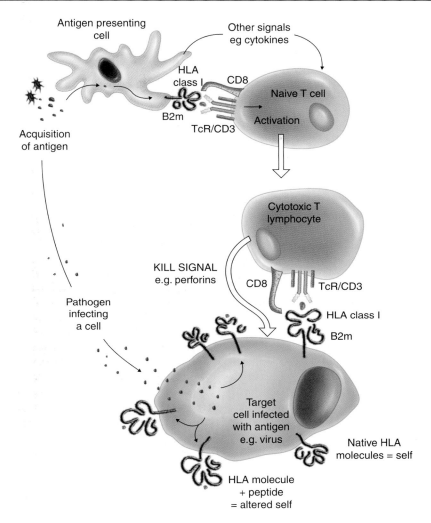

Figure 10.8 T cell–antigen-presenting cell interaction. An antigen-presenting cell and a naive T cell interact with HLA class I molecules and the TcR. The T cell CD8 molecule provides assurance that the presenting cell is self. The CD3 complex activates its zeta molecules, which initiates the final maturation of the cell and its clonal proliferation. The resultant cytotoxic T lymphocytes recognize and kill those cells presenting peptide and self-HLA class I molecules, which are therefore altered self.

practical aspects of this process, first in transplantation and then in autoimmune disease.

10.3   Transplantation

The transfer of biological material from one person to another, which includes the heart, lung (occasionally both at the same time), cornea, kidney, liver, bone marrow and blood (the latter described in Chapter 11), is transplantation. These transplants would be described as 'allogeneic', whereas transplants from one species to another is xeno-transplanation. Auto-transplantation, or perhaps autologous transplantation, is often used in bone marrow transplantation, and occasionally blood transfusion, where the patients receives their own tissues.

The long-term survival of allogeneic organs is resisted by the immune system of the recipient (the host), which clearly views the incoming tissues as non-self. An exception to this is the cornea, which is acellular, essentially being transparent connective tissue. As we have seen, at the cellular level, the major recognition system defining self is HLA, and each individual has their own particular HLA types. The process of trying to find an HLA type that is as close as possible between a person who needs a transplant and those of the potential donors is called histocompatibility. It follows that the ideal transplant material (such as a kidney) from the donor will carry the same HLA molecules as those of the recipient, but in practice this is almost impossible to achieve, unless from one identical twin to another. In the laboratory, the process of trying to match the donor and recipient HLA types is called HLA matching.

As we have seen, the function of cytotoxic T cells is to recognize and kill cells it regards as non-self. In a standard immune response, such as to a virus, a peptide from that virus is presented to the cytotoxic cell by an HLA class I molecule, which is, of course, self. The combination of a viral peptide on the HLA molecule is seen by the T cells as non-self, and so a legitimate target for destruction. This process is more accurately described as altered-self. The same process occurs in transplantation. However, the host may also make antibodies to the transplanted tissues and mobilize its NK cells and macrophages that can contribute to its destruction.

Polymorphisms in human leukocyte antigen molecules

The exquisitely fine specificity of the structure of the HLA molecules is such that the HLA molecules from another person (as will be present on a transplanted tissue) are regarded as altered-self, and this is why the tissue will be rejected. The HLA system is exceptionally polymorphic, which means that the chances of finding a donor with the same combined HLA type in the A, B and C molecules is as good as impossible. For example, there are almost 2000 different variants of HLA-A, almost 2500 of HLA-B and over 1000 of HLA-C. And of course we have two copies of the chromosomes bearing the genes for these molecules, and so usually two different copies of each HLA molecule, such as HLA A2 and HLA A24, although homozygosity (such as two copies of HLA A2) is not uncommon. The remaining class I molecules HLA-E, HLA-F and HLA-G have a very limited variability, with 3, 4 and 15 variants respectively.

In order to maximize the success of a transplant, the same matching process is required for the remaining MHC molecules; and of these, HLA-DR is the most polymorphic, with over 1000 alleles. Others, such as HLA-DQ and HLA-DP, are somewhat less polymorphic, with some 150 alleles each. A grossly simplified example of an individual's tissue type may be as follows:

- HLA A2 and A24
- HLA B5 and B14
- HLA C1 and C9
- HLA DR4 and DR12.

Human leukocyte antigen typing

In order to match as closely as possible the combined HLA types of donor and those of the recipient (and much else besides, such as the ABO blood group), we must know the HLA types of all parties. The closer they match, the greater is the chance the transplant will succeed. A perfect match is most likely to be found between siblings because they can inherit the same HLA genes from each parent. In this respect, HLA matching has a great deal in common with the cross-match at the heart of the blood transfusion laboratory. Indeed, failure to cross-match donor blood will lead to its rejection by the recipient. Chapter 15 of the Guidelines for the Blood Transfusion Services (BTS) in the UK refer to HLA typing, whilst Chapter 24 focuses on haemopoietic progenitor cells.

The nomenclature of HLA types is complex and evolving, and gives us more and more information about the particular type and how it is defined. For example, acceptable HLA assignments are HLA-B12, HLA-B44, HLA-B-44(12), HLA-B*44 and HLA-B*4409. HLA-D molecules maybe described as DRB1*12:05. A more complex nomenclature is HLA-DRB1*13:01:01:02N, which tells us it is a null allele that contains a mutation outside the coding region.

Historically, these types were defined serologically by antibodies from multiparous women, and then according to lymphocyte responses, although at present the ultimate defining process is DNA analysis of appropriate genes. The use of an asterisk is reserved for specificities defined by DNA techniques, such as HLA-A*01, whereas HLA-A1 is defined serologically.

Serological typing Chapter 11 discusses haemolytic disease of the newborn and how it comes about by foetal red cells effectively immunizing the mother with Rh D positive red cells. The same can occur with white cells, and serum of women often carries antibodies against the HLA class I types of their children (and so their father), who are non-self. This system was superseded by the introduction of highly specific monoclonal antibodies.

At the practical level, a few thousand lymphocytes are placed in the wells of a specialized Terasaki plate (or tray). Different antibodies of known HLA specificities are added to each well and the cocktail is completed by rabbit serum, which provides complement components. Should the antibodies bind cells, complement will be fixed and the cell lysed; hence 'lymphocytotoxicity testing' or 'complement dependent cytotoxicity'. Dead cells are identified by their failure to exclude fluorescent dyes such as ethidium bromide. Live cells remain unstained and, therefore, do not bind the particular antibody and so the individual is negative for the HLA specificity of the respective antiserum. Sections 1 and 3 of Chapter 15 of the UK BTS guidelines have additional details on serological typing.

Gene typing   These methods have arisen because of the limitations of the serological methods (the principal being the difficulty in HLA class II typing), and are regarded as the gold standard techniques. Indeed, it has been suggested that matching for class II locus HLA-DP can be typed reliably only by DNA methods. Using polymerase chain reaction (PCR) technology, DNA typing can be managed with primers and probes, the latter linked to a radioactive or fluorescent tag that is specific for each HLA locus or allele. The most sophisticated methods can sequence the nucleotides of the DNA itself. These methods can be summarized as follows:

• Sequence-specific primers serve as templates for DNA synthesis. Each primer pair is then amplified by PCR and tested in agarose gel electrophoresis. The method is rapid, efficient (being perfomed in a 96-well microtitre plate) and appropriate for typing for solid organ transplants, but it is not satisfactory for stem cell transplantation.
• Sequence-specific oligonucleotide probes are strands of DNA that bind in the region of a particular HLA allele and are labelled with markers such as fluorochromes. The probes can be immobilized on nylon membranes or beads. However, more recently, incorporation of biotin into amplified DNAs allows detection with streptavidin and phycoerythrin. This method is particularly suitable for typing large numbers of potential donors.

• Sequence-based typing is the gold standard, but it is costly and time consuming. It involves determining the exact sequence of nucleotides at a particular section of the gene of an individual's HLA allele.

Antibodies to human leukocyte antigen molecules

Historically, antibodies to HLA molecules were discovered in the serum of multiparous women, transfused patients or patients with rejected transplants and proves the concept of alloimmunization. It follows that the presence of anti-HLA antibodies may preclude transplantation of material from a donor with that particular HLA type. These antibodies can be detected by a number of methods:

• As per using antibodies to determine HLA type, the lymphocyte cytotoxicity test can be used, but requires a very well characterized panel of lymphocytes. It is popular as it is cheap and reproducible, but is prone to false positives and false negatives.
• Enzyme-linked immunosorbent assay (ELISA) can be used to screen serum for the presence of antibodies against particular HLA molecules, purified forms of the latter being immobilized on a microtitre plate. It is good for batch screening large numbers of samples.
• Fluorescence-activated cell scanning analysis uses microbeads coated with HLA proteins to determine the presence of anti-HLA antibodies in the blood of a recipient that would promote rejection of the transplant.
• The Luminex system with xMAP technology is a particularly focused aspect of microbead technology for detecting and defining anti-HLA antibodies and the specificities of HLA molecules themselves, and is rapidly becoming the method of choice. It is more sensitive than ELISA and the lymphocyte cytotoxicity test.

References

This area is highly complex; additional details are available from the UK BTS and from the British Society for Histocompatibility and Immunogenetics. Documents from the latter, 'BSHI/BTS Guidelines for the detection and characterization of clinically relevant antibodies in allotransplantation' and 'Guidelines for selection and HLA matching of related, adult unrelated donors and umbilical cord units for haemopoietic progenitor cell transplantation', available free on line, are so detailed that they are effectively textbooks.

The practicalities of transplantation

The decision to transplant is a clinical one, and is considered when all other medical and surgical approaches have failed. Historically, the kidney was first to be mass-transplanted, although, of course, blood transfusion is also a form of transplant. Leaving aside the latter, once an individual has been admitted to the waiting list they will be fully HLA typed and screened for any existing anti-HLA antibodies, and then the waiting begins.

HLA types are inherited from one's parents. It follows that parents must, and siblings may, share some HLA groups, making relatives prime candidates to be transplant donors. However, according to the laws of Mendelian inheritance, it is also possible that there will be siblings that fail to match, as illustrated in Figure 10.9. Nevertheless, it is likely that the best possible combination of HLA types will be settled upon, even if this includes some stark mismatches. Wherever possible, a pre-transplant cross-match is desired to detect anti-HLA antibodies in the recipient that react with the donor's type.

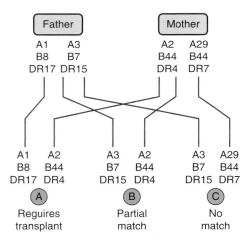

Figure 10.9 Inheritance of HLA types. Each individual inherits one HLA haplotype from their mother and a second from their father. In this very simplified example, showing only three HLA loci, sibling A, on the left, requires a transplant. Sibling B, in the middle, has inherited the same maternal haplotype, but a different paternal haplotype, so is a part match. Sibling C, on the right, has inherited a different set of haplotypes, and so is a complete mismatch. All siblings are partially matched with each parent (by definition). (After Choo SY, Yonsei Med J 2007).

Solid organ transplantation There is always a shortage of kidneys, hearts, lungs and other solid organs for transplant, most of which come from those who have recently died (cadavers). However, patients with an unusual tissue type may wait for years, and may possibly die because a suitable donor with an acceptable HLA type is never found. However, there is variability in the effect of certain HLA types on outcome: matching at HLA-DRB1 has a beneficial effect on heart and lung transplantation. In renal transplantation, the class II HLA-DR locus has a stronger impact on 1-year graft survival than does class I HLA-A and HLA-B loci; but for optimum outcome, compatibility at all three loci is desirable.

Live donors are preferable, as the time between removal of the organ (the cold ischaemic time) and its placement must be as short as possible. This is because the cold ischaemic time has a powerful effect on the survival of the organ once it has been transplanted. For example, hearts should be transplanted within 4 h of harvest. Organs taken from a dead body some distance from the recipient may well require some time in transport, during which period the cells of the organ will suffer and, if ischaemia is prolonged, will become necrotic.

The process itself is fully surgical. If a live donor, adjacent operating theatres will host two surgical teams: one to harvest and a second to place the organ into the fortunate recipient. Recovery is similar in practice to any other surgery, but the post-surgical care is profoundly different from other forms of surgery (to be described later).

Bone marrow transplantation This is used to treat life-threatening disease such as leukaemia, lymphoma and myeloma, but also non-neoplastic disease such as haemoglobinopathy and severe aplastic anaemia. The principles are broadly similar as in other transplants; that is, to match as many of the HLA types of the donor with that of the recipient as is possible.

Historically, bone marrow was harvested from donors, enriched for stem cells and depleted of mature cells, and infused into the venous circulation of the recipient. The donor cells would find their way to the recipient's bone marrow to replace and replenish their own cells. The recipient will have had their own diseased bone marrow destroyed by a combination of high-dose chemotherapy and/or total body irradiation. However, it is now possible to harvest and transplant not a simple crude mixture of bone marrow cells, but only stem cells. These can be collected from donors by the presence of the stem cell marker CD34. A further development is the

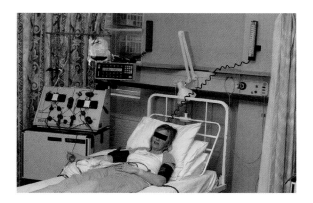

Figure 10.10 Collection of peripheral blood stem cells. The donor undergoes a modified blood transfusion donation using a cell separator. Mononuclear cells are harvested, and red cells, granulocytes and platelets are returned to the donor. (From Hoffbrand & Moss, *Essential Haematology*, Sixth Edition, 2011, Fig. 23.2, p. 300. Reproduced with permission of John Wiley & Sons, Ltd.).

harvest of CD34 positive stem cells from peripheral blood (Figure 10.10), and also from the blood in the umbilical cord of recently delivered babies, the latter being better tolerated than 'adult' stem cells.

A considerable advantage of bone marrow transplantation over solid organ transplantation is the stem cells can be manipulated *in vitro* and also stored frozen. However, cornea, skin, heart valves, tendons and ligaments can also be stored for later transplant. Chapter 24 of the Guidelines for the BTS in the UK refers to the collection, storage and use of haemopoietic progenitor stem cells.

A recurring problem with bone marrow transplantation is that of graft-versus-host disease. A crucial requirement is the need to condition the recipient by destroying their existing bone marrow, as otherwise mature leukocytes would be produced that will attack the incoming donor cells. However, this conditioning leaves the recipient immunodeficient and at risk of infections. But a further problem is that if the incoming donor stem cells are imperfectly matched to the recipient, the stem cells will produce functioning leukocytes that will, according to their function, regard their new host as non-self, and so attack the unfortunate recipient – hence graft (the incoming donor cells) versus host (the patient) disease.

Many of these problems disappear if the patient can be transplanted with their own bone marrow; that is, autologous transplantation. However, the process is still

dangerous, as absolutely all malignant cells must be destroyed before the patient's bone marrow is reintroduced. It is also crucial that the patient's bone marrow is purged of all neoplastic cells, which requires a high level of expertise in tissue culture. The disease can return if only one single malignant cell escapes this purging, as it will re-seed in the bone marrow.

A crucial aspect of bone marrow transplantation is the need to monitor levels of blood cells before and after the transplant. It is expected that the total and neutrophil counts will fall to zero by the conditioning, and then rise as the transplanted stem cells begin to repopulate the peripheral blood with new cells (Figure 10.11).

Rejection Regardless of the nature of the transplant (except acellular tissue), the constant concern is rejection by the body of its new tissue, the likelihood of which, as discussed, increases with the degree of HLA mismatch. However, the effectors of a transplant rejection process are principally cytotoxic T lymphocytes and antibodies, so that drugs (such as steroids, azathioprine and ciclosporin) aimed at suppressing T and B cells are used to prolong the life of a mismatched transplant.

However, the transplant team must find the correct balance of immunosuppression. On the one hand, too much immunosuppression to ensure the success of the transplant may also destroy the patient's ability to mount an efficient antiviral or antibacterial response; that is, iatrogenic immunodeficiency. Conversely, a low level of immunosuppression that will enable a modicum of an anti-microbe defence to be maintained may permit a response to the transplanted tissues and so lead to the rejection of the organ.

Rejection of solid organ transplants can be classified as one of three types:

• Hyperacute rejection, which happens within hours, possibly minutes. It is often caused by pre-existing anti-ABO or anti-HLA in the patient's blood, perhaps by prior blood transfusion. The principal cell attacked is the donor endothelium, and there is activation of the complement system, platelet activation and thrombosis. This type of rejection is very rare, as every attempt is made to ensure there are no existing antibodies.
• Acute rejection, which is seen to happen within a year of the transplant, but is often manifest within days. It is primarily the result of T-cell-mediated responses (inevitably cytotoxic) from the recipient that infiltrate and

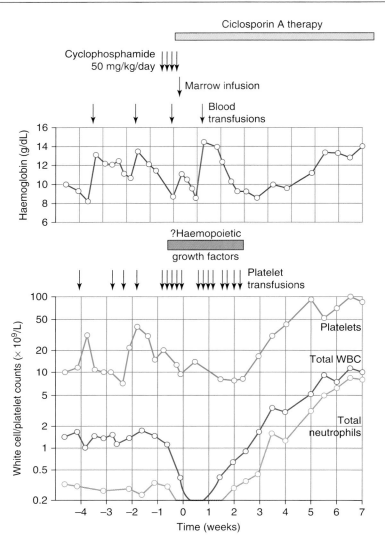

Figure 10.11 Peripheral blood counts in bone marrow transplantation. Red cell, platelet, total white cell and neutrophil counts before (on the left) and after (the right) transplantation. Note the white cell and neutophil counts fall to zero at conditioning (use of cyclophosphamide). Red cell and platelet numbers are maintained by standard transfusions. Success of the transplanted material is aided by the use of haemopoietic growth factors, such as granulocyte-macrophage colony-stimulating factor, and the use of post-transplant ciclosporin. (From Hoffbrand & Moss, *Essential Haematology*, Sixth Edition, 2011, Fig. 23.4, p. 302. Reproduced with permission of John Wiley & Sons, Ltd.).

attack the donor tissues. There may also be some anti-body-directed cellular cytotoxicity.

• Chronic rejection, which is generally caused by multiple factors with antibody, complement and cell-mediated responses. It develops months to years after the transplant, and is marked by slow deterioration in the function of the particular organ.

The ability to detect a rejection in the early stage is highly sought after. In this respect, levels of plasma markers such as soluble CD30, a T lymphocyte activation marker, show promise. The British Society for Standards in Haematology publishes guidelines for the support of graft-versus-host disease following bone marrow transplantation.

The ultimate transplant

Supposing a surgeon placed one of your father's kidneys into your mother. If fully healthy, she would be expected to reject it within weeks. So since you are a hybrid of your parent's genes, why did your mother not reject you? Indeed, we know that women can develop antibodies to the HLA types of the father (who is clearly non-self as far as the mother is concerned) of their child (who is partially self and partially non-self). However, as in haemolytic disease of the newborn, this immunological stimulation may happen after delivery. The immunology of pregnancy is a fascinating lesson in transplantation, and the foetal allograft survives for a numbers of reasons:

- The interface between the mother and her child is the foetal trophoblast, and these cells are immunologically protected and are resistant to most cytotoxic mechanisms. Placental syncytiotrophoblast and cytotrophoblast cells lack conventional HLA molecules, but express non-classical HLA-E, -F and -G proteins that may protect from NK cell attack.
- Trophoblast cells express molecules that inactivate complement component C3 convertase, and the presence of FasL may limit immunological attack.
- Immunosuppression. This includes the secretion of local nonspecific cytokines transforming growth factor-beta and IL-10 by regulatory T cells, and the generation of toxic metabolites by indoleamine 2,3-dioxygenase from trophoblast cells and macrophages. All act on T cells, B cells and NK cells.
- The uterus is an immunologically privileged site, and acts as a physical barrier. The placenta generates highly increased levels of oestrogen and progesterone that, as steroid hormones, are likely to have general immunosuppressive action.

The hypothesis of the immunosuppressive action of placental-derived hormones is supported by the observation that the pregnant woman with rheumatoid arthritis enjoys a remission of her disease. The symptoms recur once hormone levels return to normal after the delivery of the placenta. Indeed, this may be related to the fact that 75% of autoimmune diseases are found in women, and most commonly arise in the child-bearing years.

It could also be argued that the act of childbirth is in fact a form of rejection!

10.4 Autoimmunity and human leukocyte antigens

Autoimmune diseases are relatively common, affecting 5–8% of the population. Long before the development of the concepts of immunogenetics, it was recognized that rheumatoid arthritis and many other autoimmune diseases have a strong familial component (i.e. a tendency to run in families, often down the generations). We also know that the close relatives of, for example, a patient with clear Hashimoto's disease, who are asymptomatic, are at risk of having anti-thyroid autoantibodies and therefore of developing the disease. We now recognize that a large component of this is the inheritance of certain HLA types.

Aetiology

There are several theories of the aetiology of autoimmune disease. In Chapter 9 we looked at the roles of infectious agents, drugs and chemicals as precipitants of auto-immunity. The fact that considerably more women than men suffer from this group of illnesses implicates the effects of sex hormones oestrogen and progesterone, and their possible role as immunosuppressants (mim-icking as they do the structure of steroids). A recent hypothesis points to a role for CD22, present on the surface of most mature B lymphocytes, suggesting that loss of this molecule may contribute to the pathogenesis of autoimmune disease.

Loss of tolerance   One of the oldest 'immunological' mechanisms is the concept of a breakdown in the tolera-tion of self antigens by the immune system. The theory runs that some of the immature T cells developing in the thymus express a TcR that coincidentally recognizes self molecules, but that these self-reacting clones are marked out and destroyed, leaving only those cells reactive against non-self (i.e. self plus antigen) to defend us from micropathogens. The same or a similar process is likely to happen to those B lymphocytes whose BcR recognizes self tissues; these cells are eliminated, often by inducing them to go into apoptosis.

In this setting, attack of a target follows the perception by the immune system that, for example, normal pan-creatic tissue is actually non-self, leading to the destruc-tion of insulin-producing cells, and so diabetes. The reason that the pancreatic cells are attacked may be that they express aberrant HLA class II molecules that are recognized by autoreactive T cells that have escaped elimination.

Molecular mimicry   This hypothesis states that the normal immunological response to a micropathogen generates self-reactive clones that attack their body's own tissues, and so cause autoimmune disease. There are several examples that implicate the HLA system in this process:

• The amino acid sequence of the heat-shock protein of *Escherichia coli*, and certain products of *Lactobacillus lactis* and *Brucela ovis*, very closely resembles that of a section of the HLA-DRBq*04:01 molecule. This forms the basis of a hypothesis which states that T-cells reactive towards these bacterial products cross-react with the HLA type and lead to rheumatoid arthritis.

• *Chlamydia trachomatis* can trigger a reactive arthritis linked to HLA-B27. Molecular mimicry between the bowel microbe *Klebsiella* and the HLA-B27 molecule, as well as the spinal collagen types I, III and IV, indicates a pathological mechanism explains the onset of ankylosing spondylitis.

• In certain multiple sclerosis patients, T lymphocytes recognize a complex of HLA-DRB5*0101 and an Epstein–Barr virus (EBV) peptide (as they should), but also a complex of HLA-DRB1*1501 and myelin basic protein (MBP) (which they normally should not). Crystal structure determination of the DRB5*0101–EBV peptide complex reveals a marked degree of struc-tural equivalence to the DRB1*1501–MBP peptide complex, providing for molecular mimicry involving HLA molecules.

The structural details suggest an explanation for the preponderance of MHC class II associations in HLA-associated diseases.

The scope of human leukocyte antigens and autoimmune disease

There are numerous examples of links between certain HLA class I and II types and autoimmune disease (Table 10.2). Certain MHC class III loci are linked to type 1 diabetes, and another locus, close to the gene for tumour necrosis factor, is linked to rheumatoid arthritis.

Close scrutiny of Table 10.2 clearly shows that certain HLA types (principally B8, DR2, DR3 and DR4) pre-dispose to many different variants of autoimmune dis-ease across alternative aetiologies, such as of the connective tissues and endocrine system. A further com-plexity is that certain diseases are linked to more than one particular HLA type. For example, SLE is linked to HLA-B5, HLA-B8 and HLA-DR3, whilst Sjögren's syndrome has four linked types.

As it is also the case that not everyone with a particular genotype succumbs to its linked disease, in the future these associations may prompt investiga-tions leading to the factors that trigger symptoms in a susceptible individual and possibly the use of targeted therapy.

The combined haplotype of HLA-A1, HLA-B8 and HLA-DR17 is frequently associated with autoimmune disease. However, others are linked to individual HLA types. This is important as some associations are very strong, perhaps the strongest being HLA B27, which

Table 10.2 Autoimmune disease linked with HLA types.

HLA-A	HLA A1: myasthenia gravis and Sjögren's syndrome (N.B. HLA-DR3 is the primary disease link – the link with HLA-A1 is only present because of linkage disequilibrium)
HLA-B	HLA B7: Goodpasteur's syndrome
	HLA B8: Addison's disease, SLE, dermatomyositis, autoimmune haemolytic anaemia, primary biliary cirrhosis, myasthenia gravis, type 1 diabetes, Sjögren's syndrome, scleroderma, Hashimoto's thyroiditis and Graves' disease (once again, the primary association is with HLA DR)
	HLA B14: dermatomyositis
	HLA B15: type 1 diabetes and rheumatoid arthritis (both are primarily associated with HLA DR or HLA DQ)
	HLA B27: ankylosing spondylitis, postgonococcal arthritis, acute anterior uveitis
	HLA B44: rheumatoid arthritis (although primarily with HLA DR4)
	HLA B47: deficiency of 21-hydroxylase (involved in steroid metabolism)
HLA-C	HLA Cw0602: psoriasis
HLA-D	HLA DR1: Sjögren's syndrome
	HLA DR2: pernicious anaemia, Goodpasteur's syndrome and SLE
	HLA DR3: Sjögren's syndrome, autoimmune hepatitis, dermatomyositis, primary biliary cirrhosis, Addison's disease, myasthenia gravis, pernicious anaemia, Hashimoto's thyroiditis, Graves' disease, SLE and type 1 diabetes
	HLA DR4: rheumatoid arthritis, mixed connective tissue disease, Sjögren's syndrome, pemphigus vulgaris and type 1 diabetes
	HLA DR5: scleroderma
	HLA DR6: pemphigus vulgaris

SLE: systemic lupus erythematosus.

carries an astonishing 87-fold increased risk of ankylosing spondylitis. Many others are linked to HLA-D types, the strongest being:

- DQ2 and DQ8, which when together confer a 20-fold risk of type 1 diabetes; DQ8 by itself carries a 14-fold increased risk.
- DR3, which carries almost a 10-fold increased risk of Sjögren's syndrome, a sixfold increased risk of Addison's disease and a threefold risk of Graves' disease.
- DR4, conferring a four- to six-fold risk of rheumatoid arthritis.
- DR5, linked to a threefold increased risk of Hashimoto's thyroiditis.

A treatment for rheumatoid arthritis is the immunosuppressive drug ciclosporin. A retrospective study found that the presence of the HLA-DRB1*0401/*0404 genotype was strongly linked to use of this drug, providing limited evidence that this genotype confers the need for ciclosporin, and so likely to be a surrogate for more severe disease. Similarly, HLA-DRB1 is a risk factor for the requirement for orthopaedic surgery in rheumatoid arthritis, again suggesting that subjects with this type may benefit from targeted care.

However, HLA associations are not all disease promoting, as HLA-DR2 and HLA-DR5 are protective against the development of type 1 diabetes.

Blood science angle: HLA, rheumatoid arthritis and cardiovascular disease

The leading cause of death in rheumatoid arthritis (and many other inflammatory connective tissue diseases) is cardiovascular disease, supporting the hypothesis that the inflammatory component in the latter is not simply a bystander risk indicator. In a study of 182 patients with rheumatoid arthritis, increased c-reactive protein and ESR both predicted cardiovascular events and mortality. However, the presence of HLA-DRB1*0404 was a more powerful indicator of a poor outcome. The link was broadly confirmed in larger studies. Although the exact pathogenic mechanism for this is unclear (although it may relate to endothelial dysfunction), the presence of this HLA type may lead to more targeted therapy and so better disease management.

Summary

- Antibody and T lymphocytes responses to antigens are governed by genetics, and differentiate 'self' from 'non-self'.
- Combinations of V, D and J genes come together to generate millions of different binding regions.
- The BcR and TcR are combinations of cell surface molecules that recognize HLA molecules that present antigens.
- HLA class I molecules present antigens to the TcR of naive T cells bearing CD8. The latter are then transformed into cytotoxic T lymphocytes to kill those cells bearing the presenting antigen.
- HLA class II molecules present antigens to the TcR of T helper cells bearing CD4, which then presents the antigen to the BcR of B cells, resulting in the production of antibodies.
- The B cell co-receptor supports the production of antibodies independently of T cells.
- Histocompatibility is the process of matching different HLA types. It is used principally to assist the success of transplantation.
- Numerous HLA types, principally B8, DR2, DR3 and DR4, are linked to autoimmune disease. Potential aetiologies include loss of tolerance and molecular mimicry involving a pathogen.

Further reading

Choo SY. The HLA system: genetics, immunology, clinical testing, and clinical implications. Yonsei Med J. 2007;48:11–23.

Del Nagro CJ, Otero DC, Anzelon AN. *et al.* CD19 function in central and peripheral B-cell development. Immunol Res. 2005;31:119–131.

Farragher TM, Goodson NJ, Naseem H. *et al.* Association of the *HLA–DRB1* gene with premature death, particularly from cardiovascular disease, in patients with rheumatoid arthritis and inflammatory polyarthritis. Arthritis Rheum. 2008;58:359–369.

Gonzalez-Gay MA, Gonzalez-Juanatey C, Lopez-Diaz MJ. *et al. HLA–DRB1* and persistent chronic inflammation contribute to cardiovascular events and cardiovascular mortality in patients with rheumatoid arthritis. Arthritis Rheum. 2007;57:125–132.

Hütter G, Nowak D, Mossner M. *et al.* Long-term control of HIV by *CCR5*Delta32/Delta32 stem-cell transplantation. N Eng J Med. 2009;360:692–698.

Krangel MS. Mechanics of T cell receptor gene rearrangement. Curr Opin Immunol. 2009;21: 133–139.

Petersdorf EW. Optimal HLA matching in hematopoietic cell transplantation. Curr Opin Immunol. 2008;20: 588–593.

Shankarkumar U, Pawar A, Ghosh K. Implications of HLA sequence-based typing in transplantation. J Postgrad Med. 2008;54:41–44.

Sundberg EJ, Deng L, Mariuzza RA. TCR recognition of peptide/MHC class II complexes and superantigens. Semin Immunol. 2007;19:262–271.

Guidelines

Dignan FL, Scarisbrick JJ, Cornish J. *et al.* Haemato-oncology Task Force of British Committee for Standards in Haematology; British Society for Blood and Marrow Transplantation. Organ-specific management and supportive care in chronic graft-versus-host disease. Br J Haematol. 2012;158:62–78.

Dignan FL, Clark A, Amrolia P. *et al.* Haemato-oncology Task Force of British Committee for Standards in Haematology; British Society for Blood and Marrow Transplantation. Diagnosis and management of acute graft-versus-host disease. Br J Haematol. 2012;158:30–45.

Dignan FL, Amrolia P, Clark A. *et al.* Haemato-oncology Task Force of British Committee for Standards in Haematology; British Society for Blood and Marrow Transplantation. Diagnosis and management of chronic graft-versus-host disease. Br J Haematol. 2012;158:46–61.

The British Committee for Standards in Haematology: www.bcshguidelines.com.

The Guideline of the Blood Transfusion Service in the UK: http://www.transfusionguidelines.org.uk/index .aspx?Publication=RB.

Web sites

British Society for Histocompatibility and Immunogenetics: http://www.bshi.org.uk/.

Bone marrow transplantation: http://www.anthonynolan .org.

The official/standard system for naming and classifying HLA molecules: http://hla.alleles.org/nomenclature.

11 Blood Transfusion

Learning objectives

After studying this chapter, you should be able to:

- explain the glycoprotein nature of blood group molecules;
- discuss the biology of the ABO system;
- recognize the complexity of the Rh system and the importance of the D molecule;
- explain the value of different blood components;
- describe the major laboratory procedures – determination of blood group, antibody screen and the cross-match;
- highlight the role of antiglobulin testing;
- understand the needs for blood transfusion, and alternatives;
- appreciate dangers of blood transfusion.

Red Book

Almost all of the practice and management of blood transfusion in the UK is governed by the so-called 'Red Book', this being the *Guidelines for the Blood Transfusion Services in the UK*. This document is available online free of charge. Blood scientists (of whatever grade) dedicated to blood transfusion will be fully aware of, and practice in accordance with, these guidelines. The British Committee for Standards in Haematology also publishes guidelines pertinent to blood transfusion. Accordingly, this chapter cannot, nor does it intend to, replace these guidelines, but nevertheless summarizes the science and key aspects of this discipline.

The objective of blood transfusion has changed markedly over the years, from being a crude instrument to maintain haemoglobin at a certain level, to a multi-faceted therapy that saves lives. Additional changes have been the realization that a blood transfusion is far from a simple and trouble-free treatment, and that it can damage, possibly permanently, the health of the recipient.

What can be transfused?

Another development has been the move from transfusing 'whole' blood (which was once delivered in pint bottles), which therefore included plasma (with all its proteins and antibodies), white blood cells and platelets, to transfusing only specific components, usually just the red cells themselves. The latter, called 'packed cells', is not only more efficient, but also, without the white blood cells and (possibly damaging) plasma antibodies, produces fewer adverse reactions. Furthermore, the scope of transfusion has extended from red cells to include platelets and white cells. Noncellular components are blood components, including plasma, coagulation factors, and albumin. These blood components are derived from the 'leftover' plasma from a blood donation, although some (such as coagulation proteins) can also be produced by genetic engineering.

A more recent development has been the collection of bone marrow stem cells for transplantation. This is performed in specialist blood transfusion centres. Additional details of transplantation are presented in Chapter 10.

The basis of transfusion science

A fundamental process in cell biology is the ability to recognize who you are (i.e. 'self'), and so anything that is not 'self' must therefore be 'non-self'. This is important because pathogenic microbes and parasites are obviously non-self, and must be destroyed. Therefore, a transfusion of someone else's blood (non-self) into a

Blood Science: Principles and Pathology, First Edition. Andrew Blann and Nessar Ahmed.
© 2014 John Wiley & Sons, Ltd. Published 2014 by John Wiley & Sons, Ltd.

healthy body (self) will be recognized by the immune system and treated as if a pathogen, and so attacked. This is why we seek to get the incoming blood to be as close to the 'self' blood as is possible, this being called compatibility. It follows that differences between the donor and the recipient may make the transfusion incompatible, and this may be life-threatening. As is explained in Chapter 10, this incompatibility is also an issue in organ transplantation, where organs and tissues may be rejected. Indeed, blood transfusion is transplantation.

The scientific basis of self/non-self is the variation in the physical structure of certain molecules on the outside of all cells. These molecules all have a particular function, and are expressed differently on different cells. For example, GpVI on the surface of platelets bind to collagen, and so aids the adhesion of the cell to the sub-endothelium (Chapter 7). White cells express a high density of human leukocyte antigens (HLAs) to help in the recognition of self/non-self and so foreign material (Chapters 9 and 10). The genes for these molecules have different forms (alleles): GpVI varies very little between individuals, whereas HLA molecules show marked variation. The HLA system lies firmly within the field of transplantation, where considerable energy is expended in preventing the immune system of the recipient from treating the donor's transplanted organ as a giant pathogenic microbe.

This chapter focuses on variations in the structure of certain molecules on the surface of a red cell. Most of these molecules, the principals of which are called A, B, D and Kell, also have a degree of variability in their structure that lies somewhere between these two examples, and it is this variation that gives rise to problems in blood transfusion. In many cases, these differences are so marked that, if infused, the red cells are recognized as foreign, and so invoke an immune response as if they were microbial antigens on the surface of bacteria and viruses. This will induce in the patient the generation of antibodies that will attack and destroy the 'invading' red cell, thus defeating the objective of the transfusion.

In this chapter we will examine the discipline of blood transfusion by looking at the journey that the transfused material (be it red cells, white cells, platelets or blood components) takes as it leaves the donor (Section 11.1), and is then characterized and processed (Section 11.2). In Section 11.3 we will look at matching the donor blood with the recipient, and we conclude with problems in Section 11.4.

> **Blood science angle: Blood transfusion**
>
> This discipline is the practical link between haematology and immunology. Perhaps the prime reason for a transfusion is to treat a life-threatening anaemia, or perhaps haemorrhage. Blood transfusion may be regarded as a branch of immunology, considering as it does the actions of antibodies; indeed, many transfusion textbooks have entire chapters on immunology, leading to the expression 'immunohaematology'. Biochemistry does not generally have a great part to play in blood transfusion, and although the variation in blood groups is genetically determined, formal molecular genetic tests are rarely needed. However, our colleagues in microbiology may be called on if there is a suspected bacterial or viral infection.

11.1   Blood collection and processing

The blood scientist is unlikely to interact closely with the blood donor, as blood collection is generally performed in specialist units, some of which are mobile (hence the 'Blood-Mobile'). Nevertheless, awareness of the major issues regarding blood donation are a requirement. However, the blood scientist will be fully involved in the processing of the collected blood, which is performed by the NHS Blood and Transplant (NHSBT) service (formerly the National Blood Transfusion Service).

The blood donor

Naturally, we hope and expect the blood donor to be in excellent health. However, many of us have subclinical conditions, including infections. The latter can be serious, and the particular pathogenic organism could be transmitted to the unfortunate patient in a transfusion. For this reason donors are screened at the blood donation event for simple infections (such as cough, sore throat, cold, influenza, a recently completed course of antibiotics), illnesses (such as jaundice and asthma), participation in (according to the NHSBT) hazardous events or activity (such as sex with a man (if you are man) or a prostitute (of and by either sex), or having a tattoo or other skin piercing). Some exclusion criteria are permanent (such as receiving blood or blood components before 1980, or having had syphilis or being hepatitis B positive or a carrier, or being seropositive for the human immunodeficiency virus (HIV)), others are temporary (such as acupuncture in the last 4 months, or the use of needles for nonprescription drugs in the past year by a sexual partner). Additional details are provided on the NHSBT web site.

Other criteria expected of a donor are to be of weight 50 kg or greater, age 17–65 and to have a haemoglobin level of at least 125 g/L for women and 135 g/L for men. The latter is determined by specific gravity, defined by a drop of blood falling though a copper sulphate solution. This is needed to ensure a good 'rich' donation, but also to check the donor has sufficient blood to be able to give a donation. If the donor passes these tests, a donation of 400–500 mL of blood is taken into plastic bags with an anticoagulant such as sodium citrate, to which dextrose is added, the solution being kept acidic (hence ACD). Addition of phosphate (hence CPD) and/or adenine adds to the lifespan of the cells, to 28 days.

Blood processing

The donor packs of whole blood are passed to the NHSBT laboratories for characterization and processing by blood scientists. This includes determination of blood groups and screening for pathogens, and is where plasma is removed for the production of blood components.

The blood pack is then A, B and D grouped, and plasma drawn off to be processed for blood components. New donors may be additionally grouped, and special donors with rare groups may be processed individually. Donor plasma is also screened for antibodies that are already present, perhaps directed towards antigens of the Kell system (perhaps on a special 120-well microtitre plate), as these may find their way into blood components and could cause a transfusion reaction. The likelihood that the NHSBT will make an error in grouping is exceptionally remote. Similarly, great care is taken to minimize the probability of introducing a bacterial or viral pathogen into the collecting and processing of plasma. Other solutions, such as saline, adenine and glucose, possibly supplemented with mannitol, are used to prolong the lifespan of the blood and packed red cells for up to 42 days, although in many cases this is restricted to 35 days.

Pathogen screening Each collected pack is tested for a number of pathogens. These include HIV, human T-lymphocytotrophic viruses 1 and 2, hepatitis viruses and cytomegalovirus. There will also be nucleic acid testing for hepatitis C virus messenger RNA using polymerase chain reaction technology. In many cases individual samples are tested, but in others pooled plasma from up to 24 donors is collectively tested. Not all of this testing is mandatory; for example, blood drawn in the UK is not screened for West Nile virus, but blood imported from the USA is tested.

Other agents established to have caused disease in recipients include syphilis (caused by the spirochete bacterium *Treponema pallidum*), the Epstein–Barr virus, West Nile virus, parvovirus, some variants of herpes virus, malaria (the four *Plasmodium* species of *P. falciparum*, *P. vivax*, *P. malariae* and *P. ovale*), variant Creutzfeldt–Jakob disease and Chagas' disease (caused by the parasite protozoan *Trypanosome cruzeii*).

Specialized processing The most common blood pack that leaves the NHSBT service is packed red cells, generally whole blood from which most of the plasma has been removed. It therefore still includes platelet and white cells, and the latter can cause problems to patients who are immune-compromised, such as after bone marrow transplantation or with the immunodeficiency that accompanies the acquired immunodeficiency syndrome stage of HIV infection. In these cases, the donor white cells may actually attack the host, a condition called graft-versus-host disease, more details of which are presented in Chapter 10. A solution to this problem is to remove as many white cells and platelets as possible, which collectively are called the buffy coat. This can be done by the process of leukodepletion, or perhaps leukoreduction, which involves filtration, and is now standard practice so that all packs are leukodepleted by the NHSBT.

It is also possible to inactivate white cells (and many pathogens) by gamma irradiation or X-rays, although this may damage some red cell membranes. Other benefits of leukodepletion include a reduction in febrile non-haemolytic transfusion reactions and a lower rate of microbial infection (especially in cytomegalovirus). Pathogens in blood components may be inactivated with phytochemical technology, riboflavin, a solvent detergent and methylene blue.

Blood components

The major blood component is packed red cells, but there are several other blood components that are purified from the 'leftover' plasma in the donor blood. These have been known as 'blood products', but now as components. Fresh frozen plasma (FFP), cryoprecipitate (cryo) and platelets are given to stem blood loss, as described in Chapter 8. These components may also be obtained by apheresis, a specialized procedure where blood passes into a machine, the plasma and platelets are removed, and the red cells and white cells returned to the donor. Clinical indications for the use of blood components are presented in Section 11.4.

Fresh frozen plasma   FFP contains all the soluble substances found in normal blood. It also provides the patient with bulk fluid to support haemodynamics, as well as coagulation factors that help to restore haemostasis. The 'fresh' aspect refers to its need to be 'frozen' within 8 h of collection. FFP must be thawed at $37\,°C$ to ensure all the molecules have returned from the solid (frozen) phase to the liquid phase prior to transfusion. A variant of FFP is 24-h plasma, which is frozen 8–24 h after donation, and accordingly has lower levels of factor VIII. Problems with both types of FFP include allergic reactions, and there is a need for compatibility.

Each pack of FFP between 150 and 300 mL contains fibrinogen, factor VIII and other coagulation factors. Male donors are preferred, as parous women may have antibodies to the red cell and HLA molecules of the father of their children. FFP also contains normal antibodies to a variety of common pathogens, and so can be used in the treatment of immunodeficiency, and also in thrombocytopenia. The general dose is of 12–15 mL/kg, which translates to 900–1125 mL for a 75 kg patient, as may be used in major bleeding in association with fibrinolytic drugs such as alteplase, tenecteplase, reteplase and streptokinase. FFP is available commercially as Octaplas, and this component is often preferred to standard FFP in the treatment of thrombotic thrombocytopenic purpura.

Cryoprecipitate   This component is prepared from plasma by freezing (perhaps to $-20\,°C$) and a gentle thaw at $4\,°C$. At this cold temperature, certain coagulation factors remain insoluble and so can be collected as a precipitate (hence cryoprecipitate) and concentrated by centrifugation. If warmed to room temperature, these molecules that have precipitated will then go back into solution. Cryo is enriched for von Willebrand factor (vWf), and so also contains coagulation factor VIII and some fibrinogen and factor XIII. Consequently, it is most often used in von Willebrand's disease, but also in haemophilia if factor concentrates are unavailable (Chapter 8). Cryo lacks prothrombin and factors V, VII, IX, X, XI and XII, so cannot be used in patients deficient in these molecules. Like FFP, cryo is stored frozen and must be thawed at $37\,°C$ to ensure all coagulation factors are back in solution.

Both cryo and FFP are used to prevent and treat haemorrhage in haemophilia, isolated factor deficiencies, use of fibrinolytics, excessive anticoagulation with warfarin, von Willebrand's disease, hypofibrinogenaemia (where cryo is given) and disseminated intravascular coagulation. If there is massive haemorrhage, red cells are needed. The choice of FFP or cryo is made on clinical grounds and based on results from the laboratory, which can also confirm, perhaps an hour after the infusion, that it is effective.

Prothrombin complex concentrate   This product is enriched not only for prothrombin, but also for factors VII, IX and X (hence four-factor) as well as protein C and protein S. A principal use is to rapidly reverse a very high international normalized ratio (INR) in cases of warfarin overdose (if so, 2–5 mg vitamin K will also be given), and in bleeding whilst on one of the new oral anticoagulants (dabigatran, apixaban, rivaroxaban) where the patient may be at risk of gastrointestinal or intracranial haemorrhage. Indeed, a recent guideline recommended that all hospitals managing patients on warfarin should stock four-factor prothrombin complex concentrate (PCC).

If PCC is unavailable, FFP may be used to reverse warfarin-induced haemorrhage (if present) or a high INR (therefore a risk if haemorrhage). PCC is available commercially as Beriplex or Octaplex.

Platelets   These may be from a single donor or a pool from several donors, and must be collected rapidly after donation. They are stored for up to 5 days at $20–24\,°C$, and must be gently agitated by a rocking motion. However, some 7-day platelet packs are now available with additional bacterial testing of the donations. The risk of bacterial infection is higher than with red cells, most probably due to the higher storage temperature. As a fragment of the megakaryocyte, platelets bear all the same surface molecules, and also some platelet-specific molecules that have their own blood group system (details to follow) and so need to be compatible between donor and recipient.

A platelet infusion is required in order to treat a life-threatening thrombocytopenia, and ideally a single pack of 300 mL should contain over 5×10^{10} platelets, although an apheresis pack should contain over 3×10^{11} platelets. Each one of these packs is likely to raise the platelet count by $5–10 \times 10^{8}$/mL and $30–60 \times 10^{9}$/mL respectively. It follows that several packs may be required to raise the platelet count to a level considered sufficient to provide haemostasis. Platelet packs may also be called upon in a moderately thrombocytopenic patient (perhaps $<50 \times 10^9$ cells/mL) about to undergo surgery. Platelet infusion may be used in haemorrhage associated with anti-platelet drugs aspirin, clopidogrel and abciximab (the latter if the platelet count is less than 10×10^9/L). The practitioner will need to refer to national guidelines for the use of platelets.

Granulocytes These may be requested if the patient has a profound neutropenia and so is at risk of bacterial infections. Like platelets, granulocytes have their own blood group system that requires matching if a transfusion reaction is to be minimized (details are to follow).

Albumin This molecule is important because it comprises half of the total plasma protein pool, and as such is the major factor in osmolality (Chapter 17). Therefore, low levels, as are possible in severe nephrotic syndrome and after major burns, may have important repercussions. It is generally provided as a 4.5%, 5% or 20% solution, and may also be used to help treat hypovolaemia.

Blood science angle: Blood components (previously blood products)

The provision of these coagulation blood components for treating or preventing haemorrhage clearly demands knowledge of the pathophysiology of haemostasis. Similarly, the biochemistry laboratory can measure plasma albumin, and so demonstrate the presence of hypoalbuminaemia and so the likely need for replacement therapy. The laboratory will also be needed to ensure plasma levels of particular blood components have returned to normal.

Other blood components These include specific coagulation factors VIII, IX, VII, IX and X. Some of these factors (IX, VIIa, VIII) may be produced by recombinant deoxyribonucleic acid techniques. Recombinant factor VIIa should be considered if there is life-threatening bleeding associated with the use of low molecular weight heparin despite the use of protamine sulphate.

However, a caution with all these factor concentrates is the risk of thrombosis, especially in those vessels where the infusion is being given. Further details are to be found in the British National Formulary. The gamma globulin fraction (inevitably pooled from several donations) provides immunoglobulins (Igs) for the immunodeficient, whilst anti-D antibodies can be collected from those with high titres (details of these processes are in Chapter 9, and of anti-D in a section to follow).

11.2 Blood groups

The blood transfusion laboratory offers a number of components, both cells and plasma components. As red cells clearly dominate, a fuller understanding of their cell biology is demanded. All blood group systems (which may be present on all body cells or restricted to certain cells) consist of two parts. The first is molecules present at the surface of the cell; the second aspect is a series of antibodies that recognize these molecules. In some cases these antibodies are naturally occurring, whereas in others they are acquired following exposure of the particular molecule to the immune system, generally as part of a transfusion reaction. Most molecules present on red cells are also present on many other cells.

Red cell surface molecules

As indicated, all red cell surface molecules have a precise function and number of copies per cell. Most are glycoproteins, with a few being a carbohydrate, a lipoprotein or a glycolipid. Other details of the structure of the red cell membrane are to be found in Chapter 3, where Table 3.3 summarizes membrane glycoproteins in terms of the CD system. Figure 3.4 shows the association between major molecules, which come together in two clusters:

- The Ankyrin cluster brings together molecules LW, two or three copies of GpA (CD235A), RHAG, Glut-1, CD47, GpB (CD235B) and four copies of Band 3 (CD233).
- The 4.1 cluster consists of GpC (CD236C), Kell (CD238), Duffy (CD234), XK and two copies of Band 3.

Certain members of both complexes span the membrane and link (via the molecules ankyrin and 4.1 respectively) to the alpha and beta spectrins that make up the internal cytoskeleton of the cell; other molecules are expressed only on the external side of the membrane.

The synthesis of these cell surface molecules occurs in the red cell precursors (such as the erythroblast) in the bone marrow; a mature erythrocyte lacks the nucleic acids or protein synthesis apparatus to generate these molecules. The diverse functions of these molecules include adhesion, enzyme activity, regulation of the

Table 11.1 Major red cell surface molecules.

Number[a]	System names	Chromosome location of the gene	Molecules per cell ($\times 10^3$)	Molecule name/function
001	ABO	9	1000	Band 3: CD233 anion exchange
002	MNSs	4	200–1000	Glycophorins A and B: CD235
003	P	22	Unclear	Globoside 1
004	Rh	1	100–200	Facilitates Band 3/RhAG complex assembly: CD240
005	Lutheran	19	1–4	CD239: binds laminin
006	Kell	7	3–6	CD238: endopeptidase
008	Duffy	1	40–66	CD234: a receptor for chemokines
009	Kidd	18	14	Urea transporter
010	Diego	17	15	Band 3: anion exchange
011	Yt	7	Unknown	Acetylcholinesterase

RhAG: Rh-associated glycoprotein. [a] System number 007, Lewis, coded for by a gene on chromosome 19, is strictly speaking not a *blood* group system. The Lewis antigen is adsorbed onto the outside of the cell. (Modified from McCullough J., Transfusion Medicine, 3rd edition, 2012. Reproduced with permission of Wiley Blackwell).

complement system and the transport of ions and other molecules and ions across the cell membrane. These are required to maintain the physiology, size and shape of the red cell.

Table 11.1 summarizes certain aspects of the major molecules, which can be collected together into families, or groups, although some have only a single form. The most abundant molecules belong to the ABO system, and rightly bear the classification number 001, as their different varieties present the greatest challenge to the scientist. This is because they invoke particularly strong immune responses when compared with other blood group systems. However, there are very many others that frustrate the blood (transfusion) scientist.

The ABO system

The ABO system was identified and developed by Landsteiner in 1900 (who was rewarded with a Nobel Prize 30 years later), with the heritability being discovered in 1910. The system is based on the presence of two molecules, A and B, which are protein structures with five terminal carbohydrate groups, the major portions of which are part of the Band 3 molecule. These structures are widely expressed in body fluids and tissues.

Cell surface ABO molecules The common structure of the A and B structures begins with a base unit, composed of two galactose residues either side of a

molecule of N-acetyl glucosamine. A series of enzymes that effect these conversions belong to the glycosyltransferase family. The base unit of these three carbohydrates is the substrate for the enzyme L-fucose transferase, coded for by the *H* gene (also known as *FUT-1*) on chromosome 19, which adds a fucose molecule onto the terminal galactose molecule of the base unit.

If present, a second enzyme acts to define the ABO group:

• The enzyme encoded by the A gene, N-acetyl galactosamine transferase, or perhaps N-acetyl galactosaminyl transferase, adds a molecule of N-acetyl D-galactosamine to the terminal galactose.
• The enzyme encoded by the B gene, D-galactosyl transferase, adds another molecule of D-galactose to the terminal D-galactose.

The activity of the H and A genes produces blood group A. Similarly, the activity of the B and H genes leads to blood group B. The activity of both A and B genes leads to the parallel generation of their respective enzymes, and so the presence of both molecules A and B at the cell surface, and hence blood group AB.

However, if both the A and B genes are absent or malfunctioning, there is no enzyme activity and no generation of an A or B molecule. In this case the dominant membrane structure is the H molecule, and this creates blood group O. The O 'gene' is actually an A gene with a frame-shift mutation that results in failure of

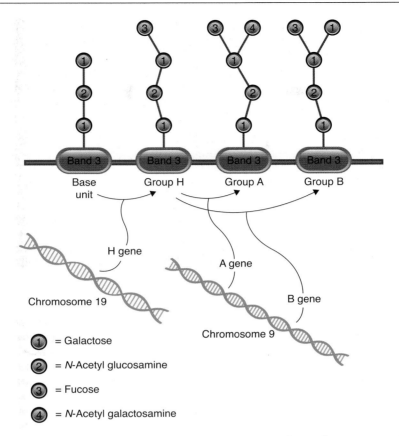

Figure 11.1 Formation of A, B and H blood group structures. The base unit is the substrate for an enzyme coded for by the H gene that adds a molecule of fucose to the terminal galactose molecule. If active, the A enzyme, coded for by the A gene, adds a molecule of *N*-acetyl galactosamine to the terminal galactose. If active, the enzyme encoded for by the B gene adds another molecule of galactose to the terminal galactose. Whilst most ABH molecules are linked with Band 3 molecule, they are also found elsewhere.

transcription. AB structures have a mass of 90–100 kDa. This process is illustrated in Figure 11.1.

Antibodies These are proteins designed to recognize a variety of molecules, generally that the immune system considers to be alien, and so should be destroyed. Details of the structure and function of these molecules, the major species of which are IgG and IgM, are found in Chapter 9.

In a fully reciprocal system, those people who are blood group A have antibodies to blood group B; that is, anti-B antibodies. Conversely, people of blood group B have anti-A antibodies in their plasma. Blood group O persons have both anti-A and anti-B antibodies, whilst blood group AB persons are free of these antibodies. AB antibodies are generally of the IgM class, and so are too large to be able to cross the placenta. The system is summarized in Table 11.2.

Interestingly, newborn infants, of whatever blood group, lack AB antibodies, which appear after several weeks of life. Therefore, ABO typing and the transfusion of neonates demands extra attention. The exact reason for the delayed appearance of these antibodies is obscure, but it implies an environmental stimulant, possibly certain bacteria and viruses.

Table 11.2 ABO blood group structures and antibodies.

Structures on the red cell surface	Antibodies in the plasma
A	Anti-B
B	Anti-A
AB	No antibodies
O	Anti-A and anti-B

Table 11.3 Racial distribution of ABO groups.

Phenotype	European	Black	Hong Kong	South Asia	Australian Aborigine
O	0.44	0.49	0.43	0.31	0.44
A_1	0.34	0.19	0.27	0.26	0.56
A_2	0.10	0.08	rare	0.03	0
B	0.09	0.20	0.25	0.30	0
AB	0.03	0.04	0.05	0.10	0

(Modified from Overfield J. *et al.*, (2008) and elsewhere).

The secretor phenotype The *Se* gene (*FUT2*, mapped to chromosome 19p13.3) controls the ability to secrete ABH molecules into the plasma and all body fluids, including faeces, saliva and semen, leading to a value in forensic science. Obeying Mendelian genetics, the *Se* gene has a silent allele, *se*, giving *Se/Se* homozygotes, *Se/se* heterozygotes (both of which secrete) and the *se/se* silent homozygote (the absence of *FUT2*), which defines non-secretors. Approximately 75% of a Caucasian population are secretors. This is also relevant to the biology of the Lewis antigen system, where *FUT2* and *FUT3* lead to Le^a and Le^b antigens.

The Bombay phenotype Just as loss-of-function mutations in the A and B genes governs the formation of the A and B structures, there is a rare phenotype resulting in the failure of the H gene. There are in fact two H alleles, the dominant *H* and the recessive *h* gene. The Bombay phenotype is due to the homozygous state of *hh*, and complete loss of the functioning gene.

The consequences of the Bombay phenotype is that only a very small amount (or even none) of the fucose molecule is grafted onto the terminal galactose of the base unit, and the overall phenotype appears as group O. However, the serum contains antibodies directed against the H molecule, which makes finding a compatible blood donor very difficult.

Subgroups of A and B The section described above is too simple; there are in fact subgroups of both A and B. More important are the subtypes of A, which include A_1, A_2, A_3, A_{el}, A_{end}, A_{int}, A_m and A_x, and are due to variability in the ABO genes, and so variability in the structure and/or the number of molecules per cell. The A_1, A_2 and A_3 variant numbers are on average 900 000, 250 000 and 35 000 copies per cell respectively. For comparison, the B molecule has about 700 000 copies per cell, whilst there are some 1.7 million H structures on a group O cell. These groups can combine, giving not simply group AB, but A_1B and A_2B.

There are also A_1A_1, A_1A_2, and A_2A_2. Blood group B variants include B_3, B_w, B_x and B_m, which are rarely found in Caucasian populations. This model may be considerably too simplistic, as up to 27 A alleles, 15 B alleles, 26 O alleles and 4 AB hybrid alleles have been described.

Genetics and the racial distribution of ABO groups

The ABO phenotypes are generated by action of ABO genes, of which all individuals have two copies, and this is why group AB exists. However, the phenotype of blood group A can exist as genotypes AA or as AO, or indeed as A_1O or A_2O. The frequency of the different ABO groups varies by racial group; these variations are summarized in Table 11.3, and although other workers may cite slightly different numbers, these will not vary more than 1 or 2%.

The most common phenotypes in most populations are groups O and A, with the combination of A_1 and A_2 being as dominant as group O. Where present, in all populations the A_1 frequency exceeds the A_2 frequency, and in the UK the frequency of the AO genotype exceeds the AA genotypes sevenfold.

The reasons for these stark differences are unclear but are presumed to be driven by natural selection. For example, group O confers a greater risk of severe infections with cholera compared with non-group O, and this may explain the high frequency of group B in certain cholera-endemic regions. Similarly, patients of group O seem to be more susceptible to gastrointestinal infections caused by *Escherichia coli*: in one study, 87.5% of patients who died of *E. coli* infection were of blood group O, the local population frequency of this blood group being 44%.

Implications of ABO blood group Decades ago, astute scientists noted that the request for blood for transfusion in certain types of surgery was linked to ABO blood group. Formal studies subsequently showed that group O subjects have a 14% reduced risk of squamous cell carcinoma and a 4% reduced risk of basal cell carcinoma compared with the other three groups combined. Group O also protects from pancreatic cancer. Group B is linked with ovarian cancer, and group A carries a greater risk of gastric cancer, whereas group O is protective. This may be linked to certain strains of *Heliobacter pylori*. Secretor status appears to confer resistance to *Haemophilus*, *Neisseria meningitidis* and *Streptococcus* species, and urinary tract infection caused by *E. coli*.

The ABO system presents a serious problem to the transfusion scientist. An incorrect transfusion would present a major clinical crisis in the recipient, and is termed an incompatible transfusion, which could prove fatal. These reactions can be noted in the laboratory, where scientists mix different samples of blood and observe the result, thus defining compatibility between blood donor and patient.

Blood science angle: von Willebrand factor

Some of the carbohydrates that define ABO groups are also present on vWf. Accordingly, perhaps 25% of the variation in vWf levels in populations can be accounted for by the ABO blood group. This has implications in haemostasis – ABO blood group influences results from the platelet analyser PFA-100 (Chapter 8), and it has been suggested that reference ranges for vWf assays should be ABO blood group specific. Since factor VIII activity relies on vWf, this may also be the case for the coagulation factor.

The Rh system

This is the second most important system, being developed from the observations of Levine and Stetson in 1939, and the experiments of Landsteiner and Weiner the following year. The former noted a case of haemolytic disease of the newborn (HDN) in a woman who had previously received a transfusion of her husband's blood, and correctly hypothesized that the two events were linked by a factor in his blood, later shown to be very likely the D antigen. Landsteiner and Weiner performed the key experiment of injecting blood from a rhesus monkey (hence Rh) into guinea pigs and rabbits, which resulted in an antiserum that reacted with the blood of most humans, hence the Rh system. With the benefit of today's technology they probably identified the LW antigen.

Molecules of the Rh system The Rh system is considerably more complicated than the ABO system, being composed of over 50 recognized glycoproteins, although on a day-to-day basis five different structures on the surface of the blood cell are commonly dealt with in the

blood bank. In practice, there is focus on the molecule known as D (i.e. RhD), as it is this structure and the antibodies that it can generate that can give rise to the strongest transfusion reaction, and also cause HDN if not correctly treated. About 85% of white Europeans are RhD positive, the molecule having a mass of some 30 kDa. However, there are a number of usual D isoforms, described as weak D, D variant, Du and partial D. Together, these forms account for 1% of all D molecules, and confound typing and attempts to find compatible blood for transfusion.

Other members of the Rh family of molecules are C, c, E and e, and each is associated with an antibody. In some circumstances the molecule C^w may be present; this is an antigen whose presence or absence may have been only partially determined. Notably, there is no 'd' molecule to complete the pairings. These molecules are not only smaller than the AB molecules, there are 5- to 10-fold fewer copies on the surface of the red cell. Very rare individuals lack all Rh molecules (hence Rh null), but this is associated with a haemolytic anaemia, demonstrating an important physiological role for these molecules.

Genetics of the Rh system Rh molecules are the product of a closely linked set of genes on chromosome 1 (at 1p36.13-p34.3), of which there are two copies. If functioning, the *RhD* gene gives rise to a D molecule. However, if the *RhD* is nonfunctional, there is no alternative product; that is, a would-be d molecule (if present). Thus, a homozygote with two functioning copies of *RhD* and a heterozygote with only one copy of *RhD* both produce the same RhD phenotype. Those homozygous for the lack of *RhD* (i.e. the theoretical 'dd' genotype) are phenotypically RhD negative.

The second gene is in fact a single long gene that, by alternative splicing, codes for E or e, and for C or c. Again, with copies of this complex on both chromosomes, a large number of combinations are possible, such as cDE or Cde, and each has a short code (in these cases, R_2 and r' respectively, a system developed by Fisher and Race). However, these combinations are not completely random: certain combinations seem to be preferred, the most common being presented in Table 11.4.

The Rh-associated glycoprotein The expression of Rh on the surface of the cell requires RhAG molecules, coded for by a gene mapped to chromosome 6p11-21. The LW molecule, coded for a gene located at chromosome 19p13.3, is also part of this complex, which includes Band 3 and ankyrin molecules.

Table 11.4 Common combination of genotypes of the Rh system.

Haplotype 1	Code	Haplotype 2	Code	Overall genotype	Combined code	Frequency (%)[a]
CDe	R_1	cde	r	CDe/cde	R_1r	31–34
CDe	R_1	CDe	R_1	CDe/CDe	$R_1 R_1$	16–18
cde	r	cde	r	cde/cde	rr	15
CDe	R_1	cDE	R_2	CDe/cDE	$R_1 R_2$	13–14
cDE	R_2	cde	r	cDE/cde	R_2r	11–13
cDE	R_2	cDE	R_2	cDE/cDE	$R_2 R_2$	2–3

[a]Frequency in a Caucasian European population.

Antibodies of the Rh system   Rh antibodies rarely occur naturally; they must be stimulated by presentation of alien red cells to a subject's immune system. The subject may be exposed to these alien red cells by an incompatible transfusion or by foetal blood passing into the maternal circulation at childbirth. The most common anti-Rh antibody recognizes the D molecule (i.e. anti-D) and is generally an IgG, and so has the ability to cross the placenta and attack the foetus (as is explained in the section that follows). Anti-C, anti-E, anti-c and anti-e are occasionally seen and may cause transfusion reactions and may also cross the placenta to attack the foetus, causing the syndrome of HDN.

Haemolytic disease of the newborn   This condition brings together several aspects of immunology and blood transfusion. It may be present in a woman who develops antibodies that cross the placenta to attack certain cells of her baby. Accordingly, this condition may also be called haemolytic disease of the foetus and newborn (HDFN). However, there are several key features that must be present for HDN to develop. In the case of HDN caused by RhD incompatibility, these are:

1. HDN can only occur in a woman who is RhD negative.
2. It can only happen if her foetus is RhD positive.
3. It cannot happen unless she has anti-D antibodies that can cross the placenta (and so are almost always IgG antibodies) and so attack the red cells of the RhD positive foetus.

These anti-D antibodies may have arisen because she has become exposed to the D antigen, perhaps on red cells that have entered her circulation at a previous childbirth or by blood transfusion(s).

The most common scenario is of an RhD negative woman carrying an RhD positive foetus. As the placenta

breaks away from the uterus after childbirth, some of the foetal red cells may enter her circulation and so (being non-self, and so foreign) will stimulate the production of anti-D antibodies. If there is a subsequent pregnancy with an RhD positive foetus, the maternal immune system will recognize the foreign D molecule and regard it as pathogenic. This will stimulate the production of anti-D IgG molecules which can cross the placenta and attack the foetal red cells, causing a haemolytic anaemia – hence HDN. Treatments include exchange transfusion, phototherapy and high-dose intravenous Igs.

For this reason, all RhD negative mothers are automatically given a dose of anti-D antibodies shortly after the birth of their baby. However, there is also a place for routine antenatal anti-D prophylaxis during pregnancy. The object of this is to destroy any foetal cells before they get the chance to provoke a genuine immune response. If necessary, the degree of this 'immunization' of the mother by her baby can be assessed by the number of foetal red cells in her blood, as these cells will express foetal variants of haemoglobin (as may be detected by the Kleihauer test, high-performance liquid chromatography (as for haemoglobinopathy) or by flow cytometry), and are larger than maternal cells (Chapters 3 and 4). If there is a large number of foetal cells present, a higher dose of anti-D may be necessary to ensure all the foetal cells have been destroyed.

Although we have focused on RhD HDN, several other blood groups can cause this condition, as is indicated in the section that follows. Let us hypothesize a theoretical blood group Z^a with two stable and mutually exclusive isoforms Z^a and Z^b. If a woman who is group Z^a is accidentally/unknowingly transfused with Z^b cells, she will develop anti-Z^b antibodies. This may occur at a first childbirth, if red cells from her child enter her own circulation. If, in a subsequent pregnancy (although it may be her first if she was exposed to Z^b by a transfusion) her child inherits the Z^b group from the father, the mother's IgG anti-Z^b antibodies are likely to cross the placenta and cause HDN. Indeed, an incompatible transfusion of this kind, and the HDN that followed, stimulated the first research in this area.

Other blood groups

There are hundreds of different molecules on the surface of cells, including red blood cells (Table 11.1). As discussed, some variants, such as A, B and RhD are very immunogenic (i.e. invoke a strong immune response and the generation of antibodies). There are also a host of minor blood groups that are weakly immunogenic and may not stimulate a strong antibody response unless there is repeated stimulus. However, if a patient is repeatedly transfused with blood that is incompatible for minor blood groups, that patient will eventually develop antibodies that will reduce the number of potential donors. The minor blood groups include the following:

- *Kell.* Incompatibility between the >30 different variants (such as K and k, Kp^a and Kp^b, and Js^a and Js^b) are, after A, B and Rh, the third most likely to cause a transfusion reaction or HDN. The molecule itself is a transmembrane endopeptidase. Antibodies (generally IgG) may cause a transfusion reaction or HDN.
- *Duffy.* This molecule (shorthand Fy, also known as CD234) is a receptor for several different cytokines. It exists as two major types, Fy^a and Fy^b, with either, neither or both being present. There are also minor molecules: Fy^3, Fy^4, Fy^5 and Fy^6. The phenotype Fya−b− is highly prevalent in Africans, less so in African Americans, and exceedingly rare in Caucasians. The IgG antibody is only moderately immunogenic, but nevertheless may cause a transfusion reaction, although it rarely causes HDN.

Blood science angle: Malaria

The Duffy molecule is the route by which parasites that cause malaria (*P. vivax* and *P. knowlesi*) enter the red cell. Therefore, people who lack this molecule (i.e. are Fy (a − b−)) are at a selective advantage. This disease is also modulated by the presence of certain abnormal haemoglobin species (notably sickle cell disease) (Chapter 4). Similarly, MN blood group molecules are a receptor for *P. falciparum*, so that those who are M^-N^- are at an advantage. However, both Fy and MN blood groups and haemoglobinopathy phenotypes are relatively easy to determine; molecular genetics are rarely called for.

- *Kidd.* This molecule (shorthand Jk), a urea transporter, is found in three isoforms: Jk^a, Jk^b and Jk^3. As with Duffy, either, neither or both a and b forms may be present. IgG antibodies tend to be weak but may also cause a transfusion reaction or HDN.
- *Lutheran.* There are 20 molecules in this system, but the dominant isoforms are Lu^a and Lu^b. This adhesion molecule may also exist in the same four isoforms as Kidd and Duffy; that is, Lu (a + b−), Lu (a + b+), Lu (a − b+) and Lu (a − b−). Most antibodies are IgM and IgA, and occasionally cause HDN.
- *MNSs.* This system comprises over 40 molecules found on glycophorins A (M and N species) and B (the S and s variants, and an isoform designated U). Genes *GYA* and *GYB* are located on chromosome 4q28-q31. Phenotypes of any combination are possible, and there can be a marked difference in racial distribution: the phenotype M + N − S − s+ is present in 8% of white populations but in 16% of blacks. Anti-S, anti-s and anti-U antibodies are clinically significant, being able to cause a transfusion reaction and, if IgG, HDN.
- *P.* This group consists of three molecules, P, P_1 and P^k, and there is also a null variant where no P molecules are present. The IgM antibodies rarely cross the placenta to cause HDN but may cause a transfusion reaction. Interestingly, paroxysmal cold haemoglobinuria is caused by IgG autoantibodies to group P on red cells.
- *Lewis.* Strictly speaking, the Lewis system of Le^a and Le^b molecules is not a blood group system, but is composed of molecules produced by epithelial tissues that reversibly bind to red cells and are linked to the complement system. The Le^a is the product of an *Le* gene (mapped to chromosome 19p13.3) and the ABO secretor gene, and may be converted into Le^b, which is a receptor for *H. pylori*. The IgM antibodies rarely cause HDN but may precipitate a transfusion reaction.
- *LW.* This system is closely associated with the Rh system, and indeed often causes confusion as antibodies may part-react with the D molecule. Three isoforms are recognized where LW molecules are present at the cell surface: LW^a, LW^{ab} and LW^b. A fourth phenotype is absence of both a and b forms; that is, LW(a − b−). Anti-LW^a and anti-LW^b antibodies are not clinically significant, and neither of them has been implicated in HDN.

Other blood group systems include those of Diego (abbreviated to Di), Cartwright (Yt), Xg (so called because the Xg^a gene is carried on the X chromosome), Dombrock (Do), Scianna (Sc), Colton (Co), Rogers (Rg)

Table 11.5 Human neutrophil antigen determinants.

HNA system	Antigen	Phenotype frequency	CD number/other classification
1	1a, 1b, 1c	46%, 88%, 5% respectively	CD16 (Fc receptor type IIIb for IgG
2	2	97%	CD177
3	3a, 3b	94.5%, 35.9% respectively	CTL2
4	4a	99.1%	CD11b
5	5a	>99%	CD11a

(Modified from Knight R. (Ed). Transfusion and Transplantation Science, Oxford University Press, 2012. Table 9.13).

and Chido (Ch), the I/i system, and several others. Readers are directed to reference textbooks held by their senior colleagues.

White cell antigens

The dominant molecules of the surface of B lymphocytes include antibodies, on T lymphocytes the T cell receptor, and both express highly polymorphic HLA molecules (Chapters 5, 9 and 10). The latter are effectively a personal blood group, unique for each individual and virtually impossible to get a perfect match for.

Neutrophils express their own system, human neutrophil antigens (HNAs), most of which are based on gene polymorphisms that code for a sub-type of Fc receptor for IgG. Neutrophils also express different varieties of CD11a and CD11b, both also associated noncovalently with CD18 (Table 11.5), as well as HLA molecules. Interestingly, neutrophils fail to express ABH, Rh and blood group K molecules.

Chapter 16 of the *Guidelines of the Blood Transfusion Services in the UK* document has more details of this aspect of neutrophil immunology.

Platelets

Platelets express many red cell molecules, but at a much lower density, and also certain HLA class I molecules. However, the platelet has its own series of molecules designated HPA (human platelet antigens), and all are the result of polymorphism in glycoproteins (Table 11.6). HPA systems 1–5 are fully established, and each has variants a and b. Alternative names for HPA-1 alleles are PI^{A1} and PI^{A2}.

There are up to 21 other HPA systems, but all these are designated 'workshop' – hence W. Most variants focus on GpIIIa and GpIIb (the most abundant platelet glycoprotein), although HPA-15 refers to polymorphisms on CD109, which is also present on certain T lymphocytes

and endothelial cells (additional details of platelet glycoproteins are in Chapter 7).

Chapter 17 of the *Guidelines of the Blood Transfusion Services in the UK* document has more details of this aspect of platelet immunology.

Clinical consequence of platelet and granulocyte incompatibility

Failure to correctly match donor platelet packs to the recipient lead to the destruction of platelets and so thrombocytopenia and post-transfusion purpura. The latter can develop 5–12 days after an incompatible transfusion, often directed against HPA-1a. A platelet version of HDN (resulting in neonatal alloimmune thrombocytopenia) develops in some neonates whose mother develops IgG alloantibodies against non-self platelet antigens resulting from paternal genes. Sections of the *Guidelines for the Blood Transfusion Service in the UK* (the Red Book) refer to platelet antigens.

Regarding white cells, as for platelets, the consequence of incompatibility is low numbers. These are forms of lymphopenia and neutropenia of the infused cells which develop once alloantibodies have been stimulated. However, if the recipient has a weakened immune system, it may not be able to attack and destroy these foreign cells, which themselves can become dangerous

Table 11.6 Human platelet antigen determinants.

System	Glycoprotein	System	Glycoprotein
HPA-1	GpIIIa (CD61)	HPA-6W	GpIIIa
HPA-2	GpIb (CD42b and CD42c)	HPA-7W	GpIIIa
HPA-3	GpIIb (CD41)	HPA-8W	GpIIIa
HPA-4	GpIIIa	HPA-9W	GpIIb
HPA-5	GpIa	HPA-10W	GpIIIa

and so cause graft-versus-host disease (developed in Chapter 10). There may also be neonatal alloimmune neutropenia, the neutrophil equivalent of HDN, with maternal antibodies often directed towards the HPA-1b molecule.

11.3   Laboratory practice of blood transfusion

The NHSBT service delivers part-characterized and micro-biologically safe components to the hospital, but more testing needs to be done before it can be given to the patient. Basic training in blood transfusion demands competency in a number of techniques: determination of blood group, antibody screening and cross-match, which is also described as compatibility testing. Other techniques include the need to identify existing antibodies in the recipient. More experienced staff will perform more complicated investigations (such as the fine specificity of an antibody) that are beyond the scope of this volume.

Determination of blood group

The request to 'Group and Save' is made by a physician or surgeon with an implication that a blood transfusion may be needed in the near future. The request is to find out the patient's blood group (*Group*), but then keep the blood sample handy (*Save*, generally in a refrigerator at 4 °C, where it will remain useable for up to 7 days). As a formal request for blood for transfusion may come at any time, the Group and Save gives the laboratory a head start in trying to match a number of packs of stored blood from those in the blood mark for the patient.

Most Group and Saves determine the ABO and Rh groups. If the ABO and RhD group of the recipient has been determined and the presence of clinically significant antibodies excluded, then the issue of matched blood is a relatively straightforward process.

The practical aspect of blood grouping is relatively straightforward, and for decades was a standard teaching tool for biomedical scientists in training, although nowadays other techniques operate. The patient's blood is washed in saline to remove all antibodies. The 'clean' red blood cells are then mixed with IgM antibodies of known identity – these being anti-A, anti-B and anti-D. Depending on the molecules present on the red cells, there may or may not be an agglutination reaction; if there is, this gives a positive identification of the presence of the particular molecule on the surface of the red cell:

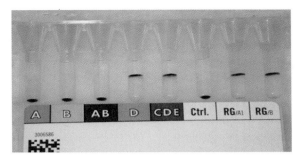

Figure 11.2 Determination of blood group in a single multi-chamber gel agglutination cuvette. Each of the eight compartments is for a separate reaction between the cells and antibody reagents present in the gel. If there is no reaction between the two, the red cells fail to agglutinate and so move to the bottom of the reaction cuvette. However, if there is a reaction, agglutinated cells remain higher up the gel in the particular column. So in this case the cells fail to react with anti-A, anti-B but do react with anti-D and so are group O, RhD positive.

The two channels on the far right are controls to ensure the reagents are working correctly. A negative tube control is also included in which patient cells are added to their plasma to detect the presence of any autoantibodies. (Image courtesy of H. Bibawi, NHS Tayside).

agglutination by the Anti-A reagent (but not the anti-B reagent) defines the cells as group A, and similarly anti-B reagent and group B. Failure of anti-A or anti-B reagents to agglutinate the cells identifies them as group O, whilst agglutination with both reagents defines group AB, as per Table 11.2.

Historically, this process is most easily demonstrated by mixing cells and reagents on a glass or porcelain tile, or in a small plastic test tube, and the agglutinated 'clots' noted. However, in current practice, grouping may be performed in a microtitre plate, or in a small plastic cuvette part-filled with gel and antisera (Figure 11.2). This is also called column agglutination technology.

Antibody screening

Often in parallel to grouping, pre-transfusion samples of plasma are screened for atypical antibodies against common red cell antigens. An indirect antiglobulin test should be used as the primary method for the screening of patients' plasma for the presence of clinically significant red cell antibodies. Other methods include those using tubes (liquid phase), microplates (liquid or solid phase), or cards/cassettes (column agglutination). If a tube method is used for antibody screening then this should involve the suspension of red cells in a low ionic strength solution.

The red cells are homozygous for all major groups, and as a minimum the following antigens should be expressed: D, C, c, E, e and K. Blood for extended phenotype and neonatal use is tested more rigorously, and also looks for antibodies against Fy^b, Jk^b, s and M. However, there are more comprehensive panels with cells bearing k, Kp^a, Kp^b, Js^a, Js^b, Fy^a, Fy^b, Jk^a, Le^a, Le^b, P1, N, S, s, Lu^a, Lu^b, and Xg^a, in addition to those already described.

Chapters 13 and 14 of the *Guidelines of the Blood Transfusion Services in the UK* document deal with issues relating to ABO and D typing, and with antibody screening.

Cross-match/compatibility testing

Not every group and antibody screen is translated into a request to cross-match. But when the call comes, scientists will retrieve the patient's blood sample that was used in the initial request to group and screen, and prepare some of the donor red blood cells and some of the patient's plasma or serum. The objective is find which packs of blood from among those in the blood bank fail to agglutinate when mixed with blood from the patient (i.e. are compatible – hence compatibility testing). The potential donor packs will already have their AB and major Rh groups defined; but as we know, there are dozens of other blood group molecules (several of which are listed in Table 11.1) that may differ between the donor and patient recipient and so may cause an agglutination.

As with many laboratory techniques, the process of cross-match/compatibility testing has evolved from a very 'hands on' manual method to one where several steps can be automated. Nevertheless, pared down to its component parts, a cross-match can be viewed as having two parts:

- Red cells from the patient (the recipient) are mixed with a series of samples of plasma or serum (often five or six) from packs of blood from different donors of the same ABO and D groups. These will have been selected from those packs in the blood bank.
- In reverse, some red cells from packs from the potential donors are mixed with the plasma or serum from the patient.

A good match is where, in both cases, the red cells are unaltered by this mixing, and therefore should not react together when in the patient. However, blood that does not match will aggregate, forming small clots, indicating an incompatibility. This process can be performed in small plastic test tubes, or even on a glass slide or porcelain plate, but nowadays the process can be developed in a 96-well microtitre plate, and the presence or absence of an agglutination detected by an autoanalyser.

If present, an agglutination is inevitably caused by a mismatch between the molecules on the red cells and the antibodies in the serum or plasma recognizing each other, and reacting together, causing blood to clump. It is presumed that the same reaction may happen pathologically in the blood vessels of the recipient and so cause a transfusion reaction.

Let us work through this with the scenario of a theoretical blood group Z^a (previously introduced to illustrate HDN) which has two isoforms at the surface of the red cells: Z^a and Z^b. These are coded for by two alleles of the Z gene, only one of which is expressed, so an individual can be only Z^a or Z^b, not both. Now suppose each isoform is regarded as foreign by those with the opposing isoform, so that someone who is Z^a will regard Z^b as non-self and so develop anti-Z^b antibodies. In the ABO system, neonates automatically develop antibodies to the AB group molecules, but in many cases the antibodies to blood groups only arise after the exposure to the foreign molecules, perhaps by a blood transfusion. Repeated blood transfusion of the same foreign material (such as Z^b red cells into a patient who is Z^a) induces the production of progressively stronger anti-Z^b antibodies, so that eventually it will be impossible to find compatible blood.

A successful cross-match will consist of mixing samples of potential donor red cells with plasma or serum from the patient. Ideally, there will be no agglutination. However, if there is agglutination, this is effectively proof of an antibody in the patient's blood that reacts with what it regards as a foreign molecule on the red cell. In our example, this may be the presence of an anti-Z^a antibody in the patient's plasma that reacts with Z^a on the surface of the red cells from the donor, rendering the donor cells incompatible and so untransfusable. These principles are illustrated in Figure 11.3.

The same principle of incompatibility also occurs in real systems, such as the presence of a different form of an Rh molecule (such as D) or to a Kell blood group molecule on the donor blood that may react with specific anti-Rh or anti-K antibodies in the patient, and so causing a transfusion reaction.

Donor cells 1	Patient's plasma	Reaction mixture	Result

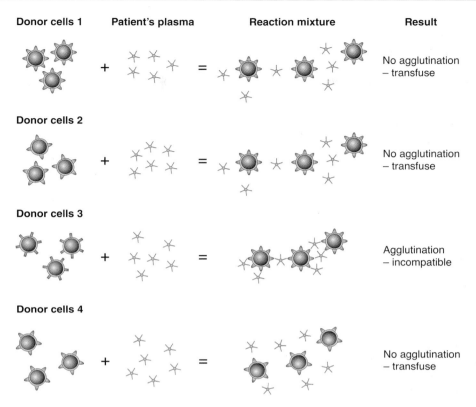

Figure 11.3 Principles of a cross-match. In this cross-match, samples of red cells from four packs of blood from potential donors are mixed with plasma from the patient. In three cases, there is no agglutination, so patient lacks antibodies to molecules on cells from the donor packs, which are therefore acceptable for transfusion. However, in one case the patient's plasma has antibodies that react with one of the potential donor cell, causing agglutination, and so these cells are incompatible and cannot be transfused.

Special investigations The scenario of excluding potential incompatibility and antibody testing described above can be criticized as being oversimplistic. There are a number of special methods that may be employed to determine the characteristic of an incompatible reaction and/or an abnormal antibody. These include:

- Whilst all cross-matching should proceed at 37 °C (body temperature), it may be that some antibodies react at lower temperatures, such as at room temperature. Such 'cold' antibodies may be clinically relevant because the vessels of the skin are cooler than those deep inside the body, and may be involved in diseases such as Raynaud's disease.
- Although rarely used, some add albumin to the antibody/red cell reaction. This provides more of a physiological medium as albumin is clearly present in the blood, as this may promote antibody–red cell reactions

- The use of low ionic strength saline is routine in many laboratories, but some supplement with added polyethylene glycol.
- The use of proteolytic enzymes to expose certain sites that may otherwise have escaped detection.

However, not all of these options are routinely used in all situations, but may be called upon to help solve a particular problem. It should be recalled, though, that not all antibody reactions found in the laboratory automatically lead to a clinical problem. Indeed, the presence of irrelevant antibodies may lead to a perfectly good pack being denied to a patient.

Antiglobulin testing

The most efficient antibodies that agglutinate red cells belong to the IgM class, because this large pentamer molecule can cross-link two adjacent cells. The other antibody classes important in blood transfusion, IgG and (to a lesser degree) IgA, although able to bind to its antigen on the red cell, are unable to cross-link different cells. However, IgG binding to red cells can still be associated with agglutination in the body if a third component binds to IgG molecules on different cells, thereby forming a cross-link.

This phenomenon can be tested in the laboratory, where the third reagent is called an antiglobulin. The antiglobulin test, developed by Robin Coombs (therefore referred to by some as the Coombs test), has two forms: direct and indirect. The anti-human globulin reagent is often raised in sheep or goats, and must be able to recognize several of the antibody classes (and so would be polyclonal), although more specific anti-IgG and anti-IgM antiglobulins are available. There are also reagents that can detect the presence of complement components on the surface of cells.

The direct antiglobulin test   In this test the patient's red cells are first washed free of plasma. The second step is to mix the washed cells with an anti-human globulin reagent to detect any antibodies that are still bound to the red cells after the washing. An alternative form of the reagent contains antibodies to complement components such as C3d, as these molecules can absorb passively onto the surface of the red cell. If present, these bound antibodies (and perhaps C3d) will be recognized by the anti-human globulin reagent and, if the reaction is strong enough, will result in agglutination. This process is illustrated in Figure 11.4.

This test therefore detects the presence of antibodies on the patient's own red cells. These may be autoantibodies, or perhaps antibodies introduced because of a prior transfusion. However, this test is not widely used in routine blood transfusion, but may be called upon in special investigations. For example, an autoantibody to the P blood group is often present in

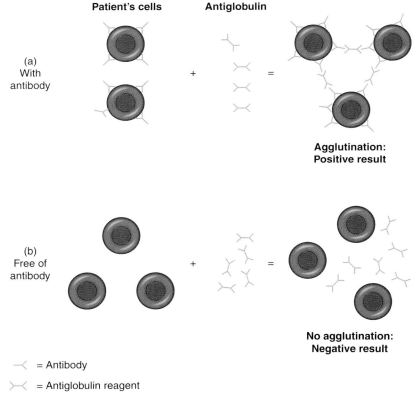

Figure 11.4   Direct antiglobulin testing. Washed cells from the patient are mixed with an antiglobulin reagent to detect antibodies on the red cells. (a) If present, antibodies will be recognized by the antiglobulin reagent, and will cross-link the cells, giving a positive result. (b) In the absence of antibodies, the antiglobulin will have nothing to cross-link, so there will be no agglutination.

chronic lymphocytic leukaemia, and can be detected with this test. These antibodies may cause a haemolytic anaemia.

The indirect antiglobulin test This test, effectively a modification of the direct antiglobulin test, is widely used in transfusion laboratories to screen for antibodies in the recipient's serum that may react with molecules on the donor's red cells. In a two-step process, the donor cells are first washed free of plasma proteins and are then incubated with serum or plasma from the potential recipient. If present, antibodies in the patient's sample will bind to the donor cells. The cells are again washed, this time to remove excess Ig molecules from the recipient. In the second step, some anti-human globulin is then added, and will detect and bind to any antibodies present on the red cell, and so form an agglutination. This process is illustrated in Figure 11.5.

This powerful technique detects antibodies in the patient's blood that are proven to react with donor cells in the laboratory. It is presumed that, if these donor cells are transfused, the same reaction will occur in the body.

However, the laboratory procedure uses an anti-human globulin reagent that is not present in the patient, but may become present due to the nature of the immune response. However, agglutination is not the only adverse effect of an incompatible transfusion. The recipient's immune system can still detect and destroy donor cells carrying an antibody by the process of complement-mediated lysis, details of which are to be found in Chapter 9.

If present, the exact nature of the antibody in a positive antiglobulin test is unknown. It is most unlikely to be an anti-A, anti-B or anti-RhD, as the status of the donor cells has already been made. It follows that the unknown antibody may react with any of the other blood group molecules listed in Table 11.1. Additional testing is required to determine the specificity of the antibody. The indirect antiglobulin test may also be used in ante-natal screening of women for IgG antibodies in potential HDN. At the practical level, antiglobulin testing is performed in the same type of gel-containing plastic cuvettes as are used in the determination of blood group (Figure 11.6), but can also be by tube or microplate.

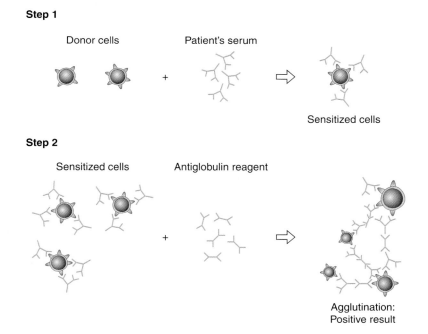

Figure 11.5 Indirect antiglobulin testing. In step 1, washed cells from the donor are mixed with serum or plasma from the patient. If present, antibodies will bind to the red cells and sensitize them. In step 2, the cells are mixed with an antiglobulin reagent to detect antibodies on the red cells. As in the direct antiglobulin test, the anti-human reagent will cross-link the cells, so that agglutination will mark a positive result.

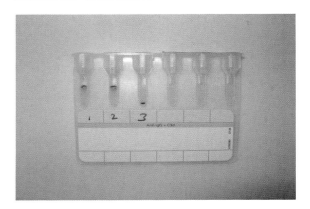

Figure 11.6 Antiglobulin testing. Testing for antibodies on red cells by the antiglobulin test can proceed in the same type of plastic cuvette with gel matrix as is used for blood group determination (Figure 11.2). In samples 1 and 2, cells have been agglutinated by the antiglobulin reagent and remain at the top of the column. Unagglutinated cells pass through to the bottom of the column, as in sample 3. The reagent shown here contains antibodies that recognize both IgG and complement component C3d, which may absorb passively onto the surface of the red cell. However, cuvettes are available that detect only the binding of IgG antibodies. (Image courtesy of H. Bibawi, NHS Tayside).

Although the most robust, this 'standard' version of the cross-match does have disadvantages. These include a lengthy and expensive process, and that the absence of antigens on the red cells may miss a rare or weak antibody. Accordingly, other variants of the cross-match have been developed, but these must be used in conjunction with an antibody screen, and provides a second check on ABO compatibility.

Immediate spin cross-match

The advantage of this method is that it is rapid, simple and will detect most ABO incompatibilities, although it should not be used where ABO grouping reveals very weak anti-A or anti-B antibodies. Other irregular antibodies must have been excluded by antibody screening. The patient's plasma is tested against a 2–3% saline suspension of donor cells with only a 2–5 min incubation, but there are several difficulties with the process. One is that the immediate spin cross-match presumes the antibody screening has not missed any antibodies (i.e. is negative); another is that it will generally not detect cold-reacting IgG antibodies or those of minor clinical significance. Some workers are concerned about a lack of

positive and negative controls, and accordingly nonspecific reactions may be reported.

Electronic issue

This is an information-based method that simply matches blood based on existing history; there is no actual mixing of samples between donor and recipient. It relies instead on the accuracy of the patient's ABO and D status, their transfusion history and antibody screen results compared with the ABO and D status of the donor cells. The practitioner is referred to current Medicines and Healthcare Products Regulatory Agency and British Committee for Standards in Haematology (BCSH) guidance.

The process is best suited for automation, and a robust laboratory information management system, and is simple, rapid and cost effective. However, there are clearly many pitfalls and problems, such as the presumption that the grouping antibody screening is accurate. There are also several instances where electronic issue is not advised, which include patients who have a history of or a current positive antibody screen, those who are not part of the computer system, patients who have had any manual cross-matching and those who have received a transfusion.

Special cases

Some rare circumstances demand special care, such as exchange transfusions and those involving the neonate and foetus.

In life-threatening emergencies, there may not be time for a full compatibility work-up. If desperate, group O and RhD negative blood may be issued before all the standard transfusion testing is complete. In order to avoid Rh sensitivity, blood for a woman under age 60 should be D negative.

Emergency issue does not render normal practice irrelevant, and all cross-matching and antibody screening should proceed as normal, but this of course depends on the human and technical resources of the laboratory at the time. Each laboratory will have its own standard operationg procedure for such circumstances.

Blood transfusion analysers

Major techniques in blood transfusion continue to evolve and, in common with other disciplines in biomedical

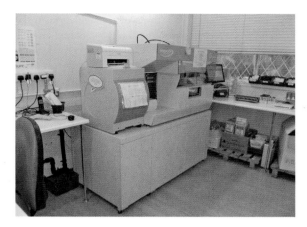

Figure 11.7 A blood transfusion autoanalyser.

science, several can be automated. Whilst reactions in gel cards (Figures 11.2 and Figure 11.6) can be read by the scientists, these can also be determined by imaging software. These have been incorporated into analysers that can perform ABO and RhD groups, and in antibody investigations. Figure 11.7 shows such an analyser, which to the uninitiated closely resembles machines found in biochemistry, haematology and immunology laboratories.

11.4 Clinical practice of blood transfusion

The requirement for a blood transfusion is made primarily on clinical grounds and is therefore unlikely to be part of the blood scientist's portfolio. Nevertheless, knowledge of the steps involved will make you a better practitioner.

Blood transfusion is far from being a simple and trouble-free procedure. First, the practitioner will ask: Does the patient *really* need it? Are there any alternatives – is autologous transfusion possible? One textbook suggests that a post-operative patient who is asymptomatic with a haemoglobin of 90 g/L probably does not require a transfusion; others would not consider a transfusion if the haemoglobin is greater than 80 g/L. As mentioned, in the past, ordering a transfusion was often simply because a physician considered it a good thing to do, that it would probably do the patient some good. This would, of course, be fine but for a number of problems, a few of which are as follows:

- In practice, whilst the ABO/Rh systems are the most important, there are many other minor blood group systems (such as Duffy and Kell) that can be a problem. The more transfusions a person receives, then the greater the likelihood that these minor incompatibilities can build up to be a real clinical and laboratory issue.
- Patients with chronic anaemia (such as severe sickle cell disease or thalassaemia) need to be repeatedly transfused. This leads to the syndrome of hyper-transfusion, and often leads to problems, as the build up of this iron can cause damage to the tissues (haematochromatosis, Chapter 4). Since one unit of blood contains some 250 mg of iron, then 15 units can more than double the body iron stores, and this is deposited in organs such as the liver, heart and pituitary, leading to liver failure, heart disease and endocrine problems respectively.
- Transfused blood can contain pathogenic organisms (viruses, bacteria, parasites), although, fortunately, owing to screening, this is becoming less of a problem. However, infection can also occur via the site of the infusion.

Therefore, not simply because of the above, the present view is that transfusion should be reserved only for those in danger of losing their life or those who will show a measurable improvement not achievable by other means. It follows that the requirement for a transfusion can only be made clinically, not in the laboratory.

Indications for transfusion

This is not simply because of haemoglobin being less than a certain number. In Chapter 4 we read that this approach led to the inappropriate transfusion of six packs of red cells that cost the recipient her life and the hospital $11.5 million. Local and national guidelines offer advice, although some may be prescriptive.

Reasons for ordering a transfusion can be many and varied, but major life-threatening indications include chronic and serious anaemia refractive to other treatment (such as severe cases of the haemoglobinopathies), life-threatening emergencies such as rupture of an aortic abdominal aneurysm, massive blood loss after a road traffic accident, and major haemorrhage (such as in haemophilia or overdoses of warfarin or heparin, although transfusion of coagulation components may also be necessary).

As mentioned, those at risk of life-threatening haemorrhage are likely to benefit from transfusion of blood components of coagulation proteins, FFP and cryo (Chapter 8). These will be stored at $-25\,°C$ or less, and are stable for 24 months. Platelets may also be transfused for severe thrombocytopenia; these are stored at $22\,°C$, with gently rocking agitation, and are stable for 5 days. Granulocytes are stored at $22\,°C$, but are stable for only 24 h. Fortunately, the UK Chief Medical Officer's National Blood Transfusion Committee has summarized action to be taken in common indications.

Red cell concentrates   In cases of acute blood loss, such as <30% loss of blood volume (<1500 mL in an adult), only crystalloid/colloid should be transfused – red cell transfusion is unlikely to be necessary. If the blood loss is 30–40% (1500–2000 mL) then both crystalloid/colloid and red cells are likely to be required in order to maintain recommended haemoglobin levels, such as 70 g/L in otherwise fit patients and >80 g/L in the elderly and those with known cardiovascular disease. Rapid volume replacement including red cell transfusion is needed if the blood loss is >40% loss of blood volume (>2000 mL in an adult). Patients who maintain their haemoglobin level after surgery will not need red cell support, but if this falls markedly, such as to the levels indicated, the haemoglobin can guide the use of red cell transfusion.

There is no evidence-based guide to the use of transfusion after chemotherapy, but most hospitals use a transfusion threshold of a haemoglobin of 80 or 90 g/L. Similarly, in the case of radiotherapy for cervical and possibly other tumours, some protocols recommend maintaining a haemoglobin above 100–110 g/L. In chronic anaemia, transfusion is required to maintain the haemoglobin to prevent the symptoms of anaemia; however, some patients may be asymptomatic with a haemoglobin >80 g/L.

Anaemia is common among critically ill patients, 60% of whom are found in intensive care units; and of these, 20–30% have a haemoglobin <90 g/L. Because of the high mortality rate of this group, and in those with traumatic brain injury or angina, the transfusion threshold is 70 g/L or below, with a target of 70–90 g/L. In acute coronary syndromes, haemoglobin should be >80 g/L.

Fresh frozen plasma   The dose of this component is generally 12–15 mL/kg body weight, equivalent to four units for an adult. Indications include replacement of single coagulation factor deficiencies, where a specific or combined factor concentrate is unavailable (e.g. factor

V), acute disseminated intravascular coagulation in the presence of bleeding and abnormal coagulation results, and thrombotic thrombocytopenic purpura (usually in conjunction with plasma exchange).

If emergency uncontrolled bleeding and massive haemorrhage is anticipated, early infusion of FFP (15 mL/kg) is recommended to treat coagulopathy. Local protocols should be followed, and the later use of FFP should be guided by timely tests of coagulation, including near patient testing. Where there is anticipated large volume blood loss associated with routine surgery, guidelines suggest the prothrombin time (PT) and activated partial thromboplastin time ratio should be maintained at <1.5. This is likely to occur after replacement of 1–1.5 times the patient's blood volume. In liver disease, there is no evidence of benefit from FFP in patients with a PT ratio of less than or equal to 1.5.

Cryoprecipitate   A standard dose is two pooled units, equivalent to 10 individual donor units, for an adult (contains approximately 3 g of fibrinogen). Cryo should be used in combination with FFP unless there is an isolated deficiency of fibrinogen. It may be used in acute disseminated intravascular coagulation (where there is bleeding and a fibrinogen level <1 g/L), in advanced liver disease (to correct bleeding or as prophylaxis before surgery, when the fibrinogen level <1 g/L) and in bleeding associated with thrombolytic therapy causing hypofibrinogenaemia. In cases of hypofibrinogenaemia secondary to massive transfusion, it should be used to maintain fibrinogen >1 g/L. It may also be used in renal failure or liver failure associated with abnormal bleeding where desmopressin is contraindicated or ineffective, and in inherited hypofibrinogenaemia, where fibrinogen concentrate is not readily available.

Platelet concentrates   A standard dose of 15 mL/kg body weight for children <20 kg; one adult therapeutic dose for adults and older children. In bone marrow failure, it may be used to prevent spontaneous bleeding in patients with reversible bone marrow failure when the platelet count is $<10 \times 10^9$/L. Prophylactic platelet transfusions are not indicated in chronic stable thrombocytopenia, but may be used to prevent spontaneous bleeding when the platelet count $<20 \times 10^9$/L in the presence of additional risk factors for bleeding such as sepsis or haemostatic abnormalities. To prevent bleeding associated with invasive procedures, the platelet count should be raised to $>50 \times 10^9$/L before lumbar puncture, insertion of intravascular lines, trans-bronchial and liver biopsy, and laparotomy, to $>80 \times 10^9$/L before spinal epidural anaesthesia and to $>100 \times 10^9$/L before surgery in critical sites such as the brain or the eyes.

Uses of platelet concentrates in critical care/surgery include in massive blood transfusion. The empirical use of platelets, according to a specific blood component ratio, is reserved for the patients with severe trauma. The aim is to maintain platelet count $>75 \times 10^9$/L and $>100 \times 10^9$/L if multiple, eye or central nervous system trauma. It may also be used in acquired platelet dysfunction (e.g. post-cardiopulmonary bypass), use of potent anti-platelet agents such as clopidigrel and with non-surgically correctable bleeding. There is also a place for platelet concentrates in acute disseminated intra-vascular coagulation in the presence of bleeding and severe thrombocytopenia, inherited platelet dysfunction disorders such as Glanzmanns thrombasthenia with bleeding or as prophylaxis before surgery.

There are three indications for the use of platelet concentrates in immune thrombocytopenia:

- Primary immune thrombocytopenia, as emergency treatment in advance of surgery or in the presence of major haemorrhage. A platelet count of $>80 \times 10^9$/L is recommended for major surgery and a count of $>70 \times 10^9$/L for obstetric regional axial anaesthesia.
- In post-transfusion purpura, in the presence of major haemorrhage.
- Neonatal alloimmune thrombocytopenia, to treat bleeding or as prophylaxis to maintain the platelet count $>30 \times 10^9$/L.

Alternatives to blood transfusion

Most of the problems with blood transfusion stem from the fact that the incoming blood is foreign (non-self). Many of these can be overcome by the process of autologous transfusion. If the requirement for blood can be predicted, such as in elective surgery, it may be possible to harvest some of the patient's own blood up to 4 weeks before their surgery, to be returned after the surgery, if required. Hence, the only hazard here is of infection during the process of removing and returning the blood. However, use of this pre-deposit of patient's own red cells is falling.

It is also possible to collect the patient's blood being lost 'normally' during surgery, concentrate it and then return it to the body. This is called autologous salvage. Further strategies to minimize blood loss in surgery are to activate the coagulation system, perhaps with recombinant (genetically engineered) factor VIIa, and the use of desmopressin and tranexamic acid to promote haemostasis.

Other opportunities to avoid transfusion include the use of iron (if the patient is iron deficient) and other supplements to stimulate red cell production. Principal among the latter is use of the growth factor erythropoietin to directly stimulate the bone marrow to produce more red cells.

11.5   Hazards of blood transfusion

Just as no drug is free of side effects or is effective in all patients, no blood transfusion event is completely safe and free of potential complications. Errors can and do occur at all places in the 'journey' from the donor to the recipient. We have already looked at hazards that may be present in the donor and the role of the NBS in testing for pathogens.

The hospital

It is generally recognized that most errors happen in the laboratory and/or once the blood has left the blood bank for its destination.

Laboratory error   Packs of blood arrive from the NHSBT already typed for ABO and Rh, and screened for major infective agents. However, the blood sample from the recipient (the patient) may be labelled incorrectly. The next source of error may be the incorrect labelling of the small portion of each potential donor pack that is taken for the cross-match.

Further errors are possible in the cross-match itself. However, these are rare because the laboratory invests heavily in the technology and reagents to ensure that if an incompatible reaction is present it is detected. There will also be positive and negative control samples. However, if the cross-match goes wrong, which is a false negative, a possible incompatible unit of blood may be issued. If several packs are incorrectly identified, the laboratory may assign an incompatible unit to the donor.

Post-laboratory error   These are inevitably the wrong blood being given to the wrong patient. The wrong pack of blood may be collected from the blood bank issuing refrigerator, or the blood may be given in error to the wrong patient. A common confusion is that two or more

Table 11.7 Key features of TRALI and TACO.

	TRALI	TACO
Key clinical features	Onset of pulmonary oedema within 6 h of transfusion	Acute respiratory distress with pulmonary oedema, and tachycardia
Temperature	Raised	No change
Blood pressure	Normal/low	Normal/raised
Central venous pressure	Normal	Increased
Improves after use of diuretics	No	Yes
White blood cell count	Transient reduction	No change

N.B. Assumes no concurrent disease such as sepsis, inflammation.
(Modified from Retter A. *et al.*, British Committee for Standards in Haematology. Br J Haematol. 2013;160:445–464. Reproduced with permission of John Wiley & Sons).

patients are to be transfused at the same place at the same time. Many of these are simply incorrect patient identification, generally by a misunderstood verbal recognition question, or by misreading the patient's ID strip at the wrist.

The responses of the body to an incompatible transfusion

As with other clinical issues, the blood scientist is unlikely to observe an incompatible transfusion at first hand, but certainly needs to know its causes, its consequences and how it will be managed. Although there is focus on reactions resulting in the destruction of red cells (i.e. haemolytic reactions), these are in the minority. The most common transfusion reactions include allergies and reactions, post-transfusion purpura, graft-versus-host disease, transfusion-related acute lung injury (TRALI) and transfusion-associated circulatory overload (TACO). Key features of the latter two conditions are summarized in Table 11.7.

Pathology From the haemolytic perspective, almost all transfusion reactions are caused by antibodies in the recipient reacting with donor cells, and can have a number of consequences. The binding of an antibody to a cell signals the immune response to attack and destroy the cell. This may be by the fixation of complement and/or antibody-dependent cellular cytotoxicity (as described in Chapter 9). Lysis of donor cells (i.e. haemolysis) will result in the release of its component haemoglobin, and if widespread it will markedly increase the plasma protein component and cause renal damage.

If there is a sufficiently large IgM response to the donor cells, agglutination and aggregation may occur, which

will cause occlusion of blood vessels, much like a thrombus of platelets. If the agglutinated cells are attacked in capillary beds, there may be collateral damage to the endothelium, leading to inflammation (vasculitis) and local tissue damage. If in the skin, this may be a rash; in the kidney, it may lead to renal failure.

Early reactions In an acute setting, symptoms and signs of a transfusion reaction vary enormously (Table 11.8). However, there can be acute non-red blood cell reactions. Chest symptoms may indicate TRALI, an uncommon condition (approximately 1 in every 5000) that generally develops within 6 h of an incompatible transfusion. In many cases there is involvement of leukocytes with degranulation and cell damage akin to an acute inflammation. If severe it may lead to acute respiratory distress syndrome.

Table 11.8 Signs and symptoms of a transfusion reaction.

Symptoms	Signs
Cough	Fever (temp. spike >40 °C)
Flushing/rash	Hypotension
Anxiety	Oozing from wounds
Chills	Haemoglobinaemia
Nausea and vomiting	Haemoglobinuria
Tremble/shakes	Tachycardia
Shortness of breath/chest pain	Failure to make urine (implying renal failure)
Headache	
Diarrhoea	
Cognitive changes/patient restless/agitated	

Upon suspicion of a reaction, the infusion should be immediately stopped. All good hospitals will have a defined protocol that must be followed. Treatment will depend on the severity of the reaction, which, if not too bad, can be rapidly reversed. But severe reactions can be life-threatening and will be treated accordingly (e.g. rapid admission to an intensive care unit for disseminated intravascular coagulation with ventilation, adrenaline and hydrocortisone).

One 'mechanical' way of treating an acute reaction is to try to 'flush' it out – this is attempted by giving fluids, but clearly requires good renal function. If there is massive donor red blood cell destruction there may well be hyperkalaemia, which may lead to arrhythmia. The donor blood pack will be collected by laboratory staff and a blood sample taken from the patient. Both will be thoroughly (re)analysed. Most errors are subsequently found to be ABO incompatibilities, often in the A and B subgroups – Rh system problems are less common, but minor blood groups still occasionally cause problems.

Late reactions Problems at 12/24 h to 3 days may be with major organs such as acute renal failure (with or without haematuria or haemoglobinuria), jaundice, congestive heart failure due to circulatory overload, and pulmonary oedema with or without adult respiratory distress syndrome.

Febrile reactions (generally with a slowly rising temperature peaking at over $40\,^\circ$C) are seen in 0.5–1% of transfusions and may be due to anti-HLA reactions. Later complications (3–14 days) after an incompatible transfusion may consist of a 'new' immunological reaction of the recipient to the donor red blood cells, causing the destruction of the latter. This time period may also see transfusion-related infections that have escaped the screening process. There may also be reactions with rare blood group molecules too weak to be detected in the laboratory.

Post-transfusion purpura (bruising) is characterized by a severe thrombocytopenia (which can last from 2 weeks to 2 months) and is caused by antibodies to molecules on the surface of platelets. In the short term a thrombocytopenia may develop in those hyper-transfused patients, as infused blood is generally platelet free. If heavily haemorrhaging, then platelet transfusion may be required.

Late reactions (apparent months or years later) include the effects of hepatitis B virus (1 in 670 000), HIV (1 in 5 million), human T-lymphotropic virus type 1 (1 in 17 million) and of the hepatitis C virus (1 in 17 million).

Parallel figures of the incidence of variant Creutzfeldt–Jakob disease (vCJD) are unavailable, but according to one authority, since its appearance in 1996, seven of those many millions who have been transfused have developed clinical disease. Three cases were linked to a donor who also later developed vCJD, and in the remaining four cases no infected donor has been identified. The mean interval from transfusion to the onset of illness is approximately 6.5 years.

Allergic reactions Although life-threatening anaphylaxis rarely occurs, allergic reactions are common. Acute urticarial reactions (e.g. hives) and anaphylaxis may be the result of the recipient responding to the donor's plasma proteins such as IgA and haptoglobin, especially in those recipients deficient in these molecules. If so, antihistamines are one possible treatment. Chemical allergens include methylene blue, used to inactivate viruses, whilst a 6-year old boy had a reaction to peanut allergen present in the donor pack. Severe urticarial reactions may be treated with steroids, and if there is hypotension then adrenaline/epinephrine may be required. Further details of allergic reactions are presented in Chapter 9.

Repercussions There will of course be an 'inquest', the findings of which will strengthen the system and so minimize the possibility of a similar error returning. A report will be forwarded to the relevant authorities; for example, the National Patient Safety Agency, and the Serious Adverse Blood Reactions and Events reporting system (the reporting unit of the Medicines and Healthcare Products Regulatory Agency), and also to the Serious Hazard of Transfusion (SHOT) group. Between 1996 and 2011 SHOT reviewed 9925 cases, with over 3000 being incorrect blood component transfused, perhaps 2300 were acute transfusion reactions, and there were over 1000 anti-D errors and over 1000 handling and storage errors.

Given that there are about 3 million components issued in the UK in 2011, the very low number of deaths (eight patients) represents a rate of 0.000 26%. There were 117 cases of major morbidity, of which the most common was an acute transfusion reaction (45%), followed by TACO (21%). Other events included transfusion-related dyspnoea (difficulty with breathing), TRALI and haemolytic transfusion reactions.

Naturally, there are many steps designed to prevent a transfusion error: generally check, check and check again. Laser bar coding is being introduced so the sample can be traced from the requesting blood sample all the way back to the patient. A guideline from the BCSH is available. Many hospitals have a policy of at least two members of staff checking the blood they are about to transfuse into one of their patients. This approach has proven to reduce mistakes and serious hazards of transfusion. Indeed, SHOT itself reports that ABO-incompatible transfusions show a 54% reduction since 2001–2002. Many wards and hospitals have their own formal policy, and professional bodies (such as the Royal College of Nursing, the BCSH and the Institute of Biomedical Science) also offer guidelines.

Summary

- Variation in the structure of key molecules of the surface of red cells frustrates the simple transfusion of blood.
- The most important group, ABO, consists of A and/or B, or neither, antigenic molecules on the red cell and the presence or absence of the opposite antibody.
- The Rh system consists of a large number of molecules, although we focus on five (D, E, C, e. c) that may be present in a variety of combinations and that may induce antibodies.
- There are perhaps eight or nine other blood groups (often described as 'minor') that often need to be addressed.
- The major techniques in blood transfusion are group, antibody screening, cross-match and the anti-human globulin test, the latter consisting of direct and indirect variants.
- Blood transfusion is requested on clinical grounds, but as it is linked to a variety of other diseases and conditions it must not be undertaken lightly. Alternatives should always be considered.
- A further reason to consider the need for a transfusion is that there is always the possibility of error, and although these have the potential to be fatal, this is rarely the case.

References

Knight R. (ed.). Transfusion and Transplantation Science. Oxford University Press, 2012.

McCullough J. Transfusion Medicine, 3rd edition. Wiley-Blackwell, 2012.

Overfield J, Dawson M, Hamer D. Transfusion Science, 2nd edition. Scion Publishing, 2008.

Further reading

Anstee DJ. The relationship between blood groups and disease. Blood. 2010;115:4635–4643.

Daniels G. Variants of RhD – current testing and clinical consequences. Br J Haematol. 2013;161:461–470.

Davies L, Brown TJ, Haynes S. et al. Cost-effectiveness of cell salvage and alternative methods of minimising perioperative allogeneic blood transfusion: a systematic review and economic model. Health Technol Assess. 2006;10: iii–iv, ix–x, 1–210.

Dodd RY. Emerging pathogens and their implications for the blood supply and transfusion transmitted infections. Br J Haematol. 2012;159:135–142.

Giangrande PLF. The history of blood transfusion. Br J Haematol. 2000;110:758–767.

Hirayama F. Current understanding of allergic transfusion reactions: incidence, pathogenesis, laboratory tests, prevention and treatment. Br J Haematol. 2013;160:434–44.

Isbister JP, Shander A, Spahn DR. et al. Adverse blood transfusion outcomes: establishing causation. Transfus Med Rev. 2011;25:89–101.

Lögdberg L, Reid ME, Zelinski T. Human blood group genes 2010: chromosomal locations and cloning strategies revisited. Transfus Med Rev. 2011;25: 36–46.

Shaz BH, Stowell SR, Hillyer CD. Transfusion-related acute lung injury: from bedside to bench and back. Blood. 2011;117:1463–1471.

Tilley L, Green C, Poole J. et al. A new blood group system, RHAG: three antigens resulting from amino acid substitutions in the Rh-associated glycoprotein. Vox Sang. 2010;98:151–159.

Zuccala ES, Baum J. Cytoskeletal and membrane remodelling during malaria parasite invasion of the human erythrocyte. Br J Haematol. 2011;154:680–689.

Guidelines

Guidelines of the Blood Transfusion Services in the UK: http://www.transfusionguidelines.org.uk/index.aspx? Publication=RB.

Makris M, Van Veen JJ, Tait CR. et al. British Committee for Standards in Haematology. Guideline on the management of bleeding in patients on antithrombotic agents. Br J Haematol. 2013;160:35–46.

Retter A, Wyncoll D, Pearse R. *et al.* British Committee for Standards in Haematology. Guidelines on the management of anaemia and red cell transfusion in adult critically ill patients. Br J Haematol. 2013; 160:445–464.

Tinegate H, Birchall J, Gray A. *et al.* Guideline on the investigation and management of acute transfusion reactions. Br J Haematol. 2012;159:143–153.

Web sites

An excellent education site: www.learnbloodtransfusion .org.uk.

British Committee for Standards in Haematology (BCSH): www.bcshguidelines.com.

NHS Blood and Transplant Service: http://www.blood .co.uk.

SABRE: http://www.mhra.gov.uk/Safetyinformation/ Reportingsafetyproblems/Blood/index.htm.

Serious Hazard of Transfusion (SHOT) group: www .shotuk.org.

The National Creutzfeldt–Jakob Disease Research and Surveillance Unit (NCJDRSU) web site www.cjd.ed .ac.uk; but see also http://www.cjd.ed.ac.uk/TMER/ results.htm and http://www.transfusionguidelines.org .uk/index.aspx?Publication=DL&Section=12&pageid =794.

The National Patient Safety Agency: www.npsa.nhs.uk.

12 Waste Products, Electrolytes and Renal Disease

After studying this chapter, you should be able to:

- appreciate the three major functions of the kidney;
- recognize the homeostatic function of the kidney;
- describe renal excretory function;
- list the endocrine aspects of the kidney;
- discuss tests of renal function;
- be aware of the causes and implications of renal disease.

We begin our exploration of the major aspects of blood science by looking at renal function. The kidney, the study of which is nephrology, is in fact (at least) three organs for the price of one. Furthermore, the close proximity of the adrenal glands also demands attention. These three functions of the kidney can be summarized as follows:

- *Homeostasis:* ensuring the correct amount of water, and the concentrations of certain ions and molecules in the blood. This also has an impact on blood volume and so blood pressure. The key analytes under homeostatic control are sodium and potassium, but we will also look briefly at the regulation of hydrogen ions and their contribution to the pH of the blood. In order to maintain blood volume, water is retained or removed, and excess ions pass into the urine.
- *Excretion:* the removal of the waste products of metabolism. These are principally urea and creatinine, which also pass into the urine, although it could be argued that the hydrogen ion is also an excretory product.
- *Endocrinology:* the secretion of, and response to, hormones. Those secreted by the kidney include the bone marrow stimulant erythropoietin and renin, the latter a key hormone in the regulation of electrolytes. Many

hormones target the kidney to mediate their homeostatic effect. For example, parathyroid hormone acts on the kidney to help regulate the amount of calcium in the blood. The kidney also has its own internal hormones.

The major blood test in renal function is urea and electrolytes (U&Es), and is the most commonly requested test in biochemistry. Missing from the abbreviation is the excretory product creatinine, and this molecule, in combination with age, sex and (if present) African origin gives a crude but useful index of renal function – the estimated glomerular filtration rate (eGFR) (Table 12.1).

Before we embark on our study of the blood science of the kidney, we must refresh our knowledge of the anatomy and physiology of this organ. Only when we have explored renal function and physiology, which we will follow in the order of homeostasis (Section 12.2), excretion (Section 12.3) and endocrinology (Section 12.4), will we be able to consider the role of the laboratory in renal disease (Section 12.5). Finally, in Section 12.6 we will explore the value of these indices by looking at some patients with different renal diseases.

12.1 Renal anatomy and physiology

The kidneys are two bean-shaped organs located at the rear of the abdominal cavity, one below the liver and one below the spleen, weighing perhaps 130–170 g each in the adult male and 10 g less in the adult female. Each is supplied with blood via the renal arteries, which branch directly off the abdominal aorta, and renal veins, which return purified blood to the inferior vena cava. A third vessel, the ureter, leaves the kidney and carries urine to the bladder.

Sitting just above each kidney are the left and right adrenal glands, which are part of the endocrine system.

Blood Science: Principles and Pathology, First Edition. Andrew Blann and Nessar Ahmed.
© 2014 John Wiley & Sons, Ltd. Published 2014 by John Wiley & Sons, Ltd.

Table 12.1 Major renal blood tests.

Analyte	Adult reference range
Sodium	135–145 mmol/L
Potassium	3.8–5.0 mmol/L
Urea	3.3–6.7 mmol/L
Creatinine	71–133 μmol/L
eGFR	>90 ml/min per 1.73 m^2

N.B. There is no right or wrong reference range; refer to the values produced by your local laboratory.

Among their secretions is aldosterone, a local hormone that acts on the kidney and is involved in the regulation of water, sodium and potassium. The crude internal structure of the kidney can be defined as a pelvis, close to the area where the artery, vein and ureter enter and leave, the cortex, which is the area closest to the outer surface, and the area between the two, known as the medulla. The entire kidney is enclosed by a capsule and is held in position in the abdominal cavity by loose connective tissues.

The nephron

The excretory and regulatory unit of the kidney is the nephron, effectively a very long tube, in which urine is formed. Each nephron, of which there are approximately 1 million per kidney in a healthy adult, is closely associated with an arterial and venous blood supply, so that excretory and homeostatic metabolism can operate in parallel (Figure 12.1). These twin processes begin at the glomerulus, a knot of arteriolar vessels at such high blood pressure that fluid is forced from the blood into the top of the nephron, the whole structure being retained within the Bowman's capsule. A key collection of cells form the juxta-glomerular apparatus, and secrete renin, which has a part to play in homeostasis. As the nephron develops, it becomes tortuous and twisted, a region known as the proximal convoluted tubule.

As the nephron leaves the proximal convoluted tubules it becomes the loop of Henle, which has descending and ascending sections. Following the ascending loop, the nephron then becomes twisted and tortuous a second time, and so becomes the distal convoluted tubule. Upon leaving this second set of tubules, the nephron merges with others into increasingly large collecting ducts, which in turn merge into the papillary ducts, and ultimately the ureter, which takes urine to the bladder.

Throughout its length, each nephron is closely associated with blood vessels, allowing water, electrolytes and molecules to be transferred between the two. The process by which these substances are regulated is homeostasis.

12.2 Homeostasis

This process of regulating water, electrolytes and other molecules in the blood happens in all parts of the nephron, although there are specific activities linked to precise regions.

Ultrafiltration

The endothelium and tissues of the glomerulus together are effectively a molecular sieve, enabling the passage of molecules with a molecular weight of less than about 67 kDa to pass into the early form of urine, or filtrate. Hence, this subprocess is described as ultrafiltration, and ideally would contain only a trace of the most abundant blood protein, albumin, as this valuable protein is too large to pass through the pores in a normal glomerulus. Thus, this early filtration is effectively blood without red and white cells, platelets, or large proteins such as immunoglobulins. This filtrate can be generated at a rate of up to 125 mL/min – this is the glomerular filtration rate (GFR), more of which we shall examine later.

Selective reabsorption

This next step, where up to 99% of the desired substances are taken back from the filtrate into the blood, leaves only those undesirable substances in the developing urine. The cells of the tubules and blood vessels are constantly interrogating each other and deciding upon which ions and molecules to reabsorb. The twin processes that regulate selective reabsorption are the laws of osmosis and electrical neutrality, and local hormones acting on different parts of the nephron.

Electrical neutrality has two parts. In the first, if there is a movement of a positive ion then it must be balanced by the reverse movement of another positive ion. In practice, this means that the movement of a hydrogen ion in one direction must be balanced by movement of potassium in the other direction. So loss of sodium will be accompanied by retention of potassium, and vice versa. The same is true for anions. Similarly, as a cation moves, so must an anion. So the movement of, for example,

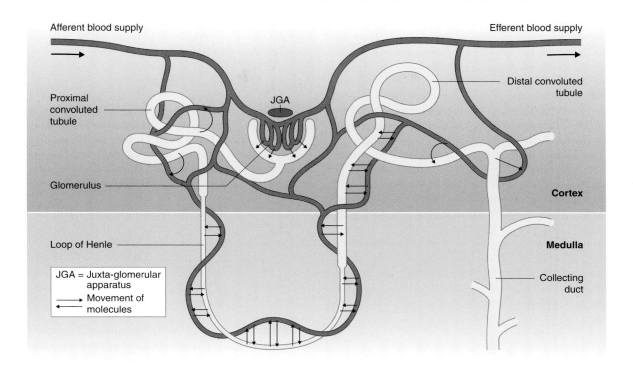

Afferent blood supply

Efferent blood supply

Proximal convoluted tubule

JGA

Distal convoluted tubule

Glomerulus

Cortex

Loop of Henle

Medulla

JGA = Juxta-glomerular apparatus

Movement of molecules

Collecting duct

Figure 12.1 The nephron. An afferent blood vessel brings blood to the glomerulus, where ultrafiltration takes place. Filtrate passes into the proximal convoluted tubule, the loop of Henle, the distal convoluted tubule and the collecting duct. The close association between the nephron and blood vessels allows the transfer of substances and water in selective reabsorption, which maintains homeostasis and mediates excretion.

sodium, must be associated by the movement, in the same direction, of an anion such as chloride. This can be extremely complicated, and, as we saw in Chapter 4, involves the pH of the blood.

Local hormones

The hormones arginine vasopressin (AVP, also known as antidiuretic hormone), aldosterone and natriuretic peptides contribute to the homeostatic regulation of water and electrolytes. Specialized cells in the hypothalamus monitor the osmolality of circulating plasma, and if too high will both stimulate thirst and the release of AVP from the adjacent posterior pituitary, which acts on the collecting ducts of the nephron to retain water. Both thirst satisfaction (i.e. drinking) and water reabsorption will change the osmolality. Therefore, AVP is effectively an 'anti-dehydration' hormone, conserving water. Conversely, if the osmolality is low, AVP and water volume fall and the plasma osmolality rises.

This feedback process is highly effective in maintaining osmolality within a very narrow range of around 285–295 mOsm/kg. In health the major contributors to osmolality are sodium, potassium and their counteranions together with the low molecular weight substances glucose and urea. The local hormone aldosterone acts independently of AVP on the distal convoluted tubule and collecting duct, promoting sodium absorption in exchange for potassium and/or hydrogen ions, leading to a rise in plasma sodium (and therefore the total ionic component of the plasma). Conversely, in the absence of aldosterone, sodium is not reabsorbed from the tubules and plasma concentrations fall.

Within the glomerulus, the juxta-glomerular apparatus monitors blood pressure (which is directly related to blood volume) and, when this rises, secretes the enzyme renin into the plasma. This enzyme acts on angiotensinogen (produced by the liver) to generate angiotensin I. This is the substrate for angiotensin-converting enzyme (produced by endothelial cells), the product being angiotensin II, which stimulates cells of the adrenal cortex to synthesize aldosterone. Angiotensin II also has other functions, such as in promoting vaso-constriction (and so possibly hypertension) and in acti-vating platelets (and so possibly thrombosis). Indeed, this pathway is the target of two important classes of drugs that lower blood pressure:– the angiotensin-converting enzyme inhibitors (ACEIs) and the angiotensin receptor blockers (ARBs).

Additional hormones include A- and B-type natri-uretic peptides (ANP and BNP). These peptides are released by the cells of the atria of the heart, probably in response to cardiomyocyte stretch, itself a conse-quence of rising blood volume. Both peptides act to reduce blood volume by promoting the loss of sodium into the urine, whoch subsequently leads to water loss in the urine. Additionally, measurement of BNP has clinical value in cardiac disease as low concentrations effectively exclude a diagnosis of heart failure. Figure 12.2 illustrates the relationships between these major hormone regula-tors of water and plasma sodium.

It is clear, therefore, that correct regulation of ion balance is highly complex, and regulated by pathways that are often antagonistic.

The electrolytes

Virtually all biomolecules, when placed in water or perhaps a buffered solution, can be viewed as being in two states: covalent and ionic. In some molecules, most of the chemical is present in the covalent form, but the ionic form dominates in most. All molecules find their balance, or equilibrium, between these two states.

The chemistry of electrolytes   The ionic form of a compound is an electrolyte; that is, a charged atom or molecule that may be involved in the passage of elec-tricity. Electrolytes are often written with a small sub-script plus or a minus, indicating their electrical charge, although, in practice, charge sign is often ignored. Consider sodium chloride, which as a solid at room temperature and pressure has a crystalline structure. Solvation of sodium chloride into aqueous solution is associated with a high dissociation to produce the cation Na^+ and the anion Cl^-. The other major cation electro-lyte is potassium (K^+). Other major electrolytes are calcium (Ca^{2+}) and magnesium (Mg^{2+}), and we will address these cations in Chapter 15. The counter-anion to the calcium cation is often phosphate (HPO_4^{2-}).

Water also has an equilibrium between its covalent form (H_2O) and its ionic form (the hydrogen ion, or proton, H^+, a cation, and the hydroxide ion, OH^-, an anion). However, water can also form a compound with carbon dioxide, namely carbonic acid (H_2CO_3), which ionizes to H^+ (again) and HCO_3^-, the bicarbonate ion. The complex interactions between water, carbon dioxide, hydrogen ions and bicarbonate ions contribute signifi-cantly to blood pH. This general area is so important that it warrants its own major section, and follows in Chapter 13. Key electrolytes are summarized in Table 12.2.

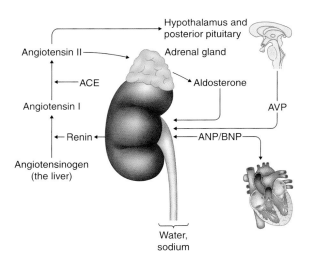

ACE = Angiotensin converting enzyme

Figure 12.2 Hormonal regulation of homeostasis. AVP from the posterior pituitary, and ANP and BNP from atrial cardio-myocytes act directly on the nephron. The juxta-glomerular apparatus secretes renin, which converts angiotensinogen to angiotensin I, which in turn is transformed into angiotensin II by angiotensin-converting enzyme. Angiotensin II acts on the adrenals to synthesize and release aldosterone, which acts on the nephron.

Table 12.2 Key electrolytes.

Cations	Anions
Na^+	Cl^-OH^-
K^+	HCO_3^-
H^+	HPO_4^{2-}
Ca^{2+}	
Mg^{2+}	

Osmolality The total mass of all the ions in a fluid (such as those in Table 12.2) contributes to the osmolality, not to be confused with osmolarity, although the two are clearly closely related (the former is mathematical, the latter a basic process in chemistry). Generally, we expect osmolality to be around 285 mmol per kilogram of fluid (inevitably water), and the major factors contributing to osmolality are sodium and glucose. Osmolality can be both measured and calculated, and a difference between the two – that is, the osmolar gap – is generally due to high concentrations of a generally unmeasured substance, such as alcohol. Hypo-osmolality is inevitably due to low sodium, as this is the major contributor to osmolality.

Concentration, mass and volume The arithmetic of the concentration [C] of any substance dissolved in water (or, actually, any fluid) is given simply by the mass [M], which may be in the unit of grams or moles of the substance, and the volume of fluid [V], generally in units of litres. This is simple to think of in terms of sugar in tea or coffee – the concentration (sweetness) can be increased by placing more sugar in the same volume of the fluid, or by putting the same amount of sugar into a smaller volume of the beverage.

However, our physiology is not a closed system of a cup or mug, but is dynamic and flexible. Accordingly, the concentration of a substance in our blood is a balance between the amount we take in by eating (i.e. mass in M_{in}) and the amount that is excreted (i.e. mass out M_{out}). Similarly, the fluid component is the product of what we drink (i.e. volume in V_{in}) and that fluid that passes out, mostly as urine (hence volume out V_{out}). This brings us to the key equation in biochemistry:

$$C = \frac{M_{in} - M_{out}}{V_{in} - V_{out}}$$

This equation immediately informs our understanding of the complex relationship between the four components, and how any combination can lead to abnormalities in the concentration of the substance in question. Consider one extreme where the volume of fluid taken as drink is exactly balanced by the fluid lost as urine. So for the concentration to rise, then M_{in} must exceed M_{out}, and vice versa for the concentration of the substance to fall. The polar extreme is where the intake and excretion of the substance are equal. If the balance in the fluid component falls, either the volume in is falling or the volume out is being increased; this will inevitably lead to

an increase in the concentration of the substance. However, if the volume of fluid leaving the body falls and exceeds that being taken in, then the fluid component of the blood will rise so the concentration of its components will fall.

In healthy adults there is voluntary control over what is eaten (M_{in}) and drunk (V_{in}), but none over what we excrete (i.e. M_{out} and V_{out}). The latter is regulated by the activities of the kidney. It follows that if the intake of food and drink is 'sensible', then abnormalities in relevant blood components are the consequence of renal dysfunction, itself inevitably driven by a disease process.

Water

We now look at the importance of the V_{in} and V_{out} components of the equation. By far the greatest source of water is in drinking – relatively little is taken in as part of solid foodstuff, despite the high water components of many fruits and vegetables. Water passes into the intestines and then into the blood. Theoretically, water is a by-product of metabolism, as in the case of the extraction of energy from the chemical bonds of glucose and oxygen by the process of respiration in the mitochondrion:

$$C_6H_{12}O_6 + 6O_2 = 6CO_2 + 6H_2O$$

Therefore, within a 180 g chocolate bar is buried 108 g of water. The contribution of this metabolic water to the general body mass of water is difficult to assess with precision. Total body water (perhaps 42 L in a 72 kg body) is to be found in three compartments – the blood (3.5 L), the tissues (10.5 L), and within cells (28 L) – and can move between these three as different conditions demand. The importance of this carbon dioxide and water will be addressed in Chapter 13, as they can combine to form first carbonic acid ($H_2CO_3^-$), which can then form the bicarbonate ion (HCO_3^-) and hydrogen ion (H^+), and so have an impact on the pH of the blood.

Physiology Increased water entering the blood has a number of consequences, which include a rise in blood volume and a fall in the concentration of blood constituents, the latter being part of the master equation of concentration = mass/volume. In this respect the role of the kidney is to maintain both blood volume and the concentration of many ions within a defined range, responding to the laws of osmosis and the action of renal hormones. The first component is important clinically because a rising blood volume will put stress on the heart, which will lose its natural rhythm, become congested (and so release ANP and BNP) and may even arrest. The kidney responds to the rising blood volume by increasing total urinary volume – the process of diuresis.

The second consequence of rising blood volume will be a reduction in the concentration of blood constituents, such as sodium, and therefore plasma osmolality. The latter will be noted by the sensors in the hypothalamus and posterior pituitary, resulting in reduced AVP and so loss of water. The kidney maintains plasma sodium within a tightly regulated range (135–145 mmol/L in health by controlling both sodium excretion (primarily through renin and aldosterone) and water (primarily through AVP).

Pathology Failure of physiology leads directly to pathology. If the kidney is unable to correct rising plasma volume through the production of urine, there is a redistribution of water between the water compartments, leading to conditions such as oedema, most often seen in the soft tissues of the ankle and lower leg. However, oedema is also found in other conditions, such as heart failure. This explains the link between the value of BNP as its activity in attempting to reduce blood volume, and so reduce stress on the heart. Once the tissues are saturated, fluid must move into cells, which therefore disrupts their function.

An uncorrected fall in blood volume leads to hypotension and reduced perfusion of the organs with blood and ultimately leads to cardiac arrest. Accordingly, the body is programmed to maintain blood volume; its does this by a number of ways, such as reducing the generation of urine. Another is by moving water from the cellular and tissue components into the blood. Depletion of the tissues of its water causes dehydration, an established physical sign.

Diabetes insipidus (N.B. unrelated to diabetes mellitus)

The patient with this condition is characterized by the production of copious watery urine (polyuria) and a compensatory excessive thirst (polydipsia), hence similarity with 'sugar' diabetes. It is caused by either the defective production of AVP or the insensitivity of the kidney to this hormone (the latter possibly due to a drug effect).

Fluid balance and insensible losses Dehydration and oedema are such clear signs of a problem with fluid balance that it is often useful to be able to assess with precision the amount of fluid a patient takes and the amount of urine they make. However, water taken in drinking often fails to balance urine production, as water can be lost through breathing and sweating (especially on hot days) and in the faeces (more so in diarrhoea). These insensible losses can be as much as 500–800 mL/day.

Sodium

This is the primary ion in the blood, as there is more of it than any other ion. Indeed, as mentioned, it is the major contributor to osmolality. Faced with increases or decreases in serum sodium, we first need to consider the equation concentration = mass/volume. The intake of both sodium and water varies, and in health the mechanisms described above maintain both, so that one good reason to have changes in serum sodium is that the diet is abnormally rich or low in this salt. Another reason is that the amount of water in the blood has increased or fallen. Since the kidney regulates both these aspects, then changes in sodium may indicate problems with this organ. As we have discussed, raised sodium may be the consequence of abnormal renal hormones.

Hypernatraemia Defined by many as serum sodium >145 mmol/L, hypernatraemia may be due to one or more of:

- a diet rich in salt (as sodium chloride, i.e. high M_{in}), which is very possible given the high salt content of processed food and as used in cooking;
- excessive resorption of sodium by the tubules (leading to low M_{out});
- low fluid (water) intake (i.e. low V_{in}) and excessive water loss in urine (i.e. high V_{out}).

Thus, hypernatraemia may be characterized by a degree of dehydration – the classical clinical sign of which is loss of elasticity of the skin (the pinch test - although unreliable in the elderly). However, the most common reason is water loss, possibly due to insufficient intake in drinking, excess loss in urine, or losses in sweat and diarrhoea.

Clinically, the simplest treatment of hypernatraemia due to dehydration is to restrict salt intake in the diet and to replace fluid orally, if possible. If this is not possible (maybe because the patient is in a coma), or if considered to be life-threatening, then isotonic fluid can be infused intravenously. However, the original cause must be detected and corrected.

Conn's syndrome

This rare condition, also known as primary aldosteronism, is characterized by the excessive secretion of aldosterone by the adrenal glands. This causes the retention of sodium and water (so rising levels in the blood) and in parallel the loss of potassium and hydrogen ions, so the blood pH rises. The increased mass of water in the blood causes hypertension. The most likely cause of the high aldosterone is oversecretion by tumour in the adrenals.

Hyponatraemia Often defined by serum sodium <135 mmol/L, this may be due to the reverse of the above; that is:

• insufficient sodium intake (i.e. low M_{in}), although this is not very likely, for reasons given above;
• failure to resorb sodium from the tubules, so it remains in the urine, ultimately being high M_{out};
• excessively high intake of water, so a high V_{in}, theoretically possible, and excessive retention of water (low V_{out}), which is quite possible;
• high concentration of proteins or lipoproteins (this is not a genuine hyponatraemia, but a laboratory artefact termed pseudohyponatraemia).

This is the most common in-hospital electrolyte disturbance, present in up to 15% of patients, and may be due to fluid retention, a recognized sign of which is oedema and reduced diuresis, although this may not always be present. However, oedema is also associated with heart failure, and also with low concentrations of plasma albumin. Sodium may also be lost from the gastrointestinal tract in diarrhoea and in vomiting.

The consequences of low sodium include low blood pressure and a corresponding high pulse rate. As with high levels of sodium, the treatment of low levels depends on aetiology, being any combination of the four bullet points. In some cases, water retention can be treated with thiazide drugs, but this can be dangerous if the kidneys are not adequately functioning.

Syndrome of inappropriate antidiuretic hormone (SIADH)

This syndrome is characterized by the excessive release of AVP by the posterior pituitary, leading to excessive water reabsorption in the collecting ducts. This results in hyponatraemia and so low osmolality, and if unchecked can lead to nonspecific symptoms such as headache, nausea and vomiting, and ultimately convulsions and coma.

Potassium

Many aspects of potassium regulation are similar to those of sodium (such as intake and outlet), but with a number of differences:

• The majority of our potassium is within cells, not in the plasma. This is the reverse of sodium, very little of which is inside cells.
• The relationship between potassium and hydrogen ions: as the concentration of hydrogen ions increases (as in acidosis) then potassium is displaced from the cell and enters the plasma. The reverse is true in alkalosis.
• Overall (unlike sodium), potassium does not vary a great deal in response to large changes in water balance.
• But crucially . . . **high potassium kills people**. So, in the face of a rising or existing high potassium, the scientist must take all necessary steps to ensure a responsible person is fully aware of the danger the patient may be in.

Hyperkalaemia High potassium, such as serum potassium >5.0 mmol/L, may be due to a number of causes, but is rarely due to excessive intake in the diet (high M_{in}). Although bananas are relatively rich in this electrolyte, it would be necessary to eat many kilograms to influence plasma concentrations. Common aetiologies include renal problems; for example, failure to excrete sufficient potassium (low M_{out}), acidosis, or release from damaged cells, as may be caused by the effect of cytotoxic chemotherapy on tumour and normal body cells, or crush injuries. Hyperkalaemia may also be present in haemolytic anaemia (as discussed in Chapter 4), as red blood cells are a rich source of this electrolyte. A high potassium may in theory be present in dehydration, when the amount of water in the blood falls (i.e. when V_{out} exceeds V_{in}), although, if present, hypernatraemia is far more likely to be present.

However, whatever the reason, hyperkalaemia is serious as high serum potassium concentrations (e.g. >7 mmol/L) can precipitate cardiac arrhythmia (as can be detected on an electrocardiogram) and then cardiac arrest. This is because potassium interferes with the cardiac cycle. Therefore, high potassium is a common and serious electrolyte emergency. Treatment includes increasing blood volume (artificial V_{in}), and insulin and glucose, as this hormone moves potassium into cells (and so out of the plasma). Other treatments include calcium gluconate, resonium A and, if extremely high, dialysis.

Addison's disease

This condition is essentially failure of the adrenals, and so the potential for low aldosterone, and thus hyponatraemia and hypotension. A consequence of the failure to resorb sodium is that potassium is retained, leading to hyperkalaemia. This disease will be discussed in greater detail in Chapter 18, which considers endocrinology.

Hypokalaemia Causes of low serum potassium (<3.8 mmol/L) include the reverse of the above, such as in alkalosis. Other causes include loss in diarrhoea and vomiting, and from a damaged kidney (high M_{out}) that may not be able to resorb potassium from the tubules, or inappropriate use of corticosteroids. A rare cause is Conn's syndrome, but this is also associated with mild hypernatraemia.

Consequences include muscle weakness and pain, and constipation. If severe, it may cause cardiac arrest. Treatment focuses on replacement, such as oral salts, or a potassium-rich drip supplement, but, as always, the primary cause must be detected and (if possible) corrected.

Drugs High blood pressure (hypertension) is a major risk factor for stroke, retinopathy and renal failure, and therefore demands attention. Treatments include reducing blood volume, and this can be done with diuretics (drugs that promote the generation of watery urine). These act in various ways to restrict the ability of the tubules to resorb water, and/or change ion absorption/resorption, so that the volume of the blood falls. Furosemide acts on the ascending loop of Henle to inhibit the resorption of sodium, so that, effectively, water stays in the filtrate. Hydrochlorothiazide acts on the distal convoluted tubule, leading to the retention of water in the urine.

The potassium sparing diuretics include spironolactone (an inhibitor of aldosterone) and amiloride (which inhibits sodium resorption by the distal convoluted tubules, connecting tubules and collecting ducts). As different drugs act on different parts of the nephron, several may be used at the same time, such as amiloride with hydrochlorothiazide. If the hypertension is resistant despite the use of these drugs, then other drugs targeting the angiotensin and aldosterone pathway can be used.

Common ACEIs include perindopril and enalapril; ARBs include irbesartan and losartan. As they target the same pathway, ACEIs and ARBs are rarely used at the same time. Hypertension can also be treated with drugs that act directly on blood vessels and the heart, not on the kidney, such as calcium channel blockers and nitrates. This subsection explains why (a) use of anti-hypertension drugs often causes electrolyte disturbances and (b) renal disease makes use of these drugs difficult.

Intravenous fluids

It follows from much of the above that loss of fluid is important, but it can be restored by intravenous fluids, usually in 500 mL or 1 L bags. Fluids may also be given to replace losses that can be predicted. Water cannot be given by itself as it will destroy red blood cells, so must be given as 5% dextrose. It will enter the blood and also moves into the tissues and cells. Saline (0.9% NaCl) rehydrates plasma and lymph, but not cells, whilst plasma expanders replace fluid deficits in the blood alone. Most clinical situations can be managed with these two fluids, plus 1.26% bicarbonate (to treat acidosis) and concentrated supplements (such as potassium, used to treat hypokalaemia). However, misuse of intravenous fluids can, of course, cause the reverse of what is being treated (such as hypertension).

Laboratory monitoring is by frequent measurement of serum electrolytes, but clinical assessment (such as pulse, blood pressure, body weight) can also be important. Measurement of fluids taken orally (by drinking) and the volume of urine produced are important, but these will not always add up to 100% due to 'insensible' losses that are not easy to calculate (from sweat, lungs and in faeces). Also, if someone is vomiting, this will also compound a possible state of dehydration.

Because the physiology of much of this opening section (summarized in Table 12.2) relies on good kidney function, then this is often presumed, if not demanded. However, this is clearly not always the case, and problems with U&Es are often the result of renal disturbances.

Therefore, we now move on to look at this organ in more detail.

Blood science angle: Hyperkalaemia and red blood cells

There are two well-known non-renal causes of hyperkalemia, and both involve the red cell and its high concentration of potassium. As soon as red cells leave the body, they start to lyse, so their potassium enters the plasma. As this is progressive, blood must be tested as soon as possible, certainly within 3 h. Red blood cells can be destroyed in the circulation in certain types of haemolytic anaemia (as is explained in Chapter 4). This may also cause serum potassium to rise.

12.3   Excretion

The regulation of electrolytes and their passage into the urine is not excretion: we reserve this word for the removal of the waste produces of metabolism. In this section we focus on urea, creatinine, the GFR and uric acid/urate. We will then also consider the pathological conditions of acute renal failure and chronic renal failure.

Waste products

The principal waste atom in all animal physiology is nitrogen. Aquatic animals can dispose of this mineral in the form of ammonia (NH_3), some reptiles do so with uric acid, but denied an excess of water in which to excrete these substances, mammals bind up their nitrogen in urea. This molecule, synthesized by liver, is efficient in that it contains two atoms of nitrogen in a relatively small molecule – $CO(NH_2)_2$. Creatinine is a breakdown product of creatine phosphate, found principally in muscles, and is a complex cyclic molecule containing three nitrogen atoms. Both urea and creatinine are excreted in large amounts of water.

A third nitrogen-rich molecule is uric acid (with four nitrogen atoms per molecule), which ionizes to the urate ion. Arising from the breakdown of deoxyribonucleic acid (DNA), urate has a low solubility and (unlike urea and creatinine) easily comes out of solution to form crystals, and so can cause severe disruption to cells and tissues.

Another major unwanted by-product of mammalian (and much other) metabolism is the hydrogen ion (H^+), which must also be excreted. Urine pH is variable depending on dietary intake and metabolic activity. Generally, the sum product of metabolic activity is an acid load that requires excretion to maintain the blood pH in a very narrow range (typically pH 7.35–7.45) and therefore urine pH is below 7, but ranges between 4 and 8. This biochemistry will be expanded upon in Chapter 13.

As discussed in Sections 12.1 and 12.2, the top part of the mini-organ of the nephron is the glomerulus – the part that interfaces directly with the blood. It is an important filter: ions and molecules with a mass less than around 65–67 kDa pass into the urinary filtrate, and most of the valuable solutes are then later reabsorbed so that those that remain make up the urine. Hence, we are interested in the quality of the function of this small area of the nephron. The most efficient and practical method for assessing the function of the glomerulus is the rate at which it cleans the blood.

Tests of glomerular function

Renal disease is frequently accompanied by changes in U&Es, but especially in serum creatinine and urea. If the former slowly rises up to 150 µmol/L and beyond, such as to 180 µmol/L, then renal function is deteriorating and the patient is probably in need of a referral to a renal physician. Creatinine itself has few adverse physiological effects, but rising urea is dangerous as it can adversely influence the function of red blood cells and other cells. Concentrations may also rise because of a decrease in effective circulating blood.

Albuminuria   An alternative test of the integrity of the kidney in general and the glomerulus specifically is the presence of protein in the urine (proteinuria). Most healthy adults can lose a small amount of protein in this way (generally less than 30 mg in a 24-h period), and this is physiologically acceptable. Nonetheless, if this loss becomes considerable then renal function must be deteriorating, and problems with the osmotic potential of the plasma may follow. Protein can be detected with dipsticks, but these have poor sensitivity for proteinuria – the laboratory utilizes analytical methods with greater sensitivity and therefore can quantitate lower levels. When glomerular damage is advanced and proteinuric loses are in the grams per litre range, then the concentration of protein, including albumin, may be low in the plasma. Hence, in the absence of liver disease, where synthesis of proteins may be compromised, hypoalbuminaemia may also indicate renal damage.

Creatinine clearance However, the primary indicator of renal function is the rate at which blood is filtered by passing through the glomerulus to produce an early form of urine; that is, the GFR. Ideally, this would be over 100 mL/min, and may rise to 125 mL/min in an elite athlete, but declines gradually with each decade of life by up to 3 mL/min per decade. This rate of filtration of 6–8 L/h, and the relatively tiny amount that ultimately reaches the bladder, is a testament to the incredible power of the nephron in reclaiming perhaps 99% of this fluid.

The GFR may be calculated by dividing the product of the urine creatinine concentration and 24-h urine by the concentration of creatinine in the serum at some time in this period, and so is also known as the creatinine clearance test. The key aspect of this test is the need for a very accurate collection of the 24-h urine, which must then be sent to the laboratory, along with a sample of serum taken any time within that 24 h. The laboratory will then measure creatinine in both fluids and apply the results into the following equation to obtain the GFR:

$$GFR = \\ \times \frac{\text{urine concentration of creatinine} \times \text{rate of urine production}}{\text{serum creatinine concentration}}$$

It is hard to overemphasize that the crucial aspect of this equation is the 24-h urine. For example, with a urine creatinine concentration of 7.5 mmol/L, a 24-h urine of 2.16 L and a serum creatinine of 150 μmol/L, then the GFR should be about 75 mL/min, which is low.

However, if the 2.16 L of urine was incorrectly taken over only an 18-h period, then a true GFR would be about 100 mL/min, which is clearly much better and closer to the normal range. Conversely, if only 1.8 L was collected, instead of the true 2.16 L, then the GFR would be only 62.5 mL/min, which is clearly worse and closer to renal failure. Both these errors could have resulted in starting or withholding crucial therapy.

Overall, therefore, this short example suggests some degree of renal impairment. However, as the serum creatinine was already raised at 150 μmol/L, then this is hardly surprising. Generally, as the creatinine rises, then the GFR falls. At the clinical level, the GFR can fall considerably (such as to 15–20% of normal), and the serum creatinine rises appreciably, before the patient becomes symptomatic. A further problem is that protein can interfere with the measurement of creatinine in urine. This is an issue because we can expect a degree of proteinuria to be present in those very patients whose renal function we seek to define.

Estimated glomerular filtration rate The inaccuracies of the 24-h urine as a method for assessing renal function prompted the search for an improved technique, based on fewer more reliable variables. Cockcroft and Gault, noting the fact that age is a major influence on creatinine clearance, derived an equation to estimate the GFR (hence eGFR) using only the patient's age, sex, weight and serum creatinine, thus removing the need for a 24-h urine collection. However, this equation fails to take account of the different renal function in those of African origin. The 'modification of diet in renal disease' (MDRD) equation does indeed require this information, in the equation

$$\begin{aligned} eGFR = 32\,788 &\times \text{serum creatinine}^{-1.154} \times \text{age}^{-0.203} \\ &\times [1.212 \text{ if African race}] \\ &\times [0.742 \text{ if female}]'\mu\text{mol/L} \end{aligned}$$

An alternative working of this equation gives

$$\begin{aligned} eGFR = 175 &\times [\text{serum creatinine} \times 0.011312] - 1.154 \\ &\times [\text{age}] - 0.203 \times [1.212 \text{ if African}] \\ &\times [0.742 \text{ if female}]'\mu\text{mol/L} \end{aligned}$$

There are several other variants, such as those that call for albumin and urea, but the local laboratory will always provide an eGFR given the required information, so (fortunately) the practitioner rarely needs to wield a calculator. In addition, various calculators are freely available on line (see 'Web sites' at the end of the chapter).

Cystatin C, a small protein with a molecular weight of perhaps 13 kDa, may be a better marker of renal function, but as yet has to displace the eGFR as the gold standard. Its value is related to a lack of association with age, sex, race and muscle mass, all of which influence creatinine, and that it may be a better marker of the risk of cardiovascular disease than creatinine is.

Urate/uric acid Strictly speaking, this is not a true marker of renal function, but is certainly to do with renal physiology and pathology.

Uric acid and its ionized form urate (dependent on pH) are derived mostly from breakdown products of purine nucleotides in the DNA, and are excreted by the kidney and in the gut. High plasma concentrations can arise from increased production (maybe 10% of cases, from the action of xanthine oxidase on xanthine and hypoxanthine) or decreased excretion (perhaps present in 90%). Thus, with poor excretion, concentrations in the blood will rise. Various drugs can also cause an increase in concentrations of this metabolite. However, a recognized cause of hyperuricaemia is the destruction of tumour cells by chemotherapy and/or radiotherapy: the increased uric acid reflects the DNA release from dead cells.

Urate is particularly insoluble, so that when concentrations rise (such as above 0.5 mmol/L) it will come out of solution and form crystals (such as in the synovial joint, which can be detected by microscopy) and stones (such as in the kidney). These lead to further problems, such as swelling, pain and inflammation, and ultimately to an arthritis (often gout, where there can also be deposits in the skin).

Blockage of the ureter or urethra by a kidney stone can lead to urinary retention and thus to several renal diseases, but increased uric acid may lead directly to renal damage by crystallizing within the nephron. Risk factors for hyperuricaemia, apart from renal disease, are obesity and alcoholism. Treatments include risk factor management, allopurinol (an inhibitor of xanthine oxidase), nonsteroid anti-inflammatory drugs and colchicine.

12.4   Renal endocrinology

The competent blood scientist will be aware that the kidney is both a producer and responder to various hormones, and so will appreciate how these impact on renal function and general pathology.

Hormones in homeostasis

We have already looked at the importance of AVP (from the pituitary), aldosterone (from the adrenals), and ANP and BNP (from the muscle cells of atria) in regulating sodium and water (Figure 12.2).

Erythropoietin

This growth factor acts on the bone marrow to promote erythropoiesis (Chapter 10). It is produced by the peritubular capillary lining cells of the renal cortex in response to the oxygenation of the blood – synthesis is higher in hypoxia. It follows that if there is renal disease then production of this hormone will be reduced. However, erythropoietin is not a marker of renal function, as some may arise from the liver.

Blood science angle: Erythropoietin

This explains why the measurement of this molecule, alongside U&Es and the eGFR, may be justified in a patient with anaemia, especially if normocytic. Details of this condition are in Chapter 4. Measurement by immunoassay may be performed in a biochemistry, immunology or haematology laboratory.

Calcium and vitamin D

As we will see in Chapter 15, the kidney is a key component in the metabolism of this ion and vitamin D.

There are several isoforms of vitamin D, and the synthesis is unusually complex. Our understanding of the metabolism is not helped by several names being used for the same molecule. Nevertheless, it is established that a precursor form is taken in the diet and is converted to cholecalciferol in the skin by the action of ultraviolet light. The next step is the conversion to 25-hydroxycalciferol by the liver, and then to 1,25-dihydroxycalciferol (vitamin D_3, also known as calcitriol) in the kidney. Parathyroid hormone (PTH) is required for this latter step.

Hormonally active vitamin D (1,25-dihydrocholecalciferol) acts to increase the tubular resorption of calcium by renal tubules, so that blood calcium rises. It follows that damage to the tubules, and/or lack of vitamin D, leads to depletion of calcium. Vitamin D deficiency at its most severe leads to bone diseases that include osteomalacia (rickets) and renal osteodystrophy. In severe hypocalcaemia, concentrations of PTH will be high, attempting to stimulate the kidney to resorb calcium. However, PTH also acts in its own right to stimulate the kidney to resorb calcium, and in addition promotes the loss of phosphates. Further details of calcium, PTH and vitamin D are present in Chapter 15, which focuses on bone.

Renal efficiency

For many, the astonishing efficiency of the kidney can be illustrated by differences in changes of various ions and molecules in the blood, post-glomerular filtrate, and in the urine (Table 12.3). In some cases, 100% of a constituent of the blood passes into the filtrate and is reabsorbed (such as valuable proteins and glucose). In other cases, little (or even none) is reabsorbed, and so remains in the urine to be excreted (waste products urea and creatinine).

Table 12.3 Constituents of blood, filtrate and urine in health.

	Plasma (%)	Post-glomerular filtrate (%)	Urine (%)	Reabsorbed (%)
Water	100	100	0.6	99.4
Protein	100	0.2	0	100
Chloride	100	100	0.8	99.2
Sodium	100	100	0.6	99.4
Bicarbonate	100	100	0.1	99.9
Glucose	100	100	0	100
Urea	100	100	47	53
Potassium	100	100	14	86
Uric acid	100	100	9.4	90.6
Creatinine	100	100	100	0

12.5 Renal disease

The whole purpose of renal testing is to determine the presence and extent of renal disease, and subsequently the monitoring of treatment.

Aetiology

The most common causes of kidney problems can be grouped into three areas (pre, true and post), based largely on anatomy.

Pre-renal disease This is characterized by factors leading to insufficient blood entering the kidney. Examples of this include narrowing of the renal artery (stenosis, generally caused by atheroma), abdominal aortic aneurysm, thrombosis within the aorta, low blood pressure, or poor cardiac output as may be present in heart failure.

True renal disease This is often seen in septic shock, glomerulonephritis (i.e. inflammation of the glomerulus and nephron), in the presence of toxins or amyloid, renal carcinoma (or secondary metastases), and in physical damage with blood loss. This cellular state of the disease may be called acute tubular necrosis.

Post-renal disease This is present if there are problems downstream of the kidney, such as with the ureter, the bladder or the urethra. Most common causes of this are kidney stones, carcinoma of the bladder or prostate, benign prostatic hyperplasia or infections. All these limit or prevent urine from flowing out, so that it will eventually back up to the kidneys themselves.

Thus, in both pre- and post-renal disease, there is nothing intrinsically wrong with the tissues of the kidney itself, or its functioning; the problem is with the tubes at either end. In the former, not enough blood is entering the kidney, so that 'cleaning' is poor. Furthermore, the kidney itself requires oxygen and energy sources (such as glucose) to perform its functions, so that deprived of these the kidney will fail. In post-renal disease, urine backing up the ureters will eventually damage the kidney. Therefore, failure to correct pre- or post-renal disease will inevitably lead to acute tubular necrosis. Management of renal disease depends on the particular aetiology, and thus the correction of pre/post-renal factors, if appropriate. This may be by, for example, antibiotics and immunosuppression for glomerulonephritis, angioplasty for renal artery stenosis, or surgery for bladder carcinoma. Figure 12.3 illustrates the anatomical classification of renal disease.

The perspective of the patient Naturally, the patient will have no idea as to the state of their nephrons, but they will be able to give useful information on a key aspect of renal function: the production of urine. A reduction in the production of urine is called oligouria (oligo = few), although this may be both a reduction in the volume passed at each urination and/or the frequency of urinations per day. Technically, oligouria may be defined as less than 400 mL/24 h. The most severe renal disease is, of course, failure to make any urine at all, or perhaps merely a dribble – anuria – generally less than 100 mL/24 h.

The patient will also be able to say if their urine has blood in it – haematuria. However, this may also arise from the bladder (as in bladder carcinoma), the prostate (as in prostatitis or prostate cancer) and the uterus. However, microhaematuria (such as a small number of red cells that they cannot be seen with the eye) may be present and may be detected by a dipstick, although centrifugation and a light microscope are preferable.

At the other end of the scale is the production of an increased volume of urine, such as over 3 L in a 24-h period. This is polyuria, and may be seen in diabetes mellitus and diabetes insipidus. Renal disease can also be classified by how rapidly it has developed.

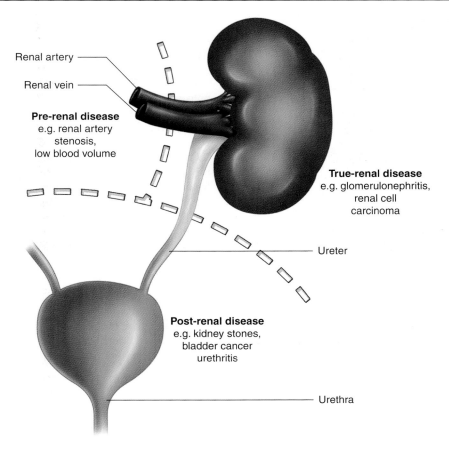

Renal artery

Renal vein

Pre-renal disease
e.g. renal artery
stenosis,
low blood volume

True-renal disease
e.g. glomerulonephritis,
renal cell
carcinoma

Ureter

Post-renal disease
e.g. kidney stones,
bladder cancer
urethritis

Urethra

Figure 12.3 An anatomical classification of renal disease. We can classify renal disease as being pre-renal, true renal, or post-renal, depending on the particular aetiology. In the first it is a problem with blood entering the kidney, and the last the problem is with urine leaving the kidney. True renal disease is where the nephrons themselves are damaged.

Acute kidney injury

This may be defined in the laboratory by the ratio of the relative rise in urea being greater than the relative rise in creatinine, not simply the concentrations themselves (as concentrations of creatinine are always lower than those of urea – perhaps 100 μmol/L versus 5000 μmol/L respectively). Other biochemical abnormalities include acidosis (because the kidney can no longer excrete hydrogen ions, which we will discuss in depth in Chapter 13) and hyperkalaemia. Of course, the patient is likely to be able to help in describing any changes in making urine. A likely story is of a sudden fall off, or even cessation, in diuresis, perhaps over several days. However, the cessation of diuresis may be due to other factors such as an inflamed urethra or acute prostatitis (inflammation of the prostate).

However, if the damage to the kidney in acute kidney injury (AKI) is excessive, and leads to acute tubular necrosis, it may well become permanently and irreversibly dysfunctional, in which case there will be deterioration to chronic renal failure. AKI is unlikely to be caused by pre-renal factors, which provides some help in confirming the aetiology.

Biochemical monitoring of renal disease will be by U&Es, and clinically by urine production. If the potassium rises dangerously, then dialysis may well be called for. Recovery from AKI will be accompanied by a marked diuresis, with a massive increase in urine production (polyuria: maybe as much as 7 L a day!), so that fluid balance may need to be checked. Soon after, a return to normal diuresis can be expected.

Chronic kidney disease

Chronic kidney disease (CKD) is the progressive and (invariably) irreversible destruction of kidney tissues, and is typically noted when the GFR falls below <60 mL/min, although it becomes critical when <15 mL/min. Using U&Es, CKD can be plotted by the relative rise in urea compared with the rise in creatinine. However, unlike AKI, in CKD there is a greater increase in creatinine and a more modest rise in urea.

The consequences of CKD are not unlike those of AKI, principally with disturbances in sodium, hydrogen, and water metabolism – there may be fluid overload or fluid depletion. If present, a metabolic acidosis may be evident with a low concentration of bicarbonate and high concentrations of hydrogen ions, and this may also contribute to hyperkalaemia. However, hyperkalaemia may independently arise from the increasingly impaired ability to excrete potassium, and may be life threatening. Low concentration of calcium, and hence raised PTH, may be present as the kidney loses its ability to promote calcium absorption in the tubules. Similarly, anaemia is likely to result as the impaired kidney will no longer be making erythropoietin. As renal disease deteriorates, protein will appear in the urine in increasing amounts, a level of 3.5 g per 1.73 m^2 in a 24-h period defining nephrotic syndrome, in which there is often hypercholesterolaemia. An alternative measure of glomerular damage is the ratio of albumin to creatinine in the urine.

Clinical features of CKD also include nocturia (urination whilst asleep, resulting from uneven diuresis) and hypertension (resulting from increased blood volume). Good management will address sodium and water intake, and diuretics may be necessary (depending on the degree of renal (dys)function). Hyperkalaemia may be managed with resonium A, as previously discussed.

Treatment and care are therefore conservative; and as renal function slowly deteriorates, the patient should be prepared physically and psychologically for dialysis, which is generally needed when the GFR falls to less than <25 mL/min. The remaining treatment is transplantation. Using the eGFR, chronic renal disease can be classified in stages (Table 12.4). As patients with the most adverse renal function are at risk of thrombotic disease, the risk factors for cardiovascular disease need to be addressed.

Renal stones (calculi)

At high concentrations, and under certain conditions, such as pH and dehydration, ions may come out of solution, form crystals and then stones in the kidney, ureter and bladder. If in the latter (and the urethra), then blockage can occur and cause post-renal disease. In some cases, stones may form around debris or aggregates of bacteria. Rarely, stones may consist of cysteine or xanthine, but most stones are composed of:

- Calcium oxalate, generally a consequence of hypercalciuria, excessive intake of oxalate or metabolic hyperoxalaemia. However, some stones may also contain a significant proportion of phosphate. A further risk factor is hyperparathyroidism, so that measurement of PTH is justified.
- Urate, inevitably caused by high serum urate and thus hyperuricuria. Thus, patients with uric-acid-linked gout may well have stones.

Table 12.4 Stages of CKD.

Stage	eGFR	Description and management
I	>90	Normal renal function, although urine analysis, structural abnormalities, family history or genetic factors may indicate renal disease. Standard monitoring and management of cardiovascular risk factors (such as blood pressure <140/90)
II	60–89	Mildly reduced renal function. Stage II CKD should not be diagnosed on eGFR alone, but urinalysis, structural abnormalities or genetic factors may support this stage. Observe and control cardiovascular risk factors as above
IIIa	45–59	Moderate decrease in renal function, with or without other evidence of kidney damage. More stringent control of cardiovascular risk factor. Consider low-dose statin regardless of serum cholesterol. Consider an ACEI regardless of blood pressure; target still <140/90
IIIb	30–44	Marked decrease in renal function, with or without other evidence of kidney damage. Statin and ACEI likely to be advisable. Blood pressure target <135/85
IV	15–29	Severely reduced renal function. Prophylactic pharmacotherapy mandatory. Planning for end-stage renal disease
V	<15	Very severe (end-stage) renal failure. Preparation for dialysis or transplant

Metabolic renal disease

As indicated in the preceding text, diabetes insipidus, Conn's syndrome, SIADH and Addison's disease all have a renal aspect. However, renal disease is very common in diabetes mellitus (i.e. diabetic nephropathy), and is almost inevitable and will be revisited in Chapter 14. About 30% of patients with end-stage renal disease have diabetes. Metabolic renal disease is a cause of bone disease, such as osteodystrophy.

Blood science angle: Inflammatory renal disease

The autoimmune disease systemic lupus erythematosus carries a strong risk of renal disease. Other inflammatory diseases includes nephritis (of the entire organ) and glomerulonephritis (of the glomerulus alone), although precise differential diagnosis between the two is difficult. In immunoblobulin A (IgA) nephropathy, IgA is deposited in the glomerulus. Goodpasteur's syndrome is caused by an autoantibody to the basement membrane of the glomerulus, which can be tested for in the immunology laboratory. These aspects are described in more detail in Chapter 9.

The genetics of renal disease

Advances in gene analysis have allowed the detection of certain mutations likely to be linked to human disease, and in this respect renal disease is no exception. The heritability of, for example, serum creatinine is between 0.41 and 0.75 (when 0 is no association and 1 is a twin-strength relationship). It follows that identification of subjects at risk of disease early in their lives may allow targeted treatments and/or lifestyle changes to perhaps avoid frank disease decades later.

One of the first major analyses of pooled data from smaller studies found significant associations between certain single-nucleotide polymorphisms and the risk of CKD as defined by concentrations of creatinine and cystatin. Key findings include mutations in *UMOD*, coding for uromodulin (also known as the Tamm–Horsfall protein), a molecule expressed specifically by cells of the ascending loop of Henle and found in abundant amounts in the urine. It may also be involved in the formation of calcium-rich kidney stones, and other mutations in *UMOD* are associated with hyperuricaemia and gout. Another gene, *GATM*, encodes an enzyme (gylcine amidinotransferase) present in the pathway

synthesizing creatine, a precursor of creatinine. Thus, mutations in this enzyme may act to influence serum creatinine independently of renal function.

A later international collaboration analysed data from over 67000 people of European ancestry, with the aim of identifying genes linked to susceptibility for reduced renal function, according to concentrations of creatinine, cystatin C and eGFR. Other genes identified were *SLC7A9*, coding for an amino acid transporter in the proximal distal tubules, mutations in which cause cystinuria, which may result in the formation of kidney stones. Another gene, *SLC34A1*, codes for a sodium/phosphate transporter. Located in cells of the proximal convoluted tubules, it mediates uptake of inorganic phosphate. Mutations in this gene also cause kidney stones.

Broadly speaking, African ancestry brings an increased risk of hypertension (and thus stroke), but also CKD. An early gene analysis in this racial group indicated a role in the angiotensinogen promoter in hypertensive renal disease, but a later one showed a much stronger link with *MYH9*, encoding non-muscle myosin heavy chain, a protein expressed in the glomerulus, peritubular capillaries and tubules themselves. Similarly, *APOL1*, coding for the apoliprotein-1, may also be involved in renal disease, although a direct pathophysiological link is obscure. Notably, both *MYH9* and *APOL1* are found close together on chromosome 22.

These findings provide new insights into the pathogenesis of CKD and underline the importance of common genetic variants likely to influence renal function and disease.

12.6 Case studies

Knowledge is nothing without practice. Now we have an understanding of renal function and disease, we can see how this information can benefit the patient. The eGFR is calculated from the MDRD equation.

Case study 11

A white European woman, aged 77, describes making less and less urine over a period of a few days, and complains to her general practitioner of something not right 'down there', with occasional pain. The doctor, feeling a hard mass in the lower abdomen, sends a venous sample of blood to the local district general hospital, with the following results:

	Result (unit)	Reference range
Na	140 mmol/L	135–145
K	5.1 mmol/L	3.8–5.0
Urea	56.5 mmol/L	3.0–6.7
Creatinine	254 μmol/L	71–133
eGFR	16.9 mL/min per 1.73 m^2 (=CKD stage IV)	>90

Clue. Compare the rise in urea to the rise in creatinine: acute or chronic?

Interpretation   Abnormal results are raised potassium, urea, creatinine and the low eGFR. The increase in potassium is very small, and so in this setting can be ignored. The increase in urea is about 10-fold over the middle of the normal range, whilst the increase in creatinine is only about four times. Thus, the relative rise in urea is over twice the relative rise in creatinine, pointing to an initial diagnosis of AKI. This is supported by the clinical history and the slowly developing anuria.

The cause is unlikely to be pre-renal, but may be true-renal (perhaps an infection of acute toxic damage), but could also be post-renal. Diagnosis may be confirmed with imaging (ultrasound), and may indicate a full bladder. A most likely cause of this is blockage of the urethra, maybe by a kidney stone or stones. There seems to be no infection, and a tumour alone would be unlikely to produce such an acute picture.

The eGFR is very low, and is close to CKD stage V – end-stage renal failure. However, the picture is clearly acute and driven by the raised creatinine. Removal of the blockage should allow the bladder to empty 'normally', although a catheter will be helpful. We expect a rapid resolution and the return of urea and creatinine to within the reference range, and an improvement in the eGFR, all of which will be helped by vigorous fluid intake (drinking and/or an intravenous infusion).

Repeat blood analysis within 8–12 h will indicate the rate of progression and the urgency of interventions.

Case study 12

A 70-year-old man from Jamaica mentions to his health-care professional that when he urinates the water in the lavatory becomes frothy. He also says he has lost about half a stone in weight in the past 4 months. A urine dipstick reveals +++ proteinuria. He denies any marked change in his pattern of urination. The practitioner orders U&Es and serum albumin.

	Result (unit)	Reference range
Na	144 mmol/L	135–145
K	4.9 mmol/L	3.8–5.0
Urea	8.9 mmol/L	3.3–6.7
Creatinine	189 μmol/L	71–133
eGRF	38.6 mL/min per 1.73 m^2 (=CKD stage IIIb)	<90
Albumin	32 g/L	35–50

Clue. Compare the rise in urea to the rise in creatinine: acute or chronic?

Interpretation   Abnormalities include raised urea, raised creatinine and low albumin. The increase in urea is less than twice the middle of the reference range, whereas the increase in creatinine is much higher, indicating chronic renal failure. The eGFR is also worryingly low. The marked proteinuria indicates nephrotic syndrome, and the severity of this case is underlined by the low serum albumin, which we presume is due to the glomerulus failing to filter this protein. The cause of the weight loss is unclear, but may be body fluid brought about by a slowly increasing diuresis that seems to have gone unnoticed by the patient.

Therefore, the problem seems to be damage to the glomerulus, which may be caused by an infection – that is, glomerulonephritis – or may be secondary to an inflammatory disease such as systemic lupus erythematosus. Damage to the glomerulus may also be present in diabetes mellitus. The precise diagnosis can only be made by renal biopsy, a procedure not to be undertaken lightly.

It is also important to repeat the blood analysis to assess progression of the CKD. Regularly repeated monitoring of eGFR changes is a key to the ongoing monitoring of GFR changes of this patient once CKD is confirmed.

Treatment of an inflammatory aetiology is with strong immunosuppressive drugs, such as azathioprine, ciclosporin and cyclophosphamide. These may be replaced with oral steroids such as prednisolone, which can slowly be tailed off as the proteinuria resolves, the creatinine falls and the eGFR rises. An ACEI may be given to provide some support to the kidney. Why Jamaica? Because African ancestry is a factor in the equation generating the eGFR. There also needs to be screening for diabetes and assessment of blood pressure, as hypertension is both a cause and consequence of renal disease.

Additional case studies where renal function is important are presented in Chapter 22.

Summary

- The main functional unit of the kidney is the nephron, the top part of which is the glomerulus, and is where ultrafiltration takes place.
- Selective reabsorption takes place in the proximal convoluted tubule, the loop of Henle, the distal convoluted tubule and collecting ducts.
- U&Es are often used to monitor renal function, the major waste products of metabolism being urea, creatinine, uric acid and the hydrogen ion.
- The efficiency of the excretory function of the kidney can be assessed by its removal of creatinine, and is a major component of the GFR.
- The estimation of the GFR by 24-h urine collection is subject to error, and has been superseded by an equation giving an estimate: the eGFR.
- Failure to excrete uric acid leads to hyperuricaemia and, ultimately, to gout.
- The kidney both releases and responds to hormones: AVP, aldosterone, ANP, BNP, erythropoietin and PTH.
- Vitamin D, the metabolism of which requires a healthy kidney, is needed for the absorption of calcium.
- Renal disease can have pre-, true- and post-renal causes.
- AKI is often associated with a greater rise in urea than the rise in creatinine.
- CKD is frequently characterized by a greater rise in creatinine than the rise in urea.
- Renal disease is frequently associated with raised uric acid, and so a gouty arthritis may develop.
- There are several metabolic and inflammatory renal diseases, and some evidence of the effects of gene mutations.

Further reading

Böger CA, Heid IM. Chronic kidney disease: novel insights from genome-wide association studies. Kidney Blood Press Res. 2011;34:225–234.
Kottgen A, Glazer NL, Dehghan A. et al. Multiple novel loci are associated with indices of renal function and chronic kidney disease. Nat Genet. 2009;41:712–717.
Levey AS, Coresh J, Greene T. et al. Using standardized serum creatinine values in the modification of diet in renal disease study equation for estimating glomerular filtration rate. Ann Intern Med. 2006;145:247–254.
Matsushita K, van der Velde M, Astor BC. et al. Association of estimated glomerular filtration rate and albuminuria with all-cause and cardiovascular mortality in general population cohorts: a collaborative meta-analysis. Lancet. 2010;375:2073–2081.
Mitra PK, Tasker PR, Ell MS. Chronic kidney disease. BMJ. 2007;334:1273.
Myers GL, Miller WG, Coresh J. et al. Recommendations for improving serum creatinine measurement: a report from the laboratory working group of the National Kidney Disease Education Program. Clin Chem. 2006;52:5–18.

Guidelines

NICE Clinical guideline CG73: Early identification and management of chronic kidney disease in adults in primary and secondary care.
NICE Clinical guideline CD114. Anaemia management in people with chronic kidney disease.
NICE Clinical guideline CD127. Hypertension: Clinical management of primary hypertension in adults.

Web sites

A calculator for eGFR: http://www.patient.co.uk/doctor/Estimated-Glomerular-Filtration-Rate-(GFR)-Calculator.htm.
National Institute for Health and Clinical Excellence (NICE): www.nice.org.uk.
The British National Formulary has information on drugs influencing renal function: www.bnf.org.

13 Hydrogen Ions, pH, and Acid–Base Disorders

Learning objectives

After studying this chapter, you should be able to:

- appreciate the importance of the pH of blood;
- understand how the kidney maintains a correct pH;
- describe the function of buffers;
- list the five clinical consequences of the failure to maintain a correct pH;
- discuss blood gases and their regulation;
- be aware of the clinical responses to problems with pH.

The topic of blood gases and pH flows naturally from renal function for a number of reasons. We must first understand how the pH of the blood is maintained, generally between pH 7.35 and pH 7.45, which is one part of the functions of the kidney. A new aspect we will examine is the importance of buffers, such as those of bicarbonate and phosphate. Only then will we be able to consider the pathology of pH and blood gases. These are at the two extremes of pH – when the pH is too low there is acidosis, whereas when the pH is too high the result is alkalosis (whose stem, alkali, is the reverse of acid). Both can have severe consequences for the blood and the body. These two extremes are inevitably related to two different pathological processes: those of metabolism and those of respiration (or, more accurately, gas exchange at the lung). These lead to five different types of clinical problems:

- *Metabolic acidosis:* present when the pH is low (perhaps <7.2). It is inevitably due to problems with metabolism, such as the excessive production of hydrogen ions, failure to excrete hydrogen ions (generally by the kidney) and the loss of bicarbonate.
- *Metabolic alkalosis:* present when the pH is high (perhaps >7.6), and is again due to metabolic problems.

The most common causes include excessive loss of hydrogen ions from the intestines and/or the kidney.
- *Respiratory acidosis:* present when the pH is low (perhaps <7.2). In this case, the problem arises due to problems with the lung, leading to an increase in carbon dioxide.
- *Respiratory alkalosis:* present when abnormalities in lung function lead to a high pH, generally in excess of 7.6. This most often happens when carbon dioxide levels are low.
- *Mixed acid–base conditions:* given the complexity of biochemistry, it is entirely acceptable that rare conditions are present that adopt different aspects of the four major conditions.

Complete knowledge of these conditions is impossible without an understanding of a relatively small number of molecules. The key analytes we will be addressing in this section are hydrogen ions (H^+, also known as protons), the pH of blood, concentrations of bicarbonate, and of the gases oxygen and carbon dioxide (Table 13.1). An additional index is to do with different numbers of charged particles (ions) in the blood: the anion gap. Once we have a basic knowledge of these indices we will then be able to explore their importance in terms of the five conditions described above (Sections 13.3–13.5).

13.1 Ions and molecules

Ionisation

As discussed in Section 12.2, virtually all biochemical substances have a molecular form and an ionic form; when dissolved in a fluid the latter allows the passage of electricity, and as such is called an electrolyte. Water is the best known example of an electrolyte, whilst oil is an

Blood Science: Principles and Pathology, First Edition. Andrew Blann and Nessar Ahmed.
© 2014 John Wiley & Sons, Ltd. Published 2014 by John Wiley & Sons, Ltd.

Table 13.1 Major tests in pH and blood gases.

Analyte	Reference range
Hydrogen ions (H^+)	35–45 nmol/L
pH	7.35–7.45
Bicarbonate	22–29 mmol/L
Carbon dioxide (P_{CO_2})[a]	4.7–6.0 kPa[b]
Oxygen (P_{O_2})[a]	12–14.6 kPa
Anion gap	15–20 mmol/L

[a]Arterial blood; others are in venous blood.
[b]The unit of pressure is the pascal.

example of a nonelectrolyte. The latter does not promote the passage of electricity as it does not ionize.

Consider a theoretical compound of two elements, A and B, which come together to form a molecule with the formula AB. If we place AB into water, some of it will split into two parts, and be ionic, becoming A^+ and B^- (the ions), whereas some will remain as one molecule; that is, in the form AB (the molecule). These two forms are interchangeable: there are times when the molecular form dominates, and in other conditions there is more of the ionic form. This balance, referred to as its equilibrium, can be represented as

$$AB \leftrightarrow A^+ + B^-$$

Note that the element A adopts a small plus sign; that is, A^+. This is an atom of element A without a negatively charged electron, and so A becomes positively charged. But as in magnetism, where you cannot have a north pole without a south pole, there must always be a negative component to balance the positive ion, hence the minus sign on the B^- (the counter ion), as it has gained an electron. We will return to this requirement for electric balance in the sections that follow.

Each compound ionizes to a different degree. In some the balance lies over to the right (ionic) side, and in some the balance is on the left (molecular) side. The degree of this balance is governed by the dissociation constant, and is represented by the letter K. This can be written as an equation of the different components:

$$K = \frac{[A^+] \times [B^-]}{AB}$$

We use square brackets to denote concentration, so that $[A^+]$ refers to the concentration of A^+.

Water and pH

In nature, there is no compound AB, but an excellent example of ionisation is the chemistry of sodium chloride (salt, NaCl), which converts to Na^+ and Cl^- in water. NaCl dissociates very freely into the ionic form with almost none remaining as a molecule. As far as water is concerned, the covalent form H_2O is in equilibrium with the ionic form (H^+ and OH^-). The dissociation constant K of water is approximately 1×10^{-14}, so putting this into the equation

$$K_{water} = \frac{[H^+] \times [OH^-]}{[H_2O]}$$

means that concentration of ionic water is $100\,000\,000\,000\,000$ ($=10^{14}$) less than that of covalent water. So that although $[H^+]$ is very, very small indeed, it makes up half of the ionic form, the other half being $[OH^-]$, and so the figure comes down to 'only' $10\,000\,000$ ($=10^7$) times less. However, use of these large numbers in everyday language is clearly difficult, so scientists have developed a way of simplifying this by taking the logarithm to the base 10, so that $10\,000\,000$, which can also be written as 1×10^7, simply becomes 7, and is why we say that (generally pure) water is at pH 7.

Acids and pH

One way that we can define an acid is as a substance that acts to increase the concentration of hydrogen ions (which might be the theoretical molecule HB, as in the example above), but another definition is of the hydrogen ion itself (i.e. H^+). It follows that you may consider water to be an acid, as it provides H^+ when it ionizes. Despite this lack of agreement between chemists, there is, however, complete consensus of the biochemical definition of acidity of a solution, being the concentration of the protons; that is:

$$pH = -\log[H^+]$$

So because of the minus sign, this means that the higher the concentration of hydrogen ions, then the lower is the pH. But for every acid (such as H^+) there has to be a conjugate base (such as B^-) to balance the electrical aspect of the equation. The concentration of B^- (i.e. $[B^-]$) in a liquid confers alkalinity, so that if $[B^-]$ is high compared with its conjugate acid, the solution is alkaline.

Acidity is a problem for the body because hydrogen ions are constantly being generated, as they are waste products of metabolism, and so need to be expelled from the body, which it does through the kidneys. Furthermore, increasing concentrations of hydrogen ions are bad because rising acidity has many adverse effects on the body, such as on the homeostasis of potassium. Whilst it is true that the reverse of acidity (i.e. alkalinity) is also to be avoided, normal body metabolism does not directly generate negatively charged ions as its does the positively charged hydrogen ion [H^+]. It does, however, generate a great deal of carbon dioxide (CO_2), and this molecule, although it does not ionize directly, does have profound effects on pH and pathology.

Ideally, the kidney should be able to regulate the hydrogen ion concentration of the blood, and so the degree of acidity and alkalinity. Therefore, renal dysfunction will naturally lead to pH problems. However, we have another system designed to keep the pH of the blood within its limits: the system of buffers. In many cases these provide essential support in maintaining acid/alkali homeostasis.

Buffers

Buffers have been called 'proton sponges', mopping up any excess hydrogen ions that have been added to a solution, and then releasing them when they are in short supply. Buffers play a major role in controlling acid–base balance in the body. A buffer is a combination of a weak acid in solution with its conjugate alkali. The major plasma buffer is bicarbonate; others include proteins, phosphate and ammonia.

Bicarbonate Almost all oxygen in the blood is carried by the haemoglobin in red cells. It is required to form water at the end of the electron transport chain that generates adenosine triphosphate (ATP). Oxygen is required as a vehicle for removing the carbon atoms in energy substrates such as glucose, in the form of carbon dioxide, which must then be excreted. A small amount of carbon dioxide is returned from the tissues to the lungs inside red cells (as carbaminohaemoglobin, described in Chapter 3); a little more is carried dissolved in the plasma. However, most carbon dioxide is carried in combination with water, in the form of carbonic acid, which is generated by the enzyme carbonic anhydrase (also referred to as carbonate dehydrase):

$$CO_2 + H_2O \leftrightarrow H_2CO_3$$

However, recall that all compounds reach an equilibrium between a molecular form and an ionic form, and carbonic acid is no exception:

$$H_2CO_3 \leftrightarrow HCO_3^- + H^+$$

Therefore, the ionisation of carbonic acid generates a hydrogen ion, but its counter-anion, the bicarbonate ion, HCO_3^-, often referred to simply as bicarb, is very important. Indeed, the bicarbonate system provides 70% of the blood's buffering capacity. Bicarbonate is so called (with a bi–) because carbonic acid can have two cations – in the example above it is two hydrogen ions, but there can also be one hydrogen and one sodium (i.e. $NaHCO_3$, sodium hydrogen carbonate), or indeed just two sodium atoms (i.e. Na_2CO_3, sodium carbonate).

Within red cells, the hydrogen ion can be absorbed by haemoglobin, but the bicarbonate ion passes out of the cell into the plasma. This therefore leaves an imbalance in the electrical charge within the cell, so that another negatively charged ion must enter the red cell to balance the bicarbonate leaving it. This ion is chloride (Cl^-), and its movement is called the chloride shift. This requirement to maintain electrical neutrality we shall meet again.

We have a relatively large amount of bicarbonate in the plasma (around 24–25 mmol/L) that is available to soak up hydrogen ions. However, if there are too many hydrogen ions present, the buffering capacity of the bicarbonate system will eventually be overwhelmed. Accordingly, concentrations of plasma bicarbonate will fall, to less than 20 mmol/L, and so what remains will have difficulty in keeping the pH within its prescribed limits. As we will see, failure to maintain an adequate concentration of bicarbonate ions in the blood can have clinical consequences.

Other buffers Proteins, especially albumin, account for 95% of the non-bicarbonate buffering capacity of plasma. They act as buffers both inside cells and in the plasma. Albumin behaves as a weak acid, owing to its high concentration of negatively charged amino acids (mainly carboxyl and aspartic acids).

Phosphoric acid, H_3PO_4, like carbonic acid, is a weak acid, and ionizes to give the phosphate ion and up to two hydrogen ions:

$$H_3PO_4 \leftrightarrow H_2PO_4^- + H^+ \leftrightarrow HPO_4^{2-} + 2H^+$$

Phosphate occurs in both organic and inorganic forms within the body. At physiological pH 7.4 most phosphate within plasma and the initial part of the glomerular filtrate exists as monohydrogen phosphate (HPO_4^{2-}). This form of phosphate can accept H^+ to form dihydrogen phosphate ($H_2PO_4^-$). Owing to its relatively low concentration, phosphate tends to be a minor component of extracellular buffering, but it plays a significant contribution in buffering urine. High concentrations of phosphate are present in bone and within cells. This is significant in some acidotic (low plasma pH) states when phosphate can be released from the bones and act as a buffer in the plasma.

Unlike plasma, urine pH may vary considerably between pH 4.8 and 7.8, but is more likely to be acidic. Urinary buffers are required in acid–base homeostasis as they provide the major mechanism by which hydrogen ions are excreted from the body and are essential in the generation of bicarbonate. Ammonia, phosphate and bicarbonate are all involved in urine buffering.

The breakdown of the amino acid glutamine by the enzyme glutaminase in cells of the nephron results in the generation of a molecule of ammonia. A second molecule is produced by the subsequent metabolism of glutamate to α-keto-glutarate. Ammonia (NH_3) can diffuse out of the renal tubular cells and into the filtrate, and easily absorbs a hydrogen ion to become the ammonium ion:

$$NH_3 + H^+ \leftrightarrow NH_4^+$$

So, fortunately, this small pathway eliminates two toxins at the same time. The remaining α-keto-glutarate within the cell can also act as a buffer by absorbing H^+, eventually resulting in the generation of glucose, and therefore an additional source of energy. This mechanism is remarkably efficient, as the ammonium ion accounts for over half of the hydrogen ion excretion derived from metabolic acids.

13.2 Blood gases

The unit of gas pressure is the pascal (Pa), and the pressure exerted by the air at standard temperature and altitude (sea level) is 101.26 kPa. Air is composed mostly of oxygen, nitrogen, carbon dioxide and water vapour, and each gas exerts an individual pressure roughly proportional to its concentration in air. Thus, the partial pressure of oxygen is 21.12 kPa, nitrogen is 79.6 kPa, carbon dioxide is 0.04 kPa and water vapour exerts 0.5 kPa. However, in arterial blood, there is far less

oxygen than in the air, with partial pressure (P_{O_2}) between 12 and 14.6 kPa. Conversely, there is far more carbon dioxide in arterial blood than in the air – exerting a partial pressure (hence P_{CO_2}) between 4.7 and 6.0 kPa.

Clearly, oxygen is crucial for life, and blood levels fall as it is consumed in respiration. The reverse happens for carbon dioxide, whose levels rise as the blood passes through capillary beds. However, carbon dioxide is not merely an excretory product that we need to expel from the body, but it is also a powerful metabolite in its own right, and counteracts acidity. Indeed, it may be argued that the pH of the blood depends on the ratio between the concentration of bicarbonate (as $[HCO_3^-]$) and the amount of carbon dioxide in the blood (as P_{CO_2}).

Measurement

In several clinical settings it is important to know the amount of oxygen and carbon dioxide in the blood. One way of measuring blood oxygen is through the skin – the transcutaneous route – with a probe on the end of a finger, which makes use of the different absorption of light when haemoglobin is oxygenated. However, this pulse oximeter is not very sensitive and depends on the thickness and hardness of the skin. Therefore, a blood sample itself is required, and this needs to be from an artery, typically one of the radial arteries in the wrist, although other sites include the brachial and femoral arteries. In physiology, we expect much less oxygen, but slightly more carbon dioxide, in venous blood (Figure 13.1), reflecting the consumption of oxygen and generation of carbon dioxide in respiration.

Arterial blood for blood gases must be taken into a syringe preloaded with a small amount of heparin (as this anticoagulant is acidic). Once obtained, care must be taken to seal the syringe with a cap, so preventing contamination with bubbles of air, and analysed in a blood gas analyser within an hour. Ideally, it will be assessed immediately, and if not, then cooled in an ice bath until processed (Figure 13.2). The blood gas analyser may also be programmed to measure pH, bicarbonate and lactate (the latter occasionally requested in certain metabolic conditions).

Generally speaking, the greatest call for blood gases is in accident and emergency, intensive care, neonatal units and surgical theatres, as these tests are often required urgently. Blood science equipment remote from the laboratory (i.e. in 'near patient testing' situations) is at risk from poor quality assurance and other laboratory

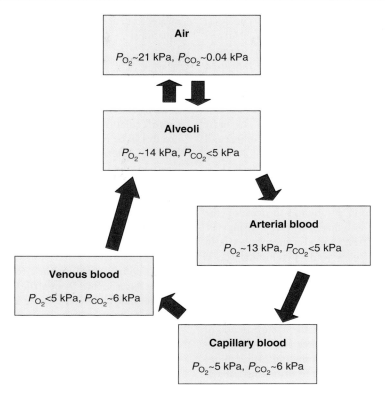

Figure 13.1 Gases in the air, the lungs and the blood. Partial pressures of oxygen and carbon dioxide in the air, the alveoli, arterial blood, capillary blood and venous blood. As blood passes round the circulation in a clockwise manner, it gradually loses oxygen and gains carbon dioxide.

Figure 13.2 A blood gas analyser, as is present in an accident and emergency unit. The front left-hand side has two reagent reservoirs; on the right are three reservoirs. The touch-sensitive screen allows programming. Results are printed out on a strip of paper.

backup, and staff may not be trained as comprehensively as are laboratory staff.

Both gases are measured by modified ion electrodes – for carbon dioxide this is a gas-sensing electrode, whilst for oxygen the chemistry involves a silver/silver chloride electrode and a platinum electrode. Notably, reactions of both gases are strongly temperature dependent, and the analyser reaction chambers are at 37 °C.

Interpretation

The chemistry of the electrodes is converted into the partial pressures of the gases in the sample, indicated by an upper case P, thus P_{O_2} and P_{CO_2}. Rarely, total carbon dioxide may be requested, being the sum of dissolved carbon dioxide, carbonic acid and bicarbonate. However, 95% of this total value is contributed by the bicarbonate. P_{O_2} is often used to monitor the efficiency of oxygen therapy in a patient with severe lung disease, delivered at pressure from a bedside cylinder. In blood gas analysis, the bicarbonate is not measured directly but is calculated from the H^+ and the P_{CO_2} alone, in contrast to 'normal' serum biochemistry, where bicarbonate is measured directly by a standard chemical technique.

A primary reason for low oxygen in arterial blood is poor ventilation and/or poor absorption across the air–blood interface of the alveoli. However, serious heart disease (anatomical abnormalities and heart failure) and profound obesity may also be responsible for highly unusual blood gas results, and in many cases this is clear. The ventilation rate (generally 11–16 breaths/min) is part governed by P_{CO_2} – a high level will stimulate an increase in the rate of breathing in an attempt to drive off the excess – but this will also happen with a rising H^+ concentration. A falling P_{O_2} will also prompt a rise in the ventilation rate, perhaps to 30 per minute (i.e. a breath every 2 s), which is close to panting.

Regulation of blood gases and pH

As we have discussed in Chapter 12, the waste products of our normal metabolism include urea and creatinine, to which we can now add hydrogen ions and carbon dioxide. The former are excreted only in urine; the latter is excreted at the lungs in our exhaled breath, but it can also be excreted in urine as bicarbonate. It follows that disease in either organ will lead to changes in these blood components. Concentrations of hydrogen ions being too high, and thus the pH being low (tending to, or lower than, pH 7.2), is acidosis. Conversely, when the pH is high (tending to, or exceeding, pH 7.7), with low concentrations of hydrogen ions, there is likely to be alkalosis. However, the picture is complex and therefore requires some explanation.

Although many references cite the 'normal' range for plasma pH as 7.35–7.45, it does not follow that pH 7.34 is always acidosis, or that pH 7.46 is always alkalosis, any more than a haemoglobin of 132 g/L (reference range 133–167 g/L) always defines anaemia. Yes, pH is important, but it takes more than the pH result to define clinical acidosis or alkalosis; the condition of the patient must also be addressed.

At the lung   Hydrogen ions can exist by themselves or as part of water, and an increased hydrogen ion concentration leads to a low pH and acidosis. However, carbon dioxide, as part of a buffering system, can combine with water to give carbonic acid (assisted by the enzyme carbonic anhydrase), and then bicarbonate in the following master equation we have already met in two separate equations:

$$H_2O + CO_2 \leftrightarrow H_2CO_3 \leftrightarrow H^+ + HCO_3^-$$

In physiology, an increase in hydrogen ion concentration, and thus a fall in pH, can be countered by an increase in the rate of ventilation to remove the carbon dioxide. The laws of equilibrium state that carbon dioxide lost in exhaled breath will be replaced by carbonic acid, which itself can only be generated by hydrogen ions combining with bicarbonate ions. What this means in practice is that the entire equation moves to the left. It also calls for sufficient plasma bicarbonate to mop up the hydrogen ions.

As discussed, carbon dioxide generated in the tissues by respiration forms carbonic acid with water. This weak acid dissociates to the bicarbonate ion, almost all of which is carried in the plasma, and a hydrogen ion, which can be carried within a red cell by haemoglobin, leaving bicarbonate to contribute to the plasma pool (Figure 13.3). However, some carbon dioxide is also carried by haemoglobin.

At the alveolus, carbon dioxide and the hydrogen ion leave the red cell, and the latter reforms carbonic acid, which in turn regenerates carbon dioxide and water, both of which can pass into the alveolar air and so be exchanged for atmospheric oxygen. The relative simplicity of this system, which is driven by laws of mass action and gas diffusion (such as Dalton's laws, which we need not consider) contrasts markedly with the complexity of the homeostasis of pH at the renal tubule.

At the kidney   The lung must (simply) adhere to the laws of gas exchange, but the kidney must adhere not only to the laws of osmosis, but also of electrical neutrality, which at times may be antagonistic and counter to the best interests of the physiology of excretion and homeostasis. One way of getting round this is to actively pump ions and molecules in and out of cells, which requires energy in the form of ATP, and also by the action of hormones such as aldosterone, arginine vasopressin (AVP) and the natriuretic peptides, acting on cells of the nephron. These hormones were discussed in Section 12.2.

Thus, regulation can be viewed as being in two parts: active (requiring energy) and passive (not requiring energy). The former consists of the functions of the three hormones, which interact with pumps to move sodium and water across the cell membrane, often against a concentration gradient. These pumps require energy as ATP, generated by respiration, which of course consumes an energy source such as glucose, but which also produces

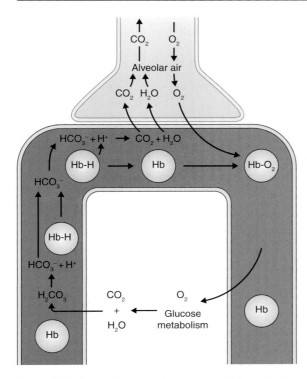

Figure 13.3 Gas exchange at the alveolus. Carbon dioxide generated in the tissues initially forms carbonic acid, which then ionizes to bicarbonate and a hydrogen ion. The latter can be carried by haemoglobin. The constituent parts reform at the alveolar surface, and water and carbon dioxide pass into the exhaled breath.

carbon dioxide and water (Figure 13.4, top part). Therefore, the act of generating energy to regulate the transport of water in itself generates water. The generation of carbon dioxide within the cell may also have metabolic consequences.

Bicarbonate can be generated from carbon dioxide and water (Figure 13.4, middle part), but this also produces a hydrogen ion. However, the latter can pass into the urinary filtrate and bind with buffering phosphate ions and ammonia. The bicarbonate can pass into the blood where it may perform its own buffering duties in absorbing hydrogen ions. As hydrogen ions leave the cell, the electrical neutrality can be maintained by the movement of sodium from the filtrate, and ultimately into the blood.

Glutamine, mostly from the muscles and liver (Figure 13.4, lower part), can pass into the tubules, where it is the substrate for glutamase (producing glutamate) and then glutamate dehydrogenase, which produces α-keto-glutarate, and the production of two molecules of

ammonia. The latter can diffuse into the filtrate and combine with (and so buffer) two hydrogen ions, thus reducing the pH of the urine.

It must be stressed that this scheme is an attempt to simplify hundreds of different pathways occurring at different places in the nephron. For example, it is difficult to simply demonstrate, in a single diagram, the generation of 'new' bicarbonate and the resorption of 'old' bicarbonate from the urinary filtrate. Nevertheless, Figure 13.4 underlines the complexity of the biochemistry, and so an indication of what might happen when these pathways break down; that is, in acidosis or alkalosis.

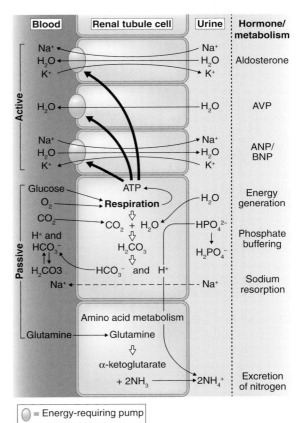

Figure 13.4 Regulation at the nephron. A cartoon of five tubule cells and the possible biochemical pathways that may be present. The top three are cells involved in the hormone-controlled regulation of sodium and water. Potassium moves to maintain electrical balance. The lower cells show potential regulation by passive movement of ions across the cell membrane, although some may be actively pumped. Note the sodium can also move passively, but this, like the potassium movement, is also to maintain the electrical balance.

Thus, the primary pathology of these two conditions depends on many renal and respiratory factors. In some, a primary pulmonary defect in ventilation affects the P_{CO_2} and thus the ability to generate bicarbonate ions. Alternatively, deranged metabolic disorders may lead to the failure of the buffering systems, with pH problems, and the loss or retention of bicarbonate. Thus, two types of acidosis and alkalosis are recognized, and follow problems with the lungs and/or the kidneys.

Compensation This is a precise biochemical term that refers to a physiological response to altered pH dynamics, and acts to neutralize that change. Broadly speaking, if the change in pH is due to respiratory reasons (such as altered P_{CO_2}), then metabolic (mostly renal) changes operate to compensate for this. Conversely, if a metabolic problem leads to change in $[HCO_3^-]$, then a respiratory compensation acts by changing the P_{CO_2}. Respiratory compensation is faster than metabolic compensation, but is less effective in the long term. Compensation will be revisited in Section 13.3.

Total CO_2 Rarely, total CO_2 can be requested, being the sum of the carbon dioxide in bicarbonate, carbonic acid, in carbaminohaemoglobin, and the carbon dioxide dissolve in the plasma. However, as 95% of total carbon dioxide is in the bicarbonate fraction, it is often used. Although relatively simple to determine, total carbon dioxide has its limitations, one of which is that it cannot alone define an acid/alkaline status as the hydrogen ion concentration and P_{CO_2} are unknown. Indeed, total carbon dioxide may be increased in both respiratory acidosis and in metabolic alkalosis.

Chloride (Cl^-) This anion is present at a relatively high concentration (95–107 mmol/L) in the plasma, second in concentration to sodium (135–145 mmol/L), its most common counter-cation. Increased chloride is likely to be found in renal failure, but it is most commonly needed in the calculation of the anion gap (to be discussed in Section 13.3).

Blood science angle: Blood gases and haemoglobin

Red blood cells can carry both hydrogen ions and carbon dioxide. However, defective haemoglobin (for whatever reason) may not be able to do this as efficiently. Conversely, problems with the carriage of hydrogen ions and carbon dioxide may adversely influence the ability of the cell to carry and deliver oxygen to the tissues, leading to the symptoms of anaemia (Chapter 4).

13.3 Acidosis (pH <7.3)

This may be defined simply by an increase in plasma hydrogen ions, and thus a low pH, and has a number of adverse effects. For example, variations in pH may affect the biological activity of certain enzymes and cations, such as the binding of calcium and magnesium to albumin. An increased hydrogen ion concentration can result in the release of these cations from proteins, increasing the concentration of the biologically active form in the plasma, which can have pathological consequences. Almost all causes of acidosis are due to two separate processes: metabolic acidosis and respiratory acidosis.

Metabolic acidosis

Metabolic acidosis is characterized by an increased production, increased ingestion and decreased excretion of hydrogen ions, or any combination of these. A second mechanism is depletion of buffering bicarbonate. Metabolic acids producing hydrogen ions are derived from three main sources: the anaerobic metabolism of glucose to lactate and pyruvate, the anaerobic metabolism of fatty acids, and the oxidation of sulphur-containing amino acids (cysteine and methionine) and cationic amino acids (arginine and lysine). From these sources, our cells release 40–60 mmol of hydrogen ions into the plasma each day, possibly more. Other factors leading to an increased plasma hydrogen ion concentration include (i) the inability to excrete these ions, (ii) failure of the buffering system and, of course, (iii) a combination of the two.

Failure to excrete sufficient hydrogen ions is clearly a renal problem, as we have discussed. Failure of the buffering system (most of which is bicarbonate) may also be due to renal disease and the failure of the renal tubule cells to reabsorb or generate this ion. Bicarbonate may also be lost in acute or chronic diarrhoea. Although the carbon dioxide that generates the bicarbonate may arise from respiration, there may also be acidosis resulting from abnormal carbon dioxide gas exchange at the alveolus (Figure 13.3).

The compensatory response to metabolic acidosis, as will be sensed by chemoreceptors, is to increase the ventilation rate. This can be marked, with hyperventilation of deep and forced breaths, which of course demands adequate pulmonary function. However, there also needs to be renal compensation, with the excretion of hydrogen ions and the increased reabsorption and generation of bicarbonate ions.

Examples of clinical syndromes associated with an increasing hydrogen ion concentration include poisoning by methanol, salicylate or the ingestion of acids. The latter extends to highly carbonated soft drinks, which have a very low pH. As discussed, renal failure will lead to failure to excrete hydrogen ions, whilst bicarbonate can be lost in severe diarrhoea.

Lactic acidosis is particularly notable as it may arise from anaerobic respiration, which can also follow shock from causes such as myocardial infarction and liver failure. It is also a recognized problem after treatment with certain hypoglycaemic drugs, such as metformin, and is commonly found in diabetic and alcoholic ketoacidosis. In the laboratory it may be defined as lactate >5 mmol/L. A common treatment of acidosis is intravenous infusion with a sodium bicarbonate solution. Blood samples from the opposing arm will be needed to monitor the resolution of the acidosis, and the fluid will help the kidney in washing out any unwanted ions.

The anion gap This is simply the difference between the major cations (sodium and potassium) and the major anions (chloride and bicarbonate); that is:

$$\text{Anion gap} = ([Na^+] + [K^+]) - ([Cl^-] + [HCO_3^-]) \\ \sim 10 - 20 \text{ mmol/L}$$

Hence, the sum of the cations should exceed the sum of the anions, the balance being unmeasured (and possibly unknown) anions. Measurement of the anion gap is useful in certain cases of metabolic acidosis, such as:

1. When a raised anion gap is present, suggesting hyperuricaemia, ketoacidosis, increased lactate or certain drugs (such as aspirin or methanol).
2. Or when a normal anion gap is present, as is present when bicarbonate ions are lost by renal or gastrointestinal disease. In such cases, the lost bicarbonate may be replaced by increased chloride ions, so there is no net anion change.

Respiratory acidosis

This condition, characterized by a raised P_{CO_2} (hypercapnia) is caused by an increase (probably due to retention) of carbon dioxide. However, it may also be related to the increased generation of carbon dioxide in muscle during exercise, although this is likely to be short-lived. If carbon dioxide remains high, or its concentration rises in the plasma, then it continues to influence the equilibrium of bicarbonate:

$$H_2O + CO_2 \leftrightarrow H_2CO_3 \leftrightarrow H^+ + HCO_3^-$$

Thus, the equilibrium of the equation will move to the right, and concentration of both bicarbonate and hydrogen ions will rise, the latter driving a reduction in the pH. The renal compensation solution to this is to excrete the excess hydrogen ions, which can be effected by the ammonium and phosphate pathways (Figure 13.4). In addition, bicarbonate can be reclaimed from the filtrate and fresh bicarbonate can be generated. However, these responses are relatively slow.

The primary reason for respiratory acidosis is hypoventilation – reducing the ventilation rate by 75% of its normal rate lowers the pH by 0.4, bringing it down, for example, from 7.4 to 7.0. Causes include drugs (e.g. narcotics and anaesthetics – which suppress the respiratory centre), nerve problems (motor neurone disease, spinal cord lesions, poliomyelitis) or muscle problems (intercostal muscles, diaphragm) and trauma, perhaps choking. The ventilation rate is part-governed by the respiration centre in the medulla, so that damage to this tissue, perhaps by stroke or trauma, may also lead to a reduction in breathing, and so acidosis. Clear respiratory disease such as bronchitis, emphysema, pulmonary oedema, asthma, fibrosis and chronic obstructive airways disease all have the potential to lead to a chronic acidosis.

13.4 Alkalosis (pH >7.5)

This may be defined as a decrease in the concentration of hydrogen ions in the plasma, and so a rise in pH. Its effects reflect those of acidosis in numerous physiological systems, and also has two aetiologies: metabolic alkalosis and respiratory alkalosis.

Metabolic alkalosis

This is characterized by raised bicarbonate and a low hydrogen ion concentration, and is most often due to loss of hydrogen ions from the gastrointestinal tract by vomiting and diarrhoea. If so, there will be clear evidence, and lost fluids may need to be replaced by intravenous infusion, which also gives the opportunity to infuse other corrective therapies, such as potassium to reverse the likely hypokalaemia (see Chapter 12). There may also be a compensatory rise in P_{CO_2}, but as lung function is normal, excess carbon dioxide can be expelled easily.

Certain drugs (such as thiazide diuretics) may cause excessive hydrogen ions to be lost in urine. Here, loss of sodium and water induces high levels of bicarbonate. Because ventilation is generally normal, P_{CO_2} levels can be unchanged (or modestly raised), and so bicarbonate rises. Another cause is excess ingestion of alkali solutions (such as sodium bicarbonate solution by athletes).

The respiratory compensatory response is to decrease the rate of breathing, so retaining carbon dioxide, although this is generally minimal since any rise in P_{CO_2} will stimulate breathing. The renal compensation is to retain hydrogen ions.

Respiratory alkalosis

This is often a consequence of hyperventilation (perhaps hysterical), with the rapid excretion of carbon dioxide, so that P_{CO_2} falls, leading to hypocapnia. The bicarbonate is generally normal but may fall a little. This may be voluntary, mechanical or by stimulation of the brain stem respiratory centre (by pain, drugs, fever, hypoxia due to high altitude, anaemia and/or pulmonary disease and oedema giving poor perfusion).

The renal compensation is to reduce the reabsorption of bicarbonate from the urinary filtrate, so that pH falls.

Major aspects of acidosis and alkalosis are summarized in Table 13.2.

13.5 Mixed acid–base conditions

These occur when different aspects of two or more of the four primary conditions are present simultaneously.

Given the complexity of acid/alkaline biochemistry, this is hardly surprising, and is relatively common. For example, in salicylate overdosing there may be stimulation of the respiratory centre in the medulla oblongata, and so a reduced P_{CO_2}, indicating a respiratory alkalosis. However, there is also often a raised hydrogen ion concentration, suggesting a metabolic acidosis. In such cases, the measurement of the anion gap will be helpful and is likely to be raised. Nevertheless, also very useful, and likely to be indispensable, would be the direct measurement of serum salicylate.

Use of diuretics often produces a confused picture, especially when compounded by an organic disease. A patient with chronic obstructive pulmonary disease may be expected to have a respiratory acidosis, but if taking a thiazide that depletes the blood of potassium, there may be indications of metabolic alkalosis.

13.6 Clinical interpretation

Once an acid–alkali abnormality is suspected, inevitably on clinical grounds, such as the patient hyperventilating (tachypnea, perhaps over 20 breaths/min) or hypoventilating (bradypnea, maybe less than 10 breaths/min), arterial and venous blood samples are obtained. The latter is likely to be sent for urea and electrolytes (U&Es), possibly with the extra request for chloride and bicarbonate (so that the anion gap can be calculated) and possibly also lactate.

The first step is to consider the hydrogen ion concentration, which defines acidosis or alkalosis. Next, look at the P_{CO_2}, and then, if necessary, the concentration of bicarbonate and PO_2. Interpretation can then follow.

Table 13.2 A summary of acidosis and alkalosis.

Type	Major chemistry	Potential causes
Acidosis: pH <7.3		
Metabolic	Increased H^+ that exceeds the body's buffering system, failure to excrete H^+, and/or loss of HCO_3^-	Ketoacidosis and lactate acidosis, acute and chronic renal failure, diarrhoea
Respiratory	Increased P_{CO_2} (hypercapnia)	Disease/depression of the central nervous system, pulmonary disease
Alkalosis: pH >7.45		
Metabolic	H^+ loss	Vomiting, loss by nephrons in renal disease, excess bicarbonate ingestion
Respiratory	Reduced P_{CO_2} (hypocapnia)	Hyperventilation

1. *Increased hydrogen ion concentration, and therefore acidosis.*
   i. Decreased P_{CO_2}, and therefore metabolic acidosis. Bicarbonate is also generally reduced.
   ii. Normal P_{CO_2}, and therefore uncompensated metabolic acidosis. This is likely to be because this is a simultaneous respiratory acidosis with retention of carbon dioxide. Plasma bicarbonate will be reduced.
   iii. If there is definitely a metabolic acidosis, but the picture is unclear, assessment of the anion gap may be helpful.
   iv. Increased P_{CO_2}, and therefore respiratory acidosis. If this is simple, bicarbonate will be increased. However, if acute, there can be a raised hydrogen ion concentration, P_{CO_2} and (slightly) bicarbonate as the renal responses have yet to develop. In a chronic setting, a markedly raised bicarbonate may be present and due to renal retention.
2. *Normal hydrogen ion concentration.* This does not always mean there are no metabolic problems, as there may be two concurrent competing problems. For example:
   i. Decreased P_{CO_2}, possibly because of a mixed respiratory alkalosis and metabolic acidosis. Bicarbonate is generally reduced.
   ii. Normal P_{CO_2}, and therefore no acid/alkali disturbance.
   iii. Increased P_{CO_2}, and therefore either a fully compensated respiratory acidosis, or a mixed respiratory acidosis and a concurrent metabolic alkalosis. Both lead to a raised concentration of bicarbonate.
3. *Decreased hydrogen ion concentration, and therefore alkalosis.*
   i. Decreased P_{CO_2}, and therefore respiratory alkalosis. If this is uncomplicated, bicarbonate will also be decreased.
   ii. Normal P_{CO_2}, and therefore uncompensated metabolic alkalosis. Plasma bicarbonate will be increased.
   iii. Increased P_{CO_2}, and therefore another complex disturbance such as metabolic alkalosis with some respiration changes (i.e. hypoventilation) in compensation. Another explanation may be combined metabolic alkalosis and respiratory acidosis. Plasma bicarbonate will also be increased.

Management

First, the aetiology must be defined and understood, which will indicate the underlying cause(s) and therefore point to possible solutions. Where this is not immediately possible, then neutralization of the pH problem should be undertaken. This will be urgent if the hydrogen ion concentration exceeds 100 nmol/L (pH <7), as is likely to be present in metabolic acidosis. The simplest resolution of the acidosis is a slow infusion of bicarbonate to buffer the hydrogen ions.

In respiratory acidosis the objective is to improve alveolar ventilation and so lower the P_{CO_2}, possibly by mechanical means, or by giving oxygen. A saline drip (possibly with potassium supplementation) may be useful to stimulate renal perfusion. A common treatment for the hyperventilation of respiratory alkalosis is to get the subject to breathe into a paper bag. This will generally force more CO_2 in inspired air, and thus into the blood.

The response of the body to intravenous infusion should be checked with blood samples taken from the opposite arm.

13.7 Case studies

Case study 13

A woman in her mid 20s is brought to the accident and emergency unit at midnight unconscious having collapsed at a night club. Her friends do not consider her to be drunk but say she may well have hit her head as she fell. The subject is poorly responsive, pale and clammy, and with a breath rate of about 20–25/min. Blood pressures are 110/70, pulse rate 90/min. Serum biochemistry and blood gases are ordered.

	Result (unit)	Reference range
Serum analytes		
Na	150 mmol/L	135–148
K	3.7 mmol/L	3.3–5.0
Urea	8.2 mmol/L	3.0–6.7
Creatinine	85 μmol/L	71–133
Glucose	5.0 mmol/L	3.5–5.5
Bicarbonate	22 mmol/L	24–29
Chloride	99 mmol/L	95–108
Blood gases		
Hydrogen ions (H^+)	29 nmol/L	35–45
pH	7.55	7.35–7.45
Bicarbonate	25 mmol/L	24–29
P_{CO_2}	3.8 kPa	4.5–6.1
P_{O_2}	12.9 kPa	12–14.6

Interpretation The abnormal serum results are slight hypernatraemia and uraemia, with serum bicarbonate at the bottom end of the reference range. The blood pressure is not too bad, but there is somewhat of an increased heart rate (tachycardia). The blood gas result is consistent with respiratory alkalosis (raised pH, low H^+ and low P_{CO_2}, lowish bicarbonate), which fits in with the hyperventilation (about twice the normal rate). Note the report of possible head injury that may have involved the brain stem, where the respiratory centre is located.

A check of the woman's handbag found an inhaler, implying a mild lung problem, and antibiotics, implying an unknown infection. Normal glucose rules out diabetic ketoacidosis causing the collapse. Electrolyte changes may be secondary to a pre-renal problem, possibly dehydration. It would also be prudent to consider the possibility of drug abuse causing a respiratory stimulation. However, a call for screening for drugs of abuse maybe unnecessary. The situation does not seem to be life threatening – the respiratory rate is not extremely high – so that rest and observation may be all that are required. Ideally, this would be in a quite side room – a formal admission would not be necessary. A saline drip may help the pre-renal uraemia and hypernatraemia.

Case study 14

A 75-year-old man is delivered to accident and emergency in a collapsed state; concerned neighbours had called 999. He is confused and uncooperative, and refuses to have blood pressure taken, or to have an electrocardiogram, but curiously consents to blood tests. His respiratory rate is 15 breaths/min.

	Result (unit)	Reference range
Serum analytes		
Na	140 mmol/L	135–148
K	4.0 mmol/L	3.3–5.0
Urea	4.3 mmol/L	3.0–6.7
Creatinine	92 μmol/L	71–133
Glucose	4.5 mmol/L	3.5–5.5
Bicarbonate	15 mmol/L	24–29
Chloride	100 mmol/L	95–108
Blood gases		
Hydrogen ions (H^+)	54 nmol/L	35–45
pH	7.15	7.35–7.45
Bicarbonate	17 mmol/L	24–29
P_{CO_2}	3.2 kPa	4.5–6.1
P_{O_2}	10.5 kPa	12–14.6

Interpretation   The serum biochemistry is normal, except for a low bicarbonate. Blood gases show a high hydrogen ion concentration and low pH, with low bicarbonate and P_{CO_2}. This is consistent with a diagnosis of metabolic acidosis. Without a blood pressure measurement, progress is difficult, and consideration of hypotension and/or dehydration will have to be made empirically, although standard U&Es are normal. The acidosis may be treatable with a bicarbonate drip.

The next step, having addressed the immediate problem, is to consider the cause. This type of acidosis may be due to ingestion of acids, increased generation of hydrogen ions and decreased excretion of hydrogen ions. The latter may be present in renal failure, but U&Es give absolutely no indication of this. Normal glucose almost completely rules out diabetic ketoacidosis, but there may be alcoholic or lactic acidosis, which can be confirmed or refuted by a blood test.

An additional investigation would be to determine the anion gap, being the sum of the cations minus the sum of the anions, and should be between 10 and 20 mmol/L. From the data given above, sodium plus potassium is 144 mmol/L, whereas chloride plus bicarbonate is 115 mmol/L, leaving an anion gap of 29 mmol/L, markedly greater than the 10–20 mmol/L that is expected. Therefore, we suspect that a missing anion is making up the difference. Candidates for this include salicylate, sulphates and phosphates, which can also be tested for in the laboratory. Other possibilities include methanol and glycols.

Summary

- Arterial blood must be analysed as soon as possible, but if a delay is inevitable, the syringe should be stored on ice.
- Acidosis (low pH) can be respiratory (such as following lung disease as in chronic obstructive airways disease) or metabolic (due to overproduction of and/or failure to excrete hydrogen ions).
- The anion gap may be useful in some cases of metabolic acidosis.
- Alkalosis can be respiratory (e.g. due to hyperventilation) or metabolic (such as loss of acidic gastric secretions).
- The body responds to acidosis and alkalosis by compensations, generally the reverse of the cause. Hence, a respiratory problem will be addressed with a metabolic response, and vice versa. Respiratory compensation is rapid (minutes or hours), whereas renal compensation may take 12–24 h to be effective.

Further reading

Koeppen BM. The kidney and acid–base regulation. Adv Physiol Educ. 2009;33:275–281.

Kraut JA, Madias NE. Serum anion gap: its uses and limitations in clinical medicine. Clin J Am Soc Nephrol. 2007;2:162–167.

Story DA. Bench-to-bedside review: a brief history of clinical acid–base. Crit Care. 2004;8:253–258.

Williams AJ. ABC of oxygen: assessing and interpreting arterial blood gases and acid–base balance. BMJ. 1998;317:1213–1216.

Web site

Interesting examples of acid–base problems: http://www.anaesthesiamcq.com/AcidBaseBook/ab9_6.php.

14 Glucose, Lipids and Atherosclerosis

Learning objectives

After studying this chapter, you should be able to:

- explain the importance of glucose and its role in metabolism;
- appreciate the consequences of low concentrations (hypoglycaemia) and high concentrations (hyperglycaemia, leading to diabetes);
- describe the biochemistry of lipids;
- explain the consequences of abnormal lipid concentrations (i.e. dyslipidaemia);
- understand how these risk factors contribute to the pathophysiology of atherosclerosis;
- describe the clinical consequences of atherosclerosis;
- outline the role of the laboratory in heart disease.

This chapter outlines the importance of two metabolites of great relevance to human disease: glucose and lipids. Increased concentrations of these molecules lead to major metabolic diseases: diabetes (resulting from hyperglycaemia) and dyslipidaemia (from abnormalities in cholesterol and triacylglycerols). Whilst the importance of high concentrations of these molecules is apparent, low concentrations can also lead to life-threatening conditions, the most crucial of which is low glucose (hypoglycaemia).

We will also discover how these diseases are related to the pathophysiology of cardiovascular disease (CVD), the major cause of mortality in the developed world. The leading cause of CVD is atherosclerosis, a disease process that attacks the inner lining of the blood vessels (the endothelium) causing high blood pressure and thrombosis. Although this process is also linked to two other risk factors – hypertension and smoking – neither of them have simple, direct and accessible blood tests that are useful in diagnosis and management.

The consequences of atherosclerosis and its risk factors are damage to the cardiovascular system, and this manifests mostly as critical limb ischaemia, stroke and heart attack. There are no reliable blood tests to demonstrate a recent stroke or the presence of critical limb ischaemia, although there are several clinical tests (such as imaging) to help with these conditions. However, the demonstration of a heart attack is an area where the laboratory can provide essential information. Essentially, the basis of this condition is damage to the muscle cells of the heart (cardiomyocytes), which release a number of analytes roughly in proportion to the severity of the heart attack.

This chapter will first address the major risk factors of glucose (in Section 14.1) and lipids (in Section 14.2). We will then be in a position to explore (in Section 14.3) the role of the laboratory in CVD in general, and heart disease specifically. The chapter will conclude with case studies (Section 14.4). The major blood tests we will be looking in this chapter are summarized in Table 14.1, although others will be described as and when relevant.

14.1 Glucose

The body's major source of energy is glucose. This carbohydrate is the principal substrate for the Embden–Meyerhof glycolytic pathway that feeds the Krebs cycle in the process of respiration, generating adenosine triphosphate (ATP). The most important organ in the body is the heart, and the constituent cardiomyocytes need a constant source of energy, such as is provided by glucose, in order to keep beating. Almost every cell in the body generates its energy in this way, so that a shortage of glucose may lead to a widespread shortage of energy that affects all organs.

Blood Science: Principles and Pathology, First Edition. Andrew Blann and Nessar Ahmed.
© 2014 John Wiley & Sons, Ltd. Published 2014 by John Wiley & Sons, Ltd.

Table 14.1 Major blood tests in the pathophysiology of atherosclerosis.

Analyte	Reference range
Glucose	3.5–5.5 mmol/L
HbA$_{1c}$ (glycated haemoglobin)	3.8–6.2%
	20–42 mmol/mol
Total cholesterol	<5.0 mmol/L
High-density lipoprotein cholesterol	>1.2 mmol/L
Low-density lipoprotein cholesterol	<3.0 mmol/L
Triacylglycerols	<1.7 mmol/L
Creatine kinase	<150 U/mL
Creatine kinase MB	<15 U/mL
Troponin[a]	Low risk <0.1 ng/mL
	Moderate risk 0.1–0.6 ng/mL
	High risk >0.6 ng/mL

[a]Reference range depends on method and the type of molecule (troponin T, troponin I). Changes in troponin are important in the assessment of myocardial infarction after onset of acute chest pain.

In this section, we will first briefly review the metabolism of glucose, and how abnormalities in this metabolism cause disease. We will then look at the clinical conditions that call for the services of the laboratory, and how they are defined. In increasing order of blood glucose and risk of CVD, the relevant conditions are:

- hypoglycaemia
- impaired fasting glucose (IFG – also known as impaired fasting glycaemia)
- impaired glucose tolerance (IGT)
- the metabolic syndrome
- diabetes (of which there are two forms – type 1 and type 2).

In this chapter we take diabetes to be diabetes mellitus, as opposed to diabetes insipidus, which can be either central diabetes insipidus, caused by insufficient production of anti-diuretic hormone (ADH) in the hypothalamus or nephrogenic diabetes insipidus due to lack of sensitivity to ADH in the kidney (as explained in Chapter 12), and in which there is (generally) no abnormality in glucose regulation. Both types of diabetes are linked by the increased production of urine, hence the

Greek term diabetes, which means 'through a siphon', but for completely different reasons.

The metabolism of glucose

Glucose may be taken in the diet, but is more likely to be the product of the breakdown of complex carbohydrates by digestive enzymes. Glucose can also be liberated from its storage form, glycogen, and concentrations in the blood are generally maintained within a relatively tight margin of 3.5–5.5 mmol/L. The key regulator of this process, the hormone insulin, acts on cells to promote the movement of glucose from the blood and into the cells. A second hormone, glucagon, effectively counters the effects of insulin, both hormones acting on the liver, muscle and adipose tissues, although there are other hormones involved in glucose metabolism.

Insulin   The pancreas has two roles: an exocrine function in the secretion of digestive enzymes and an endocrine action in the secretion of hormones. The latter are derived from cells of the islets of Langerhans, of which the dominant form (80%) are beta cells, which secrete insulin. The remainder are alpha cells, secreting glucagon, and the delta cells, secreting somatostatin.

The secretion of insulin is not steady state, but is stimulated by concentrations of glucose. This occurs in a number of steps. First, there is uptake of glucose by a specific beta cell membrane-bound transporter called GLUT2. Once inside the cell, the glucose is then phosphorylated, and downstream metabolic pathways result in the release of insulin granules and their export into the blood. Other known insulin secretagogues include leucine, arginine, certain fatty acids, ketones, sulphonylureas and other hormones, called incretins.

Insulin is produced by the cell's protein synthesis mechanisms (the ribosomes and endoplasmic reticulum) initially as pre-pro-insulin, a single chain of 86 amino acids, the pre- section then being removed to leave pro-insulin. This in turn is processed to give the mature insulin molecule of 51 amino acids, with a relative molecular mass of 5.8 kDa. The remaining pro- section (the connecting or c-peptide), is also secreted into the blood alongside the insulin. Although there must be one c-peptide molecule for each insulin molecule, in the blood, the two have different half-lives, leading to differences in their molar ratio.

Once in the blood, insulin is free to bind to specific insulin receptors on the surface of its target cells, typically the hepatocyte. This leads to the phosphorylation of the cytoplasmic tail of the receptor, which initiates cytoplasmic signalling that leads to the increased expression and translocation of GLUT4 receptor molecules to the plasma membrane that bring about the import of glucose into the cell. Binding of insulin to its receptor promotes several metabolic processes, which include:

- the conversion of intracellular glucose into stores of glycogen (i.e. glycogenesis);
- the inhibition of glycogenolysis (the process where glycogen stores are broken down into glucose – the reverse of glycogenesis);
- the inhibition of the formation of glucose from non-carbohydrate sources such as pyruvate and glycerol, primarily in the liver (gluconeogenesis);
- the inhibition of the breakdown of lipids to non-esterified fatty acids (i.e. lipolysis);
- the synthesis of fatty acids (lipogenesis, mostly within adipocytes);
- inhibition of the release of glucagon.

There are also many other functions of insulin unrelated to glucose metabolism, such as increasing the movement of potassium into the cell, increasing the secretion of acid into the stomach, and promoting the reabsorption of sodium from the urinary filtrate by tubules of the nephron (as is fully explained in Chapter 12). In general, insulin is a growth-promoting hormone which stimulates anabolic processes.

Glucagon The second major glucose hormone is glucagon, a 29 amino acid peptide with a relative molecular mass of 3.5 kDa. Released from the alpha cells of the islets of Langerhans, this hormone effectively acts in the opposite manner to insulin, and release is promoted by low concentrations of glucose, and possibly by concentrations of insulin. High concentrations of glucose, arginine, non-esterified fatty acids and ketones all inhibit the release of the hormone, whereas a high-protein meal, adrenaline, alanine and the action of the autonomic nervous system all promote glucagon release.

Glucagon acts on target cells, such as hepatocytes, via a specific glucagon receptor, and when occupied the receptor begins what are series of messages not dissimilar to that of the insulin receptor. Broadly speaking, the effects of glucagon are the reverse of those of insulin; for example:

- the promotion of glycogenolysis;
- the promotion of gluconeogenesis;
- the inhibition of glucose uptake;

Table 14.2 The effects of insulin and glucagon.

Effect	Insulin	Glucagon
Glycogenesis	Promotes	Inhibits
Glycogenolysis	Inhibits	Promotes
Lipolysis	Inhibits	Promotes
Lipogenesis	Promotes	Inhibits
Broad net result	Reduces blood glucose, increases blood lipids	Increases blood glucose, reduces blood lipids

- the promotion of lipolysis;
- the production of ketones (ketogenesis).

Therefore, in this setting, the functions of glucagon are to increase intracellular and plasma glucose, counteracting the actions of insulin. Other counter regulatory hormones, such as cortisol, adrenaline and growth hormone, are also stimulated by low blood sugar; they counter the actions of insulin, and so raise the concentration of blood glucose. The opposing functions of insulin and glucagon are summarized in Table 14.2. However, there are some functions of glucagon that are not mirrored by insulin, and vice versa.

Somatostatin and incretin hormones Somatostatin is effectively inhibitory, in the digestive system suppressing levels of other hormones, including gastrin (the hormone that stimulates gastric acid secretion), and decreases the rate of emptying of the stomach. Somatostatin also inhibits exocrine functions of the pancreas. Incretin hormones include glucagon-like peptide 1 (GLP-1) and glucose-dependent insulinotrophic peptide. The latter is released from cells of the duodenum and jejunum in response to food, and acts on islet beta cells to stimulate insulin secretion, but release of insulin is inhibited by somatostatin. Other intestinal cells (in the ileum and colon) secrete a different secretagogue, GLP-1, which also stimulates insulin secretion, but in addition suppresses glucagon secretion. However, the receptor for GLP-1 is also found on other tissues, including the brain (where it may promote satiety – being satisfied with a meal), the heart and the kidney. By acting on the stomach, GLP-1 slows gastric emptying, so delaying the delivery of food to the small intestines.

Insulin resistance and sensitivity A useful model to explain the regulation of glucose by insulin, the functioning of the beta cells and the development of hyperglycaemia, and so diabetes, is to consider the qualitative aspects of sensitivity and resistance. Insulin sensitivity is simply about how easy it is to get glucose into the cell. At times of low plasma glucose, the cell needs all the glucose it can get, and so is very sensitive to insulin in encouraging the import of the carbohydrate. It follows that key features of glucose homeostasis are:

1. the sensitivity of the beta cell to glucose;
2. the extent to which the beta cells can generate and release insulin;
3. the response of the target cell (hepatocyte, myocyte) to the insulin by the latter occupying the insulin receptor; and
4. how efficiently the target cell can import the glucose.

Although these four states exist independently, they interact. Insulin resistance seems likely to result from years of hyperglycaemia, where the beta cells are unable to generate sufficient insulin to correctly regulate sugar concentrations. However, it could be argued that some component of insulin sensitivity and resistance is due to abnormalities in glucagon metabolism. In parallel, target cells (principally hepatocytes, muscle cells and adipocytes) are increasingly swamped by high concentrations of glucose and fail to respond correctly to insulin binding to their insulin receptors. In addition, there is a reduced uptake of lipids and the mobilization of fatty acid stores, resulting in high concentrations of plasma lipids and triacylglycerols. This is why many patients with diabetes attend a lipid clinic, and vice versa.

These factors have been built into several models to assess insulin resistance and beta cell function:

• The euglycaemic clamp model notes changes in glucose, for several hours, after the infusion of glucose, but potassium must also be infused as insulin causes hypokalaemia.
• In the insulin suppression test, somatostatin (or an analogue) is first infused to suppress the body's natural insulin, and then both insulin and glucose are infused. Serial measurement of the two in blood taken from the noninfused arm provides information on insulin resistance and sensitivity.
• The homeostasis model assessment (HOMA, updated to HOMA2) uses paired fasting insulin and glucose concentrations. It has a major advantage over other methods because, using fasting blood only, it does not call for the infusion of glucose or insulin. The HOMA

technique can help determine those at greatest risk of the development of diabetes and is more suited to the study of populations rather than individuals.
• Although it may be disputed that the oral glucose tolerance test (OGTT) is a test of insulin sensitivity and beta cell function, it is nonetheless a very popular physiologically obvious test of the ability of the body to process a glucose load. We will look at this test in more detail shortly.

Impaired lipoprotein regulation

We have already noted that insulin and glucagon influence the formation and breakdown of lipids (lipogenesis and lipolysis respectively). Whilst this is physiological and therefore desirable, in the presence of abnormalities in the regulation of glucose, and abnormal glucohormones, this can lead to abnormalities in lipid metabolism, and therefore to disease. This we will return to in the section on diabetes that follows.

The laboratory and glycaemia

The laboratory offers two standard tests: glucose and glycated haemoglobin (HbA_{1c}). The OGTT (which we will examine in detail in a further section) measures glucose in two blood samples: one taken just before and one 2 h after a glucose drink. These tests are used to diagnose and monitor the management of glucose-related disease.

Glucose   Although the measurement of glucose is relatively straightforward, there are a number of factors we must address before we consider its chemistry.

1. Blood for glucose estimation must be taken into an anticoagulant such as fluoride oxalate. The latter is needed because white blood cells, whilst alive, respire and so consume glucose, so that concentrations of glucose fall with time. The fluoride effectively stops the electron transport chain in the mitochondria, and so the consumption of glucose.
2. Glucose concentrations differ in whole (venous) blood compared with plasma (where concentrations are 10–15% higher), as there will be some glucose within the red blood cells. This is therefore pertinent in the use of near-patient testing machines that generally measure capillary whole blood. It is important to bear in mind that point-of-care glucose meters may not be accurate at very low or very high blood glucose concentrations, and

the "gold standard" test is still to measure blood glucose using a laboratory analyser.

3. Glucose has a relatively short half-life, and concentrations are lowest early in the morning after an overnight fast. Therefore, levels mid morning will be higher due to the absorption of glucose from the breakfast. A random blood sugar (taken at any time), if high, may be a genuine pathological hyperglycaemia or the consequence of a large sugary drink and some chocolate bars. Consequently, many services accept only a fasting sample.

Estimations of glucose rely on the ability of the enzyme glucose oxidase (EC 1.1.3.4) to generate glucuronic acid and hydrogen peroxide. This can be assessed by a system that uses an ion-selective electrode, and is common in hand-held and near-patient testing machines. However, another method is to use a second enzyme, peroxidase, alongside 4-amino-phenazone and phenol, to generate a pink colour that can be measured in a spectrophotometer at 515 nm. This method is well suited to autoanalysers and so is used widely where there is a high demand for this measurement. In a modification of this assay, the fluorescent product resorufin is generated from the non-fluorescent product amplex red, which demands a spectrophotometer able to detect the appropriate wavelength (575–585 nm).

Glycated haemoglobin Glucose is sticky. It sticks to proteins, carbohydrates and fats, whether in the plasma or in the cell membranes. Chapter 3 explains why the membrane of the red blood cell is highly modified to allow the passage of oxygen in and out, and is therefore markedly different to all other cell membranes. Indeed, a red cell membrane must be relatively thin. This super-adaptability brings other aspects, such as that glucose passes into the cell easier than it does a nucleated cell. Once inside the cell, the glucose binds irreversibly to all forms of haemoglobin (HbA, HBA_2, HbC and HbS, if present), hence glycated haemoglobin. HbA_{1c} measures the degree to which HbA is glycated.

The presence of a large mass of glucose on the haemoglobin molecule means that the glycated haemoglobin has a different chromatographic mobility than that which has only a trace of glucose. The position that glycated haemoglobin migrates to is easily seen on standard electrophoresis and in high-performance liquid chromatography, although the latter is the preferred option as it lends itself more easily to automation and has better sensitivity and specificity (Figure 14.1).

It is entirely normal to have some of one's haemoglobin glycated. However, the degree of glycation is proportional to the average concentration of glucose in the blood. Therefore, HbA_{1c} is a surrogate marker of blood glucose, since the binding of glucose to the haemoglobin is irreversible. Hence, the HbA_{1c} result is effectively a record of the amount of glucose in blood in the preceding 10–12 weeks, as this is the lifetime of the red blood cell. This is important for several reasons:

- The patient can quite easily fast for 48 h before their clinic visit, so that a fasting blood glucose will be low. However, such a fast will not reduce their HbA_{1c} concentration.
- A blood glucose concentration is most reliable when fasting. However, as HbA_{1c} gives a long-term view of hyperglycaemia, it can be used as a general screening tool, and is useful even if the patient admits to a recent large carbohydrate-rich meal.
- Because HbA_{1c} has a long half-life, measuring it on two consecutive weeks will see little change, even if the subject has been starved of glucose, and/or has markedly changed their glucose-lowering medications. Indeed, many laboratory services will decline an HbA_{1c} request within a month or 6 weeks of a previous measurement.

Historically, the unit of HbA_{1c} was per cent. However, the unit of mmol/mol (HbA_{1c}/haemoglobin) is now used as it takes into account the concentration of haemoglobin. This was called for not only because of a need for international standardization, but also by the observation that, in some people with a low haemoglobin (and so possibly anaemic), a high degree of glycation may be overlooked. The new unit therefore demands the measurement of haemoglobin. However, in the absence of a haemoglobin result, the new units can be obtained from the percentage results according to the equation

$$HbA_{1c} \ (mmol/mol) = [HbA_{1c}(\%) - 2.15] \times 10.929$$

Other laboratory tests in glucose pathology A consequence of severe hyperglycaemia, as in diabetes due to insulin deficiency, is the presence of ketones in the blood, and also an acidosis, so that blood gases and concentrations of bicarbonate may be requested. An allied metabolism influences concentrations of potassium, so this too may be called for. Many other proteins apart from

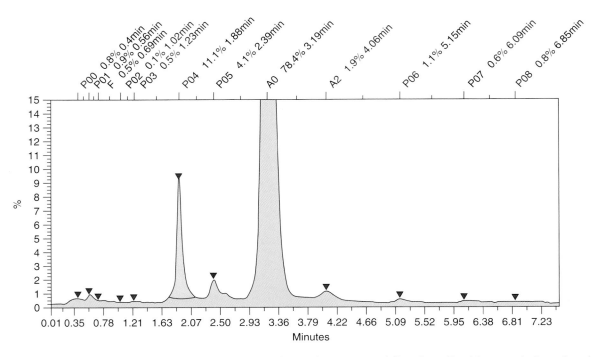

Figure 14.1 The HPLC trace of a sample of blood. The vertical axis is the percentage of all results attributable to a particular peak, and the horizontal axis is the duration of the analysis. The peak P04 at 1.88 min makes up 11.1%, and is due to glycated haemoglobin, strongly suggestive of a diagnosis of diabetes mellitus. (Courtesy of Dr S. Marwah, City Hospital, Birmingham).

haemoglobin can be glycated. Fructosamine, the result of the glycation of largely albumin, may be used to assess hyperglycaemia over the past 2–3 weeks. As diabetes progresses, it attacks certain tissues and organs, notably the kidney, causing diabetic nephropathy. If so, the assessment of renal function by urea and electrolytes (U&Es) and by the ratio of albumin to creatinine in the urine may be required.

Having developed an understanding of the physiology and homeostasis of glucose metabolism, we now move to situations where the loss of this homeostasis leads to pathology.

Hypoglycaemia

We all need glucose as a source of energy, and low concentrations may therefore lead to loss of energy. Should this be the heart, then it may result in arrhythmia (as detected by an abnormal electrocardiogram (ECG)) and cardiac arrest. Indeed, hypoglycaemia accounts for 3–4% of deaths in young diabetics. Loss of glucose to the brain may lead to personality changes and loss of consciousness, which may resemble having drunk too

much alcohol. Other symptoms include tremor, sweating, anxiety and/or tingling around the tongue, lips and limbs. Most diabetics learn to recognize the onset of hypoglycaemic symptoms and can usually self-treat at home by eating and/or drinking glucose. It is likely that hypoglycaemia in the diabetic population is underdiagnosed.

The laboratory The major national body of healthcare professionals is Diabetes UK, which in turn has links with European and international bodies such as the World Health Organization (WHO) and the International Diabetes Federation (IDF). Between them, they publish guidelines on what may be described as 'normal', and therefore 'abnormal'. Notably, the concentrations of glucose that define different conditions are not fixed, often vary between different authorities and over the decades have become more stringent.

Many reference ranges recommend a fasting glucose between 3.5 and 5.5 mmol/L. Of course, this does not mean that a concentration of 3.4 mmol/L defines hypoglycaemia any more than 5.6 mmol/L defines hyperglycaemia. Nevertheless, a commonly used point to define hypoglycaemia in the UK is a fasting blood glucose concentration of 2.5 mmol/L or below. In some cases it may be advisable to measure concentrations of c-peptide.

Pathophysiology   Hypoglycaemia may arise from several alternative pathways, some of which may be present at the same time:

- As we will see, a common treatment for the hyperglycaemia of diabetes is to inject insulin, which acts to move glucose out of the blood and into cells. However, too much insulin can move too much glucose out of the blood, and so cause hypoglycaemia.
- Similarly, the excessive use of other glucose-lowering drugs (such as sulphonylurea) may also lead to hypoglycaemia.
- In those insulin-dependent diabetics who take part in too much physical exercise (athletics, dancing), which uses up their glucose stores, a 'normal' injection of insulin may precipitate hypoglycaemia.
- A tumour of the beta cells (an insulinoma) secretes large amounts of insulin, or pro-insulin, which is cleaved in the plasma to yield insulin.
- A severely damaged liver (Chapter 17) may not be able to store glycogen and convert it into glucose.
- Hypoglycaemia may be present in premature or small-for-date neonates as their liver may be insufficiently mature to be able to store and metabolize glucose.

'Dad's gone all funny . . . '

A healthcare professional was called on the phone by her mother with the message that 'Your Dad's gone all funny . . . '. The man in question was attending hospital for an endoscopy to investigate abdominal discomfort, which required an overnight fast. Upon reaching the endoscopy suite the professional found her father partially dressed, singing, dancing and acting in a most uncharacteristic manner. He could not be pacified or argued with. Somebody recommended a blood glucose measurement, which returned a result of 1 mmol/L.

Diagnosed immediately with profound hypoglycaemia, he was treated with oral glucose and within 30 min he calmed down and his normal personality returned within the hour. Soon after, high concentrations of serum insulin were found, and he was diagnosed with an insulinoma.

In retrospect, the family realized he was always first up to breakfast, made his own lunch, was irritable prior to the evening meal and always had biscuits to take to bed. Clear signs of an unconscious adaptation to hypoglycaemia driven by the hyperinsulinaemia.

Treatment   Recovery from hypoglycaemia usually occurs within a few minutes following the intake of sugar. In serious cases, where the patient may be uncooperative or unconscious, intravenous glucose or intramuscular or subcutaneous glucagon may be given initially, which is usually sufficient to restore consciousness. Oral glucose can then be given. For the diagnosis of hypoglycaemia to be correct, the clinical picture must resolve when glucose is given to the patient. This observation of symptoms, observed at confirmed levels of hypoglycaemia, which resolve after the administration of glucose, is known as Whipple's triad.

Impaired fasting glucose

Normal fasting blood glucose generally has an upper limit of perhaps 5.5 mmol/L. The WHO defines those with a fasting blood glucose between 6.1 and 6.9 mmol/L as having IFG. However, these numbers (6.1 and 6.9) are arbitrary, with little pure science attached. Indeed, the American Diabetic Association defined IFG as being in the range 5.6–6.9 mmol/L.

However, in many cases IFG is a misdiagnosis as fasting may not have been for sufficiently long, or the subject knowingly or accidentally consumed some glucose prior to their blood test. For some, even milk in tea or coffee, or a peppermint sweet can be enough to increase blood glucose. Furthermore, even caffeine can influence carbohydrate metabolism. But does it matter?

Yes it does. IFG carries a risk of CVD and mortality above that of normal fasting glucose. Furthermore, it also brings a risk of conversion to diabetes, and so is often described as pre-diabetes. Fortunately, both are reversible upon 'treatment' of the mildly increased glucose. As IFG is often found in the overweight (body mass index [BMI] >25 kg/m^2) and in the obese (BMI >30 kg/m^2) this may be achieved by weight reduction, linked to increased exercise and diet control.

Impaired glucose tolerance

Diabetes is not simply about too much glucose in the blood, it is also about how that glucose is handled by the body, which itself demands a good insulin response. This can be determined by noting the response to a challenge of a short, sharp ingestion of glucose, and then in observing the consequences. This is called the OGTT.

The oral glucose tolerance test   The subject is brought to the laboratory in a fully fasted state, perhaps for as much as 14 h. No calories are allowed, only a modest amount of water. A blood sample is taken (time zero), and the subject then takes a drink containing 75 g of glucose within the next 5 min. Blood samples are then drawn during the morning, possibly at hourly intervals, and the rise in glucose is monitored as the sugar enters the blood.

Some of this blood will perfuse the insulin-secreting beta cells of the pancreas, which then begin to secrete insulin. This will therefore lead to a fall in blood glucose as the sugar enters various cells, and eventually the concentrations will fall, perhaps down to that when fasting. A bell-shaped plot can be drawn showing the rise and fall of the concentrations of glucose (Figure 14.2).

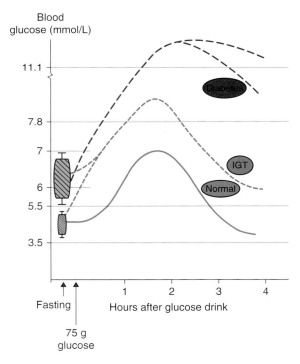

Figure 14.2   In the OGTT, blood glucose concentrations rise, and then fall as insulin mediates the passage of the sugar into cells. The continuous line represents a typical normal profile, the dotted line the profile of IGT and the dashed line the profile as expected in diabetes. The latter two conditions may also have raised fasting glucose (5.6–7.0 mmol/L). N.B. The exact cut-off points vary; refer to your own local service.

Interpretation   In any population there will be a variety in the response to a standard stimulus (in this case the 75 g of glucose): in some cases the peak in blood glucose may be 100 min, in others perhaps 150 min. Similarly, there will be some whose insulin response is very effective, the rising glucose is rapidly addressed and the peak glucose concentration is low. In others, whose insulin response is weak, the glucose is not metabolized effectively, so the peak concentration of glucose will be higher.

Although the physiology of this response is interesting, no aspect of human pathophysiology can be purely academic. Practitioners need the laboratory to give firm answers and directions as to diagnosis and treatment. Therefore, by consensus we take the concentration of glucose at 2 h to be a standard; and we take the peak concentration of glucose that defines a normal response to be less than 7.8 mmol/L.

It follows that a peak concentration of glucose above this level implies a slow or sluggish insulin response. We may say that the high concentration of glucose is in fact 'tolerated' by the body, whereas someone with a different physiology would not tolerate this degree of hyperglycaemia. As this high concentration is regarded as abnormal, we say a peak result of perhaps 8.5 mmol/L to define IGT.

Implications and treatment   Just as IFG carries a risk of CVD greater than normal glucose metabolism, then so IGT brings a greater risk of both morbidity and mortality than does IFG. Indeed, we can extend this continuum to actual diabetes, which brings greatest risk of premature CVD. Intensive lifestyle intervention and the drug metformin (to be described below) in people with IFG and IGT plus BMI >25 kg/m² both bring reduction in the conversion of these diseases to full blown diabetes.

The metabolic syndrome

This disease has been 'invented' to fit situations where a series of abnormalities seem to cluster together in certain individuals with common problems. The abnormalities themselves are:

• raised fasting plasma glucose (>5.6 mmol/L) or diabetes;
• general whole-body obesity (as defined by BMI >30 kg/m²) and/or central obesity (waist-to-hip ratio >0.9 in males or >0.85 in females);
• raised blood pressure (systolic blood pressure >130 mmHg, or diastolic blood pressure >85 mmHg, or on anti-hypertensive treatment;

- raised triacylglycerols (>1.7 mmol/L) or specific treatment;
- reduced high-density lipoprotein cholesterol (<1.03 mmol/L in males, <1.29 mmol/L in females), or specific treatment;
- microalbuminuria (such as albumin/creatinine ratio >3.0 mg/mmol).

However, the exact definition varies according to the particular authority (such as the definition of IFG as >5.6 or 6.1 mmol/L). For example, the WHO requires an established disease of hyperglycaemia and any two of the above. The International Diabetes Federation puts obesity and an abnormal waist-to-hip ratio first, and only then considers the other factors. Another authority considers not the waist-to-hip ratio, but the absolute waist circumference: ≥102-cm in men and ≥88-cm in women. Even the use of the expression metabolic syndrome has been debated, as some prefer to use 'syndrome X' or perhaps 'insulin resistance syndrome'. Other linked problems include hyperuricaemia, and therefore gout, polycystic ovarian syndrome and hyperpigmentation of the skin.

Implications The metabolic syndrome is slightly to the side of the spectrum of normoglycaemia to IFG, then to impaired glucose tolerance, and finally to diabetes. Nevertheless, the metabolic syndrome also carries a risk of CVD. It has been estimated that 25% of the population of the USA have the metabolic syndrome, which, if correct, implies a considerable future impact on the healthcare system.

Diabetes

This is the most severe form of hyperglycaemia, and carries the greatest risk of morbidity and mortality. There are two separate pathophysiological processes based on the absence of insulin (type 1 diabetes) and resistance to insulin (type 2), although both lead to hyperglycaemia and roughly the same cluster of symptoms.

Presentation of diabetes Perhaps the best known symptom is polyuria – generation of a great deal of urine. This is because as the glucose arrives at the kidney and is excreted, the laws of osmosis demand that water also be excreted. This may lead to dehydration, and almost certainly accounts for increased thirst, and so drinking water (polydipsia) in an attempt to replace the water lost in urine. There may also be weight loss

(regardless of food intake) and fatigue (possibly because body cells are deprived of the glucose they need as an energy source).

The overwhelming causes of diabetes (autoimmunity and obesity) we are about to discuss. However, diabetes may also be secondary to other disease such as pancreatitis and cancer of the pancreas, malnutrition, Cushing's disease (with high concentrations of cortisol), acromegaly (with high concentrations of growth hormone) and Conn's syndrome (with high concentrations of aldosterone). Note that all the latter diseases are selective overactivity of the endocrine system. Drugs known to cause diabetes include thiazide diuretics, beta blockers and steroids.

Epidemiology and economics Figures for the presence of diabetes vary considerably. For example, according to the WHO there were 1.76 million diagnosed diabetics in the UK in 2000, estimated to rise to 2.67 million by 2030. Many feel these data are conservative, as the charity Diabetes UK estimates that there may be a further 1 million undiagnosed diabetics. The UK government's own figures are of 2.2 million diabetics in England in 2010, whilst in March 2013 Diabetes UK estimated 4.5% of the UK to be diabetic, which is in the region of 3 million people. The UK National Audit Office calculated that from 1998 to 2008 the incidence of type 2 diabetes rose by 54%. If one adds IGT and IFG, the figure for known and unknown hyperglycaemia may exceed 10% of the population.

Naturally, this places a considerable economic burden on the healthcare industry. In 2002 the estimated total annual cost of diabetes to the National Health Service (NHS) was £1.3 billion, with the total cost to the UK economy much higher. In 2004, Diabetes UK estimated that diabetes accounted for around 5% of all NHS spending. At Birmingham and Midlands Eye Hospital, diabetes accounts for 50% of attendances, reflecting the degree of eye disease (retinopathy) that diabetes causes. All patients will of course require laboratory monitoring, and there will be other tests such as for dyslipidaemia and renal function.

The definition of diabetes We have already reviewed the causes and consequences of hyperglycaemia, which are generally loss of insulin secretion and/or the resistance of the cell to insulin. In 2006 the WHO and International Diabetes Federation defined diabetes as a fasting glucose ≥7.0 mmol/L or 2-h OGTT glucose ≥11.1 mmol/L, and the clinical symptoms described above. In the UK the blood test must normally be confirmed on at least two different occasions before a definitive diagnosis is made. However, there are several different manifestations of diabetes:

- type 1 diabetes, where the beta cells no longer make insulin;
- type 2 diabetes, where insulin production may be normal or reduced, but the target cells fail to respond;
- the diabetes that appears during pregnancy; and
- rare and unusual variants.

Type 1 diabetes

Pathophysiology   What causes the failure of the beta cell to generate insulin? Several mechanisms are established, with the most common being autoimmune attack. In some cases it may follow an infection, often by a trigger virus such as coxsackie B or influenza. Other triggers include environmental agents such as toxins. The autoimmune aspect is notable because of an established link with certain human leukocyte antigens (HLAs), notably HLA DR3 and HLA DR4, which account for 95% of HLA associations. Furthermore, the concordance of type 1 diabetes in identical twins is 40%.

Each autoimmune disease is characterized by at least one autoantibody, and in type 1 diabetes autoantibodies may be directed towards the beta cells or towards insulin. Indeed, the pancreas of many type 1 diabetics is infiltrated with lymphocytes. Autoimmune disease is explained in Chapter 9. A second cause of type 1 diabetes is steroids, themselves often used to treat inflammatory disease such as colitis and polymyalgia rheumatica. The pathophysiology is effectively that the beta cells are simply destroyed, and there is no autoimmune aspect.

Clinical aspects   Presentation of autoimmune type 1 diabetes is often acute, and before the age of 40, but with a peak incidence at around 9–13 years of age. Treatment is by replacement therapy – that is, daily injection(s) of insulin; hence the descriptor of insulin-dependent diabetes mellitus. In the past, insulin was prepared from animal sources, so that antibodies could be generated to negate the hormone. Fortunately, human recombinant insulin is now available.

Type 2 diabetes

Pathophysiology   This is certainly the dominant form of the disease, with perhaps 4–5% of the world population

(171 million people in 2000, 285 million in 2009 and estimated to be 336 million in 2030) affected, a figure rising to 10% in Europe. The underlying cause of type 2 diabetes is that target cells (of the liver, of muscles, of the kidney, leukocytes, etc.) respond poorly to insulin, or even fail to respond at all, so are described as resistant to the hormone. The resistance is likely to be the result of long-term hyperglycaemia, which may have been present for years, perhaps decades. In contrast to type 1 diabetes, type 2 diabetes can arise at any age, although in the past it was described as 'maturity onset'. However, with the increase in childhood obesity (the major risk factor for type 2 diabetes), this age effect is disappearing.

Although there is no clear genetic aspect in type 2 diabetes, there may be a familial link, because if both parents have type 2 diabetes then the lifetime risk of their child developing type 2 diabetes is increased to about 60%. However, this link may also be due to the habit of a poor diet learned in childhood.

Mechanisms of beta cell dysfunction   In contrast to type 1 diabetes, in type 2 disease the loss of beta cell function is not due to autoimmune destruction. There is growing evidence that the key pathology is excessive and/ or premature apoptosis, for which there are a number of triggers, such as inflammatory cytokines, non-esterified fatty acids and chronically elevated glucose. It is also possible that incretins have a role in maintaining beta cell function, so that changes in GLP-1 (which is decreased in diabetes) may be important. Evidence in support of this hypothesis comes from the observation that infusion of GLP-1 improves glucose control, promotes weight loss and improves insulin sensitivity and insulin secretion. Furthermore, GLP-1 suppresses glucagon, which normally acts to increase blood glucose, so that low concentrations of GLP-1 in diabetes are an additional link to hyperglycaemia via this second glucohormone.

Diabetes in pregnancy

Perhaps 4–5% of women develop diabetes during pregnancy (hence gestational diabetes mellitus) most often in their second or third trimester. The dominant cause is the inability of the maternal beta cells to deliver enough insulin to the body to cope with the extra metabolic demands of her pregnancy, so that glucose concentrations are high. The precise cause is unclear but may be due to an abnormal response to raised oestrogen, placental lactogen and adiponectin in susceptible women.

Most cases of gestational diabetes occur late in pregnancy, and if the foetal pancreas produces enough of its own insulin, there may be few major pathological consequences for the infant. However, if maternal hyperglycaemia is high early in her pregnancy, her infant may suffer with low birthweight (microsomia), high birthweight (macrosomia), malformations such as cleft lip/ palate, and birth difficulties. Furthermore, infants of mothers who had diabetes during their pregnancy are themselves at risk of becoming obese and developing diabetes later in life.

The diabetes generally resolves in the post-partum period. However, gestational diabetes may have long-term consequences for the mother, whose risk of developing mature diabetes increases to some 30%, compared with 10% for women who have normal glucose during their pregnancy. South Asian and Afro-Caribbean women not only have a higher rate of type 2 diabetes than white European women, but are more likely to develop type 2 diabetes if they have had gestational diabetes.

However, pregnancy in the established diabetic woman requires her to be more stringent with her blood glucose concentrations, with more personal testing of blood sugar. Oral agents may need to be replaced by insulin injections. This may also require dietary changes and additional exercise, such as swimming. Fortunately, these are well known, and support is available from antenatal clinics. Breastfeeding is actively encouraged in diabetes as in non-diabetes. Indeed, breast feeding an infant reduces their risk of diabetes as an adult.

The genetics of glycaemic disease

As will have been obvious from the previous text, glycoregulation is complicated and relies on the coordinated expression of at least two hormones and their receptors. Therefore, it is inevitable that gene defects will be present that have an impact on glucose regulation. There are a number of reasonably well-characterized syndromes:

- Maturity-onset diabetes of the young (MODY) is the collective term for a series of inherited disorders. Most are caused by mutations in the gene for the enzyme glucokinase, which phosphorylates glucose. Without this phosphorylation, glucose cannot be metabolized and so builds up in the cell.
- Other forms of MODY are caused by loss-of-function mutations in genes for intracellular signalling molecules

such as hepatocyte nuclear factor, of which there are several isoforms.
- One of the most common forms of diabetes in the neonate (aged up to 6 months) is one or more mutations in genes coding for cell membrane receptors or ion channels, such as the potassium channel or the sulphonylurea receptor.

Although the dominant cause of type 1 diabetes is linkage to certain HLA types, there are other susceptibility genes in the insulin region on chromosome 11, the gene coding for lymphoid protein tyrosine phosphatase, cytotoxic T lymphocyte-associated protein 4, and the interleukin-2 receptor.

In common with other pathologies, the use of molecular genetics in a clinical setting is often to confirm a particular type of disease, and is generally of little use in directing treatment, which remains control of blood glucose. However, although genome-wide association studies have identified some mutations that predispose to type 2 diabetes, by far the most important predictors are (increased) BMI and (lack of sufficient) physical activity.

Clinical aspects of diabetes

Many of us are aware of the risk of CVD that diabetes confers. However, there are also numerous acute and chronic metabolic states that cause other types of illness.

Diabetic ketoacidosis This syndrome is the consequence of uncontrolled hyperglycaemia, and may even be the precipitating event that leads to the diagnosis of diabetes. Diabetic ketoacidosis (DKA) is more frequently seen in type 1 diabetes, and carries a marked mortality (perhaps as high as 5%). In 25% of cases the precipitating factor or factors are unknown, but it is known to develop from a background of pneumonia, urinary tract infections, heart attack, trauma and stroke.

From the patient's perspective, there may be a 24–48-h history of polydipsia, polyuria, weakness, nausea and drowsiness – in fact, all the classic signs and symptoms of diabetes. If unaddressed this may lead to dehydration, and so low blood pressure and a tachycardia (high pulse rate) as the heart tries to maintain cardiovascular haemodynamics. A deep and rapid breathing (Kussmaul ventilation) may also be present. Failure to perfuse the brain with blood (carrying oxygen) may lead to personality changes and the appearance of being intoxicated with alcohol, as in hypoglycaemia. Up to 10% of subjects may fall into a coma.

Much of this can be explained by abnormal biochemistry. Blood high in glucose will reach the kidney, and so the sugar will pass into the urine. The ability of the nephrons to reabsorb the valuable commodity of glucose is finite and rapidly exceeded, so that glucose remains in the urine. The laws of osmosis demand that water follows the glucose, leading to a great deal of 'sweet urine', easily detectable with a urinary dipstick, and so the polyuria. As water leaves the blood, then blood volume will fall, leading to low blood pressure, hypotension and failure to perfuse organs such as the brain and kidneys. If the latter cannot clear the blood of the high glucose, concentrations remain high and so a vicious circle can develop.

If there is severe insulin deficiency (as in type 1 diabetes), glucagon is not inhibited, which also results in hyperglycaemia, lipolysis and the release of fatty acids from adipocytes. Deprived of glucose, the cell must find energy from elsewhere, and the fatty acids are a good substrate, but metabolic by-products include acetoacetate and beta-hydroxybutyrate, both of which are ketones. These leave the cell and enter the blood, where their overall chemical effect is to lower the pH, so that the blood becomes acidotic; hence the term ketoacidosis. If present at a sufficiently high blood concentration, these ketones are excreted via the lungs, giving the characteristic odour of pear-drops.

Chapter 13 deals with blood gases and pH, and the importance of the bicarbonate molecule (HCO_3^-) as a buffer against a fall in pH and so an increasing acidosis. However, if the ketoacidosis is marked, then concentrations of bicarbonate will be consumed and blood levels will be reduced. As the pH falls, the concentration of potassium rises, and this may contribute to clinically important hyperkalaemia.

Treatment is by controlled fluid replacement (not merely to provide haemodynamic support, but also to help flush out the glucose and other toxins) and intravenous insulin. Naturally, all of this requires frequent monitoring. Figure 14.3 summarizes the biochemistry and the clinical aspects of DKA.

Hyperosmolar hyperglycaemic syndrome Chapter 12 explains the concept of osmolality, being the total mass of all the ions and molecules in a fluid, which we expect to be around 285 mmol per kilogram of fluid. Glucose is a major contributor to osmolality, so that in hyperglycaemia the osmolality will be increased, perhaps to 320–340 mmol/kg. Ketones can also contribute to this hyperosmolar state, which may be present in DKA, but hyperosmolar hyperglycaemic syndrome is often found in the absence of ketones. Indeed, hyperosmolar hyperglycaemic syndrome may also be known as hyperosmolar non-ketotic state.

The clinical consequences are part of the general diabetes/hyperglycaemia picture; that is, polyuria and dehydration, hypotension, and the possibility of coma. Treatment is controlled rehydration, perhaps with up to 9 L of saline in 48 h. Inappropriate fluid replacement can cause brain damage and can sometimes be fatal. Insulin is likely to be needed to gently reduce the hyperglycaemia at a rate of perhaps 3–4 mmol/L of glucose per hour.

Impaired lipoprotein regulation The later consequences of the loss of insulin and glucagon homeostasis extend to dyslipidaemia, especially raised triacylglycerols and reduced high-density lipoprotein (HDL) cholesterol. Evidence that the latter is a risk factor for CVD is strong, but a direct role for high triacylglycerols is unclear. Nevertheless, high fatty acids are undesirable for a number of reasons. These matters will be developed below and in Section 14.2.

Long-term complications of diabetes

Diabetes influences many tissues and organs, inevitably to cause dysfunction and damage, and so disease. We have already noted the glycation of haemoglobin (HbA_{1c}), which results in less oxygen transport and a short life of the cell. Glucose also binds to a range of plasma proteins and lipids, thus generating advanced glycation endproducts (AGEs) that themselves may be toxic. The binding of AGEs to a cell membrane receptor may promote inflammatory changes and perhaps apoptosis.

The primary target for the disease process is the endothelial cells that line blood vessels. Damage to this tissue is common in diabetes, possibly because of the effects of AGEs. A damaged endothelium is no longer able to participate in the regulation of blood vessel tone and in anticoagulation, thereby leading to thrombosis and hypertension, both common in diabetes. It is also likely to lose its resistance to lipids, enabling them to perfuse the blood vessel wall and so promote atheroma.

Microvascular disease Damage to the endothelium in small blood vessels is common in diabetes, and causes several distinct conditions:

• Damage to the blood vessels of the kidney causes nephropathy. This may be due to high blood pressure

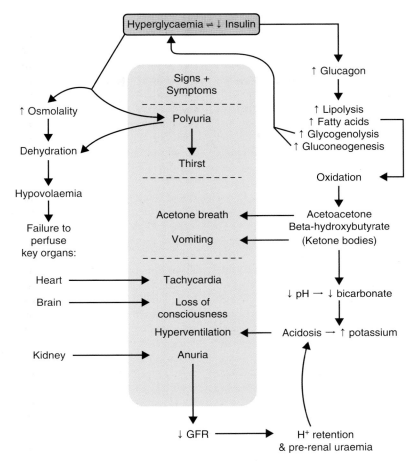

Figure 14.3 The biochemistry and clinical aspects of diabetic ketoacidosis. Hyperglycaemia and low insulin lead to physiological changes (left), signs and symptoms (central box) and biochemical changes (right).

within the kidney or in the circulation as a whole, and/or to damage to the glomerulus, possible by AGEs. This leads to excess albumin in the urine (microalbuminuria) and, if uncontrolled, to nephrotic syndrome (as explained in Chapter 12). To help with the hypertension, many patients will be prescribed an angiotensin-converting enzyme inhibitor or an angiotensin II receptor antagonist.
• Damage to the blood vessels of the eye causes retinopathy. The damaged vessels become leaky and tortuous, and also blocked by thrombosis. Accordingly, diabetics require annual eye checks.
• Damage to blood vessels feeding nerves causes neuropathy, but the nerves themselves can also be damaged, causing loss of function. This can manifest itself as tingling and loss of sensation.

• Damage to blood vessels of the skin causes venous ulceration, skin breakdown and poor wound healing.

Several of these can interact. For example, loss of sensation in the foot will make the patient unaware of skin breakdown and blisters, for example, which will fail to heal properly, causing ulceration and infections. In its most severe form, blockage of arteries feeding the toes will lead to gangrene and so the need for amputation. Damage to the blood vessels feeding the genitals of both sexes leads to sexual dysfunction.

Macrovascular disease As the disease progresses, more and more blood vessels, and in particular the larger arteries, are attacked, causing heart attack, stroke and critical limb ischaemia. These will be discussed in greater detail in Section 14.3.

Other disease   Diabetes is also characterized by an increased frequency of connective tissue disease such as limited joint flexibility, carpel tunnel syndrome and Charcot joint and foot. Diabetics are at increased risk of clots in veins (venous thromboembolism). In the leg this causes deep vein thrombosis and in the lung there is pulmonary embolism (both of these are discussed in depth in Chapter 8). People with diabetes are more likely to suffer from depression than the general population are.

Treatment of diabetes

Clinical trials in both type 1 (Diabetes Control and Complications Trial) and type 2 diabetes (UK Prospective Diabetes Study) have shown that intensive management of hyperglycaemia, weight and blood pressure are effective in reducing the progression of the disease and in the frequency of major end points such as myocardial infarction and stroke. There are several types of treatment, many of which are undertaken in parallel. Lifestyle changes are the first line of treatment in the newly diagnosed, and consist of education about the disease, weight reduction (certainly in type 2, where overweight is inevitable), attention to diet, and exercise.

Beta cell stimulation   Insulin secretion can be promoted by the sulphonylurea class of drugs, which bind to a receptor (SUR1) that is linked to potassium channels. This is the mode of operation of drugs such as tolbutamine, although more recently developed drugs such as gliclazide are becoming more popular. Exenatide is an analogue of the incretin hormone GLP-1, and although it must be injected, it also suppresses glucagon and slows the emptying of the stomach after a meal. Dipeptidyl peptidase IV (DPP IV) inhibitors inhibit the enzyme DPP IV which metabolizes GLP-1 to its inactive form; thus, they prolong the action of GLP-1.

Insulin sensitizers   These drugs do not directly act to release insulin, but they do make the beta cell more responsive. The first in this class, metformin, can be taken orally and so is very widely used. It also inhibits gluconeogenesis. A recent powerful class of drugs are the thiazolidinediones (the glitazones), which bind to intracellular peroxisome proliferator-activated receptors to reduce insulin resistance and increase insulin sensitivity.

Insulin   This is required by all type 1 diabetics and an increasing number of those with type 2 disease. Most patients self-inject up to three times in a 24-h period; others use a pump to deliver a constant supply of the hormone. Other methods of delivering insulin include an aerosol for nasal use, and high-pressure jets for use across the skin where needles are contra-indicated. Many different forms of insulin are available, arising from various sources and with different half-lives and potencies. Some are appropriate after a meal, whilst other are more suited for overnight use.

Other treatments   If dyslipidaemia is present (such as hypercholesterolaemia), then a statin may be prescribed. High blood triacylglycerols are generally treated with a drug of the fibrate class. Hypertension will be managed by a combination of different agents. The absorption of carbohydrates by the intestines can be inhibited by acarbose and other alpha-glucosidase inhibitors. Amylin is a small peptide product of beta cells that slows gastric emptying and promotes satiety. Its effects are mimicked by the drug pramlintide. A newer therapy emerging is the use of sodium glucose co-transporter 2 (SGLT2) inhibitors. These compounds inhibit the SGLT2 co-transporter in the kidney which reduces the amount of glucose that is reabsorbed from the urine, thus reducing blood glucose concentrations. Surgeons offer gastric banding and gastric by-pass surgery to help reduce weight in the obese (i.e. bariatric surgery), and in diabetes this can be very effective in weight loss and in resolving hyperglycaemia.

Monitoring and management of diabetes

Most diabetics will be managed by their family physician, supported by the local hospital. The UK Department of Health has published a national service framework for diabetes (http://www.nhs.uk/NHSEngland/NSF/Pages/Diabetes.aspx) outlining key steps for patients and practitioners. The National Institute for Health and Clinical Excellence (NICE) has published a large number of documents on precise aspects such as diabetes in pregnancy, foot care in diabetes and on the best use of drugs, such as exenatide. These are available free of charge from the NICE website.

It is certainly the fact that patients who take ownership of their disease do better. Accordingly, where possible, diabetics are encouraged not only to address their own diet, but also monitor their blood glucose and treat if abnormal. The latter can be achieved with a portable hand-held point-of-care machine. However, for a better long-term view of blood glucose, fructosamine and HbA$_{1c}$ are preferable.

Clinical management is likely to address blood pressure, weight and BMI, with referral to specialists such as chiropodists and podiatrists for foot care, ophthalmologists for retinopathy and psychiatrists for depression.

Diabetes and haematology

Diabetes clearly interests the haematologist because of the glycation of haemoglobin, and if there is an infected venous ulcer, there may also be a raised white cell count and erythrocyte sedimentation rate. The immunology laboratory can provide proof of autoantibodies to insulin and to the insulin receptor, and as the disease develops, nephropathy can be assessed with U&Es, and the urinary albumin/creatinine ratio.

14.2 Dyslipidaemia

The relationship between cholesterol and CVD was established unequivocally in the second half of the 20th century, giving rise to the expression of 'hypercholesterolaemia', and so the general implication that high cholesterol is bad. This view is inappropriate, in that we now know that 'total' cholesterol is in fact the sum of two independent components that can be separated in the laboratory by their density, hence HDL cholesterol and low-density lipoprotein (LDL) cholesterol. Epidemiology and other studies demonstrated that increased concentrations of LDL are a strong predictor of CVD, and are therefore to be avoided, whilst high concentrations of HDL are protective of CVD, and so are to be promoted. Consequently, the preferred term to use is 'dyslipidaemia'.

Indeed, several lipids are truly essential, in that they contribute to the integrity of the cell membrane, form the substrate for several hormones, such as those regulating sexual function, and are also involved in complex metabolic pathways such as in the synthesis of vitamin D. Fats can also be a source of energy. However, it is certainly an accepted view that the body will store those lipids it cannot immediately use in adipocytes, which build into deposits of fat, and ultimately to obesity and other disease such as diabetes.

Lipids, as a class of biochemicals, consist of a number of families, which include:

- fatty acids
- triacylglycerols (also known as triglycerides)

- cholesterol
- lipoproteins
- phospholipids, sphingolipids and glycolipids.

In this section we will first examine the chemistry of the major lipids, and how abnormal concentrations contribute to disease. We will then look at the clinical conditions that call for the services of the laboratory, and how they are defined.

Lipid families

Fatty acids   A fat is essentially a biological molecule of very low solubility in water, perhaps even completely insoluble. However, the attachment of certain chemical groups makes fat more soluble, and an acid group is one way of doing this, hence fatty acid. In our setting fatty acids are long chains of carbon atoms with an acid group at one end.

The presence of at least one double bond in the chain makes the molecule unsaturated, but if there are no double bonds we say it is saturated. If there are numerous double bonds, the molecule is polyunsaturated. The most common fatty acids are described in Table 14.3. Notably, an unsaturated fat such as oleic acid can be converted to a saturated fat by adding hydrogen (i.e., hydrogenation), in which case it becomes stearic acid.

The presence of a double bond has two major implications: first, other atoms and molecules can be added to the double bond; second, there can be structural implications with the possibility of two isomers – the *cis* form and the *trans* form (Figure 14.4). Generally, the *cis* forms have a beneficial effect, as the isoform provides relative rigidity to the molecule and, when part of the membrane, provides support. Furthermore, unsaturated fatty acids

Table 14.3 Common fatty acids.

Name of fatty acid	Number of carbon atoms	Double bonds	Nature
Lauric	12	None	Saturated
Palmitic	14	None	Saturated
Stearic	18	None	Saturated
Palmitoleic	16	One	Monounsaturated
Oleic	18	One	Monounsaturated
Linoleic	18	Two	Polyunsaturated
Linolenic	18	Three	Polyunsaturated
Arachidonic	20	Four	Polyunsaturated

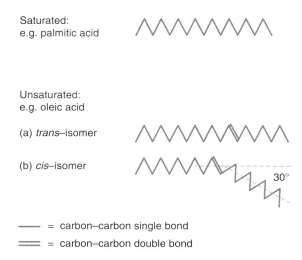

Saturated:
e.g. palmitic acid

Unsaturated:
e.g. oleic acid

(a) *trans*–isomer

(b) *cis*–isomer

30°

——— = carbon–carbon single bond
═══ = carbon–carbon double bond

Figure 14.4 Structure of fatty acids. The presence of a double bond allows two isomers of a particular molecule: the *cis* form, when the molecule is effectively straight, and the *trans* form, where it has a bend of 30° in the middle.

produced industrially have double bonds in the *trans* form, which is not found in a natural healthy diet.

Fatty acids are the starting point for the synthesis of a number of molecules involved in platelet and vascular biology. For example, arachidonic acid is the base unit for the synthesis of prostaglandins, leukotrienes and thromboxanes, molecules important in many areas, such as in platelet function. Eicosapentaenoic and docosahexaenoic acids, fatty acids extracted from fish oils, are often prescribed after a heart attack.

In many cases, free fatty acids are essential metabolic fuels, and provide energy to the heart. Indeed, this may be why there are modest deposits of fat (adipose tissue) around a healthy heart. Cells that store fatty acids are called adipocytes.

Triacylglycerols Blood concentrations of these lipids are made up of that which is taken in as part of the diet and that which is synthesized.

There are several classes of triacylglycerols. All have a common structure of a backbone of glycerol, a three-carbon molecule, each of which is linked to a long-chain fatty acid (an acyl – typically of 12–20 carbon atoms in length), so that triacylglycerols are esters (a molecule from an acid and an alcohol). In theory, almost any combination of three fatty acids listed in Table 14.3 may combine with glycerol, and the particular fatty acids give the triacylglycerol its name (Figure 14.4). For example,

three palmitic acids combining with a glycerol gives tripalmitin, or perhaps tripalmitoylglycerol. There may also be two or even three different fatty acids combining with the glycerol; for example, 2-stearodipalmitoylglycerol, which tells us of a stearic acid in the middle of the three carbon atoms, flanked by a palmitic acid on each side. Synthesis begins with a molecule of glycerol-3-phosphate (itself derived from glucose), to which a fatty acid is added by acyl transferase to create a monoacylglycerol. This process, occurring within an adipocyte, is repeated to give a diacylglycerol, requiring a diacylglycerol transferase, and again to a triacylglycerol (Figure 14.5).

Triacylglycerols can be a significant source of energy. They can be hydrolysed back to their component fatty acids and glycerol by lipase enzymes, whose synthesis may be part-regulated by hormones, notably insulin and glucagon.

Cholesterol The structure of cholesterol is complex and radically different from that of triacylglycerols. Like triacylglycerols, concentrations in the blood are the sum of that which is absorbed from the diet (both plant and

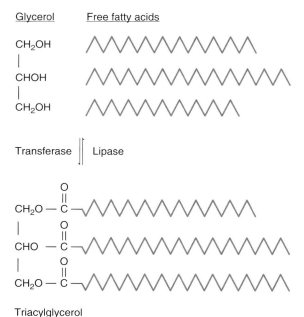

Glycerol Free fatty acids

CH_2OH

$CHOH$

CH_2OH

Transferase Lipase

CH_2O — C (=O)

CHO — C (=O)

CH_2O — C (=O)

Triacylglycerol

Figure 14.5 Metabolism of triacylglycerols. Each molecule is synthesized by linking fatty acids to a glycerol by transferase enzymes. If required as a source of energy, the triacylglycerol can be digested back to its component parts by lipase enzymes.

animal material) and that which is synthesized in the liver.

In the liver, cholesterol is synthesized by a series of complex biochemical steps based on isoprene and acetyl-coenzyme A (CoA) (Figure 14.6). Several isoprene units can be fused to give a ring structure, a feature also found in steroids, certain sex hormones and vitamin D. A key early step is the biosynthesis of hydroxymethylglutaryl-CoA (HMG-CoA) from acetyl-CoA and acetoacetyl-CoA, followed by its reduction to form mevalonate. This step is regulated by the enzyme HMG-CoA reductase, and as the rate-limiting step is the target of an important class of drugs that reduce the production of cholesterol. These drugs, the HMG-CoA reductase inhibitors, also known as statins, are very widely prescribed for people with high cholesterol and those with CVD (post-myocardial infarction, stroke) regardless of their serum cholesterol. We shall refer to these drugs in sections that follow.

In the intestines, cholesterol and a fatty acid may be fused by the enzyme acyl-CoA, cholesterol acyltransferase, to form a cholesterol ester. An alternative name for this enzyme is lethicin-cholesterol acyltransferase (LCAT). Cholesterol taken in foodstuffs is absorbed mainly in the upper part of the small intestines (the duodenum and jejunum), and depends on the solubilization, with other fats and bile salts, to form minute droplets called mixed micelles. Cholesterol and triacylglycerols pass across the enterocytes of the intestines, and enter the blood as chylomicrons.

Lipoproteins A key feature of fats is that they are insoluble in water. This is why fats and acids combine to make fatty acids more soluble. A further example of this is the formation of lipoproteins from a lipid and a specialized protein – an apoprotein, of which there are several types. These apoproteins include apolipoprotein a, (apo(a), with subtypes A1, A2, A4 and A5), apoprotein B (apoB, of which there are two types, $apoB_{48}$ (made by intestinal cells) and $apoB_{100}$ (by hepatocytes), apoprotein C (apoC, of which there are several types), apoprotein D and apoprotein E (apoE, of which there are three isotypes). Some apoproteins combine: apo(a) with apoB and some lipids to form lipoprotein A (Lp(a)).

Phospholipid, sphingolipids and glucolipids Phospholipids, which are important as they can make up almost half of the cell membrane, are related to triacylglycerols, and often contain a choline group as well as the obligatory phosphate group. Phosphatidic acid is glycerol with two of the three carbon groups linked to a fatty acid, with a phosphate group on the third carbon atom. Attach a molecule of the amino acid serine to the phosphate group and it becomes phosphatidylserine; add choline instead of serine and it becomes phosphatidylcholine (lecithin).

Sphingolipids are based on a serine molecule, to which can be added a fatty acid to yield sphingosine, whilst a second fatty acid makes it ceramide. Addition of a phosphate group and choline yields sphingomyelin. The complicated nature of these chemicals is such that sphingomyelin may also be classified as a phospholipid, and can be seen as being analogous to phosphatidylcholine. The presence of glucose or galactose defines a lipid as a glycolipid. Addition of galactose to ceramide

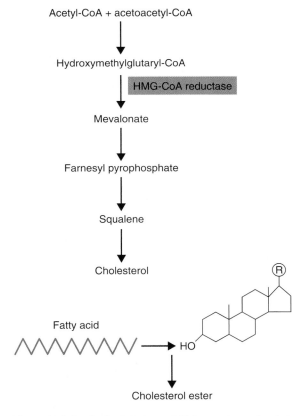

Figure 14.6 Synthesis of cholesterol. This complex molecule is constructed initially from acetyl groups, which are fused to form hydroxymethylglutatyl-coenzyme A (HMG-CoA). This is converted to mevalonate by the enzyme HMG-CoA reductase, which is inhibited by the statin group of lipid-lowering agents. Later steps see squalene being converted to cholesterol, which can then join with a fatty acid to form cholesterol ester.

generates a cerebroside, whilst addition of several carbohydrates makes it a ganglioside.

Whilst the structure and metabolism of these molecules is interesting, they rarely occupy a great deal of a biochemist's workload. However, deficiencies of enzymes involved in these pathways lead to precisely defined disease. Gaucher's disease, the build-up of glucocerebrosides, is caused by lack of beta-glucocerebosidase, whilst Fabry's disease, characterized by excess ceramide trihexose, is an X-chromosome-linked condition of the failure to produce alpha-galactocerebrosidase. Biochemically similar conditions include Tay–Sach's disease and Niemann–Pick disease.

Lipoproteins in the blood and tissues

The major lipids come together to form a (frequently confusing) series of molecules of varying size and composition. A common definition of lipoprotein microparticles is to use their size and density, which gives:

- chylomicrons
- very low density lipoproteins (VLDLs)
- intermediate density lipoproteins (IDLs)
- LDLs
- HDL.

Chylomicrons With a diameter of 80–500 nm, these particles are effectively fats that have been absorbed across the gut wall, and that have been 'proteinated' by enterocytes to enable ease of transport in the blood. The first major protein to be associated with chylomicrons is apoB$_{48}$, and the dominant lipid is triacylglycerol, with small amounts of phospholipids, cholesterol and cholesterol esters. However, this first type of chylomicron, a spheroid of perhaps 75–1200 nm diameter, becomes modified by the addition of more lipoproteins, apoC2 and apoE. ApoC2 is a cofactor for the enzyme lipoprotein lipase, and this and other enzymes slowly digest the triacylglycerols of the chylomicrons, which therefore get smaller, ending up as chylomicron remnants, with a diameter of some 40 nm. Remnants are eventually absorbed by the liver via apoE with receptors on the hepatocyte membrane.

Low, intermediate and very low density lipoproteins

In contrast to chylomicrons, VLDL (diameter 30–80 nm) is constructed in the liver from triacylglycerols, cholesterol and the apoproteins apoB$_{100}$, apoC1, apoE, and picks up apoC2 and more apoE whilst in the blood.

The acquisition of these additional apoproteins, and the digestion of triacylglycerols by lipases, results in an increase in density and its transformation into IDL.

Markedly smaller IDLs, with a diameter of 25–30 nm, are characterized by little (if any) apoC1 and apoC2, but by high concentrations of apoE. The latter is recognized by receptors on hepatocytes, so that IDLs may be absorbed and recycled. As IDLs mature they lose their apoE, so that the dominant lipoprotein becomes apoB$_{100}$, at which point they are recognized as LDL.

LDLs, with a diameter of 19–25 nm, are the final particle in the lipoprotein pathway, effectively IDLs that have been stripped of triacylglycerols and apoE, and thus are rich in cholesterol and apoB$_{100}$. Thus, LDLs are the major blood carrier of cholesterol, and accordingly are important in the pathogenesis of atherosclerosis, the subject of Section 14.3. Ultimately, they are absorbed by the hepatocyte via an interaction between apoB$_{100}$ and the LDL receptor, which also recognizes apoE. The consequences of loss-of-function mutations in the gene for the receptor include failure of the hepatocytes to absorb LDL, so that concentrations in the blood remain high. We will return to this mutation shortly.

However, whilst in the blood, LDLs can be modified by oxidation or glycation, and so can be recognized by nonspecific scavenger receptors (such as CD36) on the surface of hepatocytes, macrophages and other cells.

High-density lipoprotein These particles, the smallest of the lipoproteins (diameter 6–11 nm), are composed almost entirely of cholesterol, and phospholipids, and with apoproteins; mostly apoA1 and apoA2, but with smaller amounts of apoC2 and apoE. They are the most dense particle because of their lack of triacylglycerols. As they circulate, HDL particles pick up cholesterol from the blood and tissues, and so become larger. The collection of cholesterol from the tissues is mediated via a molecule called ABC transporter, whilst apoA1 binds to the LCAT enzyme, which fuses free cholesterol with a fatty acid to form an ester, and so enters the particle.

Eventually the HDL particle passes into the liver, where hepatocytes bearing an HDL receptor will absorb the particle and digest the cholesterol. Thus, HDL can be seen as a general scavenger, which collects 'left over' cholesterol and delivers it back to the liver, so that it is often called the reverse transport of cholesterol. It is likely that this pathway is an important protective mechanism for minimizing the amount of free cholesterol that may otherwise promote atherosclerosis, as we will discuss in

Table 14.4 Major features of lipoproteins.

	Density (g/cm^3)	Major apoproteins	Cholesterol (%)^a	Triacylglycerol (%)	Phospholipids (%)	Protein (%)
Chylomicrons	<0.94	A, B, C	1–4	86–94	3–8	1–2
VLDL	0.94–1.006	B, C, E	18–22	55–65	12–18	8–15
IDL	1.006–1.019	B, E	27–46	25–40	15–22	12–19
LDL	1.019–1.063	B	41–55	6–12	20–25	20–25
HDL	1.063–1.210	A	10–20	1–4	25–40	35–55

^{a}Includes cholesterol esters and free cholesterol.

Section 14.3. However, HDL is also probably useful in delivering cholesterol to other organs, such as those generating steroid hormones (i.e. the adrenals and gonads).

Further analysis has identified sub-fractions of HDL: HDL$_2$ and HDL$_3$. The former is larger, with more cholesterol but less protein than the latter. Accordingly, HDL$_3$ is more dense than HDL$_2$.

Overall, therefore, lipoprotein particles evolving from chylomicrons to LDL, lose triacylglycerols and gain proteins. Table 14.4 summarizes these and other features of lipoproteins.

Lipoprotein (a)   The Lp(a) particle, with a diameter of 25–40 nm, is closely related to LDL in that it is cholesterol rich and contains apoB$_{100}$. However, Lp(a) also contains apo(a), and has a density similar to that of a heavy LDL particle (perhaps 1.052–1.063 g/cm^3). It exists in a number of isoforms, and concentrations vary considerably between individuals, being governed by the apo(a) gene on chromosome 6. The precise function of Lp(a) is unclear, but it is notable that its structure resembles plasminogen and tissue plasminogen activator, molecules involved in the dissolution of thrombus (as explained in Chapter 7).

Cell surface receptors   Some ions and molecules pass through the cell membrane relatively easily, but for others the membrane is impassable. However, the membrane offers several routes to the cell, such as receptors. An excellent example from Section 14.1 is the insulin receptor. The LDL receptor recognizes and binds to apoB$_{100}$ and apoE, thus enabling the passage of LDL into the cell. The LDL-receptor-related protein, expressed by hepatocytes, can bind several ligands, including apoE and lipase. CD36 can bind both normal and oxidized LDL, whilst scavenger receptor class B member 1 is a scavenger that can bind to HDL, LDL,

VLDL and oxidized LDL. Loss-of-function mutations in the genes for these receptors can lead to a variety of clinical conditions, as we will see.

Enzymes, transporters and transfer molecules   A small number of non-lipids and non-lipoproteins are needed to complete our picture of the metabolism of this group of biochemicals:

- Lipoprotein lipase (LPL: EC 3.1.1.34) is the major enzyme in the catabolism of lipids, and can, for example, digest triacylglycerols back to their constituent fatty acids and glycerol. It can be found on the surface of endothelial cells, within hepatocytes and may be produced by the pancreas.
- Cholesterol ester transfer protein (CETP) promotes the transfer of cholesterol esters and triacylglycerols between HDL and VLDL.
- LCAT mediates the transfer of a fatty acid from phosphatidlycholine (lecithin) to cholesterol to create a cholesterol ester.
- Phospholipid transfer protein is able to transfer proteins between various lipoprotein particles, particularly triacylglycerols and HDL.

There are several ways in which cholesterol disease can be classified. The most popular system focuses on abnormalities in total cholesterol and its sub-fractions, and on triacylglycerols. Chylomicrons are very rarely requested in routine clinical biochemistry.

Disorders of cholesterol

Cholesterol in the blood has two sources: the diet and the body, where almost all of it comes from the liver, with possibly some being produced by intestinal cells. Put simply, in some people, the liver makes a lot of cholesterol; others make much less, often under genetic control that may have a heritable aspect. Some lipid disease is secondary to other conditions, such as diabetes.

Dietary hypercholesterolaemia   A poor diet, rich in cholesterol, is a clear cause of hypercholesterolaemia. Such a poor diet is likely to have been present for years, and so the individual is likely to be obese. Thus, resolution of hypercholesterolaemia by diet alone may take years to be effective, as even in a very low fat diet cholesterol can leach out of stores to keep blood concentrations high. Nevertheless, a low-fat diet is a key requirement for the treatment of obesity, and often of hypercholesterolaemia. A modest intake of alcohol results in an increase in HDL.

Primary hypercholesterolaemia   Cholesterol is synthesized by the liver, and in many cases is regulated by several genes. Both loss-of-function and gain-of-function mutations exist that directly influence concentrations in the blood. These include:

• Familial hypercholesterolaemia (FH), caused by failure of the LDL receptor to remove LDL from the circulation, so that concentrations remain high. Literally hundreds of mutations in the gene for LDL have been described, and are present in 1 in 500 of the population (i.e. are heterozygotes), leading to a serum cholesterol of perhaps 7 or 8 mmol/L. In homozygotes, serum cholesterol is often in double figures. Diagnosis is based on increased LDL, family history of premature CVD and subcutaneous deposits of cholesterol, generally on the tendons (xanthoma) or eyelids (xanthelasma).
• The LDL receptor recognizes apolipoprotein B_{100} in LDL, so absorbing it into the hepatocyte. Familial defective apoprotein B_{100} is the disease so named because of an abnormality in the gene, so that its mutated $apoB_{100}$ product fails to be recognized by the receptor. This results in raised serum LDL and a phenotype resembling FH.
• The intracellular domain of the LDL receptor interacts with other intracellular molecules involved with endocytosis. Abnormalities in these molecules result in poor receptor recycling and ultimately impaired clearing of LDL from the plasma. Patients have a serum cholesterol concentration between that of heterozygous and homozygous FH, the condition being described as autosomal recessive hypercholesterolaemia.
• Stanol esters, such as sitosterol, are plant products that resemble human cholesterol and are absorbed like cholesterol. However, a transporter molecule exports these plant sterols back into the intestines and also into bile. An abnormality in the transporter reduces the passage of cholesterol and plant sterols into the bile, resulting in sitosterolaemia and xanthoma.

In contrast to these single-gene-defect conditions (i.e. monogenetic), there are many other genetic abnormalities leading to raised cholesterol. It is likely that several are present in an individual and act in concert, leading to the concept of polygenic (i.e. many genes) hypercholesterolaemia.

Secondary hypercholesterolaemia   This can be classified as being due to either disease or to a drug effect. The former include:

• Curiously, in anorexia nervosa, where calorific intake is deliberately restricted, there is increased VLDL and LDL.
• In diabetes mellitus, as discussed in Section 14.1, insulin resistance leads to the overproduction of VLDL as insulin normally suppresses the release of fatty acid stores in adipocytes.
• Hypercholesterolaemia is a common consequence of stages IV and V chronic renal failure (Chapter 12). A likely mechanism is the overproduction of apoB and VLDL by the liver.
• The mechanism by which hypothyroidism causes hypercholesterolaemia (and possibly raised triacylglycerols) is likely to be reduced uptake of LDL by the LDL receptor. The link is so strong that hypothyroidism should always be considered in an abnormal lipid profile.
• Monoclonal gammopathies and paraproteinaemias (as in the white blood cell malignancy of myeloma) may interfere with the uptake of lipoproteins by their receptors, resulting in raised IDL. The abnormal gammaglobulins may also form complexes with lipoproteins, and so reduce the rate of their removal. These complexes may form xanthomas.
• Cholestasis is associated with increases in serum cholesterol due to accumulation of cholesterol and phospholipids in the liver which eventually pass into the blood. Of course, if there is cholestasis there is likely to be jaundice and abnormal liver function tests, as described in Chapter 17.

Numerous medications are linked to hypercholesterolaemia. These include those used to treat acquired immunodeficiency syndrome and hypertension (thiazide diuretics, beta blockers), in addition to synthetic oestrogen (in the oral contraceptive pill and as hormone replacement therapy), ciclosporin, glucocorticoids (mimicking Cushing's syndrome) and retinoic acid.

Increased HDL In contrast to total cholesterol and LDL, which, from an epidemiological perspective, are to be avoided, for the same reasons, high HDL is to be encouraged. Where present, this may be due to increased apoA1 or a reduced absorption of HDL by hepatocytes. Additional mechanisms include increased LCAT, reduced CETP activity (the latter possibly related to the overexpression of apoC1) and reduced hepatic lipase activity. Despite the view that raised HDL is beneficial, this is not always the case, as it may not necessarily result in improved reverse cholesterol transport.

Hypocholesterolaemia Just as some hypercholesterolaemia may be the upper extreme of the reference or normal range, so hypocholesterolaemia may reflect the low extreme. The extent to which the latter may be regarded as a disease is questionable. Many cases of cancer are associated with hypocholesterolaemia, but this is a likely response to the malignancy. An intensive low-fat diet and excessive treatment may of course lead to low cholesterol, but there are known biochemical syndromes.

- Hypobetalipoproteinaemia may be due to genetic defects that result in a familial form of the disease. Heterozygotes, who are generally asymptomatic, have lower total cholesterol than 'normal' (perhaps 2.5 mmol/L), but concentrations in homozygotes may be as low as 1.3 mmol/L. The most common mutations lead to a shortened form of $apoB_{100}$, giving variants of length 31, 37 and 41.
- An extreme (and very rare) form of the above is a complete lack of apoB, described as abetalipoproteinaemia, leading to failure to produce VLDL and LDL. It is caused by a mutation in the gene for microsomal triacylglycerol transport protein, so that chylomicrons and VLDL particles are incorrectly packaged. This also leads to malabsorption of vitamin E, an eye disease and unusually shaped red blood cells, which are called acanthrocytes.
- Tangier disease is caused by a mutation in a gene regulating apoA1, hence hypoalphalipoproteinaemia. This leads to an inability to transport cholesterol out of the cell and so to its accumulation within the cell. A further consequence is grossly reduced concentrations of HDL, as this relies on apoA1.

- Low concentrations of HDL may be genetic, secondary to other disease (such as diabetes and raised triacylglycerols), end-stage renal disease and severe inflammation. In the case of liver disease, low HDL results from inability to generate apoA1, a key component of the lipid. A variety of drugs also cause low HDL, such as steroid hormones and the blood pressure lowering agents thiazide diuretics and beta blockers.

Disorders of triacylglycerols

Like cholesterol, blood triacylglycerols may arise from diet and the liver, and the ability of the liver to synthesize these lipids also varies from person to person. The parallel with cholesterol extends to various genetic causes of raised concentrations, and may also be due to secondary causes, one of the strongest being pancreatic disease.

Dietary hypertriacylglycerolaemia Clearly, a high-fat diet is most likely to lead to high plasma triacylglycerols, and like cholesterol is closely linked to obesity. In particular, central obesity is linked to this problem, as central adipocytes are more likely to release fatty acids. Alcohol is a major independent influence on concentrations of triacylglycerols, and also VLDL.

Primary hypertriacylglycerolaemia These all have a genetic, and so an inherited, basis. The most common are:

- Familial endogenous hypertriglyceridaemia, which is most likely to be caused by mutations in genes for hepatic lipase, apoCIII and apoA5.
- Lipoprotein lipase deficiency, as this enzyme digests acylglycerols, so that in its absence the concentrations of the fat remain high.
- Familial hypertriacylglycerolaemia, an autosomal dominant condition, which results from overproduction of VLDL, and elevated chylomicrons are common.

Secondary hypertriacylglycerolaemia Probably the most common pathological state causing raised triacylglycerols is pancreatitis, and of course in Section 14.1 we learned that diabetes also causes high concentrations of this fat. Increased triacylglycerols are also found in metabolic syndrome, although this does not *cause* the high concentrations.

Other conditions causing high triacylglycerols include the autoimmune disease systemic lupus erythematosus, alcohol, nephrotic syndrome and hypothyroidism.

Combined raised cholesterol and triacylglycerol

Familial combined hyperlipidaemia (FCH), estimated to be present in 0.5–2.0% of the population, is characterized by raised triacylglycerols and raised cholesterol (as VLDL and LDL), and there is also the likelihood of low HDL. There are also raised levels of $apoB_{100}$. The precise cause is abnormalities in apoC and CETP. The genetics of FCH are unclear; in some communities it is linked to a site on chromosome 1, although others have linked it to multiple loci on chromosomes 9, 11 and 16. Diagnosis is more certain if similar dyslipidaemia can be demonstrated in family members.

Both hypercholesterolaemia and hypertriacylglycerolaemia can also arise at the same time from a poor (lipid-rich) diet, but may also be found in other disease, such as diabetes, and often as part of Cushing's syndrome (Chapter 18).

Clinical genetics and dyslipidaemia

The clinical molecular genetics of these (and other) lipid disorders seeks to precisely define the nature of the abnormality. However, in high concentrations of the particular lipid this is often only of partial interest as treatment focuses on diet and drugs to reduce plasma lipoproteins, as we shall see in the subsection that follows. Nevertheless, a genetics service may be called upon to determine the gene status of a subject who is (so far) asymptomatic, but whose family members have a clear genetic abnormality.

Despite this, there is interest in three genetically driven isoforms of apoE (apoE2, apoE3 and apoE4). ApoE3 is 'normal', whereas apoE2 is associated with lower LDL and apoE4 with higher LDL. Consequently, the latter may be a risk factor for atherosclerosis.

The clinical consequences and treatment of dyslipidaemia

The outward signs of lipid disease are the deposition of yellowish nodules of lipids in certain anatomical sites, such as the eyelids (xanthelasma) and the tendons, notably of the heel, and joints (xanthoma). There may also be small cutaneous nodules on the hands. Together, although possibly unsightly, these signs have no adverse clinical consequences.

However, the same process of the deposition of lipids into the wall of blood vessels has a major impact on health as it leads to atherosclerosis, to be discussed in Section 14.3. Therefore, maintaining a low lipid profile in the blood is desirable if this disease, and its many manifestations, is to be resisted. The most effective method of ensuring low serum lipids is pharmaceutical intervention, although of course a low-fat diet in parallel is strongly encouraged.

Treatment of hypercholesterolemia Quite possibly the most successful class of drugs in the 20 years from 1990 has been the statins. These drugs inhibit the rate-limiting enzyme in the synthesis of cholesterol – that is, HMG-CoA reductase – in hepatocytes (Figure 14.6). Thus, inhibition of this enzyme (i.e. by an HMG-CoA reductase inhibitor) leads to depletion of intracellular cholesterol and an increased uptake from the plasma.

Numerous trials have unequivocally demonstrated that statins (notably pravastatin, simvastatin and atorvastatin) reduce the risk of heart attack and stroke in both primary and secondary care, and that the mechanism for this is by the reduction in serum cholesterol. Indeed, the success of this class of drug is such that it is routinely given to those at high risk of CVD (typically, diabetics), regardless of their particular cholesterol result. In these subjects, the target concentrations for cholesterol-lowering therapy are total cholesterol of less than 4.0 mmol/L and an LDL of less than 2.0 mmol/L. However, these figures change with time, and are often more stringent in those at the highest risk of CVD, such as obese diabetics who have already had a heart attack. The treating practitioner may feel that the high risk justifies a target of less than 3.5 mmol of total cholesterol and/or 1.5 mmol of LDL cholesterol, and an HDL result greater than 2.0 mmol/L.

However, no drug is free of any side effect. The most commonly reported problem with this class of drugs is muscle pain. Biopsies from painful muscles show damage and the breakdown of muscle cells: rhabdomyolysis. In severe cases there may be so much muscle damage that increased concentrations of the enzyme creatine kinase appear in the blood. In some cases, patients can be transferred to a different statin, and in others a supplement of ubiquinone can be given. Other concerns are whether or not statins are safe in those with acute or chronic liver disease, as some statins can cause increased concentration of transaminase enzymes (implying hepatocyte damage).

Although statins dominate lipid-lowering therapy, several other drugs are available should the need arise (such as intolerance to statins). These include:

● Drugs acting on the absorption of lipids by the intestines. Colestipol and cholestyramine are bile acid sequestrants, preventing their reabsorption so that the liver synthesizes the bile acids instead of cholesterol. A further consequence is the increased expression of the LDL receptor, resulting in increased LDL absorption from the blood.

- Ezetimide lowers dietary and biliary cholesterol by inhibiting its absorption by intestine brush-border enterocytes. The uptake of triacylglycerols and fat-soluble vitamins is unaffected. The most efficient use of this agent is in combination with a statin, and a combined tablet is available.
- Niacin (vitamin B_3) blocks the breakdown of fats, leading to a decrease in free fatty acids and so decreases VLDL and cholesterol release by hepatocytes. A further bonus (from the cardiovascular perspective) is that HDL concentrations rise.

Treatment of hypertriacylglycerolaemia Fibrates are the cornerstone of the treatment of high triacylglycerols. Several are available, the most commonly used being bezafibrate and gemfibrozil. They act on intracellular peroxisome proliferator-activated receptors, which then upregulate various genes involved in carbohydrate and lipid metabolism. The net result is an increase in HDL and a reduction in triacylglycerols. Interestingly, the structure of fibrates resembles that of thiazolidinedione drugs (the glitazones) used in diabetes to improve insulin resistance. In many cases statins can be used in parallel. A diet rich in fish oils (of which the active component is omega-3 fatty acids) is also effective in reducing hypertriacylglycerolaemia.

Treatment of combined hyperlipidaemia Since the metabolism of the two major lipid classes is so different, then the common approach is to use two different drugs, such as a statin and a fibrate. However, the effectiveness of such combinations must be monitored and may not always be as successful as hoped.

Other treatments of dyslipidaemia Raised triacylglycerols and cholesterol are often present in obesity, which has a number of potential treatments, although, overall, regrettably their success rate is poor. Aids to weight loss such as intestinal lipase inhibitors and centrally acting appetite suppressors may help by restricting fat intake, and increased exercise is clearly of benefit. In the pre-statin era, plasmaphoresis (a type of dialysis) was the most effective method of reducing cholesterol in homozygous FH, but today is rarely needed. Surgical approaches include partial gastrectomy, gastric banding and intestinal bypass surgery.

Screening for elevated Lp(a) in those at intermediate or high risk of atherosclerosis has been advocated, a desirable concentration being <50 mg/dL. One way to attain this may be use of niacin for Lp(a) and atherosclerosis risk reduction.

The laboratory and lipids

Unsurprisingly, lipid analysis is a major part of laboratory work, and focuses on total cholesterol, HDL-cholesterol and triacylglycerols. LDL-cholesterol is not measured directly in a routine setting. Lipids are more commonly measured by a variety of chemistries, most of which are based on enzymes (many of which are purified from bacteria), but some may also be assessed by electrophoresis. Immunoassay (such as enzyme-linked immunosorbent assay) is popular in the quantification of apoproteins.

Total cholesterol Measurement of cholesterol is dominated by the enzymatic method, which has several steps. Perhaps the most popular technique is first to convert cholesterol esters to cholesterol by adding the enzyme cholesterol esterase to the sample of serum or plasma being tested. The reaction also generates fatty acids, but these play no part in the analysis.

The second step sees the conversion of the cholesterol to cholest-4-en-3-one by the action of cholesterol oxidase and oxygen. However, the important aspect is not the cholesterol metabolite but the hydrogen peroxide that is a further product. In the presence of phenol and peroxidase, this molecule oxidizes the colourless 4-aminophenazone to give a red-coloured product. The reaction vessel and conditions are set up so that the rate-limiting step in the amount of colour is the concentration of hydrogen peroxide, which itself had to have been generated directly from cholesterol (Figure 14.7).

This method can accurately measure cholesterol in the range 0.08–20.7 mmol/L, which is more than enough for all human disease. However, it cannot be performed on blood anticoagulated with citrate, oxalate or fluoride.

High-density lipoprotein cholesterol A number of methods are available for measuring this particle. Historically, the dominant method is precipitation, where molecules containing apoB are separated. As this apoprotein is absent from HDL, whatever remains is taken to

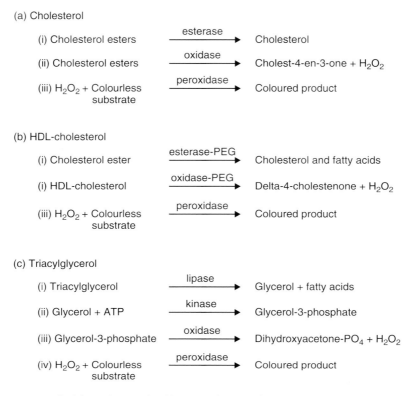

Figure 14.7 The enzymatic method for each major lipid has a number of defined steps and requires several different reagents, and notably all conclude with the generation of hydrogen peroxide. Nevertheless, each is amenable to automation and has excellent reproducibility.

be the HDL. The precipitant itself is based on magnesium, with a variety of cofactors such as heparin, dextran sulphate, phosphotungstate and polyethylene glycol (PEG). However, the precipitation method may be influenced by VLDL and IDL in high concentrations.

One of the most common methods, like the measurement of total cholesterol, has a number of well-defined steps. In the first, an aliquot of the serum or plasma reacts with the enzyme cholesterol esterase conjugated to PEG. This generates HDL-cholesterol and fatty acids. In the second step, the HDL-cholesterol is oxidized by the action of PEG-cholesterol oxidase to generate delta-4-cholestenone and hydrogen peroxide. Finally, the latter reacts with 4-amino-antipyrine and a complex sodium aniline compound to produce a purple–blue pigment that is quantified photometrically. As with the total cholesterol method, the reaction conditions are such that the density of the coloured product is directly proportional to the concentration of HDL in the plasma

or serum. The method is accurate over the range 0.08–3.1 mmol/L.

Triacylglycerols   The first step in one of the more frequently used methods is to separate the glycerol from its three fatty acids by lipoprotein lipase. The glycerol is then phosphorylated by ATP and a kinase enzyme, generating adenosine diphosphate and glycerol-3-phosphate. The next step is the oxidation of the latter to dihydroxyacetone phosphate and hydrogen peroxide by glycerol phosphate oxidase. The final step is similar to that of total cholesterol and HDL, in that the hydrogen peroxide enables the formation of the red-coloured product 4-(p-benzoquinone monoimino)-phenazone from 4-aminophenazone and 4-chlorophenol. The technique is accurate over the range 0.05–11.3 mmol/L.

Assay characteristics   Each of these three methods has excellent assay characteristics, such as very low coefficients of variation. Generally, there is no or minimal interference from bilirubin, haemolysis or lipaemia (i.e. high triacylglycerols or chylomicrons), unless exceptional. Very high concentrations of paraproteins, as may be present in myeloma and similar illnesses, may interfere with the techniques.

Low-density lipoprotein cholesterol This is one of the strongest predictors of the risk of CVD, and in this respect is much more sensitive and specific than is total cholesterol. Regrettably, LDL is very difficult to measure directly, and so is not offered routinely. Fortunately, concentrations of LDL can be determined indirectly by the following equation, also called the Friedewald equation:

$$LDL \text{ cholesterol} = \text{total cholesterol} - HDL \text{ cholesterol} - \frac{\text{triacylglycerol}}{2.2}$$

However, the equation is imperfect because:

1. It assumes most circulating triacylglycerols are in VLDL.
2. It assumes the relationship between the concentrations of VLDL and triacylglycerols is consistent up to a triacylglycerol concentration of 2.5 mmol/L. Above this level, according to some, the calculated LDL result may not be reliable.
3. The equation is only valid up to a triacylglycerol result of 4.5 mmol/L.
4. Blood triacylglycerol concentrations show a strong diurnal variation, especially after a lipid-rich meal; it should only be measured in a fasting (perhaps 12-h) blood sample.
5. The calculated LDL result relies on three other results, and small errors in each (which could in isolation be tolerated) may be additive, leading to an unacceptable error in the LDL result.

Apoproteins Proteins are particularly suited to measurement by immunoturbidimetric methods as they can be readily automated. However, the distribution of each apoprotein in Table 14.4 is such that the measurement of any individual apoprotein is relatively uninformative as they can be found in several disease lipoproteins. Perhaps the only apoprotein-based immunomethod used in a semi-routine manner is for Lp(a). However, there are circumstances where measurement of apoproteins may be of value in a routine setting, as in the determination of the apoB/apoA1 protein ratio, which may have advantages over using the LDL/HDL ratio when assessing cardiovascular risk.

Blood science angle: Dyslipidaemia

Unlike diabetes, although lipoproteins are a major component of the red cell membrane, dyslipidaemia in itself is of little interest to haematologists, or to immunologists. Perhaps the only indication would be a grossly elevated hypertriacylglycerolaemia that interferes with other tests.

14.3 Atherosclerosis

A problem with the 'scientific' approach to hyperglycaemia and dyslipidaemia is the tendency to treat the particular laboratory result, and not necessarily the patient. A modern view of health is to promote several different aspects of the risk of CVD (and other disease, such as cancer) with the aim to reduce the risk factors to as low as possible. Advice is to not smoke, maintain BMI between 18 and 25 kg/m^2, eat a diet containing less than 30% fat, of which saturated fats constitute less than 10% of total fat, and less than 300 mg of cholesterol daily, eat five portions of fresh fruit or vegetables daily, eat unbattered fatty fish twice a week (ideally grilled), restrict alcohol intake to less than 21 units (1 unit equals 10 g) for men and 14 units for women weekly, restrict salt (NaCl) intake to less than 6 g per day,– restrict or avoid the use of salt in cooking and if added to food at the table – and take regular aerobic exercise, such as fast walking for half an hour or swimming on most days.

Failure to address these points places the individual at increased risk of atherosclerosis, the disease which attacks the arterial circulation. The consequences of the disease, attacking first the inner lining of the vessel (the endothelium), is failure to deliver oxygenated blood to the tissues, leading to ischaemia. Atherosclerosis of the arteries feeding the heart results in angina and myocardial infarction. The same process attacking the blood supply to the brain causes transient ischaemic attack and stroke, whilst damage to the vessels of the leg lead to intermittent claudication, critical limb (leg) ischaemia and possibly amputation. These diseases almost never develop in the absence of one or more risk factors.

The risk factors for atherosclerosis

The pathophysiology of this disease has been developing for over a century. Victorian pathologists recognized that deposits within the walls of arteries that caused their narrowing were rich in cholesterol and other fats. In the 20th century, these observations were extended, and other factors were discovered. It was subsequently shown that acute myocardial infarction (a heart attack) is the consequence of these cholesterol-rich deposits. Epidemiology studies also unequivocally demonstrated that tobacco smoking, diabetes and hypertension are also important risk factors. Other possible risk factors, such as stress, hyperhomocysteinaemia, and the lack of certain vitamins, have failed to gain widespread acceptance. However, it is clear that platelets have a major role in this disease, and there is evidence that inflammation may also be involved.

There are no clear single gene defects that lead directly and independently to atherosclerosis. Those genetic defects that are linked to this disease operate through diabetes, hypertension and dyslipidaemia.

The pathogenesis of atherosclerosis

It has long been known that the coronary arteries of many who die of a heart attack are infiltrated with lipids, so much so that the arteries are distended and narrowed. Early researchers found that most of this lipid resembled gruel, or, in Greek, *atherae*. This root combines with 'sclerosis', meaning thickening, to give atherosclerosis. The mass of cells and other material is called an atheroma. However, the damaged arteries of those with heart disease represent only the final stages of the disease.

The dominant theory of the pathogenesis of atherosclerosis is of 'response to injury'. This theory states the damage to, and/or dysfunction of, the endothelium (i.e. a lesion) is the first step from which all vascular pathology arises. Once this damage occurs, other steps follow that culminate in symptomatic, and then serious, disease. Although a continuum, the pathogenesis of atherosclerosis can be illustrated in a series of steps that, to some degree, follow an ordered pathway.

The initial phase   The target organ for the disease process is the endothelium, and we can find clear evidence of damage to these cells in all of the major risk factors in the absence of actual clinical disease. Cigarette smoke is an excellent example of a noxious factor toxic to many cells, including the endothelium. Another injurious feature may be high blood pressure, and there is evidence that the high blood glucose present in diabetes damages endothelial cells.

A damaged endothelium loses its barrier function, and so becomes unable to resist the movement of lipids from the plasma into the subendothelium and the media of the artery. This process is more rapid in hyperlipidaemia, and more so if there is also diabetes and hypertension. Lipid infiltration (especially LDL-cholesterol, which may become oxidized) results in the appearance of fatty streaks.

The development of atheroma   Monocytes are programmed to enter the tissues to scavenge for dead cells and damaged tissues – at which point they are described as macrophages, and when they ingest lipids are described as foam cells. A damaged or activated endothelium is more likely to promote the passage of monocytes into the subendothelium by increasing the expression of adhesion molecules, and this may accelerate the transformation of the fatty streak into an atheroma.

As the abnormality in the vessel wall develops, it expands to a degree where it starts to narrow the lumen of the blood vessel, which then restricts the blood flow, which then flows more rapidly. As the lipid deposits increase, there are also changes to the smooth muscle cells of the media, a consequence of which is failure to control blood pressure (often described as hardening of the arteries). In some cases there may be infiltration of calcium, which causes additional hardening. At this point a lipid- and platelet-rich lesion may be described as an atherosclerotic plaque.

The late stages   As the atherosclerotic plaque develops there is an increased narrowing (stenosis) and irregularity of the lumen of the vessel. As the atheroma within the plaque grows, its lipid-rich centre becomes necrotic, and this places strain on the outer margins of the lesion. This process may be promoted by enzymes released by macrophages within the plaque. As the endothelium and the nervous system fail to control the contractions of the smooth muscle cells in the media, arterial spasm can occur, causing irregular and temporary narrowing of the blood vessel. This may further reduce blood flow down an already partially occluded artery.

The final stages see physical pressure being placed on the weakened atherosclerotic plaque, which may be described as 'vulnerable', eventually leading to rupture, and spillage of atheroma and thrombotic material into the arterial blood stream. The pressure of blood flow will drive this material downstream until it becomes lodged in arterioles and capillaries. The devastating consequences of plaque rupture we shall discuss in the coming section.

The consequences of atherosclerosis

The purpose of arterial blood is to deliver oxygen and glucose to the tissues. Should it fail to do so, because the arteriole is blocked (occluded) by fragments of atheroma or by thrombosis, then the tissue that would be normally fed with oxygen and glucose is denied these nutrients and, unless blood supply is rapidly restored, will die. This process is called an infarction. Although infarction resulting from plaque rupture and/or thrombosis is certainly the most serious pathology, there can be other clinical consequences of atherosclerosis.

At rest, such as whilst sleeping, the oxygen demand of most muscles (including the heart) is low and can be satisfied by a relatively small amount of blood, so long as it is adequately oxygenated. However, at times of high activity of the muscle, such as in strenuous exercise, the same stenosed artery will not be able to satisfy the increased demand for oxygenated blood.

If deprived of sufficient oxygen, muscles can still obtain energy from glucose via anaerobic respiration. Unfortunately, this leads to the build-up of a metabolic by-product, lactic acid, high concentrations of which cause a cramp-like pain. Indeed, this also happens to athletes involved in long-term exercise when their muscles outstrip the ability of the cardiovascular system to supply oxygen.

The heart

The blood supply to the muscles of the heart (the myocardium) consists of coronary arteries running on the outer surface of the heart (the epicardium) and which then branch and burrow to deliver oxygen and glucose to the cardiomyocytes of the myocardium. If these arteries become narrowed by atherosclerosis, then at times of high oxygen demand there is insufficient delivery of oxygen and this results in ischaemia; that is, ischaemic heart disease. However, cardiomyocytes can generate some energy from glucose in the absence of oxygen (anaerobic respiration), but this inefficient process also generates lactic acid. The resulting pain associated with the excess lactic acid is angina, and is an established warning sign of more severe disease to come.

Pathophysiology of myocardial infarction   If the arterial supply is severely compromised, perhaps by several regions of stenosis, or an occlusion, then an infarction can occur. Because the downstream tissues that are damaged are cardiomyocytes, then the event is a myocardial infarction, or heart attack. The consequence of damage to the cardiomyocytes is that they will cease to contract and will eventually die. If the occlusion in the particular coronary artery is large, then a large mass of the myocardium will be affected, which can lead to irregularities in the heart beat, and possible death. Should the myocardial infarction be the result of the sudden rupture of an atherosclerotic plaque, then the symptoms of angina-like pain will come on rapidly, and if so we can further qualify the process as an acute myocardial infarction. However, there are other consequences of a myocardial infarction apart from the chest pain. These include changes in the pattern of the heart beat (as detected by the ECG) and changes in the concentrations of certain molecules in the blood. These changes include enzymes and other proteins found within cardiomyocytes.

Cardiac enzymes   The heart is clearly a very metabolically active organ, demanding the activity of numerous enzymes, such as aspartate aminotransferase (AST), lactate dehydrogenase and creatine kinase (CK).

- AST (EC 2.6.1.1) is widely distributed, and indeed is a liver function test (Chapter 17). Concentrations start to rise perhaps 12 h after a myocardial infarction, reaching a peak at 36 h.
- Lactate dehydrogenase (EC 1.1.1.27), catalysing the conversion of lactate to pyruvate, is very widely distributed throughout the body. It is present in cardiomyocytes, red blood cells and cells of other organs, so that raised concentrations in the blood may be the result of, for example, red cell destruction in haemolytic anaemia. Nevertheless, rising levels following the symptoms of a heart attack add to the confidence of the diagnosis of myocardial infarction. However, the peak concentration is generally much later than that of AST, often 36 h after the event.
- CK (EC 2.7.3.2) catalyses the phosphorylation of creatine, so consuming a molecule of ATP and generating phosphocreatine, a key step in an actively metabolizing cell such as a cardiomyocyte. However, CK is also found in skeletal and smooth muscle, and damage to muscles, such as of the legs after having run a marathon, or of the heart after a myocardial infarction, or in other muscle disease such as polymyositis; all result in increased concentrations

of this enzyme. Concentrations of CK begin to rise 4–6 h after a myocardial infarction, peaking at 24 h.

• Fortunately, we can distinguish damage to the myocardium from damage to other muscles, as a subtype of CK (i.e. CK-MB) is found almost exclusively in heart muscle. Other types of CK are CKMM, found mostly in skeletal muscle, and CKBB, found predominantly in smooth muscle. Thus, raised plasma CK-MB is specifically indicative of a heart attack. Broadly speaking, there is a clear relationship between increased CK and CK-MB and the mass of heart muscle that has been damaged, so that this blood test effectively tells us of the severity of the heart attack.

Muscle protein Myoglobin is effectively a small (18 kDa) version of haemoglobin within the cell, transporting oxygen through the cytoplasm to the mitochondrion. When muscle cells are damaged (as in myocardial infarction), myoglobin leaves the cell and enters the blood, raised concentrations being detected within 2–3 h of an infarction, peaking after 12 h.

Troponin is a complex of three isoforms (C, I and T) that bind to the actin–myosin complex of the muscle cell, and can be measured by immunoassay. However, unlike all the other molecules in this subsection, troponins are considerably more specific for cardiomyocytes, and leave the damaged cell after an infarction. Unfortunately, the best window for troponin measurement is 10–12 h after the presumed event, and normal concentrations after 12–24 h effectively rule out a myocardial infarction. A further problem is that there are many cardiac and non-cardiac causes of raised troponins, so that caution is required and alternative diagnoses considered (Table 14.5).

A recent potential addition to the panel of cardiac damage markers is heart-type fatty acid binding protein (hFABP). This 15 kDa protein product of the *FABP3* gene is released from cardiomyocytes after ischaemic damage, and so be may used in a similar manner to troponin, CK and CK-MB. However, levels of hFABP rise sooner after the ischaemic event, and so may be used in concert with other markers. However, like troponin, hFABP may be increased after pulmonary embolism, so caution is required.

The diagnosis and treatment of myocardial infarction None of the signs, symptoms and investigations we have discussed are, by themselves, good enough to make a firm diagnosis of a heart attack. This is needed so that the correct treatment is undertaken. The WHO

Table 14.5 Causes of a raised troponin.

Cardiac disease and interventions	Non-cardiac diseases
Cardiac amyloidosis, cardiac contusion, cardiac surgery, cardioversion, closure of atrial septal defect	Critically ill patients, high-dose chemotherapy, primary pulmonary hypertension
Dilated cardiomyopathy, heart failure, hypertrophic cardiomyopathy, myocarditis	Pulmonary embolism, renal failure, stroke, subarachnoid haemorrhage
Percutaneous coronary intervention, post-cardiac transplantation, supraventricular tachycardia	Scorpion envenoming, sepsis and septic shock, ultra-endurance exercise

recommends that a diagnosis be made only after a typical history of chest pain (such as behind the sternum, radiating to the neck, between the shoulder blades and to the left arm), and an increase in cardiac enzymes (such as CK and CK-MB), and typical ECG changes (such as ST segment elevation). It may be that the diagnosis is unclear, and that cardiac markers are obtained every 4- 6 hours to provide crucial confidence.

However, in contrast to WHO guidelines, that of NICE Clinical Guideline 95 on chest pain of recent onset focuses on history, ECG changes and troponins, measured initially and 10–12 h after onset of symptoms. Similarly, NICE Clinical Guideline 94, on unstable angina and non-ST elevation myocardial infarction, calls for blood tests for troponin I or T, creatinine, glucose and haemoglobin, and fails to mention CK or CK-MB.

These NICE guidelines also describe preferred treatment, which should begin as soon as possible, the objective of which is to restore blood flow, generally by dissolving the clot (fibrinolysis), or by angioplasty (which involves passing a wire into the coronary artery carrying the stenosis). If the disease is very severe, and not amendable to angioplasty, it may be necessary to create a 'new' artery to bypass the diseased or occluded artery. This 'new' artery is often a section of one of the patients' own veins, and so the procedure is called a coronary artery bypass graft.

Further treatments of a myocardial infarction whilst in hospital include anticoagulants (such as heparin), anti-platelet drugs (such as aspirin and clopidogrel) and drugs to relieve blood pressure (such as nitrates, beta blockers and calcium channel blockers). An additional treatment is a statin, regardless of the patient's cholesterol concentration. When discharged from hospital, the patient is likely to attend a cardiac rehabilitation class where further education takes place, such as on a healthy life style, with advice on exercise. Drug treatment, especially aspirin, is likely to continue for life. The patient will most likely return to the hospital for follow-up outpatient appointments with a cardiologist.

Heart failure The major driving force of the heartbeat is the contraction of the left ventricle, which ejects perhaps 60–80% of its volume of blood to the aorta each beat. If this ejection fraction falls, perhaps to 40% and below, then the heart becomes unable to provide the body with the blood it needs, leading to the diagnosis of heart failure. The major causes of the damage to the left ventricle include hypertension (which may have been undiagnosed for years) and myocardial infarction.

The heart responds to structural stress by releasing a variety of peptides that act on the kidneys to reduce blood volume by failing to reabsorb sodium. One of these natriuretic peptides, B type (hence BNP), is produced by atria and ventricles, and accordingly increased concentrations are released when the muscles of these chambers are under excessive strain. Isotypes of BNP include a prohormone (proBNP) and an amino terminal fraction, NTproBNP. Therefore, raised concentrations of BNP or NTproBNP can be used to diagnose heart failure, which is particularly valuable as heart failure is very difficult to define from symptoms alone.

NICE Clinical Guideline 108, on chronic heart failure, underlines the value of BNP and NTproBNP, as these molecules also predict poor prognosis. Patients with suspected heart failure and a BNP concentration >400 pg/mL (116 pmol/L) or an NTproBNP >2000 pg/ml (236 pmol/L) must be referred urgently to a specialist centre.

Peripheral artery disease

Atherosclerosis manifests itself in damage to arteries feeding the brain and arteries feeding the legs. As regards the brain, atheroma in the carotid arteries may rupture and shed thrombus into the blood which can flow in to the brain and cause a stroke. But a stenosed carotid artery will reduce blood flow, leading to a transient ischaemic attack, which has a parallel with the angina of ischaemic heart disease. However, stroke can be the result of clots arising from the heart.

Similarly, atherosclerosis of the iliac, femoral and popliteal arteries restricts blood flow to the leg and foot, causing intermittent claudication. Thrombus occluding arterioles causes critical limb ischaemia, which may lead to the amputation of toes and even the foot and ultimately the entire leg.

There are no reliable laboratory tests for the presence of peripheral atherosclerosis. However, risk factors can provide some surrogates of the presence or extent of disease. For example, a major cause of stroke is hypertension, and another is atrial fibrillation, neither of which can be assessed in the laboratory. Nevertheless, the disease process in diabetes inevitably causes occlusive disease of the leg and foot, with ulceration and gangrene.

14.4 Case studies

Case study 15

An overweight 45-year-old woman presents to her general practitioner with a history of increasing tiredness and lethargy. She also complains of vaginal infections with thrush and says she often needs a large cool drink when having a hot flush, which happens at least once, often twice a day. The practice has a near-patient testing device, which gives a blood glucose result of 12.0 mmol/L.

Interpretation The blood glucose is far above the reference range for a fasting sample of 3.5–5.5 mmol/L, but do we know this sample is fasting? If the sample is indeed fasting, it strongly implies diabetes. But given the history, and the importance of the potential diagnosis, the latter must be formally confirmed. For this, a fasting sample is required, and possibly also a full blood count with HbA_{1c} should be obtained. An OGTT may also be advised. An additional point to consider is the accuracy of the near-patient testing device, which can be prone to error, and which needs to be regularly checked.

Case study 16

A 25-year-old woman goes to a well-woman clinic because her 59-year-old father has just had a heart attack. On questioning, she never knew her grandfather as he died of a heart attack before she was born. The practitioner takes blood for a lipid profile. The results are as follows:

	Result (unit)	Reference range
Total cholesterol	7.8 mmol/L	2.5–5.0
HDL-cholesterol	1.9 mmol/L	>1.0
Triacylglycerols	4.9 mmol/L	<1.7
LDL-cholesterol	Uncalculable	<3.0

Interpretation   The sample clearly shows hypercholesterolaemia. However, there is also hypertriacylglycerolaemia, and LDL-cholesterol cannot be calculated. The latter is because the formula for LDL-cholesterol is only valid up to a triacylglycerol result of 4.5 mmol/L. This very high result suggests either a true clinical hypertriacylglycerolaemia or a recent meal rich in fats. Accordingly, for an accurate LDL and triacylglycerol result, a fasting concentration is required. However, this is not strictly necessary as neither the total cholesterol nor the HDL-cholesterol result should be influenced by fasting.

The picture is strongly suggestive of FH, which probably accounted for the father's heart attack and the premature death of the grandfather. Genetic testing will confirm if the disease is due to a mutation in the gene for the LDL receptor, for apoB, or another molecule. However, this is largely academic as the woman will almost certainly be treated with a statin regardless of the actual cause of the abnormality. NICE Clinical Guideline 71, on the identification and management of FH, describes a set of criteria, such as total cholesterol >7.5 mmol/L, that define the disease.

Summary

• Diabetes is the result of failure of glucose metabolism, caused by either a complete lack (type 1) or a relative lack and/or defective action (type 2) of insulin.
• In many cases, diabetes is preceded by IFG and IGT.
• The major risk factor for type 2 diabetes is obesity.
• The primary laboratory tests are fasting blood glucose, HbA_{1c} and the OGTT.
• A major crisis in hyperglycaemia is DKA.
• The long-term consequences of diabetes include atherosclerosis of small arteries (causing retinopathy, neuropathy and nephropathy) and large arteries (leading to

myocardial infarction, stroke, and critical limb ischaemia).
• The major groups of lipids include fatty acids, triacylglycerols, cholesterol and lipoproteins.
• Routine laboratory tests are total cholesterol and HDL-cholesterol, and triacylglycerols. LDL-cholesterol is calculated from an equation.
• Hypercholesterolaemia and hypertriacylglycerolaemia may arise from a fat-rich diet and/or the production of high amounts by the liver.
• Increased blood lipids, especially LDL-cholesterol, are a major risk factor for atherosclerosis, but are treatable with statins.
• Atherosclerosis is initiated by attack on the endothelium, causing thrombosis, hypertension and atheroma within the vessel wall.
• Thrombosis, perhaps from a ruptured atherosclerotic plaque, may cause an infarction, which if in the arteries of the heart leads to a heart attack.
• After a myocardial infarction, damaged heart muscle cells release CK and troponins, and so are essential in diagnosis. An important subtype of CK is CK-MB.

Further reading

Brown TM, Bittner V. Biomarkers of atherosclerosis: clinical applications. Current Cardiol Rep. 2008;10: 497–504.

Buijsse B, Simmons RK, Griffin SJ, Schulze MB. Risk assessment tools for identifying individuals at risk of developing type 2 diabetes. Epidemiol Rev. 2011;33: 46–62.

Charlton-Menys V, Durrington PN. Human cholesterol metabolism and therapeutic molecules. Exp Physiol. 2008;93:27–42.

Horejsí B, Ceska R. Apolipoproteins and atherosclerosis. Apolipoprotein E and apolipoprotein(a) as candidate genes of premature development of atherosclerosis. Physiol Res. 2000;49(Suppl 1):S63–S69.

Imamura M, Maeda S. Genetics of type 2 diabetes: the GWAS era and future perspectives. Endocrinol J. 2011;58:723–739.

Kassi E, Pervanidou P, Kaltsas G, Chrousos G. Metabolic syndrome: definitions and controversies. BMC Med. 2011;9:48.

Kitabchi AE, Umpierrez GE, Fisher JN. et al. Thirty years of personal experience in hyperglycemic crises: diabetic ketoacidosis and hyperglycemic hyperosmolar state. J Clin Endocrinol Metab. 2008;3:1541–1552.

Krentz AJ. Lipoprotein abnormalities and their consequences for patients with type 2 diabetes. Diabetes Obesity Metabol. 2003;(Suppl 1):S19–S27.

Lewis GF, Rader DJ. New insights into the regulation of HDL metabolism and reverse cholesterol transport. Circ Res. 2005;6:1221–1232.

Lippi G, Mattiuzzi C, Cervellin G. Critical review and meta-analysis on the combination of heart-type fatty acid binding protein (H-FABP) and troponin for early diagnosis of acute myocardial infarction. Clin Biochem. 2013;46:26–30.

Little RR, Rohlfing CL, Sacks DB. Status of hemoglobin A1c measurement and goals for improvement: from chaos to order for improving diabetes care. Clin Chem. 2011;57:205–210.

Miller M, Stone NJ, Ballantyne C. et al. Triglycerides and cardiovascular disease: a scientific statement from the American Heart Association. Circulation. 2011;123:2292–2333.

Thygesen K, Alpert JS, White HD. Universal definition of myocardial infarction. J Am Coll Cardiol. 2007;50:2173–2195.

Van Belle TL, Coppieters KT, von Herrath MG. Type 1 diabetes: etiology, immunology, and therapeutic strategies. Physiol Rev. 2011;91:79–118.

Wallace TM, Levy JC, Matthews DR. Use and abuse of HOMA modelling. Diabetes Care. 2004;27:1487–1495.

Wendland EM, Torloni MR, Falavigna M. et al. Gestational diabetes and pregnancy outcomes – a systematic review of the World Health Organization (WHO) and the International Association of Diabetes in Pregnancy Study Groups (IADPSG) diagnostic criteria. BMC Pregnancy Childbirth. 2102;12:23.

Guidelines

Catapano AL, Reiner Z, De Backer G. et al. ESC/EAS guidelines for the management of dyslipidaemias: The Task Force for the Management of Dyslipidaemias of the European Society of Cardiology (ESC) and the European Atherosclerosis Society (EAS). Atherosclerosis 2011;217(Suppl 1):S1–S44.

JBS2: Joint British Societies' guidelines on prevention of cardiovascular disease in clinical practice. Heart. 2005;91:1–52.

NICE Clinical Guideline CG15: Type 1 diabetes: Diagnosis and management of type 1 diabetes in children, young people and adults.

NICE Clinical Guideline 48: Secondary prevention in primary and secondary care for patients following a myocardial infarction.

NICE Clinical Guideline CG63: Diabetes in pregnancy: Management of diabetes and its complications from pre-conception to the postnatal period.

NICE Clinical Guideline CG67: Lipid modification: Cardiovascular risk assessment and the modification of blood lipids for the primary and secondary prevention of cardiovascular disease.

NICE Clinical Guideline 71: Identification and management of familial hypercholesterolaemia.

NICE Clinical Guideline CG107: Chronic heart failure.

Sinclair AJ, Paolisso G, Castro M. et al. European Diabetes Working Party for Older People 2011 clinical guidelines for type 2 diabetes mellitus. Diabetes Metab. 2011;37(Suppl 3):S27–S38.

Web sites

Diabetes UK website. www.diabetes.org.uk.

International Diabetes Federation, The IDF consensus worldwide definition of the metabolic syndrome, 2006. www.idf.org.

National institute for Health and Clinical Excellence (NICE) www.nice.org.uk.

National Service Framework for diabetes: http://www.nhs.uk/NHSEngland/NSF/Pages/Diabetes.aspx.

World Health Organization/International Diabetes Federation, The definition and diagnosis of diabetes mellitus and intermediate hyperglycaemia. Report of a WHO/IDF consultation. World Health Organization, 2006. www.WHO.org.

15 Calcium, Phosphate, Magnesium and Bone Disease

Learning objectives

After studying this chapter, you should be able to:

- describe the physiology of calcium, phosphate and magnesium homeostasis;
- list the roles of parathyroid hormone in regulating calcium concentrations;
- explain the synthesis of vitamin D and its roles in bone metabolism;
- understand the laboratory assessment of these molecules;
- discuss the causes and consequences of hyper- and hypocalcaemia;
- describe the causes and consequences of hyper- and hypomagnesaemia;
- be aware of different types of bone disease;
- understand the contribution of the laboratory to the diagnosis and treatment of bone disease.

Bone is a complex structure of calcium and phosphates, formed into a quasi-crystalline structure by specialized cells that provides the body with overall strength, key organs with protection, and attachment points for the lever action of muscles. A key cofactor in healthy bone formation is vitamin D, the production of which itself requires good renal and liver function, and adequate sunlight. Bone is a living tissue that also hosts the progenitor stem cells that give rise to mature red cells, white cells and platelets. Together, the two most prevalent bone diseases, osteoporosis and osteoarthritis, cause a considerable degree of morbidity (in terms of pain and disability) and so have a high cost burden for the National Health Service and the UK economy in general, but are rarely a direct cause of excess mortality.

Calcium, phosphate and magnesium are inorganic minerals and key components of this tissue. Bone contains almost all of the body's calcium and the greater proportion of its phosphate and magnesium. In addition to its mechanical role in locomotion and protection of organs (such as the brain, lungs and heart), bone also acts as a reserve for these minerals. Indeed, the concentration of calcium in the plasma is maintained within its reference range by absorption of dietary calcium and also by the exchange of calcium between blood and bone. Changes in plasma calcium concentrations above or below its reference range can result in clinical disorders that have characteristic clinical features, and in extreme cases may be fatal. Similarly, homeostatic control of phosphate and magnesium concentrations is essential to prevent disorders associated with changes in their plasma concentrations.

Those routine blood tests most applicable to calcium and bone are listed in Table 15.1. Albumin is present in this panel because it binds to, and carries, some calcium. The inclusion of alkaline phosphatase (ALP) in this chapter is justified because concentrations in the blood may arise from bone, with high levels in certain bone diseases. Indeed, many consider a 'bone panel' of these tests, in the same way that the liver function tests (LFTs) are valuable in hepatic disease, and where urine and electrolytes (U&Es) mark renal function.

This chapter will first discuss the physiology, biochemistry and homeostasis of calcium, phosphate and magnesium in Sections 15.1–15.3, whilst Section 15.4 looks at the laboratory measurement of these ions and related molecules. In Sections 15.5–15.7 we will look at the pathology of these metabolites. How calcium and phosphate come together to form bone, and its own diseases, will be examined in Section 15.8 and Section 15.9 respectively. The roles of the four members of the 'supporting cast' of vitamin D, parathyroid hormone (PTH), albumin, and ALP will be discussed as the particular text demands. We will conclude with case studies in Section 15.10.

Blood Science: Principles and Pathology, First Edition. Andrew Blann and Nessar Ahmed.
© 2014 John Wiley & Sons, Ltd. Published 2014 by John Wiley & Sons, Ltd.

Table 15.1 Major blood tests in the biology of bone.

Analyte	Reference range
Total calcium	2.2–2.6 mmol/L
Free calcium	1.2–1.37 mmol/L
Phosphate	0.8–1.4 mmol/L
Magnesium	0.8–1.2 mmol/L
Vitamin D[a]	20–150 nmol/L
Alkaline phosphatase	30–130 IU/L
Albumin	35–50 g/L
Parathyroid hormone	1–6 pmol/L

[a]As we will see, the biochemistry of vitamin D is complex, and there are several isoforms, often (adding to the confusion) with more than one name.

15.1 Calcium

Calcium is the most abundant mineral in the body. An average adult weighing 70 kg has approximately 1 kg of calcium, distributed around various body compartments of the blood, the extracellular fluid (ECF) and bone. Almost all body's calcium (99%) is present in the bone in the form of complex crystals similar to those of calcium hydroxyapatite $(Ca_{10}[PO_4]_6[OH]_2)$. The total calcium content of ECF is approximately 23 mmol, of which 9 is present in the plasma. About 100 mmol of calcium in the bone is exchangeable with the ECF; and indeed, a total of 500 mmol is exchanged daily.

The biology of calcium

In addition to its role in bone, calcium is also required for release of neurotransmitters and the initiation of muscle contraction. As an intracellular signal it acts as a second messenger for the action of hormones, growth factors, and so on following their interaction with receptors on target cells. Calcium also acts as a cofactor for many enzymes, such as those of the coagulation pathway. Indeed, sequestration of calcium prevents clot formation, and is why it is the object of anticoagulants sodium citrate and ethylenediaminetetraacetic acid (EDTA). This is why calcium cannot be measured in most samples of anticoagulated blood, and so we refer to serum calcium, although calcium can be measured in blood anticoagulated with heparin. However, even measurement of calcium in serum is controversial as it has been argued that some calcium in the serum from clotted blood may have arisen from stores within platelets.

A normal healthy diet provides approximately 25 mmol of calcium per day, and is the recommended daily allowance (perhaps 1000 mg/day), often achieved through dairy products, nuts and vegetables. Calcium is absorbed across the intestinal wall and enters the blood as the free, divalent ionized cation (Ca^{2+}), and approximately half of it binds nonspecifically with albumin. However, this binding is dependent on the pH, a fact that may have (as we shall see) pathological significance. The difference between free and bound calcium (together giving total calcium) is important and will be returned to more than once. For example, at the healthy glomerulus, albumin does not pass through into the urinary filtrate, whereas free calcium does, and is therefore subject to reabsorption.

Only the ionized form of calcium is physiologically active, and it is this fraction which is regulated by homeostatic mechanisms. Perhaps 240 mmol of calcium is filtered by the kidneys every day, where the bulk of this is reabsorbed in the tubules, and the remainder, some 5 mmol per day, is lost in the urine. Gastrointestinal secretions also contain calcium, and some of this is reabsorbed together with dietary calcium. Around 20 mmol of calcium is lost in the faeces daily, as that which is not absorbed from the diet and that present in shed enterocytes (Figure 15.1).

The regulation of calcium in the blood is determined by homeostatic mechanisms of the body in response to free calcium, and concentrations are maintained by the action of two regulators: PTH and vitamin D. As indicated, the metabolism of the latter is complex, and for the present we shall refer to it simply as vitamin D. Full details will follow.

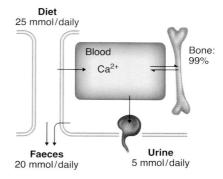

Figure 15.1 Distribution and turnover of calcium. Calcium in the diet passes through the intestines into the blood, where some binds to albumin. It may then pass into the bone or be excreted in the urine.

Parathyroid hormone

PTH is released from the parathyroid glands, four or more small bodies found behind the thyroid gland in the neck. When the concentration of blood calcium falls below a certain threshold, this are detected by calcium-sensing receptors in the parathyroid glands that are stimulated to release PTH into the bloodstream.

PTH is a polypeptide hormone consisting of 84 amino acid residues generated by a gene on chromosome 11 as a large precursor molecule (preproPTH) of 115 amino acid residues. In the rough endoplasmic reticulum of parathyroid cells, 25 residues are cleaved to produce proPTH and then further processing in the Golgi apparatus results in cleavage of six residues to give mature PTH.

The sequences of amino acids removed are thought to be necessary in the intracellular transport of the hormone. The biological activity of PTH is situated in the amino terminal 1 to 34 amino acid residues. It is stored in granules until its release and has a half-life of 5–10 min in the blood before being rapidly metabolized in the liver and excreted by the kidneys.

Following metabolism, fragments of PTH are produced and can be detected in the blood together with intact PTH. These fragments include an amino terminal fragment which is biologically active and exists for 5–10 min, a carboxyl terminal fragment with a half-life of 2–3 h and other smaller fragments. Immunoassays for PTH had the limitation that they detected some of these fragments that are biologically inactive, thus making interpretation of the results difficult. However, more recent immunoassays measure only the intact PTH and provide more reliable results.

After its release into the bloodstream, PTH has a number of actions:

• It acts on the bone to stimulate bone resorption by cells residing within the matrix of the bone, a process that therefore releases calcium (and phosphate) and so increases the plasma calcium concentration.
• It acts on the kidneys to stimulate 1-hydroxylation of vitamin D, which in turn acts on the gastrointestinal tract (GIT) to promote the absorption of calcium.
• A second action on the kidney is to act on the tubules to stimulate the reabsorption of free calcium from the filtrate.
• PTH decreases bicarbonate reabsorption by the kidneys so that more bicarbonate is lost in the urine. This produces more acidic conditions in the blood, which in turn helps to raise the ionized calcium concentration.

• The action of PTH on the kidneys also reduces reabsorption of phosphate, thus promoting its loss in the urine.

Eventually, these actions will bring calcium back into the reference range, but will continue to add to blood until concentrations exceed a second (upper) preset level, at which point the parathyroid glands will stop releasing PTH. As concentrations of the hormone fall, the five bullet points are reversed, such as calcium is not reabsorbed from the urinary filtrate or absorbed from the intestines. Thus, concentrations will eventually fall, until they reach the lower preset level that triggers PTH release, leading to rising concentrations. This system of feedback regulation is very common in endocrinology: the end product (calcium) regulates and is regulated by the mediator (PTH). These mechanisms are illustrated in Figure 15.2.

Structurally related to PTH is PTH-related protein (PTHrP), a molecule with roles in breast, bone and tooth development. If excess concentrations are produced, as perhaps by a tumour, then it can adopt true PTH-like

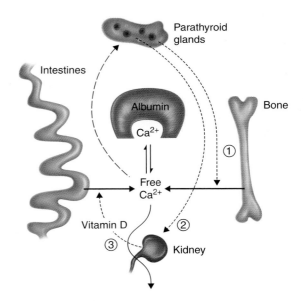

Figure 15.2 Role of PTH in regulating blood calcium. Falling concentrations of free calcium induce the release of PTH from the parathyroid glands, with three major consequences: (1) stimulation of cells within the bone, (2) reabsorption of calcium from urine by the renal tubules and (3) stimulation of the generation of active vitamin D that promotes the absorption of calcium in the intestines.

characteristics and lead to raised concentrations of calcium.

Calcitonin

This hormone of 32 amino acid residues is secreted by the parafollicular or C cells of the thyroid gland. As secretion is prompted by high concentrations of calcium, its main functions therefore oppose those of PTH. These are:

- reduction of calcium absorption by the intestine;
- inhibition of bone resorption by bone cells; and
- reduction in the renal tubular reabsorption of calcium, so allowing it to remain in the filtrate and so the urine.

Therefore, the effects of all three are to reduce the concentration of calcium in the blood.

Vitamin D

The synthesis of this molecule is complex, with different forms having different names. Historically, when vitamin D was first identified, and it was recognized that various forms were present, one particular preparation was named vitamin D_1. This notation is no longer used. Over the years, consensus on naming has been poor and subject to changes over time (in that old textbooks will not reflect modern usage) and according to geography (USA, UK, Europe). Nevertheless, there is a great deal of agreement, such as that a good place to start is with cholesterol, as follows:

- Ergosterol (a type of cholesterol in vegetable matter) is converted in the skin to vitamin D_2 (also known as ergocalciferol) by the ultraviolet wavelengths of sunlight.
- In the liver, 7-dehydrocholesterol (provitamin D) is produced as a component of the synthesis of mature cholesterol, but in the skin is also acted upon by the ultraviolet light in sunshine, the product being vitamin D_3 (cholecalciferol, or perhaps simply calciferol).
- In the liver, both species are hydroxylated in a reaction catalysed by 25-hydroxylase to form 25-hydroxycholecalciferol (25-HCC), or calcidiol.
- A second hydroxylation of 25-HCC occurs due to the activity of 1-alpha-hydroxylase in the kidneys, to form the active form of the vitamin: 1,25-dihydroxycholecalciferol (1,25-DHCC) or calcitriol.

When plasma calcium concentration is low, release of PTH occurs and can stimulate activity of the enzyme

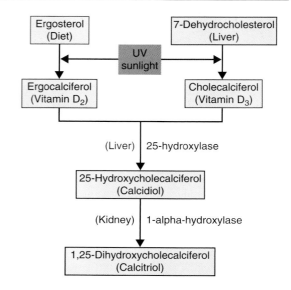

Figure 15.3 Synthesis of Vitamin D. Various isoforms of vitamin D are metabolized, often by hydroxylation, in the skin, liver and kidney. The most active form of vitamin D is the 1,25-dihydroxy species known as calcitriol.

1-hydroxylase in the kidneys, increasing formation of calcitriol. However, when plasma calcium is normal or high the enzyme 24-hydroxylase then becomes active and converts some of the 25-HCC to 24,25-DHCC, which has very weak activity with regard to calcium reabsorption from the gut. These steps are summarized in Figure 15.3.

Biologically active vitamin D (i.e. calcitriol) has a number of functions, which include:

- Stimulation of the increased absorption of calcium across the intestinal wall. However, in order to maintain electrical neutrality, the transport of positively charged calcium ions by the intestinal epithelial cells must be balanced by the import of negatively charged anions. This is most efficiently achieved by the accumulation of inorganic phosphate ions (HPO_4^{2-}); thus, calcitriol also stimulates the intestinal absorption of phosphate.
- Like PTH, calcitriol also stimulates an increase in the renal tubular reabsorption of calcium, therefore reducing losses of calcium in the urine and helping to increase plasma calcium concentrations.
- It also stimulates the release of calcium from bone by acting on osteoclasts in the bone, causing bone resorption, which increases plasma calcium concentrations.

- Calcitriol also inhibits the release of calcitonin, the hormone that reduces plasma calcium by inhibiting release of calcium from bone.

15.2 Phosphates

From a pure chemistry viewpoint, the root molecule is phosphoric acid, H_3PO_4, which ionizes in the blood to $H_2PO_4^-$ and then to HPO_4^{2-}, with the concurrent generation of protons (H^+). The phosphate anion is one of the dominant species (alongside chloride, hydroxide and bicarbonate) in human biochemistry, and is often balanced electrostatically by Ca^{2+}. However, the overriding feature of this relationship is that the chemistries of the two ions complement each other and form the hard crystalline structure of calcium hydroxyapatite, the mineral component of bone and teeth.

The biology of phosphate

This mineral is required in many aspects of the biochemistry of the body to, for example to maintain cell wall integrity and participate in metabolic processes such as glycolysis and oxidative phosphorylation. As regards the former, the importance of phospholipids was discussed in Chapter 14. Phosphates are a key component of molecules such as adenosine monophosphate (AMP), adenosine diphosphate (ADP) and adenosine triphosphate (ATP), and function as a urinary buffer for excretion of H^+ ions in the kidneys (as discussed in Chapter 13). It is also needed for synthesis of 2,3-diphosphoglycerate, which regulates dissociation of oxygen from oxyhaemoglobin (further details in Chapter 5). Phosphate is required for phosphorylation and dephosphorylation, which controls activity of many enzymes and intracellular second messengers, such as the activation of glycogen synthetase kinase-1. Some describe 'free' phosphate in the blood as inorganic, in order to distinguish it from that fraction of phosphate that is a constituent of molecules such as ATP.

The majority of body phosphate (i.e. about 80–85%) is in the bone, the remainder being 15% in cells and 0.1% in the ECF. Most of the intracellular phosphate is attached covalently to lipids and proteins. The daily intake of phosphate is around 40 mmol (derived mainly from dairy products and green vegetables), of which 14 mmol is lost in the faeces. Perhaps 26 mmol of phosphate is lost daily in the urine.

Regulation of plasma phosphate

As we have already noted, phosphate concentrations in the blood are part-controlled by PTH and the vitamin D isoform calcitriol. PTH decreases phosphate reabsorption in the kidneys, causing loss of phosphate in the urine and a fall in plasma phosphate concentrations. Calcitriol increases phosphate absorption in the gut and, therefore, raises the concentration of plasma phosphate.

15.3 Magnesium

This is the body's fourth most common cation and eleventh most common element. However, 99% of it is in the bone, muscles and other tissues. The fact that only 1% of total body magnesium is in the blood means that (like calcium) measurement of this electrolyte does not necessarily reflect body stores. The reference range for magnesium is shown in Table 15.1. Magnesium acts as a cofactor for some 300 enzymes and is required for maintenance of structures of ribosomes, nucleic acids and certain proteins. It is also required for normal neuromuscular function and for the synthesis and release of PTH.

Regulation of magnesium

A normal healthy daily diet contains perhaps about 15 mmol of magnesium, of which some 30% is absorbed in the GIT, the remainder being lost in the faeces. Free magnesium crosses the enterocyte, enters the plasma, and some is adsorbed nonspecifically on to carrier proteins. The free magnesium crosses the glomerulus, and regulation is achieved mainly by reabsorption in the proximal tubules and loop of Henle in the kidneys. It follows that renal disease (to be described below) has an impact on blood concentrations. Furthermore, several factors can influence the rate of excretion of magnesium. Both hypercalcaemia and hypophosphataemia decrease renal reabsorption of magnesium promoting its loss in the urine. PTH increases renal reabsorption of magnesium.

15.4 The laboratory

Laboratory determination of calcium

As discussed, serum is the preferred sample for the measurement of calcium, although heparin is an acceptable anticoagulant, but it can also be measured in urine. Measurements of calcium can determine both free (ionized) and total (free plus that fraction bound to plasma proteins, notably albumin). As it is the free, ionized form that is biologically active, it is the preferred measure, but is more difficult to perform analytically.

Plasma proteins will become insoluble in acid conditions, and so can be removed by centrifugation, free calcium being measurable in the supernatant. Most routine biochemistry methods rely on the reaction of calcium with δ-cresolphthalein at a high pH, giving a red colour measurable at a wavelength of 570–580 nm.

The albumin issue The most common methods for measuring calcium used to determine its total concentration. However, measurements of total calcium are affected by changes in albumin concentrations, which can give rise to misleading results. Changes in the concentrations of albumin, for whatever reason, can affect that of total calcium because approximately half of the calcium in blood is bound to albumin. For example, when albumin concentrations increase, total calcium measurements are also increased. Similarly, when albumin concentrations decline, then total calcium concentrations also decline. However, the concentrations of free unbound, ionized calcium are *not* affected by changes in albumin so there is no physiological or pathological change. Two formulae are used to modify measured calcium concentrations to give a *corrected* calcium value. For albumin concentrations less than 40 g/L, the corrected calcium can be determined by

$$Corrected[calcium] = measured\ total[calcium] + 0.02(40 - [albumin])$$

For albumin concentrations greater than 45 g/L, the corrected calcium is determined by

$$Corrected[calcium] = measured\ total[calcium] - 0.02([albumin] - 45)$$

The alternative is to measure the ionized calcium directly, perhaps using an ion-selective electrode, which is currently the method of choice.

The effect of pH

The binding of calcium to albumin is not covalent but relies on electrostatic forces. Thus, changes in acid–base homeostasis can influence the balance between free and bound calcium but have no effect on total calcium concentrations. Ionized calcium ions compete with hydrogen ions (H^+) for negative binding sites on albumin, so that changes in the ionized fraction may occur in acute acid–base disorders.

In alkalosis, the concentration of H^+ in the plasma is low and so more calcium ions bind to albumin as H^+ dissociates from albumin, thus increasing protein-bound calcium. Thus, the concentration of ionized calcium may decline below its reference range and produce the associated symptoms of hypocalcaemia.

However, in acidosis, the concentration of plasma H^+ increases and more of these will bind to the negative sites on the albumin, displacing the calcium ions; therefor, the effect is to decrease protein-bound calcium but increase free plasma ionized calcium. When the ionized calcium concentration rises above its reference range, then symptoms typical of high plasma calcium can be seen in affected patients.

Laboratory measurement of phosphate

A common method for the measurement of phosphate relies on its reaction with ammonium molybdate under acid conditions. Although this forms a colourless phosphomolybdate complex, it can be measured at 340 nm in the ultraviolet part of the spectrum. However, a further reaction with reducing agents gives a blue-coloured complex which can be measured at 600–700 nm.

In common with calcium, anticoagulants such as EDTA, sodium citrate and fluoride oxalate cannot be used, but in the case of phosphate it is because they interfere with the formation of molybdate complexes.

Laboratory measurement of magnesium

In common with many other divalent cations (such as calcium), magnesium is present in the blood mostly (55%) in a free ionized form as Mg^{2+}, the remainder being bound to albumin and other carriers, and also complexed with other ions such as phosphate.

Blood must be collected into heparinized plasma, or no anticoagulant, thereby providing serum. This is because, as with some other electrolytes, the anticoagulants EDTA, citrate and oxalate will chelate the magnesium, rendering it unmeasurable. In the laboratory, three methods are available:

• The reaction between magnesium and dyes such as formazan and methylthymol blue, providing a colour change that can be easily quantified by a spectrophotometer.

- Atomic absorption spectrophotometry, which measures the emission of magnesium-specific wavelengths of light. However, the sample must be cleared of protein, perhaps by lanthanum hydrochloride.
- There are also specific ion electrodes that measure free magnesium.

Vitamin D isoforms

The complex nature of the different isoforms of vitamin D make accurate assessment of the biologically active form (1,25-DHCC) difficult. Radioimmunoassay and the ability of vitamin D to complex with binding proteins are available, but the most popular methods include isotope-dilution liquid chromatography–mass spectrometry and tandem mass spectrometry (Chapter 2). These methods rely on complex analysers that are generally to be found in reference laboratories.

In one such complex method, liquid–liquid extraction is used to prepare serum or plasma samples for analysis, and zinc sulphate used to enhance the response and aid protein precipitation. Methanol is used to deproteinate samples and release vitamin D from its binding protein. The freed vitamin D escapes to hexane, an extremely hydrophobic solvent. After centrifugation, the vitamin D rich *n*-hexane layer is transferred to an analysis vial, dried down under nitrogen gas to concentrate the 25-hydroxy vitamin D, reconstituted in mobile phase and is analysed by liquid chromatography–mass spectrometry. The tandem mass spectrometry system ionizes these samples and detects specific parent daughter ion transitions for 25-hydroxy vitamin D_3 and D_2.

Parathyroid hormone and calcitonin

As small peptides, probably the most convenient method is immunoassay, of which enzyme-linked immunosorbent assay (ELISA) is perhaps the most popular. However, a problem is that the particular monoclonal antibody may recognize a part of the target molecule (the epitope) that is shared with other isoforms. This may be a problem in differentiating PTH, preproPTH, proPTH and PTHrP. Close attention to the specificity and sensitivity of the particular method is required.

15.5   Disorders of calcium homeostasis

Disorders of calcium homeostasis occur when the free ionized or corrected calcium concentration falls outside its reference range. Hypercalcaemia describes high plasma calcium concentrations above the reference range, whereas hypocalcaemia is characterized by calcium concentrations below its reference range. Abnormalities of calcium homeostasis are often detected by accident, when blood is tested for another reason, or when calcium is monitored in diseases associated with abnormal calcium homeostasis or because of the presence of clinical features of hypo- or hypercalcaemia.

The causes of hypercalcaemia

Raised calcium is often an asymptomatic laboratory finding, but care must be taken as hypercalcaemia is often indicative of other diseases. In general, most causes of hypercalcaemia can be classified as being due to malignancy, an overactive parathyroid gland (with or without vitamin D abnormalities), and renal issues (including renal failure), but there are of course dozens of rare and unusual causes. The most common causes of hypercalcaemia are the presence of malignant tumours and hyperparathyroidism.

Tumours of breast, prostate and certain types of lung cancers are known to secrete PTHrP, which has an activity similar to that of PTH in that it can increase the concentration of plasma calcium. However, unlike PTH, the release of PTHrP is not controlled by the concentration of ionized calcium. Furthermore, cytokines and prostaglandins produced by tumours that metastasize to bones may increase bone resorption, which, in turn, releases calcium and contributes towards the raised plasma concentrations.

A second aspect of the hypercalcaemia of malignancy is a cancer within the bone itself. This may be a cancer of the bone's own tissues, such as a myeloma, or deposits of distant cancer that have metastasized to bone. In both cases, in order for the tumour to grow, it must make space within the matrix of the bone. One way of achieving this is to trick the bone-resorbing cells into dissolving bone; another is for the tumour to directly secrete ALP, which is involved in bone remodelling. Both lead to increased release of calcium from the bone; hence the hypercalcaemia. In theory, the parathyroids are unaffected by the cancer, and so continue to act normally. If so, they are likely to respond to the hypercalcaemia by reducing their release of PTH, so that concentrations will be low or even absent.

Hyperparathyroidism, an overactivity of the parathyroid glands, is the second most common cause of hypercalcaemia. Its occurrence is unrelated to age and sex, but is found most commonly in post-menopausal women. It results in excessive production of PTH, leading to hypercalcaemia. Two distinct types of this condition are recognized. In primary hyperparathyroidism, there is an increased production of PTH, possibly due to the development of a parathyroid adenoma that secretes PTH and/or PTHrP. Thus, there will be both a high calcium and a high PTH.

The kidney has a special place in calcium metabolism, as is evident from Figure 15.1, Figure 15.2 and Figure 15.3, but mostly from the viewpoint of low calcium. However, in hypertension, the use of thiazide diuretics can reduce renal calcium excretion and can lead to hypercalcaemia, although this rarely occurs in people with otherwise normal calcium metabolism. It is more likely to occur in patients with increased bone resorption or hyperparathyroidism. Hypercalcaemia is occasionally seen in acromegaly, probably due to stimulation of 1-hydroxylase activity in the kidneys (and so raised vitamin D stimulating intestinal calcium absorption) by excess growth hormone.

There are many rare and unusual causes of hypercalcaemia. Some of these are as follows:

• An excess intake of vitamin D supplements.
• Granulomatous diseases such as tuberculosis or sarcoidosis. The granulomas contain macrophages in which there is increased conversion of 25-HCC to 1,25-DHCC, which in turn increases plasma calcium.
• Idiopathic hypercalcaemia of infancy is a condition associated with hypercalcaemia due to increased sensitivity to vitamin D by bone and gut, but the precise mechanism by which this occurs is unknown.
• The ingestion of large amounts of milk together with alkali antacids (hence 'milk–alkali syndrome'), such as bicarbonate to relieve symptoms of peptic ulceration. This can cause an alkalosis and reduce renal calcium excretion, giving rise to hypercalcaemia.
• Patients with bipolar disorder and on lithium therapy may develop mild hypercalcaemia. Lithium probably raises the cut-off point for PTH inhibition by calcium.
• Although in hyperthyroidism there are often increased thyroid hormones, they have no direct effect on calcium homeostasis but can increase bone turnover by increasing osteoclast activity, giving rise to mild hypercalcaemia.
• A second aspect of hyperthyroidism is that the innocent parathyroid glands may be stimulated to overproduce PTH.
• When patients are immobilized for significant periods of time, there is decreased bone formation but continual resorption of bone and release of calcium that is lost in the urine.
• Hypercalcaemia may occur in such patients who also suffer from increased bone turnover; for example, as in Paget's disease.

The consequences of hypercalcaemia

The clinical diagnosis of mild hypercalcaemia is difficult (if not impossible) because it is either asymptomatic, or symptoms are vague and nonspecific. However, as the calcium concentrations rise, more disease-specific symptoms appear. Amongst these are changes resulting from interference with nervous system function. These include tremor and palpitations, the latter of which can be detected by ECG. These arrhythmias may result in cardiac arrest. Hypercalcaemia may also depress neuromuscular excitability, and this may present as constipation and abdominal pain. Further neurological effects include an adverse effect on the central nervous system, causing depression, nausea, vomiting and often dehydration in affected patients. Gastrin release can also be stimulated by hypercalcaemia, leading to excessive gastric acid secretion in the stomach, which in turn can cause peptic ulceration. An additional serious consequence of prolonged hypercalcaemia is renal damage.

Investigation and management of hypercalcaemia

A number of biochemical tests can assist in the diagnosis of hypercalcaemia and its causes. Tests used to investigate suspected hypercalcaemia include measurement of concentrations of ionized or total calcium. If total calcium is measured, then albumin concentrations need to be determined in case corrected calcium measurements are required. Measurements of PTH are also valuable in diagnosis of primary hyperparathyroidism, one of the common causes of hypercalcaemia.

If a patient is suspected of hypercalcaemia, then usually the corrected calcium is determined; and if greater than 2.6 mmol/L, this indicates hypercalcaemia. Patients who have very high concentrations of calcium equal to or in excess of 3.5 mmol/L require urgent treatment to correct the hypercalcaemia. Measurement of PTH will help to determine whether the cause is malignancy, where there will be a low or undetectable PTH concentration, or primary hyperparathyroidism, which is characterized by a high PTH concentration.

How hypercalcaemia is treated depends on both its severity and cause. Patients with high concentrations of calcium, perhaps over 3.5 mmol/L, need to be treated urgently and may require dialysis or emergency parathyroidectomy. A common approach is to identify the underlying cause of the hypercalcaemia and treat this wherever possible. For example, surgical removal of a parathyroid adenoma may be necessary in patients with hypercalcaemia due to primary hyperparathyroidism. Intravenous saline should be administered in dehydrated patients to restore the glomerular filtration rate (GFR), which not only improves hydration but also enhances calcium excretion. In some patients, drugs such as frusemide may be used to inhibit the renal reabsorption of calcium and promote calcium excretion. Other drugs, such as bisphosphonates, lower plasma calcium concentration by inhibiting bone resorption and may be of value in treating hypercalcaemia.

The causes of hypocalcaemia

Recall that low calcium (<2.2 mmol/L) by itself is not necessarily true hypocalcaemia; it may be related to a low albumin, and corrected calcium may rise to be within the normal range. Investigation centres on abnormal intake and regulation. Low concentrations of calcium, like high concentrations, are generally due to one of a small number of causes, but there are also many infrequent causes.

As vitamin D is required for calcium absorption across the intestinal walls, then calcium deficiency can arise in an individual due to dietary deficiency of vitamin D or its malabsorption perhaps in inflammatory bowel disease. It can also occur because of inadequate synthesis of vitamin D in the skin due to lack of exposure to sunlight, as may be present in individuals in certain communities who cover their skin for religious, medical and/or cultural reasons.

In chronic renal failure there may be a decrease in the absorption of calcium from the intestines because of the decreased synthesis of vitamin D isotype 1,25-DHCC from 25-HCC in the kidney. If so, there may be increased output of PTH due to the hypocalcaemia, which may cause metabolic bone disease because of its effect on osteoclastic activity. However, low blood calcium may simply be the result of the failure to collect calcium from the filtrate, so that concentrations in the urine will be high.

Hypocalcaemia may occur in patients suffering from hypoparathyroidism, a condition characterized by decreased output of PTH. Hypoparathyroidism can be divided into two types: congenital hypoparathyroidism, where there is congenital absence of the parathyroid glands, and acquired hypoparathyroidism, which can be idiopathic or autoimmune in nature or arise following surgery of the parathyroid glands (i.e. a parathyroidectomy). There may also be secondary damage to the parathyroid glands due to disease of the thyroid and its treatment, such as poor surgical technique and radioactive iodine (as discussed in Chapter 18).

Like hypercalcaemia, there are several uncommon reasons for hypocalcaemia. These include:

- Pseudohypoparathyroidism, which is a condition where excessive secretion of PTH occurs because the target tissues fail to respond to this hormone (resistance) and hypocalcaemia persists. This condition is more common in males than females. Patients present with skeletal abnormalities, including a short stature, mental retardation, cataracts and testicular atrophy.
- Hungry bone syndrome. This can cause hypocalcaemia following surgical treatment of hyperparathyroidism in patients who have had prolonged secondary or tertiary hyperparathyroidism. In this condition, calcium from the blood is rapidly deposited in the bone, causing hypocalcaemia.
- Striated muscle cell damage (rhabdomyolysis). A consequence of this is the release of intracellular phosphates into the blood. In response to this, calcium may be moved out of the blood and become deposited in the bones and extraskeletal tissues. Similarly, intravenous phosphate administration may cause hypocalcaemia.
- Magnesium deficiency. Magnesium is required for both secretion and action of PTH, and a deficiency of this cation produces hypocalcaemia.
- Acute pancreatitis. During this, calcium can precipitate in the abdomen, causing hypocalcaemia. There is often decreased PTH secretion and increased calcitonin release, which may also contribute towards the hypocalcaemia.

Artefactual indications of often profound hypocalcaemia are common. These arise when blood is accidentally collected into tubes containing the anticoagulant EDTA or citrate.

Investigation and management of hypocalcaemia

Like hypercalcaemia, common features of hypocalcaemia include neuromuscular dysfunction, as calcium is an essential requirement at the synapse. Therefore, lack of calcium leads to poor message conduction and 'confusion' of muscles, which becomes manifest as tremor. Typical symptoms include peripheral neuropathy such as parasthesia ('pins and needles' sensation) affecting the hands, feet and perioral regions of the body. In some patients, fatigue and anxiety may occur, along with muscle cramps which can be painful and sometimes may progress to tetany affecting, face, hand and feet muscles. A second common symptom of hypocalcaemia is petechiae. These are small red spots on the skin that arise due to broken capillaries. Sometimes, petechiae may develop into a rash. In some extreme cases, patients may have bronchial or laryngeal spasms in addition to life-threatening complications such as cardiac arrhythmias.

Hypocalcaemia may also cause depression, memory loss and hallucinations. In chronic hypocalcaemia due to hypoparathyroidism, the high phosphate concentration leads to a precipitation of calcium phosphate in the lens of the eye, and may precipitate the development of a cataract.

If a patient is suspected of hypocalcaemia then usually the corrected calcium is determined and if less than 2.2 mmol/L confirms hypocalcaemia. These patients should be assessed for their renal function with U&Es as this is a common cause of hypocalcaemia. If they do not have renal disease then measurement of PTH can reveal possible causes of the hypocalcaemia. Low or undetectable PTH often occurs in patients with magnesium deficiency, post-surgery or those with idiopathic hypocalcaemia. PTH concentrations are high in patients with hypocalcaemia due to vitamin D deficiency, pseudohypoparathyroidism and in other rare causes of hypocalcaemia.

The main approach to managing hypocalcaemia is to treat the underlying cause wherever possible. Mild cases of hypocalcaemia are often treated by oral calcium supplements. Patients with hypocalcaemia due to vitamin D deficiency may be placed on calcitriol and its precursors. Magnesium supplements may be prescribed for patients with hypocalcaemia due to magnesium deficiency. However, eventually, the root pathological cause must be determined and treated. Table 15.2 summarizes calcium metabolism.

15.6   Disorders of phosphate homeostasis

Disorders of phosphate homeostasis occur when the concentration of phosphate falls outside its reference range. Hyperphosphataemia describes high concentrations of plasma phosphate above its reference range,

Table 15.2 Causes of abnormal calcium.

Hypercalcaemia	Hypocalcaemia
Hyperparathyroidism Cancer: a. PTH-related peptide b. bone lysis	Hypoparathyroidism Renal disease: a. failure to generate vitamin D b. failure of tubular calcium resorption
Many other causes: vitamin D supplementation, drugs (e.g. lithium, diuretics), thyroid disease	Dietary and/or sunlight insufficiency in vitamin D, malabsorption, hyperphosphataemia

N.B. Features are not in order of frequency or importance.

whereas hypophosphataemia refers to phosphate concentrations below its reference range.

Causes of hyperphosphataemia

Renal failure is one of the commonest causes of increased blood phosphate. The GFR falls as the function of the kidney deteriorates, and therefore loss of phosphate in the urine declines and its plasma concentration increases. Hypoparathyroidism is characterized by low PTH secretion, which causes reduced renal excretion of phosphate, giving rise to hyperphosphataemia. In acromegaly there is increased activity of 1-hydroxylase, which results in an increased synthesis of calcitriol. This, in turn, increases absorption of dietary phosphate from the GIT. In addition, the excess growth hormone acts directly to reduce renal excretion of phosphate and both effects contribute towards the hyperphosphataemia. In pseudohypoparathyroidism, there is resistance to PTH action, causing decreased renal excretion of phosphate and, therefore, hyperphosphataemia.

Excessive dietary or intravenous intake is a rare cause of hyperphosphataemia and is more likely in patients suffering from concurrent renal failure. Excessive intake of vitamin D may increase calcitriol formation, which may cause hyperphosphataemia because calcitriol increases phosphate absorption from the intestines.

Increased release of phosphate from cells can cause a shift of phosphate from the cytoplasm to ECF, causing hyperphosphataemia. For example, phosphate is released from red cells into the plasma during intravascular haemolysis and this could give rise to hyperphosphataemia. During diabetic ketoacidosis, deficiency of insulin prevents uptake of phosphate by various cells, causing hyperphosphataemia. In catabolic states, that is any condition where there is increased turnover of cells (e.g. when treating malignancy with chemotherapy), increased catabolism results in release of phosphate, which in turn causes hyperphosphataemia.

Artefactual causes of hyperphosphataemia include delayed separation of serum from blood before analysis and/or haemolysis of blood, as red cells have relatively high levels of phosphate. In both cases, measurement of the ion will reveal high concentrations even though the patient has no phosphate disorder.

Consequences of hyperphosphataemia

High concentrations of phosphates can affect calcium metabolism as the calcium is precipitated out as calcium phosphate, causing hypocalcaemia. As a consequence, these patients can suffer from tetany, particularly if the plasma phosphate rises rapidly. Hyperphosphataemia can also cause metastatic calcification; that is, deposition of calcium phosphate in soft tissues, which often occurs in severe hyperphosphataemia, as in patients with renal failure.

Excess free phosphate can be taken up into vascular smooth muscle cells and activate a gene *cbfa-1* which promotes calcium deposition in the vascular cells, causing calcification. Hyperphosphataemia-induced resistance to PTH is believed to contribute towards renal osteodystrophy.

The first line of biochemical tests useful for investigating hyperphosphataemia includes measurement of phosphate, calcium and U&Es. Later, urinary phosphate and plasma PTH measurements may be informative. When hyperphosphataemia is detected, laboratory artefacts are considered and excluded first. Increased phosphate intake is then considered; if excluded, the renal function is then assessed to exclude renal failure. Conditions which cause a shift in phosphate from cells into the ECF can be considered; if excluded, then tests are done to assess whether the hyperphosphataemia is due to endocrine causes such as hypoparathyroidism. If these are excluded, then the hyperphosphataemia is probably due to a rare cause which will need investigation.

Hyperphosphataemia is managed by identifying and treating the underlying cause wherever possible. An active form of treatment is the oral intake of aluminium, calcium or magnesium salts, as they can bind to phosphate in the gut and so reduce phosphate absorption. However, caution is required when using calcium salts as they can promote calcium phosphate deposition in the blood vessel wall. Resin binders such as sevelamer (Renagel) promote phosphorus excretion without affecting calcium concentrations. Sevelamer decreases vascular calcium deposition in patients with renal failure.

Phosphate binders that contain aluminium should not be used in patients with renal failure as they can cause aluminium toxicity. Haemodialysis may be required in severe cases of hyperphosphataemia, and particularly for patients suffering from renal failure. Infusion of insulin will promote cellular uptake of phosphate and should be used in patients with diabetic ketoacidosis. In cases of phosphate toxicity, gastric lavage and oral phosphate binders are used to prevent further absorption.

Hypophosphataemia

Low concentrations of phosphates are less common than hyperphosphataemia, but more damage is caused when this does occur. Common causes include a decreased intake of phosphate, increased renal loss and increased uptake by cells.

Inadequate intake of phosphate is a rare cause of hypophosphataemia. The mineral is common in many foodstuffs, so that dietary deficiency in unlikely in those taking a healthy, balanced diet. Vitamin D deficiency results in decreased synthesis of calcitriol, and therefore decreased absorption of phosphate from the gut, giving rise to hypophosphataemia. Phosphate binding agents such as certain antacids that contain aluminium hydroxide can bind phosphate in the gut, thus preventing its absorption into the blood.

Increased renal loss of phosphate may cause hypophosphataemia. For example, in primary hyperparathyroidism there is increased secretion of PTH, which causes excessive loss of phosphate via the kidneys. In addition, certain diuretics can cause increased loss of phosphate via the kidneys, resulting in hypophosphataemia. A number of congenital renal tubular defects cause hypophosphataemia because the mechanism for phosphate reabsorption is defective and so much of the phosphate is lost in the urine. Examples of this include Fanconi's syndrome and vitamin-D-resistant rickets.

Increased cellular uptake of phosphate may cause hypophosphataemia. This may happen in the recovery phase of diabetic ketoacidosis; as patients are administered insulin, this in turn promotes cellular uptake of phosphate, giving rise to hypophosphataemia. The hypophosphataemia develops because of depletion of total body phosphate as a consequence of osmotic diuresis.

Table 15.3 Causes of abnormal phosphates.

Hyperphosphataemia	Hypophosphataemia
Increased intake	Decreased intake of phosphate and/or vitamin D
Renal disease (failure to excrete)	Renal disease(excess excretion)
Cellular release: tissue destruction haemolysis, diabetic ketoacidosis	Cellular uptake, alkalosis, alcoholism

N.B. Features are not in order of frequency or importance.

Respiratory alkalosis can cause hypophosphataemia by stimulating the enzyme phosphofructokinase, and therefore use of phosphate to form intermediates in glycolysis, such as glucose-6-phosphate. Chronic alcohol intake is a rare but growing cause of hypophosphataemia, but the pathogenesis is complex and multifactorial. Reduced absorption, poor diet, vomiting and diarrhoea probably all play a role.

In common with abnormalities of calcium, the clinical effects of hypophosphataemia include paraesthesia, ataxia and coma. If there is a profoundly low phosphate concentration, muscle weakness can occur, causing respiratory dysfunction. Often, patients with hypophosphataemia have an increased susceptibility to infections (possibly due to defective phagocytosis). Hypophosphataemia may cause osteomalacia or softening of the bones.

When hypophosphataemia is detected in a blood sample, it is crucial to first exclude medication that may cause hypophosphataemia. Possible cause for movement of phosphate from ECF to intracellular fluid should be investigated; and if excluded, then measurement of urinary phosphate may reveal causes that are either due to inadequate phosphate intake or excessive renal losses. A major aim of management is to treat the underlying cause wherever possible as this will resolve the hypophosphataemia. Oral or parenteral administration of phosphate may be required to correct a phosphate deficit. Table 15.3 outlines the causes of hyper- and hypophosphataemia.

15.7 Disorders of magnesium homeostasis

Disorders of magnesium homeostasis arise when blood magnesium concentrations fall outside the reference range. Concentrations of blood magnesium above the reference range are referred to as hypermagnesaemia, whereas those below the reference range are referred to as hypomagnesaemia.

Hypermagnesaemia

Causes of high magnesium are easy to classify:

- increased intake (oral, parenteral, antacids, laxatives);
- decreased excretion (renal failure, mineralocorticoid deficiency, hypothyroidism);
- release from cells (post-cytotoxic chemotherapy necrosis, crush injuries, diabetic ketoacidosis, tissue hypoxia).

Renal disease is the most common cause of hypermagnesaemia, and often is found alongside abnormalities in calcium. Mildly raised concentrations may be present in the principal mineralocorticoid deficiency – Addison's disease – where there is a decline in the blood volume.

Clinical consequences and management of hypermagnesaemia

The major problems of hypermagnesaemia, like that of calcium, are neuro-muscular, such as respiratory weakness and paralysis, hypotension, arrhythmia and cardiac arrest, with nausea and vomiting. The common causes of hypermagnesaemia are usually evident from clinical examination of patients. Severe hypermagnesaemia tends to occur in patients with chronic renal failure, especially if they take laxatives or antacid preparations containing magnesium. Treatment is according to aetiology. In normal renal function, this can be assisted by promoting excretion by drinking water or by intravenous administration of diuretics. However, severe hypermagnesaemia due to renal failure may require dialysis.

Hypomagnesaemia

Low magnesium is more common than hypermagnesaemia and arises owing to either decreased intake of magnesium or increased losses of magnesium from the body via the renal and/or gastrointestinal routes. The first two causes are effectively the reverse of the causes of increased plasma magnesium, as follows:

- decreased intake (starvation, malabsorption, parenteral, prolonged gastric suction);
- increased renal loss (osmotic diuresis, alcoholism, hypercalcaemia, hypoparathyroidism, hyperaldosteronism, drug effects);

• increased losses via the intestines (prolonged diarrhoea, laxative abuse, intestinal fistula).

The use of loop and thiazide diuretics by renal and hypertensive patients frequently present with hypomagnesaemia, as do those using certain antibiotics. The latter often act on the loop of Henle to reduce absorption of magnesium, thus promoting its loss in the urine and causing hypomagnesaemia. Patients taking cytotoxic chemotherapy (such as cis-platinum) are at risk of low magnesium (and much else) because of the nonspecific damage to the nephron. In alcoholism, hypomagnesaemia may be due to a number of factors, such as increased renal excretion, inadequate dietary intake, vomiting and diarrhoea. Hypoparathyroidism causes renal wasting, and this may account for increased loss of body magnesium in the urine, whereas hyperaldosteronism increases renal flow and promotes loss of magnesium in the urine.

Clinical consequences and management of hypomagnesaemia

The clinical effects of hypomagnesaemia are very similar to those seen in hypocalcaemia and arise largely due to the role of magnesium in neuromuscular function. In extreme cases, there will be general muscular paralysis, cardiac arrhythmia and arrest. However, pathological situations where a single electrolyte abnormality is present occur very rarely, and in many cases disturbances of sodium, potassium and calcium are also likely to be present.

Management of a patient with hypomagnesaemia should (as with all diseases) begin with identifying and treating the underlying cause(s) (as above) wherever possible. Oral magnesium supplements may be adequate for mild cases of hypomagnesaemia, but in severe magnesium deficiency, such as in malabsorption or chronic renal failure, intravenous infusions may be required.

15.8   Bone physiology

Bones are rigid parts of the skeleton. Their functions include movement, support and protection of various body organs. The strength comes from the outer surfaces of compact bone; the inner part of bone is spongy and houses the haemopoietic tissues that produce red and white blood cells and platelets. Bone also serves as the storage site of minerals, particularly calcium and phosphate. There are two types of bone: cortical, which is sometimes referred to as compact bone, and trabecular, which is often referred to as spongy bone. Most of the bones in the body are made from both these types, although there is a variation within the bones at different sites. The hard outer surface of bone is cortical and accounts for 80% of the bone mass in a healthy adult. The bone interior consists of the trabecular type and accounts for the remaining 20% of bone mass.

The biology of bone

The technical name for the matrix of bone is osteoid, an organic matrix containing deposits of hydroxyapatite. Osteoid contains mainly type I collagen with some nonfibrous proteins. Hydroxyapatite is a mineral composed of calcium, phosphate and water and gives the bone its hardness and weight. The collagen fibrils are arranged in sheets, with the fibrils in one particular layer lying at 90° to those on either side. Bone is metabolically active and undergoes continuous remodelling that comprises two processes: bone formation and bone resorption.

Bone remodelling is mediated by two types of cells. Osteoblasts synthesize the hard bone matrix; some remain at the surface of the bone whilst others are embedded in the bone itself, and so are effectively trapped. These trapped cells (osteocytes) occupy a small cavity (a lacuna), whilst canaliculi radiate and contact other cells. Eventually, the osteoblasts and osteocytes form concentric rings of osteoid that give the compact bone its characteristic microscopic structure. At the centre of each set of rings is a Haversian canal that carries a blood vessel. A single unit of concentric rings is an osteon (Figure 15.4).

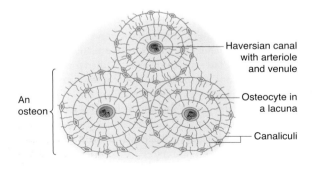

Figure 15.4 Structure of compact bone. An osteon consists of a series of concentric rings of canaliculi with a Haversian canal at the centre.

Whilst bone matrix is deposited by osteoblasts, it is eroded by osteoclasts, large multinucleated cells that originate, like macrophages, from the haemopoietic tissues of the bone marrow. Degradation of bone is by the secretion of acid to break down the hydroxyapatite, and of enzymes such as proteolysis cathepsin and matrix metalloproteinases that digest the collagens. Osteoblasts and osteoclasts work together when remodelling bone. Indeed, resorption of old bone occurs first by osteoclasts and then osteoblasts appear at the site to fill the cavity with osteoid, which becomes calcified.

The two processes of bone formation and resorption are tightly linked: when a certain amount of bone is resorbed, an equivalent amount is formed to replace it. The proliferation of osteoblasts is stimulated by some growth factors; for example, transforming growth factor-β and growth hormone. In contrast, PTH and some cytokines arrest osteoblast proliferation and indirectly stimulate bone resorption. The activity of osteoclasts is inhibited by calcitonin, but is promoted by the cytokine interleukin-6. The process of bone formation and resorption is carefully balanced, and any imbalance will result in disease.

Markers of bone turnover

Several biochemical markers can be used to assess bone formation and resorption and are of use when investigating metabolic bone disease. Those that can indicate bone formation are ALP activity and osteocalcin, whereas those for bone destruction are urinary hydroxyproline and urinary pyridinoline. These markers may have a place in investigating osteoporosis, bone metastases and Paget's disease, and in the use of drugs acting on bone, such as bisphosphates.

Bone contributes 15–30% of the total serum ALP activity; the remainder comes from the liver, the intestines and some other sources. In the bone, ALP activity is found in the membranes of osteoblasts and is released during osteoblast turnover and is therefore a good marker of bone formation. Children who have active growth have high concentrations of plasma ALP. However, specialized electrophoretic techniques are available to differentiate bone ALP from other isoenzymes of ALP.

Osteocalcin is a non-collagenous protein that is synthesized by osteoblasts and released into the plasma during bone formation; hence its use as a bone marker. Hydroxyproline, an imino acid, is a component of collagen that is released and excreted in the urine during bone resorption. Urinary hydroxyproline measurements

can be significantly influenced by the ingestion of dietary gelatin. Pyridinoline and deoxypyridinoline are collagen cross-links not influenced by diet and found in the bone and cartilage. They are released during bone resorption and their urinary concentrations can be used as markers for bone resorption.

Point of nomenclature

Proline and hydroxyproline are amino acids because they have both an amino group and an acid group. However, both of these groups are part of a ring structure, so the entire molecules are better described as *imino* acids.

Other promising bone markers include two telopeptides: N-terminal telopeptide (also known as amino-terminal collagen cross-links (NTX)) and the closely related C-terminal telopeptide (carboxy-terminal collagen cross-links). The latter is a preferred marker as it has superior sensitivity and specificity, and is cleaved from mature type 1 collagen by osteoclasts. Procollagen I N-terminal propeptide is a product of the cleavage of newly synthesized collagen and, since concentrations reflect synthesis by osteoblasts, it can be used as a marker of diseases where there is overactivity of these cells. All these peptides can be measured by immunoassay, principally ELISA and radioimmunoassay.

Blood science angle: Bone

Haematologists have a special interest in spongy bone (which fills the hollow centre of large bones) as it is where blood cells are produced, but is also where the white blood cell tumours of leukaemia and myeloma are found. Bone does not impact a great deal into routine immunology, except perhaps in the differential diagnosis of osteoarthritis and rheumatoid arthritis.

15.9   Bone disease

Metabolic bone disease causing abnormal bone function arises from a number of disorders that are often associated with abnormal calcium, vitamin D and phosphate homeostasis. They are often detected and monitored using bone markers, and most are clinically reversible once the underlying defect has been rectified. The major metabolic bone diseases are osteoporosis, osteomalacia and rickets, and Paget's disease. However, there are several other diseases of bone, the basis of which is inflammation. A common feature is thinning of the bone (osteopenia), leading to weakness and fractures.

Osteoporosis

This is the most prevalent metabolic bone disease and is particularly common in women after the menopause. In the UK, one in four women and one in twenty men after the age of 60 years will have osteoporosis. About one-third of hospital orthopaedic beds in the UK are occupied by patients with this disease.

Osteoporosis is characterized by equal losses of osteoid and minerals, resulting in a decrease in bone mass. The rate of bone formation is usually normal but the rate of resorption increases and there is greater loss of trabecular than cortical bone. With normal ageing there is a progressive decline in osteoblast activity compared with osteoclast activity. Therefore, a progressive decline in bone mass occurs, which translates to an increased risk of fractures. In its early stages it is asymptomatic, but as the disease progresses bone pain (for example severe backache), spontaneous fractures and collapse of vertebrae and fractures of ribs and hips occur readily with minor trauma.

Established causes of osteoporosis include:

• ageing – especially post-menopausal (primary osteoporosis);
• endocrine – thyrotoxicosis, Cushing's syndrome, diabetes mellitus, oestrogen deficiency (women), androgen deficiency (men), hyperparathyroidism;
• malignancy, such as multiple myeloma;
• drugs (such as prolonged treatment with unfractioned heparin), alcoholism;
• miscellaneous – prolonged immobilization (bed, wheelchair), malabsorption of calcium, weightlessness.

The diagnosis of osteoporosis depends largely on clinical examination and radiological investigations: bone mineral density, which is measured by dual-energy X-ray absorptiometry. Biochemical tests are of little use in the investigation of primary osteoporosis as concentrations of calcium, phosphate and ALP are within reference range, although the ALP activity may rise after a fracture. Urinary hydroxyproline excretion may be increased if there is rapid bone loss, but is often within reference range. Urinary excretion of pyridinium cross-links are increased in patients with osteoporosis, but also in other conditions associated with increased bone

resorption. Biochemical tests may, however, be of use in detecting the underlying causes of secondary osteoporosis.

Hormonal replacement therapy may be of value in preventing primary osteoporosis after the menopause. Patients at risk of developing osteoporosis should be counselled to avoid known risk factors; for example, excessive alcohol intake. Adequate dietary intake of calcium, vitamin D and regular exercise are necessary in individuals at risk of developing osteoporosis. Treatment with bisphosphonates (drugs which suppress bone resorption) and oestrogens may be beneficial in individuals suffering from osteoporosis.

Osteomalacia and rickets

Both these conditions (the former in adults, the latter in children) are characterized by softening of bones due to a defect in mineralization of the bone matrix or cartilage, or both. They are rare conditions, but are more common in certain groups, such as South Asians and the elderly.

A common cause is calcium or phosphate deficiency, or both (both are required for bone mineralization). Calcium deficiency is usually due to deficiency of vitamin D, which may be dietary or due to reduced exposure to sunlight. The addition of adequate vitamin D in most foods has virtually eliminated rickets in children in the developed world. Vitamin D deficiency may also occur in patients with malabsorption; for example, those suffering with coeliac disease.

Renal disease can result in reduced formation of calcitriol required for absorption of dietary calcium. Phosphate deficiency occurs in renal tubular defects (reduced reabsorption of phosphate) such as the Fanconi syndrome giving rise to vitamin-D-resistant rickets or osteomalacia. Tumour-induced osteomalacia is uncommon but causes increased renal phosphate excretion, resulting in hypophosphataemia and osteomalacia.

The most common symptoms are bone pain, muscular weakness and increased susceptibility to fractures even after minor trauma. Children with rickets may suffer from bone pain and have bowed legs and cranial defects. The biochemical findings typically include hypocalcaemia, hypophosphataemia and elevated concentrations of PTH. Concentrations of vitamin D and its metabolites are often low in osteomalacia or rickets, whilst ALP is usually raised.

When managing patients with osteomalacia or rickets, a common approach is to identify the causal factors and then aim to treat the underlying cause wherever possible. Oral or parenteral administration of vitamin D may be required in patients with vitamin D deficiency.

Paget's disease

Also known as osteodystrophia deformans, this disease is named after Sir James Paget, the British surgeon who described this condition in 1877. It is characterized by excessive osteoclastic bone resorption followed by formation of new bone of abnormal structure laid down in a disorganized manner, resulting in deformed bones. The disease, more common in men, is rarely diagnosed in those aged under 40. As many individuals with Paget's disease are asymptomatic, it is difficult to estimate the prevalence of this condition, but may affect about 5% of the UK population over the age of 50 years.

Paget's disease may be asymptomatic in some patients, but clinical features can occur in other sufferers. These include bone pain and bone deformities, such as bowed tibia, kyphosis (hunching of the back), increasing skull size and bone fractures (Figure 15.5). Diagnosis is usually made on the basis of clinical investigation supported by radiological investigations. Paget's disease is diagnosed using X-rays and imaging of the affected bone. Calcium and phosphate concentrations in serum samples tend to be within their reference ranges, but hypercalcaemia may occur in immobilized patients. The abnormal bone turnover causes increased activities of ALP, and there is increased urinary excretion of NTX.

The aim of managing Paget's disease is to give relief from bone pain and to prevent disease progression. Analgesics may be used to give relief from the pain. In severe cases, bisphosphonates may be used to reduce osteoclastic activity. Patients with bone deformity may require supports such as heel lifts or specialized footwear. Corrective surgery may be required where joints are damaged, for fractures or severely deformed bones, or where nerves are being compressed by enlarged bones. Serial measurements of ALP and urinary NTX may be used when monitoring treatment.

Osteomyelitis

Technically, this is inflammation of the inside of the bone (i.e. spongy bone). Since lone autoimmune disease of the bone marrow is exceptionally unusual, the osteomyelitis must be due to an infective agent, almost always bacteria, such as a staphylococcus. In theory, therefore, diagnosis cannot be complete without the demonstration of a

Figure 15.5 Pretty as a picture. The National Gallery in London has a portrait, perhaps unkindly titled 'A Grotesque Old Woman'. Note the increased distance between the upper lip and the base of the nose, and the expanded bone of the front of the skull. The clavicles are also prominent, and it is likely that some fingers are misshapen. On reflection, these are all signs of bone deformity; and indeed, we now recognize that she is likely to have suffered from Paget's disease. (Quinten Massys, An Old Woman ('The Ugly Duchess') © The National Gallery, London).

pathogenic microbe, but in practice this is generally presumed. Alternatively, there may be a skin wound that is so deep it goes down to the bone. Other changes to the bone consistent with osteomyelitis are made by X-ray. If the infection is profound there may be increases in certain inflammatory markers in the blood (raised white count, erythrocyte sedimentation rate) but changes in calcium, phosphates, PTH and vitamin D are most unusual.

Osteoarthritis

This disease causes a great deal of morbidity but is rarely a cause of mortality. In its most severe form, the major risk factor for osteoarthritis is being overweight or obese, and joint replacement may be required. However, patients with osteoarthritis have an increased risk of cardiovascular disease and cancer, but this is inevitably because of being overweight, which brings a risk of diabetes.

As is implied, the pathological basis is inflammation, but in most cases the focus of this remains within one particular synovial joint, although in other patients two, or at most three joints are affected. Because the disease process remains local to the joints, there is no systemic disease and so no abnormalities in the blood. Indeed, if there are abnormalities in blood results, the disease is unlikely to be osteoarthritis, and may actually be rheumatoid arthritis (discussed in full in Chapter 9).

Renal osteodystrophy

This disease is generally only found in severe chronic kidney disease (CKD), and is the consequence of long-term hypocalcaemia due to failure to resorb calcium from the filtrate. Allied to poor absorption of calcium from the intestines (itself a result of the inability of the kidney to generate vitamin D), calcium leaches out of the bone in an attempt to maintain blood calcium concentrations, leading initially to osteopenia, and then fractures. In response to the low calcium, a raised PTH is expected, and it is also likely that there is hyperphosphataemia.

The scaffolding hypothesis

This likens bone to a highly complex scaffold, as may be required for extensive building work (Figure 15.6), where the strength of the scaffolding relies on the correct placement of long metal poles by a gang of scaffolders. The poles are fixed together with clamps, tightened by spanners. Should the scaffolders have too few poles, clamps or spanners, the scaffolding will be weak. This is not necessarily the fault of the scaffolders.

This can be used as a partial metaphor for bone and its diseases, where the long metal poles may be the calcium and phosphates, and the scaffolding gang may be the osteoblasts that build up bone. Should the scaffolders (osteoblasts) have the wrong plans for their scaffold (bone), no amount of extra metal poles simply being thrown at the scaffolding (calcium/phosphate) will put it right. Perhaps there is a problem with the delivery of the metal poles to the scaffolding site (lack of vitamin D in rickets/osteomalacia, perhaps lack of PTH?). If the scaffolders are not motivated correctly, the scaffold they

Figure 15.6 Bone as scaffolding. If we see bone as simple scaffolding, the long metal poles represent calcium and phosphates, but these poles need to be secured by clamps, which themselves must be place by scaffolders with special tools. The scaffolding will be unsafe if any of these components are defective.

build may be weak (perhaps lack of oestrogen in post-menopausal women leading to osteoporosis).

When the building work is complete, the scaffolding needs to be removed by a second gang of scaffolders. But in bone we have two groups of 'scaffolders', the osteoblasts and osteroclasts (the latter breaking down bone), operating at the same time to remodel our skeleton. Perhaps Paget's disease is a lack of coordination between these two cell types (scaffolders) who have failed to receive the correct or conflicting instructions from their respective foremen, so resulting in abnormal scaffolding (bone). It could also be argued that osteoarthritis is the scaffolding of the body wearing out, perhaps from over-use, or maybe from the effects of rust.

Table 15.4 summarizes the value of blood tests and other investigations.

The genetics of bone, calcium and phosphates

Although we have noted several alternative aetiologies of bone, calcium and phosphate disease, there are conditions that seem likely to have a genetic component.

Familial hypocalciuric hypercalcaemia This auto-somal dominant condition, which develops from childhood, is characterized by chronic hypercalcaemia, and although usually asymptomatic on outset, symptoms of hypercalcaemia are inevitable. Most cases are due to mutations in the *CASR* gene that codes for calcium-

Table 15.4 Blood tests in different bone diseases.

	Osteoporosis	Osteomalacia, rickets	Paget's disease	Osteoarthrtitis	Osteomyelitis
Calcium	Normal	Decreased	Normal	Normal	Normal
Phosphates	Normal	Decreased	Normal	Normal	Normal
Vitamin D	Normal	Decreased	Normal	Normal	Normal
PTH	Normal	Raised	Variable	Normal	Normal
ALP	Normal	Raised	Raised	Normal	Normal
Basis of the disease	Decreased bone mass	Soft bones	Abnormal bone structure	Inflammatory: within a synovial joint	Inflammatory: within bone itself
Other investigation	X-ray, bone scan	X-ray	X-ray, bone scan	X-ray	X-ray
Leading treatment(s)	HRT (but risks venous thrombosis)	Vitamin D/phosphate	Bisphosphonates, calcitonin	Analgesia	Antibiotics

sensing receptors in cells of the parathyroid glands and kidneys. Since these defective receptors cannot detect calcium concentrations, the parathyroid gland produces inappropriately high concentrations of PTH which causes the hypercalcaemia. In addition to hypercalcaemia, these patients present with hypocalciuria. The reduced loss of calcium in the urine is due to the failure of kidney cells to recognize and excrete excessive calcium from the body.

Familial tumoral calcinosis This condition refers to a group of disorders inherited in an autosomal recessive fashion characterized by hyperphosphataemia. The latter is due to increased reabsorption of phosphate through the renal proximal tubule, resulting in elevated phosphate concentration and deposition of calcified deposits in cutaneous and subcutaneous tissues, as well as, occasionally, in visceral organs. The cause seems likely to be due to mutations in at least three genes: *GALNT3*, encoding a glycosyltransferase termed ppGalNacT3, *FGF23* encoding a potent phosphaturic protein, and *KL* encoding Klotho, a co-receptor for FGF23, thereby integrating the genetic data into a single physiological system.

Paget's disease Advances in molecular genetics suggests that most cases of this disease are caused by a combination of (a) rare alleles that exert a large effect that cause autosomal dominant inheritance of the disease and (b) more common alleles that exert a smaller effect on the bone. Mutations of *SQSTM1* (on chromosome 5, coding for a molecule called sequestosome-1) are the most common cause of classical Paget's, being present in about 10% of patients, but mutations in *RANK* (receptor activator of nuclear factor kappa B, involved in osteoblast

function) on chromosome 18 are also common. Seven predisposing loci have been identified by genome-wide association studies, and these increase the risk of the disease individually by 1.3- to 1.7-fold, but have combined effects that account for about 86% of the population-attributable risk of *SQSTM1* negative patients.

In common with many conditions, knowledge of the precise genetic lesion(s) in bone disease does not necessarily drive management. However, it does provide confidence that the diagnosis is correct, and it may also lead to new treatments, perhaps aimed at gene products such as receptors.

15.10 Case studies

Case study 17

A 58-year old woman presents with muscle weakness, tiredness and lethargy, confusion, alternate polyuria and thirst, intermittent loss of memory and nausea. A urine dipstick for glucose is negative. Fasting blood is taken for routine biochemistry tests.

	Result (unit)	Reference range
Na	149 mmol/L	135–145
K	4.9 mmol/l	3.8–5.0
Urea	7.1 mmol/L	3.3–6.7
Creatinine	129 μmol/L	71–133
Glucose	3.9 mmol/L	3.5–5.5
Albumin	33 g/L	35–45
Calcium	3.0 mmol/L	2.2–2.6
eGFR	39 mL/min per 1.73 m^2	>90

Interpretation This woman's symptoms are vaguely suggestive of diabetes, hence the urine test for glucose. Although a useful test in many cases, it can be negative in diabetes, hence the fasting blood sample, which in many cases confirms absence of diabetes. However, the abnormal results of hypernatraemia and uraemia point to renal disease. Although potassium and creatinine are both within their reference ranges, they are certainly high, leading to a degree of concern. There is also hypercalcaemia and hypoalbuminuria. It is tempting to presume that the latter problems are the consequence of chronic renal failure. This is supported by the low estimated GFR (eGFR), which translates to CKD stage IIIb, and warrants specialized investigation and treatment. A further aspect is dehydration, which may classify the hypernatraemia as pre-renal. A further blood test for PTH is required.

Case study 18

A 65-year-old man presents with a 6-month history of progressive pain in the left leg, the right leg being asymptomatic. His level of d-dimer is raised, a finding often associated with deep vein thrombosis. However, as the pain seems to be more due to bone than to soft tissue, U&Es and LFTs were ordered, and he was sent for an X-ray.

	Result (unit)	Reference range
Na	139 mmol/L	135–145
K	4.2 mmol/l	3.8–5.0
Urea	4.7 mmol/L	3.3–6.7
Creatinine	89 μmol/L	71–133
eGFR	79 mL/min per 1.73 m^2	>90
Bilirubin	14 μmol/L	<21
Alanine transaminase	26 IU/L	<50
Aspartate aminotransferase	45 IU/L	<60
Gamma-glutamyl transpeptidase	54 IU/L	<70
ALP	196 IU/L	20–130

Interpretation The raised d-dimer is of little value – it may be due to several different disease processes – and so does not define a deep vein thrombosis, a possible cause of the leg pain. U&Es and LFTs are justified as two simple screening tests, and these tell us that renal function is acceptable (stage II CKD), and so is liver function, but with a single exception – that of ALP. If there are clear signs of abnormal bone formation on the X-ray, this gives quite a firm diagnosis of Paget's disease. An alternative explanation is invasion of bone by metastatic cancer, such as of the prostate and bowel. Additional tests may be calcium and PTH, although these are expected to be within the reference range if the patient has Paget's disease, and for cancer markers PSA (prostate-specific antigen) and CEA (carcinoembryonic antigen). These are described in detail in Chapter 19.

Case study 19

A 73-year-old woman was admitted to hospital complaining of muscular weakness, abdominal discomfort and constipation. She also suffered from vomiting and diarrhoea. Her blood tests gave the following results:

	Result (unit)	Reference range
Na	141 mmol/L	135–145
K	4.1 mmol/l	3.8–5.0
Urea	4.6 mmol/L	3.3–6.7
Creatinine	93 μmol/L	71–133
eGFR	73 mL/min per 1.73 m^2	>90
Calcium	1.5 mmol/L	2.2–2.6
Albumin	40 g/L	25–45
Magnesium	0.4 mmol/L	0.8–1.2

Interpretation This patient has a low calcium, normal albumin and suffers from hypocalcaemia. Renal function is unremarkable (the eGFR equates to CKD stage II). She also has hypomagnesaemia, which may have caused the hypocalcaemia. As the albumin concentration is normal, this is not a pesudohypocalcaemia. The likely pathology is because a low serum magnesium reduces action of PTH and also inhibits its secretion from the parathyroid glands. She should be treated by using calcium and magnesium supplements.

Summary

- The greater proportion of body calcium is in bone; blood concentrations are regulated by PTH, calcitriol (vitamin D) and calcitonin.
- Disorders of calcium homeostasis present as either abnormally high plasma calcium (hypercalcaemia) or abnormally low plasma calcium (hypocalcaemia).
- Phosphate is present largely in the bones, and its plasma concentrations are controlled by PTH and calcitriol (vitamin D).

- Disorders of phosphate homeostasis cause hyperphosphataemia or hypophosphataemia.
- Disorders of magnesium homeostasis cause hypermagnesaemia and hypomagnesaemia.
- The major metabolic diseases affecting the bone are osteoporosis, osteomalacia and Paget's disease. Other bone diseases are osteoarthritis and osteomyelitis, both of which are inflammatory.
- Markers of bone disease have a role in diagnosis and in the monitoring of treatment.

Further reading

Cooper MS, Gittoes NJ. Diagnosis and management of hypocalcaemia. Br Med J. 2008;336:1298–1302.

Cremers S, Bilezikian JP, Garnero P. Bone markers – new aspects. Clin Lab. 2008;54:461–471.

Guerrera MP, Volpe SL, Mao JJ. Therapeutic uses of magnesium. Am Fam Physician. 2009;80:157–162.

Holick MF. Vitamin D status: measurement, interpretation, and clinical application. Ann Epidemiol. 2009; 19:73–78.

Lau WL, Pai A, Moe SM, Giachelli CM. Direct effects of phosphate on vascular cell function. Adv Chronic Kidney Dis. 2011;18:105–112.

Peacock M. Calcium metabolism in health and disease. Clin J Am Soc Nephrol. 2010;5:523–530.

Ralston SH, Albagha OM. Genetic determinants of Paget's disease of bone. Ann N Y Acad Sci. 2011;1240:53–60.

Ralston SH, Langston AL, Reid IR. Pathogenesis and management of Paget's disease of bone. Lancet. 2008; 372:155–163.

Scharla S. Diagnosis of disorders of vitamin D metabolism and osteomalacia. Clin Lab. 2008;54:451–459.

Shoback D. Clinical practice: hypoparathyroidism. New Engl J Med. 2008;359:391–403.

Guidelines

NICE Clinical Guideline CG146. Osteoporosis: assessing the risk of fragility fracture.

NICE Clinical Guideline, CG59. Osteoarthritis: The care and management of osteoarthritis in adults.

NICE Technology appraisal, TA160. Alendronate, etidronate, risedronate, raloxifene and strontium ranelate for the primary prevention of osteoporotic fragility fractures in postmenopausal women (amended).

Web sites

National Association for the Relief of *Paget's Disease*: www.paget.org.uk.

Home site of the National Osteoporosis Society: www.nos.org.uk.

Patient orientated site on renal osteodystrophy: www.edrep.org/pages/textbook/osteodystrophy.php.

16 Nutrients and Gastrointestinal Disorders

Learning objectives

After studying this chapter, you should be able to:

- discuss the key nutrients and their function in the human body;
- outline the consequences of nutritional disorders;
- explain the role of the laboratory in nutrition;
- describe the gastrointestinal tract and its function;
- appreciate the value of good gastrointestinal function and the consequences of disease of this super-organ.

Adequate nutrition is vital for normal functioning of the human body. This chapter will first review, in Section 16.1, the major nutrients, their functions and disorders associated with their deficiencies or excesses. Absorption of nutrients occurs after digestion of foods in the gastrointestinal tract. In Section 16.2 we will review the key features of the gastrointestinal tract and discuss the associated pathology and laboratory investigation.

16.1 Nutrients

Nutrients, inevitably derived from foodstuffs, can be divided into two broad categories: macronutrients, which are required in amounts greater than 1 g per day and can be seen as energy providers, and micronutrients, which are required in amounts of less than a gram a day and include minerals and vitamins (Table 16.1). It follows that other minerals and elements (calcium, magnesium, sodium, potassium, chlorine, phosphorus) are not part of this chapter, being discussed elsewhere.

Macronutrients

The major foodstuffs can be classified into one of three groups: carbohydrates, proteins and lipids. These major foodstuffs provide energy. Carbohydrate and proteins provide about 4 kcal energy per gram, but lipids provide 9 kcal energy per gram. In comparison, ethanol provides 7 kcal energy per gram, although the latter has a considerable side-effect profile. The average body requires 2000–3000 kcal per day, depending on sex and physical activity; if the daily calorific intake exceeds this, then the excess is converted to fat and leads to obesity. Conversely, a diminishing energy intake leads to starvation.

Carbohydrates This group of nutrients are heterologous formations of different proportions of monosaccharide monomers such as glucose, fructose and galactose. They form dimers (disaccharides such as sucrose and lactose), and by extension oligosaccharides (3–10 monomers) and so polysaccharides (>10 monomers) such as starch (in plants) and glycogen (in animals). In the intestines, enzymes such as amylase and lactase digest complex carbohydrates into simple carbohydrates that can be absorbed from the gut wall.

Proteins Similarly, this group consists of irregular polymers of any combination of 20–30 monomers (amino acids), although only eight amino acids are truly essential (as they cannot be synthesized) for a healthy physiology. Additional classification depends on the presence of functional groups: hydrophobic, hydrophilic, and extra carboxylic and/or amino side groups. Proteins in the diet are digested by proteases such as pepsin and trypsin, the products being short chains of amino acids (peptides).

Lipids These may be defined as molecules that are insoluble (or weakly soluble) in water, but are more soluble in a polar solvent. The major groups as far as human biology is concerned are triacylglycerols (also known as triglycerides), free fatty acids and cholesterol. The former two contain long chains of carbon atoms,

Blood Science: Principles and Pathology, First Edition. Andrew Blann and Nessar Ahmed.
© 2014 John Wiley & Sons, Ltd. Published 2014 by John Wiley & Sons, Ltd.

Table 16.1 Nutrients.

Macronutrients	Carbohydrates, lipids, proteins
Micronutrient vitamins	Lipid-soluble: vitamins A, D, E and K
	Water-soluble: vitamins B and C
Micronutrient minerals	Copper, iodine, iron, selenium, zinc

some linked by double bonds, and if so are described as unsaturated.

Micronutrients

These are substances needed in amounts of less than a gram per day. They can be summarized as vitamins and minerals (the latter also called trace elements). Of the 13 vitamins, four are lipid-soluble and the remainder are water-soluble.

Lipid-soluble vitamins After absorption, vitamins A, D, E and K are transported in the blood bound to carrier proteins and stored either in the liver or in fat tissue, from where they can be mobilized as required.

• Vitamin A, also known as retinol, has three main functions in humans. These are in promoting healthy rods and cones of the retina (deficiency leading to night blindness), control of differentiation and proliferation of certain cells such as epithelial and bone cells, and in glycoprotein metabolism. Newly absorbed vitamin A needs to be carried in the blood by a specific transport protein, retinol-binding protein.
• Vitamin D is described fully in Chapter 15, and will not be reproduced in detail here. The metabolism is complex, intermediates including ergocalciferol and cholecalciferol, and deficiency leads to the bone diseases rickets in children and osteomalacia in adults.
• Vitamin E (tocopherols) functions as an antioxidant and free-radical scavenger, principally protecting lipids (particularly those in cell membranes from oxidation). A low intake had been linked with an increased risk of cardiovascular disease, but supplements have not shown any benefit. As vitamin E is fat-soluble, its concentration is usually expressed relative to the total cholesterol concentration.
• Vitamin K also has a complex biochemistry; intermediates including phylloquinone, phytomenadione and menaquinone. Discovered in Germany, the 'K' comes

from koagulation, and as such is a required cofactor in the synthesis of certain coagulation factors, as is explained in Chapter 8.

Water-soluble vitamins These remaining nine B and C vitamins are easily absorbed and move freely around the body to their sites of action. The importance of vitamins B_6, B_{12} and folate is explained in Chapters 3 and 4.

• Vitamin B_1 (thiamine) is a cofactor for a number of metabolic processes, most notably the conversion of pyruvate to acetyl-coenzyme A. Deficiency leads to impaired carbohydrate metabolism, the condition beriberi, and potentially to life-threatening lactic acidosis.
• Vitamin B_2 (riboflavin) is also required for numerous enzymes, such as flavoproteins. Although there is no specific deficiency disease, low levels are associated with angular stomatitis (inflammation of the side of the mouth), glossitis (inflammation of the tongue) and anaemia.
• Vitamin B_3 (niacin), part of nicotinamide adenine dinucleotide and nicotinamide adenine dinucleotide phosphate, is involved in many redox reactions in human metabolism. Severe deficiency results in pellagra (leading to dermatitis, diarrhoea and dementia), but in milder deficiencies there may be nonspecific signs, such as glossitis.
• Vitamin B_5 (pantothenic acid) is part of coenzyme A and plays a fundamental role in metabolism. However, it is so common in food that there is no precise deficiency disease.
• Vitamin B_6 (pyridoxine) is also so prevalent in foodstuffs that deficiency is rare, but if so then anaemia, glossitis and neuropathy may be present.
• Vitamin B_7 (biotin) is a cofactor in some carboxylase reactions involved with fatty acid metabolism and gluconeogenesis. Specific deficiency is rare, the most common sign being dermatitis. As it happens, biotin binds to avidin (found in raw egg white), a factor exploited in laboratory methods such as enzyme-linked immunosorbent assay.
• Vitamin B_9 (folic acid, in the form of tetrahydrofolic acid) is a cofactor in DNA synthesis. Deficiency results in anaemia, and has been associated with neural tube defects in the foetus, so that supplementation is often recommended.
• Vitamin B_{12} (cobalamin) is a cofactor for methionine synthase and methylmalonyl coenzyme A, enzymes required for the healthy production of red blood cells, so that deficiency leads to anaemia.

- Vitamin C (ascorbic acid) does not have a role as a specific cofactor, but acts instead as a general antioxidant. It is also required in the absorption of iron and the synthesis of amino acids and proteins such as collagen. Deficiency leads to scurvy, signs and symptoms of which include poor wound-healing, gum disease and haemorrhage.

Other requirements

A healthy diet also calls for water and inorganic macronutrients to replace those lost in the urine and faeces, and indigestible material (roughage, mostly plant matter). Electrolytes such as sodium and potassium, and calcium and magnesium are discussed in Chapter 12 and Chapter 15 respectively.

Specific nutrients: inorganic micronutrients

- Copper is a cofactor for numerous enzymes involved in a wide range of metabolic processes, such as oxygen usage and respiration. It is relatively easily absorbed and carried in the blood bound to albumin and caeruloplasmin. Deficiency leads to a diverse mixture of signs and symptoms, such as anaemia, bone abnormalities and increased risk of infections. However, two genetic conditions are recognized: Menkes disease results from copper deficiency, whilst Wilson's disease is the consequence of high concentrations of the metal. Copper and caeruloplasmin increase in pregnancy, in those taking oral contraceptives and during the acute-phase response.
- Iodine is an essential constituent of thyroid hormones tri-iodothyronine and thyroxine. The iodide anion is rapidly absorbed and circulates free (unbound to protein), is taken up by thyrocytes (where it is also stored) and is easily excreted by the kidney. Deficiency leads to hypothyroidism and, possibly, to the development of a goitre (swelling of the thyroid, Chapter 18).
- Iron is required principally as a component of haem in haemoglobin. Deficiency is a cause of anaemia (with signs and symptoms such as tiredness and angular stomatitis). The body is programmed to retain iron, so that excess intake leads to iron deposits in various organs, such as the heart and brain. In some cases this iron overload (haemochromatosis) has a genetic basis. Full details of iron metabolism are presented in Chapters 3 and 4.
- Selenium is a component of many enzymes, such as those involved in redox reactions including glutathione

peroxidase, iodothyronine 5-deiodinase, and thioredoxin reductase. In high concentrations it is relatively toxic, and intake of >3.2 mg/day may cause alopecia and abnormalities of the skin and nervous system. Selenium deficiency has been known to result in cardiomyopathy (Keshan disease) and a childhood osteoarthropathy (Kashin–Beck disease), which have been observed in parts of China where the soil has a low selenium content.
- Conversely, zinc is relatively nontoxic, and required by many different types of enzymes, deficiency leading to depressed growth, anorexia (loss of appetite), loss of taste and smell, night blindness behaviour disturbances, poor wound-healing, impaired immune function, skin lesions and diarrhoea. Acrodermatitis enteropathica is a rare defect in zinc absorption and causes dermatitis, alopecia (loss of hair) and diarrhoea and treated using zinc supplementation. High concentrations of zinc (>100 mg/day) may interfere with copper and iron metabolism and lead to potential deficiency and anaemia respectively.

In general, micronutrients are necessary as cofactors for specific enzymes in metabolic pathways. Excess amounts can lead to toxicity, whereas deficiencies of individual micronutrients can lead to classical deficiency diseases such as scurvy. Some trace elements have other metabolic effects when taken in quantities higher than usual dietary amounts, an example being lithium in the treatment of bipolar disorder.

Laboratory assessment of nutrients

Assessment of specific vitamin or mineral status is required only if a deficiency is suspected, or to confirm adequacy of supply in long-term artificial nutrition. Vitamins are generally measured by high-performance liquid chromatography (HPLC), folate and B_{12} are measured by immunoassay, and vitamin D isotypes by immunoassay, HPLC and mass spectrometry linked to liquid chromatography. Vitamin K is often assessed by its role as an essential cofactor in the synthesis of coagulation proteins, as the prothrombin and activated partial thromboplastin times.

Measurement of some trace elements is confounded by their binding to carrier and storage proteins (such as iron by transferrin and ferritin, and copper by caeruloplasmin), which are generally measured by immunoassay. However, these proteins may be influenced by the acute-phase response for eg ferritin will increase but transferrin is a negative acute-phase reactant, so this may need to be addressed, perhaps by c-reactive protein measurements (Chapter 9). Metals may be assessed by atomic absorption or emission spectroscopy, but these are being superseded by the superior technology of newer methods. For example, inductively coupled plasma mass spectrometry can assess several elements at the same time, unlike atomic absorption, which can only scan a single element at a time. These and other methods are outlined in Chapter 2.

Nutritional disorders

The desire to eat is fed by the satiety centre in the arcuate nucleus in the hypothalamus, which acts via, and responds to, a variety of hormonal and neural stimuli, although there are many other factors, several of which are psychological. A decreased desire to eat is anorexia, the polar extreme of which is an increased desire to eat, leading to polyphagia (meaning many eating). The ultimate consequences are death by starvation for the former and obesity for the latter.

The mathematical model for defining these conditions is the body mass index (BMI), given by the weight in kilograms divided by the square of the height in metres (i.e. kg/m^2). This is driven by the recognition of the adverse health effects of overweight and underweight, and precise categories are recognized (Table 16.2).

However, the International Association for the Study of Obesity, the International Obesity Task Force and the World Health Organization propose BMI cut-off points that are more stringent for adult Asians, where 23.0–24.9 kg/m^2 is defined as overweight and $\geq$25.0 kg/m^2 is defined as obesity. BMI is used differently for children. It is calculated in the same way as for adults, but then compared with typical values for other children of the same age.

Table 16.2 Categories of BMI.

BMI (kg/m^2)	Descriptor
<15	Very severely underweight
15.1–16.5	Severely underweight
16.6–18.5	Underweight
18.6–25.0	Normal (healthy) weight
25.1–30.0	Overweight
30.1–40.0	Obese
>40.1	Morbidly obese

The pathology of underweight   The dominant pathophysiology of underweight is anorexia nervosa. It is dominated by psychological factors, although there may be some relevant gene lesions. There is no clear consensus as to a precise BMI that defines this condition, although BMI <17.5 kg/m^2 is commonly cited. In addition, these patients suffer from rapid weight loss, fear of gaining weight and loss of menstrual cycle (amenorrhoea) in females.

The pathology of overweight   The most common clinical condition of overweight is as part of the Prader–Willi syndrome, a multifactorial condition linked to mutations in a number of genes on chromosome 15. It can be characterized clinically by overeating in childhood, which continues into adolescence and adulthood, and which, unless addressed, leads inevitably to morbid obesity. With a frequency of 1 in 10 000–25 000 it cannot possibly account for more than a tiny fraction of overweight in today's society.

The major causes of overweight and obesity are of course high calorific intake alongside a reduction in physical activity. Consequences of obesity include diabetes, venothromboembolism (both deep vein thrombosis and pulmonary embolism), osteoarthritis and obstructive sleep apnea. The molecular basis of the extremes of BMI may be related to levels of hormones such as leptin, ghrelin and obestatin, although measurement of these molecules (by immunoassay) is far from the routine in the blood science laboratory, but may become so.

Nutritional intervention

For those patients with life-threatening nutritional disorders, direct intervention can take several forms: oral supplementation of specific nutrients, enteral nutrition (perhaps via a nasogastric tube) and a percutaneous endoscopic gastrostomy tube (passing food through the skin directly into the stomach). However, another role for the blood scientist is to cooperate with dieticians, pharmacists and other professionals in the management of total parenteral nutrition (TPN).

The typical patient requiring TPN will be one where all or part of the gastrointestinal tract is nonfunctional. This may be because of a pathology, such as very severe Crohn's disease or another inflammatory bowel disease, bowel obstruction, prolonged icterus, radiation enteritis, hypermetabolic states (such as in repair after severe burns) or that so much of the intestines have had to be removed surgically that not enough are present to

Table 16.3 Components of a typical TPN infusion.

Component	Concentration/amount (per kg of patient's weight/day)
Water	30–40 mL
Energy (generally as dextrose)	Medical patient: 30 kcal Post-surgical patient: 60 kcal
Amino acids	Medical patient: 1 g Post-surgical patient: 2 g
Minerals	Chromium 15 µg, copper 1.5 mg, iodine 120 µg, manganese 2 mg, phosphorus 300 mg, zinc 5 mg
Vitamins	Absorbic acid 100 mg, biotin 60 µg, cobalamin 5 µg, folate 400 µg, niacin 40 mg, pantothenic acid 15 mg, pyridoxine 4 mg, riboflavin 3.6 mg, thiamine 3 mg, vitamin A 4000 IU, vitamin D 400 IU, vitamin E 15 mg, vitamin K 200 µg

enable adequate absorption. Accordingly, the patient must obtain all their nutrition by another route, generally a large-bore indwelling intravenous catheter, often in the vena cava. Strict sterile technique must be used during insertion and maintenance of the central line, and external tubing should be changed every 24 h with the first infusion of the day. Infections and thrombosis are a common problem, and in the long term, fatty liver may develop, which may progress to liver failure.

Standardized TPN packs exist leading to a decrease in the number of on-site manufactured packs. Commercial products are available, and these are tailored for the specific needs of the particular patient with supplements. The constituents of a typical solution are shown in Table 16.3.

Most calories are supplied as carbohydrate, but the exact amount and concentration depend on other factors, such as metabolic needs. For example, in a fever, energy requirements need to increase by 12% for each degree centigrade increase in temperature. Lipids are also a source of energy, and essential fatty acids and triacylglycerols may make up 20–25% of total calories.

The volume of fluid delivered varies between 1 and 2.5 L, and the concentration of electrolytes and other elements adjusted. For example, a 2.5 L pack should generally contain 60–80 mmol of sodium, 50–60 mmol of potassium, 4.5–5.5 mmol of magnesium, 4–5 mmol of calcium and 20–25 mmol of phosphate. Micronutrient requirements include fluoride 1 mg/day, molybdenum 0.3 mg/day, selenium 0.07 mg/day, chromium 0.2 mg/day and vanadium 0.002 mg/day. Other special aspects include:

• Reduced protein content and a high percentage of essential amino acids in renal insufficiency not being treated with dialysis or for liver failure.

• For heart or kidney failure, limited volume (liquid) intake in heart or renal failure.
• A lipid emulsion that provides most of the non-protein calories to minimize CO_2 production by carbohydrate metabolism in respiratory failure.

The laboratory will be called on to perform analysis of the factors in a TPN pack providing dieticians and pharmacists with the tools needed to fine-tune the components of the infusion to suit the individual patient. Naturally, the laboratory will also be called upon to determine levels of numerous analytes several hours after an infusion is complete. This is needed to allow an equilibrium between the blood and the tissues to develop. It is also often useful to analyse the urine of a patient on TPN.

16.2 The intestines

The often-overlooked intestines are a 7–8 m long super-organ, whose full and effective function demands the participation of numerous other organs and systems. The main functions of the intestines are the ingestion, digestion and absorption of nutrients, conservation of secretions, and elimination of unabsorbed waste products. Although this is not a physiology textbook, it is timely to review the entire process of digestion. However, we will exclude the liver and its diseases as these are the subject of Chapter 17.

Intestinal physiology

The intestinal tract is a highly modified hollow tube running from the mouth to the anus. Pavlov's dogs and other experiments have taught us of the importance of the higher centres of the brain in anticipating feeding and the effects this has on subsequent digestion.

The upper intestines   Digestion begins with the mastication of food by the teeth, and the action of salivary enzymes, one of which is amylase. The senses (sight of food, taste and smell) and chewing stimulate parts of the brain to prepare intestinal organs for digestion. The bolus of food passes down the oesophagus, under the influence of gravity and peristalsis, to the stomach.

This organ is a J-shaped structure lying directly under the diaphragm where food is stored, but digestion continues with the action of enzymes such as pepsin and lipase. Hydrochloric acid from parietal cells brings the local pH down to 5–6. There are also mechanical factors to mix the food, which becomes chyme. The stomach also secretes mucus and intrinsic factor, the latter required for the absorption of vitamin B_{12} further down the intestines. Secretions are regulated by local hormones (such as gastrin) and by innervation via the vagus nerve.

The pancreas   This soft oblong organ is sited between the spleen and the diaphragm on the left side of the abdomen, and in close contact with the stomach. It consists of a loose amalgam of cells and tissues, with a draining tube (the pancreatic duct) that (in most people) merges with the bile duct, forming the ampulla of Vater, which then empties into the duodenum.

The pancreas has two functions. Endocrine roles in the blood regulation and metabolism of glucose are explained in Chapter 14. Exocrine roles include the production of enzymes by glandular epitheial cells that make up 99% of the mass of the pancreas. Each day these tissues generate approximately 1200–1500 ml of pancreatic juice, consisting of amylase, trypsin, chymotrypsin, carboxypeptidase, lipase, ribonuclease and deoxyribonuclease. In many cases these enzymes are produced as zymogens (such as trypsinogen) that minimize autodigestion, and these become activated in the duodenum. A further component of pancreatic juice is bicarbonate, which raises the pH and partially counteracts the low pH of the chyme leaving the stomach. Parasympathetic impulses from the vagus nerve and the hormones secretin and cholecystokinin regulate pancreatic secretions.

The liver, gall bladder and bile   Chapter 17 has extensive details of this large organ, which secretes approximately 900 ml of bile daily. This alkaline fluid (pH 7.6–8.6) consists of water, bile salts, cholesterol, lecithin, bile pigments and ions. The principal component of bile salts and pigments is bilirubin, a metabolic product of haem (Chapter 3). Bile salts are crucial in the emulsification of fat globules, breaking them down into droplets to allow further digestion by lipases.

Within the liver, the biliary tree is a network of tubes originating in the lobules. Small biliary vessels (canaliculi) carry bile, and merge into larger vessels (ductules) that eventually form the bile duct. The cystic duct leads to and from the gall bladder, a small sac that can store bile. Under certain conditions (such as insufficient bile salts or lecithin) the cholesterol in the bile in the gall bladder may come out of solution and precipitate as gallstones.

The small intestines   Partially digested food (chyme) enters the duodenum (perhaps 25 cm long), which stimulates the release of hormones secretin and cholecystokinin that in turn stimulate the release of pancreatic and biliary secretions. Phospholipases, cholesterol esterases and nucleases hydrolyse their relevant substrates to release fatty acids, cholesterol and oligonucleotides.

Inorganic iron is absorbed mainly in the duodenum, aided by the acid from the stomach. The reduced ferrous (Fe^{2+}) form is more readily absorbed than the oxidized ferric (Fe^{3+}) form, and both need to be liberated from haem, although this too can cross the enterocytes.

The duodenum transforms slowly into the jejunum (some 2.5 m in length), where absorption of the small molecular products of digestion is facilitated by enterocytes. This section also includes mucus-secreting goblet cells and exocrine glands secreting proteases such as carboxypeptidases, trypsin, chymotrypsin and elastase. Fats are emulsified into droplets by the mechanical action of peristalsis and the detergent action of both bile acid conjugates (bile salts) and phospholipids. Key fat-digesting enzymes include pancreatic lipase. Normal jejunal secretion is stimulated by distension of the duodenum, and modulated by neural and humoral mechanisms. Most water-soluble vitamins are absorbed in the duodenum and jejunum. The latter merges into the ileum, which is about 3.6 m long. This section is particularly relevant as it is where the vitamin B_{12}/intrinsic factor and almost all of the bile acid conjugates are absorbed.

The large intestines   This consists of the caecum (a very short 6 cm blind pouch), the colon (1.3 m long) and the rectum (20 cm long), the intestines terminating with the anus. As the intestines develop, bacterial commensals become important, such as involvement in the formation of secondary free bile acids deoxycholic and lithocholic acids in the distal ileum and colon, and in vitamin K metabolism. The bacteria also ferment the chyme, leading to gas production (flatus) and its release (flatulence).

In these two sections, water and electrolyte (sodium, potassium and chloride) uptake and secretion are regulated, partially under the control of aldosterone and osmosis, although hydrogen and bicarbonate ions are important. Slow and steady removal of water from the chyme leads to the formation of faeces, effectively the indigestible food, generally vegetable fibre. Failure to regulate these factors leads to diarrhoea.

Other functions of the intestines The intestines have an important barrier function, minimizing the entry of potential pathogens, partially achieved by the acid of the stomach. Any breakdown in the mucosa can result in bacterial invasion and systemic infection. Within the intestinal walls are aggregates of immunological tissue (Peyer's patches, as is explained in detail in Chapters 5 and 9), whilst endocrine tissues secrete hormones directly into the circulation to local and distant organs that regulate fluid and enzyme secretion, gut motility, sphincters and also appetite.

Intestinal disease

Disease of the mouth, throat and oesophagus All three areas are at risk of cancer, but if present it rarely causes direct abnormality in the physiology of digestion; the major problems are loss of appetite, difficulty in eating and the inevitable surgery. The oesophagus may become inflamed (i.e. oesophagitis), and can be part of diseases such as the autoimmune condition scleroderma (Chapter 9).

Diseases of the stomach These include:

• Weakness of the cardiac sphincter, the valve formed by a ring of muscle that regulates the passage of food from the oesophagus to the stomach, and its retention. A weak sphincter allows part-digested food and gastric secretions to reflux up into the oesophagus where it may cause disease.
• Peptic ulcers, often caused by excessive acid secretion and/or loss of protective mucus. In 80% of cases this may be due to infection with an inflammatory response to *Helicobacter pylori*. This bacterium metabolizes urea to ammonia, so providing protection from gastric acidity. Consequently, the urea breath test may be useful in diagnosis. Patients drink a fluid with C^{13}-urea, and if

present the organism will metabolize the urea and C^{13}-labelled carbon dioxide will be detected in expired air. Fragments of *H. pylori* can be detected in faeces.
• Cancer, of which there are several variants. A gastrinoma secretes high levels of gastrin, which overstimulates the release of acid and so may cause peptic ulceration and have effects on the small intestines, such as the inhibition of 'alkaline' enzymes of the pancreas, leading to malabsorption. All stomach cancers are amenable to surgery, but this may also remove the cells secreting intrinsic factors, leading to vitamin B_{12} deficiency.
• Autoimmune disease, such as autoantibodies directed towards the cells that generate intrinsic factor, destruction of which also leads to vitamin B_{12} deficiency. The immunology laboratory can help with the detection of antibodies to gastric parietal cells and to intrinsic factor (Chapter 9).
• Gastric atrophy and/or gastritis, one of the most common causes of which is alcohol abuse, and which once more leads to micronutrient malabsorption.

Disease of the biliary tree Although liver disease is the object of Chapter 17, it cannot be ignored in this section. Briefly, gallstones forming in the gall bladder may leave it and pass down the bile duct, where they may become lodged, and so obstruct the flow of bile. This leads to the condition of obstructive cholestasis. Alternatively, the bile duct may become inflamed, leading to inflammatory cholestasis. The consequences of either form of cholestasis are (a) that bile fails to reach the duodenum, which has repercussions for digestion, and (b) bile remains in the liver and ultimately causes jaundice and liver damage with raised liver function tests. Should a gallstone become lodged in the ampulla of Vater it will also block the pancreatic duct.

In biliary atresia, the bile duct is absent. In a congenital form it is a cause of neonatal jaundice. In the adult, it may arise as a result of the progressive destruction of the bile duct, possibly by an autoimmune process.

Disease of the pancreas The key biochemical marker is amylase, although some may arise from salivary gland disease. The isoform *p*-isoamylase is a more sensitive and specific test, with values >10 times the upper limit of normal considered almost diagnostic for acute pancreatitis, whereas fivefold increases can be present in mesenteric artery infarction (occlusion of blood supply to tissue), diabetic ketoacidosis, and acute biliary tract disease. A second marker is pancreatic elastase, which is resistant to digestion by other enzymes, and can be measured in faeces: low concentrations reflect pancreatic insufficiency. Blood lipase and lactate dehydrogenase may also be raised. Indeed, blood lipase is more sensitive and specific than amylase, and its use is increasing. Pancreatic diseases include:

- Acute pancreatitis; the two main causes are gallstones and alcohol abuse, in addition to others such as biliary disease, viral and bacterial infection, trauma and certain medications (e.g. oestrogens, corticosteroids). However, the growing public health issue is the pancreatitis resulting from alcohol abuse. The tissue inflammation leads to cell damage and auto-digestion by the pancreas's own enzymes.
- Chronic pancreatitis. By definition, this is a slowly developing and insidious inflammatory condition leading to failure of the organ to secrete its enzymes. Causes are as above, but rarely may also be genetic and autoimmune.
- Blockage of the pancreatic duct, which clearly leads to a reduced enzyme secretion from an otherwise healthy pancreas, which will therefore become damaged in a setting that has parallels to obstructive cholestasis. One of the leading causes of blockage is cancer of the head of the pancreas at the point where it meets the duodenum.
- Pancreatic cancer. This may develop *de novo* or from a background of other pancreatic disease, and which has a particularly poor prognosis. Patients presenting with jaundice but no pain often results in an earlier diagnosis. Diagnosis is often delayed because of few space occupying effects.

Pancreatic disease may be present in other conditions, such as cystic fibrosis. The consequences of pancreatic disease include malabsorption of proteins, fats and carbohydrates, with steatorrhoea (excessive fat in the faeces). However, the major metabolic consequence of pancreatic disease is diabetes.

Disease of the small Intestine   There are a number of disease processes leading to malabsorption of chyme by the duodenum, jejunum and ileum; these include non-specific bacterial overgrowth and failure of peristalsis, although there are a number of known pathologies:

- Crohn's disease is an umbrella term for an inflammatory disease of any part of the small and large intestines, although the most common site is the ileum.
- A diverticulum is a pouch or in-tucking of the intestines. These can be present in the small or large intestine; if inflamed, the resulting condition is diverticulitis.

- Coeliac disease is an autoimmune condition caused initially by hypersensitivity to gliadin, a gluten protein found in wheat and other cereals. With an incidence of perhaps 1 per 100 Caucasians, there is a strong linkage with the HLA-DQ2, and is often associated with other autoimmune diseases such as type 1 diabetes mellitus and thyroid disease. Additional details are provided in Chapters 9 and 10.
- Tropical sprue. The aetiology of tropical sprue is unclear, but may be the result of an excessive immune response to a pathogenic organism. There are no specific blood tests, but as in all malabsorption syndromes, eventually concentrations of micronutrients will fall.
- Lactose intolerance (inability to digest lactose, also present in up to 5% of the population). This is caused by inadequate concentrations of lactase. As lactose is a disaccharide of galactose and glucose, the condition is also called disaccharidase deficiency. In the absence of a reliable and simple test for the enzyme, the potential patient consumes a meal of 50 g of lactose. A diagnosis is made 3 h later if blood concentrations fail to exceed 1.1 mmol/L. An alternative is a hydrogen breath test, which operates under similar conditions. Other tests include measurement of reducing substances in faeces semiquantitatively using the Benedict's test and then identifying the sugars present using thin-layer chromatography.

Bile acid metabolism   Although the formation, storage and excretion of bile are discussed above and in Chapter 17, we will now look at its role in digestion. Entering the duodenum from the confluence of the bile duct and pancreatic duct, bile (principally glycocholic, deoxycholic, lithocholic and taurocholic acids, structurally all based on cholesterol) promotes the activity of certain enzymes and the absorption of lipids. Although perhaps 20–30 g of bile acids enter the intestines daily, perhaps 95% are reabsorbed and recycled through the liver.

However, various conditions lead to the failure to reabsorb these bile salts, which can therefore be seen as malabsorption. Causes include mucosal disease or resection of the terminal ileum, bile acid transporter mechanism, or disordered ileal signalling affecting hepatic synthesis and bile acid pool size. An additional cause is inappropriate bacterial colonization.

Disease of the large Intestine   The major diseases of the large intestine are Crohn's disease, ulcerative colitis, irritable bowel syndrome (IBS), diverticular disease, polyps and carcinoma. The first we have already met, as it may present in the small intestines, and with ulcerative colitis forms the inflammatory bowel diseases (IBDs).

As the name implies, ulcerative colitis is characterized by inflammation and ulceration of the colon, leading to bloody diarrhoea and lower abdominal pain. However, whilst the aetiology of Crohn's disease is autoimmune, the exact cause of the inflammation in ulcerative colitis is unclear. IBS is a condition in search of a clear pathophysiology, it being characterized by symptoms such as abdominal pain, gastroesophageal reflux and bloating. Diverticulitis is found more often in the large intestines than in the small intestines, symptoms including abdominal pain, fever, nausea, vomiting and constipation. Of course, these are all nonspecific and can be present in the IBDs.

There are few truly useful blood tests in bowel disease. If an active inflammation is present, there may be increased erythrocyte sedimentation rate, white cell count and an acute-phase response, all nonspecific. However, faecal calprotein may be valuable. This 36 kDa protein, resistant to enzymatic digestion, is a neutrophil product and has bacteriostatic activity. Raised concentrations are present in the faeces of those suffering from Crohn's disease, ulcerative colitis and certain tumours, but levels are normal in IBS.

Cancer of the large and small intestines

All sections of the intestines are prone to cancer, and in most cases treatments are surgery and chemotherapy. In the small intestines these may be adenocarcinoma, sarcoma and leiomyoma (i.e. of the muscles of the intestinal walls) and lymphoma (from the intestinal lymph nodes). As is discussed in Chapter 8, tumours of the intestines may be part of the spectrum of multiple endocrine neoplasia.

The ileum is the preferred site for a carcinoid tumour. The pathological basis is neuroendocrine, with the secretion of serotonin and other vasoactive peptides. These pass directly, via the hepatic-portal circulation, to the liver and the body in general, leading to dermal vasodilation (flushing) and diarrhoea. Serotonin is metabolized to 5-hydroxyindole acetic acid, and can be measured in urine. Most reference ranges are <40 μmol/24 h; values >120 μmol/L are diagnostic of carcinoid tumour, and concentrations correlate well with tumour size and can be used to monitor therapy.

Colorectal cancer is one of the leading causes of morbidity and mortality in the UK. Serum concentrations of carcinoembryonic antigen (CEA) are useful in monitoring the effectiveness of treatment, generally

surgery to remove the tumour, followed by chemotherapy. However, CEA may be increased in cancer of the stomach, pancreas, lung and breast, and also in non-malignant conditions of these organs, such as ulcerative colitis, pancreatitis, chronic obstructive pulmonary disease, and in tobacco smokers. Most cases of colorectal cancer arise from polyps, but these are generally asymptomatic and are only detectable by colonoscopy and sigmoidoscopy. However, faecal occult blood is a useful screening tool in selected high-risk groups.

> ### Blood science angle: Nutrition and intestinal disease
>
> Haematologists take a close interest in the intestines. Vitamin K is the target of the anticoagulant warfarin, and disease of the stomach and ileum can lead to insufficient malabsorption of vitamin B_{12}, and so megaloblastic anaemia. Nutritional lack of iron can lead to microcytic anaemia. These points are discussed in Chapters 3 and 4.

Diarrhoea A simple definition of diarrhoea is passage of >200 g faeces a day, but this fails to account for diet, the body mass of the subject, or the degree of water present. Increases in stool mass of 500–1000 g/day are highly significant of disease and require medical attention; however, diarrhoea has to be more severe to result in dehydration. A number of different pathologies are recognized.

- Perhaps the most common in the UK is that associated with IBS, Crohn's disease and ulcerative colitis, accounting for up to half of all visits to gastroenterology outpatient departments. This diarrhoea is described as exudative, and the pathology is inflammatory.
- If the intestines consider foodstuffs to be a hazard, they will move it rapidly through and then expel it, the state being called hypermobility. This is common in a change in diet, perhaps whilst on holiday. Alternatively, the food may be expelled because it is infected. There may also be increased motility in opiate withdrawal (as these generally cause constipation) and thyrotoxicosis.
- Viral infections, such as norovirus, rotavirus and adenovirus, but also bacterial infections such as campylobacter, salmonella and shigella.
- Self-induced diarrhoea due to the excessive use of laxatives.
- Secretory diarrhoea follows the increased secretion of, or reduced absorption of, water and ions. An example of

this, still pertinent in the developing world, is the effects of the cholera toxin, which opens membrane channels leading to the export of water, sodium, potassium and bicarbonate into the intestinal lumen, resulting in water diarrhoea and potentially fatal dehydration.

• Osmotic diarrhoea is the consequence of a solution of very high osmotic potential (generally sugar or salt), or perhaps malabsorbed sugars. The laws of osmosis demand the movement of water from the body and wall of the intestines into the lumen to equilibrate the high ionic content of the chyme. However, it may also be the result of malabsorption or maldigestion, as may occur in IBD and pancreatic disease respectively.

• Mixed secretory and osmotic diarrhoea may be present in coeliac disease, the presence of unabsorbed bile acids and fatty acids, longstanding diabetes, and abnormal bacterial colonization.

There are no specific blood tests for diarrhoea per se, but hypernatraemia may be present in dehydration.

However, analysis of the diarrhoea itself may be useful, such as the faecal osmotic gap, which requires the measurement of faecal sodium and potassium. In health, the sum of these ions, in millimoles per litre multiplied by 2, should be around 290 arbitrary units, and a discrepancy (if present) between the measured osmolality and the calculated osmolality is called the faecal osmotic gap. A gap of <75 mOsm/kg, excludes osmotic diarrhoea, whereas a gap >75 mOsm/kg excludes secretory diarrhoea. Figure 16.1 summarizes major aspects of diseases of the gastrointestinal tract.

Malabsorption

Malabsorption refers to inadequate absorption of nutrients following digestion. Although aspects of malabsorption have already been mentioned, it deserves a section on its own. Malabsorption should not be confused with maldigestion, which is a failure of digestion itself. However, undigested molecules may not be

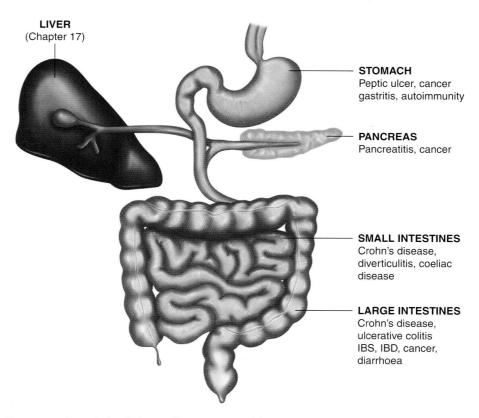

Figure 16.1 Key aspects of intestinal pathology. Different sections of the intestines are host to particular diseases, but some diseases are present in more than one location. Most diseases are inflammatory and neoplastic in nature.

absorbed, and thus the clinical features that present may be similar. Malabsorption can be general, affecting a range of different nutrients, or more specific, where only certain nutrients are affected. The clinical features observed in a patient with malabsorption arise because of two broad reasons:

- deficiency of nutrients, such as due to anaemia (iron, folate, vitamin B_{12} deficiency), osteomalacia (vitamin D deficiency), bleeding tendency (vitamin K deficiency), oedema (protein deficiency, such as of albumin) and growth failure/weight loss (generalized deficiency of nutrients);
- retention of nutrients within the bowel lumen, which can cause steatorrhoea, diarrhoea, abdominal pain and flatulence.

Malabsorption of nutrients may occur for one or more reasons and some of the common causes are:

- deficiency of pancreatic enzymes, as in chronic pancreatitis;
- bile salt deficiency, such as in liver disease and biliary obstruction;
- diseases of the intestine causing failure in absorption of nutrients, as in coeliac disease and Crohn's disease;
- overgrowth of bacteria, such as in internal fistulae and jejunal diverticulosis.

The purpose of investigation is to confirm that malabsorption of nutrients is indeed occurring and to identify the possible cause(s). A clinical history is crucial and may suggest malabsorption if the patient presents with the clinical features described above. Clinical examination of the abdomen and evidence of weight loss or deficiency of a particular nutrient can aid diagnosis. Presence of steatorrhoea is characteristic of malabsorption and confirms the diagnosis. A number of simple chemical tests can indicate the possibility of malabsorption, and these are listed in Table 16.4.

More specialized biochemical tests for malabsorption include those such as the faecal fat test. This test is rarely used nowadays, as it is an unpleasant test to undertake for all concerned. It involves collection of faeces for 3 days and estimation of their fat content, a high value indicating the presence of steatorrhoea. Other tests include the triolein breath test, where the patient consumes triolein labelled with either ^{13}C or ^{14}C isotope. This triolein is digested, absorbed in the gut and then appears in the breath as carbon dioxide where the labelled component

Table 16.4 Simple tests for initial investigation of malabsorption.

Test	Deficiency in malabsorption
Reduced full blood count	Anaemia
Low serum iron, folate or vitamin B_{12}	Anaemia
Low serum urea	Protein deficiency
Low serum albumin	Protein deficiency (hypoalbuminaemia)
Low serum calcium	Hypocalcaemia
Low serum phosphate	Hypophosphataemia

can be measured using an appropriate technique. In malabsorption, the amount of label in expired air is reduced.

Other tests include the xylose absorption test, where the patient fasts overnight, empties their bladder in the morning and then consumes 5 g of xylose dissolved in 500 mL of water. Urine samples are collected over the next 5 h and a blood sample taken after 1 h. Low values are present for xylose absorption in blood and excretion in urine in patients with malabsorption. However, all of these tests can be problematic and are rarely used. The approach nowadays is that if there is clinical evidence of malabsorption, then the simple tests outlined in Table 16.4 are performed and, if necessary, further supported by measurements of faecal elastase, endomysium and tissue transglutaminase antibodies (Chapter 9), biopsy of the small intestine and other imaging techniques. Management of malabsorption involves treating the underlying cause and ensuring the patient is adequately nourished.

16.3   Case studies

Case study 20

A 46-year-old man was taken to hospital complaining of severe abdominal pain that radiated through to his back. This pain had developed suddenly a few hours earlier. On clinical examination he was in mild haemodynamic shock (systolic and diastolic blood pressure 115/69 mmHg) and had a tender abdomen. He had no previous history of gastrointestinal disease and there was no evidence of intestinal obstruction on radiographic examination. However, he did confess to heavy alcohol intake.

	Result (unit)	Reference range
Na	144 mmol/L	135–145
K	4.7 mmol/L	3.8–5.0
Urea	9.6 mmol/L	3.3–6.7
Creatinine	86 μmol/L	44–133
eGFR	88 mL/min per 1.73 m^2	>90
Glucose	11.0 mmol/L	3.1–6.0
Amylase	5000 IU/L	<100

Interpretation The clinical presentation is typical of acute pancreatitis, and this is confirmed by the high serum amylase concentration. These patients often have hyperglycaemia which is transient in nature and secondary to the pancreatic disease. The high urea and normal creatinine are likely to be due to a combination of dehydration and shock, which causes reduced blood volume (or renal hypoperfusion). The eGFR equates to chronic kidney disease stage II.

Case study 21

A 6-year-old child presented with the symptoms of anaemia, weight loss and abdominal distention. The stools obtained from the child for faecal fat test were noted to have an offensive smell and were loose, bulky and pale in colour.

	Result (unit)	Reference range
Albumin	28 g/L	32–50
Iron	6 μmol/L	14–29
Faecal fat	39 g/3 days	<21

A biopsy of the small intestine showed atrophy of the villi.

Interpretation The clinical presentation and test results indicate coeliac disease, a key feature of which is atrophy of the villi and malabsorption of nutrients. The child has malabsorption of fats as they are lost in the faeces producing the characteristic fat-rich faeces (steatorrhoea). Malabsorption of protein and iron gives the hypoalbuminaemia and anaemia respectively. The child

was placed on a gluten-free diet and the symptoms improved and the child began to gain weight within 4 weeks.

Summary

- Macronutrients include carbohydrates, proteins and lipids; micronutrients include vitamins and minerals.
- Disorders of nutrition include overweight and obesity, and anorexia. Blood scientists collaborate with pharmacists, dieticians and others to ensure correct total parenteral nutrition.
- Diseases of the intestines are often neoplastic, autoimmune or inflammatory. They often lead to diarrhoea, and the malabsorption of macro- and micronutrients, the latter being detectable in the laboratory.

Further reading

Asquith M, Powrie F. An innately dangerous balancing act: intestinal homeostasis, inflammation, and colitis-associated cancer. J Exp Med. 2010;207:1573–1577.

Plauth M, Cabré E, Campillo B. *et al.* ESPEN guidelines on parenteral nutrition: hepatology. Clin Nutr. 2009;28:436–444.

Scaldaferri F, Correale C, Gasbarrini A, Danese S. Mucosal biomarkers in inflammatory bowel disease: key pathogenic players or disease predictors? World J Gastroenterol. 2010;16:2616–2625.

Shenkin A. The key role of micronutrients. Clin Nutr. 2006;25:1–13.

Vilela EG, Torres HO, Martins F.P. *et al.* Evaluation of inflammatory activity in Crohn's disease and ulcerative colitis. World J Gastroenterol. 2012;18:872–881.

Guidelines

NICE CG32 : Nutrition support in adults: oral nutrition support, enteral tube feeding and parenteral nutrition. www.nice.org.uk.

17

Liver Function Tests and Plasma Proteins

Learning objectives

After studying this chapter, you should be able to:

- appreciate the importance of the liver;
- understand how the structure of the liver relates to its functions;
- describe the major liver function tests (LFTs);
- list the major plasma proteins formed in the liver;
- explain the function of albumin and other major plasma proteins;
- discuss the uses of these blood tests in liver and related diseases;
- be aware of the clinical responses to problems with the liver.

The liver is the largest single discrete organ in the body, with a mass of 1.4–1.8 kg (perhaps 3–4 lbs) in the adult male, 1.2–1.4 kg in the female, and is located in the right upper part of the abdominal cavity, immediately below the diaphragm. The liver is composed of millions of liver cells (hepatocytes) that play a major part in many metabolic and excretory mechanisms. One of these is to synthesize a large number of proteins for export into the plasma; indeed, it has been suggested that the liver produces 90% of the plasma proteins, and is why this chapter will also look at plasma proteins. The major plasma protein is albumin, a critical regulator of osmosis and transporter of many compounds. Like the kidney, the functions of the liver can also be classified into a small number of areas, but the function of the liver is more diverse:

- *Metabolic activity.* This includes the synthesis (anabolism) of a large number of different proteins, lipids and carbohydrates, many destined for 'export' to the blood.

Good examples of these are cholesterol and the iron transport molecule transferrin.

- *Storage.* The liver converts glucose and other carbohydrates into the storage compound glycogen in times of plenty, and then back to glucose in times of need. It also stores iron (in ferritin), vitamin B_{12} and copper.
- *Detoxification.* This is the catabolic removal of dangerous and unwanted substances; not only natural toxins, but also complex drugs.
- *Excretion.* Under this heading we consider the generation and removal of the waste products of metabolism, such as urea and bile.

As regards the latter, bile is a collection of molecules and ions generated by hepatocytes that is transported to the duodenum by the system of vessels called the biliary tree. Bile can also be stored in the gall bladder.

Therefore, assessment of liver function can be followed by observing various aspects of the above. However, direct access to the liver (indeed, as for any organ) can only be achieved by biopsy, a procedure not to be undertaken lightly. Therefore, as for many organs and processes, we fall back on blood tests, in this respect referred to as LFTs. Together, we use these tests to investigate conditions that include acute liver failure, chronic liver failure, cholestasis, liver cancer and jaundice.

Most scientists recognize some key LFTs, all measurable in plasma, although serum is preferred (Table 17.1). The first, bilirubin, is a waste product, whereas the remainder are enzymes involved in the various anabolic and catabolic processes present in this organ. It could be argued that concentration of albumin, the major single factor in plasma proteins, may also reflect liver function.

As we did for renal function, before we address the value of these tests in pathology, we must refresh our knowledge of the structure of this organ, and then its

Blood Science: Principles and Pathology, First Edition. Andrew Blann and Nessar Ahmed.
© 2014 John Wiley & Sons, Ltd. Published 2014 by John Wiley & Sons, Ltd.

Table 17.1 Major LFTs and plasma proteins.

Analyte	Reference range
Bilirubin	<17 µmol/L
Alkaline phosphatase (ALP)	30–130 IU/L
Gamma-glutamyl transpeptidase (GGT)	5–55 IU/L
Alanine aminotransferase (ALT)	5–42 IU/L
Asparate aminotransferase (AST)	10–50 IU/L
Total protein	60–80 g/L
Albumin	35–50 g/L

There are dozens of other molecules synthesized by the liver, and so could be considered LFTs. The most important will be discussed as they arise.

function (i.e. its physiology). This we will do in Section 17.1; in Section 17.2 we will examine each LFT in detail, and in Section 17.3 we will look at different types of diseases of the liver. In Section 17.4 we will look at plasma proteins, and the chapter will conclude with two case studies in Section 17.5.

An alternative to the word 'liver' that is often used to describe this organ is 'hepatic', and from this we get:

- hepatitis – inflammation of the liver;
- heparin – an anticoagulant first purified from this organ;
- hepatocyte – a liver cell;
- hepatoma – a mass or growth within the liver, possibly (not always) a cancer;
- hepatic artery and hepatic vein – blood vessels to and from the liver;
- hepatectomy – surgical removal of all or part of the liver.

17.1 Anatomy and physiology of the liver

Anatomy

The liver is formed from two lobes, the largest (being five-sixths of the total mass) being on the right. Tucked underneath the liver is the gall bladder. The liver receives blood from the hepatic artery, being a branch of the aorta, and from the intestines via the hepatic portal vein. The latter carries nutrient-rich blood loaded with amino acids, peptides, carbohydrates and lipids. The hepatic vein carries blood from the liver to the inferior vena cava, and a lymphatic vessel takes protein-rich lymph to the thoracic duct. The liver is innervated by both sympathetic

and parasympathetic fibres. A most unusual feature of this organ is its ability to regenerate after parts of it have been removed surgically.

Although a single unit of the kidney is the nephron, there is no true single cell of the liver. However, most of the workings of the liver are handled by the hepatocytes, although these cells make up perhaps only two-thirds to three-quarters of the liver, the remainder being the endothelial cells lining the blood and other vessels, specialized macrophages (Kupffer cells) and cells such as stellate cells (also known as Ito cells) and fibroblasts that produce supporting connective tissues. There is evidence that Kupffer cells eliminate and detoxify micro-organisms, endotoxin, degenerated cells, immune complexes and toxins such as alcohol. In addition, they are likely to participate in immune responses and may be involved in tumour surveillance.

The gross architecture of the liver can be described in terms of a lobule of tissues, a circular or perhaps orthogonal structure at the centre of which is a branch of the hepatic vein (Figure 17.1). On the periphery of each

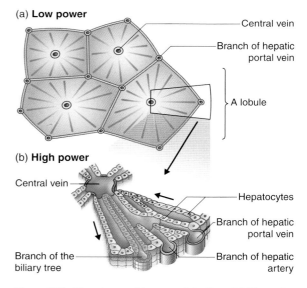

Figure 17.1 The micro-architecture of the liver. (a) The major functional unit of the liver is the lobule, with branches of the hepatic portal vein, the hepatic artery and biliary tree at the periphery, and a branch of the hepatic vein at the centre. (b) Blood passes along sinusoids of merged venous and arterial blood, and perfuses tissues composed of hepatocytes and Kupffer cells, supported by fibroblasts and connective tissues. Bile is generated and passes into the biliary vessels that lead to the gall bladder and bile duct.

lobule is a branch of the hepatic artery, the hepatic portal vein, and a biliary vessel, that is itself a branch of the bile duct. Blood flowing from the hepatic portal vein and hepatic artery merge, and the resultant mixture passes into the central vein, then the hepatic vein and so to the inferior vena cava. A branch of the biliary tree moves the developing bile in the opposite direction. Hepatocytes are located between the different vessels.

The gall bladder and biliary system Chapter 16 has details of this small organ and its related vessels. Together they store and conduct bile from the lobules of the liver to the duodenum. The biliary system consists of a network of fine vessels, with a wall often only one or two cells thick, and with similarities to the lymphatics. Bile is a complex mixture of glycocholic, deoxycholic, lithocholic and taurocholic acids (structurally all based on cholesterol, often conjugated to amino acids), lecithin, mucus, pigments and inorganic salts. As we shall see, and as also explained in Chapter 16, various pathological conditions arise from disease of the gall bladder and biliary tree.

Physiology

The functions of the liver can be described broadly in four groups: metabolism, excretion, storage and detoxification. The greater part of the excretory function of the liver is the removal of bilirubin, which contributes to bile, and also of nitrogenous waste in the form of urea.

Metabolism Although all functions of the liver are vital, the greater proportion of the workings of this organ is to do with metabolism. Indeed, the liver may be described as a giant factory and recycling centre.

The liver synthesizes and then exports a large number of proteins into the blood. Perhaps the only major plasma proteins not produced by this organ are the immunoglobulins (Igs), which are produced by B lymphocytes. Thus, if the concentrations of these non-Ig proteins (such as albumin, fibrinogen and prothrombin) fall, we might consider failure of production. Alternatively, low concentrations may arise from excess consumption, certainly possible for the coagulation proteins, as may be present in the severe clotting condition disseminated intravascular coagulation (Chapter 8). However, another explanation for low concentrations of plasma proteins in general, and defined molecules specifically, is malnutrition. Concentrations of albumin may be low because of loss due to damage to the glomerulus, as in nephrotic syndrome (Chapter 12), or in burns.

Enzymes involved in both anabolic and catabolic pathways include the aminotransferases alanine aminotransferase (ALT) and aspartate aminotransferase (AST), alkaline phosphatase (ALP) and gamma-glutamyl transpeptidase (but also called gamma-glutamyl transferase, both abbreviating to GGT) (Table 17.1). Between them, these enzymes are the tools the liver cells use to break up large molecules into their component parts (catabolism). It is also possible that some of these parts will be used to synthesize new molecules (anabolism). Unused parts are excreted (e.g. as urea).

The liver also has roles in hormone metabolism, such as the synthesis of transcortin, the protein that carries the hormones cortisol, progesterone and aldosterone. The liver is also required for the 25-hydroxylation of a precursor of vitamin D, so that it becomes 25-hydroxyvitamin D. The latter is then hydroxylated again in the kidney to give 1,25-dihydroxyvitamin D – the most active form of the hormone. These molecules are described in Chapter 15.

The liver also synthesizes fatty acids, triacylglycerols, cholesterol, phospholipids and lipoproteins. Indeed, the root of the word 'cholesterol' is long established, and gives us many other words, several of which have a degree of medical relevance (such as cholera and cholestasis), whereas links with others (such as colic and melancholy) are less obvious. Nevertheless, lipid metabolism is complicated, and as it is linked with cardiovascular disease, it is discussed in detail in Chapter 14. Here, it is sufficient to say that lipoproteins are not regarded as LFTs.

The acute-phase response Many of the protein products of the liver are synthesized in a steady state. However, the liver is sensitive to the acute-phase response. This process is generally initiated by an external factor or factors (such as an infection) and has many repercussions, one of which is the generation of inflammatory cytokines such as interleukins. Hepatocytes respond to the stimuli of interleukins by increasing or decreasing the synthesis of various molecules collectively described as acute-phase reactants (Table 17.2).

For this (and other) reasons, the acute-phase proteins are not considered to be LFTs. Nevertheless, the importance of the acute-phase response ise discussed in Chapter 9, on immunopathology.

Excretion of bilirubin and bile When red blood cells come to the end of their life, the proteins, carbohydrates and iron are recycled. The haem molecule is broken down by the enzyme haemoxygenase, which liberates

Table 17.2 Acute-phase reactants.

Increased	Decreased
C-reactive protein, serum amyloid A, complement components, fibrinogen, prothrombin, coagulation factor VIII, plasminogen, ferritin, hepcidin, caeruloplasmin, haptoglobin, orosomucoid, alpha-1-antitrypsin, alpha-1-antichymotrypsin, Igs	Albumin, transferrin, retinol-binding protein, antithrombin, transcortin

the atom of iron (which is also recycled), and partially breaks down the porphyrin ring to carbon monoxide and biliverdin. The latter is converted to bilirubin by the enzyme biliverdin reductase. The precise biochemistry of the family of haem-type molecules is complex and beyond the scope of this chapter, but is clearly relevant to red blood cell biology (Chapters 3 and 4).

This destruction of red blood cells (haemolysis), generally by macrophages, can happen in the bloodstream (where it is described as intravascular haemolysis), or in organs such as the liver (by Kupffer cells), bone marrow and spleen (i.e. extravascular haemolysis). This can generate as much as 500 mmol/275 mg in a volume of 800–1000 ml daily. 'Normal' bilirubin, as it is having left the site of production, is relatively insoluble in water at physiological pH and so is not readily excreted, and passive binding to albumin prevents or minimizes uptake by non-liver cells and enables it to be transported to the liver.

In the liver, the albumin–bilirubin complex dissociates and bilirubin enters the hepatocyte by a carrier-mediated process where it binds to cytosolic proteins, mainly glutathione *S*-transferase B. Once inside the cell, bilirubin is conjugated with glucuronic acid by the action of transferase enzymes to form mono- and di-glucuronides (25% and 75% respectively) that are water-soluble and are thus more likely to be excreted in urine in this form. From the hepatocyte, bilirubin conjugates then pass to the biliary system (so making up bile, which has a pH between 7.6 and 8.6), and so sequentially to the gall bladder (where it is stored), the bile duct and then into the intestines.

Under certain circumstances, the constituents of bile can come out of solution and form crystals, and eventually stones. Most of these gallstones are rich in cholesterol, and can leave the gall bladder, and, if small enough,

pass safely through to the intestines and so are excreted. If large, fragments can sometime become lodged in the bile duct, and so cause acute pain and disease.

Bile pigments reaching the large intestine are degraded by bacteria to urobilinogen and stercobilinogen, and some of these return to the systemic circulation and are subsequently excreted in the urine. Approximately 95% of bilirubin in the plasma is unconjugated. The bile salts that fail to be reabsorbed are excreted in the faeces. Some urobilinogens oxidize spontaneously to stercobilin, an orange–brown pigment responsible for the characteristic brown colour of the faeces. Thus, if any of the liver, gall bladder and bile duct are unable to perform the function of sequestering bilirubin into bile and then passing it to the intestines, the concentration of bilirubin in the blood rises, leading to the yellow colour of the skin known as jaundice, to be discussed in a section that follows.

The production and secretion of bile by the liver is essential for normal digestion and particularly for the absorption and metabolism of fats in the diet. The complex mixture that is bile contains not only bilirubin, but also cholesterol, cholesteryl esters, proteins and bile salts. Collectively, these chemicals are required for the absorption from the small intestine of the free fatty acids and monoglycerides formed by the action of pancreatic lipases.

Different forms of bilirubin in the plasma are also filtered by the glomerulus, but are not reabsorbed, so are concentrated in the urine. Indeed, this is why our urine is a straw colour. If there is glomerular damage, and albumin is lost, then so will the bilirubin that is passively bound to the protein. The metabolism and excretion of bilirubin are summarized in Figure 17.2.

Excretion of urea   Through the production of urea, the liver plays a significant role in the removal of excess nitrogen from the body in the urine – 75% of our nitrogen is excreted in this form. Amino acids are first transaminated to glutamate and then deaminated to ammonia; the nitrogen then passes via citrulline and arginine to generate urea in the urea cycle (also called the ornithine cycle). This process, which operates partially in the cytoplasm and partially in the mitochondrion, also generates fumarate, and requires the energy of three molecules of adenosine triphosphate (ATP) for each molecule of urea that is generated (Figure 17.3).

Failure of the kidney to excrete urea results in high concentrations of this metabolite (as discussed in Chapter 12). High plasma urea (i.e. uraemia) is never

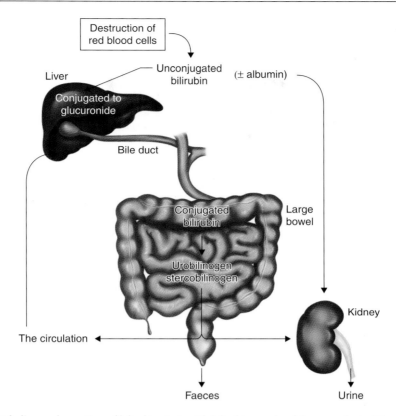

Figure 17.2 The metabolism and excretion of bilirubin starts with bilirubin as a breakdown product of the red blood cell. In the plasma it is transported to the liver by albumin, where it is conjugated to glucuronide, becomes part of the bile and passes into the intestines. Some bile metabolites are acted upon by intestinal bacteria, and some may be reabsorbed. The remainder is excreted in the faeces.

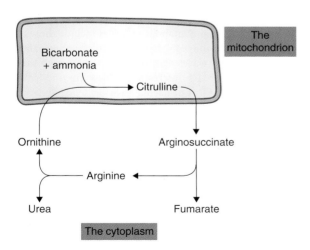

Figure 17.3 The urea/ornithine cycle. This pathway operates partly in the cytoplasm and partly in the mitochondria of hepatocytes.

due to increased synthesis by the liver. The other major nitrogen metabolite, creatinine, may arise from the liver, but the vast majority is the result of catabolic processes in muscles. An 80 kg person needs to be able to excrete about 12–22 g of nitrogen daily from those amino acids that are surplus to requirements.

Storage Plants store the energy in carbohydrates as starch; animals do so as glycogen, which is generally about 500 g in the average body. However, only 20% of glycogen is in hepatocytes, the remainder being in muscle cells (myocytes). At times of excess, glucose is converted first to glucose-6-phosphate, and then to the polysaccharide glycogen by the process of glycogenesis, a process promoted by insulin. However, in gross excess, glucose is converted to fat (i.e. lipogenesis), and this can occur within the liver and elsewhere in the body. The adipocyte is a cell specially adapted to store fat.

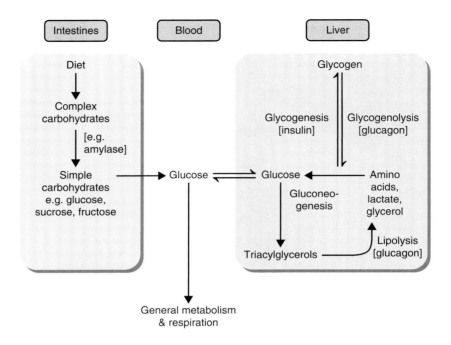

Figure 17.4 The liver stores carbohydrate as glycogen (glycogenesis). Glucose can be liberated from the glycogen stores (glycogenolysis) or can be generated from amino acids and glycerol from fatty acids (gluconeogenesis).

Conversely, at times of need, glucose is liberated from glycogen by the process of glycogenolysis, itself stimulated by the hormone glucagon. However, glucose can be generated by standard metabolic pathways (i.e. gluconeogenesis), which can be stimulated by the hormones cortisol and thyroxine. This process can use lactate, glycerol and amino acids such as alanine, cysteine, glycine and serine to first generate pyruvate, and then glucose. Glucose can also be used for fatty acid synthesis via the glycolytic pathway and the citric acid cycle, and this sugar can also be generated from other carbohydrates such as galactose. The role of the liver in carbohydrate metabolism is summarized in Figure 17.4. The liver also stores iron (in ferritin and haemosiderin), vitamins A, B_{12} and D, and copper.

Detoxification

As a busy factory, the liver degrades as well as synthesizes and recycles. This includes plant, animal and fungal toxins (such as the fungal aflatoxin), as well as drugs we either take by choice (such as paracetamol and alcohol) or are given as medications (antibiotics, chemotherapy, anaesthetics). In many cases, this detoxification is complex and expensive (i.e. demands the energy in the form of ATP). Many catabolic enzymes include those of the large cytochrome P450 family (so called because they absorb light at a wavelength of 450 nm), although they may also be anabolic. Interestingly, these enzymes contain iron in the form of haem.

Blood science angle: The liver and the haematologist

This organ is of interest to haematologists for broadly two reasons: iron and coagulation. For the former, liver failure is likely to lead to anaemia, as the liver is the major site of iron storage, and of the iron overload disease of haemochromatosis. For the latter, several coagulation proteins (such as prothrombin and fibrinogen) are synthesized by the liver, so that haemorrhage is also a consequence of chronic liver failure. These aspects are expanded upon in Chapters 4 and 8.

17.2 Liver function tests

Section 17.1 has shown that the liver has many diverse functions. Consequently, it is most unlikely that one single blood test could mark all these different aspects, and this is why we have standard LFTs. Four are enzymes; one (bilirubin) is an excretory product (Table 17.1). However, some consider albumin to be an LFT, although this is not a widely held view.

Aminotransferases

These enzymes, previously designated as 'transaminases', are responsible for the transfer of an amino group from an α-amino acid to an α-oxo acid. They can also be described as part of the tools required to break up and rebuild other molecules, so can be seen as hammers, spanners and saws in a metal workshop. In the real world, these tools belong in the workshop, and likewise liver enzymes belong in hepatocytes. It follows that increased concentrations of these enzymes in the blood (where they seem unlikely to serve a physiological purpose) indicate damage to liver cells. The two aminotransferases are

- AST (EC 2.6.1.1), found in liver, heart, skeletal muscle, kidney, brain and red cells, and
- ALT (EC 2.6.1.2), which has a similar distribution, but its concentrations are lower in non-hepatic tissues.

Bilirubin

As discussed, this yellow–orange-coloured pigment is a breakdown product of haem, and is cleared from the blood by the liver. If the liver is unable to clear bilirubin, it remains in the blood and becomes deposited in the skin, causing jaundice. Differentiation of conjugated from unconjugated hyperbilirubinaemia is often a valuable tool in determining the cause of jaundice.

The daily destruction of red cells produces perhaps 250–350 mg of bilirubin, which in health is more than adequately disposed of, as clearance can reach perhaps 400 mg/day in adults (5 mg/kg per day) if necessary. The half-life of unconjugated bilirubin is less than 5 min, whereas that which is bound to albumin has a half-life of 17–20 days. The day-to-day variation is surprisingly high (15–30%), whilst those of African descent have levels 15–30% lower than in Caucasians.

Alkaline phosphatase

This enzyme (EC 3.1.3.1), a hydrolase, may be defined by the enzymatic activity of whole plasma in removing phosphate groups from many different molecules, including proteins and carbohydrates. Thus, it is defined functionally, and accordingly may include several distinct proteins with the same enzyme activity (i.e. isoenzymes), which may arise from different tissues such as the placenta and intestines. Four genes on different chromosomes encoding alternative ALPs have been identified and sequenced. In the circulation, the plasma ALP activity is mainly derived from liver and bone in approximately equal amounts.

As liver-derived ALP isoenzyme is located on the outside of the bile canalicular membrane, increased plasma concentrations are a sign of intra- or extrahepatic bile duct obstruction. However, the variant due to bone is dominant in the young, almost certainly because of the increased activity of this diffuse organ as it is constantly remodelling and growing. In contrast to bilirubin, ALP is 10–15% higher in African-Americans, and increased concentrations after a meal are presumed to arise from intestinal cells. ALP activity can rise markedly in the third trimester of pregnancy owing to the presence of the placental isoenzyme.

Gamma-glutamyl transpeptidase

This enzyme (EC 2.3.2.2) transfers glutamyl groups from gamma-glutamyl peptides to other peptides or amino acids. It is present throughout the liver and biliary tract, but is also present in smaller amounts in other organs, such as the heart, kidneys, lungs, pancreas and seminal vesicles, although the liver is responsible for 95% of plasma concentrations. Concentrations of the enzyme decrease shortly after a meal, then return to its starting value – so that measurements taken mid morning may be artefactually low because of breakfast.

Albumin

This 67 kDa molecule is a specific protein product of the liver and has many diverse nonspecific functions, several of which have pathophysiological consequences. It may be thought of as an LFT, as low plasma concentrations may be because of reduced synthesis, although there are several other reasons. We shall revisit this molecule in the coming sections.

17.3   Diseases of the liver

As we have seen, the liver is a complex organ, with numerous functions, many of which are linked to specific or nonspecific tests. The aetiology of many liver diseases can also be complicated, but for many the most obvious sign of liver disease is jaundice.

Jaundice

We begin exploring liver disease with this clear indicator, present when concentrations of bilirubin are so high in the blood, generally over 40 μmol/L, that they enter the skin and other tissues. This is most apparent when the sclera (white) of the eye takes a yellow coloration (Figure 17.5), and is also described as icterus, the patient being icteric. The yellow colour can develop on any external surface.

We recognize three ways in which jaundice can develop:

- haemolysis, following red cell destruction;
- damage to liver cells; and
- failure to excrete bile, most likely due to obstruction.

We will examine each of these in turn. With so much bilirubin in the blood, this eventually finds its way to the

Figure 17.5 Jaundice is the yellowing of the skin and mucous membranes due to the deposition of bilirubin. It is often most obvious in the eye, where the white sclera gives good contrast. (Photo credit: CDC/Dr. Thomas F. Sellers/ Emory University; http://commons.wikimedia.org/wiki/File: Jaundice_eye.jpg).

kidney, and passes through the glomerulus into the filtrate. As it is not reabsorbed, the urine becomes dark yellow and possibly orange/brown, an additional clinical sign. A further sign is a change in the colour of the faeces from brown to grey, a likely indication of obstruction of the bile duct so that the bile does not enter the intestinal tract. There are many varied reasons why jaundice is present, but this sign is by no means the only aspect of problems with this organ.

Diseases of the liver may be classified according to their aetiologies, but also by anatomy and by the speed of presentation. These have parallels with renal disease (Chapter 12), and we will begin with pre–, true–, and post-hepatic disease (Figure 17.6).

Pre-liver disease

This is characterized by problems outside the liver, 'before' the blood reaches this organ. Perhaps the only aspect of this is in excessive red cell destruction. As mentioned, when red blood cells come to the end of their lifespan they are removed from the circulation, often in the liver and spleen, and broken up for recycling. The unrecycled part, bilirubin, remains in the blood and is eventually cleared, mostly by the liver, and some by the kidney. In health, the liver is perfectly able to process this bilirubin.

However, if there is excessive haemolysis, then the liver is unable to keep up with purifying the blood of the high concentrations of bilirubin, so concentrations rise and can cause jaundice. Thus, there is nothing intrinsically wrong with the liver itself, it is simply unable to keep up with the high concentrations of bilirubin resulting from excessive red cell destruction. Jaundice is often seen in haematology disease, such as acute or chronic haemolytic disease (such as a serious infection with malaria, or in a haemoglobinopathy or membrane defect disease, respectively). These are fully explained in Chapter 4.

However, bilirubin (and bile) itself is not simply a marker of liver disease; it is toxic, causing a toxaemia. Consequently, high concentrations in jaundice may be responsible for damage to other cells and tissues, including endothelial cells that line the blood vessels, the brain and indeed the liver itself, in the form of hepatocyte necrosis and apoptosis. This can be so severe that dialysis of the blood is required. Consequently, jaundice of this nature may cause a vicious cycle of hepatocyte damage in 'true' liver disease, as the high concentrations destroy the ability of the hepatocyte to

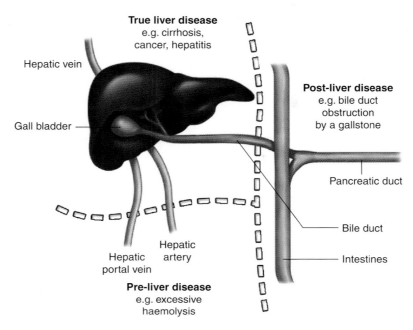

Figure 17.6 Anatomical classification of liver disease. We can consider liver disease in terms of pre–, true– and post-factors. The principal pre-liver cause of jaundice is excessive haemolysis, whilst most post-liver disease is caused by partial or complete obstruction of the bile duct, leading to cholestasis. True liver disease is characterized by damage to hepatocytes by a defined toxin or condition such as cancer or cirrhosis.

excrete this toxin. Furthermore, bile acids may also behave as cancer promoters.

True liver disease

In true liver disease there is actual damage to the cells of the liver itself (i.e. principally hepatocellular damage), so it cannot partially or fully perform its functions. It follows that true liver disease is accompanied by extreme hyperbilirubinaemia described in the previous paragraph. Damaging factors include chemical toxins (such as in cases of poisoning by inhaled solvents in paint spray) or viral hepatitis. In the case of industrial poisons, or self-inflicted or accidental overdose, recall that the liver is site of the removal of toxins, which it is fully able to do should levels of such toxins be modest (as in nature). However, in the modern age, the liver may be presented with supra-natural levels of toxins that it cannot cope with, leading to the destruction of the hepatocyte.

Damage to the hepatocyte is initially characterized biochemically in the short term by increases in the aminotransferases ALT and AST. Later, there is likely to be rising bilirubin as the excretory function of the liver is influenced. True liver disease can also be the result of a long-standing low level of damage to the liver, as is present in alcoholism and several other situations.

Post-liver disease

Vessels leaving the liver are the hepatic vein, the hepatic lymphatic and the bile duct. Thrombosis of the hepatic vein is rare, and if present blood cannot leave the liver, which then becomes congested and so can suffer. The hepatic vein may also be compressed from the outside by other disease, such as a tumour, in which case the same congestive disease is present. Disease of the hepatic lymphatic system itself is exceptionally rare, and if present will be secondary to other disease.

Cholestasis The major aspect of post-hepatic disease is obstruction or stenosis of the bile duct by factors such as gallstones, or a tumour of the head of the bile duct or pancreatic duct where they meet the duodenum. Thus, bile cannot be excreted, and so remains within the liver and biliary system, leading to obstructive jaundice.

When tissues are inflamed, they become engorged and swollen. Thus, the gross diameter of the inflamed tube will increase, but the lumen of that vessel will be reduced. Consequently, it cannot carry as much fluid. Inflammation of the bile duct leads to inflammatory cholestasis, and the reduced ability of the bile to pass to the intestines, so that the gall bladder becomes engorged, leading ultimately to jaundice. Furthermore, if the lumen is reduced, gallstones leaving the gall bladder are more likely to cause an obstruction.

The hepatocyte is of course unaware of this problem, so continues to generate bile, which fills the gall bladder, leading to congestion. Once the gall bladder is full, bile remains in the liver (causing hepatocellular damage) and then passes into the blood, thereby causing jaundice. Thus, again, as in post-renal disease, the root problem is not the tissue of the liver itself, but with the associated plumbing. However, as with pre- and post-renal disease, failure to correct pre- and post-liver disease will eventually lead to true liver disease.

Liver disease can also be characterized by how rapidly it develops, which may be rapidly, perhaps over days or weeks, or chronic, which may take years to become evident.

Acute liver disease   It is inevitable that acute liver disease, generally leading to liver failure, is caused by a precise object or situation. From the pre-liver perspective, this may follow from a sudden haemolytic event such as malaria. True liver disease may be caused by dozens of factors, such as accidental exposure to an industrial toxin or poison, or perhaps a shock to the body following a different condition, such as after major abdominal surgery or septicaemia. A good example of an acute post-liver disease is obstructive cholestasis caused by a gallstone.

Hepatitis is inflammation of the liver. Classically, this is precipitated by an infective agent, but there may also be autoimmune or other causes. Thus, acute inflammation of the liver, associated with hepatocyte damage, may be caused by viruses, bacteria or parasites. Whilst some viruses actually attack and destroy the hepatocyte, in many cases the cause of the liver disease is the normal immune response (often T lymphocyte mediated) to that virus (such as the hepatitis A, B and C viruses), which in these cases is clearly excessive.

A cautionary tale: Acute liver failure

Several weeks after marriage, a young woman developed an irreversibly rapid and fulminant jaundice that progressed to acute liver failure and her death. The post-mortem cause was hepatitis B virus. The husband subsequently remarried, only to find his second wife suffering almost exactly the same fate. The unfortunate widower identified a third possible spouse, but during their courtship she also became acutely jaundiced, but this time the hepatitis was not fatal.

Investigations into the man found him to be an asymptomatic hepatitis B virus carrier with all LFTs mildly abnormal. Although clearly (with hindsight) inoculating his three sexual partners with semen loaded with hepatitis B virus, the compassionate nature of the case was clear and the authorities decided against charging him with manslaughter.

Doubtless by now he will have died from one of the established consequences of chronic hepatitis B virus infection, such as hepatocellular carcinoma and terminal cirrhosis.

From a letter to a learned journal during the 1980s

The immune system is also primed to destroy and remove damaged cells, tissues and debris, and this may lead to inflammation and attack on innocent cells, in a form of autoimmune disease. This is why hepatitis may be the consequence of damage to the liver in the absence of microbial infection. If severe, acute hepatitis may deteriorate to frank acute liver failure. Should this damage be only partially healed, or the immediate destructive process persists, then chronic liver disease may develop.

Chronic liver disease

By definition this develops slowly, and a clear causative agent may never be found, as in inflammatory disease. However, the distinction between acute and chronic liver disease can be difficult, and so often depends on clinical signs (such as jaundice) and symptoms (such as haemorrhage). The autoimmune condition primary biliary cirrhosis, alcoholic liver disease and the iron storage disease haemochromatosis are examples of diseases that have no acute phase, so that diagnosis may be made years after the pathological start of the disease. Almost all chronic liver disease is 'true' (as opposed to pre– or post–) liver disease. However, an example of chronic post-liver disease is the progressive blocking of the bile duct by a tumour at the point where it joins the intestine.

Ascites Ascites is the excessive accumulation of plasma-like fluid in the peritoneal cavity. In up to 80% of patients it is a consequence of advanced cirrhosis (other causes include heart failure, cancer and inflammation), and conversely 50–60% of patients with cirrhosis develop ascites within 10 years of their diagnosis. The primary aetiology is portal hypertension, such as in Budd–Chiari syndrome (caused by occlusion of the portal vein by thrombosis, and therefore primarily a thrombotic disease), that leads to the export of extracellular fluid which accumulates in the abdomen (oedema). Further pathology involves renin and aldosterone, and thus the kidney, with the retention of sodium and water, so that little sodium is excreted. The laboratory may be called upon to assess the concentrations of sodium, albumin and markers of inflammation with the oedema.

Autoimmune hepatitis Acute infection with a hepatitis A or B virus may precipitate an autoimmune hepatitis, but most cases are the consequence of long-standing disease, and a causative agent may not be found. However, in contrast to acute autoimmune hepatitis, most cases (80%) of chronic autoimmune disease are characterized by autoantibodies such as anti-nuclear antibodies, anti-smooth muscle cell antibodies, and antibodies to the microsome and to mitochrondria. There may also be increased concentrations of IgG. A diagnostic problem is the nonspecific nature of these autoantibodies.

Cirrhosis Once damaged, such as after a brief viral infection or exposure to a poison, the liver has a remarkable capacity to regenerate, as is demonstrated by renewal after the need to remove entire lobes. However, if the cause of the damage is long-lived, such as in a chronic infection, the renewal and repair processes may not be able to cope, leading to the deposition of collagenous fibrous tissues in place of functioning hepatocytes. When appropriately stained and viewed by light microscopy, these fibrous tissues are said to resemble cirrus clouds; hence the term cirrhosis.

Thus, cirrhosis is also the end result of several lengthy pathological processes. These include autoimmune diseases such as primary biliary cirrhosis, chronic hepatitis B infections, long-standing biliary obstruction, primary sclerosing cholangitis (inflammation of the biliary system) and alcoholism. However, perhaps 30% of cases of cirrhosis have no clear causative agent. Eventually, with more and more cirrhosis and the loss of functioning tissues, the disease process causes liver failure.

Fatty liver (steatosis) Triacylglycerols are a very effective form of storing energy. When correctly packaged they have a higher calorific density than do carbohydrates or proteins (9 kcal/g versus 4.5 kcal/g versus 4 kcal/g respectively) and, being hydrophobic, have few osmotic or colloidal consequences. The fat is stored in adipocytes, with large deposits in various parts of the body, such as around the heart, and also in the liver. However, in excessive fat intake, in certain rare metabolic conditions and in alcoholism, fat is laid down within the liver. Steatosis is defined as greater than 55 mg of triacylglycerols per gram of liver, for which a tissue biopsy may be necessary. Initially, small droplets of lipids form within hepatocytes; but as these enlarge, the function of the liver cell suffers.

The aetiology of fatty liver disease can be classified as alcoholic or not, the latter being non-alcoholic steatohepatitis (NASH). The damage caused by lipid overloading attracts a low-grade inflammatory response and/or collagen deposition, leading to cirrhosis in 10–30% of NASH patients within 10 years, and over a quarter of NASH-cirrhotic patients subsequently develop hepatocellular carcinoma. NASH may be secondary to a number of conditions, such as diabetes mellitus, malnutrition, hypertension and obesity.

As with all chronic liver diseases, fatty liver is accompanied by a low and steady increase in the LFTs, a fall in albumin and loss of other hepatic functions, such as the storage of micronutrients.

Gilbert's syndrome This condition is the most common cause of an apparently unexplained modest increase in plasma bilirubin. Its frequency has been estimated at 3–10% of the population with a male/female ratio of 3:1. Bilirubin is relatively insoluble, which leads to problems in clearance by the liver. A solution to this is to link bilirubin to glucuronide, thus creating conjugated bilirubin, which enables its passage into the bile.

The enzyme responsible for this conjugation, UDP-glucuronosyl transferase, has a number of isoforms, several of which have low activity, leading to poor conjugation and failure to move bilirubin into the liver, so that concentrations remain high. The condition is generally relatively benign, but may be a problem if compounded with other disease such as haemoglobinopathy and its accompanying haemolytic anaemia.

Blood science angle: Gilbert's syndrome

The pathophysiology of Gilbert's syndrome, first described in 1901, is perhaps a classic example of science revealing the precise cause of the condition. Molecular genetics has found a stable mutation in the TATA promotor region of the gene (on chromosome 2) coding the enzyme that is responsible for most cases of the syndrome. Accordingly, this mutation can be used to help confirm or refute the diagnosis of Gilbert's syndrome and persistent hyperbilirubinaemia.

Primary liver cancer The primary malignancy of the tissues of the liver itself is hepatocellular carcinoma (HCC). It is the sixth most common malignancy in the world and the third greatest cause of liver-related death. In the UK it makes up 1 in 100 cancers, and well over 3000 new cases arise each year. An alternative descriptor is hepatoma, but technically the latter is simply a mass within the liver that may not necessarily be a malignancy.

The vast majority of cancers of the liver are of the hepatocytes; less common is cancer of the bile ducts, called cholangiocarcinoma. The remaining cells of the liver – fibroblasts, stellate cells, endothelial cells and Kupffer cells – are very rarely malignant. The major causes of liver cancer are known. These are chronic viral infection (hepatitis B and C virus), alcoholism and cirrhosis.

The risk and incidence of HCC varies considerably around the world. In China and South Asia, hepatitis B virus is endemic and is the leading cause, whereas in the developed world, where vaccination against hepatitis B virus is common, alcoholism and hepatitis C virus are the most common causes, although cirrhosis will be an intermediate stage. The liver can also be the site of secondary cancer (i.e. metastases) that originate in other tissues, such as cancer of the intestines, breast, ovary, lung, kidneys and prostate.

The key laboratory test is alpha-fetoprotein (AFP). This molecule is a normal component of foetal physiology, being coded by a gene on chromosome 4, where it acts as a variant of albumin, binding metals and bilirubin. In the non-pregnant adult, high concentrations initially imply HCC, although alternative diagnoses suggested by a raised AFP include ataxia telangiectasia, where LFTs are expected to be normal.

Secondary liver cancer The liver is a common site for secondary deposits of metastatic tissues from other cancers, such as those of the colon, breast, lung, kidney and prostate. These may form 90% of cases of liver malignancy, with 'true' liver disease (i.e. of the hepatocytes and biliary tree) comprising only 10%.

Blood science angle: The genetics of liver disease

We have already looked at the role of genetics in Gilbert's syndrome, but there are many other such examples. Wilson's disease is caused by a mutation in *ATP7B* that normally transports copper into bile and mediates its incorporation into caeruloplasmin. The closely related Menkes disease is linked to a mutation in *ATP7A* that also leads to a copper pathology. The most common glycogen storage disease is caused by a mutation in the gene for acid alpha-glucosidase found on chromosome 17. There is evidence that mutations in the apolipoprotein C gene may be important in NASH; another is a missense mutation in *PNPLA3*, which codes for a phospholipase. The genetics of cholesterol are discussed in Chapter 14 and of haemochromatosis in Chapter 4.

Other liver disease The liver is the subject of infection by several protozoan and metazoan parasites, the liver fluke (*Fasicola* species) being perhaps one of the best known. There are also numerous other diseases and conditions involving this organ. We have already mentioned the iron storage disease haemochromatosis; other storage diseases include that of glycogen, in which there are abnormal deposits of this carbohydrate in hepatocyte, and in muscle and nerve cells; another is Gaucher's disease. Wilson's disease is caused by the build-up of copper in the liver and brain. Serum copper is low, but is increased in the urine. Some factors leading to liver disease are summarized in Table 17.3.

Table 17.3 Some selected factors leading to liver disease.

Viral	Autoimmune
Hepatitis viruses A to C	Autoimmune hepatitis
Epstein–Barr virus	Primary biliary cirrhosis
Cytomegalovirus	Primary sclerosing cholangitis
Arboviruses	*Bacterial and protozoal*
Parasites	Tuberculosis
Liver flukes	Malaria
Schistosomes	Amoebiasis
Toxocara	Leishmaniasis
Tapeworms	*Anatomical and vascular*
Metabolic	Gall stones
Haemochromatosis	Biliary atresia
Wilson's disease	Venous thrombosis
Gilbert's disease	Budd–Chiari syndrome
Fatty liver	

Disease of the biliary tree

These have been mentioned in passing in the preceding section, but are worthy of bringing together. The biliary system is a network of vessels that collect bile from liver lobules and deliver it to the intestines. There are three major disease of this system, and all are found with abnormal LFTs.

Primary sclerosing cholangitis As is evident from the suffix –itis, this is inflammation. 'Ang' refers to a vessel, and 'chol' to the liver; hence inflammation of vessels of the liver. The inflammation of the cells that comprise these vessels inevitably leads first to their thickening (sclerosis) and ultimately their destruction and so inability to drain the bile that the hepatocytes synthesize. The disease can strike at the micro-vessels within the liver, or larger vessels that merge to form the bile duct, but both lead to cirrhosis and liver failure. Notably, a major risk factor for this disease is ulcerative colitis, and many patients have increased levels of autoantibodies such as p-anti-neutrophil cytoplasmic antibodies and anti-smooth muscle cell antibodies. These can be detected by immunopathological methods (see Chapter 9 on immunopathology), whilst human leukocyte antigens A1, B8 and DR3 are linked to this disease (see Chapter 10).

Primary biliary sclerosis This, too, is characterized by a slow and progressive destruction of the biliary tree with cirrhosis, fibrosis and obstruction. Commonly found autoantibodies include anti-nuclear, anti-centromere and anti-mitochondrial antibodies, and although the former two are relatively nonspecific, the latter is much more specific, being present in 90% of patients. Unfortunately, definitive diagnosis is by biopsy. Nevertheless, other antibodies, such as to glycoprotein 210 show promise not only in assisting diagnosis but also in prognosis.

Cholangiocarcinoma This is a relatively rare cancer of the epithelial cells of the biliary system. It may arise from other liver diseases, such as those listed above, but also after exposure to carcinogens and liver parasites (including viruses). From a liver function perspective, the anatomical site of the tumour defines severity: if it develops in the wall of a large vessel it will occlude a larger number of biliary vessels (and so influence a larger mass of the liver) than if the tumour arises in a small vessel. Blood tests are unhelpful, although concentrations of carcinoembryonic antigen (CEA) may be increased.

Other disease Xanthogranulomatous cholecystitis is a rare form of inflammation of the gall bladder, which may resemble cancer of the gall bladder, although the latter is extremely rare. Gallstones may lead to cholecystitis, which is severe and may require surgery.

Liver function tests in liver disease

The interpretation of LFTs is complicated; there are few examples of clinical situations where a single blood test has sufficient sensitivity and specificity to reliably define a precise disease. Accordingly, practitioners will need to gather as much information as possible before making a clinical judgement.

Aminotransferases An increase in concentration of AST and ALT, alongside a relatively smaller increase in other tests, indicates damage to the hepatocyte, of which there are many potential causes, including infective agents, autoimmune disorders and toxins. Infections with microbial pathogens (Table 17.3) all induce increased AST and/or ALT, although the latter is better than AST for monitoring viral activity in chronic hepatitis B and C as it is released from damaged hepatocytes more easily than is AST. Measurement of ALT is preferred as it is much less sensitive to interference than AST is, and some laboratories no longer measure AST.

Unfortunately, AST and ALT are found in many cells other than hepatocytes, including skeletal and cardiac muscle. However, damage to the latter is likely to precipitate increase in creatine kinase, an enzyme absent from the liver. Thus, raised AST and ALT after a myocardial infarction does not necessarily imply liver disease. Similarly, raised AST is commonly found in haemolytic anemia as this enzyme is present in red blood cells. Another blood test, lactate dehydrogenase (LDH), can be helpful in differential diagnosis, but once more the competent blood scientist will recognize that forms of LDH can arise from non-red cell sources such as cardiomyocytes damaged after a myocardial infarction and liver disease itself. The role of blood tests in cardiology is discussed in Chapter 14.

Many drugs and toxins can induce raised AST and/or ALT. These include alcohol, industrial solvents, toxins from certain mushrooms, and the cholesterol-lowering drugs of the statin class. It is debatable whether or not the anticoagulant warfarin is a toxin, as it is widely used as a rat poison. In our own species, it is given in a controlled manner, acting on the liver to interfere with the synthesis of several proteins involved in the coagulation pathway and so reduces the risk of thrombosis in conditions such as atrial fibrillation and after a pulmonary embolism (Chapter 8).

Paracetamol deserves special mention, not merely because it is one of the major self-inflicted poisons, being the most common cause of acute liver failure in the UK. A metabolite of paracetamol is N-acetyl-parabenzoqui-none imine (NAPQI), which in recommended doses the normal metabolic pathways of the hepatocyte can safely address. However, ingestion of over 10 g of paracetamol (200 mg/kg body weight) leads to high concentrations of NAQPI that overwhelm the protective metabolism of the cell (such as glutathione) and cause symptoms within 48–72 h. Concentrations of glutathione can be replenished by N-acetylcysteine infusion to restore the sulphation and glucuronidation pathways and so minimize levels of NAPQI.

Bilirubin This is regarded as the only true LFT, as the clearance of this potential toxin is a major function of this organ. As we have discussed, the primary non-hepatic cause of raised bilirubin is excessive haemolysis. However, true liver disease, such as Wilson's disease, may also cause red blood cell destruction and an anaemia.

Bilirubin in the neonate deserves special mention. The neonatal liver is relatively underdeveloped and shortly after birth may not be mature enough to process its own bilirubin once independent of the mother, and so may lead to hyperbilirubinaemia and jaundice. This should normalize within 2–3 days after birth as the liver matures. However, there are a large number of pathological conditions causing jaundice that may become evident in the neonatal period, such as haemolytic disorders, hypoxia, galactosaemia and fructosaemia.

Alkaline phosphatase The most common cause of raised ALP is obstruction of the bile duct or its tributaries (i.e. cholestasis), as may be caused by gallstones. Increased ALP is also found in primary biliary cirrhosis or in biliary or pancreatic malignancy. Increased blood pressure in the portal vein or reduced blood flow due to heart failure also result in raised plasma ALP.

The major non-hepatic source of plasma ALP is bone. One particular isoform of ALP is produced by osteoblasts, so that raised ALP may be present in metabolic bone disease such as Paget's disease. Increased ALP may also be present in other bone disease, such as a primary or secondary bone cancer (typically metastases from prostate cancer and breast cancer).

Gamma-glutamyl transpeptidase Once more, there are numerous hepatic and non-hepatic causes of raised GGT. Increased levels are often present in any form of cirrhosis or hepatitis, haemochromatosis, HCC and secondary cancer, tuberculosis, sarcoid and Wilson's disease.

One of the problems with GGT is that plasma concentrations are influenced by several commonly used drugs, such as phenytoin, barbiturates, carbamazepine and alcohol. Concentrations are also associated with the body mass index, especially when this exceeds 30 kg/m^2. This may be due to fat deposits in the liver (steatosis) which may well be present (possibly in a sub-clinical manner) in diabetes mellitus, non-alcoholic steatohepatitis and non-alcoholic fatty liver disease.

Comparing the liver function tests In paracetamol poisoning, AST needs to be considered alongside bilirubin in assessing prognosis using serial results, perhaps over a period of a week. This is because if the AST and bilirubin fall in parallel this suggests hepatic recovery and good prognosis. Conversely, a falling AST with a rising bilirubin indicates critical loss of hepatocytes and a poor prognosis.

The ratio of the aminotransferases can be helpful. An AST/ALT ratio greater than 2 implies alcohol misuse because of the release of mitochondrial AST due to damage from alcohol metabolites. A rise in AST may also be compared with the rise in ALP, where a ratio in favour of AST is suggestive of hepatitis.

ALP and GGT often rise and fall together in many conditions. However, the former is a better marker of biliary and bone disease, and the latter of alcoholic disease.

The complexity of liver function tests It is clear from this section that virtually any (if not all) liver disease(s) may be accompanied by any combination of abnormalities in the standard LFTs. Furthermore, levels come and go at different rates under different conditions, such as in an acute inflammation versus a chronic damaging attack. Accordingly, the practitioner needs to have a full awareness of the significance of abnormalities in the LFTs. Although incomplete, a simple synopsis of the LFTs is presented in Table 17.4.

17.4 Plasma proteins

The laboratory offers both a global estimate of the proteins in the blood (total proteins) and also individual proteins, of which the most notable is albumin (with a concentration of perhaps 40 g/L). Other proteins,

Table 17.4 A crude synopsis of LFTs.

LFT	Pathology	Caveat
AST, ALT	Hepatocyte damage	Nonspecific
Bilirubin	Hepatocyte damage, cholestasis	Haemolysis
ALP	Biliary damage, cholestasis	Bone
GGT	Biliary damage, cholestasis	Alcohol, drugs
Albumin	Hepatocyte damage	Acute-phase reactant, renal damage

N.B. Pathology leads to increased concentrations of the LFTs, but to low concentrations of albumin.

collectively called globulins, and together with a concentration of approximately 30 g/L, include cancer markers, enzymes, transport proteins, hormones, cytokines, coagulation factors and complement proteins. This section will examine only a selection of the individual proteins in which the measurement of their concentration has proved useful in the diagnosis and management of diseases. Others will be discussed in other chapters.

Protein may also be measured in the urine and cerebrospinal fluid, and this gives us some important information, perhaps regarding renal function, a malignancy, or a disorder of the central nervous system.

Total proteins

As is clear, this is the sum of hundreds of different proteins. Nowadays, total protein measurements are not considered very useful as specific protein assays are available. However, the major interest in this index is proteins as a global score, and so we focus (with respect to the reference range) on high and low concentrations.

Hyperproteinaemia In theory, an increase in any one of the hundreds of individual proteins could cause an isolated hyperproteinaemia, but in practice there are only a few major causes. One is a general increase in many plasma proteins as part of the acute-phase response, although concentrations of some actually fall (Table 17.2). In an established infection, there may be increased Igs (principally IgG and IgM) in addition to those of the acute phase. The third is an increase in one particular protein (a paraprotein), as may be present in myeloma or a related condition. Hyperproteinaemia may also be due to dehydration.

Hypoproteinaemia In contrast, there are several possible reasons for low concentrations of total proteins. One of these, malnutrition, is rare in the developed world, but the principal alternative reason for hypoproteinaemia is low concentrations of albumin. We will return to this important molecule in a subsection that follows.

Measurement of proteins One of the most common techniques for the measurement of total proteins is the Biuret method. This colorimetric technique relies on the formation of a purple colour when copper ions (provided by, for example, copper sulphate) react with the peptide bonds that link individual amino acids in the proteins at an alkaline pH (possibly due to sodium or potassium hydroxide). This can be detected in a spectrophotometer at a wavelength of 540 nm and the result from the patient's sample compared with a standard of known protein concentration.

However, this relatively crude technique cannot be used for the individual plasma proteins. The concentrations of specific plasma proteins can be measured using a variety of different methods:

- Immunological methods rely on the ability of a specific antibody (often a monoclonal antibody) to recognize the particular protein. Variants of this method include enzyme-linked immunosorbent assay, chemiluminescent and fluorescent immunoassay, immunoturbidimetry (where light is absorbed) and immunonephelometry (where light is scattered). In these latter two methods, a buffer containing polyethylene glycol is used to enhance the reaction. Another method, radial immunodiffusion, is rarely used.
- Albumin can be measured very simply by its ability to bind with certain dyes, principally bromocresol purple and bromocresol green. However, the latter also reacts with non-albumin proteins.
- As discussed above, there are many enzymes in plasma (ALP, GGT, ALT, AST, amylase, lipase), and these may be assessed by their catalytic activity, given an appropriate substrate.
- Different proteins can be separated by electrophoresis, of which there are two major variants: gel electrophoresis and capillary electrophoresis.
- There can also be a combination of these different methods, such as immunoelectrophoresis (a sub-technique of gel electrophoresis), although some have fallen out of popularity. For example, rocket immunoelectrophoresis is rarely used nowadays.

Protein electrophoresis Although plasma proteins are a very heterologous group of molecules, they can be subdivided into groups depending on a key aspect of their physicochemical make-up. Each protein is the sum of its individual amino acids, and as all amino acids have a particular acid/alkaline nature, then so too does a particular protein. This is exploited during electrophoresis, of which the most common types are gel and capillary. These powerful techniques allow the separation of a large number of molecules according to their overall electrical charge and molecular weight, and as such have applications in several aspects of blood science (such as the characterization of different haemoglobin molecules in the investigation of sickle cell disease in Chapter 4).

In gel electrophoresis, a small volume of plasma is absorbed onto a solid phase, such as a sheet of cellulose acetate or agarose. In the case of the former, when dry, the solid phase is moistened with a buffer and an electrical charge applied to the two ends. The component proteins of the sample are driven across the solid phase by the electric change and come to rest at a point dependent on their own electrical nature. The electric charge is then stopped and the solid phase is processed (often with a stain such as Coomassie blue) to allow the identification of the different species of proteins in the sample.

The strength of this technique is that it breaks down the total plasma proteins into two groups: albumin and the globulins. Furthermore, the latter can be subdivided into a small number of bands: alpha, beta and gamma. Typical results are shown in Figure 17.7.

Each band contains a number of individual proteins, often referred to as globulins, so called because of their globular (spherical) three-dimensional shape. Members of this family include the globulins of haemoglobin and myoglobin, although these should be found only in red blood cells and muscle cells respectively. Globulins present in the plasma are:

- Alpha globulins, which include alpha-1-lipoprotein, alpha-1-antitrypsin (AAT), orosomucoid, and serum amyloid A. The dye bromocresol green reacts with alpha-globulins such as AAT and alpha-2-macroglobulin as well as with albumin.
- Beta globulins, which include alpha-2-antiplasmin, alpha-2-lipoprotein, alpha-2-macroglobulin, angiotensinogen, caeruloplasmin, haptoglobin, protein C and thyroxine-binding globulin.

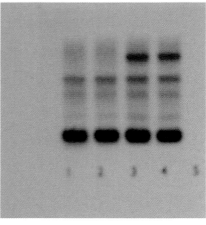

Gamma globulins
Beta globulins
Alpha-2-globulins
Alpha-1-globulins
Albumin

Figure 17.7 Protein electrophoresis. The separation of groups of proteins according to their overall electrical change. Dependent on the pH of the buffers, and their starting point, different molecules migrate towards the anode or cathode. The figure shows four samples of two patients in duplicate. The heavy blue band at the bottom is albumin, the different globulin bands being easily characterized. Note that samples 3 and 4 have a much heavier staining of the gamma-globulin region. This is because of an abnormality, and is called a paraprotein, the origin of which may be malignant. (Image courtesy of H. Bibawi, NHS Tayside).

- Gamma globulins, which are antibodies, also called Igs. This group of molecules can be further classified according to their structure, hence (in order of concentration), IgG, IgM, IgA, IgE and IgD. Further details of the gamma globulins are present in Chapters 5 and 9, which deal with white blood cells and immunology.

The density of staining of the proteins in Figure 17.7 gives us an idea of their concentrations. By converting the density into a number we can quantify the different alpha, beta and gamma globulin fractions. This then gives us the ability to compare concentrations of the different proteins in health and disease, and with knowledge of the total protein result allows a quantification of the amount of protein in each band (Figure 17.8).

Gamma globulins are synthesized in B lymphocytes, some which may be in the liver, although the vast majority are in lymph nodes, the spleen and the bone marrow. The liver is the site of albumin production, and of the many alpha and beta globulins. Increased concentrations of these molecules may well be present in liver

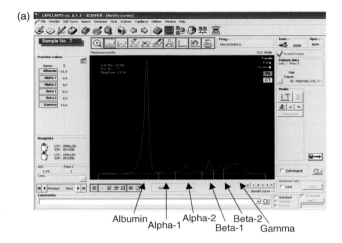

Albumin
Alpha-1
Alpha-2
Beta-1
Beta-2
Gamma

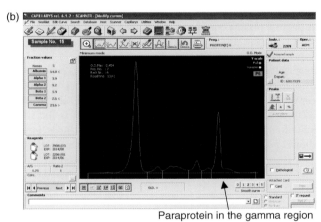

Paraprotein in the gamma region

Figure 17.8 Quantification of proteins by electrophoresis. These figures are screenshots of densitometer plots. A beam of light passes over the electrophoresis gel (Figure 17.7) and plots the density of the stain, which is proportional to the concentration of protein present in the gel.

(a) A normal plot. The large peak on the left is albumin, others (alpha, beta and gamma) are marked. The software draws an orange line between each peak and then calculates the proportions of each peak as a percentage of the whole. This is present on the left-hand side of the plot. So with the total protein result (such as 70 g/L), then the albumin result is 61.4% of this; that is, 43 g/L. Similarly, the gamma-globulin result of 14.4% translates to 10 g/L.

(b) Abnormal plot. The dominant peak is still the albumin, at 54.8%. But the total protein result is higher, perhaps 78 g/L, which gives an albumin result of 42.7 g/L, very close to that of the normal plot. However, the major difference between the plots is the large gamma-globulin band, which is 23.6% of the total protein result, translating to 18.4 g/L. Thus, this second plot has a far larger gamma-globulin component (in fact, 8.4 g = 84% more), which we call a paraprotein. This result would trigger further investigations to determine the exact nature of the paraprotein, which is dealt with in Chapter 9.

disease such as cirrhosis and inflammatory disease. However, as the disease becomes increasingly chronic, then concentrations of all proteins will fall as the organ also fails.

As the liver is sensitive to the acute-phase response, then so too are the proteins synthesized by this organ. Most are increased in the acute-phase response, but concentrations of some fall, such as albumin. The greatest relative increase is in concentrations of c-reactive protein (CRP), but this is most unlikely to show up on an electrophoresis plot as the plasma concentration is so much lower than other specified proteins.

The scope of plasma proteins   Table 17.5 summarizes those plasma proteins with the highest concentration. We will now look at several of these plasma proteins in depth, although the Igs (IgG, IgA and IgM) are discussed in depth in Chapters 5 and 9. Chapter 9 also discusses two other important plasma proteins missing from Table 17.5: complement components C3 and C4. However, we will also address two important molecules present at low concentrations: CRP, a marker of acute and chronic inflammation, and ferritin, important in iron metabolism.

Missing from Table 17.5 are enzymes, such as amylase and those of the LFTs. This is because plasma concentrations are reported as units per litre, not as milligrams or micromoles, as are many other proteins. One of the reasons for this is that, taking ALP as an example, we measure the enzymatic activity of the plasma, not concentrations of a particular protein that just happens to have enzyme activity. Indeed, the total ALP activity of the plasma may be the combination of different enzyme molecules from the liver, the bone and possibly elsewhere.

Albumin

This molecule begins our exploration of specific plasma proteins not merely because of the alphabet. A simple polypeptide of 584 amino acids giving a molecular weight of 66–69 kDa, it is the most abundant single protein in blood plasma and has a plasma half-life of perhaps 21 days. A specific product of the liver, enough is produced daily to replace that consumed in metabolism, degraded by plasma enzymes, and that which is lost in urine. It is important for at least three reasons: transport, osmosis and as a disease marker.

Transport   Several important constituents of the plasma are largely hydrophobic (such as bilirubin and certain lipids) and so are relatively insoluble. Binding to albumin helps their transport and distribution. This binding also helps take toxins (once more, such as bilirubin) out of the circulation. Several ions and drugs are also carried by albumin: these include calcium (which has further biochemical implications – Chapter 15), iron (in addition to that carried by transferrin and ferritin – Chapter 3), iodine and phenytoin. This binding is influenced by the pH of the blood (as noted in Chapter 13).

Osmosis   This process is one of the fundamental forces in biology, which effectively seeks to maintain a level of 'thickness' and 'thinness' of a fluid. As far as blood is concerned, this is the sum of all the substances in the blood, and includes ions and much larger molecules. Indeed, in Chapter 12 we read about the concept of osmolality. Albumin is a major contributor to the osmotic potential of the blood, and this oncotic pressure prevents fluid leaking out of capillaries and into the tissue fluid.

Albumin as a disease marker   The liver very rarely makes more albumin than it needs to, so hyperalbuminaemia is most unusual and often reflects dehydration.

Table 17.5 Major plasma proteins.

Plasma protein	Typical plasma concentration (g/L)	Function	Acute-phase reaction
Albumin	40	Many	Decrease
IgG	12	Immunity	Increase
Fibrinogen	3.0	Coagulation	Increase
Transferrin	2.5	Iron transport	Decrease
IgA	2.0	Immunity	Increase
AAT	1.25	Enzyme regulation	Increase
IgM	1.1	Immunity	Increase
Haptoglobin	1.0	Binds haemoglobin	Increase
Caeruloplasmin	0.5	Copper metabolism	Increase
Thyroxine-binding globulin	0.05	Binds thyroxine	Increase
CRP	0.003	Binds bacteria	Increase
Ferritin	0.0002	Stores iron	Increase

However, there are often low concentrations in the blood – hypoalbuminaemia. A potentially large number of pathological conditions give rise to this, and these include the following:

- Malnutrition. The liver is a factory, and as such needs raw material to produce its products. Thus, in times of malnutrition, the liver is unable to synthesize albumin. However, if present, this will also cause the reduced synthesis of other products, and so a generalized hypoproteinaemia. Severe protein deficiency leads to marasmus, a syndrome likened to starvation. The closely linked kwashiorkor relates to protein wasting.
- The acute-phase response. In health, this process follows an acute inflammatory or other physiological shake-up, but it can also be the consequence of chronic inflammation, as in autoimmune disease such as rheumatoid arthritis. The liver responds by the increased synthesis of many proteins, but also by reducing the concentrations of albumin (and transferrin) it generates (Table 17.2). This may be because the liver diverts protein synthesis to acute-phase reactants such as CRP and ferritin.
- Malabsorption and protein-losing enteropathy. This can occur despite a good diet, and is most likely to be due to inflammatory bowel disease such as coeliac disease and Crohn's disease. Other causes include drug effects, such as nonsteroidal anti-inflammatory drugs.
- Liver disease. We have already seen how conditions such as cirrhosis lead to hypoalbuminaemia, and is a consequence of the disease process.
- Renal disease. Recall from Chapter 12 that the top of the nephron (the glomerulus) is a molecular sieve, and that the threshold of the sieve is the size of albumin. Thus, in health, a small amount of this protein may pass into the glomerular filtrate, and subsequent parts of the nephron can reabsorb some of this protein. However, where there is damage to the glomerulus, as in glomerulonephritis, the loss of albumin is likely to be excessive, and greater than the capacity of the nephron to reabsorb this protein, and then plasma concentrations fall. High concentrations of protein in the urine (proteinuria) define the nephrotic syndrome (to be detailed in a section that follows).
- Low albumin is also a consequence of burns to the skin, possibly because of the need to regenerate the damaged tissues and loss of albumin in the exudate.

The consequences of hypoalbuminaemia Plasma albumin can fall to as little as 15 g/L before there are major pathological effects. However, low plasma albumin has implications for osmosis. The laws of osmosis demand a certain osmotic pressure, so that if the plasma is too 'thin', then if the protein concentration cannot be increased, then water must be exported to the interstitial fluid in the tissues, leading to oedema. This fluid may lead to swollen ankles, fluid in the lungs (pulmonary oedema) and around the heart, and, in severe cases, to fluid in the peritoneal cavity – ascites.

Alpha-1-antitrypsin

This molecule, of relative molecular mass 52 kDa and constructed from 394 amino acids, was originally characterized by its ability to inhibit the protease enzyme trypsin, but it is now known to inhibit other enzymes such as elastase. For this reason, it may also be described an alpha-1-proteinase inhibitor. As release of AAT occurs when neutrophils are activated at the location of an infection, it therefore serves to restrict the activity of elastase to those sites where this aspect of an inflammatory response requires control. Indeed, this is probably why it is an acute-phase reactant. The referral to 'alpha' indicates its electrophoretic status.

AAT is in fact one of several protease inhibitors; others include antithrombin and C1 esterase inhibitor, and may arise from the liver and other tissues. Collectively these are known as serpins, the 'ser' because they inhibit enzymes that have a serine amino acid residue near the binding site (hence **ser**ine **p**rotease **in**hibitors), although not all members of the serpin superfamily are inhibitors.

AAT is important principally because of the lung disease emphysema. It may also be required to help differential diagnosis of neonatal jaundice. In the absence of AAT, neutrophil elastase is unrestricted, and so continues to degrade the structural protein elastin. This has several consequences, such as pulmonary disease, and also cirrhosis. Accordingly, one of the most important laboratory uses of AAT is to define the deficiency of the enzyme, as an aid to diagnosis. However, there are several genetic variants of AAT: the M, S and Z types which are inherited in a Mendelian manner. The homozygous form of the latter species (i.e. ZZ) is associated with very low plasma AAT (10–15% of normal) and so a risk of liver disease (possibly jaundice) and/or lung disease (where excessive elastase activity effectively destroys the alveoli).

The exact M–S–Z phenotype of AAT cannot be determined by measuring its concentration in blood. Accordingly, isoelectric focusing is often required. Furthermore, attention must be paid to the acute-phase response, which may increase concentrations of this molecule.

Caeruloplasmin

This large (relative molecular mass 132 kDa) copper-containing protein enables the uptake of iron. This metal is transported through the intestinal cell wall in the ferrous form (Fe^{2+}), but can only be picked up and carried to the bone marrow by transferrin in the ferric form (Fe^{3+}). Caeruloplasmin catalyses this transformation, and so is described as a ferroxidase. A second function of caeruloplasmin is as a carrier and possibly a storage molecule for copper, as this molecule contains 90% of the metal. The clinical requirement for the measurement of caeruloplasmin centres on two diseases: Wilson's disease and Menkes syndrome.

Wilson's disease   As already indicated, this condition results from a mutation in *ATP7B* on chromosome 13 that normally codes for a molecule that transports copper into bile and mediates its incorporation into caeruloplasmin. In the presence of the mutation, which has a heterozygous carrier frequency of 1/100, copper builds up in the liver, leading to cirrhosis and cancer. The caeruloplasmin that is secreted lacks copper and has an abnormally short half-life, leading to low plasma concentrations (often <0.2 g/L). However, concentrations may be higher in inflammatory disease, as caeruloplasmin is an acute-phase reactant. Free copper is still exported by the liver, but accumulates in the kidneys, eyes and brain, causing organ damage. A haemolytic anaemia is also often present. The disease is treated by the drug penicillamine, which helps to remove copper.

Menkes syndrome   This disease is caused by a mutation in *ATP7A*, found on the X chromosome, which results in impaired uptake of copper by the intestines, with a poor and uneven distribution throughout the body. Like Wilson's disease, caeruloplasmin concentrations are low and what is present is low in copper. Differential diagnosis therefore requires a tissue biopsy (ideally the liver), reporting low copper in Menkes syndrome but high copper in Wilson's disease.

C-reactive protein

This molecule, with a relative molecular mass of 25 kDa, and formed from 224 amino acids, is the gold standard marker of inflammation, and is therefore also of considerable interest to immunologists. Coded for by a gene on chromosome 1, it is so-named because it binds to the capsular (c) polysaccharide (now known to be phosphocholine) of pneumococcus bacteria, and is released by the liver in response to inflammatory cytokines. The magnitude of this response can be as much as 1000-fold within 24 h, but with a short half-life of 18 h it falls back to normal just as quickly.

CRP is raised in acute and chronic inflammatory disease, whether or not related to an infection, and is therefore a useful aid not only in diagnosis but also in assessing the severity of the disease. However, there are a (small) number of caveats regarding this molecule:

- CRP is often raised after an acute severe shock to the body that is unrelated to inflammation or infection. Examples of this include after a myocardial infarction, a stroke or after trauma, such as major surgery. The likely reason for this is that the body predicts the likelihood of an infection, so upregulates CRP as a precaution.
- Its nonspecific nature does not provide a great deal of help in confirming an exact inflammatory diagnosis or even an infection. This can be a severe limitation. Indeed, raised CRP is often noted in cancer, possibly related to the generalized changes to physiology in a malignancy.
- Unlike other markers of disease (such as cholesterol), there are no specific medications that act directly on CRP – concentrations fall after successful immunosuppression (such as by steroids), although statins may reduce concentrations independently of cholesterol. Fortunately, raised CRP by itself does not have major implications for cell pathobiology (unlike, for example, bilirubin or urea, which are toxic).

Nonetheless, there is increasing evidence that CRP may be useful in stratifying the risk of cardiovascular disease, but this has yet to have a great effect on management where the major risk factors for this disease remain dominant (Chapter 6). As an inflammatory marker, it is preferred to erythrocyte sedimentation rate (ESR) as the latter is also abnormal in anaemia, which should not influence the CRP.

The very low plasma concentrations in health (perhaps 3 mg/L, at least 1000 times less than albumin – Table 17.5) demand a high-performance assay; accordingly, high-sensitivity CRP (hsCRP) immunoassay is the method of choice.

Ferritin

The importance of iron to our ancestors is demonstrated by the complex and expensive methods the body has for absorbing and retaining this metal. Indeed, we have no specific method for eliminating iron. Once absorbed, iron is transported to storage sites in various organs (but mostly the liver) by transferrin, where it is incorporated first into ferritin and then into haemosiderin.

Genes on chromosome 19 code for monomer units that combine to form a 450 kDa molecule, each being able to carry thousands of atoms of iron. Thus, ferritin in the plasma can transport and deliver a large amount of iron, often to the bone marrow. Measurement of ferritin is often used to give an approximation of whole body iron stores. Additional details are present in Chapters 3 and 4.

Fibrinogen

This liver-derived molecule (coded for by a gene on chromosome 4) is cleaved by the enzyme thrombin (itself derived from prothrombin) to give fibrin, an essential component of clot formation (haemostasis). With a relative molecular mass of some 340 kDa, low concentrations (as may be found in the life-threatening coagulopathy disseminated intravascular coagulation) impair the ability of the blood to clot, whilst high concentrations promote thrombosis.

An important medication for the treatment or prophylaxis of venous thrombosis is the drug warfarin, which inhibits the ability of the liver to synthesize this protein. Warfarin is most often used to treat patients at high risk of thrombosis (such as after orthopaedic surgery) and those who already have a clot in their veins (such as a deep vein thrombosis or a pulmonary embolism). An acute-phase reactant, additional details are present in Chapters 7 and 8 on blood coagulation.

Haptoglobin

When red blood cells are destroyed whilst in the blood (intravascular haemolysis, Chapter 4), the haemoglobin spills out in the plasma, where it binds to haptoglobin. This molecule is a tetramer of two α and two β polypeptides (coded for by genes on chromosome 16) that can come together in different combinations giving variants of molecular weights from 86 to up to 900 kDa. Different variants of haptoglobin offer protection from

diabetic renal disease, whilst others are linked to an increased incidence.

The precise purpose of haptoglobin is unclear. Possible functions include easing the recycling of the haemoglobin via the macrophage scavenging receptor (CD163), and minimizing potentially damaging effects on the kidney. As the haptoglobin–haemoglobin complex is cleared from the blood, concentrations of haptoglobin fall, so that low levels imply intravascular haemolysis. If there is red cell destruction within cells (liver, spleen; i.e. extravascular haemolysis) then the haemoglobin does not enter the blood and so haptoglobin concentrations remain normal.

As with so many plasma proteins, interpretation of the plasma concentrations depends on the acute-phase response, which in the case of haptoglobin may increase by some three- to eight-fold. Of course, many of the conditions leading to intravascular haemolysis are likely to have an inflammatory component.

Thyroxine-binding globulin

This is one of the molecules that transport thyroid hormones tri-iodothyronine and thyroxine from the thyroid to the tissues. The other carriers are albumin and transthyretin. A 65 kDa product of the liver, coded from the X chromosome, thyroxine-binding globulin is not to be confused with thyroglobulin. Rarely, a request for measurement of thyroxine-binding globulin is part of an investigation of hypothyroidism. Additional details are present in Chapter 18.

Transferrin

Iron needs to be transported from the intestines and storage pools (such as in the liver) to the bone marrow to be inserted into developing red cells. Transferrin performs this function. With a molecular weight of 77–80 kDa, and coded for by genes on chromosome 3, each transferrin molecule can carry two atoms of ferric (Fe^{3+}) iron. However, some iron may be carried by albumin and lactoferrin. Notably, transferrin is a reverse acute-phase reactant – concentrations fall in inflammatory disease, possibly as a mechanism to deny iron to bacteria and so minimize their growth.

Lack of transferrin leads to iron-deficiency anaemia, therefore justifying its measurement, although many other iron-related tests will probably be ordered as the metabolism of this metal is complex. Further details are presented in Chapters 3 and 4.

Other plasma proteins

There are hundreds of other plasma proteins with a variety of functions. To name but a few, these include pro-thrombin (coagulation), amylase (in pancreatitis and pancreatic cancer) and the glucoregulatory hormones insulin and glucagon. Each is addressed in its own particular chapter. Many plasma proteins are also cancer markers, as are addressed in Chapter 19. These markers include:

- prostate-specific antigen (the prostate)
- AFP (liver)
- CA-125 (ovary)
- paraproteins (myeloma)
- CEA (principally colorectal cancer, but it may also be raised in gastric, pancreatic, lung and breast cancer).

> ### Blood science angle: Plasma proteins
>
> Many plasma proteins may be 'claimed' by the different disciplines that make up blood science. The Igs and CRP are of undoubted importance in immunology (Chapter 9), whilst fibrinogen, transferrin, haptoglobin and ferritin are all essential to the understanding of haematology (Chapters 3 and 4). The endocrinology of the thyroid requires a knowledge of thyroxine-binding globulin (Chapter 18).

Analysis of urine

The molecular sieve of the glomerulus does not allow the passage of proteins greater in mass than albumin (therefore, Ig, caeruloplasmin, von Willebrand factor, and others). Many of those smaller proteins that do pass through the sieve are reabsorbed in the tubules. However, it is therefore normal for a trace of protein to be found in the urine.

However, in renal damage (especially glomerulo-nephritis), more proteins are lost and fewer are reab-sorbed, leading to protein in the urine; that is, proteinuria. This can be detected with urine dipsticks, but these are of poor sensitivity and specificity, so that formal laboratory analysis is required. The dipstick method relies on bromophenol blue, which reacts with albumin, so that increased amounts of other proteins will not be recorded, thus giving a false negative.

A technical problem with the analysis of urine is that all molecules of interest are at very low concentration compared with the blood, so that many methods are inappropriate. However, urine can be concentrated, and the proteins can be precipitated by a strong acid solution,

such as trichloroacetic acid. Dyes such as pyrogallol red-molybdate and Coomassie brilliant blue may be used to give an overall urine protein concentration. Most requests for the analysis of urine are one of two types: Bence–Jones protein (BJP) and albumin.

Bence–Jones protein   Antibodies are constructed from 'heavy' and 'light' chains, generally in equal amounts, the latter having a relative molecular mass of 25 kDa proteins. Consequently, the light chain molecule is small enough to pass through the glomerular sieve, and most is likely to be reabsorbed further down the nephron. However, the presence of a small amount of free light chain in urine (perhaps 10 mg/L) can be present in healthy individuals.

In certain circumstances too many light chains are produced, and so may be excreted. However, if a large number are produced by B lymphocytes, then their concentration in the blood will rise. As the ability of the kidney to reabsorb free light chains is finite, much of this increased protein will appear in the urine, perhaps at a concentration of 30 mg/L, and is called BJP.

The most serious disease where a great deal of BJP is found is in the white blood cell malignancy myeloma. The exact nature of the BJP can be ascertained by immunoelectrophoresis: it may be excess light chains or small fractions of heavy chains. The aetiology and importance of BJP is explained in Chapters 5 and 9, where diseases of white blood cells are discussed.

Albumin   Increased amounts of protein in the urine (proteinuria) are a general sign of renal damage. However, in some circumstances it is helpful to know the amount of albumin in the urine (i.e. albuminuria). This can be achieved by its measurement using an immunoassay.

The dipstick method is unable to detect lows of albumin, less than 250 mg/L. Detection of lower levels – that is, less than 100 mg/L (microalbuminuria) – demands special testing, and is commonly requested to check for an early indication of renal disease, as may be present in diabetes (hence diabetic nephropathy, as discussed in Chapter 14). However, analysis of urine is influenced by the amount of water taken, and so excreted. Thus, a patient with a low water intake is likely to generate a more concentrated urine than a patient who drinks a lot of water. To get round this, a better indication of renal disease is the ratio of albumin to creatinine (abbreviated to the ACR). As is common, the reference range varies: for some, a result less than

3 mg/mmol in women and less than 2 mg/mmol in men carry the best outcome, whilst others seek a result less than 2.3 mg/mmol regardless of sex. As renal disease progresses, so does the increase in the ACR.

Possibly the most accurate determination of albuminuria and proteinuria is to measure them in urine collected over a 24-h period. However, even this can be problematical and is of course unpopular with the patient.

Other analyses   The standard urine dipstick can offer many analyses, but none exceed the accuracy, sensitivity and specificity of 'standard' laboratory testing, be it biochemistry, haematology, immunology or microbiology. The most common dipstick analytes are leukocytes, nitrites, urobilinogen, protein, pH, specific gravity, ketones, bilirubin and glucose.

Wilson's disease and Menkes disease can be differentiated by measuring urine copper: concentrations will be high in the former but low in the latter. However, urine copper may be raised in other liver disease.

17.5   Case studies

Case study 22

A male, aged 68, reports being increasingly tired and lethargic over a 6-month period, with weight loss, despite his wife's attempts to 'feed him up'. More recently, he complains of abdominal discomfort. Blood results are as follows:

	Result (unit)	Reference range
Na	138 mmol/L	133–148
K	4.0 mmol/L	3.3–5.6
Urea	6.1 mmol/L	3.0–8.3
Creatinine	89 μmol/L	44–133
Total protein	59 g/L	60–80
Albumin	28 g/L	35–50
ALP	390 IU/L	30–130
AST	89 IU/L	10–50
Bilirubin	22 μmol/L	<17
ALT	72 IU/L	5–42
GGT	205 IU/L	5–55
Prothrombin time	15 s	11–14
Amylase	25 IU/L	<110
Fibrinogen	1.3 g/L	1.5–4.0
ESR	15 mm/h	<10
hsCRP	2.0 mg/L	<5

Interpretation   There are no abnormalities in the tests of renal function, so we can discount such problems with reasonable surety. However, all liver function tests are abnormal, so this organ is clearly under suspicion. We also have a prolonged prothrombin time, low fibrinogen and abnormal ESR but normal amylase and hsCRP. The abnormalities in prothrombin time and fibrinogen, with low total protein and albumin also indicate liver dysfunction, but the raised ESR provides little extra help. No increase in hsCRP points to a lack of an ongoing systemic inflammatory response.

Additional tests would include hepatitis virus screening (A, B and C), AFP (to test the possibility of liver cancer), and ultrasound imaging to check the integrity of the gall bladder. Since there is an ESR result then we presume there must be a full blood count somewhere. If not, it should certainly be ordered, as there may be a question of anaemia. The immunology laboratory will help with considering primary biliary cirrhosis by searching for autoantibodies. Additional questions to the patient may reveal use of alcohol (although, of course, this may not always be reliable). A biopsy may help, although this may be dangerous in view of low coagulation proteins.

Case study 23

A 45-year-old female solicitor complains of a 10-day period of increasing itchiness, lethargy, nausea and right abdominal pain. On further questioning, she reports a 6-month history of short bursts of acute pain stabbing on the right side. Her body mass index is 33.5 and her complexion is fair. Her urine is quite a dark yellow and her faeces are lighter brown than usual. The practitioner orders urea and electrolytes, CRP, total protein and albumin, and LFTs. Blood results are as follows:

	Result (unit)	Reference range
Na	141 mmol/L	133–148
K	4.5 mmol/L	3.3–5.6
Urea	7.1 mmol/L	3.0–8.3
Creatinine	76 μmol/L	44–133
Total Protein	72 g/L	60–80
Albumin	39 g/L	35–50
ALP	512 IU/L	30–130
AST	63 IU/L	10–50
Bilirubin	36 μmol/L	<17
ALT	40 IU/L	5–42
GGT	154 IU/L	5–55
hsCRP	1.2 mg/L	<5

Interpretation There are again no abnormalities in the tests of renal function, so once more we can discount problems for this organ. However, although total protein and albumin are normal, four out of five LFTs are abnormal, indicating liver disease. The aetiology seems to be chronic, but there is clearly an acute exacerbation. CRP is normal, therefore excluding a systemic inflammatory response. Excessive itchiness (pruritis) is a common sign of liver abnormalities.

The condition that best fits this picture is obstructive cholestasis, for which obesity is a risk factor, as may be the fair complexion. Note that ALP and GGT are relatively higher than AST and bilirubin. The patient reports dark urine, which could be the excretion of bilirubin, and light faeces because the bile is not being released into the intestines. Further studies are needed to confirm, and then remove, the obstructing gallstone, quite possibly by the process of endoscopic retrograde cholangiopancreatography (generally referred to as ERCP).

Summary

- The functions of the liver can be classified as metabolism, storage, excretion and detoxification.
- The LFTs include aminotransferases AST and ALT, bilirubin, ALP and GGT. Some consider albumin to be an LFT.
- The classic sign of liver disease is jaundice, although investigation of the aetiology includes consideration of pre-liver, true liver and post-liver causes. Disease may also arise acutely, or be chronic.
- Common liver diseases include cirrhosis, hepatitis, liver cancer and obstructive or inflammatory cholestasis
- All proteins can be classified as albumin or globulins – the latter include alpha, beta and gamma varieties.
- Electrophoresis (gel or capillary) is the major laboratory technique for quantifying these subsets of total proteins.
- The most important single plasma protein is albumin, making up perhaps half of the total protein pool. Low concentrations (hypoalbuminaemia) may be caused by liver disease or loss of albumin from a damaged kidney.

- There are many other plasma proteins, some of which are involved in coagulation (fibrinogen, prothrombin), are hormones (insulin, growth hormone), enzymes (ALP, GGT, amylase) or cancer markers (AFP, CEA).
- Increased protein in the urine defines the nephrotic syndrome.
- BJPs are abnormal Ig proteins (mostly light chains) in the urine, invariably due to a myeloma or a related malignancy.

Further reading

American Gastroenterological Association. Medical position statement: evaluation of liver chemistry tests. Gastroenterology. 2002;123:1364–1366.

Bogdanos DP, Invernizzi P, Mackay IR, Vergani D. Autoimmune liver serology: current diagnostic and clinical challenges. World J Gastroenterol. 2008;14: 3374–3387.

Camper WD, Osicka TM. Detection of urinary albumin. Adv Chronic Kidney Dis. 2005;12:170–176.

Carrell RW, Lomas DA. Alpha-1-antitrypsin deficiency – a model for conformational diseases. New Engl J Med. 2002;346:45–53.

Fontana RJ. Acute liver failure including acetaminophen overdose. Med Clin North Am. 2008;9: 761–794.

Frith J, Jones D, Newton JL. Chronic liver disease in an ageing population. Age Ageing. 2009;38:11–8.

Radu P, Atsmon J. Gilbert's syndrome – clinical and pharmacological implications. Israel Med Assoc J. 2001;3:593–598.

Schuppan D, Addhal NH. Liver cirrhosis. Lancet. 2008;371:838–851.

Tarantino G. From bed to bench: which attitude towards the laboratory liver tests should health care practitioners strike? World J Gastroenterol. 2007;13:4917–4923.

Web sites

www.britishlivertrust.org.uk
http://www.nlm.nih.gov/medlineplus/ency/article/ 000205.htm

18 Hormones and Endocrine Disorders

Learning objectives

After studying this chapter, you should be able to:

- appreciate the physiology of the organs and hormones of the endocrine system;
- be aware of the structure and function of the pituitary gland and hypothalamus;
- explain the functions of anterior and posterior pituitary hormones, and the causes and consequences of abnormal hormone production;
- describe and explain the pathophysiology of diseases of the thyroid and parathyroid glands;
- understand the structure of the adrenals, their products and how abnormalities lead to disease;
- discuss the endocrine basis of fertility and subfertility;
- recognize the complexity of steroid chemistry and laboratory methods for their analysis.

The endocrine system regulates numerous bodily functions by the action of small molecules called hormones, the study of which is endocrinology. A key feature is that a hormone is released by an endocrine gland (a gland that has no duct) directly into the blood, which carries it to its target organ or organs elsewhere in the body on which the hormone acts.

Each endocrine organ produces its own hormone or hormones, and in many cases the organs are stimulated by releasing factors, which themselves could be described as hormones (Table 18.1). The locations of the endocrine organs are shown in Figure 18.1. In this chapter we will examine each of these organs, and how each of the hormones they release will influence the functions of their target tissues and organs. We will also consider the pathology of the organs, and how they lead to the diseases from which the patients suffer. However, one pair of hormones (insulin and glucagon) are fully described in

Chapter 14, so will not be discussed in depth at present. Although parathyroid hormone is mentioned in Chapter 15, it will be detailed in this chapter.

However, before we address specific aspects of the endocrine system, we will in Section 18.1 first discuss general factors regarding these organs and their hormones, and then move to the physiology of the endocrine system. Once this has been achieved, in Section 18.2 we will look at diseases of the hormones, inevitably due to the organs that produce them. The chapter will conclude in Section 18.3 with case studies.

18.1 Endocrine physiology

Although seemingly complex (sometimes with justification), the endocrine system consists of a number of discrete organs secreting a number of well-defined molecules (Table 18.1, Figure 18.1). Complexity arises because of the fact that many hormones, although having a major role in a particular physiology, also have one or more minor effects in other areas. For example, the major role of prolactin in human physiology is to promote milk formation in the lactating female, but it also has a minor effect on sperm generation, although the major hormone here is testosterone.

Hormones

These small molecules (relative molecular mass $<25\,\mathrm{kDa}$) can be classified as being lipid-soluble or water-soluble.

Lipid-soluble hormones These include steroids (derived from cholesterol) and thyroid hormones (derived from the amino acid tyrosine). Being lipid soluble (and so water insoluble), steroid and thyroid hormones must be carried in the blood by specific

Blood Science: Principles and Pathology, First Edition. Andrew Blann and Nessar Ahmed.
© 2014 John Wiley & Sons, Ltd. Published 2014 by John Wiley & Sons, Ltd.

Table 18.1   The endocrine system.

Organ	Major hormones
Anterior pituitary	Adrenocorticotrophic hormone (ACTH), growth hormone (GH), prolactin, thyroid stimulating hormone (TSH), luteinizing hormone (LH) and follicle stimulating hormone (FSH)
Posterior pituitary	Oxytocin and arginine vasopressin (AVP; also known as anti-diuretic hormone, ADH)
Thyroid	Tri-iodothyronine (T3) and tetra-iodothyronine (T4, also known as thyroxine)
Parathyroid	Parathyroid hormone (PTH)
Adrenals	The adrenal cortex: cortisol, androgens and aldosterone. The adrenal medulla: adrenaline (epinephrine), noradrenaline (norepinephrine), dopamine
Pancreas	Insulin and glucagon
The gonads	The ovaries: oestrogens and progesterone The testes: testosterone

transport proteins, generally produced by the liver. For example, cortisol must be carried by cortisol-binding globulin or transcortin, whilst thyroxine (T4) is carried by thyroxine-binding globulin. The receptors for lipid-soluble hormones are located inside their target cells (the cytoplasm or the nucleus), so must pass through the cell membrane, which is permissible because of the lipid nature of the membrane.

Water-soluble hormones   These include protein/peptide hormones (such as insulin and GH), catecholamines and eicosanoid hormones (derived from arachidonic acid, such as prostaglandins and leukotrienes). They do not require a carrier protein, and they interact with their target cells via specific receptors on the cell membrane, an interaction that stimulates a cascade of intracellular second messengers (such as cyclic adenosine monophosphate) that result in changes in the cell, such as the activation of certain genes.

Control of hormone secretion   As mentioned, some endocrine organs are stimulated to produce their hormone products by other small molecules called releasing factors, and some hormones act on other endocrine organs. The release of some hormones is also promoted by nervous stimuli. A further aspect of regulation is a

feedback mechanism, of which there are two types: positive and negative. In positive feedback, the hormone stimulates the organ to release more of the hormone. A negative mechanism is present where increasing concentrations of the hormone or second molecule act to reduce levels of the hormone (Figure 18.2).

We begin our examination of endocrinology with the pituitary. This organ has been described as the master control endocrine organ, as it acts on several other endocrine organs. However, we must first address the anatomy of the pituitary, and its relationship with the hypothalamus. The pituitary gland lies beneath the hypothalamus, to which it is connected by the pituitary stalk or infundibulum. The pituitary itself has anterior and posterior lobes.

The anterior pituitary

Regulation of the secretion of hormones from the anterior lobe of the pituitary is by specific releasing factors or inhibiting factors that arise from the hypothalamus. The anterior pituitary hormones can be classified into one of two groups: glycoproteins (FSH, LH and TSH) and single-chain polypeptides (ACTH, GH and prolactin). Reference ranges are shown in Table 18.2. Each glycoprotein hormone consists of two subunits (α and β). The α subunit is common to FSH, LH and TSH (and also human chorionic gonadotrophin, a product of the embryo and placenta), while the β subunit is unique to each hormone and confers its biological activity.

Adrenocorticotrophic hormone   This hormone, a protein of 39 amino acids and molecular mass of 4.5 kDa, acts on the adrenal glands to stimulate the release of cortisol. Notably, a 13-amino acid cleavage product of ACTH is melanocyte-stimulating hormone. The release of ACTH is under the control of a releasing factor called corticotrophin-releasing hormone derived from the hypothalamus.

ACTH is not released in a steady state: there is a circadian or diurnal rhythm, with concentrations being low at midnight and rising in the final hours of sleep to peak shortly after awakening, and therefore so does cortisol. The concentrations of ACTH and cortisol then steadily decline throughout the day. Concentration of ACTH is also influenced by fear, illness, extreme emotions, hypoglycaemia and surgery, leading to the concept that ACTH, and therefore cortisol, is a stress hormone. However, this is viewed with suspicion by

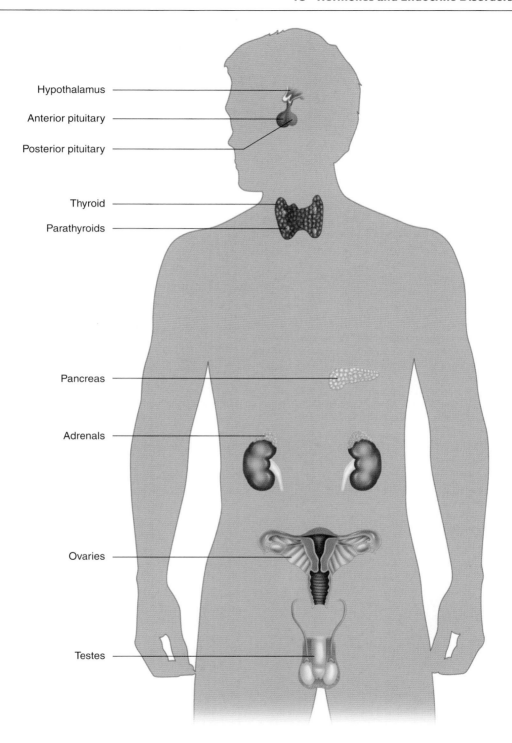

Figure 18.1 The location of endocrine organs, variously positioned in the skull, then neck, the abdomen and at the base of the trunk.

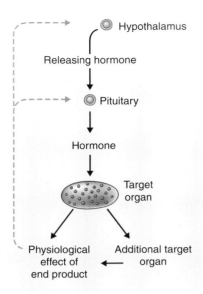

Figure 18.2 Feedback inhibition. In almost all cases, the hypothalamus secretes a releasing hormone that stimulates the pituitary, which then releases a second hormone which acts on a target organ. The products of the latter feed back to the hypothalamus and pituitary to suppress this pathway. The target organ may act on a second target organ which itself may release a factor or factors that have a physiological effect.

Table 18.2 Reference ranges for hormones of the anterior pituitary.

Hormone	Reference range
ACTH	5–20 pmol/L, 7–51 ng/L (dependent on time of day)
GH	<10 mU/L (<3.4 μg/L)
Prolactin	Male: 86–324 mU/L Female 103–497 mU/L
TSH	0.2–3.5 mU/L
LH	Male: 0.8–8.0 IU/L, Pre-menopausal female: 1–96 IU/L (dependent on stage of the menstrual cycle)[a] Post-menopausal female: 8–59 IU/L
FSH	Male: 0.7–11.0 IU/L Pre-menopausal female: 1.8–22 IU/L (dependent on stage of the menstrual cycle)[a] Post-menopausal female: 26–135 IU/L

[a]See Table 18.9 for levels throughout the menstrual cycle.

scientists, as 'stress' is very difficult to accurately define and so to quantify.

Growth hormone   GH is a 191-amino protein with a molecular mass of 22 kDa. Produced by somatotroph cells (and so occasionally described as somatotrophin), the most abundant cells in the anterior pituitary, it acts on many tissues to regulate their metabolism. For example, in the muscle cell it promotes glucose uptake, and in the adipocyte it decreases glucose uptake but increases lipolysis. In the liver it has no effect on glycogenolysis, but increases gluconeogenesis and ketogenesis, but a major effect on this organ is to stimulate the release of insulin-like growth factor-1 (IGF-1). This second hormone accounts for a major part of the biological activity of GH, by promoting the growth and development of skeletal muscle, cartilage, bone, liver, kidney, skin, haemopoietic tissues and cells of the lung.

Secretion of GH occurs in pulses, most frequently at night during the early hours of sleep (especially deep sleep), and is part controlled by the release of growth-hormone-releasing hormone (GHRH) from the hypothalamus. However, another product of the hypothalamus, somatostatin, inhibits the release of GH. During the day the plasma GH levels are considerably lower, whilst levels in teenagers can be double those of adults. Notably, both GH and IGF-1 part-regulate the production of GH by negative feedback at the hypothalamus and anterior pituitary.

Prolactin   Prolactin, the major species of which is a single-chain polypeptide of 198 amino acids and molecular mass of 22.5 kDa, is unique in that its secretion is controlled by an inhibitory factor (dopamine). The concentration of prolactin in the plasma is increased during pregnancy and lactation, but its physiological role in non-lactating women and in men remains unknown, although it may have a role in regulating other hormones such as the gonadotrophins, and the prolactin receptor is widely distributed. By itself prolactin has only a weak effect in stimulating lactation, but it will bring about milk secretion after the breast has been primed by factors such as oestrogens, progesterone, glucocorticoids, GH, insulin and T4, and, of course, stimulation of the nipple by a suckling infant. Prolactin also has a role in breast development.

Thyroid stimulating hormone   Secretion of this hormone, a peptide of 205 amino acids and a molecular mass of perhaps 25 kDa, is largely under the control of thyrotrophin-releasing hormone (TRH) from the hypo-thalamus. However, the release of TSH can be inhibited by somatostatin. The end organ (the thyroid) is stimulated to synthesize and release two other hormones: T4 and T3. Both of these hormones then exert a negative feedback effect on the pituitary and hypothalamus to reduce TSH and TRH secretion respectively.

Luteinizing hormone and follicle-stimulating hormone

These two hormones, each with a relative mass of some 24 kDa, are collectively termed the gonadotrophins because (a) they both act in parallel on the gonads to the two sexes (testes in the male and ovaries in the female) and (b) their release from the pituitary is stimulated by a common factor – gonadotrophin-releasing hormone (GnRH), a decapeptide released from the hypothalamus. In the male, LH (composed of 112 amino acids in two chains) promotes the maturation of the testes, and in the female it stimulates the development of the ovary and ovulation. In the male, FSH (composed of 110 amino acids, also in two chains) supports the production of sperm, and in the female it acts to regulate the maturation of the ovarian follicle and so the production of oestrogens.

The gonads synthesize testosterone, oestrogens and progesterone, and increasing concentrations of these hormones feed back to the pituitary and hypothalamus to inhibit the release of GnRH. In both sexes, GnRH release is also inhibited by inhibin, itself under the control of FSH. Conversely, activin (closely chemically related to inhibin) stimulates FSH production by the pituitary.

The posterior pituitary

This is essentially a collection of nervous tissues. Technically, it could be argued that the posterior pituitary is not a gland as it does not itself actively synthesize the hormones it releases (oxytocin and arginine vasopressin), but instead is where they are stored (in the terminal regions of specialized axons) and from where the hormones are released into the venous microcirculation. Reference ranges for the two hormones are shown in Table 18.3.

Oxytocin This hormone, of nine amino acids and molecular mass 1 kDa, is synthesized in the paraventricular nucleus of the hypothalamus, and travels along nerves to the pituitary. It is most active during childbirth, where its release is stimulated by the contractions of the

Table 18.3 Reference ranges for hormones of the posterior pituitary.

Hormone	Reference range
Oxytocin	Men: 215–390 pg/mL Women: 250–400 pg/mL Lactation: >600 pg/mL
AVP	<1.7 pg/mL (in platelet-free plasma)

smooth muscle cells of the uterus and the distension of the cervix by the infant. Its function is to promote contraction of the uterine wall. Indeed, oxytocin is Greek for 'swift birth'.

After delivery, oxytocin stimulates milk ejection from glands in the breast in response to mechanical stimulation from the suckling infant, and continues to act to reduce the size of the uterus. The function of oxytocin in males and non-pregnant females is unclear but may relate to sexual activity.

Arginine vasopressin This hormone, also of nine amino acids and mass 1 kDa, is also known as vasopressin, or anti-diuretic hormone (ADH – diuresis is the process of generating urine). It is synthesized in the hypothalamic supraoptic nucleus, and, like oxytocin, also travels along nerves to the pituitary, from where it is released.

As discussed in Chapter 12, arginine vasopressin (AVP) acts on the kidneys to increase water absorption by the tubules, so decreasing urine output, which results in the production of concentrated urine. It also has a role in maintaining blood pressure, hence the vaso- part and the -pressin part. Secretion is sensitive to blood pressure – AVP concentrations rise when blood volume and blood pressure are low. A second major influence on AVP is osmolality. The blood vessels of the hypothalamus have osmoreceptors that sense a change in the sum concentration of all ions and other small molecules in the blood that contribute to osmolality. If these receptors consider osmolality to be rising, it can be countered by increasing the amount of water in the blood by inducing the release of AVP.

The thyroid

The thyroid gland is located at the front of the neck, below the larynx. It is a butterfly shaped organ whose two lobes ('wings') lie to the left and right of the trachea, and are linked by an isthmus. Weighing perhaps 30 g in the

Table 18.4 Reference ranges for thyroid hormones.

Hormone	Reference range
T3	Free: 4.0–6.5 pmol/L
	Total: 1.2–2.3 nmol/L
T4	Free: 10–25 pmol
	Total: 60–160 nmol/L

Table 18.5 Reference range for parathyroid hormone.

Hormone	Reference range
PTH	1–6 pmol/L

adult, it is composed microscopically by spherical thyroid follicles that synthesize the two thyroid hormones T3 and T4. Reference ranges are shown in Table 18.4. Between the follicles are parafollicular cells or C-cells, which produce calcitonin, a participant in the homeostasis of calcium (Chapter 15).

Disorders of thyroid function can result in either inadequate or excess production of thyroid hormones causing altered cellular metabolism and development of associated clinical features.

Thyroid hormones   T4 and T3 are synthesized in the follicular cells and follicle colloid of the thyroid. Intracellular thyroglobulin is a required cofactor for the fusion of two tyrosine amino acids to form a tyrosyl group, to which three or four atoms of iodine are sequentially added. A key enzyme in this process is thyroperoxidase. T4 (making up 95% of the thyroid hormones) and T3 (5%) leave the thyroid, but for most efficient carriage they must bind to thyroxine-binding globulin (which carries about 70% of T4 and 80% of T3), albumin (25%) or transthyretin (5%). The remaining 0.05% of T4 and 0.2% of T3 circulate free. Upon arrival and passage into the target cell, T4 is de-iodinated of one atom to yield T3, the biologically active form. It follows that T4 is a pro-hormone. T3 crosses the cytoplasm and interacts with receptors on the nuclear membrane.

Regulation of tri- and tetra-iodothyronine   As discussed, TSH stimulates the thyroid to synthesize and release the thyroid hormones. The level of TSH is itself regulated by free T3 and T4 (i.e. that which is not being bound to carriers) which feed back to the pituitary. TSH passes through the bloodstream and interacts with specific receptors on the surface of thyroid cells to initiate hormone production.

Function of thyroid hormones   The thyroid hormones have numerous influences on metabolism on almost every cell in the body. Effects include increasing basic

metabolic rate, production of blood cells (especially red blood cells), regulation of fat, carbohydrate and protein synthesis and storage, regulation of intracellular transcription and translation bringing about changes in cell size, number and differentiation. They also promote cellular differentiation and growth, influence heart rate and the cardiac cycle, and even the rate of hair growth and density of subdermal fat.

The parathyroids

The parathyroid glands consist of four glands located on the posterior surface of the thyroid gland and together have a mass of 25–40 mg in humans. Their function is to synthesize and release PTH, an 88-amino acid peptide with a molecular mass of perhaps 9.4 kDa. The reference range for PTH is shown in Table 18.5.

As discussed in Chapter 15, PTH acts in three independent ways to increase the concentration of calcium in the blood:

• indirectly motivating osteoblasts to break down bone (where 99% of body calcium is found);
• promoting the reabsorption of calcium from the distal convoluted tubules and ascending limb of the loop of Henle (see Chapter 12 for renal function); and
• increasing the absorption of calcium across the gut wall by promoting the release of vitamin D.

The release of PTH depends not on the pituitary but on the concentration of free calcium in the blood (i.e. the fraction which is not bound to albumin). When free calcium is low, levels of PTH rise to redress the balance. However, there is a diurnal variation of PTH, being highest in the early hours of the morning and lowest at 9 a.m. This is generally not an issue as blood for the analysis is rarely taken at, for example, 2 a.m. The activity of PTH is countered by calcitonin (a 32-amino acid molecule with a relative mass of 3.6 kDa produced by parafollicular cells of the thyroid). The action of calcitonin is to reduce the synthesis of vitamin D, and so calcium absorption by the intestine, to inhibit osteoclasts, and to reduce the tubular reabsorption of calcium, so allowing it to remain in the filtrate and so in the urine.

The adrenals

These glands, each weighing 4–6 g, are physically joined to the kidneys by connective tissues and have defined regions. The middle is called the medulla and the outer part is the cortex, which is further divided into three parts. The zona glomerulosa makes up 10% of the cortex, the zona fasciculata 75–85% and zona reticularis 5–15%. The hormone-producing cells of the medulla may also be referred to as chromaffin cells, a histological expression defined by their affinity for chromium salts. These cells are modified neurones, and the medulla itself is supplied with nerves from the sympathetic nervous system. Each part of the adrenal produces its own set of hormones

(Table 18.6) that leave the gland to influence their respective target tissues (Figure 18.3). Reference ranges are shown in Table 18.7.

Adrenaline, noradrenaline and dopamine These catecholamines are produced by chromaffin cells of the adrenal medulla, often in response to exercise, trauma and perceived danger. Adrenaline and noradrenaline, also known as epinephrine and norepinephrine, have a host of activities, such as raising blood pressure and heart rate, dilation of pupils and pulmonary and bronchial airways (thereby increasing gas exchange) and generally mimicking the stimulation of the sympathetic nervous system (as both can be neurotransmitters). They are

Table 18.6 The adrenal gland and its hormones.

Region	Hormones	Target
Medulla	Adrenaline (epinephrine), noradrenaline (norepinephrine), dopamine	Various cells and tissues
Cortical zona reticularis (adjacent to the medulla)	Cortisol, adrenal androgens	Various cells and tissues
Cortical zona fasciculate	Cortisol, adrenal androgens	Various cells and tissues
Cortical zona glomerulosa (the outermost layer)	Aldosterone	The kidney

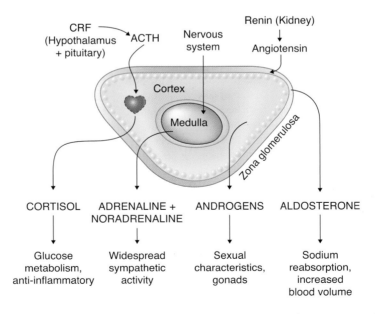

Figure 18.3 Functional anatomy of the adrenal gland. The adrenal consists of three areas: the innermost medulla, the cortex and the outer zona glomerulosma. Each has specific functions. CRF: corticotrophin releasing factor; ACTH: adrenocorticotrophic hormone.

Table 18.7 Reference ranges for adrenal hormones.

Hormone	Reference range
Aldosterone	100–500 pmol/L
Cortisol	200–800 nmol/L at 8 a.m.
	50–200 nmol/L at midnight
Dehydroepiandrosterone sulphate (DHEAS)	Men: 2–10 μmol/L Women: 3–8 μmol/L
Dehydroepiandrosterone (DHEA)	Both sexes: 1–5 nmol/L
Androstenedione	Men: 2–12 nmol/L Women: 2–8 nmol/L
Urine:	
adrenaline	<200 nmol/day
noradrenaline	<680 nmol/day
metanephrines	<2.1 μmol/day
dopamine	<3 μmol/day

synthesized from tyrosine in a pathway that includes L-dopa and dopamine (shortened from dihydroxyphenethylamine), generating molecules of some 183 Da mass. The latter is also an important neurotransmitter in the brain, but in the peripheral circulation its activities resemble those of adrenaline.

Catecholamines have a very short half-life (measurable in minutes) and are degraded by transferase enzymes to metanephrines, which are stable. All catecholamines and their metabolites are generally measured in a 24-h urine sample.

Cortisol Cells of the adrenal cortex have receptors for low-density lipoprotein cholesterol, enabling the uptake of cholesterol. This lipid is the substrate for a series of different metabolic pathways, one of which results in the production of cortisol (Figure 18.3). The first and rate-limiting step in this pathway, the generation of pregnenolone, is the delivery of cholesterol to the mitochondria, which is facilitated by ACTH. Further steps involve 21-hydroxylase enzymes.

Cortisol, with a relative mass of 362 Da, has a large number of actions, such as facilitating an increase in blood sugar by promoting glycogenolysis (and so may be described as a glucocorticoid), and is also involved in fat and protein metabolism. In this respect it acts as a counter to insulin. There is also a role for this hormone in the actions of other hormones (adrenaline, noradrenaline and glucagon), and in muscle, nervous system and intestinal function. As regards the kidney, it has weak

effects in inhibiting sodium loss and also acts as a diuretic, thereby countering the effects of AVP.

A further important function of this molecule is in immunosuppression, in that it may inhibit inflammatory responses to injury and limit the acute-phase response. Indeed, this property is developed by the synthesis of artificial versions of cortisol that have much stronger immunosuppressive and anti-inflammatory effects – these drugs being hydrocortisone, prednisolone and dexamethasone.

In the bloodstream, in common with other steroid-based hormones, cortisol must be carried by a transporter protein, corticosteroid-binding globulin (or transcortin, an alpha globulin product of the liver) and albumin. Levels of cortisol vary during the 24-h period: levels are highest early in the morning (perhaps 8–9 a.m.) and lowest between midnight and 4 a.m. This has important clinical consequences.

Adrenal androgens Androgen is the generic name for a steroid-based sex hormone, most of which, of course, arises from the gonads. These small molecules (typically with a mass of some 288 Da) include DHEA and DHEAS, and androstenedione and testosterone. The latter is converted to dihydrotestosterone, which is four times more potent than testosterone. However, the production of testosterone by the adrenals is trivial compared with that generated by the testes. In the adult male, androgens and FSH cooperate to act on testicular Sertoli cells to support spermatogenesis. In the female, androgens can act as precursors for oestrogens. These will be discussed more fully in sections to come.

Aldosterone This important regulator of homeostasis is also called a mineralocorticoid, and is also discussed in Chapter 12, which deals with renal function. When blood volume rises, the renal juxta-glomerular apparatus secretes renin, which acts on angiotensinogen to generate angiotensin I. This is the substrate for angiotensin-converting enzyme, the product being angiotensin II, which stimulates the zona glomerulosa cells of the adrenal cortex to synthesize aldosterone.

Aldosterone, which has a mass of 360 Da, acts on the proximal distal tubules and collecting ducts of the nephron to reabsorb sodium, which is followed by water. In order to balance the increase in tubular sodium, potassium passes into the urinary filtrate. Indeed, hyperkalaemia stimulates aldosterone production, as does blood loss, and there may also be a role for ACTH. Thus, the overall position is to increase blood volume, and so possibly blood pressure.

Table 18.8 Reference ranges for testosterone.

Hormone	Reference range
Testosterone	Men aged 15–40: 15–30 nmol/L
	Men aged >40: 10–25 nmol/L
	Women: 0.5–2.5 nmol/L

The gonads

The target organs for LH and FSH are the gonads: testes and ovaries, which secrete testosterone, oestrogens and progesterone. Although the natural focus is on sex, in both men and women, oestradiol (the major oestrogen) has nonsexual functions in bone, endothelial and lipoprotein metabolism. In normal women, both testosterone and androstenedione are produced and secreted in measurable quantities by both the ovary and the adrenals. Reference ranges for testosterone are shown in Table 18.8.

Sex-hormone-binding globulin Approximately 98–99% of androgens and oestrogens circulate bound to (mostly) liver-derived glycoproteins, the aptly named sex-hormone-binding globulin (SHBG), a molecule of 90 kDa coded for by a gene on chromosome 17. However, a proportion of these sex hormones may also be carried by albumin and transcortin (corticosteroid-binding globulin, which also carries cortisol, testosterone and progesterone). The remaining 1–2% of these sex hormones are 'free' and are able to enter their target cells, this fraction being described as the free androgen index (FAI). In the male, testosterone is also carried by androgen-binding protein, a product of the Sertoli cells of testes, and bears a striking structural homology to SHBG.

SHBG is an important regulator of sex hormones concentration in the plasma. Falling concentrations allow the FAI to increase and so may lead to hyperandrogenaemia. However, the biology of SHBG is compounded by its regulation by oestradiol and testosterone, and by insulin and T4. Consequently, for full understanding of the physiological and pathological roles of SHBG, knowledge of the levels of several hormones and nonspecific carriers is required.

Male sexual development In the foetal hypothalamus, GnRH stimulates the pituitary to secrete LH, which acts on testicular Leydig cells instructing the production of testosterone. This continues in childhood, but up to puberty levels are low. Pulses of LH and FSH occur and initiate puberty: the testes grow, and spermatogenesis begins as the testosterone levels rise. The hormone also induces the production of androgen-binding protein by Sertoli cells, the action of which is to concentrate levels of testosterone near sperm-producing areas.

Testosterone induces further changes in the adolescent, which include the development of the secondary sex characteristics: growth of the external genitalia, development of a muscular body shape, deepening of the voice, facial and body hair growth, and development and maintenance of the sex drive. In addition to testosterone, Leydig cells also produce androstenedione and DHEA, whilst locally produced testosterone from the testicular interstitial cells is essential to spermatogenesis.

Spermatozoa originate in the germinal epithelium and gradually mature and develop over two and a half months, assisted by Sertoli cells. If the secretion of LH and FSH is interrupted, spermatogenesis is impeded and stops.

FSH and LH are regulated by negative feedback inhibition by levels of testosterone. In addition, FSH acts on Sertoli cells to cause the secretion of inhibin, which also negatively regulates the production of FSH by the pituitary. Sertoli cells also produce activin, which enhances FSH biosynthesis.

Concentrations of sex hormones rise at puberty and remain reasonably constant until late middle age (perhaps age 50 years), at which point testosterone slowly falls, often described as the andropause, the male equivalent of the female menopause. As testosterone falls, the pituitary responds by increasing the synthesis of FSH (Figure 18.4).

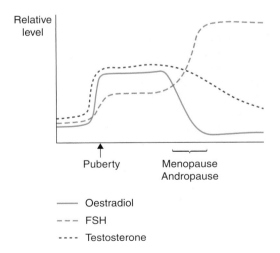

Figure 18.4 Fluctuations in concentrations of oestradiol, FSH and testosterone in the life cycle. Concentrations rise in puberty, and change again at the menopause and andropause.

Female sexual development As in the male, foetal and pre-pubertal levels of LH and FSH are low. The foetal ovary is established by 10 weeks and by mid-term some 4 million primordial follicles are present, although only a tiny proportion (considerably less than 1%) of these will transform into mature ova. The same hypothalamus–pituitary axis releases LH and FSH, the former acting on ovarian theca cells to produce androstenedione and testosterone. FSH acts on follicular cells to induce the enzyme aromatase to convert testosterone to oestradiol, the principal oestrogen (the others being oestrone and oestriol). LH acts on the corpus luteum to produce both progesterone and oestradiol. These hormones produce in the adolescent female parallel changes as occur in the male; that is, secondary sexual characteristics of breast development, reforming the pelvis, fat deposition, and maturation of the uterus which leads to menstruation.

Three other molecules are of note:

- Anti-Mullerian hormone (AMH) is secreted by ovarian granulosa cells to (a) inhibit the development of primary ovarian follicles and (b) inhibit the actions of FSH. Low levels of AMH predict an early menopause, and this is useful in cases of assisted conception as AMH can be used to assess a woman's ovarian reserve in response to controlled ovarian stimulation.
- Inhibin (of which there are two isoforms, A and B) is also produced by ovarian granulosa cells, and may also have a role in inhibiting FSH secretion.
- In the ovarian follicle, activin increases the activity of LH and FSH.

The menstrual cycle The process by which the hypothalamus, pituitary and ovaries orchestrate changes in the uterus and elsewhere to prepare the body for pregnancy is established. Full details are available elsewhere, and accordingly a brief sketch only will be presented here. Although the menstrual cycle can vary greatly from perhaps 23 to 33 days, it is convenient to illustrate with a 28-day cycle. Reference ranges for these hormones by stage of the cycle are presented in Table 18.9.

Taking day 1 to be the first day of menstrual bleeding (which may last 3–5 days), days 1–12 are described as the follicular phase, days 13–14 the ovulatory phase and days 15–28 as the luteal phase.

- In the follicular phase, oestradiol concentrations are low and FSH is high, the latter causing follicle development and the secretion of oestradiol by the ovary. Increasing concentrations of oestradiol induce the cervix

Table 18.9 Reference ranges for LH, FSH, oestradiol and progesterone according to the menstrual cycle.

Hormone	Reference range (stage of cycle)
LH	Follicular: 3–13 IU/L
	Mid cycle: 14–96 IU/L
	Luteal: 1–11 IU/L
FSH	Follicular: 4–13 IU/L
	Mid cycle: 5–22 IU/L
	Luteal: 1.8–7.8 IU/L
Oestradiol-17-β	Follicular: 110–180 pmol/L
	Mid cycle: 550–1650 pmol/L
	Luteal: 370–700 pmol/L
Progesterone	Follicular: <10 nmol/L
	Mid cycle: <10 nmol/L
	Luteal: >10, peak 30 nmol/L

to produce thinner mucus, which is easily penetrable by sperm. As concentrations continue to rise it feeds back to inhibit FSH levels, and also promotes the secretion of LH.
- The mid-cycle LH surge triggers release of the ovum from the follicle (ovulation) which passes into the fallopian tube where it may be fertilized. The remaining tissue, the corpus luteum, begins to secrete progesterone, and at this stage the expression of oxytocin receptors is at its highest.
- Rising concentrations of progesterone from the corpus luteum (hence luteal phase), alongside high levels of oestrogens, act on the inner lining of the uterus (the endometrium) to prepare it to receive an embryo. High concentrations of the hormones feed back to the pituitary and hypothalamus to reduce the release of LH and FSH by feedback inhibition.
- As the corpus luteum becomes exhausted it fails to produce progesterone, required by the endometrium. Thus deprived of progesterone, the endometrium breaks down and leaves the uterus as menstrual blood, which is day 1 of the next cycle.

Pregnancy The only physiological reasons for cessation (amenorrhea) of the menstrual cycle are pregnancy, lactation and the menopause. The embryo becomes embedded in the endometrium, and the outer layer (the trophoblast) secretes the hormone human chorionic gonadotrophin (hCG). The latter has the same action as LH, and acts on the corpus luteum to maintain progesterone and oestradiol secretion, thus maintaining the endometrium and the pregnancy. Once the placenta is established, it also secretes oestrogens and progesterones.

Lactation Rising concentrations of oestrogens, progesterone and (possibly) hCG act on the breast to prepare it to feed the infant. The placenta releases the hormone human placental lactogen, which has the same effect. Prolactin rises during pregnancy and is the major hormone promoting lactation. Following birth, oxytocin promotes uterine contraction, but also supports lactation. Infant suckling of the nipple induces the pituitary to secrete both prolactin and oxytocin. Notably, dopamine has an inhibitory effect on prolactin, so that concentrations of the former will be low or absent in lactation.

The menopause After perhaps 40 years of fertility (such as ages 13–53) the store of ova is depleted, and the ovaries cease the production of oestrogens so that ovulation stops. Without a post-ovulation corpus luteum, concentrations of progesterone also fall. However, oestrogens are still produced by other tissues, such as the adrenals, and the lower concentrations of oestrogens and progesterone are detected by the pituitary, which consequently continues to secrete FSH and LH, possibly at increased concentrations.

Many of the symptoms of the menopause can be treated with oral female sex hormones – hormone replacement therapy. However, long-term use, such as for osteoporosis, is clouded by the increased risk of breast and endometrial disease, and venous thromboembolism.

Changes in oestradiol, testosterone and FSH over the life cycle are illustrated in Figure 18.4. Concentrations of oestradiol and testosterone are highest in the middle, actively reproductive section. Concentrations of the former fall markedly after the menopause; levels of the latter fall more slowly. With a functioning pituitary, FSH rises with age as fertility falls.

Laboratory issues In the male, testosterone production is maximal from the early 20s to the late 30s and then gradually declines, accompanied by rising LH and FSH, leading to the andropause (Figure 18.4). Consequently, age-specific reference ranges may be called for. In young and middle-aged men, there is a diurnal rhythm of testosterone secretion, with higher circulating concentrations in the morning than in the afternoon or evening. The diurnal rhythm is lost or reduced beyond middle age.

Clearly, there is marked variation in all endocrine hormones at different stages of the female life cycle, and reference ranges for all such stages are established.

Steroid chemistry

The close structural association between the cortical hormones deserves special attention. The root molecule for the synthesis of these hormones is cholesterol, which is the substrate of enzymes such as mitochondrial cytochrome P450scc, which removes side chains and performs hydroxylations, to generate pregnenolone. This process is accelerated primarily by ACTH, but also by FSH and LH.

Pregnenolone is the substrate for 17α-hydroxylase, which generates 17-α-hydroxy-pregnenolone, and for 3-β-hydroxysteroid dehydrogenase, which produces progesterone. At this point it gets complicated, as both 21-α-hydroxylase and 11-β-hydroxylase enzymes can act on different metabolites (Figure 18.5). However, order can be sought in a number of pathways:

- Progesterone can be engineered to give 11-deoxycorticosterone, then corticosterone, then 18-hydroxycorticosterone, and ultimately, via the action of aldosterone synthase, the mature aldosterone molecule.
- Other enzymes can act on progesterone to give 17-hydroxyprogesterone, then deoxycortisol, and finally cortisol.
- Both 17-α-hydroxypregnenolone and 17-α-hydroxyprogesterone can be substrates for enzymes generating DHEA and androstenedione, the latter by different routes.
- Androstenedione can be converted to oestrone and to testosterone by different enzymes. Both can be converted to oestradiol, the most potent of the oestrogens, and testosterone and can be reduced to the far more active dihydrotestosterone.

The very close association between the structures of these molecules is illustrated by Figure 18.6, which in many cases differ by very small substitutions. For example, oestradiol and testosterone differ by an —OH and a =O at position 'A'.

Steroids are degraded mainly in the liver by transferases to inactivated metabolites, some of which are sulphated; others are conjugated to glucuronide. Clearance from the body is mostly in the urine, but some are lost through the intestinal tract.

18.2 The pathology of the endocrine system

Having looked at the physiology of the endocrine system, we will now look at pathology. This can be

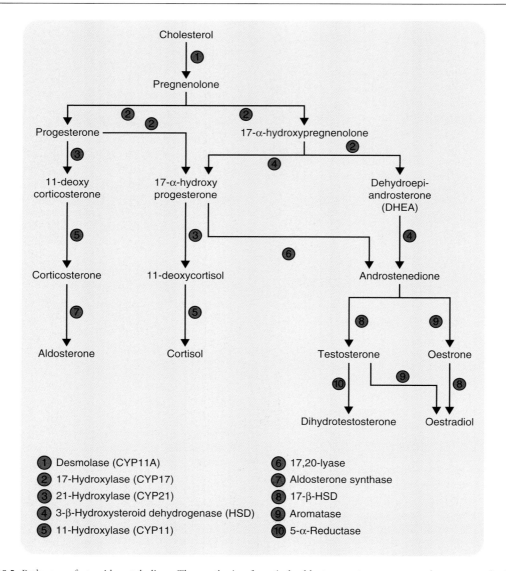

Figure 18.5 Pathways of steroid metabolism. The synthesis of cortisol, aldosterone, testosterone and oestrogens begins with cholesterol, with many complex intermediates which are the substrate for a host of enzymes: (1) desmolase; (2) 17-hydroxylase; (3) 21-hydroxylase; (4) 3-β-hydroxysteroid dehydrogenase; (5) 11-hydroxylase; (6) 17,20-lyase; (7) aldosterone synthase; (8) 17-β-hydroxysteroid dehydrogenase; (9) aromatase; (10) 5-α-reductase. The activity of the enzyme aldosterone synthase is promoted by angiotensin, the end product of a pathway initiated by the kidney, involving angiotensin-converting enzyme and the enzyme renin (Chapter 12). The aromatase enzyme is the objective of inhibitors used to treat breast and ovarian cancer, and also gynaecomastia.

classified simply as overactivity or underactivity. Most of the former will be adenoma tumours – cancers that secrete or release a molecule – although inflammation can also cause disease. However, the complexity of the endocrine system, such as that two hormones often act serially to produce an effect, means that explanation based on organs will be awkward. Instead, we will look at disease causes by defects in the production of the hormones themselves, regardless of their anatomical origin.

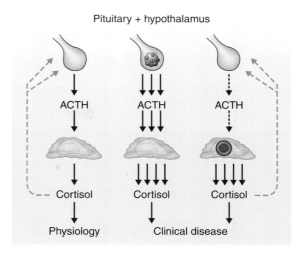

	A	B	C
Cholesterol	—OH	—CH₃	Long chain
DHEA	—OH	—CH₃	=O
Cortisol	=O	—CH₃	—C(O)–CH₂OH
Aldosterone	=O	—CHO	—C(O)–CH₂OH
Testosterone	=O	—CH₃	—OH
Oestradiol	—OH	—CH₃	—OH

Figure 18.6 Steroid structure. Very minor differences in the structures of cholesterol and five steroid hormones. N.B. This is very simplified, and with apologies to card-carrying steroid biochemists.

Abnormalities in adrenocorticotrophic hormone and cortisol

These two hormones are linked, in that ACTH released from the anterior pituitary acts on the adrenals to promote the formation of cortisol. Thus, abnormalities in cortisol may arise from malfunction of the pituitary and/or the adrenal glands.

Cushing's syndrome This is probably the most well-known adrenal disease, discovered by the American surgeon Harvey Cushing in 1932, and is the consequence of high levels of cortisol. Clinically, it is characterized by a number of features that include round or 'moon' face, with central obesity, fat deposits around the shoulders (both possibly leading to diabetes), hypertension, polyuria, stretch marks on the abdomen (striae) and, in females, hirsutism (mostly manifesting as facial hair).

There are a number of different types, depending on the exact site of the cellular basis of the disease (Figure 18.7 and Figure 18.8).

- Overactivity of the anterior pituitary, perhaps by a pituitary adenoma, leading to high levels of ACTH. This in turns acts on an 'innocent' adrenal gland, which produces high levels of cortisol. This particular abnormality may also be referred to as Cushing's disease.
- Similarly, overactivity of the adrenal cortex due to the presence of an adrenal adenoma or carcinoma that

Pituitary + hypothalamus

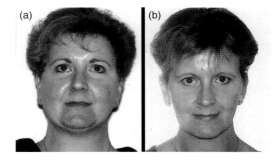

Figure 18.7 Pathogenesis of Cushing's syndrome. On the left, ACTH arising from the pituitary acts on the adrenals, which secrete cortisol. The latter in turn feeds back to the pituitary to regulate ACTH production. In the middle, a pituitary adenoma secretes high levels of ACTH that flood the adrenal, which responds by secreting high concentrations of cortisol. On the right, an adrenal tumour secretes excess cortisol, which feeds back to a normal pituitary so concentrations of ACTH fall. Ectopic production of ACTH and clinical disease caused by steroid therapy are not shown.

secretes cortisol. This process is independent of ACTH, concentrations of which fall due to negative feedback on the pituitary.

Figure 18.8 Clinical features of Cushing's syndrome. (a) The patient exhibits the round ('moon') face characteristic of this disease, a consequence of the high concentrations of cortisol secreted by her adrenocortical adenoma. (b) The same patient 6 months after removal of the tumour. (From Holt & Hanley, Essential Endocrinology and Diabetes, 6th edition, 2012, Figs. 6.9a, b, p. 112. Reproduced with permission of John Wiley & Sons, Ltd.).

- An ACTH-producing tumour outside the endocrine system (an ectopic tumour), such as cancer of the lung or bronchus.
- Cushing's syndrome may also arise as a long-term consequence of the use of steroids to treat chronic inflammatory diseases such as rheumatoid arthritis or inflammatory bowel disease.

Addison's disease   This is effectively the reverse of Cushing's syndrome, and was discovered by British physician Thomas Addison in 1855. It also has a number of distinct aetiologies which are characterized by low concentrations of cortisol, but the dominant pathology is destruction of the adrenal cortex. This may, in turn, be due to factors such as autoimmune disease (being the most common form), tuberculosis, infiltration by a malignancy and the response to an infection, perhaps by a bacterial infection. Consequently, as a result of cortical cell destruction (i.e. a generalized hypoadrenism), it is also likely that there are also abnormalities due to loss of aldosterone and adrenal androgens.

Many of the clinical aspects are also the reverse of Cushing's disease, and include hypotension (leading to hypoperfusion and syncope) and weight loss. There is often hyperpigmentation, nausea and vomiting, muscle and joint pains, and hypoglycaemia (due to the loss of the effect of cortisol in regulating blood sugar) which may, in acute crises, induce a coma. The hyperpigmentation is often due to increased concentrations of melanocyte-stimulating hormone, which shares a common metabolic pathway to ACTH.

Investigation of abnormalities in adrenocorticotrophic hormone and cortisol   Clearly, the approach is to measure both molecules, but knowledge of various other analytes, especially sodium, potassium and calcium, can be informative. However, attention must be paid to the time of day, as ACTH and cortisol have a diurnal variation. Nevertheless, a low cortisol (<100 nmol/L at 8 a.m., reference range 200–800 nmol/L) suggests adrenal insufficiency, although this may be tested by the synacthen provocation test with 250 μg/0.25 mg of ACTH delivered by intramuscular or intravenous routes. Blood is taken at intervals (such as 30 and 60 min) and any response in cortisol noted.

A high concentration of cortisol (such as 1500 ng/mL, reference range <200 ng/mL) alongside low or even undetectable ACTH strongly suggests Cushing's syndrome due to an adrenal tumour. The low ACTH supports the view that pituitary and hypothalamic function is normal and that it is simply responding to feedback inhibition. However, if there is a pituitary adenoma or an ectopic tumour, ACTH concentrations will be within or above the reference range. In cases of ectopic generation of ACTH, there is often hypokalaemia (low potassium) and alkalosis (high pH). Failure of aldosterone is likely to lead to low serum sodium alongside high potassium and high calcium.

Despite the reasonably well-defined laboratory abnormalities and clinical signs and symptoms, a more complete diagnosis is often desired. One way to do this is to challenge the endocrine system to determine the presence of a functioning or nonfunctioning feedback mechanism.

The diagnosis of ACTH-dependent Cushing's syndrome can be supported by the results of the dexamethasone suppression test. This drug is a powerful artificial mimetic of cortisol, and if injected in sufficient quantities should feed back to the pituitary and hypothalamus to inhibit the release of ACTH and so reduce concentrations of cortisol. At the practical level, 1–2 mg dexamethasone is given late at night, and blood taken 8–9 h later for the measurement of cortisol. A high dose challenge with 8 mg can also be helpful, and in both doses the cessation of certain other drugs (such as barbiturates and oral oestrogens) may be necessary. Cortisol can also be measured in the urine.

If the patient has ectopic ACTH secretion or an adrenal tumour, there should be no suppression of concentrations of cortisol. In this case, an ACTH concentration will be diagnostic, in that it is raised in ectopic ACTH secretion and suppressed in patients with an adrenal tumour, due to negative feedback.

In some cases it may be necessary to sample blood from vessels leading into and out of the pituitary and hypothalamus (the petrosal sinus), and compare this blood with that taken from a peripheral vein. Patients with pituitary-driven Cushing's disease will show a central:peripheral gradient of ACTH of ≥3:1, which is diagnostic of a pituitary source. An alternative is to challenge the production of ACTH by an infusion of corticotrophin-releasing hormone that bypasses the hypothalamus.

Management of Cushing's syndrome   The management of these patients depends on the precise aetiology. An ectopic ACTH-secreting tumour may be amenable to surgery and/or chemotherapy, and there is also a place for surgery on the pituitary. Drugs such as metyrapone, which inhibits the synthesis of cortisol by the adrenal, have a place but limited efficacy.

Abnormalities in growth hormone

Overactivity of the pituitary results in excess production of GH, which leads to a series of clinical signs collectively called acromegaly. However, as with other endocrine disorders, high GH may be due to ectopic secretion by non-pituitary tumours such as of the lung, adrenals or pancreas. Rarely, there may be increase in the secretion of GHRH. Acromegaly is rare with an incidence of three or four cases per million per year.

In the adult, the consequences of raised GH, and so increased IGF-1, include a large jaw (prognathism) and heavy bone growth over the eyebrows, with large hands and feet. Indeed, an early symptom is the need for a larger shoe. Other signs and symptoms are a large tongue, gaps between the teeth, headaches, sweating, with visual and sleep problems. In children, excess GH simply increases the growth rate, leading to tall and large children.

The pathophysiology of the disease follows the abnormal pituitary, with numerous non-GH functions. For example, there may also be changes that can be ascribed to abnormal sex hormones – loss of libido in both sexes, amenorrhea and erectile dysfunction. Other common problems later in the disease include hypertension, diabetes (both possibly related to endocrine abnormalities), an enlarged liver, with heart failure and kidney failure.

Investigation of acromegaly   Clearly, measurement of GH is key, but there can also be a place for the assessment of IGF-1 (although normal liver function is required) and GHRH. Medical imaging with magnetic resonance imaging of the head can also be useful in looking for a pituitary mass. But for blood scientists, GH release has a diurnal pattern, with most secretion during sleep and very little when awake. It follows that high concentrations during the day strongly suggest acromegaly.

GH has a number of functions, and participates in glucose metabolism. Accordingly, the definitive test for acromegaly is the response of GH to the oral glucose tolerance test (explained fully in Chapter 14). In health, GH concentrations fall to undetectable levels 2 h after a drink of 75 g of glucose. High concentrations that fail to fall very strongly support the diagnosis of acromegaly. As indicated above, there may also be incidental changes in other anterior pituitary products: TSH, FSH, LH, ACTH and prolactin, and so associated changes in their target organs. Plasma IGF-1 concentrations are also increased in patients with acromegaly. Measurements of IGF-1 are also of value in monitoring therapy as they do not fluctuate as much as GH values do.

Management of acromegaly   If increased GH is due to a tumour, normal cancer treatment may be effective, being a combination of surgery, chemotherapy and radiotherapy. Surgery to the pituitary is of course extremely difficult and not without risk (such as of meningitis), one route being through the nasal cavity and another through the mouth.

Chemotherapy includes drugs such as octreotide and lanreotide to suppress GH production, which can be very effective, with injections every 2–4 weeks. An alternative agent is to use a dopamine agonist, the most common being bromocriptine, which is taken orally.

The laboratory will be called upon to monitor blood concentrations of GH, and possibly IGF-1 during active treatment, and annually thereafter.

Low concentrations of growth hormone   GH deficiency is a consequence of problems with the pituitary. However, as we will see, pituitary dysfunction is rarely specific for a single GH measurement, so that other changes, such as hypothyroidism, are likely to be present.

In the child, GH deficiency is suspected if the growth rate fails to approach the 95th percentile of the normal range, most so as the rate comes closer to the 99th percentile. In the absence of other causes, GH deficiency has an incidence of 1 in 4000 live births, and may be due to many causes, such as congenital malformations. In the adult, GH deficiency is present in perhaps 10 per million, and is associated with reduced bone mass and osteoporosis, increased body fat and raised low-density lipoprotein cholesterol.

A problem with defining GH deficiency is that concentrations are very low most of the day. A better method is to provoke the pituitary with an agonist that, in health, will result in a burst of GH production. Insulin is a convenient stimulant. An increase in IGF-1 completes the picture and demonstrates a functioning liver. However, in true pituitary-based GH deficiency, provocation causes minimal concentrations (if any) of GH to appear in the blood.

Treatment of GH deficiency in children is with daily injections, but replacement therapy in the adult is not recommended for the 'natural' decline in GH concentrations, but may be appropriate in adult-onset disease.

Abnormalities in prolactin

Hyperprolactinaemia   Of all the benign pituitary adenomas, a prolactinoma is the most common, generating high concentrations of the hormone; that is, hyperprolactinaemia. If the tumour has a diameter less than 10 mm (often according to magnetic resonance imaging, which is preferable to computed tomography), it is described as a microprolactinoma, while a macro-prolactinoma is greater than 10 mm in diameter.

Prolactin is unusual among the anterior pituitary hormones as it is regulated by suppression – by dopamine. Hence, failure of dopamine from the hypothalamus to act on the anterior pituitary leads to unregulated release of prolactin. However, there is a long list of drugs that can cause hyperprolactinaemia, such as certain tranquilizers, drugs to treat hypertension, and oestrogen. Raised prolactin is also associated with Cushing's disease, renal failure and polycystic ovary syndrome (PCOS).

The clinical consequences of this depend on the patient's age, sex and duration of the hyperprolactinaemia. Men and post-menopausal women often present with symptoms of a pituitary mass, such as headache and visual disturbance due to compression of the optic chiasm. Both sexes can present with loss of libido, breast pain, headache and visual disturbances, and men present with erectile dysfunction and gynaecomastia. Pre-menopausal women usually present with amenorrhoea, subfertility and galactorrhoea (inappropriate release of breast milk, although this is not unheard of in men and post-menopausal women). Abnormalities in the pituitary may result in deficiency of other pituitary hormones; for example, GH, TSH and ACTH. Indeed, abnormal LH and FSH may be responsible for some of the reproductive changes.

Investigation of hyperprolactinaemia   The blood test is relatively straightforward, but timing is important as prolactin has a diurnal variation (high during sleep and lowest between 9 a.m. and noon) and a menstrual variation. Interpretation must consider physiological factors of pregnancy, lactation and breast manipulation, and pathological factors outlined above. Concentrations of prolactin in various conditions are shown in Table 18.10.

Treatment of raised prolactin is generally with dopamine agonists such as bromocriptine (N.B. as in acromegaly), and regular blood tests for efficacy. However, if an adenoma becomes so large that it causes problems such as pressing on the optic nerve, surgery may be necessary.

Every laboratory method has a range of values over which it provides an accurate result. A problem with some cases of hyperprolactinaemia is that concentrations defined by some immunoassays can be so high that they

Table 18.10 Increased concentrations of prolactin.

Cause	Prolactin (mU/L)
Macroprolactinaemia	9000–120000
Microprolactinaemia	3000–9000
Pregnancy	<8000
Dopamine antagonists	<5000
Renal failure	<5000
Hyperthyroidism	<2500
Stress	<2000
Reference range	Women <500, men <450

exceed the top of this range, and can interfere with the assay so that concentrations appear normal. Thus, when the laboratory result conflicts with the clinical picture, this phenomenon, known as a hook effect, or the prozone effect, may be present. The solution is to dilute the plasma two- or three-fold in buffer, which will reduce the concentration and may well bring it down into the working range of the assay. A false hyperprolactinaemia may be present if prolactin is bound to an immunoglobulin G molecule, the complex being described as macroprolactin. Fortunately, it can be removed from a sample of plasma by polyeythelene glycol.

Hypoprolactinaemia   Failure of the pituitary to produce prolactin may be due to excess dopamine, extreme weight loss and drugs designed to treat Parkinson's disease. Clinical associations include ovarian dysfunction, erectile dysfunction, abnormal spermatogenesis and other reproductive problems. Diagnosis calls for low or absent serum prolactin.

Thyroid disease

Disease of the thyroid is easily classified into overactivity (hyperthyroidism, which may also be called thyrotoxicosis), underactivity (hypothyroidism, also known as myxoedema) and cancer.

Hyperthyroidism is associated with significant short- and long-term morbidity and mortality, and although its incidence is less than that of hypothyroidism at 800 per million, this figure exceeds many other endocrine diseases. In both diseases, 10 times as many women are affected as are men. Hypothyroidism is a common insidious condition with significant morbidity and the subtle and nonspecific signs that are often incorrectly associated with other conditions. With an incidence of 3500 per million, hypothyroidism is one of the most prevalent of the endocrine diseases with a frequency second only to diabetes. However, it has been estimated that 3–5% of women have thyroid disease, with up to 10% of those aged over 75 years having subclinical hypothyroidism.

Hyperthyroidism The laboratory will be called upon to confirm or refute increased TSH, T3 and/or T4, which will be requested after clinical assessment. The causes of these abnormalities can be:

- overactivity of the pituitary, which generates increased TSH that in turn acts on an 'innocent' thyroid gland, or
- overactivity of the thyroid gland itself with a normal pituitary.

The most likely cause of pituitary hyperthyroidism is an adenoma, which secretes inappropriate amounts of TSH. However, tumours secreting TSH are very rare. A second diagnostic arm, as in all suspected pituitary masses, is imaging. The high concentrations of TSH cause release of T3 and T4, so is relatively easy to define in the laboratory.

A mass in the thyroid is a goitre, and is effectively a focus of overactive cells, which may be malignant (Figure 18.9). These masses are often quite firm, and so are described as nodules, and several such nodules can be present (i.e. multi-nodular goitre). These nodules are the site of the excessive production of free T3 and/or T4, easily measurable in the laboratory. If there is a normally functioning pituitary, the physiological feedback

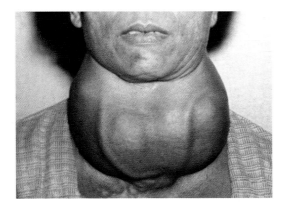

Figure 18.9 Goitre. An exceptionally large goitre due to profound and chronic iodine deficiency in rural Africa. This case would never develop to such an extent in the UK. (From Holt & Hanley, Essential Endocrinology and Diabetes, 6th edition, 2012, Fig. 8.5, p. 170. Reproduced with permission of John Wiley & Sons, Ltd.).

mechanism results in low TSH. Imaging is a further diagnostic aid, where thyroid scintillation scanning often shows patchy uptake of isotope with multiple hot and cold areas being noted throughout the gland.

Pathophysiology of hyperthyroidism These can be listed with relative simplicity.

- Graves' disease (named after the Irish physician Robert Graves, who described it is 1835) is the consequence of autoimmune attack on the thyroid, with the inevitable inflammation, leading to thyroiditis. It is the most common form of hyperthyroidism, accounting for perhaps 75% of cases. The relevant autoantibodies are directed towards the receptor for TSH on the surface of thyroid cells. The binding of the antibody mimics the effects of genuine TSH binding, so that the cell is inappropriately stimulated to generate T3 and T4. If this occurs in a precise location there may be a nodular goitre.
- An alternative type of thyroiditis is that first described by Hakaru Hashimoto in 1912. The pathological basis of the disease is antibodies to thyroid peroxidase, an enzyme required for iodine metabolism and its incorporation into thyroglobulin. A later consequence is the invasion of the thyroid tissue with leukocytes, the classic sign of inflammation. However, many cases of hypothyroidism are linked to Hashimoto's disease.
- A tumour of thyroid that secretes T3 and/or T4 (i.e. an adenoma).
- Sub-acute thyroiditis is characterized by a rapid presentation and often follows a viral infection, such as by the Coxsackie virus. However, there are various other forms, such as postpartum thyroiditis, sub-acute lymphocytic thyroiditis and a thyroiditis following mechanical damage.
- Use of certain drugs, such as amiodarone, as is used in atrial fibrillation. This drug coincidentally structurally resembles T4, which may be why it leads to thyroid disease. Hyperthyroidism may also be due to the excessive use of T4.

Clinical features of hyperthyroidism The common feature is a swollen neck, which may be a single goitre, multiple goitres or generalized soft tissue swelling. An established feature of Graves' disease is exophthalmia, where an excess presentation of the sclera (the whites) of the eye because of eyelid retraction gives the classic staring expression.

There are very many effects of high concentrations of T3 and T4, most of which can be described in terms of a

Table 18.11 General features of thyroid disease.

	Hypothyroidism	Hyperthyroidism
Common signs and symptoms	Lethargy, dry coarse and pale skin, slow speech and mental function, pallor, hoarse voice, constipation	Increased irritability and sweating, tremor, lethargy, breathlessness, muscle weakness
The heart, temperature control, and weight	Bradycardia	Tachycardia, palpitations, arrhythmia
	Cold intolerance and cool extremities	Heat intolerance and warm extremities
	Weight gain	Weight loss
Thyroid hormones	Low T3 and/or T4 (underactive thyroid)	High T3 and/or T4 (overactive thyroid)
Changes in TSH	Pituitary failure: low TSH	Pituitary failure: low TSH
	Pituitary overactivity: high TSH	Pituitary overactivity: high TSH
Treatment	Replacement therapy(oral T4)	Suppression of the thyroid (drugs, surgery, radioactive iodine)
Allied clinical conditions	Myxodema, goitre, Hashimoto's thyroiditis	Exopthalmia, goitre, Hashimoto's thyroiditis, Graves' disease

'speeding up' of metabolism that often follow the overactivity of the sympathetic nervous system. Nonspecific symptoms include weight loss (despite a good appetite), increased sweating, difficulty in sleeping, thinning of the skin and hair, muscle weakness, nervousness and irritability (Table 18.11).

Management of hyperthyroidism There are three treatments of this condition:

• Anti-thyroid drugs, such as carbimazole and propylthiouracil. Cardiac symptoms may be treated with beta-blockers.
• Radioactive iodine, typically ^{131}I, causes damage to the thyroid cells. However, this therapy is difficult to control and reverse, and in many patients their disease converts to hypothyroidism.
• Surgery for those with an enlarged thyroid and clearly identifiable nodular goitre(s). This thyroidectomy can be total or subtotal. However, there is danger of damaging the phrenic nerve and removing some parathyroid tissue.

A problem with all of these treatments is that too much may lead to hypothyroidism.

Hypothyroidism This disease is characterized by insufficient or even absent concentrations of TSH, T3 and/or T4, and can be due to hypothalamic failure, pituitary failure or thyroid failure. The first (tertiary hypothyroidism), accounting for perhaps 5% of all cases of hypothyroidism, is due to failure to generate TRH. More common (secondary hypothyroidism, 5–10%) is pituitary failure, which manifests biochemically with low or

absent TSH, and it may be part of hypopituitarism. However, the most common cause (primary hypothyroidism, 80–90% of cases) is failure of the thyroid gland itself.

Pathophysiology of primary hypothyroidism As with hyperthyroidism, these can be listed with ease:

• The most common cause of primary hypothyroidism in the developed world is Hashimoto's thyroiditis (as in hyperthyroidism), where the autoantibodies cause progressive destruction of thyroid tissue.
• In the developing world, the major cause is lack of iodine in the diet.
• As indicated above, hypothyroidism may be the result of the overtreatment of hyperthyroidism.
• Drugs that cause hypothyroidism include lithium and amiodarone (which both reduce T3 and T4 secretion), dopamine and glucocorticoids (decrease TSH secretion) and radioactive iodine (destroys thyroid tissue).

Other causes include congenital hypothyroidism in infants due to failure of the thyroid to develop. Lack of treatment leads to cretinism. As discussed, diseases or injuries affecting the hypothalamus or pituitary can result in reduced production of TRH and TSH respectively, causing reduced T3 and T4.

Clinical features of hypothyroidism As with hyperthyroidism, there are a large number of nonspecific signs and symptoms, such as weight gain, depression and cold intolerance. Later in the disease, a goitre maybe present, especially in iodine deficiency, and is due to hyperplasia of thyroid tissue. An additional sign is myxoedema, characterized by heavy facial flesh with enlarged cheek folds, eyebrows and lips (Table 18.11).

Investigation and diagnosis of hypothyroidism Once again, in common with hyperthyroidism, this involves measurement of T3, T4 and TSH. In primary hypothyroidism, failure of the thyroid gland to generate T3 and T4 is recognized by a functioning pituitary, which should therefore release high concentrations of TSH. However, in secondary and tertiary hypothyroidism, failure of the hypothalamus and pituitary to generate TSH will also lead to low T3 and T4. Thus, the key analysis is of concentrations of TSH.

In order to further classify the particular type of disease, determination of autoantibodies to thyroid peroxidase (present in 95% of patients with Hashimoto's thyroiditis) and the TSH receptor may be helpful. If a firm diagnosis is still unclear, the response of the pituitary–thyroid axis to an infusion of TRH may be helpful.

Management of hypothyroidism Regardless of aetiology, most cases are treated with replacement therapy of oral T4, often initiated with 50 μg daily, and then increased as symptoms and blood results demand, aiming at T4 in the reference range. Also, treatment should reduce TSH to the lower part of the reference range. In primary hypothyroidism, high TSH should fall as the pituitary detects rising plasma T4.

The laboratory and thyroid disease For the most complete picture of thyroid pathophysiology, measurements of T3, T4 and TSH are required. However, it has been argued that concentrations of T3 and T4 generally correlate together so closely that measuring both is a poor economy. As concentrations of T4 exceed that of T3, the former gives better sensitivity and specificity, so that many services no longer offer measurement of T3 routinely.

This can be taken one step further, in that it may be assumed that if the TSH is within the reference range then there should be no abnormality in T3 and T4. Consequently, a growing trend is to first measure TSH and then, if this is normal, to report that the patient is euthyroid (i.e. true thyroid) and that therefore measurement of T3 and T4 is unjustified. However, if TSH is outside the reference range, the move is to measure T4, as it may be that there is an abnormality. This effectively turns TSH into a screening test. However, a TSH measurement on its own should be treated with

caution as it may miss hypothyroidism due to a hypopituitary cause.

The laboratory may also be called upon to measure other analytes, as hyperthyroidism can be associated with raised serum calcium, angiotensin-converting enzyme, and liver function tests (especially gamma-glutamyl transpeptidase). Similarly, hypothyroidism is linked to raised cholesterol, creatine kinase, creatinine and thyroxine-binding globulin. SHBG is decreased in hypothyroidism and increased in hyperthyroidism, potentially leading to changes in sexual physiology.

The thyroid and pregnancy In physiology, the increased oestrogen in pregnancy drives an increase in thyroxine-binding globulin and may also be responsible for increased TSH; and hCG has a weak effect on T3 and T4. Many women experience nausea and vomiting early in pregnancy, and this may mask or be taken as a symptom of thyroid disease, and there may be a need for iodine supplements.

The foetal thyroid is slow to develop, and so the foetus relies on maternal T3 and T4 crossing the placenta. Subclinical hypothyroidism is present in 2–3% of pregnancies, whilst clear hypothyroidism develops in 0.3–0.5%, The latter can bring serious complications for the mother and her child, such as miscarriage, premature delivery, pre-eclampsia and low birth weight. Treatment is with oral T4, and the hypothyroid woman attempting pregnancy will, like the diabetic woman, need to have regular blood tests.

An overactive thyroid is likely in 0.2–0.4% of pregnancies, and is often detected by the development of exophthalmia (suggesting Graves' disease) and/or a diffuse goitre. However, care in diagnosis is required as hCG may stimulate the thyroid. Hyperthyroidism leads to a risk of pre-eclampsia and low birthweight, and if due to certain autoantibodies these may cross the placenta and influence foetal development. Surgery is a possible treatment, but chemotherapy with propylthiouracil is recommended in the first trimester, transferring to carbimazole for the remainder of the pregnancy. Thyroid problems occur in perhaps 5% of women within the year after their pregnancy, and is generally a thyroiditis that resembles Hashimoto's disease. Both hypothyroid and hyperthyroid phases are recognized, and of these some 20% will develop permanent hypothyroidism.

Thyroid cancer This disease accounts for 4% of female cancer and <3% of male cancers. In the UK this translates to a rate of 4.8 per 100 000 women and

1.3 per 100 000 men, with a common mean age of diagnosis of 50 years. The presenting feature is often a nodule or 'stringy' fibrosis with symptoms of hyper- or hypothyroidism.

The cancers themselves can be classified by histology as papillary (the most common, about 80%), follicular (15%) and medullary (6.5%). Other rare types include thyroid lymphoma, sarcoma and carcinoma. Unsurprisingly, the laboratory will be called on for concentrations of T3, T4 and TSH, although most patients will be biochemically euthyroid. However, concentrations of T3 and T4 may be raised if the tumour is an adenoma. Measurement of calcitonin is also justified, especially in actual or family history of medullary cancer, a cancer as part of the multiple endocrine neoplasia (MEN) syndrome, or hyperparathyroidism. Regrettably, calcitonin alone cannot differentiate benign from malignant disease.

The place of these molecules as cancer markers is also discussed in Chapter 19.

Blood science angle: The thyroid

The products of this organ influence many other organs, and so do its diseases. Haematologists may order thyroid function tests to investigate a normocytic anaemia, common in hypothyroidism. Immunologists can help with determining autoantibodies to thyroid peroxidase and to the TSH receptor and thus aid the diagnosis of Hashimoto's and Graves' diseases.

Anterior pituitary hypofunction

The anterior pituitary is a small mass of tissue, and is responsible for producing ACTH, TSH, LH and FSH. If disease such as an adenoma develops in one particular anatomical area, perhaps responsible for producing a defined hormone, there is a strong likelihood that other areas producing other hormones will also be influenced. A deficiency in one or more hormones is described as hypopituitarism, and has an incidence of approximately 40 per million per year. Panhypopituitarism is reserved for failure of all hormone production and is rare and is generally the consequence of exceptional pathology.

We have already discussed the large number of causes of pituitary disease, which include tumours, infarction, granuloma, traumatic brain injury, radiation, hypothalamic disease and infection. Clinical presentation depends on which hormones are abnormal. Indeed, upon clinical suspicion of hypopituitarism, blood tests for ACTH, TSH, FSH and LH (bearing in mind the possibility of diurnal variation) may be called for.

To aid diagnosis, the pituitary may be challenged with insulin, as we have already mentioned. However, this provocation should result in increased GH and in increased ACTH/cortisol, and must be associated with hypoglycaemia to stimulate the pituitary. Management is by replacement therapy of whatever hormones are deficient.

Posterior pituitary dysfunction

In contrast to the anterior pituitary, which receives local hormones from the hypothalamus, the posterior lobe responds to nervous impulses to release two hormones: oxytocin and AVP (also referred to as ADH by some). As with anterior pituitary disease, abnormal posterior pituitary function can be idiopathic or follow malignancies of hypothalamus and pituitary, head trauma, or following surgical or radiation therapy.

There are a few conditions reporting abnormal serum oxytocin: concentrations are low in anorexia nervosa, a low level post-menopause is associated with osteoporosis, and anti-oxytocin autoantibodies have been reported.

Decreased levels of arginine vasopressin The primary role of this hormone is to regulate water excretion by the aquaporin-2 channels in the collecting ducts of the nephron. When AVP binds to its receptor, aquaporin channels are active, and move water from the filtrate, which moves into the blood. A homeostatic model sees increasing blood osmolality (of which sodium is a major component) being recognized by the hypothalamus, which instructs the pituitary to release AVP, which results in more water in the blood, and so normalization of the high osmolality. Increased water in the blood leads to increased blood volume and generally higher blood pressure.

It follows that reduced concentrations of AVP (and/or its receptor) lead to failure to recover water from the filtrate, leading to excess urine production (as much as 20 L over 24 h), a syndrome described clinically as diabetes insipidus. A consequence of this is hypernatraemia and dehydration, with increased thirst (polydipsia), a feedback response of the body attempting to increase blood volume.

The laboratory in diabetes insipidus Clearly, the key measurement is of AVP, but this is not performed routinely due to sampling issues. However, diabetes insipidus can generally be ruled out if plasma osmolality is within its reference range (275–295 mmol/kg), and more so if urine osmolality exceeds 750 mmol/kg. Conversely, diabetes insipidus is more likely in the presence of hypernatraemia (sodium >145 mmol/L), plasma osmolality >300 mmol/kg and urine osmolality <750 mmol/kg.

However, as in many endocrine disorders, a provocation or challenge can be useful, and in this case requires the patient to not drink for up to 7 h. A normal response is to retain water, thus concentrating the urine to >700 mmol/kg, which maintains a normal osmolality. In diabetes insipidus, water is not retained, plasma osmolality rises and urine osmolality fall, typically to less than 270 mmol/kg. A further test is to infuse desmopressin (also known as DDAVP, a synthetic analogue of AVP), which should cause the urine to become more concentrated and reduce plasma osmolality. An important point in this test, and the treatment of diabetes insipidus with desmopressin, is that it assumes the kidney can respond to the AVP or desmopressin, which it may not if there is renal disease. If so, the disease may be described as nephrogenic diabetes insipidus to contrast it from a pituitary cause.

Increased concentrations of arginine vasopressin

This condition leads to the syndrome of inappropriate antidiuretic hormone, characterized by retention of water, leading to hyponatraemia, rising blood volume and hypertension. Treatment is to antagonize the release of AVP (which may arise from an adenoma) by chemotherapy directed towards the pituitary, or the use of a vasopressin receptor antagonist of the 'vaptan' class, such as conivaptan.

Parathyroid dysfunction

As with other disease of endocrine organs (and, of course, every other organ!) we focus on overactivity (hyperparathyroidism) and underactivity (hypoparathyroidism). However, the parathyroid is unusual in endocrine terms as it has no other organ or hormone or releasing factor molecule directly regulating it. Instead, it is regulated by calcium.

Hyperparathyroidism The consequence of overactivity of the parathyroids is high concentrations of PTH. Several types are recognized, the most common being:

- Primary hyperparathyroidism, which occurs when parathyroid tissue is overactive, most likely because of

a malignant transformation into an adenoma or carcinoma. This type of disease is inevitably the cause of high concentrations of serum calcium (hypercalcaemia).
- Secondary hyperparathyroidism, which is present when a healthy parathyroid reacts physiologically to pathologically low concentrations of calcium (hypocalcaemia, perhaps the consequence of renal disease and/or lack of vitamin D), mild decrease in serum magnesium, and/or high concentrations of phosphates (hyperphosphataemia) by secreting high concentrations of PTH.

Other forms of hyperparathyroidism are those that result from parathyroid hyperplasia, itself the product of long-term secondary disease, and from changes to the glands after surgery.

Hyperparathyroidism is dangerous for a number of reasons. The PTH itself can cause bone disease by activating osteoblasts, which in turn stimulate osteoclasts to strip calcium from bone, and so a risk of developing renal osteodystrophy. Increased calcium can interfere with nervous conduction and the cardiac cycle, potentially leading to cardiac arrhythmias and arrest. Treatment depends on aetiology, such as surgery and/or chemotherapy for malignancies, and the treatment of the kidney in renal disease. A calcium mimetic, cinacalcet, may also be useful in suppressing the release of PTH.

Hypoparathyroidism Failure of the parathyroids to release PTH results in low concentrations of calcium. As in hyperparathyroidism, aetiology is restricted to a small number of causes:

- The consequence of thyroid disease, such as removal of healthy parathyroid tissue during the removal of diseased, possibly malignant, thyroid tissue.
- Autoimmune attack of the parathyroid gland, either directly or indirectly, as the result of thyroiditis.
- Deposition of iron as part of the disease haematochromatosis.

Other causes include magnesium deficiency, defective calcium sensing and the anatomical disease DiGeorge syndrome. The direct treatment of the organ itself is difficult, as in many cases the parathyroid tissues are effectively dead and cannot be stimulated. Inflammation may in theory be treated with immunosuppression, but delivering this specifically to the parathyroid can be difficult. Damage caused by haematochromatosis may be reversible upon iron chelation. Replacement therapy consists of daily injections of recombinant PTH.

Blood Science: Principles and Pathology

Adrenal pathology

We have already discussed two major abnormalities of this organ from the perspective of the pituitary: the result of overstimulation of the adrenals by ACTH leads to high cortisol (Cushing's syndrome), whilst lack of ACTH and adrenal underactivity leads to Addison's disease. However, both syndromes, and others, can be found in the presence of a normal pituitary, as the adrenal glands themselves may fail and/or be the subject of attack by a number of pathologies, such as autoantibodies or bacterial or viral infection. Where this leads to adrenal insufficiency, concentrations of ACTH are likely to be raised.

Congenital adrenal hyperplasia As will be evident from Figure 18.4, the metabolic pathways underlying the synthesis of steroid hormones are complex and interwoven. Consequently, abnormalities in the genes for the various enzymes are the basis of several disease and syndromes, collectively known as congenital adrenal hyperplasia (CAH). Congenital, because they are evident at birth or shortly thereafter; and hyperplasia, because they are associated with enlarged adrenals. CAH may be a cause of Addison's disease, resulting from low cortisol, aldosterone, adrenaline and noradrenaline.

The most common form of CAH (exceeding all others by a factor of 10) is deficiency of 21-hydroxylase, occurring in perhaps 1 in 12 000 infants in the UK annually. The enzyme is rate limiting for the synthesis of various molecules, and since other pathways are present there is a build-up in other metabolites, such as 17-hydroxypregnenolone, which will appear in large amounts in the urine. However, some becomes the substrate of other enzymes, often leading to the accumulation of excess androgens (male hormones). This has consequences for the female foetus, as high androgens disrupt the development of normal female sexual anatomy with ambiguous genitalia, which can appear almost male with no testes. Some female babies may be incorrectly assigned the male gender on the basis of a quick inspection of the genitalia.

The neonate with 21-hydroxylase deficiency is likely to show signs of electrolyte disturbance (hyperkalaemia and hyponatraemia) from day 5 and can die. However, a mild form of the deficiency has no effect until later in childhood with the development of precocious puberty, often before the age of 10 and accompanied by the early appearance of secondary sexual characteristics, due to high levels of androgens.

A further consequence of lack of the enzyme is reduced synthesis of adrenaline because the development of the adrenal medulla relies on cortisol production from the adrenal cortex. Lack of cortisol also leads to hypoglycaemia. In the likelihood of a normal hypothalamus and pituitary, low concentrations of cortisol will be detected and concentrations of ACTH will rise.

The second most common enzyme deficiency leading to CAH, accounting for 5% of cases, is of 11-β-hydroxylase, which would normally act on 11-deoxycortisol to generate cortisol. There is also reduced synthesis of androgens and aldosterone, leading respectively to abnormal sexual development and to the retention of sodium and water, and so to hypertension and oedema.

The laboratory in congenital adrenal hyperplasia The approach is the same regardless of the particular enzyme defect, focusing on urea and electrolytes, ACTH, renin, cortisol, aldosterone and sex hormones. Further investigations are of concentrations of particular enzymes and intermediate metabolites such as 17-hydroxyprogesterone and pregnanetriol.

Conn's syndrome This condition, first described by American endocrinologist Jerome Conn in 1955, is the consequence of an aldosterone-secreting tumour, and so hyperaldosteronaemia (typically 1500 pmol/L, reference range 100–500 pmol/L). This in turn acts on the nephron, leading it to retain sodium and excrete potassium; and a further consequence is loss of hydrogen ions, leading to metabolic alkalosis and so the possibility of hypocalcaemia.

These changes lead to muscle weakness, headache and hypertension, the latter a consequence of the increase in blood volume, itself due to the increased uptake of water in an attempt to normalize the hypernatraemia. This increased blood pressure leads the juxta-glomerular apparatus to secrete renin. Accordingly, the ratio of aldosterone to renin can be used to confirm or refute a particular diagnosis: a high ratio implies primary aldosteronism.

Although interesting, Conn's syndrome is exceedingly rare; considerably more common is secondary hyperaldosteronism resulting from renal and/or liver disease.

Catecholamine abnormalities Increased concentrations of adrenaline and/or noradrenaline, produced by the adrenal medulla, are characterized by symptoms such as tachycardia, palpitations, hypertension, headaches and weight loss. All of these effectively mimic overactivity of the sympathetic nervous system. Increased glucose may be present as a result of the action of the catecholamines on carbohydrate metabolism – the promotion of lipolysis and inhibition of the cellular uptake of glucose.

Adrenaline and noradrenaline themselves have a short half-life and are broken down by enzymes such as monamine oxidases to stable metanephrines, which can be measured in blood and urine, but require the patient to be at physical rest and psychologically calm. Indeed, concentrations of plasma adrenaline and noradrenaline rise 10-fold during heavy exercise or athletics. Perhaps the greatest example of stress is myocardial infarction (a heart attack), where levels of adrenaline may rise 20-fold and concentrations of noradrenaline by sevenfold.

The principal catecholamine abnormality is a phaeo-chromocytoma – a tumour of the chromaffin cells – and can be present in one or both adrenals. If sufficiently large it may be imaged by computed tomography or magnetic resonance. However, perhaps 15% of these tumours are outside the adrenals. Approximately five new cases per million population present annually, generally in the young and middle aged. Typical plasma adrenaline concentrations in phaeochromocytoma may be 280 pg/mL (compared with 40 pg/mL at rest), whilst noradrenaline may exceed 5000 pg/mL (200 pg/mL at rest).

Treatment of adrenal disease   As with other disease, management is at two levels: chemotherapy and surgery. The patient's initial physiology must first be addressed, an example of this being replacement therapy, such as cortisol in Addison's disease. High concentration of molecules may be treated by suppression, an example being adrenal suppression by cyproterone acetate or spironolactone. However, the root cause must be determined, and if this is a tumour then cytotoxic chemotherapy and/or surgery are options.

Abnormalities in male reproductive endocrinology

In this section, to be followed by female reproductive disease, we will focus on LH, FSH, testosterone, oestrogens and progesterone, having already discussed the physiology and pathology of prolactin. GnRH, a product of the hypothalamus, acts on the anterior pituitary to induce the synthesis of LH and FSH. These in turn act on the testes and ovaries to direct the production of testosterone, and of oestrogens and progesterone respectively. It follows that abnormalities in any one of these stages may lead to subfertility, and possibly infertility.

Generally speaking, endocrine causes of male infertility are quite rare. However, it may be present as a severe manifestation of thyroid disease, CAH and of hyper-prolactinaemia, in which case treatment of the underlying condition is appropriate. Failure of the hypothalamus to produce GnRH or of the pituitary itself to generate FSH or LH will lead to defective spermatogenesis, although this can be induced by exogenous gonadotrophin treatment or pulsatile administration of GnRH, FSH or LH as the particular case requires. Although testosterone is the high-visibility hormone, its activity is increased fourfold by its conversion to dihydrotestosterone by the enzyme 5-α-reductase. It follows that deficiency of this enzyme leads to incomplete sexual development.

Primary testicular failure (perhaps after mumps virus infection) involving the germinal epithelium results in impaired or absent spermatogenesis, and FSH and LH are usually raised, the latter also if there is damage to the Leydig cells, and so reduced testosterone. Treatment of cancer with cytotoxic drugs frequently damages the germinal cells, and this can be irreversible. For the same reason, the testes must be lead screened during local radiotherapy, such as for bladder cancer.

Spermatogenesis   The quality of spermatogenesis may be assessed in terms of count, morphology and motility of sperm within an ejaculate. A low sperm count is oligospermia; the absence of sperm is azoospermia. It is very unlikely that blood scientists will be performing these analyses; they are inevitably the province of our colleagues in microbiology. However, if sperm generation is abnormal, the blood scientists will be called upon to assist with the endocrine analyses we have studied.

> **Blood science angle: Testosterone**
>
> Men whose concentration of testosterone falls (whether by pituitary, adrenal or testicular disease) are at risk of also having a reduction in haemoglobin and the red cell count, and this may contribute to an anaemia. This is because the hormone is an erythropoietic growth factor.

Testicular cancer   Malignancy of these organs is most common over the ages 20–39, with a lifetime frequency of 1 in 250. However, there are many different types, several of which are benign. The blood scientist is unlikely to be fully involved in diagnosis or management, which is generally clinical (testicular swelling, pain), but rarely a tumour may secrete hCG. In those who have lost both testes to surgery, treatment with testosterone supplements may require serum monitoring.

Male endocrinology as target for therapy The majority of prostate cancers are responsive to therapy aimed at reducing testosterone by influencing the pituitary secretion of FSH and LH secretion. Anti-androgens, such as flutamide, bind to but do not activate the androgen receptor. Finasteride and dutasteride inhibit the testosterone metabolic pathway. A clear consequence of this approach is that spermatogenesis is inhibited and there is a loss of erection and sex drive. There may also be gynaecomastia.

Abnormalities in female reproductive endocrinology

The cyclical nature of the female sexual endocrinology and reproduction, although more complex, automatically makes it more susceptible to pathology than in the male. Nonetheless, the sexes have key features in common, these being the hypothalamus, the pituitary, LH, FSH, the adrenals and paired gonads.

Pathophysiology It is expected that around the ages of 10–12 years girls will start to show signs of puberty. Failure to do so, especially into the teenage years, the most obvious sign being the absence of menstruation (amenorrhoea), will prompt investigation. The same investigations are applicable if otherwise unexplained amenorrhoea (e.g. in the absence of pregnancy) develops in the adult. Amenorrhoea, subfertility or irregular cycling may in part be due to abnormalities in SHBG and the FAI. Since SHBG binds testosterone, low concentrations of the globulin may lead to high free testosterone and so hyperandrogenaemia, causing the symptoms.

Abnormalities in all stages are recognized. The hypothalamus may fail to produce GnRH, possibly because of abnormal developmental processes involving GnRH-secreting neurones. This is often described as hypogonadotrophic hypogonadism, and there is absent secretion of LH and FSH by the pituitary.

The most common causes of pituitary disease include tumours, and LH and FSH secretion may still be present, although at reduced levels. Other tumours causing amenorrhea include those secreting prolactin, ACTH and GH. Malignancies elsewhere in the body may also give rise to secondary tumour deposits (metastases) in the pituitary, leading to reduction in LH and FSH. Rare causes of pituitary failure include ischaemia, resulting from poor blood supply, iron overload (whether by hypertransfusion or haemochromatosis), cranial radiotherapy and

autoimmune disease. Idiopathic hypopituitarism, when no obvious cause can be found, is also recognized, and of course all of these may be present in men. The consequences of these endocrine disorders are failure to stimulate the ovary, and so the absence of oestrogens, progesterone and ovulation (hence ovulatory failure). However, in theory the adrenals are still functioning and so should produce some steroid hormones.

Before the menopause, ovarian failure (defined by failure to produce viable ova), like testicular failure, rarely presents without an allied pathology, such as those listed above. When this happens before the age of 42 it is generally defined as premature menopause. Autoimmune processes can also cause premature ovarian destruction, and both radiation and cytotoxic therapy used in the treatment of common malignancies have this effect, as does hysterectomy. However, in many cases there is no clear cause (i.e. idiopathic).

Numerous other chronic processes can cause amenorrhoea and disrupt ovulation. These include extreme weight loss and low body fat (anorexia nervosa and bulimia, excessive athletics), chronic renal failure, poorly controlled diabetes, both hyperthyroidism and hypothyroidism, and chronic liver disease.

Endometriosis This condition is characterized by the growth of endometrial cells outside (ectopic to) the uterus, the most common site being attached to the ovaries, probably because endometrial cells are dependent on oestrogens. However, endometrial cells may be found anywhere in the pelvic cavity. Should the endometriosis be extensive, it may influence ovarian function and fertility may suffer. Although there are no specific blood tests for endometriosis, downregulation of LH and FSH secretion by a long-acting GnRH analogue will eventually suppress oestrogens and, hence, the proliferation of the endometrial cells.

Polycystic ovary syndrome This is a complex condition, with a broad range of abnormalities (leading to a difficulty in definition) that occur in a variety of combinations (leading to a difficulty in diagnosis). It is one of the most common female endocrine disorders, possibly present in 5–10% of women of reproductive potential, although it may be twice as frequent in South Asians. PCOS is the leading cause of anovulatory infertility, and is perhaps best characterized by factors such as:

• grossly reduced or absent ovulation (generally defined by biochemistry);

- excess androgens (defined by biochemistry, but also implied by symptoms);
- polycystic ovaries (defined by ultrasound).

Despite these changes, PCOS is not always associated with infertility, although the precise diagnosis of PCOS is difficult because of the poor sensitivity and specificity of these three points. As we have seen, many factors cause ovulation failure, whilst androgen-secreting adrenal, pituitary and ovarian tumours can all cause abnormal concentrations of protein and steroid hormones. The cysts in PCOS are not true cysts, but are immature ovarian follicles, but these can be undetectable in a fraction of PCOS women; and conversely, cysts may be present in women who do not have PCOS. Other differential diagnoses include late-onset CAH (11-hydroxylase and 21-hydroxylase deficiency), hypothyroidism, hyperprolactinaemia and Cushing's syndrome.

The precise pathophysiology is also unclear because of the potential for metabolic pathways to overlap and influence each other. Nevertheless, an early crucial step seems to be an abnormal release of GnRH by the pituitary, which leads to increased LH and reduced FSH (leading to a high LH/FSH ratio). There may also be a place for the suppression of FSH by increased inhibin. These changes can explain the reduced follicular maturation, and so chronic anovulation. Increased LH can stimulate the ovaries to increase the secretion of ovarian and adrenal androgens (testosterone, androstenedione and DHEA), leading to hirsutism (male characteristics) and acne. In this setting there may also be a place for measurement of SHBG as the FAI has better sensitivity and specificity than testosterone alone.

There may also be a place for AMH; high concentrations may arise from certain follicles, and this may inhibit the early stages of follicular recruitment and inhibit the actions of FSH, and form a link with amenorrhoea. Conversely, low AMH is often present in women whose amenorrhoea is due to premature ovarian insufficiency.

A parallel pathophysiology involves obesity, with up to 25% PCOS patients in secondary care being overweight (body mass index, BMI $>25 \, kg/m^2$) and approximately 50% being obese (BMI $>30 \, kg/m^2$). This overweight and obesity are of course risk factors for insulin resistance and diabetes, and the latter are present in 40–50% of patients with PCOS. This links with hypertension and with specific atherogenic abnormalities of lipoprotein metabolism (raised triacylglycerols and low high-density lipoprotein cholesterol) and so the metabolic syndrome.

Furthermore, insulin, in concert with LH, acts on the ovary to increase the release of androstenedione and testosterone, and can also downregulate SHBG, which effectively increases the bioavailability of these androgens.

PCOS is also associated with other disease, such an increased risk of gestational diabetes, pregnancy-associated hypertension and pre-eclampsia. Babies tend to be delivered early and are more likely to be admitted to a neonatal intensive care unit, and in later life PCOS is associated with long-term health consequences in terms of diabetes and cardiovascular disease.

Investigations in female reproductive endocrinology

Abnormal sexual development, amenorrhea, irregular menstruation and failure to conceive will all be investigated by measuring a panel of hormones and other molecules. However, the cyclical nature of some of these analytes demands correct timing of the blood sample.

- In a regularly cycling woman, FSH should be measured on days 3–5, expecting a result of $<9 \, IU/L$ in those aged under 30, with concentrations rising with increasing age. A concentration $>15 \, IU/L$ in the follicular phase suggests a perimenopausal state, whilst concentrations $>20 \, IU/L$ in amenorrhea strongly imply ovarian failure or an established menopause.
- In a normal menstrual cycle, LH can be measured as a predictor of ovulation. When measured on a daily basis, a doubling in LH indicates the start of the mid-cycle LH surge. In amenorrheaic women, LH and FSH may be measured at any time, and low concentrations (LH and FSH $<2 \, IU/L$) suggest a hypothalamic or pituitary cause. The ratio between LH and FSH can be useful, as a ratio >2 is suggestive of PCOS.
- In the change to the menopause, days 3–5 oestradiol is unchanged and bears little relation to ovarian reserve. After the menopause, oestradiol values overlap with those encountered in the follicular phase of the normal cycle, whilst FSH values are elevated. Oestradiol has an important use in the monitoring of post-menopausal hormone replacement therapy when this is being given via oestradiol implants.
- In regularly cycling women, a concentration of progesterone greater than $15 \, nmol/L$ in a blood sample 5–9 days before menstruation can be taken as confirmation of ovulation. However, in women with irregular cycles, samples may be taken two or three times a week until menstruation occurs and interpreted retrospectively.

• Measurement of testosterone and SHBG is justified in amenorrhoea, irregular menstruation, subfertility or with signs of androgen excess (the latter as in PCOS). The ratio between testosterone and SHBG gives the FAI, and (except where SHBG is low) gives a measure of free testosterone. This index is preferable to testosterone alone in androgen excess and PCOS. However, the FAI may be raised in androgen-secreting ovarian and adrenal tumours, and late-onset CAH.

• Other useful measurements include androstenedione, 17-hydroxyprogesterone and DHEAS, and are advised if extracted testosterone is >5 nmol/L. Cortisol and ACTH, perhaps with a dexamethasone suppression test, will help clarify the position regarding Cushing's syndrome. AMH has value in estimating ovarian reserve of follicles, and has no cyclical variation. Concentrations are elevated in PCOS but are undetectable after the menopause.

Limitations in assays for steroid and protein hormones

LH, FSH, prolactin and AMH are almost universally measured by immunoassay, perhaps enzyme-linked immunosorbent assay. In almost all cases the reagent manufacturers will provide control samples, but good laboratory practice demands that each laboratory establishes its own reference range.

Oestradiol and testosterone circulate mainly bound to SHBG and albumin with a small amount in the free form; accordingly, the steroid may need to be extracted from its carrier by an organic solvent, such as diethyl ether. Most analysis of these hormones is in unextracted serum, and is amenable to automated immunoassay, although various oestrogen-containing medications contain compounds which cross-react variably with antisera used in oestradiol assays. Unfortunately, solvent extraction can itself lead to problems, a solution to which includes techniques such as liquid chromatography–mass spectrometry.

Multiple endocrine neoplasia

This syndrome is defined by the concurrent presence of more than one endocrine tumour, and is so well established that three types are recognized: MEN 1, MEN 2A and MEN 2B. Several genes are implicated in the aetiology of MENs.

MEN 1 This variant is characterized by the presence of hyperplasia and neoplasia in any combination of parathyroid (present in 95% of cases), pancreatic (30–80%) and pituitary tissues (15–90%), although other endocrine and non-endocrine tumours may rarely be present. It generally has an equal sex distribution, and a frequency of 1 in 30 000 individuals, generally appearing in the fifth decade, with parathyroid disease usually being the first to appear. Strong familial links and an apparent Mendelian inheritance pattern have been strengthened by the discovery of a predisposing gene.

MEN 2 This type is as rare as type 1 (frequency 1 in 30 000), but presentation focuses on thyroid carcinoma, phaeochromocytoma and tumours of other endocrine tissues. There are two subtypes: MEN 2A and MEN 2B. The former is characterized by the addition of primary hyperparathyroidism resulting from parathyroid cells hyperplasia or adenoma. Medullary thyroid carcinoma is generally the first manifestation, and biochemical manifestations of medullary thyroid carcinoma appear between 5 and 25 years of age. MEN 2B is more aggressive, but accounts for only 5% of all cases of MEN 2. There is a greater likelihood of tumours of the nervous system and intestines, but not of the parathyroids, and also somatic disease, such as a decreased upper/lower body ratio and skeletal deformations. Accordingly, morbidity and mortality are higher in patients with MEN 2B than in patients with MEN 2A.

The laboratory in multiple endocrine neoplasia

Blood scientists are likely to help in the differential diagnosis of MEN by providing evidence of abnormal release of the hormone products of different tumours.

Molecular genetics of endocrine disease

Many of the conditions we have been examining arise from neoplasia, inflammation or other pathologies. However, there are many whose aetiology can be traced to a defined genetic lesion, which will have been present from birth (but not always recognized). Cases of the adult acquisition of genetic disease (therefore, a somatic mutation) are very rare, and almost all are associated with a malignancy.

Chromosomal abnormalities More fundamental abnormalities leading to gonadal failure are of chromosomes. In males, the most prominent is Klinefelter's syndrome, due to an extra X chromosome, giving the karyotype 47-XXY. Testicular interstitial cells are reduced in number and testosterone secretion is reduced, secondary sex characteristics are less well developed and gynaecomastia may be present. Seminiferous tubules are abnormal and sperm production is absent. Partial deletions of the Y chromosome lead to low sperm counts.

In women, the leading chromosomal abnormality is Turner's syndrome, in which there is absence of an X chromosome, leading to the karyotype 45-XO. There are several phenotypic manifestation of this condition, such as a short stature and a webbed neck. As regards reproduction, there is complete ovulatory failure, and so amenorrhea, but oestrogen replacement therapy can be used to assist secondary sexual development in the teenager. The uterus is effectively normal, and some women with Turner's syndrome can produce a healthy child by assisted reproduction and donated ova.

Genetic mutations There is a considerable catalogue of mutations associated with particular endocrine disorders.

- Addison's disease has been linked to mutations in the gene for steroidogenic factor 1 (SF1), which (curiously) controls reproduction, the *DAX-1* gene, possibly related to glycerol kinase, and to mutations in the genes for the ACTH receptor. However, the most common and strongest link is with genes coding for enzymes involved in steroid synthesis (CAH, more on which is to follow).
- Cases of acromegaly have been described where the disease is caused by abnormalities in the promotor region for GH, leading to high concentrations in the blood and so the clinical signs.
- A genetic component to endometriosis is implied by a 10-fold increased risk with an affected first-degree relative (mother or sister). Candidate genes lie on chromosome 10 and on chromosome 7.
- Similarly, familial clustering implies an autosomal dominant genetic lesion in PCOS, buts its precise location remains obscure.
- Mutations in several genes, notably the von Hippel–Lindau tumour suppressor, and the four genes (SDHx) coding for subunits of succinate dehydrogenase, have been linked to phaeochromocytomas. However, perhaps the best characterized lesion is that of the *RET* proto-oncogene.
- Most cases (40–45%) of papillary thyroid cancer are characterized by mutations in the *BRAF* gene, leading to an altered V600E molecule that participates in cell signalling. Follicular and papillary carcinomas and follicular adenomas may be associated with point mutations in *RAS*, which encodes for another signalling molecule. These may be used to guide post-surgical management and risk stratification.

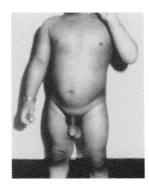

Figure 18.10 Male precocious puberty. Molecular genetic techniques discovered a gain-of-function mutation in the gene encoding the LH receptor in this patient. The consequences of this are the hyper-reactivity to high testosterone from Leydig cells, and so the development of pubic hair and genitals. The patient is 2 years old but is the size of a 4-year-old. (From Holt & Hanley, Essential Endocrinology and Diabetes, 6th edition, 2012, Fig. 3.13, p. 39. Reproduced with permission of John Wiley & Sons, Ltd.).

- A gain-of-function mutation in the gene encoding the LH receptor can lead to increased levels of high testosterone secretion from Leydig cells, and so the development of pubic hair and genitals (Figure 18.10).

The genetics of multiple endocrine neoplasia MEN has been linked to *MEN1*, *RET* (as in phaeochromocytomas) and *TRK1*, which codes for a kinase receptor. *MEN1*, restricted to MEN type 1, has been located on chromosome 11, cloned and sequenced. Its product is the Menin protein, and reputed to be a tumour suppressor. A gain in function mutation in *RET* and an unknown mutation in *TRK1* are more likely to be found in MEN 2A and 2B. These mutations may be sought in combined tumours and/or hyperplasia of the parathyroids, pituitary, thyroid and pancreas.

Intriguing as these finding are, they are most unlikely to occupy a great deal of the workload of the blood scientist working in molecular genetics. However, our final section may well provide some diversion.

Genetics of steroid production Returning once more to Figure 18.4, the synthesis of steroid intermediates is controlled by genes for certain enzymes. The most common abnormality results from a mutation in *CYP21A2* on chromosome 6 that codes for 21-hydroxylase, an enzyme of relative mass 52 kDa. A defect in this gene causing 21-hydroxylase deficiency leads to 95% of cases of CAH.

The genes for 11-hydroxylase (*CYB11B1*) and for aldosterone synthase (*CYP11B2*) are close together on chromosome 8 and have 95% homology. Deficiency in the former is found in many cases of CAH (5% of cases), whilst the latter leads to aldosterone deficiency. Mutations in the genes for 3-β-hydroxysteroid dehydrogenase (on chromosome 1), desmolase (chromosome 8) and for 17-hydroxylase (chromosome 10) are linked with CAH very rarely.

However, these mutations are likely to be of interest in defining the exact type of MEN or CAH. Generally, the diagnoses of these diseases are made using conventional clinical and laboratory methods.

18.3 Case studies

Case study 24

A 46-year old woman, weighing 70 kg, presents to her general practitioner (GP) early in the morning on her way in to work with a history of abdominal pain and vomiting, irregular urination (sometimes little flow, at other times great flow) and weight loss. On examination her blood pressure was 115/67 mmHg, pulse rate 80 beats per minute. Her urine was negative for glucose and ketones. Blood was taken for routine biochemistry on the GP's bench-top analyser.

	Result	Reference range
Na	126 mmol/L	135–145
K	5.7 mmol/L	3.8–5.0
Urea	11.9 mml/L	3.3–6.7
Creatinine	146 µmol/L	71–133
Bicarbonate	20 mmol/L	24–29

Interpretation There is hyponatraemia, hyperkalaemia and a degree of acute renal disturbance (raised urea greater than raised creatinine), and the estimated glomerular filtration rate (eGFR) is a worrying 36 mL/min per $1.73\,m^2$, giving chronic kidney disease (CKD) stage III. Low bicarbonate implies an acidosis. The low blood pressure suggests hypotension and dehydration, and the GP immediately started her on a saline drip, and sent blood to the local hospital for cortisol. After the infusion, the woman was kept in the practice, and 6 h later bloods were repeated. All results had improved, and some had resolved.

Next day the hospital e-mailed the cortisol result of 50 nmol/L (reference range 200–800 nmol/L), which strongly implies adrenal failure, quite possibly Addison's disease. The electrolyte disturbances are very likely to be due to failure of the adrenals to generate aldosterone, which can also be confirmed with a blood test. The GP referred the woman to the local endocrinologist for additional investigations (such as the synacthen test), determination of the cause (such as autoimmune adrenal attack) and treatment (such as replacement of the absent cortisol with oral hydrocortisone).

Case study 25

A 63-year-old man presents with his wife to their GP, who between them describe a vague 6-month history of an assortment of symptoms that include bone pain, impatience at mealtime, thirst and constipation. The wife also said he easily lost his temper and was often forgetful. Routine biochemistry tests were sent to the local hospital.

	Result (Unit)	Reference range
Na	136 mmol/L	135–145
K	4.7 mmol/L	3.8–5.0
Urea	6.9 mmol/L	3.3–6.7
Creatinine	106 µmolL	71–133
Bicarbonate	25 mmol/L	24–29
Calcium	2.88 mmol/L	2.2–2.6
Albumin	39 g/L	35–50

Interpretation The urea is slightly raised, the creatinine level gives an eGFR of 65 ml/min per $1.73\,m^2$ and CKD stage II. The calcium is high, prompting an investigation of hypercalcaemia. On their own initiative, the laboratory performed a PTH measurement, and found a result of 15 pmol/L (reference range 1–6 pmol/L), strongly implying hyperparathyroidism, which scanning suggested was due to an adenoma, confirmed by biopsy.

This diagnosis alone did not fully satisfy the GP, who asked the patient to return in the morning fasted for blood glucose. This test returned a result of 1.8 mmol/L (reference range 3.5–5.5 mmol/L), prompting the measurement of insulin, which as 18 mU/L (reference range 2–10 mU/L) led to the diagnosis of an insulinoma. Thus, with two separate endocrine tumours, the patient was ultimately diagnosed by the consultant endocrinologist with MEN type 1.

Summary

- Hormones are chemical messengers produced by endocrine glands that circulate in the blood and whose concentrations are regulated by positive and negative feedback.
- The hypothalamus releases factors that in turn stimulate the pituitary to release other hormones that act on other endocrine glands.
- Common pathological processes leading to overactivity or underactivity include neoplasia, inflammation and autoimmunity. When more than one endocrine tumour is present, a diagnosis of MEN is possible.
- Treatment of endocrine gland overactivity includes surgery and chemotherapy; treatment of underactivity is to replace the missing hormone.
- Anterior pituitary overactivity generally results in high concentrations of prolactin, GH and ACTH causing hyperprolactinaemia, acromegaly and ACTH-dependent Cushing's syndrome respectively. Underactivity (hypopituitarism) leads to reduced concentrations of these hormones, and therefore other disease.
- Posterior pituitary dysfunction can result in low AVP, which in turn causes diabetes insipidus.
- TSH from the anterior pituitary acts on the thyroids, promoting T4 and T3, which are required for normal cellular metabolism.
- Hyperthyroidism, resulting from high T3 and T4, leads to thyrotoxicosis, which can manifest as exophthalmia, whilst low concentrations lead to hypothyroidism, the most extreme clinical feature being myxoedema.
- The adrenal cortex (outer region) produces aldosterone, cortisol and androgens. The inner region, the medulla, produces the catecholamines: adrenaline, noradrenaline and dopamine.
- Adrenal overactivity leads to Cushing's syndrome, whilst underactivity leads to Addison's disease. Conn's syndrome is the product of an aldosterone-secreting tumour, whilst phaeochromocytomas secrete catecholamines.
- Reproduction in the male and the female have common features: hypothalamic GnRH acts on the pituitary, which releases LH and FSH that act on the ovary and the testis.
- Ovaries secrete oestrogens and progesterone that together drive sexual development, ovulation, promote fertilization and the successful implantation of a zygote. This continues until the exhaustion of the ovarian follicle pool at the menopause.
- Under the influence of testosterone, testes produce a continuous supply of very large numbers of healthy motile sperm.
- Ovulation is the result of a roughly monthly cycle, whereas spermatogenesis is a continuous process, although in both sexes the pathological processes involving the hypothalamus and pituitary lead to cessation of gonadal function.
- Common laboratory blood and urine tests for hormones include immunoassay, tandem mass spectrometry and chromatography.

Further reading

Almeida MQ, Stratakis CA. Solid tumors associated with multiple endocrine neoplasias. Cancer Genet Cytogenet. 2010;203:30–36.

Bernichtein S, Touraine P, Goffin V. New concepts in prolactin biology. J Endocrinol. 2010;206:1–11.

Boelaert K, Franklyn JA. Thyroid hormone in health and disease. J Endocrinol. 2005;187:1–15.

Dauber A, Kellogg M, Majzoub JA. Monitoring of therapy in congenital adrenal hyperplasia. Clin Chem. 2010;56:1245–1251.

Fung MM, Viveros OH, O'Connor DT. Diseases of the adrenal medulla. Acta Physiol. 2008;192:325–335.

Ghayee HK, Auchus RJ. Basic concepts and recent development in human steroid biosynthesis. Rev Endocrin Metabol Disorders. 2007;8:289–300.

Findling JW, Raff H. Cushing's syndrome: important issues in diagnosis and management. J Clin Endocrinol Metab. 2006;91:3746–3753.

Hammond GL. Diverse roles for sex hormone-binding globulin in reproduction. Biol Reprod. 2011;85:431–441.

Karasek D, Shah U, Frysak Z. *et al.* An update on the genetics of pheochromocytoma. J Hum Hypertens. 2013;27:141–147.

Melmed S. Update in pituitary disease. J Clin Endocrinol Metab. 2008;93:331–338.

Melmed S. Acromegaly pathogenesis and treatment. J Clin Invest. 2009;119:3189–3202.

Nygaard B. Hyperthyroidism. Am Fam Physician. 2007;76:1014–1016.

Roberts CG, Ladenson PW. Hypothyroidism. Lancet. 2004;363:793–803.

Teede H, Deeks A, Moran L. Polycystic ovary syndrome. BMC Med. 2010;8:41.

Young WF Jr. Minireview: primary aldosteronism – changing concepts in diagnosis and treatment. Endocrinology. 2003;144:2208–2213.

Whiting MJ. Simultaneous measurement of urinary metanephrines and catecholamines by liquid chromatography with tandem mass spectrometric detection. Ann Clin Biochem. 2009;46:129–136.

Web sites

http://www.gpnotebook.co.uk/simplepage.cfm?ID=-1831862252

http://endocrinology.com/

19 Cancer and Tumour Markers

Learning objectives

After studying this chapter, you should be able to:

- discuss key aspects of the biological basis of cancer;
- explain the key features of tumour markers and their use;
- describe the main tumour markers measured in clinical practice;
- review the role of the laboratory in diagnosis and management of cancer patients;
- explain the role of molecular genetics and pharmacogenomics in cancer.

In this chapter, we will first describe the nature of cancer, its biological basis and explain some of the terminology used in Section 19.1. This will lead, in Section 19.2, to a discussion of the value of measuring tumour markers in management of cancer patients and describe the major tumour markers encountered in clinical practice. We conclude in Section 19.3 with a brief look at the molecular genetics of cancer markers, and then some case studies in Section 19.4.

19.1 General concepts in cancer biology

After cardiovascular disease, cancer is the second most common cause of death. In the UK, each year, perhaps 1.25 million people have a diagnosis of cancer, and well over 100 000 will die of the disease. However, these crude figures mask a huge variety in the frequency and adverse consequences of individual cancers over a lifespan (Table 19.1).

Nomenclature and classification

The suffix-oma tells us of an abnormal growth. Some describe an abnormal growth as a tumour, others reserve *tumour* for a mass of cells that will ultimately cause ill health. If the cellular basis of this is of fat cells (adipocytes, perhaps loaded with triacylglycerols), we would describe it as a lipoma. There are (generally) no major adverse health consequences of an isolated lipoma, but growths in other tissues may ultimately be fatal. We can classify cancer according to the cellular basis of the tumour; for example:

- Carcinoma – a tumour based on epithelial cells. These are common in the lung and intestines as these organs are effectively modified epithelium.
- Sarcoma – a (considerably less common) tumour of connective and structural tissues (bone, muscle). If the cancer is of the bone, we describe it as an osteosarcoma, if of striated muscle it is a rhabdomyosarcoma.
- Adenoma – a tumour of glandular tissue that therefore has the ability to secrete excess levels of its healthy product.

Further classifications look at potential pathology; if of minor importance it is benign, whilst others directly leading to disease are malignant. Transformation from a benign tumour to a malignant tumour is common. Indeed, many an adenoma transforms to a malignant form, in which case it may be described as an adenocarcinoma. Tumours that spread from their original location (such as the breast or prostate) to another organ or tissues (such as bone) are said to be metastatic.

Tumour biology

All cancers ultimately arise from abnormalities in the DNA of the individual. These may in turn arise from a number of factors, such as:

- Radiation, as clearly demonstrated by the events in Japan after 1945, and in workers in the nuclear industry.

Blood Science: Principles and Pathology, First Edition. Andrew Blann and Nessar Ahmed.
© 2014 John Wiley & Sons, Ltd. Published 2014 by John Wiley & Sons, Ltd.

Table 19.1 Cancer deaths in 2010.

	Age band	Men	Women	Men/women difference
All neoplasia	All ages	74 267	67 179	1.10
(C00–D48)	35–44	1106 (1)	1396 (1)	0.79
	55–64	10 967 (10.8)	9567 (6.8)	1.15
	75–84	24 885 (22.5)	20 834 (14.9)	1.19
Breast	All ages	63	10 290	0.006
(C50)	35–44	0	490 (1)	—
	55–64	9	1815 (3.7)	0.005
	75–84	13	2545 (5.2)	0.005
Colorectal	All ages	7700	6402	1.20
(C18–C21)	35–44	82 (1)	83 (1)	0.99
	55–64	1174 (14.3)	661 (8.0)	1.78
	75–84	2575 (31.4)	2153 (25.9)	1.20
Testis/ovary	All ages	62	3676	
(C62/C56)	35–44	11 (1)	58 (1)	
	55–64	6 (0.54)	695 (12.0)	
	75–84	3 (0.27)	1028 (17.7)	
Prostate/cervix	All ages	9638	816	
(C61/C53)	35–44	6 (1)	106 (1)	
	55–64	563 (93.8)	123 (1.16)	
	75–84	3943 (657.1)	145 (1.37)	

Data beneath tumour type is the International Classification of Disease-10 code. Numbers are numbers of deaths. Data in parentheses is proportion of the death in the two older groups compared with the youngest ($=1$). Considering all neoplasia, the incidence clearly rises with age, but younger women have a higher incidence than younger men, which is reversed later in life. Note also that the incidence of cancer of the testes falls with age, but is far less frequent than ovarian cancer. However, the incidence of prostate cancer far exceeds that of cervical cancer.

• Oncogenic viruses, an excellent example being the Epstein–Barr virus in Burkitt's lymphoma (Chapter 6) and the retrovirus HTLV-1 causing adult T lymphocyte leukaemia.
• Carcinogenic hydrocarbons, such as the high prevalence of certain cancers in particular industries.
• Curiously, physical inactivity, which causes 10% of breast cancer and 10% of colon cancer.

However, several of these may interact. The cellular basis of cancer includes uncontrolled growth, which is a type of neoplasia. This is unrelated to hyperplasia, which is generally the noncancerous expansion of an organ, inevitably the result of a particular disease process. Problems arise principally from two areas. One is that the abnormal growth interferes with the functions of other organs, such as the occlusion of a tube, an example of which is a liver tumour that may press on the bile duct and so cause, and/or contribute to cholestasis and so jaundice.

A second aspect of cancer is that the particular tissue becomes overactive or underactive. Since the common basis of many neoplasias is genetic, this translates to gain-of-function or loss-of-function gene mutations respectively. Where the organ secretes or otherwise produces a particular product that has biological activity (such as a hormone), this may have important pathophysiological repercussions. However, it may be that the gene mutation produces a molecule that has no biological activity, and in these cases the abnormalities simply mark the cancer. These molecules are described as tumour markers.

Clinical oncology

A further aspect of tumour biology is treatment, which has three arms: surgery, chemotherapy and radiotherapy. Clearly, the former involves physical removal of the tumour. If the tumour was secreting a particular molecule, concentrations should fall. Much chemotherapy is simply sophisticated poison, which aims to poison the tumour (ideally, to its death). However, it can be difficult to target these cytotoxic drugs to the tumour, so that healthy tissue is also poisoned. Examples of damage done by cytotoxic drugs include:

- Jaundice, as assessed by levels of bilirubin, and a consequence of damage to liver. Other liver function tests are also likely to be increased
- Anaemia, bruising and bleeding, and infections. These are the consequences of suppression of the bone marrow and so reduced numbers of red cells, platelets and white cells respectively.
- Diarrhoea, the result of destruction of the epithelial cell line of the intestines, although there are no specific blood tests.
- Renal failure caused by attack on the cells of the nephron, principally the glomerulus, with raised urea and electrolytes, and so a high estimated glomerular filtration rate (eGFR).

Fortunately, the pharmaceutical industry is providing better, more targeted drugs with much reduced toxic side effects. Some (such as monoclonal antibodies) are directed to molecules specifically on the surface of tumours. Others, such as tyrosine kinase inhibitors (potent in certain types of leukaemia) target metabolism. In modest doses, targeted radiotherapy, which uses X-rays to destroy cancerous cells, is relatively nontoxic. However, at high doses, nearby healthy cells can be damaged. In all these situations, the laboratory provides key information regarding the cancer molecule(s) that is likely to have implications for diagnosis and/or management.

Tumour markers

The concept of raised concentrations of a particular molecule feeds the notion of a tumour marker. This considers whether or not the molecule is found only in the cancer and is not found in health, and whether or not it is found in other conditions. These two requirements are called sensitivity and specificity.

Sensitivity This quality refers to the degree to which the test can correctly identify those people who have the disease and those who have not got the disease. The former are called true positives and the latter true negatives. However, there are no tests that are 100% successful; there will inevitably be a number of people who have the disease but are not detected as such by the test – the false negatives. Hence:

$$\text{Sensitivity} = \frac{\text{true positives}}{\text{true positives plus false negatives}}$$

So if a new test correctly identies 17 of 20 people with a certain cancer, then the sensitivity is 85%. This may have serious consequences for the three people incorrectly labelled, as they will be denied potentially life-saving treatment.

Specificity This test is the reverse – it refers to the ability to identify those people who have not got the disease, the true negatives. But once more, no test is perfect, and there will inevitably be some people who are incorrectly given a clean bill of health when they actually have the disease. These people are false negatives. Hence:

$$\text{Specificity} = \frac{\text{true negatives}}{\text{true negatives plus false positives}}$$

Similarly, the test may correctly identify 19 out of 20 people are healthy (so that the specificity is 95%). So if we used this test alone to guide management, we may be giving a dangerous cytotoxic drug to one person who has not got the cancer.

Clinical value On balance, we want a test with the highest sensitivity and specificity. However, the results of almost all laboratory tests are continuously variable, such as a number between 20 and 200 mmol/L. It follows that the choice of a particular number that decides yes or no (such as 50 mmol/L) will be crucial. Generally, the lower we set this number, the more likely we are to include all the subjects with the disease, but we will also incorrectly label so many people as having the disease (false positives). Conversely, the higher we set the yes/no cut-off point (such as 90 mmol/L), the more likely we are to exclude those free of the disease, but we will incorrectly label some who have the disease as being healthy (false negatives). So the hard part is in deciding which particular cut-off point is best in minimizing false positives and false negatives. However, further mathematics gives other useful information about a test, such as the likelihood ratio, the positive predictive value and the negative predictive value.

The ideal tumour marker could be used in screening, diagnosis, prognosis, staging, management and monitoring of cancer. No tumour marker currently in use fulfils all of these criteria. This is why few practitioners will make a diagnosis based on the laboratory result alone. Figure 19.1 provides an overview of the use of tumour markers in clinical practice.

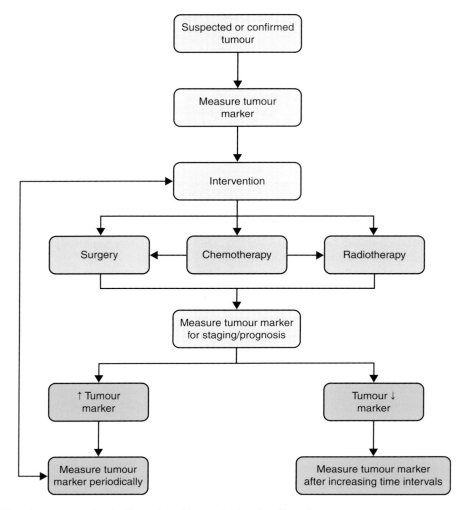

Figure 19.1 Use of a cancer marker in diagnosis and in monitoring the effect of treatment.

19.2 Blood science and cancer

There are numerous examples of changes in blood cells and molecules in a patient with cancer. In some cases, these are tumour specific, and these generally have precise pathological consequences. In other situations, changes are nonspecific for either the particular cancer or particular consequences. We can consider these changes in a number of ways, such as:

- changes with no clear biological effect
- changes that do have an effect on physiology.

Markers with no clear biological effect

An established feature of many cancers is the production of 'nonsense' molecules, and molecules that had a role early in life but which are irrelevant in the adult, the latter being called oncofoetal proteins. Examples of this are carcinoembryonic antigen (CEA) and alpha-fetoprotein (AFP). Other molecules are expressed specifically on the surface of cancer cells, and can be shed or cleaved off by enzymes so that the molecule appears in the blood. Often, concentrations in the blood reflect the size of the tumour. Examples of these tumour-specific antigens include those of the carbohydrate antigen (CA) series, and several others.

Carbohydrate antigen 15.3 (CA 15.3, also known as MUC1) is a glycoprotein expressed by epithelial cells in the bladder, stomach, pancreas, ovary, respiratory tract and breast. In normal breast tissue, MUC1 is shed into breast milk. In breast tumours, the loss of normal tissue organization causes MUC1 to be shed into the circulation, where it can be detected by immunoassay. However, poor sensitivity and specificity makes CA 15.3 unsuitable for screening for breast cancer. Up to 5% of healthy individuals may have slightly raised CA 15.3 concentrations; raised levels may also be seen in liver disease and in patients with other types of advanced adenocarcinomas. Nevertheless, CA 15.3 may have value as a prognostic marker in breast cancer.

Carbohydrate antigen 19-9 (CA 19-9), a form of the Lewis blood group antigen, may be useful in earlier detection of pancreatic cancer, but its value is limited by low sensitivity and specificity. Raised concentrations are seen in obstruction of the bile duct and in chronic liver disease. In addition, 5–10% of the population are Lewis genotype negative and, therefore, do not express CA 19-9. The low prevalence of pancreatic cancer also ensures a low positive predictive value for the test, meaning it has little use as a screening tool. However, once the diagnosis has been made, it may be useful in monitoring recurrence of cancer after surgery.

Carbohydrate antigen 125 (CA-125) is a glycoprotein present on the peritoneum, pleura, gastrointestinal tract and female reproductive tract. Modestly raised serum concentrations are present in menstruation and pregnancy, endometriosis, peritonitis or cirrhosis. However, high concentrations are often found in women with intra-abdominal malignancies, both ovarian and non-ovarian. The lack of specificity limits its role in aiding differential diagnosis and in screening for specific cancers. Despite this, once the disease is known to be present, CA-125 may be helpful in monitoring treatment (surgery and chemotherapy) and in predicting relapse. Regarding the latter, increasing CA-125 may predict clinical relapse in perhaps 70% of patients, enabling salvage chemotherapy on the assumption that enhanced clinical outcomes can then be achieved. But caution is required as concentrations may rise within 2 weeks of treatment due to release of CA-125 from damaged cancer cells. CA-125 measurement in primary care is now recommended by the National Institute of Health and Clinical Excellence for women with some symptoms of ovarian cancer – targeted screening.

CEA is a highly glycosylated cell surface glycoprotein originally found on foetal colon and adenocarcinomas but not in healthy colon. However, raised serum concentrations are also found in breast, lung, gastric and ovarian cancer, as well as alcoholic liver cirrhosis, gastrointestinal inflammatory diseases and in smokers. Indeed, low sensitivity and specificity in detecting the earliest stages of colorectal cancer (Dukes' stages I and II) limits its use as a screening test and as a diagnostic aid.

Despite these potential problems, CEA may be useful as a prognostic indicator, as patients with a high preoperative concentration have a worse outcome. After successful surgical treatment of cancer, CEA concentrations should return to normal levels within 4–6 weeks. Failure to fall to normal concentrations is associated with early recurrence of disease at the original site, but rising concentrations may indicate distant metastatic disease or recurrence at the same site after incomplete removal.

Prostate specific antigen (PSA), a serine protease with a relative molecular mass of perhaps 34 kDa, was originally thought to be synthesized solely by epithelial cells in the prostate, but is now also recognized to be present in sweat glands, endometrium, salivary glands and breast tissue. Serum PSA is raised in patients with prostate diseases such as benign prostatic hypertrophy (BPH) and prostate cancer, but also in acute urinary retention, acute prostatitis and prostatic ischaemia. However, although the role of PSA as a screening tool is controversial, it may be useful in plotting potential disease, such as the transformation of BPH to prostate cancer, and the potential return of disease after surgical treatment (radical prostatectomy).

AFP is a large glycoprotein synthesized by foetal yolk sac, liver and intestine, but concentrations are undetectable after the first year of life. AFP may be measured in pregnancy, as high concentrations may indicate foetal neural tube defects. However, the major use of AFP is in the diagnosis and management of hepatocellular carcinoma, although it may also be raised in cirrhosis, tumours of Sertoli and Leydig cells of the testes (seminomas), Wilms' tumour, colitis and ataxia telangiectasia.

Human chorionic gonadotrophin (hCG) is a 45 kDa glycoprotein secreted by placental trophoblasts. It is related to the hormones LH, FSH and TSH, as all share their α-subunit with hCG. In the absence of pregnancy, hCG is raised in ovarian and testicular germ-cell tumours, extragonadal germ-cell tumours and choriocarcinoma. The latter is a rare (1 in 50 000) and highly malignant tumour that originates in the outermost of the membranes surrounding a foetus (the chorion). Fortunately, treatment with chemotherapy is highly successful in such tumours, but hCG concentrations are measured serially for a year to guard against recurrence.

Neither AFP nor hCG are useful as population-screening tools for germ-cell tumours as their sensitivity and specificity are too low. They are both raised in a variety of other conditions, thus decreasing their specificity for germ-cell tumours. Both may aid in diagnosis, but again their sensitivity and specificity are not good enough for use as sole diagnostic tools, where biopsy is far more important, and both AFP and hCG can be used to predict prognosis, and the elevation of AFP immediately before chemotherapy is especially useful.

Paraproteins B lymphocytes construct and release antibody (immunoglobulin) molecules, a process which is normally very efficient. However, in several circumstances the production of the antibody is aberrant, leading to abnormal sections of antibody molecules (heavy chains and light chains) being present in the blood. These abnormal proteins are called paraproteins, and increased concentrations are often found in malignancies of B lymphocytes, the dominant being myeloma. They inevitably cause increased plasma viscosity and a raised erythrocyte sedimentation rate (ESR), although the latter is not normally thought of as a tumour marker.

Paraproteins can appear in the urine as Bence–Jones protein, and these inevitably mark a malignancy. Full details of B cell cancers and paraproteins, and the ratio between different types of light chains, are presented in Chapters 6 and 9.

Blood science angle: Paraproteins

These abnormal proteins may be measured in biochemistry, immunology or haematology laboratories, and each sub-discipline has its own interest. For example, patients with myeloma may also have raised serum calcium and low haemoglobin. The high protein load in paraproteinaemia may lead to renal failure, which can be monitored by urea and electrolytes.

Markers with a biological effect

Tumours, especially those of the endocrine system (Chapter 18), can secrete increased amounts of their normal product. These can have profound, often catastrophic, pathological consequences. In other instances, the effects of tumours can also have biological effects. These substances include:

- *Insulin.* High concentrations secreted by an insulinoma can produce a profound, often fatal hypoglycaemia. However, an autoantibody to beta cells of the islets can also stimulate the production of the hormone, and this is an important differential diagnosis that only the laboratory can make. Occasional tumours may secrete insulin-like factors that have the same effect as the normal molecule.
- *Prolactin.* This hormone is increased in patients with prolactinomas. As it promotes milk production after childbirth, inappropriate milk secretion (galactorrhea) in either sex may result from hyperprolactinaemia and so a pituitary tumour.
- *Growth hormone.* Raised in, and causing, acromegaly, excessive levels arise from an overactive pituitary, which may be malignant.
- *Catecholamines adrenaline and noradrenaline.* Concentrations of the hormones are high in patients with phaeochromocytomas. The latter is a tumour of the adrenal medulla releasing increased adrenaline and noradrenaline, which have numerous biological effects such as high pulse rate, sweating and palpitations. Notably, several of these symptoms are also found in the mutual differential diagnosis of hyperthyroidism.
- *Adrenocorticotrophic hormone (ACTH).* Ectopic and pituitary tumours may also be active in the secretion of increased ACTH by the adrenal glands, which will drive increased cortisol and so a Cushingoid syndrome.
- *Arginine vasopressin (AVP, also known as anti-diuretic hormone).* High levels may be secreted by a pituitary adenoma or an ectopic tumour (such as may occur in small cell lung cancer). Excess AVP leads to water retention and low serum sodium, and so to the syndrome of inappropriate antidiuretic hormone, but it can also occur in patients without malignancy or by certain drugs.
- *Calcium, leading to hypercalcaemia.* This may be caused by primary hyperparathyroidism, but may also be caused by mobilization of calcium in bone by a primary bone cancer (such as myeloma) or by secondary cancer (metastases) from many different cancers, such as those of the lung, breast, ovary and prostate. Thus, in respect of the latter two cancers, raised calcium may be present with a raised CA-125 and PSA respectively. Hypercalcaemia has a biological effect in that raised calcium can cause cardiac arrhythmia and arrest. Concentrations of parathyroid hormone (PTH) should be measured to exclude primary hyperparathyroidism. Additional details are present in Chapter 15.

Table 19.2 Bowel cancer.

We have already noted the value of levels of CEA in the blood in colon cancer and adenocarcinomas but not in healthy colon, but that levels are also raised in other cancers. There are three other tests of note, although these are not blood tests.

Faecal occult blood
Blood in the faeces is likely to arise from an intestinal pathology such as a malignancy. Indeed, this relationship is so strong that faecal occult blood has been promoted as a tumour marker, and may have a value in population screening of selected populations, such as the elderly, for colorectal cancer. However, the test is not recommended in younger patients. If symptomatic, colonoscopy is recommended owing to low sensitivity and specificity of test.

Faecal pyruvate kinase isoenzyme type M2 (faecal M2-PK)
This molecule, an isoenzyme of a key protein in glycolysis, is overexpressed in tumour cells and can be readily detected in faeces. A meta-analysis indicated superior sensitivity and specificity for colorectal cancer when compared with faecal occult blood, leading to possible use as a community screening tool.

Mucus
Goblet cells of the large intestines (mostly the rectum) secrete mucopolysaccharides (mucus) to help lubricate the passage of the faeces. In certain cases an intestinal adenocarcinoma will oversecrete this lubricant, leading to a great deal of this mucus being passed out via the anus. The blood scientist may be called on to analyse secretions such as this.

Combining markers

Many patients with cancer often have several signs and symptoms of a developing tumour, and the laboratory can often help with confirming or refuting a preliminary diagnosis, as can imaging such as X-ray and ultrasound scanning. However, researchers are constantly providing additional tools to help refine this process, an excellent example of which is ovarian cancer, the fifth most frequent female cancer that kills over 140 000 women worldwide each year.

As indicated above, the gold standard laboratory marker of this disease is CA-125. A cut-off point of 35 IU/mL may be used to help direct further analyses (such as ultrasound of the abdomen and pelvis) to refine a diagnosis in a woman with signs and symptoms of this cancer. The problem is that this cut-off point is bound to include some false positives and false negatives: raised concentrations of CA-125 are found in only 50% of women with early (stage 1, limited to the ovaries) disease. This has led to the search for other tumour markers.

Human epididymis protein-4 (HEP4, a small peptide coded for by a gene on chromosome 20q12–13.1), is expressed by a large number of normal tissues but is upregulated in ovarian cancer. The development of an enzyme-linked immunosorbent assay (ELISA) to HEP4 has been used to show raised concentrations in the serum of women with ovarian cancer and to demonstrate its superiority over CA-125 as a marker of this neoplasia. However, as both markers are raised in women with epithelial ovarian cancer, the combined use of both markers is superior to the use of either marker alone, and identifies over 93% of women with this disease. Blood tests can also be combined with other signs and symptoms to help diagnose cancer (Table 19.2).

Haemato-oncology

This small group of important diseases are characterized by the abnormal production of blood cells. Chapter 4 describes erythroleukaemia and polycythaemia rubra vera, Chapter 6 has details of malignancies of white blood cells (such as the leukaemias), whilst the pathology of essential thrombocythaemia is part of Chapter 8. We have already looked at serum paraproteins and urinary Bence–Jones protein as markers of myeloma.

In each case, it could be argued that high numbers of red cells, white cells and platelets act as their respective cancer marker. As regards functionality, leukaemic cells fail to provide immunological defence, whereas the increased platelet count may lead to both thrombosis and (if the platelets are nonfunctional) haemorrhage. The appearance of plasma cells in the blood is a marker of a poor prognosis and rapid disease progression in myeloma. Likewise, blast transformation in chronic leukaemia marks a more aggressive acute phase that demands equally aggressive treatment.

Table 19.3 Selected cancer markers.

Marker (unit)	Reference range	Equivocal result	Unequivocal result
CA-125 (U/mL)	<35	35–50	>75
CEA (ng/mL)	<2.5	4–10	>15
PSA (ng/mL)[a]	<4	4–10	>15
AFP (μg/L)	<20	25–50	>75
hCG (mIU/mL)	<5	10–25[b]	>35[b]
Prolactin (mU/L)	Men <350, women <500	1000–9000[c]	>12 000
Insulin (pmol/L)	50–300[d]	100–450	>500
(mU/L)	2–10	12–15	>20
Calcium (mmol/L)	2.2–2.6	2.8–3.2	>3.5
CA 15.3 (kU/L)	<25	30–40	>50

N.B. In many cases, reference ranges are specific for only one particular method (often an ELISA), so that care is required in interpretation. Attention must be paid to details of the expected values provided by the manufacturer of the method.
[a]PSA reference range increases with age.
[b]Concentrations when pregnant are often >1000.
[c]May reach 8000 in pregnancy.
[d]Strongly influenced by recent feeding: note alternative reference ranges.

In theory, a high red cell count may provide increased capacity to carry oxygen. Indeed, there is the case of the Finnish cross-country skier Eero Antero Mäntyranta who was reputed to be able to take advantage of his erythropoietin-secreting tumour and the increased exercise capacity it provided as a high red cell count.

Interpretation

The use of cancer markers by themselves is rarely recommended as many do not have a fully adequate sensitivity and specificity profile. Other information, such as imaging and histology, is required to make a robust diagnosis. Nevertheless, there are instances where the tumour marker is so grossly abnormal that differential diagnoses are unlikely. It is for the practitioner to decide for themselves, often aided by local or national guidelines. However, Table 19.3 may provide some guidance.

19.3   Molecular genetics

Cancer is characterized at the level of the chromosome by changes to genes. In many cases these alterations in DNA can be used to aid diagnosis, and the mutations responsible can be detected by a number of methods, such as Southern blotting, fluorescence *in-situ* hybridization, cytogenetics, polymerase chain reaction and deoxyribonucleic acid (DNA) microarrays, some of which are described in Chapter 2. There are numerous examples of this, such as:

• The use of cytogenetics in the diagnosis of chronic myeloid leukaemia, where part of chromosome 9 is translocated to chromosome 22. This brings together two sections of DNA that coincidentally create a functioning fusion gene which is the basis of the disease. Our understanding of this led directly to the development of a specific drug that has revolutionized the treatment of this cancer. This is developed in more detail in Chapter 6.
• *BRCA-1* (found on chromosome 17) codes for a protein important in the repair of DNA. Certain loss-of-function mutations in *BRCA-1*, and another, *BRCA-2*, are associated with an increased risk of breast cancer (hence the abbreviation BR and CA). Presence of one of these mutations increases a woman's risk of breast cancer fivefold, but the mutation also carries an even higher risk of ovarian cancer. Thus, sensitivity and specificity of the tests for these mutations is high and often guide treatment, especially if there are other risk factors, such as obesity and a family history.
• The presence of the V617F mutation in *JAK2* is effectively a cancer marker for polycythaemia rubra vera (Chapter 4), although it is also present in other myeloproliferative disorders.
• Certain variants in the gene for thymidine kinase are prognostic markers of overall survival in patients with epithelial ovarian cancer.

Individual chapters each refer to examples where molecular genetics has informed both diagnosis and management, such as in myeloma and lymphoma, as described in Chapter 6.

Pharmacogenomics

This is often described as personalized genetics, and closely related to pharmacogenetics. The practical aspects of the work of Darwin and Mendel state that there will be variation in different genes and entire genomes that confer advantage and disadvantage. Molecular genetics is able to define the presence of particular variation of genes (polymorphisms) that influence responses to various treatments, so enabling the practitioner to tailor the treatment to the individual depending on their genetic make-up.

Many of these polymorphisms are present in genes coding for cytochrome P450 enzymes that influence the metabolism of dozens of commonly prescribed drugs. The genomes of some individuals include genes for enzymes that have weak or slow activity, whilst others have genes coding for enzymes with strong or fast activity. Therefore, a standard dose of, say, 100 mg may in one person result in 80 mg of active drug, whereas in another the activity may be only 20 mg. However, the link between genomics and enzyme activity may not be linear, as there will be variations in the rate of absorption of the drug across the intestinal wall.

There are several instances where variability in response to cancer chemotherapy is driven by variant of certain genes:

- Deficiency of the enzyme dihydropyrimidine dehydrogenase (DHD), which is involved in the metabolism of uracil and thymine, is not generally thought of as a problem. However, it is if the patient needs to take the drug 5-fluorouracil, in which case a seemingly moderate dose may be dangerously toxic and possibly lethal.
- Similarly, thiopurine methyltransferase (TPMT) is an enzyme involved in the metabolism of the drug azathioprine, commonly used as an immunosuppressant and cancer chemotherapeutic. Those with TMPT deficiency caused by mutations in the TPMT gene fail to metabolize azathioprine at the normal rate, so that once more the drug will be more active. In both DHD and TMPT deficiency, lower doses of the particular drug are used.

- The survival of patients with advanced urothelial cancer or epithelial ovarian cancer when treated with platinum-based chemotherapy (cisplatin/carboplatin) is linked to certain alleles in the gene 'excision repair cross-complementation group 1'.

Blood science angle: Cancer

Although cancers start in a single organ, their effects are often systemic and nonspecific. A recognized effect of many tumours is a normocytic anaemia (as explained in Chapter 4), despite the absence of metastases that may interfere with erythropoiesis in the bone marrow. Indeed, anaemia may be the first symptom of cancer. Furthermore, cancer is a pro-thrombotic disease – patients are at risk of deep vein thrombosis (DVT) and pulmonary embolism. Similarly, many tumours promote a nonspecific acute-phase response, despite the absence of a clear infective agent or an autoimmune process (details in Chapters 9 and 14). This is likely to be a generalized response of the body to the developing cancer.

19.4 Case studies

Case study 26

An 84-year-old man with a recent (5-year) history of lower urinary tract infections complained of discomfort in the lower abdomen, and alternating inability to urinate and then incontinence. The practitioner took venous bloods for urea and electrolytes, and for PSA.

	Result (unit)	Reference range
Na	139 mmol/L	135–145
K	4.3 mmol/L	3.8–5.0
Urea	6.9 mmol/L	3.3–6.7
Creatinine	136 µmol/L	71–133
PSA	13.5 g/L	<4.0

Interpretation   There is a degree of renal disease as both urea and creatinine are slightly above the top of their reference range, the latter giving an eGFR of 46 ml/min per 1.73 m^2, and so chronic kidney disease stage IIIa. However, PSA is raised markedly, even accounting for the age of the patient, strongly suggesting prostate disease, which is supported by the symptoms, although they may also be due to bladder disease. Nevertheless, the high PSA is a fact and demands further investigation. This is likely to be a biopsy to determine an inflammatory or neoplastic pathology, which will then guide management. It may also be prudent to measure serum calcium and PTH as surrogates of bone disease.

Case study 27

A 56-year-old woman, presents to the accident and emergency unit with a painful and swollen leg, and is diagnosed with a DVT. She has none of the standard risk factors for venous thromboembolism (such as smoking, diabetes, varicose veins, obesity and family history – Chapter 8) and was treated with low molecular weight heparin, followed by warfarin for 6 months.

Several months later, at a visit to check her international normalized ratio, she complains of becoming tired and listless, needing to sleep each afternoon. The blood scientist orders a full blood count and ESR, which finds a haemoglobin of 109 g/L (reference range 118–148 g/L) and an ESR of 13 mm/h (reference range <10 mm/h). With a mean cell volume of 89 fL (reference range 80–98 fL), a diagnosis of normocytic anaemia is made and the general practitioner is informed.

At the final visit to the oral anticoagulant clinic, the patient complains of vomiting (with some blood present), increasing tiredness, loss of appetite and stomach pains. The blood scientist recommends she refers to the general practitioner, who sends her for an intestinal endoscopy. This finds a growth in her stomach, which biopsy proves to be a carcinoma.

Interpretation For many people, unexplained weight loss is an early sign of cancer. In this woman's case, the first symptoms were the otherwise unexplained DVT, followed by a normocytic anaemia. It is established that many cancers are pro-thrombotic, and also cause an anaemia of chronic disease. It does not necessarily follow that the tumour has metastasized to the bone marrow.

Summary

- Cancer, the second most common cause of death, is characterized by abnormal cell growth. The most commonly affected organs include the breast, lung, colon/rectum and prostate.

- Cancers may be marked by certain molecules in the blood, but few have such good sensitivity and specificity that they can be used alone.
- Cancer markers with no clear biological effect include CA 15.3, CA 19-9, CA-125, CEA, PSA, AFP and paraproteins.
- Cancer markers with a clear biological effect include insulin, prolactin, catecholamines, ACTH, AVP and calcium.
- The laboratory combines with other services, such as imaging, to provide a robust diagnosis and help direct and monitor management

Further reading

DeBerardinis RJ, Thompson CB. Cellular metabolism and disease: what do metabolic outliers teach us? Cell. 2012;148:1132–1144.

Kraljevic PS, Sedic M, Bosnjak H. *et al.* Metastasis: new perspectives on an old problem. Mol Cancer. 2011;10:22.

Moore RG, Jabre-Raughley M, Brown AK. *et al.* Comparison of a novel multiple marker assay vs the risk of malignancy index for the prediction of epithelial ovarian cancer in patients with a pelvic mass. Am J Obstet Gynecol. 2010;228:e1–e6.

Smith RA, Cokkinides V, Brooks D. *et al.* Cancer screening in the United States, 2010: a review of current American Cancer Society guidelines and issues in cancer screening. Cancer J Clin. 2010;60:99–119.

Soreide K, Berg M, Skudal BS, Nedreboe BS. Advances in the understanding and treatment of colorectal cancer. Discov Med. 2011;12:393–404.

Sturgeon CM, Duffy MJ, Stenman UH. *et al.* National Academy of Clinical Biochemistry Laboratory Medicine Practice Guidelines for Use of Tumor Markers in Testicular, Prostate, Colorectal, Breast, and Ovarian Cancers. Clin Chem. 2008;54:e11–e79.

Tian C, Markman M, Zaino R. *et al.* CA-125 change after chemotherapy in prediction of treatment outcome among advanced mucinous and clear cell epithelial ovarian cancers. Cancer. 2009;115:1395–1403.

Tonus C, Sellinger M, Koss K, Neupert G. Faecal pyruvate kinase isoenzyme type M2 for colorectal cancer screening: a meta-analysis. World J Gastroenterol. 2012;18:4004–4011.

Guidelines

The National Institute for Health and Clinical Excellence (NICE) has a number of 'pathways' for different, cancers. Although these often focus on treatment, some refer to the laboratory. http://www.nice.org.uk/Search.do?x=0&y=0&searchText=cancer&newsearch=true#/search/?reload

NICE Clinical guideline CG122: The recognition and initial management of ovarian cancer.

Web sites

www.cancerresearchuk.org/Cancer
www.dh.gov.uk/health/category/policy-areas/nhs/cancer/

20 Inherited Metabolic Disorders

Learning objectives

After studying this chapter, you should be able to:

- outline the modes of inheritance for metabolic disorders;
- list the consequences of an enzyme deficiency and how this gives rise to an inherited metabolic disorder (IMD);
- describe examples of key IMDs encountered in clinical practice;
- discuss the laboratory investigation of IMDs;
- explain the value of antenatal diagnosis and prenatal screening.

Markers that enable us to detect IMDs (but often referred to as inborn errors of metabolism) are crucial because many of these conditions are due to a genetic defect and present in childhood and have lifelong consequences. Although individually rare, IMDs together may be present in up to 1 in 1000 live births, and over 4000 single gene defects have been identified. However, these figures mask enormous variability in frequency, severity (ranging from benign to fatal) and the particular metabolic pathway(s) involved.

In this chapter, we will first remind ourselves, in Section 20.1, of the genetics of inheritance, most of which operate at the level of the chromosome. Section 20.2 will follow with details of how these operate at the biochemical level, and Section 20.3 on the IMDs of organelles. These sections will also look at specific abnormalities. In the penultimate Section 20.4 we look at which of these IMDs are the object of screening processes, and the chapter concludes with case studies in Section 20.5.

20.1 The genetics of inheritance

By definition, all disorders must have arisen from a faulty gene or genes from one or both parents. Several different forms of this are recognized, based on which gene carries the particular abnormality.

- *Autosomal recessive inheritance.* This is present when both parents carry a single copy of an abnormal gene (i.e. they are heterozygotes) on an autosome (i.e. on one of the 22 pairs of non-sex chromosomes). In almost all cases the abnormality may be described as a 'loss of function', so that effectively the gene does not work and there is no protein product. However, the normal gene on the other copy of the chromosome pair continues to function normally and so makes up for the deficit. The independent assortment of Mendelian genetics guarantees that, on average, 25% of potential offspring will inherit a copy of the mutated gene from both parents, and so the particular disease will be evident. The vast majority of IMDs are of this type.
- *Autosomal dominant inheritance.* In this group, the effect of the mutated copy of a gene on one chromosome is so strong that its effect is always present (hence dominant). The abnormality may be described as a 'gain of function'. If this is the case, then the parent is very likely to know about it, and will pass their gene (and so the disease) to 50% of their children.
- *X-linked inheritance.* This is present when a mutation is present on the X chromosome. As females have two X chromosomes, the normal gene on the other X chromosome is likely to be able to make up the deficit. However, as males have no second X chromosome,

Blood Science: Principles and Pathology, First Edition. Andrew Blann and Nessar Ahmed.
© 2014 John Wiley & Sons, Ltd. Published 2014 by John Wiley & Sons, Ltd.

Table 20.1 Examples of IMDs.

Pattern of inheritance	Examples (further detail in Chapter x)
Autosomal recessive	Simple haemoglobinopathy (Chapter 4)
	Wilson's disease (Chapter 17)
	Chronic granulomatous disease (Chapter 9)
Autosomal dominant	Huntingdon's disease (formerly chorea)
	Renal polycystic disease (Chapter 12)
X-linked	Haemophilia (Chapter 8)
	Ornithine transcarbamylase deficiency
	X-linked agammaglobulinaemia (Chapter 9)
	Glucose-6-phosphate dehydrogenase (Chapter 4)
	Duchenne and Becker muscular dystrophy

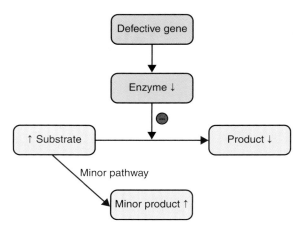

Figure 20.1 Consequences of an enzyme deficiency in a metabolic pathway. An enzyme deficiency causes reduction of product, together with accumulation of substrate, and minor products of metabolic pathway.

but a Y chromosome, the mutated gene is always present, and so the disease inevitably appears. However, in females the process of lyonization randomly inactivates one of the X chromosomes, so the mutant gene may become active. In this setting it could be argued that the Y chromosome is a 'loss of function' abnormality of an X chromosome. Examples of diseases from each of these are presented in Table 20.1.

Although sperm have mitochondria (which have their own deoxyribonucleic acid (DNA)), these do not enter the ovum, so that all mitochondria in the zygote, and so ultimately the adult, are derived from the mother. Mitochondrial genes include those for enzymes of the electron transport chain, such as for cytochromes.

Inherited metabolic disorders

Most inherited diseases are due to a genetically determined absence or modification of a key protein. For example, in agammaglobulinaemia there is an absence of antibodies, whereas modification of haemoglobin occurs in sickle cell anaemia. Other examples include defective receptor synthesis as in familial hypercholesterolaemia, where there is a defect in low-density lipoprotein receptors, and defective carrier proteins as in cystinuria, where renal reabsorption of cystine is impaired. In many cases the defective or absent protein is an enzyme, and this

results in an IMD as enzymes catalyse steps in metabolic pathways.

The genetic defect may result in absence of an enzyme catalysing a step in a metabolic pathway. This can have a number of consequences, which may underlie the clinical features observed in affected patients as outlined in Figure 20.1.

Absence of the enzyme means its substrate cannot be converted to a product and so there is lack of product of this pathway. This product may have had vital functions in the body and its deficiency will give rise to abnormal function and development of clinical features. The enzyme deficiency will also cause a build-up of the substrate as it is no longer being converted to product. Accumulation of high concentrations of the substrate may have detrimental effects if it is toxic when in excess. A small amount of the substrate may normally be channelled through a minor pathway giving rise to minor products. When in excess, more of the substrate may enter this minor pathway, increasing production of these minor products which in turn may have detrimental effects if they are toxic.

The strategies used for treating IMDs in general terms could include replacement of product of metabolic pathway or removal of excess substrate and minor products if any of these are responsible for the clinical consequences. More ambitious approaches might include replacement of the missing enzyme or the faulty gene which codes for this enzyme.

20.2 Molecular inherited metabolic disorders

Although an imperfect system, IMDs may be classified by the chemical and molecular basis of the abnormality.

Amino acid disorders

Methionine and homocysteine These amino acids are interrelated in a metabolic pathway that involves several other amino acids and enzymes, two of which are of note (Figure 20.2). Cystathionine-β-synthase fuses homocysteine and serine into cystathione, so that lack of this enzyme leads to the build-up of homocysteine in the blood; hence hyperhomocysteinaemia. As the blood perfuses the kidneys, the homocysteine passes easily into the urine and is not reabsorbed, leading to high concentrations in the urine – homocystinuria.

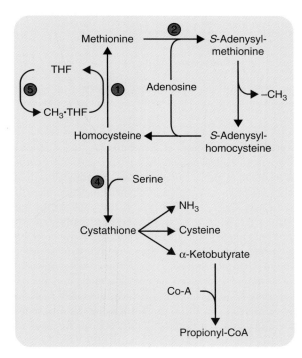

Figure 20.2 Metabolism of methionine and homocysteine. This grossly simplified metabolic pathway serves as a useful illustration of the interrelationships of certain amino acids, their linked enzymes and vitamin cofactors. The numbers refer to enzymes: (1) methionine synthase (requires vitamin B_{12}), (2) adenosyl transferase, (3) methyl transferase, (4) cystathionine synthase (requires vitamin B_6), (5) methylene tetrahydrofolate reductase (MTHFr).

A second enzyme methylenetetrahydrofolate reductase (MTHFr) is required for the conversion of homocysteine to methionine. Lack of functioning MTHFr also means that concentrations of homocysteine rise, also causing homocystinuria. This leads to a multisystem disorder whose clinical features include bony abnormalities, osteoporosis, psychiatric disorders, mental retardation, dislocated lenses, developmental delay and thrombosis.

Ornithine and citrulline Ornithine and carbamoyl phosphate are converted to citrulline in the mitochondrion by ornithine transcarbamoylase as part of the urea cycle. Lack of this enzyme (which is X-linked) leads to a build-up of ornithine, but more crucially to the build-up of carbamoyl phosphate, which is converted to orotic acid and so an orotic aciduria. However, more sinister is the build-up in ammonia, which can be fatal. Other aspects of the urea/ornithine cycle are discussed in Chapter 17, where Figure 17.3 is pertinent.

Phenylalanine and phenylketones Phenylalanine is an essential amino acid and readily available in the diet. It is converted to another nonessential amino acid, tyrosine, by the enzyme phenylalanine hydroxylase. Deficiency of this enzyme causes phenylketonuria (PKU), a condition which has an autosomal recessive mode of inheritance and an incidence of 1 in 10 000 in the UK.

Thus, PKU is associated with lack of tyrosine, required for synthesis of tissue proteins, melanin, thyroid hormones and catecholamines. However, failure of phenylalanine hydroxylase may also be due to poor provision of hydrogen from tetrahydrobiopterin and leads not only to reduced concentrations of tyrosine, but to increased concentrations of phenylalanine in the blood known as hyperphenylalaninaemia (Figure 20.3). This high phenylalanine becomes the substrate for other enzymes, such as ketoglutamate aminotransferase, the product ultimately being a phenylketone, which is excreted in the urine; hence PKU.

Clinical features of PKU are absent at birth but develop soon afterwards in untreated children. They include irritability, poor feeding, vomiting, eczema, evidence of melanin deficiency (pale skin, fair hair, blue eyes). However, the most serious and irreversible consequence of PKU is mental retardation which develops within 3–6 months following birth in untreated children. It is believed that the mental retardation arises because the high phenylalanine is toxic to the developing brain. The precise mechanism behind this toxicity is unclear, but possibilities include interference with brain amino acid

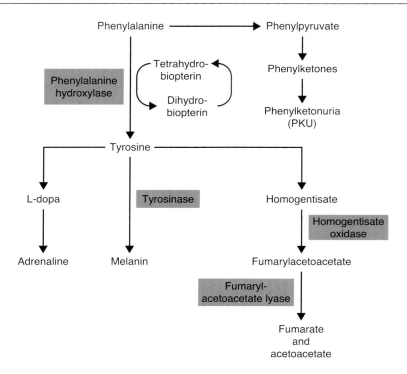

Figure 20.3 Metabolism of phenylalanine and tyrosine. Phenylalanine is the substrate for phenylalanine hydroxlase, the product being tyrosine, which is the substrate for at least three pathways. (Image from Blann AD, *et al.* (2009) Haematology Morphology. Training CD-ROM. © IBMS/Sysmex UK).

metabolism or inhibition of neurotransmitter release by the high phenylalanine concentration.

Diagnosis of PKU involves demonstration of a high serum phenylalanine concentration in excess of 1 mmol/L (normal <0.1 mmol/L). Affected children are treated by placing on a diet low in phenylalanine but with adequate amounts of tyrosine. Maintenance of this diet is particularly important in the early stages of life as there is rapid growth of the brain and central nervous system. However, the diet can be more relaxed as the child gets older. Children develop normally if they receive treatment shortly after birth. In the past, cases of PKU were missed as there was no screening programme for this condition and untreated children rarely had an IQ >70 and required life-long institutionalization. Regular monitoring and control of blood phenylalanine concentrations is also important in pregnant females who have PKU. This is because hyperphenylalaninaemia can cause damage to the developing foetus even if the foetus itself does not have PKU. Indeed, pregnant women with uncontrolled PKU often give birth to infants suffering from congenital malformations and mental retardation.

Tyrosine This amino acid is the substrate for at least three pathways (Figure 20.3). It can generate neurotransmitters L-dopa, noradrenaline and adrenaline, and also, via tyrosinase, the skin pigment melanin. Lack of tyrosinase leads to albinism and the potential for an increased risk of skin cancer. The third pathway leads first to homogentisate, increased concentrations of which are present in deficiency of homogentisate oxidase, leading to the relatively benign condition of alkaptonuria.

However, beyond this step is the enzyme fumarylacetoacetate lyase, deficiency in which ultimately leads to tyrosinaemia. This becomes evident as liver failure (hence abnormal liver function tests), renal tubular dysfunction (raised urea and electrolytes) and rickets. Chronic liver disease may also be present, and if so there may be increased alpha-fetoprotein in addition to abnormal liver function tests.

Organic acid disorders

Common causes of these disorders begin with amino acids, such as valine, leucine and isoleucine, which are converted to carboxylic acids by removal of the amino groups, and which lead to acetyl- and propionyl- metabolites. Downstream abnormalities lead to propionic, malonic and methylmalonic aciduria. Metabolites are measurable in serum and urine using mass spectrometry.

Purine and pyrimidine disorders

Purine and pyrimidine metabolism is complex, with pathways for synthesis, catabolism and salvage. Three major IMDs are recognized:

• Loss-of-function mutations in the genes for adenosine deaminase and purine nucleoside phosphorylase are among the more common causes of immunodeficiency and the inability to defend against microbes; additional details are presented in Chapter 9.
• Xanthine oxidase, a large enzyme of relative mass 270 kDa, catalyses the oxidation of hypoxanthine to xanthine. Deficiency of the enzyme results in failure to generate xanthine, a downstream consequence of which is low uric acid concentrations in plasma and urine. A clinical note: this enzyme is the target of allopurinol, a treatment for gout, generally caused by high uric acid, and easily measured in the laboratory.
• Hypoxanthine–guanine phosphoribosyl transferase, coded for by a gene on the X chromosome, is a component of the 'salvage' pathway of purine metabolism. Lack of the enzyme causes Lesch–Nyhan syndrome, the consequences of which include a high plasma and urine uric acid, which may in turn cause gout and renal stones. Treatment is with allopurinol, but this does not help other manifestations, such as mental retardation, neurological symptoms and self-mutilation.

Some disorders can be detected by urine organic acid analysis, but most purine and pyrimidine disorders will need purine and pyrimidine analysis in blood or urine, followed up by specific enzyme or DNA analyses.

Lipid disorders

Metabolic pathways release the energy in fatty acids by oxidation. Perhaps the most important is deficiency of the enzyme medium-chain acyl coenzyme A dehydrogenase, the consequences of which include inability to generate ketones when fasting. This leads to diarrhoea and vomiting, and although mostly apparent in infants, there are cases of presentation in adulthood. Treatment is by avoiding fasting.

The dominant lipid IMDs lead to hypercholesterolaemia, the most common forms being familial hypercholesterolaemia (FH) and hypertriacylglyceridaemia. The former is present in its heterozygous state at a frequency of 1/400 of the UK population. In its homozygous form, FH leads to a serum total cholesterol in excess of 10 mmol/L, and statin therapy is a necessary life-saving treatment. Because of their importance in the pathogenesis of atherosclerosis, lipid disorders are described in detail in Chapter 14.

Carbohydrate disorders

These can be life-threatening as carbohydrates are the major source of energy. Several major aspects are recognized.

Disorders of galactose metabolism These include deficiency of the enzymes galactose-1-phosphate uridyl transferase, galactokinase and uridine diphosphate galactose-4-epimerase. Lack of the first leads to a build-up of its substrate (galactose-1-phosphate), and so classical galactosaemia, which has an incidence of 1 in 60 000 in the UK. The symptoms in classical galactosaemia arise due to accumulation of galactose-1-phosphate and affected babies present with vomiting, failure to thrive, jaundice and liver failure. In addition, cataracts may develop due to build-up of galactitol and its consequent osmotic effects in the eye lens of these patients.

Galactosaemia is suspected with the detection of non-glucose reducing sugars in the urine, and further investigations to confirm diagnosis include measurement of erythrocyte galactose-1-phosphate uridyl transferase activity. Treatment involves excluding galactose and lactose from the diet and is monitored by measuring erythrocyte galactose-1-phosphate.

Disorders of fructose metabolism Deficiency of enzymes in fructose metabolism are rarer than those of galactose; the most common is lack of aldolase B, an autosomal recessive mutation in *ALDOB* on chromosome 9. It leads to hereditary fructose intolerance, but only when the sugar is present in the diet.

Disorders in glycogen metabolism Carbohydrates are stored in intracellular granules of glycogen; abnormalities in its formation lead to glycogen storage disease (GSD). One authority suggests that there are 11 distinct abnormalities, most of which are defined by lack of a particular enzyme, such as GSD type 1 (von Gierke's

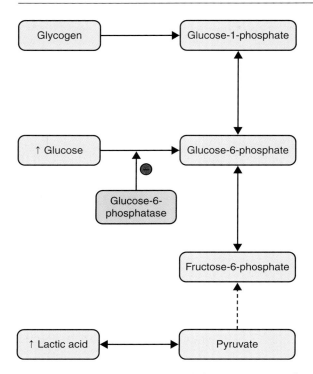

Figure 20.4 Glucose and glycogen metabolism. Maintenance of fasting blood glucose concentrations by glycogenolysis and gluconeogenesis in the liver. Deficiency of the enzyme glucose-6-phosphatase blocks the conversion of glucose-6-phosphate to glucose.

disease) caused by lack of glucose-6-phosphatase. This disorder is also known as glucose-6-phosphatase deficiency, and an example of a disorder where there is lack of a product due to enzyme deficiency.

Owing to deficiency of this enzyme, both glycogenolysis and gluconeogenesis, the two reactions required to maintain blood glucose concentrations during fasting, are blocked (Figure 20.4).

> N.B. Do not confuse glucose-6-phosphatase deficiency with glucose-6-phosphate dehydrogenase (G6PD) deficiency.

Children affected by glucose-6-phosphatase deficiency suffer from hypoglycaemia, this occurs during fasting and can be severe. They may also have hepatomegaly due to accumulation of glycogen in the liver and lactic acidosis, a consequence of blood lactate accumulation as the lactate cannot be utilized via gluconeogenesis.

Affected individuals may also have hyperlipidaemia due to increased fat synthesis. Investigation of glucose-6-phosphatase deficiency includes administration of glucagon and adrenaline, both of which fail to stimulate a rise in blood glucose concentrations in affected patients. The diagnosis can be confirmed by demonstrating absence of the enzyme glucose-6-phosphatase in a sample obtained following liver biopsy. Treatment of this condition involves maintaining blood glucose concentrations via constant nasogastric infusion, for example, during the night and frequent ingestion of starch and glucose.

When the GSD is focused on hepatic cells, then there may be abnormal liver function tests, and where muscle cells are the target (as in McArdle's disease), raised creatine kinase is present in the serum. Other established enzyme deficiency diseases include those of fructose-1,6-bisphosphatase, phosphoenolpyruvate carboxykinase and pyruvate carboxylase.

As discussed in Chapter 4, G6PD deficiency leads to a haemolytic anaemia. The mutation on the X-chromosome is particularly prevalent in populations originating in West Africa.

Cystic fibrosis

Cystic fibrosis is a relatively common abnormality, being present in 1/1600 of a Caucasian population as an autosomal recessive condition, so that 1 in 20–25 of the population are carriers. Onset of cystic fibrosis may occur at birth or later in childhood. Cystic fibrosis is characterized by exocrine secretions of high viscosity, as these secretions have an altered ion and water content. The correct viscosity is crucial for protein secretions with protective, digestive and lubricant functions. The defect in cystic fibrosis is believed to be chloride transport across membranes. This in turn is due to a defective gene that codes for a protein known as the cystic fibrosis transmembrane conductance regulator, which normally functions as a chloride ion channel in the membrane.

Patients with cystic fibrosis suffer from recurrent respiratory infections, malabsorption of nutrients due to inadequate pancreatic secretions, intestinal obstruction (also known as meconium ileus) and diabetes mellitus, which may occur as a late feature. Those suspected of cystic fibrosis are investigated using the sweat chloride test, which involves collection of skin sweat, the flow of which is stimulated using pilocarpine, and this sweat is then analysed for its chloride concentration. A concentration of sweat chloride >60 mmol/L is diagnostic of cystic fibrosis.

Affected patients are treated using physiotherapy and antibiotics to improve respiratory function and prevent infections respectively. Pancreatic enzymes may be added to food to counteract the effects of inadequate pancreatic secretions. Although prognosis of cystic fibrosis has improved, many children nevertheless still die in early adulthood. Screening involves the demonstration of a high plasma immunoreactive trypsin concentration, and a positive finding is followed by the sweat chloride test to confirm diagnosis. Molecular genetic analysis is used to identify which of the common mutations are responsible.

20.3 Organelle inherited metabolic disorders

IMDs recognized to cause disease can be found as abnormalities in intracellular organelles – lysosomes, peroxisomes and mitochondria (technically speaking from a cell biology viewpoint, many take the stance that storage granules are not organelles).

Lysosomes

These organelles contain hydrolytic enzymes that break down a specific substrate. Lack of the enzymes leads to build up of the substrate, which may lead to the bursting of the lysosome and so cell damage. Examples, all of which are autosomal recessive, include:

- Tay–Sachs disease, caused by a mutation in the gene for hexosaminidase A on chromosome 15, which digests gangliosides.
- Gaucher's disease, caused by lack of glucosylceramidase (also known as beta-glucosidase). A large number of mutations have been described, making exact molecular diagnosis difficult.
- Pompe's disease, caused by deficiency of acid alpha-glucosidase. This condition can also be classified as GSD type 2, as above.

Peroxisomes

These bodies contain a variety of enzymes involved in a range of reactions, involving often fatty acids and amino acids. The most common include:

- Zellweger syndrome, the consequences of one or more mutations in *PEX* genes, required in the correct assembly of the organelle.
- Adrenoleukodystrophy, a consequence of a mutation in *ABCD1*, which codes for a membrane component required for the oxidation of very long chain fatty acids.

Mitochondria

Mitochondria have their own DNA (mtDNA), of which perhaps a third codes for genes for enzymes of the respiratory chain. Defects in these genes lead to a number of problems, the most common being in muscles; that is, myopathy. Other defects lead to porphyria, a disease of disordered haemoglobin synthesis. A key enzyme is aminolevulinate synthetase, which acts within the mitochondria, and loss-of-function mutations in the coding gene cause inadequate synthesis of haem. Additional details are presented in Chapters 3 and 4.

20.4 Antenatal diagnosis and neonatal screening

It is advantageous to be able to detect many of these inherited disorders so that treatment can begin as soon as possible. Indeed, in many cases couples with a family history that are planning a child often seek guidance. The frequency of IMDs is higher in children whose parents are cousins, as this increases the co-inheritance of recessive genes that are unlikely to be present if the parents are unrelated. Consequently, certain populations have a higher frequency of IMDs.

When an inherited disorder cannot be treated, then early prenatal diagnosis allows the option of terminating the pregnancy. The following are indications for prenatal diagnosis:

- disease is very serious and justifies termination;
- there is no treatment available for the disease;
- there is a significant risk that the child will be affected by the disease;
- parents are prepared to consider termination of pregnancy if the foetus is affected.

Testing of a foetus *in utero* is usually carried out by chorionic villus sampling as early as 10 weeks' gestation, whereas the safer amniotic fluid sampling (amniocentesis) is an alternative method usually carried out at 16–18 weeks' gestation. In both cases, maternal contamination must be excluded.

Screening for a disease is an attempt to detect the condition before it presents clinically. A number of criteria have to be considered before a screening programme can be established for a particular condition. These criteria are:

- Does the disease have a relatively high incidence? If not, then it is unlikely a screening programme will be set up to detect such few cases.
- Will it be possible to detect the disease within a few days of birth?
- Do we have an easily measurable marker which can be used to identify affected subjects? Will the results for this test be available before any damage occurs to the affected child?
- Is it possible to miss the disease clinically and would this cause irreversible damage if left untreated?
- Can the disease be treated?

For screening in neonates, a 5–8 days of age capillary specimen is collected from a heel prick, and blood is spotted onto cards made of absorbent filter paper to form dried blood spots. The cards are sent to the screening laboratories by post. Most newborn screening laboratories in England test >50 000 babies per year. Common screening includes:

- PKU, generally tested on dried blood spots. Most laboratories use tandem mass spectrometry to detect increased phenylalanine.
- Congenital hypothyroidism is defined by increased TSH (typically >200 mU/L, reference <10 mU/L) and/ or lack of thyroxine, and can be determined on dried blood cards. Further details of thyroid diseases are present in Chapter 18.
- Haemoglobinopathies (sickle cell disease and thalassaemia, of which there are numerous variants) can be detected by high-performance liquid chromatography (HPLC) or isoelectric focusing, as is described in Chapter 4.
- Cystic fibrosis, for which screening is carried out by the measurement of immunoreactive trypsin in dried blood spots. It is also possible to obtain DNA and so probe for the relevant mutation(s), such as in *CFTR* on chromosome 7 that leads to abnormal movement of chloride and sodium ions across epithelial membranes, which is the basis of the pathology.

The laboratory

IMDs may be detected by a variety of techniques. Some are best determined by enzyme assay, whereas others require more complex techniques such as HPLC and genetics, possibly in series. However, in many cases the gold standard requires DNA, so that the involvement of the local molecular genetics unit will be needed. An example may be the finding of increased creatine kinase in a young man with symptoms suggestive of a diagnosis of muscular dystrophy.

Molecular genetics are likely to offer the definitive diagnosis by confirming the presence of a mutation in the dystrophin gene at Xp21. Both forms, Duchenne and Becker, are recessive X-linked diseases: the former has a frequency of 30 in 100 000 males, the latter a rate of 3–6 per 100 000.

Similarly, haemoglobinopathy is initially diagnosed by HPLC. However, the exact nature of the lesion (of which there are scores of variants, as noted in Chapter 4) can be determined at the level of the DNA by standard molecular genetics techniques. The parents and siblings of the affected patients are also likely to be tested.

20.5   Case studies

Case study 28

The parents of a 6-day-old baby girl are concerned that her general behaviour is markedly different from her three siblings when they were babies – she is listless and fractious, does not sleep well and takes to the breast poorly. A blood specimen collected from her heel was sent to the pathology laboratory as a dried spot for measurement for screening of common IMDs, primarily hypothyroidism, cystic fibrosis and PKU.

The sample yielded a phenylalanine result of 2075 μmol/L (reference range <100 μmol/L).

Interpretation   This child clearly has a very high serum phenylalanine concentration. This is likely due to classical PKU, where there is deficiency of phenylalanine hydroxylase. The high serum phenylalanine is toxic to the developing brain and can cause irreversible mental retardation in untreated children. Children with PKU are managed by placing them on a low phenylalanine diet that contains sufficient tyrosine.

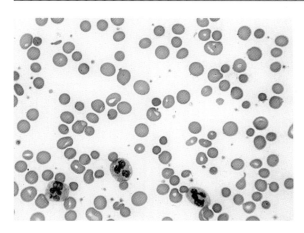

Figure 20.5 Blood film from case study 29. There is marked anisocytosis with microcytes (possibly some microspherocytes) and many reticulocytes (leading to polychromasia) and some schistocytes. Chapters 3 and 4 have parallel figures of red cell morpohology. (Image courtesy of H. Bibawi, NHS Tayside).

Case study 29

The Jamaican parents of a 10-day-old baby boy bring their first-born son to their general practitioner, stating he cries a lot and feeds poorly, much more so than his older sister when she was a neonate. The general practitioner finds him to have mild sclerotic jaundice and to be pale. Bloods are taken for full blood count, direct antiglobulin test and liver function tests.

These reveal haemoglobin 50 g/L (reference range for a 2-week-old, 98–134 g/L), mean cell volume 96 fL (reference range 88–100 fL), bilirubin 38 μmol/L (reference range 5–25 μmol/L). The blood film (Figure 20.5) shows changes consistent with haemolytic anaemia. The direct antiglobulin test is negative.

Interpretation The blood results and history point to an inherited metabolic disease that is causing red cells to be destroyed. In view of the sex and racial background of the case, he is tested for G6PD. The G6PD enzyme level was markedly reduced at 0.75 units/g haemoglobin (reference range 4.6–13.5 units/g haemoglobin).

The mother is subsequently found to be a heterozygote for G6PD deficiency, her son having inherited the X chromosome carrying the mutated gene. The genetic basis of the disease may arise from one of several mutations, which can be confirmed by standard molecular genetic testing. This in turn leads to variability in presentation and the abnormalities in the red cells, which in this case is manifest as a reticulocytosis; in other cases there may be a marked schistocytosis and Heinz bodies. Additional details of G6PD deficiency are present in Chapter 4.

Summary

- IMDs arise from adverse mutation in the genes for a variety of enzymes. Most are autosomal recessive; other types include autosomal dominant and X-chromosome linked.
- Common IMDs (perhaps 1 in 4500 live births) that call on the laboratory include PKU, hypothyroidism, lactic acidosis and galactosaemia.
- Antenatal diagnosis and neonatal screening are employed to detect IMDs at an early stage.
- Many IMDs can be diagnosed by standard laboratory methods, but ultimate diagnosis is by molecular genetics.

Further Reading

Aerts JM, Kallemeijn WW, Wegdam W. *et al.* Biomarkers in the diagnosis of lysosomal storage disorders: proteins, lipids, and antibodies. J Inherit Metab Dis. 201; 34:605–619.

Endo F, Matsuura T, Yanagita K, Matsuda I. Clinical manifestations of inborn errors of the urea cycle and related metabolic disorders during childhood. J Nutr. 2004;134(6 Suppl):1605S–1609S.

Fouchier SW, Rodenburg J, Defesche JC, Kastelein JJ. Management of hereditary dyslipidaemia; the paradigm of autosomal dominant hypercholesterolaemia. Eur J Hum Genet. 2005;13:1247–1253.

Pandor A, Eastham J, Beverley C. *et al.* Clinical effectiveness and cost-effectiveness of neonatal screening for inborn errors of metabolism using tandem mass spectrometry: a systematic review. Health Technol Assess. 2004;8:1–121.

21 Drugs and Poisons

Learning objectives

After studying this chapter, you should be able to:

- recognize the major issues in toxicology;
- discuss the toxicology of specific poisons encountered in clinical practice;
- describe the major drugs of abuse (DOAs);
- be aware of the importance of therapeutic drug monitoring (TDM);
- describe the major drugs that are monitored during therapy.

In this chapter we will look at drugs and other chemicals. Many of these can directly cause disease, and the study of these poisons (toxicology) we will review in Section 21.1. We will examine the major specific toxins in Section 21.2. However, in some cases the drugs we healthcare professionals give our patients may be dangerous, so that their effect must be monitored; hence, TDM will be focused upon in Section 21.3. The chapter will conclude in Section 21.4 with case studies.

21.1 Toxicology

Toxicology is the scientific study of poisons and poisoning, requiring the measurement of the particular toxin, but also of its effects on the body. Toxins themselves are often thought of as drugs or chemicals (xenobiotics), but high concentrations of the body's own molecules, such as bilirubin, potassium and urea can also be toxic. Indeed, many substances can be poisonous if there is excessive exposure.

However, let us not forget that we healthcare practitioners routinely give drugs to susceptible individuals, and of course this can go wrong. When this happens, the drugs are effectively poisons. Poisoning can be defined as an interaction between a foreign chemical and a biological system which results in damage or harm to a living organism. There are many different ways in which poisoning can occur, such as the practitioner prescribing the wrong dose, or patient failure to understand and follow instructions. Other disease may turn a relatively harmless drug into a poison, and this is particularly applicable in those with concurrent renal and/or hepatic disease. These poisonings are called iatrogenic. Figure 21.1 outlines what happens to a drug prescribed for medication purposes. When its plasma concentration is higher than the therapeutic range (e.g. following an overdose), then toxicity may occur.

Poisoning is a common reason for emergency hospital admission, and the reasons for this might be accidental, suicidal or criminal. The role of the blood science laboratory in toxicology is crucial and may involve monitoring of vital functions; for example, by assessing liver and renal function. These may include:

- demonstrating the presence of a suspected poison in the blood to confirm diagnosis;
- monitoring therapy by measuring the concentration of a suspected poison at regular intervals, which should decrease if the therapy is effective.

Before looking at specific poisons, we will look at and examine general concepts in toxicology.

General biochemical features of poisoning

Some toxins poison the entire body, whereas others are organ specific. However, several key organs are more susceptible than others.

The kidney This organ suffers more than its share of damage because of the complex nature of its metabolic

Blood Science: Principles and Pathology, First Edition. Andrew Blann and Nessar Ahmed.
© 2014 John Wiley & Sons, Ltd. Published 2014 by John Wiley & Sons, Ltd.

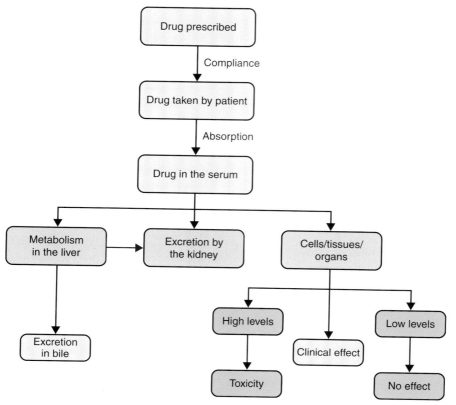

Figure 21.1 Outline of drug metabolism and clinical effect at target sites. This is far from the case that the amount of drug interfacing with cells and tissues is that which is prescribed. Factors such as compliance, absorption, metabolism and excretion all influence drug activity.

functions, including excretion. Abnormalities in acid–base homeostasis (acidosis and alkalosis) are common, especially in aspirin and ethylene glycol poisoning, so the kidney is under increased metabolic pressure as it is a key determinant of blood pH. This may be compounded if there is lactic or diabetic acidosis, or if the respiratory centre in the brain stem is damaged, causing a respiratory alkalosis, and there may also be a hyperkalaemia.

As is explained in detail in Chapter 12, increased plasma concentrations of urea and creatinine indicate renal impairment. A ratio of urea to creatinine of >100 when both are measured in millimoles per litre is suggestive of pre-renal uraemia due to hypovolaemia.

A particularly useful index in biochemical toxicology is the osmolar gap. In health, direct measurement of osmolality correlates very closely with the sum of several major ions, generally sodium, glucose and urea. If there is a difference between osmolality and the sum of these

molecules, then there must be another molecule, which in the setting of toxicology is likely to be a poison. Similarly, the anion gap is the difference between the measured anion component of the blood and the sum of several key anions that include chloride and bicarbonate.

The liver This organ is often damaged because one of its roles is to collect and then detoxify poisons such as paracetamol; this it is perfectly able to do if concentrations are modest. But in excessive concentrations, the liver is overwhelmed, and is itself poisoned. If present, hepatocyte damage is implied by impaired clotting (increased prothombin time), low albumin and high concentrations of the transaminases (alanine aminotransferase and aspartate aminotransferase). The key clinical sign is jaundice, being the consequence of raised bilirubin, but this (as is explained in Chapter 17) may be caused by a number of factors, such as excessive haemolysis, pathogen-induced damage (the hepatitis virus) and cholestasis.

Pharmacokinetics and pharmacodynamics

Pharmacokinetics considers the absorption, distribution, metabolism and excretion of drugs, including toxins, whilst pharmacodynamics refers to the mechanisms of the effects of drugs on the body, and the relationship between concentration and biological effect.

The rate by which exposure to a toxin translates to damage to organs depends on factors such as route. These are transdermal (by absorption, injection, animal bites and stings), the lungs (when inhaled) or the intestines (ingested as food or drink). Some DOAs may be absorbed through the buccal and nasal mucosa. Drugs also vary in their distribution around the body: some are fat-soluble, and therefore home to adipose tissues, whereas others may bind strongly to plasma proteins, from which they may be released very slowly.

The metabolic effects of enzymes in the liver and plasma may be to modify a particular molecule to render it less toxic. However, the reverse is also the case: relatively harmless molecules may become toxic once processed. Since there is variability in the activity of liver enzymes, a dose that is safe in one individual may be pathogenic in another.

Almost by definition, toxic events are atypical, and toxins themselves are not usually encountered by a routine laboratory. Accordingly, specialized techniques may be required, such as liquid–liquid solvent extraction and chromatography for nonvolatile organic compounds, fluorescence and atomic absorption spectrometry and induction-coupled plasma mass spectrometry. However, there is a place for routine methods such as spectrophotometry, ion-selective electrodes and immunoassay. Fine details of these are introduced in Chapter 2.

21.2 Toxicology of specific compounds

For convenience, poisons can be divided into several different groups that vary in their chemical structures and properties, and the ways in which they influence (and damage) physiology.

- Toxic anions include cyanide, fluoride, nitrite and oxalate. They often bind to metal ions in metalloproteins such as enzymes and thus interfere with their function. For example, cyanide causes tissue hypoxia by binding to iron in cytochrome c reductase and inhibiting electron transfer.
- Corrosives cause destruction of body tissues on contact. They can be strong acids or bases (such as sulphuric acid and sodium hydroxide), organic acids (such as acetic acid), heavy metals (strontium) and strong detergents.
- Gaseous and volatiles include products of combustion such as carbon monoxide, solvents and gases used in manufacturing or the home, and alcohols, the principal one being ethanol.
- Metals include iron, lead and arsenic. They can cause acute toxicity; for example, ingestion of an overdose of iron tablets can cause liver damage. Alternatively, toxicity may follow the accumulation of the metal as a result of exposure to relatively low amounts over a prolonged period of time, an example of this being criminal arsenic poisoning.
- Toxins are biological compounds produced by plants, animals, bacteria and fungi; for example, digitalin from the foxglove, snake venom, botulinum toxin and the phallotoxins and amatoxins of the death cap mushroom.
- Pesticides are chemicals used to kill pests and include organophosphate insecticides and the herbicide paraquat.
- Drugs and other substances used to produce a desired effect that can also have unwanted toxic effects. This is especially likely in nonaccidental self-poisoning, when medicines are taken in excessive amounts, and in drug misuse.

Naturally, a combination of poisons is increasingly toxic, especially if from different classes and/or absorbed by different routes. However, in practice, only a very small proportion of fatal toxicology events are iatrogenic, and most clinically significant poisonings are caused by only 20–30 substances. The most common are carbon monoxide, ethanol, paracetamol, aspirin and DOAs (Table 21.1).

Carbon monoxide

This is the most common fatal toxic gas. In household and industrial settings, high levels are likely to be due to incomplete combustion of hydrocarbon gases. Another major source is the internal combustion engine. The toxic effect is due to the binding of carbon monoxide to haemoglobin, forming carboxyhaemoglobin, which reduces oxygen delivery to tissues so that the patient effectively suffocates. Other consequences include metabolic acidosis.

Table 21.1 Common causes of fatal toxic events.

Poison	Number of deaths/year
Drugs, medicaments and biological substances	1783
Substances chiefly of non-medicinal source, including carbon monoxide and alcohol	610
Opioids, including heroin and methadone	376
Carbon monoxide	257
Alcohol	164
Tricyclic antidepressants	145
Paracetamol	134
Other antidepressants and antipsychotic drugs	109
Cocaine and other psychostimulants	61
Benzodiazepines	32

(Modified from Ahmed N (Ed.), Clinical Biochemistry, Oxford University Press, 2011).

Carboxyhaemoglobin can be measured by spectrophotometry, gas chromatography and high-performance liquid chromatography (HPLC), whilst treatment is to drive the carbon monoxide from the haemoglobin by giving 100% oxygen via a mask or nasal tube. Further aspects of carbon monoxide and carboxyhaemoglobin are to be found in Chapters 3 and 4.

Paracetamol

This over-the-counter analgesic, the second most common cause of death by poisoning in the UK, is readily absorbed by the intestines, which in recommended doses peaks in blood after 1–2 h with a half-life of perhaps 2 h. However, in higher doses, concentrations may peak at 4 h with a half-life of 12 h. It is metabolized in the liver to sulphate (20–40%) and glucuronides (40–60%), and the toxic metabolite N-acetyl-p-benzo-quinoneimine (NAPQI). Glutathione can be protective, but in overdose the levels are soon consumed, leaving the remaining NAPQI to damage hepatocytes and cells of the nephron. However, it is the destruction of hepatocytes that most directly causes death, although renal failure may be present in 15% of cases. Figure 21.2 outlines the metabolism of paracetamol and its toxicity following an overdose.

Relevant laboratory tests, therefore, include prothrombin time and liver function tests (LFTs), focusing on raised serum alanine and aspartate transaminases in the acute phase, with raised bilirubin, alkaline phosphatase and gamma glutamyl-transferase over the next 24 h. Beyond this there will be changes consistent with chronic liver failure, such as a prolonged prothrombin time leading to haemorrhage. Renal damage will be assessed by raised urea and creatinine. Treatments include acetylcysteine and methionine, which provide sulphydryl groups to counter the effects of NAPQI. Paracetamol is usually measured spectrophotometrically following chemical or enzymatic reaction. It can also be measured by immunoassay, gas–liquid chromatography or HPLC.

Salicylate

Aspirin (acetylsalicylic acid) is another over-the-counter analgesic, with similar pharmacokinetics in recommended doses and overdoses as paracetamol. It is metabolized in the liver to salicylate, which is the active form of the drug, as shown in Figure 21.3.

Clinical features of overdose (perhaps plasma salicylate >350 mg/L in children and >450 mg/L in adults), like that of paracetamol, include nausea and vomiting, vasodilatation, sweating, tinnitus, impaired hearing, pulmonary oedema, convulsions and arrhythmias.

An increase in blood salicylate following an overdose stimulates the respiratory centre in the brain, causing hyperventilation and ultimately producing a respiratory alkalosis. The kidneys respond by retaining H^+ ions but losing bicarbonate and potassium ions. Salicylates also uncouple oxidative phosphorylation, thereby decreasing adenosine triphosphate production. This is accompanied by increased heat production with fever and sweating with resultant loss of fluid and dehydration. In addition, the salicylates cause increased breakdown of lipids, producing ketones, and inhibition of the tricarboxylic acid cycle, causing accumulation of pyruvic and lactic acids, all of which contribute towards a latent metabolic acidosis.

The laboratory may be called on to consider the anion gap (likely to be raised), hypo- or hyperkalaemia and hypoglycaemia. If the liver is damaged there will be raised LFTs. If there is a severe acid–base disturbance (mixed respiratory alkalosis with metabolic acidosis), then alkalinization of the urine with intravenous sodium bicarbonate may be called for. Plasma salicylate >700 mg/L generally calls for dialysis.

Drugs of abuse

This is a collective term used for several classes of drugs used for recreational purposes (i.e. not as therapy for a

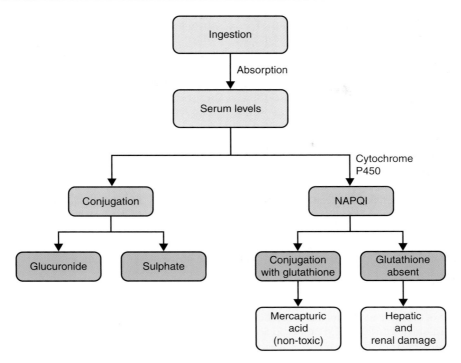

Figure 21.2 Pathways involved in the metabolism of paracetamol. Conjugation to glucuronide, sulphate or glutathione renders the drug nontoxic. However, in their absence, paracetamol causes hepatic and renal cell damage. NAPQI: *N*-acetyl-*p*-benzoquinoneimine.

particular health issue), although in some cases there are established therapeutic uses (analgesia) or health benefits (cardiovascular physiology). Several DOAs are manufactured illicitly and may contain contaminants that may be more toxic than the drugs themselves, and most are generally measured in urine.

Ethanol Almost certainly the most widely used and abused drug globally, and with many industrial uses, in small doses the healthy benefits are established, just as overdose is the cause of considerable morbidity and mortality, not only to the user. Generally ingested orally, concentrations peak after 0.5–3 h, dependent on dose and if taken with food. Perhaps 90% of alcohol is metabolized by hepatic alcohol dehydrogenase to acetate and thence to the tricarboxylic acid cycle, the remainder being excreted by the lungs (hence the 'breathalyser') and kidneys.

The social aspects and consequences of alcohol abuse are established, and exceed all other DOAs combined. Common features of overdose include dizziness, ataxia, nausea and vomiting. Later there may be drowsiness, hypotension, hypothermia, respiratory depression, convulsions and coma. Suppression of reflexes can lead to aspiration (inhalation) of vomit, which may be fatal. Laboratory features include raised plasma osmolality, mild metabolic acidosis, ketosis and hypoglycaemia in children and some adults.

Chronic misuse of alcohol (alcoholism) leads to dependency, and in the laboratory there will be increased gamma-glutamyltransferase, triacylgylcerols and erythrocyte mean cell volume.

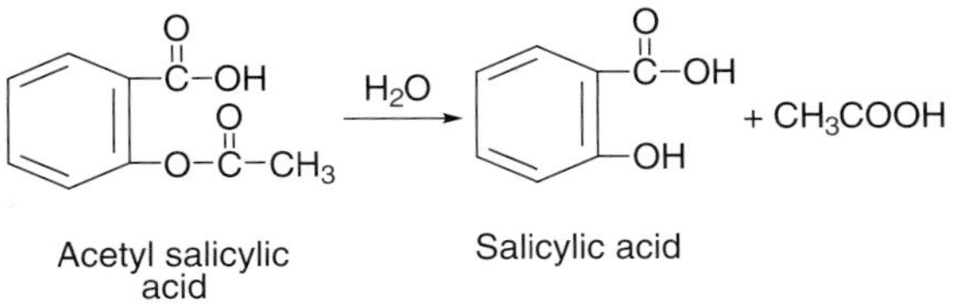

Acetyl salicylic acid → Salicylic acid + CH_3COOH

Figure 21.3 Metabolism of acetylsalicylic acid. The liver converts acetylsalicylic acid to its active salicylic acid (salicylate) by hydrolysis and deacetylation, and so the generation of a molecule of acetic acid.

Amphetamines These drugs (including ecstasy: 3,4-methylenedioxymethamphetamine, MDMA), can be taken orally, nasally, as smoke, or intravenously. They are central nervous system stimulants and common clinical features in overdose include tachycardia, hyperthermia and hypertension that may be fatal. Biochemical features include metabolic acidosis with increased alanine and aspartate transaminase, the latter due to hepatic damage. Use of ecstasy may cause hyponatraemia due to inappropriate arginine vasopressin secretion and self-induced water intoxication resulting from hyperthermia.

In 2009 there were 30 deaths due to ecstasy, compared with 100 deaths associated with horse riding. Mysteriously, there are few calls for the criminalization of the latter.

Cocaine In modest doses, this drug is a central nervous system stimulant, but high doses cause physiological (not psychological) depression, especially of the respiratory centre. Co-abuse of cocaine and alcohol may result in the formation of cardiotoxic ethylcocaine. Features in acute overdose are those similar to that of amphetamines, but more intense, and methods of ingestion are the same. However, cocaine-induced vasospasm is well known and can cause myocardial infarction with raised creatine kinase and its MB isoform. In the laboratory there may be metabolic acidosis (which may need treatment with intravenous bicarbonate) with hypokalaemia or hyperkalaemia. Diagnosis can be confirmed by detection of the stable metabolite benzoylecgonine. Chronic cocaine use leads to cardiomyopathy and psychiatric disorders, including paranoia and intense depression when use is ceased.

Opiates These drugs include morphine, heroin (diamorphine) and codeine, which belong to the alkaloid class of drugs that derive from the sap of the opium poppy, *Papaver somniferum*, and synthetic opioids such as methadone. Being taken intravenously, orally or smoked, they interact with opioid receptors in the brain to provide very effective analgesia, but unfortunately opiates are strongly addictive. Interaction with receptors in the intestines leads to reduction in motility and constipation.

A common consequence of overdose is suppression of the respiratory centre in the brain stem that may lead to death, but there is also drowsiness and coma. This can be noted in the laboratory with a low P_{O_2} with a high P_{CO_2}. Treatment aims to assist ventilation and provide oxygen; the effects of the opiate may be controlled with the antagonist naloxone. Opiates can be measured with relative ease by immunoassay in the serum and urine. These immunoassays may not directly detect synthetic opioids such as methadone (used to treat heroin addiction), but methadone-specific immunoassays are available.

Barbiturates and benzodiazapines These are central nervous system depressants widely used as sedatives, hypnotics, anaesthetics and anti-epileptics, but in many cases barbiturates have been replaced by benzodiazepines. They also have a place in alleviating withdrawal from opiate addition, but, like opiates, long-term use is associated with dependency and addiction. Overdose is associated with drowsiness, ataxia and slurred speech, measurement being by immunoassay.

Cannabis This drug is usually smoked, but can be eaten when baked into cakes or biscuits. It is derived from *Cannabis sativa*, the active chemicals being lipophilic tetrahydrocannabinols. It produces feelings of relaxation and contentment, sleepiness and lack of concentration, features interpreted as reflecting an induced apathy and low energy. The active metabolite stimulates the appetite centre, so promoting feeding and thus overweight. This latter feature prompted the development of a formal pharmaceutical drug which sought to block the cannabinoid receptor, thus reducing the desire to eat and so treating overweight and obesity.

In the USA, cannabis is used to help for the nausea and vomiting associated with cancer chemotherapy, and in the anorexia of acquired immunodeficiency syndrome. This view is spreading to other developed nations, with frequent calls from academics for a degree of decriminalization. Cannabinoids are often included as part of a DOAs screening panel but are seldom of importance in acute overdose.

Generally speaking, DOAs do not generate a great deal of work for the routine laboratory. However, large institutions are likely to have (at least) the ability to measure some DOAs, most preferring to send samples to a specialist toxicology service that is equipped with the relevant equipment (mass spectrometer, etc). Such samples are likely to derive from a need to define a particular DOA that may be causing a particular set of symptoms so that treatment can begin. Other samples may be provided by the police and prison services seeking evidence to pursue a criminal charge, such as driving under the influence of drink or drugs.

Heavy metals

Poisoning with heavy metals tends to occur as a result of industrial or environmental exposure or use of heavy-metal-containing cosmetics and rarely because of suicidal attempts. The symptoms of heavy metal poisoning depend on the amount ingested and the duration of the exposure. Diagnosis of heavy metal poisoning is made by measuring concentrations of the metal in blood, urine or hair and by observing the associated abnormalities related to its toxicity. The general approach used for treatment includes removal of the source and increased removal of the heavy metal from the body. Common heavy metals encountered in clinical practice include lead, aluminium, arsenic and cadmium.

Lead   Poisoning with this agent can be acute (which is rare) or chronic (which is more common). Sources of lead exposure in children include old paint, toys or eye cosmetics, the latter often imported from the Indian subcontinent. In adults, the sources of lead exposure tend to be occupational, especially in organizations involved in manufacturing batteries or smelting. Lead poisoning can cause a characteristic anaemia (Chapter 4) as well as liver, renal and neurological damage. Very severe cases can develop encephalopathy, seizures and coma.

Investigation of lead poisoning involves measurement of blood lead and a value higher than 0.5 μmol/L should be investigated further. The red cells of the affected individual may have high protoporphyrin concentrations owing to the inhibitory effect of lead on certain enzymes involved in haem synthesis. Management of lead poisoning involves removing the patient from the source of exposure and enhancing excretion of lead from the body using chelating agents such as dimercaptosuccinic acid.

Aluminium   This metal is found in water, although its concentrations can vary. It is a potential hazard for patients on renal dialysis as it can enter the body across the dialysis membrane. Indeed, the water used in dialysis is treated to remove any contaminating metals. Acute aluminium toxicity is very rare. However, aluminium toxicity in patients with renal failure can cause bone disease. Aluminium toxicity is investigated by measurement of its concentration in the plasma. Affected individuals are treated by removal of the source and by use of chelating agents such as desferrioxamine (also used to remove iron).

Arsenic   Acute ingestion of arsenic causes vomiting, shock and severe gastrointestinal pain. Chronic exposure causes persistent diarrhoea, renal failure, dermatitis and polyneuropathy. Exposure to arsenic can be assessed by measuring its content in hair or its excretion in urine. Treatment of affected individuals is by enhancing its excretion using a chelating agent or by dialysis in patients with renal failure.

Cadmium   Toxicity due to this metal may occur in individuals working in industries where there is exposure to cadmium fumes. It is also found in tobacco smoke, and accordingly many smokers have concentrations of blood cadmium twice that of nonsmokers. Individuals affected by severe cadmium toxicity suffer from kidney and liver damage and bone disease. Measurements of blood and urinary cadmium can give an indication of exposure. Treatment involves removal from the cadmium source.

21.3   Therapeutic drug monitoring

The variation in the human phenotype is obvious, and from the blood science perspective the concentrations of analytes in the blood and tissues also vary considerably from person to person, and this is the basis of the reference range for blood test results.

The same variability is present in how different people absorb, metabolize and excrete drugs: some process their drugs rapidly and efficiently, others take longer – there is nothing 'wrong' with this. However, there are several instances where we need to know how much of a drug, inevitably given orally, ends up biologically active in the bloodstream. This may be important because 90% of a standard dose may end up in the blood in one person, whereas in another only 30% of the same standard dose may be active, so that the former may effectively be overdosing, or the latter underdosing, both of which could be fatal. The whole area of studying how much of a drug ends up in the blood, how rapidly it does and how it is cleared from the blood is called pharmacokinetics. The relationship between blood concentration and effect is called pharmacodynamics.

General concepts

The process of checking the concentration of a drug in the blood is called TDM. By definition, TDM is measurement of drug concentrations to optimize drug dose (i.e. maximum benefit with least toxicity) for an individual. The aim of this is to ensure the oral dose translates to the 'therapeutic window' in the blood; that is, the range of concentrations where the drugs is effective, but not overeffective. Factors that influence the therapeutic window include:

• Route of administration, being oral (swallowed), sublingual (under the tongue), intravenous, subcutaneous, intramuscular, rectal, vaginal and transdermal.
• Absorption, where appropriate, across the gut wall (which is modified skin) or the skin itself. There may be an effect of the stomach and the acid it secretes. It also assumes healthy intestines.
• Metabolism in the plasma (where proteases and other enzymes may act on the drug) and the liver (where metabolic processes may alter the structure of the drug). Indeed, in some cases a therapeutic agent taken by the patient is an inactive pro-drug and we rely on their metabolism to convert it to an active drug. This is called first-pass metabolism.
• Excretion by the kidney and lungs, which again demands the health of these organs. Should there be renal disease, and the drug is not excreted at its normal rate (i.e. in health), then levels will build up in the blood, effectively increasing the concentration.

The opening phase of the use of a particular drug, such as warfarin and digoxin, may see an initial high dose (a loading dose) followed by a lower dose (the maintenance dose) for a longer period.

Once in the body, drugs may move into different anatomical regions, dependent on their particular chemistry. For example, some lipophilic drugs will naturally be more at home in fatty tissues, may need carrier proteins and move easily across the cell membrane. All of these factors come together to give us features such as the half-life of a drug, which effectively tells us how long it stays in the blood. However, the biological effect of a drug may be different from its half-life.

A further complication is that the drug may be modified during its absorption, in the plasma, or by the liver, in which case it is described as a prodrug. An example of this is the prodrug aspirin being modified to the active drug salicylate, and in some cases it may be worthwhile measuring the active and prodrug components. This pro/active principle is present in physiology, where thyroxine and testosterone are effectively pro-hormones for tri-iodothyronine and dihydrotestosterone.

Which drugs to monitor?

There are a number of instances where TDM is necessary, but almost all are because of the dangers of underactivity or overactivity, and therefore in finding the correct dose for that patient. TDM may also be necessary in order to demonstrate that the patient is actually taking the drug; that is, that they are compliant.

Anticoagulants   Warfarin (in common with many chemotherapeutics) is effectively a sophisticated poison, acting on the liver to suppress the synthesis of coagulation factors that form thrombi. Effective though it is, overdose leads to oversuppression and insufficient coagulation proteins, resulting in potentially fatal haemorrhage. TDM for warfarin is also necessary because of the wide range of effectiveness (some patients need 1 mg/day, others 15 mg/day), and also because of the many other drugs (such as antibiotics) that interfere with its bioavailability. In certain circumstances, TDM of low molecular weight heparin may be needed as the dose used to prevent a thrombus (prophylaxis) is less than that for treating a thrombus. Further details are presented in Chapter 8.

Anticonvulsant drugs   TDM is needed as the severity of the convulsions of epilepsy varies greatly from patient to patient. Underdosing will not prevent these convulsions, whilst overdosing will not only be toxic but will also severely restrict quality of life. This picture is compounded as patients may be taking more than one type of drug at the same time, and the metabolism of one drug may influence that of another. Carbamazepine, for example, may be rapidly or slowly metabolized by different isoforms of the hepatic microsomal enzyme cytochrome P3A4, and this underlies the wide adult recommended doses, ranging from 800 mg to 2000 mg daily for epilepsy. Notably, the rectal dose is higher (125 mg corresponds to 100 mg taken orally), with plasma concentration monitoring recommended. A loading dose of phenytoin is generally in the region of 20 mg/kg (maximum 2 g) and a maintenance dose of perhaps 100 mg/day. However, with a relatively narrow therapeutic window, the laboratory may be called on to check serum concentrations are within the reference of perhaps 7–20 mg/L.

Antimicrobial drugs   TDM of aminoglycoside antibiotics such as gentamicin and tobramycin is established. Isoniazid, rifampin, pyrazinamide and ethambutol, alongside anti-retroviral agents, are examples of drugs whose pharmacokinetics vary markedly in different individuals, so that measurement in serum overcomes variation in absorption and half-life. TDM of anti-fungals, such as fluconazole, itraconazole, posaconazole and voriconazole, in the prophylaxis and treatment of pathogens such as aspergillus has been advocated in high-risk groups such as the immunosuppressed.

Cancer chemotherapy Even more so than warfarin, many of these drugs (such as 5-flurouracil, 6-mercaptopurine and azathioprine) are sophisticated poisons, so that the active and correct dose can be crucial in avoiding the effects of overdose, which can be jaundice, rash and bone marrow suppression. In addition, several agents are often given together, making a link between plasma concentrations and biological activity difficult. Furthermore, part of the toxicity of these agents is gastrointestinal, so that absorption of these drugs (and much else) is likely to be abnormal. Monitoring levels of methotrexate is important as high levels produce severe side effects, some of which are ameliorated with folinic acid.

Many of these toxicities are less relevant for more targeted drugs, such as tyrosine kinase inhibitors (e.g. imatinib) in treating chronic myeloid leukaemia. TDM is required to ensure the oral dose of perhaps 400 mg daily is sufficient to ensure that a trough plasma imatinib concentration of 1 mg/L is achieved, as this is associated with an increased likelihood of maintaining a good response.

Blood science angle: Chemotherapy

An apparently appropriate dose of cancer chemotherapy in one patient may be toxic in another. In some cases this is due to variation in genes for key enzymes involved in metabolism. Chapter 19 discusses how deficiencies of the enzymes dihydropyrimidine dehydrogenase and thiopurinemethyltransferase can be defined by molecular genetics.

Cardiovascular drugs The effectiveness of lipid-lowering agents is monitored by serum total and high-density lipoprotein cholesterol and triacylglycerols, as required, and doses increased if required to hit National Institute of Health and Clinical Excellence targets. TDM of digoxin and amiodarone are established, although their clinical relevance is often unclear as monitoring is mostly by heart rate and electrocardiogram. Nevertheless, the daily oral dose of digoxin (as in the treatment of some cases of heart disease such as atrial fibrillation) is likely to be around 125–250 μg/day in order to achieve plasma concentrations of perhaps 0.8–2.0 μg/L. TDM is also useful in assessing dose requirements for flecainide usage in children.

Psychoactive drugs TDM of lithium carbonate (used to treat bipolar disorder) is important not merely to ensure it is effective, but also because it influences calcium metabolism (as is explained in Chapter 15). The optimal plasma concentration range is 0.4–1.0 mmol/L taken 12 h after an oral dose, with mild to moderate toxicity often occurring between 1.5 and 2.0 mmol/L. Lithium is excreted in urine, and can also be measured in serum using an ion-selective electrode or by colorimetry. Apart from its use in epilepsy, carbamazepine can be used in the prophylaxis of bipolar disorder unresponsive to lithium, generally as 400 mg daily in divided doses increased until symptoms are controlled, usually in the range 400–600 mg daily, to a maximum of 1.6 g daily.

Of the antipsychotic drugs, monitoring of clozapine is required as there is a 50-fold inter-patient variation in the rate of metabolism. However, haematologists need to take blood for a full blood count as clozapine is known to cause neutropenia. Furthermore, a low leukocyte or neutrophil count is a caution in the use of another antipsychotic, olanzapine. Both antipsychotics are influenced by smoking, and dose adjustment may be necessary if smoking started or stopped during treatment.

Other drugs that may need therapeutic monitoring include anti-malarials (such as quinine and allied agents) and immunosuppressants (such as ciclosporin). Major drugs requiring therapeutic monitoring are presented in Table 21.2.

Table 21.2 Common TDM target ranges.

Drug	Plasma target range
Aminoglycoside antibiotics	Peak 5–10 mg/L
Gentamycin/tobracycin	Trough <2 mg/L
Carbamazipine	4–10 mg/L, 17–42 μmol/L
Ciclosporin	120–300 nmol/L
Digoxin	0.8–2.0 ng/mL, 0.6–1.0 mmol/L
Lithium	0.6–1.0 mmol/L
Phenytoin	10–20 mg/L, 40–80 μmol/L
Theophylline	10–20 mg/L, 55–110 μmol/L

(Modified from Smith AF. *et al.*, Clinical Biochemistry, Blackwell Publishing, 1998).

21.4 Case studies

Case study 30

A 21-year-old male was brought to hospital in a state of confusion by a friend who found him at home with an empty bottle of aspirin. He had a high temperature of 39 °C, was sweating and hyperventilating (respiratory rate 20 breaths/min).

	Result (unit)	Reference range
Serum biochemistry		
Na	132 mmol/L	135–148
K	3.2 mmol/L	3.3–5.0
Urea	10 mmol/L	3.0–6.7
Creatinine	90 μmol/L	71–133
Glucose	3.3 mmol/L	3.5–5.5
Bicarbonate	11 mmo/L	24–29
Salicylate	520 mg/L	100–250
Blood gases		
Hydrogen ions (H^+)	61 nmol/L	35–45
P_{CO_2}	3.2 kPa	4.5–6.1

Interpretation In the serum sample, there is slightly low potassium, raised urea, slightly low glucose, low bicarbonate and high salicylate. The increased urea with a creatinine within the reference range implies an acute renal syndrome. The markedly low bicarbonate (being the major plasma anion) implies acidosis. This preliminary finding is confirmed by the blood gases, with high hydrogen ions and a low partial pressure carbon dioxide level.

The findings are typical of what you might see in a salicylate overdose. The patient suffers from a metabolic acidosis and has respiratory compensation (hence the increased rate of breathing). A salicylate overdose is an example of a mixed acid–base disorder where the initial disturbance is a respiratory alkalosis due to stimulation of the respiratory centre by salicylate. This initial acid–base disturbance becomes overwhelmed by the developing metabolic acidosis.

Additional details of blood gases, acidosis and alkalosis are found in Chapter 13.

Case study 31

A 61-year-old woman with bipolar disorder, although stable on 600 mg lithium carbonate once daily, complains of nausea, vomiting, constipation and myalgia. The practitioner orders urea and electrolytes, albumin and calcium.

	Results (unit)	Reference range
Na	142 mmol/L	135–145
K	4.7 mmol/L	3.8–5.0
Urea	4.3 mmol/L	3.3–6.7
Creatinine	142 μmol/L	135–145
Calcium	2.77 mmol/L	2.2–2.6
Albumin	39 g/L	35–50

Interpretation The only abnormality is hypercalcaemia, which may account for the symptoms. However, this cannot be caused by renal disease as urea and electrolytes are all within range. Upon receipt of these results, the practitioner rings the laboratory to see if they can perform an assay for parathyroid hormone (PTH) and lithium on the original serum sample. This is possible, and the PTH result is 1.1 pmol/L (reference range 1–6 pmol/L), whilst the lithium was sent to the local toxicology reference laboratory, which returned a result of 1.5 mmol/L (target range 0.6–1.0 mmol/L).

The serum concentration of lithium is 50% higher than the top of the reference range, and therefore seems to be the cause of the hypercalcaemia, especially as the PTH is at the low end of the reference range. Management seems likely to be reducing the daily dose of lithium and then see if this resolves the symptoms. Hopefully, the patient's psychological state will remain stable.

Summary

- Toxicology often requires specialist knowledge and equipment, both found in reference laboratories.
- Common poisons include aspirin, paracetamol, carbon monoxide and DOAs. The most useful analyses are of the liver function tests.
- The therapeutic monitoring of several drugs is recommended if there are concerns about a narrow therapeutic window, compliance and interference by other drugs or pathologies.
- Drugs requiring therapeutic monitoring include certain anticoagulants, antibiotics, drugs used in psychiatry and digoxin.

References

Ahmed N (ed). Clinical Biochemistry. Oxford University Press, 2011, ISBN 978-0-19-953393-0.

Smith AF, Beckett GJ, Walker SK, Rae PWH. Clinical Biochemistry. Blackwell, 1998, ISBN 0-632-04834-4.

Further reading

Alapat PM, Zimmerman JL. Toxicology in the critical care unit. Chest. 2008;133:1006–1013.

Flanagan RJ. Developing an analytical toxicology service: principles and guidance. Toxicol Rev. 2004;23:251–263.

Nieuwlaat R, Connolly SJ, Mackay JA. et al. Computerized clinical decision support systems for therapeutic drug monitoring and dosing: a decision-maker-researcher partnership systematic review. Implement Sci. 2011;6:90.

Walker DK. The use of pharmacokinetic and pharmacodynamic data in the assessment of drug safety in early drug development. Br J Clin Pharmacol. 2004; 58:601–608.

22 Case Reports in Blood Science

The need to merge different disciplines into a new structure, blood science, by definition admits the overlap of biochemistry, immunology and haematology (and possibly much more). We conclude this textbook with generated examples, some of which are based on actual cases, where the different sciences all contribute to the complete study of disease; that is, pathology.

Abbreviations

ALP	alkaline phosphatase
APTT	activated partial thromboplastin time
ALT	alanine aminotransferase
AST	aspartate aminotransferase
CEA	carcinoembryonic antigen
CRP	c-reactive protein
DKA	diabetic ketoacidosis
ECG	electrocardiogram
ESR	erythrocyte sedimentation rate
FBC	full blood count
eGFR	estimated glomerular filtration rate
GGT	gamma glutamyl transferase
GP	general practitioner
Hct	haematocrit
hsCRP	high-sensitivity CRP
ICU	intensive care unit
INR	international normalized ratio
K	potassium
LFT	liver function test
MCV	mean cell volume
Na	sodium
PSA	prostate specific antigen
PTH	parathyroid hormone
PT	prothrombin time
RCC	red cell count
T3	tri-iodothyronine
T4	thyroxine
TSH	thyroid-stimulating hormone
uACR	urinary albumin/creatinine ratio
U&Es	urea and electrolytes
WCC	white cell count

Blood Science: Principles and Pathology, First Edition. Andrew Blann and Nessar Ahmed.
© 2014 John Wiley & Sons, Ltd. Published 2014 by John Wiley & Sons, Ltd.

Case report 1

History

An 82-year-old man presents to his GP with a 3-month history of excessive and progressive tiredness and lethargy, and occasional pain in some bones, but not his joints. During the recent winter he twice needed antibiotics to treat a chest infection. On examination he was thin, pale and with two small lumps in the front of the neck. The blood test results are given in Table 22.1.

Table 22.1 Case report 1.

Analyte	Result (unit)	Reference range	Analyte	Result (unit)	Reference range
Haemoglobin	98 g/L	133–167	WCC	8.0×10^9/L	4.0–10.0
RCC	4.1×10^{12}/L	4.3–5.7	MCV	82.5 fL	77–98
Platelets	157×10^9/L	143–400	ESR	85 mm/h	<10
PT	13 s	11–14	APTT	28 s	24–34
Hct	0.34 L/L	0.35–0.55			
Neutrophils	3.0×10^9/L	2.0–7.0	Lymphocytes	4.0×10^9/L	1.0–3.0
Monocytes	0.3×10^9/L	0.2–1.0	Eosinophils	0.15×10^9/L	0.02–0.5
Basophils	0.05×10^9/L	0.02–0.1	Atypical cells	0.5×10^9/L	<0.02
Na	140 mmol/L	133–144	K	5.3 mmol/L	3.4–5.1
Urea	7.2 mmol/L	3.0–8.3	Creatinine	125 μmol/L	44–133
eGFR	51 mL/min per 1.73 m^2	>90	Calcium	2.82 mmol/L	2.2–2.6
Albumin	39 g/L	35–50	Total protein	85 g/L	60–80
Bilirubin	15 μmol/L	<21	ALP	155 IU/L	20–130
AST	26 IU/L	10–50	ALT	22 IU/L	5–42
GGT	49 IU/L	<70	CRP	6.5 mg/L	<5

Interpretation

There are many abnormalities. These are low haemoglobin, haematocrit, red cells and eGFR, with high ESR, lymphocytes, atypical leukocytes (which exceed the number of monocytes), total protein, calcium, ALP, plasma viscosity and CRP.

With low red cell indices and a history of tiredness and lethargy, the patient is anaemic, and an MCV in the reference range makes this a normocytic anaemia. The greatest abnormality is the ESR, one of the primary reasons for which is rheumatoid arthritis. However, with no history of overt joint pains, we can probably exclude this. Although the lymphocyte count is raised, this could conceivably be due to an infection, but few infections can, by themselves, cause such a grossly abnormal ESR. The increased number of atypical leukocytes demands attention.

Of the biochemistry, the key abnormalities are the low eGFR, and the raised ALP, calcium and total protein. These results would prompt a second round of investigations; the raised calcium may be due to mobilization from bone or may be due to primary hyperparathyroidism, which would be helped by knowledge of the PTH concentration. An alternative is that the hypercalcaemia is due to bone metastases, a possible source being the prostate, hence the request for the measurement of PSA. The hyperproteinaemia would prompt a request for plasma viscosity and serum protein electrophoresis.

Additional analyses

Analyte	Result (unit)	Reference range
PSA	2.2 ng/mL	0–5
Urinalysis	++ proteinuria	Trace
Plasma viscosity	1.92 mPa	1.5–1.72
PTH	1.2 pmol/L	1–6
Serum protein electrophoresis	Major band in the gammaglobulin region	

Discussion and diagnosis

The low PTH makes primary hyperparathyroidism a most unlikely cause of the hypercalcaemia, and similarly, although an often imprecise and nonspecific marker, the low PSA also effectively rules out prostate cancer. The key result is the major gammaglobulins band (a para-protein) detected by electrophoresis, which is likely to be responsible for the high plasma viscosity, which is likely to be the cause of the grossly abnormal ESR. The low eGFR indicates a degree of renal failure, and this may be related to the moderate proteinuria, which may be a Bence–Jones protein. Together these form the firm diagnosis of myeloma, supported by high calcium and ALP, indicative of bone resorption (and so the bone pain). The

normocytic anaemia is an established aspect of this malignancy, whilst the raised lymphocyte count and atypical cells may be some plasma cells which have escaped the bone marrow. The small lumps in the neck may well be lymphadenopathy, evidence of the movement of the neoplasia from bone marrow and into the lymph nodes, and is a sign of a poor prognosis. Other investigations would be X-rays and bone scans to assess the extent of the infiltration of the tumour into the skeleton.

Case report 2

History

A 63-year-old man with type 2 diabetes, a body mass index of 28.5 kg/m^2, mildly reduced renal function, and a below-knee amputee, is seen at his home by the district nurse. She is concerned that his remaining foot feels cool and is pale, and takes some blood. Whilst doing so, the patient recalls a blood test was done 'months and months ago'. On examination

today his systolic/diastolic blood pressure is 142/82 mmHg. His medications are amlodipine 5 mg once a day, frusemide 40 mg twice a day, ramipril 5 mg once a day, simvastatin 20 mg once a day and metformin 1000 mg once a day. A week later the district nurse telephones and asks for an opinion. You review his blood results (Table 22.2) and compare them with bloods taken 9 months ago.

Table 22.2 Case report 2.

Analyte	Reference range	9 months ago	Now
Haemoglobin	133–167 cg/dL	132	129
MCV	80–98 fL	85.5	86.4
RCC	$(4.3–5.7) \times 10^{12}$/mL	4.9	4.7
Hct	0.35–0.55 L/L	0.42	0.41
Platelets	$(143–400) \times 10^9$/mL	296	301
ESR	<10 mm/h	8	12
HbA$_{1c}$	3.8–6.2%	8.7	9.2
Total WCC	$3.7–9.5 \times 10^9$/L	7.6	7.9
Neutrophils	$1.7–6.1 \times 10^9$/L	5.0	5.6
Lymphocytes	$1.0–3.2 \times 10^9$/L	2.0	1.8
Na	133–144 mmol/L	138	141
K	3.4–5.1 mmol/L	4.4	5.0
Urea	3.0–8.3 mmol/L	9.1	9.3
Creatinine	44–133 μmol/L	112	168
eGFR	>90 mL/min per 1.73 m^2	61 (Stage II)	38 (Stage IIIb)
Bilirubin	<21 μmol/L	9	12
GGT	<70 IU/L	35	45
AST	<60 IU/L	43	38
ALP	40–260 IU/L	120	112
Total protein	63–84 g/L	68	65
Albumin	35–50 g/L	33	29
uACR	<2.5 mg/mmol	3.6	8.3
Total cholesterol	<5.0 mmol/L	4.9	5.1
HDL cholesterol	>1.2 mmol/L	1.4	1.3

Interpretation

The low haemoglobin and normal MCV point to a normocytic anaemia at both time points. The HbA_{1c} is high in the initial sample and has increased further in the second sample, indicative of deteriorating glycaemic control. Nine months ago the marginally acceptable total cholesterol has gone out of range. Regarding renal function, the urea is consistently high, but the creatinine and eGFR have deteriorated markedly. The albumin has fallen and the uACR has become considerably worse.

Discussion and diagnosis

The health of this diabetic seems to have deteriorated markedly in the 9 months between the samples. Haemoglobin is definitely low and the patient is likely to be anaemic, and the HbA_{1c} has become more adverse. Anaemia is not uncommon in diabetes, but a clear link is often uncertain. However, the most concerning aspect is the rise in creatinine and parallel fall in eGFR. This gives a clear diagnosis of chronic renal failure, and the stage of this has moved from stage II, straight through stage IIIa, to stage IIIb, and is mirrored by the adverse uACR. The latter may be linked to the reduction in serum albumin, which may be because of loss into the urine through a damaged glomerulus.

Risk factor management

Traditionally, blood scientists have not been involved directly in the management of cardiovascular risk factors. However, we are in the key position of being able to take an overview of several factors, as in this case. There are several aspects where comment is justified: the deteriorating renal disease, the rising hyperglycaemia, and the lipid profile. Although the latter is barely within reference values, this range is for a general population, and for high-risk groups needs to be more severe. Accordingly, this patient's target needs to be a lower total cholesterol, certainly <4.5 mmol/L, maybe even <4 mmol/L, as should be achieved by increasing the simvastatin to 40 mg a day.

Similarly, the blood pressure target in general health of <140/90 is lower in high-risk groups, and should be <135/85. This may be achieved by increasing anti-hypertensive treatment, but close attention needs to be taken in pharmacotherapy in the presence of chronic renal failure. This patient is at risk of suffering a major cardiovascular event (myocardial infarction, stroke) and would certainly benefit from 75 mg aspirin daily. Finally, the hyperglycaemia needs to be addressed, and there are a number of options, including insulin, normally reserved for type 1 diabetics. Certainly, the district nurse is likely to recommend a visit to the day hospital, where consultants can meet in a multidisciplinary team to discuss options.

Case report 3

History

A 25-year-old female intravenous drug abuser presents to her GP with progressive 24–48-h history of malaise, headaches and nausea. There is fever (38.5 °C), hyperventilation (10 breaths/minute) and tachycardia (pulse rate 90 beats per minute), and upon further questioning she reports thirst and frequent diuresis. Urgent routine biochemistry and haematology are ordered, and bicarbonate is written on the form (Table 22.3). You telephone the results to the GP, who asks your opinion.

Interpretation

There is a leukocytosis and neutrophilia with a raised ESR, indicative of a bacterial infection. There is no indication that this is related to an obvious cause, such as recent surgery. The urea is high, and although the creatinine is in the reference range the eGFR is a little low for someone so young. The level of bicarbonate is low, the calcium is very slightly raised, and the sodium is right at the top of the reference range. However, the latter

three results are of minor importance compared with the renal and white cell abnormalities.

Discussion and diagnosis

The raised urea with a normal creatinine points to acute renal failure, and with the white cell picture, and the absence of a clear infected wound, we seem to have the early stages of septicaemia. The low bicarbonate implies an acidosis, which needs to be confirmed with formal blood gases. Likely advice would be immediate referral to accident and emergency where industrial doses of intravenous antibiotics may be commenced. Technically, a blood culture would be required to prove the bacteria in the blood, but this case is urgent, so that antibiotics are likely to be started promptly. Fluids need to be kept up, in view of the report of frequent diuresis and the possibility of pre-renal hypernatraemia, and this may be easiest with a saline drip.

The case is of course high risk, and this will be a good opportunity to screen for viruses and educate/reinforce the dangers of intravenous drug abuse.

Table 22.3 Case report 3.

Analyte	Result (unit)	Reference range	Analyte	Result (unit)	Reference range
Haemoglobin	126 g/L	118–148	WCC	15.9×10^9/L	4.0–10.0
RCC	4.1×10^{12}/L	3.9–5.0	MCV	87.5 fL	77–98
Hct	0.36 L/L	0.33–0.47	ESR	25 mm/h	<10
Platelets	288×10^9/L	143–400	APTT	32.5 s	24–34
PT	11.9 s	11–14			
Neutrophils	12.6×10^9/L	2.0–7.0	Lymphocytes	2.0×10^9/L	1.0–3.0
Monocytes	1.0×10^9/L	0.2–1.0	Eosinophils	0.2×10^9/L	0.02–0.5
Basophils	0.1×10^9/L	0.02–0.1	Atypical cells	0.01×10^9/L	<0.02
Na	144 mmol/L	133–144	K	4.9 mmol/L	3.4–5.1
Urea	12.9 mmol/L	3.0–8.3	Creatinine	91 μmol/L	44–133
eGFR	69 mL/min per 1.73 m^2	>90	Calcium	2.62 mmol/L	2.2–2.6
Albumin	42 g/L	35–50	Total protein	71 g/L	60–80
Bilirubin	12 μmol/L	<21	ALP	83 IU/L	20–130
AST	35 IU/L	10–50	ALT	23 IU/L	5–42
GGT	49 IU/L	<70	Bicarbonate	22 mmol/L	24–29

Case report 4

Part 1

History

A female, aged 64 years, with a body mass index of 33.5 kg/m^2, reported progressive weight loss and tiredness over 6–9 months to her GP. Over the last 3 months she also noted periodical constipation and diarrhoea with abdominal discomfort and bloating, and even more recently was becoming 'unwell'. The GP requests FBC, ESR, U&Es, LFTs, a lipid screen, faecal occult blood, thyroid function tests and CEA. Results are given in Table 22.4.

Table 22.4 Case report 4: part 1.

Analyte	Result (unit)	Reference range	Analyte	Result (unit)	Reference range
Haemoglobin	120 g/L	118–148	WCC	6.6×10^9/L	4.0–10.0
RCC	4.5×10^{12}/L	3.9–5.0	MCV	81.7 fL	77–98
Hct	0.37 L/L	0.33–0.47	ESR	12 mm/h	<10
Platelets	318×10^9/L	143–400			
Neutrophils	3.8×10^9/L	2.0–7.0	Lymphocytes	1.9×10^9/L	1.0–3.0
Monocytes	0.5×10^9/L	0.2–1.0	Eosinophils	0.3×10^9/L	0.02–0.5
Basophils	0.06×10^9/L	0.02–0.1	Atypical cells	0.01×10^9/L	<0.02
Na	142 mmol/L	133–144	K	4.7 mmol/L	3.4–5.1
Urea	7.9 mmol/L	3.0–8.3	Creatinine	99 μmol/L	44–133
eGFR	51.5 mL/min per 1.73 m^2	>90	Calcium	2.65 mmol/L	2.2–2.6
Albumin	34 g/L	35–50	Total protein	68 g/L	60–80
Bilirubin	8 μmol/L	<21	ALP	606 IU/L	20–130
AST	32 IU/L	10–50	ALT	28 IU/L	5–42
GGT	49 IU/L	<70	TSH	7.3 mU/L	0.2–3.5
T4	4 pmol/L	10–25	CEA	125 mg/mL	<2.5
HDL cholesterol	1.4 mmol/L	>1.2	Total cholesterol	4.5 mmol/L	2.5–5.0
Triacylglycerols	1.9 mmol/L	<1.7	Faecal occult blood	Positive	Negative

Interpretation

The haematology is unremarkable, except for a very slightly raised ESR. The eGFR is low for a woman of this age, and equates to stage IIIa chronic kidney disease. There is a small hypercalcaemia, a small hypertriglyceridaemia, markedly raised ALP and evidence of abnormalities in thyroid metabolism. However, by far the most abnormal result is that of CEA, which is raised 50-fold. The faecal occult blood test is positive, so there is blood being lost somewhere in the intestines.

Discussion and diagnosis

The raised CEA and the history give a fairly clear preliminary diagnosis of colorectal cancer, which would be

confirmed with scanning and/or colonoscopy. There is also raised ALP, hypercalcaemia and hypothyroidism. However, the latter three findings pale into insignificance compared with the strong likelihood of colorectal cancer. The lowish haemoglobin may reflect blood lost in the intestines, or is part of the general biology of cancer. It is possible that the isolated raised ALP (and possibly the calcium) is secondary to the malignancy – it is not indicative of liver disease. Nevertheless, the immediate steps include a rapid referral to the oncology service with a view to surgery. Later, the thyroid question can be addressed.

Part 2

She has returned 18 months later, complaining to her GP of feeling unwell, more tired and having lost more weight. She wonders if these are some late effect of her cancer and its treatment. More tests are ordered, the results of which are given in Table 22.5.

Interpretation

Since her previous presentation, her haemoglobin has come down, and so she may be anaemic, and this may be

why her ESR has become more abnormal. However, why has this change occurred? The WCC has increased, with more neutrophils and lymphocytes, but this could be spurious. As regards the kidneys, renal function has improved markedly, with possible stage II chronic kidney disease. Does this imply the removal of a previous ill effect of the tumour? The previous hypercalcaemia and raised ALP have both resolved, but there is now a raised GGT, and all remaining LFTs have increased, although all are within their reference range. Total and HDL

Table 22.5 Case report 4: part 2.

Analyte	Result (unit)	Reference range	Analyte	Result (unit)	Reference range
Haemoglobin	110 g/L	118–148	WCC	7.5×10^9/L	4.0–10.0
RCC	4.0×10^{12}/L	3.9–5.0	MCV	90.9 fL	77–98
Hct	0.36 L/L	0.33–0.47	ESR	15 mm/h	<10
Platelets	293×10^9/L	143–400			
Neutrophils	4.2×10^9/L	2.0–7.0	Lymphocytes	2.2×10^9/L	1.0–3.0
Monocytes	0.9×10^9/L	0.2–1.0	Eosinophils	0.1×10^9/L	0.02–0.5
Basophils	0.1×10^9/L	0.02–0.1	Atypical cells	0.01×10^9/L	<0.02
Na	139 mmol/L	133–144	K	4.1 mmol/L	3.4–5.1
Urea	3.2 mmol/L	3.0–8.3	Creatinine	84 μmol/L	44–133
eGFR	62.9 mL/min per 1.73 m^2	>90	Calcium	2.45 mmol/L	2.2–2.6
Albumin	32 g/L	35–50	Total protein	62 g/L	60–80
Bilirubin	17 μmol/L	<21	ALP	93 IU/L	20–130
AST	37 IU/L	10–50	ALT	33 IU/L	5–42
GGT	143 IU/L	<70	Total cholesterol	4.0 mmol/L	2.5–5.0
HDL cholesterol	1.3 mmol/L	>1.2	Triacylglycerols	3.3 mmol/L	<1.7

cholesterol have fallen, whilst triacylglycerols have increased. The thyroid tests are not present, but perhaps there should have been a repeat of the CEA.

Discussion and diagnosis

Are we looking at the late effects of her cancer? Possibilities include metastatic disease in the liver, or the consequences of chemotherapy and/or radiotherapy.

An alternative mechanism may simply be the effect of alcohol, known to increase both GGT and triacylglycerols. This may also account for the rises in the remaining LFTs. Perhaps the fall in haemoglobin and total cholesterol reflect a degree of self-neglect that is a factor in alcoholism. Close attention to the FBC indicates that the MCV has increased by over 10%, unlikely to be a coincidence, so may also be the effect of alcohol.

Case report 5

Part 1

History

A 75-year-old woman with a body mass index of 29.5 kg/m^2 presents to accident and emergency tearful and in distress. Attended by her family, they describe an episode of collapse several hours ago during a country picnic, and which followed a game with grandchildren. The pains in her chest had not resolved, and she now has shortness of breath, with very shallow and rapid breathing. There is also a tachycardia, with a pulse rate of 95 beats per minute. An ECG and chest X-ray are performed, but these are normal. Blood results are given in Table 22.6. She is kept in overnight, and in the morning is far better, and although her breathing is better she still complains of pain in her chest.

Interpretation

The haematology is unremarkable, except for a minor basophilia. The biochemistry is also normal. The immediate concerns of the cause of the chest pain are of myocardial infarction and pulmonary embolism. Fortunately, these are effectively excluded by the normal ECG, CK, CK-MB and troponin and by the normal d-dimers respectively.

Discussion and diagnosis

There is no clear diagnosis, only a number of possibilities, and the patient is discharged with a letter to the GP.

Table 22.6 Case report 5: part 1.

Analyte	Result (unit)	Reference range	Analyte	Result (unit)	Reference range
Haemoglobin	120 g/L	118–148	WCC	8.3 × 10^9/L	4.0–10.0
RCC	4.5 × 10^{12}/L	3.9–5.0	MCV	92.5 fL	77–98
Hct	0.42 L/L	0.33–0.47	ESR	8 mm/h	<10
Platelets	313 × 10^9/L	143–400	D-dimers	150 units/mL	<500
Neutrophils	5.1 × 10^9/L	2.0–7.0	Lymphocytes	2.0 × 10^9/L	1.0–3.0
Monocytes	0.8 × 10^9/L	0.2–1.0	Eosinophils	0.1 × 10^9/L	0.02–0.5
Basophils	0.3 × 10^9/L	0.02–0.1	Atypical cells	0.02 × 10^9/L	<0.02
Na	136 mmol/L	133–144	K	4.5 mmol/L	3.4–5.1
Urea	4.5 mmol/L	3.0–8.3	Creatinine	92 μmol/L	44–133
eGFR	54.9 mL/min per 1.73 m^2	>90	Albumin	39 g/L	35–50
Bilirubin	9 μmol/L	<21	ALP	63 IU/L	20–130
AST	32 IU/L	10–50	ALT	27 IU/L	5–42
GGT	56 IU/L	<70	CK	95 IU/L	55–170
CK-MB	12 IU/L	<25	Troponin T	0.01 μg/L	<0.01

Part 2

Since her admission 6 months ago, the woman's pulmonary problems have not resolved. The symptoms seem to come and go much like attacks, and she admits that they are less worrying now she has cut down on smoking. Suspecting allergen-precipitated asthma, her GP advises her to come in for a blood test next time she has an attack. This she does, and the blood is sent to the immunology laboratory. The results are given in Table 22.7.

Interpretation

The raised IgE supports the diagnosis of asthma, and she is again advised to stop smoking and lose weight. She is sent for lung function tests, which support the diagnosis, and she is started on a salbutamol inhaler. These seem to be effective in treating the severity of the attacks, and seem to reduce their frequency. In retrospect the 'attack' she suffered may well have been asthma, possibly precipitated by allergens in the location of the picnic.

Table 22.7 Case report 5: part 2.

Analyte	Result(unit)	Reference range	Analyte	Result (unit)	Reference range
Total protein	69 g/L	60–80	Immunoglobulins	16.2 g/L	7.5–22.5
IgG	12.5 g/L	6.0–16.0	IgA	2.1 g/L	0.8–4.0
IgM	1.1 g/L	0.5–2.0	IgD	25 kU/L	0–100
IgE	110 kU/L	0–81	CRP	3.5 g/L	<5.0

Part 3

History

The woman returns 3 months later with a complaint that the inhaler has caused a cough that is unresponsive to cough mixture. She also says the inhaler makes her tired, even though she has lost weight. She admits that she has found it hard to completely give up smoking, and is now down to five cigarettes a day. She denies a fever or sweats and has a normal temperature. The GP orders an FBC and ESR (Table 22.8).

Interpretation

The FBC reveals a low haemoglobin and raised ESR. Notably, the former has fallen by 5 g/L since the initial presentation. This may conceivably be a drug effect, and she may therefore be a little anaemic. A chest infection seems unlikely, as this may explain the increase in the WCC, but the GP nevertheless orders another chest X-ray. This finds a shadow on the lung consistent with cancer. On review of the X-ray on presentation (9 months ago) an equivocal change is noted.

Table 22.8 Case report 5: part 3.

Analyte	Result (unit)	Reference range	Analyte	Result (unit)	Reference range
Haemoglobin	115 g/L	118–148	WCC	9.1×10^9/L	4.0–10.0
RCC	4.0×10^{12}/L	3.9–5.0	MCV	90.1 fL	77–98
Hct	0.36 L/L	0.33–0.47	ESR	13 mm/h	<10
Platelets	296×10^9/L	143–400			
Neutrophils	6.1×10^9/L	2.0–7.0	Lymphocytes	1.7×10^9/L	1.0–3.0
Monocytes	1.1×10^9/L	0.2–1.0	Eosinophils	0.05×10^9/L	0.02–0.5
Basophils	0.15×10^9/L	0.02–0.1	Atypical cells	0.01×10^9/L	<0.02

Discussion and diagnosis

In hindsight, it could be argued that much of this woman's chest problems were an early indication of her cancer, which may have been due to smoking. The slight basophilia in accident and emergency and the subsequent raised IgE are hard to ignore and support the asthma, but equally may reflect a cancer. Similarly, the poor lung function test may be linked to an early neoplasm, or may indeed have been asthma. A normocytic anaemia is not uncommon in cancer, and the raised ESR cell count and the rise in white cells (although within the reference range) may also be an early indicator. The woman's response to the inhaler could be psychosomatic. However, the fine historical and pathological points of the case are now irrelevant.

Case report 6

History

A 78-year-old woman complains of weight gain (a stone over a year) despite attention to her diet and doing more exercise, feeling cold (often wearing coat, hat and gloves to go out, even in mild weather) and her hair becoming brittle and thin. More recently she adds that she is tired, needing a brief snooze after lunch and again at about 6 p.m. There is no swelling, nor nodules in the neck, and this examination is painless. Her GP requests FBC, ESR, U&Es and thyroid function tests (Table 22.9).

Table 22.9 Case report 6.

Analyte	Result (unit)	Reference range	Analyte	Result (unit)	Reference range
Haemoglobin	125 g/L	118–148	WCC	5.4×10^9/L	4.0–10.0
RCC	4.5×10^{12}/L	3.9–5.0	MCV	88.5 fL	77–98
Hct	0.40 L/L	0.33–0.47	ESR	12 mm/h	<10
Platelets	233×10^9/L	143–400			
Neutrophils	3.1×10^9/L	2.0–7.0	Lymphocytes	1.5×10^9/L	1.0–3.0
Monocytes	0.5×10^9/L	0.2–1.0	Eosinophils	0.25×10^9/L	0.02–0.5
Basophils	0.05×10^9/L	0.02–0.1	Atypical cells	0.02×10^9/L	<0.02
Na	139 mmol/L	133–144	K	4.0 mmol/L	3.4–5.1
Urea	5.2 mmol/L	3.0–8.3	Creatinine	76 μmol/L	44–133
eGFR	67.9 mL/min per 1.73 m^2	>90	Albumin	41 g/L	35–50
Bilirubin	12 μmol/L	<21	ALP	56 IU/L	20–130
AST	28 IU/L	10–50	ALT	21 IU/L	5–42
GGT	42 IU/L	<70	TSH	6.6 mU/L	0.2–3.5
T4	1 pmol/L	10–25	T3	<1 pmol/L	4–6.5

Interpretation, discussion and diagnosis

The haematology profile is normal, so her tiredness seems not to be due entirely to an anaemia. All the other blood results are normal – except the raised TSH and low T4. This makes the diagnosis clear – hypothyroidism. One recent study found 8% of women over the age of 50 to be 'biochemically' hypothyroid. The fact that there is a raised TSH proves that the pituitary is responding to the low thyroid hormones. It follows that the aetiology is of the failure of the thyroid gland itself, but the treatment is the same regardless of the aetiology: daily oral thyroxine, probably starting at 50 μg, and then increasing at 25 μg increments until the unwanted symptoms resolve. This is likely to be accompanied by an increase in serum T4 and a fall into the reference range for TSH.

Although unlikely to influence treatment, it may be useful to determine the exact aetiology with an additional blood test sent to the immunology laboratory. It is almost impossible to be hypothyroid due to iodine deficiency in the UK whilst taking a healthy mixed diet, other common causes being the anti-arrhythmia drug amiodarone, or lithium to treat bipolar disorder.

Analyte	Result (unit)	Reference range
Anti-thyroid peroxidase antibodies	Positive	Negative
Anti-TSH receptor antibodies	Negative	Negative
Anti-thyroglobulin antibodies	Negative	Negative

Many define Hashimoto's thyroiditis by the presence of autoantibodies to thyroid peroxidase or to thyroglobulin, with raised TSH, so this makes the diagnosis. This is relevant because many with one autoimmune disease are at risk of another. There is also a familial aspect, as Hashimoto's thyroiditis is linked to HLA DQA1 (itself linked to DR4), and with certain types of DQB1 (that are themselves linked to DR5 and DR3). One of the DR3-DQB1 haplotypes contributes to the genetic susceptibility to type 1 diabetes and autoimmune poly-glandular syndrome. However, at this stage, complex tissue typing of the patient and her close relatives is not advised.

Case report 7

History

A 40-year-old woman with a body mass index of 25.5 kg/m^2 complains of generalized aches and pains (which are getting worse), with some early morning stiffness, especially in the legs and the back. On examination, there is some eczema. She is advised to take simple analgesics, take regular exercise, and a blood test is taken for rheumatoid factor, which comes back as negative.

Six months later she returns, saying the pains have got worse, and that she is on the maximum doses of paracetamol. On examination there is also tenderness in various muscles, and an increase in her eczema, with a spotty face. The rheumatoid factor is repeated, which comes back weakly positive. All this prompts a full haematology, biochemistry and immunology screen aimed at a connective tissue disease (Table 22.10).

Table 22.10 Case report 7.

Analyte	Result (unit)	Reference range	Analyte	Result (unit)	Reference range
Haemoglobin	121 g/L	118–148	WCC	5.4 × 10⁹/L	4.0–10.0
RCC	4.5 × 10¹²/L	3.9–5.0	MCV	88.5 fL	77–98
Hct	0.40 L/L	0.33–0.47	ESR	22 mm/h	<10
Platelets	343 × 10⁹/L	143–400			
Neutrophils	3.1 × 10⁹/L	2.0–7.0	Lymphocytes	1.5 × 10⁹/L	1.0–3.0
Monocytes	0.5 × 10⁹/L	0.2–1.0	Eosinophils	0.25 × 10⁹/L	0.02–0.5
Basophils	0.05 × 10⁹/L	0.02–0.1	Atypical cells	0.02 × 10⁹/L	<0.02
Na	139 mmol/L	133–144	K	4.0 mmol/L	3.4–5.1
Urea	5.2 mmol/L	3.0–8.3	Creatinine	96 μmol/L	44–133
eGFR	59.3 mL/min per 1.73 m²	>90	Albumin	41 g/L	35–50
Bilirubin	12 μmol/L	<21	ALP	56 IU/L	20–130
AST	28 IU/L	10–50	ALT	21 IU/L	5–42
GGT	42 IU/L	<70			
Rheumatoid factor	Titre 1/60	≤1/40	Anti-nuclear antibodies	Titre 1/80	≤1/40
Anti-dsDNA	70 IU/mL	<30, negative; 30–75, borderline; >75, positive	Anti-ENA antibodies	Weakly positive for Ro (SSA), negative for LA (SSB)	Negative
hsCRP	7.5 g/L	<5	Complement component 3	0.5 g/L	0.65–1.75
Complement component 4	0.15 g/L	0.1–0.5			

Interpretation and discussion

The initial present symptoms are vague, but on hindsight were clear harbingers of something more sinister, and it is likely that this woman is suffering from an inflammatory connective tissue disease. The only very slightly raised body mass index is likely to exclude osteoarthritis (overweight/obesity being the greatest risk factor). This exclusion is supported by the raised ESR, as osteoarthritis is a disease whose inflammation is limited to the joint and with minimal systemic effects. The most common disease in this group is rheumatoid arthritis, commonly attacking middle-aged women, and the slightly raised rheumatoid factor supports this diagnosis. However, of greater significance are the increased anti-dsDNA antibodies and the weakly positive antibodies to Ro (SSA). The high CRP, alongside the ESR, simply marks inflammation, but the reduced levels of C3 could be significant.

Diagnosis

The history and positive nuclear autoantibodies point to systemic lupus erythematosus. Note the qualification of 'point to', as the sensitivity and specificity of all these symptoms, history and autoantibodies tests (including rheumatoid factor) are far from 100%. Accordingly, the ultimate diagnosis demands both clinical and laboratory features. As there is ongoing inflammation, there is also likely to be an acute-phase response, and this may drive increased C3 and C4. Therefore, the fact that one of these is low implies the consumption of this molecule, possibly of both.

It is likely that aggressive immunosuppression will begin, and over the coming years the woman will be no stranger to the hospital or to venepuncture. Management will focus on reducing the laboratory markers, which are assumed to be surrogates of disease severity. A key index will be U&Es, as lupus nephritis is a common complication and also a marker of poor outcome. In this respect the eGFR already places her in stage IIIa chronic kidney disease.

Case report 8

A 34-year-old male postgraduate received an unrelated renal transplant 8 years ago. He had previously rejected a kidney donated by his father, and since then had been on dialysis every third day. He is currently maintained on 30 mg steroids daily and an angiotensin-converting enzyme inhibitor. Being fully aware of his condition, he regularly monitors his production of urine. He notes that over the last 2 months his urine production has been falling, and requests blood tests (Table 22.11). He is otherwise in good health, with blood pressure well within target, and a trace of proteinuria.

Table 22.11 Case report 8.

Analyte	Result (unit)	Reference range	Analyte	Result (unit)	Reference range
Haemoglobin	138 g/L	133–167	WCC	4.4×10^9/L	4.0–10.0
RCC	4.9×10^{12}/L	4.3–5.7	MCV	87.6 fL	77–98
Platelets	297×10^9/L	143–400	ESR	12 mm/h	<10
Hct	0.43 L/L	0.35–0.53			
Neutrophils	2.5×10^9/L	2.0–7.0	Lymphocytes	1.2×10^9/L	1.0–3.0
Monocytes	0.4×10^9/L	0.2–1.0	Eosinophils	0.25×10^9/L	0.02–0.5
Basophils	0.05×10^9/L	0.02–0.1	Atypical cells	0×10^9/L	<0.02
Na	142 mmol/L	133–144	K	4.8 mmol/L	3.4–5.1
Urea	7.9 mmol/L	3.0–8.3	Creatinine	106 μmol/L	44–133
eGFR	73.7 mL/min per 1.73 m^2	>90	Albumin	37 g/L	35–50
Bilirubin	10 μmol/L	<21	ALP	62 IU/L	20–130
AST	38 IU/L	10–50	ALT	32 IU/L	5–42
GGT	48 IU/L	<70	CRP	<5	<5

Interpretation

The haematology profile is normal, although the ESR is minimally increased. Similarly, the biochemistry is unremarkable. However, the patient has a firm grasp on his past laboratory history and in the past year has plotted a slowly falling haemoglobin with slowly rising creatinine. Naturally his fears are of early signs of rejection, and he is only slightly reassured by the normal CRP and relatively good eGFR.

Discussion and diagnosis

Although a one-off, this blood test is still valuable. The normal red cell indices imply sufficient erythropoietin, and therefore a functioning kidney, and all the white cell indices are within reference ranges, but are lowish, and this may be an effect of the steroids. The U&Es are within reference range, pointing to reasonable renal function.

However, although the creatinine is within 'normal' range, this applies to a general population, and more care will be needed in this high-risk group (as with Case report 7 and the risk of lupus nephritis). At present the eGFR places him in possible stage II chronic kidney disease, which is good news.

These blood results give little indication of an impending rejection crisis, but the gold standard of this is tissue biopsy, a procedure not lightly undertaken. The question here is the extent to which such a crisis may be assumed to be developing, and if so, whether or not immunosuppression should be increased (perhaps ciclosporin), which will by definition reduce the ability of the leukocytes to defend against microbial pathogens. The unfortunate expression is 'watchful waiting', and it may only be a matter of time before action needs to be taken. The renal physician can expect close questioning by the patient.

Case report 9

Day 1

The relatives call 999 on behalf of a 45-year-old woman, saying she has collapsed and has pains in her chest. In accident and emergency she is in distress, in general pain (specifically not of the chest, for which she is given paracetamol), is poorly communicative, and as she cannot stand her body mass index cannot be ascertained, although she is clearly morbidly obese. Her temperature is 39.8 °C and she has bilateral suppurative sacral sores on both buttocks with oedema on both ankles. Systolic/diastolic blood pressures are 137/85 mmHg, breathing is shallow and laboured (respiratory rate 20/min) and there are crackles in both lungs; oxygen saturation is suboptimal. An ECG is normal, and a near-patient testing device for troponin is negative.

She is admitted to a medical ward on oxygen, and being considered to be at high risk of venothromboembolism she is started on oral anticoagulation with warfarin, but in the short term is given subcutaneous unfractionated heparin. This will continue until the INR rises into the therapeutic range. The sacral sores are swabbed for microbiology and then dressed. Bloods are taken and the results are presented in Table 22.12.

Interpretation

With low red cell indices, the patient may be anaemic. There is nothing here that is grossly abnormal, but the eGFR is low, giving moderate chronic kidney disease (stage IIIa). Clinically, although the total WCC and neutrophil count are within reference ranges, there is a strong suspicion of a chest infection, and so oral antibiotics are started. She is sent for a chest X-ray (which turns out to be clear), but refuses physiotherapy. Low cardiac enzymes and troponin rule out myocardial damage and may be present in an acute coronary syndrome. D-dimers were not ordered as it would be likely that levels would be high even in the absence of a pulmonary embolism, and so would be of little value in diagnosis or management.

Day 2

The general medical condition of the patient has stabilized. Her temperature has come down a little. The haematology picture has also broadly stabilized, and the heparin is becoming effective in prolonging the APTT and so minimizing the risk of venous thrombosis. There are little major changes in the biochemistry profile. The microbiology of the sacral sores shows normal skin bacteria.

Table 22.12 Case report 9.

	Reference range	Day 1	Day 2	Day 3	Day 4	Day 5
Haemoglobin	118–148 g/L	105	104	99	75	69
MCV	80–98 fL	92.3	91.2	91.6	92.0	91.8
RCC	$(3.9–5.0) \times 10^{12}$/mL	4.1	4.3	3.9	3.0	2.8
Hct	0.33–0.47 L/L	0.38	0.39	0.36	0.28	0.26
Platelets	$(143–400) \times 10^9$/mL	245	240	233	178	125
PT	11–14 s	13	15	17	24	30
INR	2–3	1.1	1.2	1.4	1.9	2.5
APTT	24–34 s	32	40	44	70	36
APTT ratio	1.5–2.5	1.1	1.4	1.5	2.4	1.2
Fibrinogen	1.5–4.0	n.d.	n.d.	n.d.	n.d.	1.2
ESR	<10 mm/h	12	15	24	30	35
Total WCC	$3.7–9.5 \times 10^9$/L	9.0	9.0	9.9	12.2	20.8
Neutrophils	$1.7–6.1 \times 10^9$/L	5.8	6.0	6.9	9.0	16.9
Lymphocytes	$1.0–3.2 \times 10^9$/L	2.1	1.8	1.7	1.9	2.1
Monocytes	$0.2–1.0 \times 10^9$/L	0.7	0.9	1.1	1.0	1.4
Eosinophils	$0.02–0.5 \times 10^9$/L	0.3	0.2	0.3	0.3	0.1
Basophils	$0.02–0.1 \times 10^9$/L	0.07	0.1	0.05	0.04	0.1
Atypical cells	$<0.02 \times 10^9$/L	0.01	0.01	0.1	0.01	0.2
Na	133–144 mmol/L	139	140	136	134	130
K	3.4–5.1 mmol/L	4.8	4.5	4.7	4.5	4.0
Urea	3.0–8.3 mmol/L	7.9	7.7	8.1	8.5	15.5
Creatinine	44–133 μmol/L	96	99	105	125	145
eGFR	>90 mL/min per 1.73 m^2	57.9	55.9	52.2	42.7	36.0
Bilirubin	<21 μmol/L	12	10	10	12	15
GGT	<70 IU/L	60	58	65	93	101
AST	<60 IU/L	45	48	50	49	55
ALT	5–42 IU/L	30	28	35	40	38
ALP	20–130 IU/L	120	130	140	170	195
Total protein	63–84 g/L	65	66	63	65	60
Albumin	35–50 g/L	34	35	32	30	25
CRP	<5 mg/L	n.d.	n.d.	8.8	15.5	36.7
Tropinin	<0.6 U/mL	<0.6				
Creatine kinase	55–170 IU/L	123				
CK-MB	<25	21				
Blood gases (day 5 only)						
pH	7.35–7.45					6.92
P_{CO_2}	4.7–6.0 kPa					2.31
P_{O_2}	12–14.6 kPa					13.5
Bicarbonate	24–29 mmol/L					15.5

Day 3

Over the next 24 h there is little change, but with her morbid obesity and difficulty in movement she is placed in nappies, refusing a urethral catheter. She continues to refuse physiotherapy. Topical antibiotics are applied to the sacral sores. Upon review the next day, the sacral sores are more inflamed and painful. Bloods are taken, and CRP is also requested.

Interpretation

Red cell indices have deteriorated and coagulation indices reflect anticoagulation. However, the ESR has changed markedly (doubling since admission), and this cannot be due to the red cell indices alone, but may reflect a developing infection, and despite the antibiotics. This is supported by a greater than 10% increase in both white cells and neutrophils. The antibiotics are changed to intravenous, and are supplied in a single saline drip of 500 mL. Renal function has also deteriorated, giving stage IIIa chronic kidney disease, and the albumin has fallen (possible as a consequence). Three of the LFTs have increased, although the significance of this is unclear. Raised CRP confirms active inflammation.

Day 4

The patient spent a restless night, and called nursing staff several times, requesting fluids, the need to be cooler, and requests for a bedpan, which was managed with difficulty. Temperature, which had fallen since admission, is now raised at 39.1 °C. She also complained of pain whilst urinating. A fourth set of bloods was taken.

Interpretation

There has been as notable drop in haemoglobin, the RCC and Hct, leading to concerns of a haemorrhage. If this is the case, as there is no bruising, it may be internal. The anticoagulants are effective, but the platelet count has also fallen markedly, and both of these add to concerns of a haemorrhage. However, the low platelet count may be heparin-induced thrombocytopenia, so this is stopped.

The ESR, white blood cell count and neutrophils (making up 74%) have increased further, and it seems there is a full-blown bacterial infection in progress. With concerns of septicaemia, she is transferred to the ICU, where she is catheterized, and intravenous antibiotics are increased. Blood cultures are taken.

Day 5

In ICU, the patient becomes increasingly distressed and comes in and out of consciousness. Temperature rises above 40 °C, respiratory rate and pulse rate also rise. Blood pressure is 97/66 mmHg, and the saline drip is resumed, which aims to support general haemodynamics. Oxygen is now delivered by a mask, and shortly after this blood gases are taken. Routine bloods are also ordered, to which fibrinogen is added. These show a further fall in red cell indices and platelets, and hypofibrinogenaemia. The INR is now in range, but now heparin has been stopped the APTT is only marginally raised. Blood is starting to ooze from the sacral sores.

The ESR, total white cells, neutrophils (now making up 81% of the former), monocytes and atypical cells are all raised. The autoanalyser identifies some of the latter as metamyelocytes. All LFTs are rising, total protein and albumin have fallen (possibly reflecting an acute-phase response), renal function has deteriorated and CRP has risen markedly. The blood culture is positive with *Escherichia coli*. Blood gases point to an acidosis.

Interpretation

The patient may well have an internal bleed, and four units of packed red cells are ordered. The blood bank is warned of the possibility of a need for packs of platelets and possible need of fresh frozen plasma. Vitamin K is given to reverse the effects of the warfarin, and blood is oozing from the vagina, which may be menstrual. The picture is compounded by the presence of an *E. coli* septicaemia (despite intravenous antibiotics) with an acidosis. Accordingly, bicarbonate is added to the saline drip.

In the evening of the fifth day, the packed red cells infusion begins, but the patient has now lost consciousness. She is noted to be anuric (reflecting renal failure). A deepening rash develops, tachycardia reaches 150 beats per minute and there is hyperventilation. She suffers a cardiac arrest which cannot be reversed. Upon post-mortem there was indeed a bleed into the peritoneal cavity, cause of death being *E. coli* septicaemia secondary to infected sacral sores. A contributing factor was morbid obesity: her body mass index was found to be 43.4 kg/m², although she was infused with 1500 ml of fluid in her final 36 h.

Case report 10

Day 1

A 23-year-old woman presents to accident and emergency with a 3-day history of shortness of breath, chest tightness, epistaxis, and coughing up white and then brown phlegm. Her family provided a past medical history of systemic juvenile onset arthritis (diagnosed at age 14 as Still's disease), autoimmune hepatitis and pulmonary hypertension. On examination there were decreased breath sounds, a respiratory rate of 28/min (normally about 15/min), a pulse rate of 100 beats/min (therefore, tachycardia), and +++ dipstick proteinuria.

Table 22.13 Case report 10.

	Reference range	Day 1	Day 2	Day 3
Haemoglobin	118–148 g/L	98	97	129
MCV	80–98 fL	69.2	70.6	71.0
RCC	$(3.8–5.0) \times 10^{12}$/mL	4.1	4.0	4.5
Hct	0.36–0.44 L/L	0.28	0.28	0.32
Platelets	$(143–400) \times 10^9$/mL	76	53	18
ESR	<10 mm/h	25	30	40
PT	11–14 s	15.5	14.8	15.0
PTT	24–34 s	45.1	40.5	39.5
Fibrinogen	1.5–4.0 g/L	not done	2.7	not done
Total WCC	$3.7–9.5 \times 10^9$/L	13.8	11.7	28.4
Neutrophils	$1.7–6.1 \times 10^9$/L	3.6	2.5	2.6
Lymphocytes	$1.0–3.2 \times 10^9$/L	7.4	6.9	23.8
Monocytes	$0.2–1.0 \times 10^9$/L	2.5	2.0	1.6
Eosinophils	$0.02–0.5 \times 10^9$/L	0.2	0.2	0.3
Basophils	$0.02–0.1 \times 10^9$/L	0.05	0.1	0.05
Atypical cells	$<0.02 \times 10^9$/L	0.1	0.05	0.1
Na	133–144 mmol/L	140	139	143
K	3.4–5.1 mmol/L	4.1	4.0	4.8
Urea	3.0–8.3 mmol/L	7.9	7.5	8.0
Creatinine	44–133 μmol/L	92	94	100
eGFR	>90 mL/min per 1.73 m^2	70	68	63
Bilirubin	<21 μmol/L	26	25	20
GGT	<70 IU/L	195	150	145
AST	10–50 IU/L	214	175	162
ALT	5–42 IU/L	267	204	226
ALP	20–130 IU/L	563	488	383
Albumin	35–50 g/L	25	17	13
CRP	<5 mg/L	244	215	72
Amylase	<300 IU/L	not done	125	not done

There was no joint pain, rash, fever or lymphadenopathy. Blood was taken and the results are shown in Table 22.13.

Interpretation and discussion

The red cell indices are low and the patient probably has microcytic anaemia, perhaps secondary to low iron. There is also thrombocytopenia with prolonged PT and APTT, which together account for the epistaxis. The leukocytosis and lymphocytosis imply a viral infection. There is also clear liver pathology, which may account for the microcytic anaemia as this organ stores iron, and also the prolonged clotting times and hypo-albuminaemia. The d-dimer was probably requested in an attempt to clarify the likelihood of a pulmonary embolus, but with so much other pathology the high level adds little to the overall picture.

The clinical features and laboratory data point to a lung disease, potentially broncho-pneumonia, possibly related to a flare up in juvenile inflammatory arthritis. She is admitted to a medical ward and is started on intravenous fluids, antibiotics and oral prednisolone.

Overnight her notes are retrieved, which recorded numerous recurrent disease flares, with malaise, high fever (>40 °C), myositis (creatine kinase >1000 IU/L, reference range 55–170 IU/L), lymphadenopathy and some joint pain, but responds to steroids. Previous liver enzyme abnormalities included ALT 700 IU/L (reference range 5–42 IU/L) and ALP 400 IU/L (reference range 40–260 IU/L). There were no autoantibodies to smooth muscle, liver, kidney or skin, but she was ANA positive, anti-Ro positive and anti-DNA negative. Viral serology was negative for hepatitis B and C viruses, cytomegalovirus, Epstein–Barr virus and parvovirus. Two liver biopsies found no significant abnormalities other than those related with systemic onset juvenile inflammatory arthritis. Although steroids resulted in a rapid improvement in disease, there were side effects of striae, weight gain and mood swings. Steroid dose reduction resulted in disease flares. Methotrexate resulted in pruritis and nausea; ciclosporin was not tolerated. This is clearly a very poorly young woman probably suffering a flare-up in her inflammatory connective tissue disease and a potential viral infection.

Day 2

A second set of bloods is obtained mid morning. At lunchtime, the patient complains of abdominal pain, thirst, nausea and the desire to vomit. There is haematuria, a temperature of 39.1 °C, and the systolic blood pressure falls to 85 mmHg, and a saline drip is put up to support haemodynamics. On suspicion of septicaemia, blood is taken for blood culture, and oral prednisolone is changed to intravenous prednisolone.

Interpretation and discussion

The blood picture has stabilized, if not marginally improved, with many indices, including white cells, moving towards their reference range, although the Hct is still low. However, the platelet count has fallen markedly, and may lead to severe haemorrhage, especially in view of the prolonged APTT since day 1. Fortunately, the fibrinogen is within the reference range, implying the prolonged clotting times are due to deficiencies in other coagulation factors. Normal amylase minimizes the possibility of pancreatic disease. Consideration is given for a possible need for blood products, so blood is sent for Group and Save.

Although renal function is not great (chronic kidney disease stage II), it is acceptable in the short term. Happily, all LFTs, although still high, have moved towards their reference range. CRP has fallen (a positive finding), but so has albumin (an adverse finding, possibly reflecting liver pathology).

Day 3

The patient still complains of abdominal discomfort, prompting urgent gastrointestinal and surgical opinions, which leads to a CT scan of the abdomen to exclude bowel ischaemia and cholangitis. The CT scan is normal. However, the patient deteriorates, being less responsive, and is admitted to ICU on suspicion of sepsis, and bloods were taken. The management plan was to stabilize on ICU, continue antibiotics and fluids, and increase the dose of hydrocortisone. Bloods were taken for polymerase chain reaction testing, which if positive would direct antiviral therapy.

Interpretation and discussion

Although red cell indices have improved notably, the lymphocyte count has more than doubled and the platelet count has more than halved. This dangerous thrombocytopenia prompts the call for the transfusion of platelet concentrates, and this is set in motion. Although LFTs seem to be resolving, the albumin has continued to fall. In the presence of normal renal function (so it seems) this hypoalbuminaemia is likely to be due to liver failure. However, the patient was suddenly unrousable, vomited 'coffee grounds' (indicative of blood in the stomach) and went into cardiac arrest. Cardiopulmonary resuscitation was unsuccessful.

This case illustrates the difficulty in trying to suppress what seems to be the response to viral infection with a powerful immunosuppressant, in the face of liver disease. The initial management seemed to be working, but on the last day the rapid worsening of the general inflammatory pathology, marked by the lymphocytosis, may have been too much.

Case report 11

Case report modified from Lemyze *et al.* (2010). A 73-year-old diabetic female presented to accident and emergency with progressive shortness of breath and a 2-week history of diarrhoea. Her medications included aspirin, blood pressure lowering drugs irbesartan and hydrochlorothiazide, and metformin. On examination she was oliguric, disorientated and confused, with a respiratory rate of 32 breaths/min, systolic/diastolic blood pressure 76/44 mmHg, and a pulse rate 125 beats/min. Bloods were as shown in Table 22.14.

Interpretation

The haematology shows a minimally increased ESR, but a markedly abnormal HbA_{1c}, implying months of hyperglycaemia. As for the biochemistry, there is

Table 22.14 Case report 11.

Analyte	Result (unit)	Reference range	Analyte	Result (unit)	Reference range
Haemoglobin	129 g/L	118–148	WCC	7.2×10^9/L	4.0–10.0
RCC	4.5×10^{12}/L	4.3–5.7	MCV	83.2 fL	77–98
Platelets	297×10^9/L	143–400	ESR	12 mm/h	<10
Hct	0.37 L/L	0.35–0.55	HbA$_{1c}$	9.5%, 80 mmol/mol	3.8–6.2, 20–42
Neutrophils	4.5×10^9/L	2.0–7.0	Lymphocytes	1.8×10^9/L	1.0–3.0
Monocytes	0.6×10^9/L	0.2–1.0	Eosinophils	0.3×10^9/L	0.02–0.5
Basophils	0.04×10^9/L	0.02–0.1	Atypical cells	0.01×10^9/L	<0.02
Na	133 mmol/L	133–144	K	6.8 mmol/L	3.4–5.1
Urea	22.0 mmol/L	3.0–8.3	Creatinine	279 µmol/L	44–133
eGFR	15.3 mL/min per 1.73 m^2	>90	Albumin	39 g/L	35–50
Bilirubin	8 µmol/L	<21	ALP	47 IU/L	20–130
AST	39 IU/L	10–50	ALT	29 IU/L	5–42
GGT	53 IU/L	<70	Lactate	17.4 mmol/L	0.7–1.9
CRP	<5 mg/L	<5	Glucose	9 mmol/L	3.3–5.5
pH	6.72	7.35–7.45	P_{CO_2}	1.87 kPa	4.7–6.0
P_{O_2}	14.13 kPa	12–14.6	Bicarbonate	12 mmol/L	24–29

hyperkalaemia, raised urea and creatinine, the latter giving a low eGFR, synonymous with stage IV chronic kidney disease. LFTs are normal, but there is markedly raised lactate and hyperglycaemia. All blood gas indices are low, except the oxygen partial pressure.

Diagnosis

The picture is of clear diabetic lactic acidosis. It may possibly also be described as DKA, but ketones have not been directly measured. Precipitating factors include the diarrhoea, predisposing to dehydration (hence the low blood pressure) leading to the impaired perfusion of organs. The normal WCC and CRP rule out septicaemia, which commonly causes an acidosis. Indeed, factors precipitating this condition include infections, clearly not present in this case.

Urgent treatment is demanded, and is likely to be led by the need to vigorously rehydrate, probably with intravenous fluids, but also supplements of bicarbonate to reverse the acidosis. This is also likely to help the hyperkalaemia, but an additional treatment is likely to be insulin; but if so, there must be sufficient plasma glucose. The borderline low sodium may be due to the osmotic diuresis caused by the hyperglycaemia

Once stable, a complete overhaul of the patient's care is required.

Case report 12

Case report modified from Tournoy et al. (2006). A 24-year-old woman, 22 weeks' pregnant with her first child, presents to accident and emergency with fever (temperature 39.8 °C), nausea, vomiting and pain on the right of the abdomen. Abdominal ultrasound was normal and there were no signs of foetal distress. Systolic and diastolic blood pressures were 120/62 mmHg, pulse rate was 110 beats/min. Routine bloods were as shown in Table 22.15.

Interpretation

The haemoglobin is low, but this is not uncommon in pregnancy, as is an MCV towards the top of the reference range and a marginally increased ESR. However, the thrombocytopenia is unexpected and demands an explanation, as does the prolonged PT. A raised d-dimer may also be expected in pregnancy, as in this case, but here, the increase far exceeds that expected of a normal

Table 22.15 Case report 12.

Analyte	Result (unit)	Reference range	Analyte	Result (unit)	Reference range
Haemoglobin	101 g/L	118–148	WCC	6.4×10^9/L	4.0–10.0
RCC	4.8×10^{12}/L	4.3–5.7	MCV	97.2 fL	77–98
Platelets	60×10^9/L	143–400	ESR	13 mm/h	<10
Hct	0.47 L/L	0.33–0.47	D-dimers	6400 μg/L	<250
PT	17 s	11–14	APTT	32 s	24–34
Neutrophils	3.5×10^9/L	2.0–7.0	Lymphocytes	1.9×10^9/L	1.0–3.0
Monocytes	0.6×10^9/L	0.2–1.0	Eosinophils	0.3×10^9/L	0.02–0.5
Basophils	0.04×10^9/L	0.02–0.1	Atypical cells	0.02×10^9/L	<0.02
Na	138 mmol/L	133–144	K	4.8 mmol/L	3.4–5.1
Urea	5.5 mmol/L	3.0–8.3	Creatinine	79 μmol/L	44–133
eGFR	70 mL/min per 1.73 m²	>90	Albumin	43 g/L	35–50
Bilirubin	40 μmol/L	<21	ALP	147 IU/L	20–130
AST	101 IU/L	10–50	ALT	71 IU/L	5–42
GGT	75 IU/L	<70	Lactate dehydrogenase	1579 U/L	266–500
CRP	110 mg/L	<5	Glucose	9 mmol/L	3.3–5.5

pregnancy. All five LFTs are raised, as is the CRP and lactate dehydrogenase.

Discussion and diagnosis

The combination of low haemoglobin, elevated liver enzymes and low platelets is a recognized syndrome (HELLP). The H is really for haemolysis, for which an addition test such as haptoglobin may be requested (it should be low or absent). However, haemolysis is implied by the raised lactate dehydrogenase, released in this case from haemolysed red cells, although it may also arise from the liver.

HELLP is not uncommon in pregnancy, but is unusual before the 28th week, with less than 3% presenting in weeks 17–20. It is serious as it may lead to possible fatal pre-eclampsia. Consequently, another cause may be present. In this case there would be a blood film, but the blood analyser failed to report anything unusual in the red cell morphology.

However, a blood scientist looking at the blood film (Figure 22.1) quickly identified the ring-form trophozoites of *Plasmodium vivax*, one of the organisms that causes malaria. Treatment with quinine was commenced,

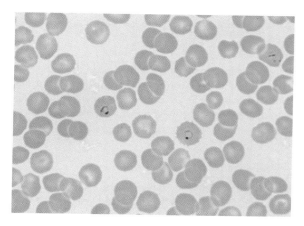

Figure 22.1 Blood film from case report 12 (Blann AD, *et al.* (2009) Haematology Morphology. Training CD-ROM. © IBMS/ Sysmex UK).

to which clindamycin was later added. It later transpired that the woman worked in the freight handling unit of an airport used by aeroplanes from equatorial Africa, from where it is presumed a mosquito fly was a stowaway. A healthy daughter was subsequently delivered.

Case report 13

Modified from Prejbisz A. *et al.*; citing Viera and Neutze (2010). In the latter, see also Viera and Neutze (2010).

A 22-year-old male reported with an increasing frequency of headaches and worsening shortness of breath. He also

Table 22.16 Case report 13.

Analyte	Result (unit)	Reference range	Analyte	Result (unit)	Reference range
Haemoglobin	148 g/L	133–167	WCC	6.0×10^9/L	4.0–10.0
RCC	5.2×10^{12}/L	4.3–5.7	MCV	91.3 fL	77–98
Platelets	302×10^9/L	143–400	ESR	6 mm/h	<10
Hct	0.47 L/L	0.35–0.53			
Neutrophils	3.5×10^9/L	2.0–7.0	Lymphocytes	1.8×10^9/L	1.0–3.0
Monocytes	0.3×10^9/L	0.2–1.0	Eosinophils	0.3×10^9/L	0.02–0.5
Basophils	0.1×10^9/L	0.02–0.1	Atypical cells	0×10^9/L	<0.02
Na	149 mmol/L	133–144	K	2.6 mmol/L	3.4–5.1
Urea	8.9 mmol/L	3.0–8.3	Creatinine	134 μmol/L	44–133
eGFR	61 mL/min per 1.73 m^2	>90	Albumin	32 g/L	35–50
Bilirubin	10 μmol/L	<21	ALP	62 IU/L	20–130
AST	38 IU/L	10–50	ALT	32 IU/L	5–42
GGT	48 IU/L	<70	CRP	<5	<5

said he was often very aware of the heart pounding in his chest. His systolic and diastolic blood pressures were 220/ 140 mmHg, with a pulse rate of 85 beats/min. This prompted the prescription of the angiotensin-converting enzyme inhibitor captopril and the beta blocker bisoprolol, which reduced his blood pressure to 170/118 mmHg, and his pulse rate fell to 75 beats/min. Blood was obtained for routine haematology and biochemistry; results are given in Table 22.16.

Interpretation

The haematology is unremarkable, but there are several abnormalities in the biochemistry. There is hypernatraemia and hypokalaemia, with very slightly raised urea and creatinine with reduced albumin. These results indicate renal dysfunction, prompting urinalysis and, as suggested by the high blood pressure, the measurement of aldosterone and renin in early morning samples. The urine analysis found albumin excretion of 60 mg/day (explaining low plasma levels) and aldosterone 45.3 μg/24 h (upper limit of reference range 30 μg/24 h) with increased levels of

potassium. The blood test reported increased aldosterone at 61.2 ng/dL (upper limit of reference range 34 ng/dL) and renin 0.1 ng/mL/24 h (reference range 4.8–48 ng/mL/ 24 h), giving an aldosterone/renin ratio of >60.

Discussion and diagnosis

The high aldosterone/renin ratio far exceeds that required (>30) for a firm diagnosis of primary aldosteronism, also known as Conn's syndrome, and as discussed in Chapter 12. The excessive secretion of aldosterone by the adrenal glands causes the retention of sodium and water (so rising levels in the blood) and in parallel the loss of potassium and hydrogen ions, so the blood pH rises (although in this case we have no blood gases). The increased mass of water in the blood causes the increased blood pressure.

A leading cause of primary aldosteronism is an adrenal adenoma, which was duly found on ultrasound, removed and confirmed on histopathology. The patient subsequently made a full recovery.

Case report 14

Case report is modified from McDonald (2012), with permission. A 12-year-old boy collapsed whilst playing during lunchtime at school. By the time the school nurse arrived he had lost consciousness. She called an

ambulance, which took him to the local accident and emergency unit. Upon arrival his blood pressure was 121/77 mmHg and he had partially regained consciousness. Bloods were as shown in Table 22.17.

Table 22.17 Case report 14.

Analyte	Result (unit)	Reference range	Analyte	Result (unit)	Reference range
Haemoglobin	148 g/L	115–158	WCC	6.9×10^9/L	4.9–13.7
RCC	5.2×10^{12}/L	4.0–5.2	MCV	91.3 fL	76–92
Platelets	302×10^9/L	150–400	ESR	6 mm/h	2–8
Hct	0.47 L/L	0.34–0.47			
Neutrophils	3.5×10^9/L	1.4–7.5	Lymphocytes	2.0×10^9/L	1.9–7.6
Monocytes	1.0×10^9/L	0.06–1.3	Eosinophils	0.3×10^9/L	0–0.75
Basophils	0.1×10^9/L	0–0.2	Atypical cells	0×10^9/L	<0.02
Na	121 mmol/L	135–146	K	3.0 mmol/L	3.5–5.0
Urea	13.6 mmol/L	2.5–6.5	Creatinine	54 μmol/L	30–60
eGFR[a]	65 mL/min per 1.73 m^2	>90	Glucose	58.1 mmol/L	3.0–6.0
Chloride	80 mmol/L	98–110	Bicarbonate	15 mmol/L	20–26
Calcium	2.47 mmol/L	2.2–2.7	Phosphate	3.15 mmol/L	1.0–1.8
Albumin	48 g/L	30–45	Total protein	89 g/L	62–82
HbA$_{1c}$	12.7%	<7	ALP	369 IU/L	60–300

[a]eGFR based on a paediatric formula that requires age, sex, height and serum creatinine.

Interpretation

There is hyponatraemia and hypokalaemia with hyper-uraemia and marked hyperglycaemia. There is also a low bicarbonate, and increased phosphates, total protein, ALP and HbA$_{1c}$. These prompted additional analyses: for blood gases (that confirmed the acidosis implied by the low bicarbonate), for beta-hydroxybutyrate (5 mmol/L, reference range 0.04–0.18 mmol/L) and for cortisol (3661 nmol/L, reference range after midday 80–350 ng/mL). The haematology is unremarkable.

Discussion and diagnosis

Together, these provide strong evidence in favour of a diagnosis of paediatric DKA. Although this case differs somewhat from Case report 11, of DKA in an elderly woman, there are broad similarities in the biochemical pathophysiology. The similarities are that the root abnormality is the failure of insulin to address rising blood glucose, from which everything follows. There will also be metabolic changes in the liver, and possibly in the muscles. Measurement of beta-hydroxybutyrate, alongside glucose, is required to assess the extent of the diabetic coma and for the exclusion of hyperosmolar non-ketotic diabetic coma.

As in adult DKA, treatment consists of insulin and intravenous fluids to correct the acid–base, electrolyte and glucose abnormalities. Bicarbonate was given to correct the acidosis, and electrolytes were measured hourly.

References

Blann AD, Clarke R, Gurney D. *et al.* Haematology Morphology Training CD-ROM. London: IBMS, 2009.

Lemyze M, Baudry JF, Collet F, Guinard N. Life threatening lactic acidosis. Br Med J. 2010;340:c857.

McDonald S. Management of a diabetic emergency: the role of biomedical science. Biomed Scientist. 2012;(July):416–417.

Prejbisz A, Klisiewicz A, Januszewicz A. *et al.* 22-year-old patient with malignant hypertension associated with primary aldosteronism. J Hum Hypertens. 2013;27:138–140.

Tournoy J, Dapper I, Spitz B. *et al.* Haemolysis, elevated liver enzymes, and thrombocytopenia in a 34-year-old pregnant woman. Lancet. 2006;368:90.

Viera AJ, Neutze DM. Diagnosis of secondary hypertension: an age-based approach. Am Fam Physician. 2010;82:1471–1478.

Appendix

Reference Ranges

Analyte	Reference range	Units
Alanine aminotransferase	5–42	IU/L
Albumin	35–50	g/L
Alkaline phosphatase	20–130	IU/L
Ammonia	<100	μmol/L
Amylase	<300	IU/L
Aspartate aminotransferase	10–50	IU/L
Basophils	0.02–1.0	10^9/L
Bicarbonate (hydrogen carbonate)	24–29	mmol/L
Bilirubin (total)	<17	μmol/L
Bilirubin (direct)	<5.2	μmol/L
Blast/atypical cells	<0.02	10^9/L
Calcium (ionized)	1.2–1.37	mmol/L
Calcium (total)	2.2–2.6	mmol/L
Chloride	95–108	mmol/L
Cholesterol (total)	<5.0	mmol/L
Cholesterol (low-density lipoprotein)	<3.0	mmol/L
Cholesterol (high-density lipoprotein)	>1.2	mmol/L
Cortisol (9 a.m.)	140–690	nmol/L
Cortisol (12 noon)	80–350	nmol/L
C-reactive protein	<10	mg/L
Creatine kinase	55–170	IU/L
Creatine kinase isotype MB	<25	IU/L
Creatinine	71–133	μmol/L
D-dimers	<500	Units/mL[a]
eGFR	>90	ml/min per 1.73 m^2
Eosinophils	0.02–0.5	10^9/L
Erythrocyte sedimentation rate	<10	mm/h

Analyte	Reference range	Units
Ferritin (male)	25–380	μg/L
Ferritin (post-menopausal female)	28–365	μg/L
Ferritin (pre-menopausal female)	7.5–224	μg/L
Fibrinogen	1.5–4.0	g/L
Folate (serum)	4–30	nmol/L
Folate (red blood cell)	360–1460	nmol/L
Follicle stimulating hormone		
follicular	4.0–13.0	IU/L
mid-cycle	5.0–22.0	IU/L
luteal	1.8–7.8	IU/L
post-menopausal	26.0–135	IU/L
male	0.7–11.0	IU/L
Gamma-glutamyl transferase	5–55	IU/L
Glucose (fasting)	3.5–5.5	mmol/L
Glucose (random)	<10	mmol/L
Glucose (urine)	negative	
Growth hormone	<10	mU/L
Haematocrit (female)	0.33–0.47	L/L
Haematocrit (male)	0.35–0.53	L/L
Haemoglobin (female)	118–148	g/L
Haemoglobin (male)	133–167	g/L
Iron	10–37	μmol/L
Insulin (fasting)	2–10	mU/L
International normalized ratio	2–3 or 3–4	
Lactate	0.7–1.9	mmol/L
Lactate dehydrogenase	240–480	IU/L

Blood Science: Principles and Pathology, First Edition. Andrew Blann and Nessar Ahmed.
© 2014 John Wiley & Sons, Ltd. Published 2014 by John Wiley & Sons, Ltd.

Analyte	Reference range	Units
Luteinizing hormone		
follicular	3–13	IU/L
mid-cycle	14–96	IU/L
luteal	1–11	IU/L
post-menopausal	8–59	IU/L
male	0.8–8.0	IU/L
Lymphocytes	1.0–3.0	10^9/L
Magnesium	0.8–1.2	mmol/L
Mean cell haemoglobin	26–33	pg
Mean cell haemoglobin concentration	330–370	pg/L
Mean cell volume	77–98	fL
Methylmalonic acid	0.08–0.28	μmol/L
Monocytes	0.2–1.0	10^9/L
Neutrophils	2.0–7.0	10^9/L
Osmolality	275–295	mOsm/kg
Parathyroid hormone	1–6	pmol/L
Partial thromboplastin time	24–34	s
P_{CO_2}	4.7–6.0	kPa
P_{O_2}	12–14.6	kPa
pH	7.35–7.45	$-\log_{10}[H^+]$
Phenylalanine	<100	μmol/L
Phosphate	0.8–1.4	mmol/L
Plasma viscosity (at 25 °C)	1.5–1.72	mPa/s
Plasma viscosity (at 37 °C)	1.16–1.33	mPa/s
Platelets	143–400	10^9/L
Potassium	3.8–5.0	mmol/L
Prolactin (female)	103–497	mU/L
Prolactin (male)	86–324	mU/L
Protein (total)	60–80	g/L
Protein (urine)	<0.25	g/L
Prothrombin time	11–14	s
Red blood cell count (female)	3.9–5.0	10^{12}/L
Red blood cell count (male)	4.3–5.7	10^{12}/L
Red cell distribution width	10.3–15.3	% coefficient of variation
Reticulocytes	25–125	10^9/L
Sodium	135–145	mmol/L

Analyte	Reference range	Units
Testosterone (female)	0.5–2.5	nmol/L
Testosterone (male)	10–31	nmol/L
Thyroid		
free T4	10–25	pmol/L
free T3	4.0–6.5	pmol/L
total T4	60–160	nmol/L
total T3	1.2–2.3	nmol/L
Thyroid stimulating hormone	0.2–3.5	mU/L
Total iron binding capacity (male)	54–72	μmol/L
Total iron binding capacity (female)	55–81	μmol/L
Triacylglycerides	<1.7	mmol/L
Transferrin	2–4	g/L
Transferrin saturation (male)	18–40	%
Transferrin saturation (female)	13–37	%
Troponin	<0.6	IU/L[a]
Urate	0.14–0.43	mmol/L
Urea	3.3–6.7	mmol/L
Vitamin B_6	5–30	μg/L
	23–129	pmol/L
Vitamin B_{12}	160–925	ng/L
	120–680	pmol/L
White blood cells	4.0–10.0	10^9/L

General notes on reference ranges:

- These can vary according to manufacturer and local conditions – this is entirely appropriate.
- Ranges can change over time.
- Do not use these figures in your daily practice.
- Some provide sex-specific reference ranges on certain analytes, others do not – this is also appropriate.

An inter-professional and interdisciplinary group are working towards harmonizing reference intervals (www.pathologyharmony.co.uk). See also De la Salle B, Pathology harmony moves on: progress on implementation in haematology. Br J Haematol. 2012;158:804–805.

[a]Reference range strongly dependent on method.

Further Reading

Ahmed N (ed.). Clinical Biochemistry. Oxford: Oxford University Press, 2011.

Ahmed N, Dawson M, Smith C, Wood E. Biology of Disease. Abingdon: Taylor and Francis, 2007.

Ayling R, Marshall W. Nutrition and Laboratory Medicine. London: ACB Venture Publications, 2007.

Beckett GJ, Ashby P, Walker SW. Lecture Notes on Clinical Biochemistry, 7th edition. Oxford: Blackwell Publishing, 2005.

Delves PL, Martin SJ, Burton DR, Roitt IM. Roitt's Essential Immunology, 12th edition. Chichester: Wiley-Blackwell, 2011.

Gaw A, Murphy M, Srivastava R, Cowan RA. O'Reilly DJ. Clinical Biochemistry: An Illustrated Colour Text, 5th edition London: Churchill Livingstone, 2013.

Hall A, Yates C (eds). Immunology. Oxford University Press, 2010.

Knight R (ed.). Transfusion and Transplantation Science. Oxford University Press, 2013.

Lamb E, Delaney M. Kidney Disease and Laboratory Medicine. London: ACB Venture Publications, 2009.

Marshall WJ, Bangert SK, Lapsley M. Clinical Chemistry, 7th edition. London: Mosby/Elsevier, 2012.

McCullough J. Transfusion Medicine, 3rd edition. Chichester: Wiley-Blackwell, 2012.

Moore G, Knight G, Blann A. Haematology. Oxford: Oxford University Press, 2010.

Reid ME, Lomas-Francis C, Olsson ML. The Blood Group Antigen Facts Book. Academic Press, 2012.

Salway JG. Medicinal Biochemistry at a Glance. 3rd edition. Chichester: Wiley-Blackwell, 2012.

Shils ME, Shike M, Olson J, Ross AC (eds). Modern Nutrition in Health and Disease. Lippincott, Williams and Wilkins, 2005.

Wild D. The Immunoassay Handbook, 3rd edition. Amsterdam: Elsevier, 2005.

Blood Science: Principles and Pathology, First Edition. Andrew Blann and Nessar Ahmed.
© 2014 John Wiley & Sons, Ltd. Published 2014 by John Wiley & Sons, Ltd.

Glossary

Abetalipoproteinaemia: inherited lack of betalipoproteins leading to low cholesterol and abnormal red cells.

ABO: a set of different antigens on the surface of red cells, the expression of which varies between individuals but which must be compatible in a blood transfusion.

Accuracy: the ability of a test for an analyte to give results comparable to the true value. Compare with precision.

Acenocoumarin: a vitamin K antagonist and anticoagulant.

Achlorhydria: low or absent secretion of acid in the stomach.

Acidosis: situation of low pH in the blood – the reverse of alkalosis.

Acid phosphatase: a metabolic enzyme found in various cells – isoforms can be used in leukaemia typing and in the study of prostate cancer.

Acquired immunity: defence against pathogens that has to be learned; it is not inbuilt.

Acromegaly: disease caused by high growth hormone resulting in increased height with large hands, feet, jaw and skull.

Activated partial thromboplastin time: screening test for use in unexplained haemorrhage, and to assess the effectiveness of heparin.

Acute: diseases or clinical events that have a rapid, often severe onset, but of short duration (compare with chronic).

Acute coronary syndrome: sudden pain in the chest whose origin is ischaemic heart disease – includes angina and myocardial infarction (heart attack).

Acute intravascular haemolysis: destruction of red cells within blood vessels due to factors such as an incompatible blood transfusion.

Acute-phase proteins: a set of plasma proteins associated with an acute-phase response; for example, C-reactive protein and fibrinogen.

Acute-phase response: a systemic response of changes in the composition of the blood, including an increased neutrophil count and the acute-phase proteins.

Acute rejection: of an allograft is one that occurs usually within a few weeks following the transplant.

Acquired immunity: the development of true immunity arising from a specific immune response to an immunogen.

Adaptive immunity: a defence that changes over time to accommodate changes in the infectious agents.

Addison's disease: condition resulting from failure of the adrenal cortex to release cortisol; may be autoimmune in aetiology.

Adenoma: a tumour that secretes an excess amount of its 'normal' product.

Adipocyte: a cell that synthesizes and stores fat.

Adrenal cortex: outer part of the adrenal gland that secretes aldosterone and cortisol.

Adrenal medulla: inner part of the adrenal gland that secretes adrenaline and noradrenaline.

Adrenocorticotrophic hormone: secretion of the anterior pituitary that regulates cortisol production by the adrenal cortex.

Aerobic: requiring oxygen.

Aetiology: biological basis of how a disease arises.

Agammaglobulinemia (or hypogammaglobulinaemia): clinical conditions in which the γ-class of immunoglobulins is deficient.

Agglutination/aggregation: the clumping together of particles such as latex, red cells or platelets.

Blood Science: Principles and Pathology, First Edition. Andrew Blann and Nessar Ahmed.
© 2014 John Wiley & Sons, Ltd. Published 2014 by John Wiley & Sons, Ltd.

Agglutinin: an antibody that causes particle and cells to agglutinate.

Albumin: the single most abundant blood protein, produced by the liver, and making up perhaps half of all plasma proteins.

Alanine aminotransferase: an enzyme found in cells of the liver and other organs; released into the blood after cell damage.

Aldosterone: steroid hormone from the adrenal cortex that acts on the nephron to retain sodium and allow potassium to pass into the urine.

Alkaline phosphatase: a metabolic enzyme and liver function test, but levels in the blood may also arise from other tissues such as bone.

Alkalosis: situation of high pH in the blood – the reverse of acidosis.

Alleles: different forms of the same gene that have arisen by mutation and found at the same place on a chromosome.

Allergens: immunogens that cause an allergy.

Allergy: a term often used to describe a Type I hypersensitivity, which results in inflammation on exposure to an allergen.

Alloantibodies: antibodies formed between different people.

Allogeneic: transplants that involve donors and recipients who are genetically nonidentical (see also autologous and syngeneic).

Allograft: a transplant between two genetically different people (see also isograft and xenogeneic transplants).

Alpha granules: intracellular bodies in platelets containing numerous molecules with roles in coagulation and wound repair.

Alternative pathway: activation of complement proteins in the absence of antibody by cell wall components of bacteria and yeasts.

Amenorrhea: an absence or stopping of the menstrual cycle.

Aminotransferase: a group of metabolic enzymes, increased blood levels of which imply cell damage.

Amyloid: insoluble fibrous proteins deposited in various tissues and organs that lead to loss of function; one species causes Alzheimer's disease.

Anaerobic: in the absence of oxygen.

Anaphylaxis: acute pathological state often precipitated by an allergic reaction to factors such as an insect sting.

Analytes: substances that are measured to assist diagnosis and monitor the treatment of disease.

Analytical sensitivity: the smallest quantity or concentration of an analyte that can reliably be detected by an analytical method. Compare analytical specificity.

Analytical specificity: the ability of an analytical method to detect only the analyte in question.

Anaemia: a disease caused by reduced amount of haemoglobin *and* the presence of symptoms such as tiredness and lethargy.

Aneuploidy: situation in which the number of chromosomes is fewer or greater than an exact multiple of the haploid number.

Aneurysm: a bulge that usually develops in weak areas of an arterial wall.

Angina: pain in the chest caused by insufficient delivery of oxygen to the muscles of the heart.

Angiotensin-converting enzyme: enzyme involved in vasoconstriction; target of a group of inhibitors (hence ACEIs) to control hypertension.

Anion gap: difference between the sum of the levels of major cations and major anions. If large, implies the presence of increased levels of an unknown molecule.

Anisocytosis: differences in the size of red blood cells.

Ankryn: a component of the cytoskeleton of the red blood cell.

Anoxia: the absence of oxygen.

Antibiotics: drugs used to treat bacterial and other infectious diseases, but which do not affect viruses. Most antibiotics are derived from microorganisms.

Antibodies: glycoproteins produced by B lymphocytes that bind to antigens (see immunoglobulins).

Anticoagulant: substance that prevents blood clotting. Includes sodium citrate and EDTA, and can be therapeutic (e.g. heparin, warfarin).

Anti-diuretic hormone (or vasopressin): a product of the posterior pituitary that acts on the nephron to retain water.

Antigen: a molecule that provokes an immune response and binds specifically to an antibody or the T cell receptor.

Antigen-presenting cells: white blood cells that take up an immunogen and process it to form a peptide that can be recognized by lymphocytes.

Antinuclear antibodies: antibodies against nuclear components that occur in some autoimmune conditions, such as rheumatoid arthritis.

Antioxidant: a substance that counteracts the toxic effects of oxygen and its metabolites.

Antiserum: a blood serum containing antibodies.

Antithrombin: one of two major inhibitors of the coagulation pathway; regulates the activity of thrombin. Requires heparin for full effectiveness.

Anti-Xa assay: used to monitor the effect of low molecular weight heparin.

Anuria: failure to pass urine; implies renal disease.

Apheresis: isolation and concentration of components of the blood for clinical use (see plasmapheresis).

Apoferritin: ferritin that does not carry iron.

Apotransferrin: transferrin that does not carry iron.

Aplasia/aplastic: refers to the failure of cells to grow and develop (as in aplastic anaemia), leading to pancytopenia.

Apoptosis (programmed cell death): also called cell suicide. A programmed release of enzymes and other proteins that eventually bring about the death of a cell.

Arrhythmias: disturbances to the heart's rhythmic contractions.

Ascites: an abnormal accumulation of fluid ('ascites fluid') in the abdomen.

Asthma: inflammatory lung disease caused by a hypersensitivity reaction.

Asparate aminotransferase: metabolic enzyme present in many cells, and raised upon cell damage; a liver function test.

Aspirate: to obtain a sample of cells or tissues, such as from the bone marrow.

Asymptomatic: to be free of symptoms (does not guarantee being free of disease).

Atheroma: a hard yellow plaque rich in cholesterol and surrounded by fibrous tissue that gradually builds up on the inside of medium-sized arteries.

Atherosclerosis or arteriosclerosis: the simultaneous development of an atheroma in an artery wall and its sclerosis (abnormal hardening or fibrosis).

Atopy: a genetic disposition to produce high levels of IgE when presented with common allergens.

Atrioventricular node: one of the two 'pacemakers' that generates electrical impulses that regulate the rate of contraction of the heart. The other is the SA node.

Atrophy: wasting away of a tissue or organ due to lack of use or a pathological process.

Audit: examination of a process aiming to identify ways in which it can be improved.

Auer rod: intracellular body found in certain blasts in myelocytic leukaemia.

Autoanalyser: machine that can process many samples in a short period of time, often involving several technical steps.

Autoantibody: antibody directed towards one's own cells or tissues.

Autoantigen: molecule that stimulates the production of an autoantibody.

Autoimmune disease/autoimmunity: situation in which the immune system mounts an immune response against 'self' tissues.

Autoimmune haemolytic anaemia: destruction of one's own red cells by an autoantibody.

Autologous stem cell transplants: use of the patient's own stem cells to restore a diseased or damaged bone marrow.

Autosomes: the non-sex chromosomes which comprise 22 homologous pairs of the 46 chromosomes found in human diploid cells.

Band form: immature granulocyte with a horseshoe-shaped nucleus.

B lymphocytes: lymphocytes that develop into antibody-producing cells.

Bacteriocidal: an antibiotic or other drug that kills bacteria.

Basophil: one of the three types of polymorphonuclear leukocyte, characterized by purple/black granules.

Basophilia: basophilic count above the top of the reference range.

BCR: B cell receptor for antigen.

BCR–ABL: a fused product of two different genes (*BCR* and *ABL*) that is the basis of the Philadelphia chromosome in CML.

Bence–Jones protein: fragments of disordered immunoglobulin proteins produced by a myeloma that pass into the urine.

Benign: a growth (often a tumour) that does not exert a major pathological effect.

Bernard–Soulier syndrome: a bleeding disorder characterized by defects in the GpIb-IX-V platelet surface receptor.

Beta cell: a cell from the islets of Langerhans in the pancreas that produces insulin.

Bicarbonate: a crucial anion that carries carbon dioxide and also provides buffering activity; low levels are found in acidosis.

Biochemistry: the chemistry of biological materials.

Bile: complex product of the liver stored in the gall bladder and excreted into the intestines below the stomach. Aids the absorption of dietary fats.

Bilirubin: a later breakdown product of haem, a component of bile, and a liver function test. Increased levels cause jaundice.

Biliverdin: the primary breakdown product of haem: converted to bilirubin.

Bipolar disorder: a psychiatric condition characterized by mood swings between mania and depression.

Bioavailability: that fraction of a drug that reaches the systemic circulation.

Biomedical science: the application of science in the biology of medicine.

Biopsy: the removal of a piece of tissue from a patient for clinical analysis.

Blast cell: a precursor cell that in health ultimately gives rise to mature cells. However, in myeloproliferative diseases, blast cells fail to differentiate.

Blood film: a drop of blood spread onto a glass slide and stained to allow the morphology of blood cells.

Bone marrow: soft tissues in the middle of hollow bones responsible for the production and maturation of blood cells.

Blood products: noncellular products of blood (albumin, coagulation proteins) that are infused into patients.

Blood science: the application of science to blood: a subspeciality of biomedical science.

Blood transfusion: the science of ensuring infused blood and blood products are compatible and safe for the patient.

Body mass index: index for the weight of an individual compared to their height, where BMI = weight (kg)/ height $(m)^2$.

Bone resorption: the release of Ca^{2+} from bone under the direction of osteoblasts.

Bradycardia: an abnormally low heart rate.

Buffer: a complex of ions that resists changes in pH

Buffy coat: combined layer of white cells and platelets that develops between the plasma and red cells when anticoagulated blood is left to stand.

Bundle of His: a group of modified cardiac muscle fibres, called Purkinje fibres, that carry electrical impulses in the muscle of the heart.

Bursa of Fabricius: organ in birds required for the maturation of B lymphocytes; no human equivalent.

C3 convertase: a proteolytic enzyme consisting of complement components C4b and C2a that cleaves C3 into two fragments – C3a and the larger C3b.

C5 convertase: a proteolytic enzyme consisting of complement proteins C3b and C3 convertase involved in the formation of the membrane attack complex.

Calcitriol: a form of vitamin D that promotes calcium absorption.

Cancer: a general term for a number of diseases in which the growth of certain body cells becomes uncontrolled, forming a tumour that may be benign or malignant.

Carcinogens: environmental agents (compounds, radiation) that cause cancers.

Carcinoma: a tumour arising from epithelial tissues.

Carbaminohaemoglobin: haemoglobin that is carrying carbon dioxide.

Carbohydrates: a group of organic compounds used as a source of energy; these include sugars and starch.

Carboxyhaemoglobin: Carbon monoxide bound to haemoglobin

Cardiac arrest: when the pump action of the heart stops, thereby causing loss of blood circulation.

Cardiolipin: phospholipid used as a capture antigen for the detection of anticardiolipin antibodies in the diagnosis of antiphospholipid syndrome.

Cardiomyocyte: a heart muscle cell.

Cardiovascular disease: diseases of the heart and circulation.

Cardioversion: procedure of converting the pathological heart rhythm atrial fibrillation back to normal (sinus) rhythm.

Carditis: general inflammation of the heart.

Carriers: individuals who harbour a pathogen, or have one effective gene and one defective gene, but are symptom free.

Catecholamine: a hormone produced by the adrenal medulla, of which there are three – adrenaline, noradrenaline and dopamine.

CD34: molecule on the surface of stem cells that allows its identification and collection, possibly for use in bone marrow transplantation.

CD4: a molecule expressed predominantly on T helper lymphocytes, but also present on some monocyte/macrophages, that binds to MHC class II molecules.

CD8: a molecule expressed predominantly on cytotoxic T lymphocytes that binds to MHC class I molecules.

Cell-mediated immunity: defence against pathogenic microbes by leukocytes, although if CMI is defective then the body itself is attacked.

Centrifuge: crucial laboratory apparatus to separate cells from plasma or serum.

Cerebral haemorrhage: rupture of a blood vessel in the brain resulting in impaired oxygen supply and subsequent brain damage.

Cerebrovascular accident: loss of function of tissues and/or organs caused by lack of oxygen to the brain; synonymous with stroke.

Chemotactic: the ability of substances (such as C3a and C5a) secreted by cells to attract other cells to the area.

Chemotherapy: literally, therapy with chemicals, but taken to imply drugs.

Cholestasis: the failure of bile to reach the small intestine.

Cholesterol: a sterol produced by the liver, but can also be taken in the diet. It is a precursor to steroids, and is also a component of the cell membrane.

Chromogenic substrate: chemical that mimics the natural substrate of an enzyme and so allows an analyte to be detected or quantified.

Chromatography: an essential laboratory method for separating and then quantifying compounds.

Chronic: diseases or clinical events that develop gradually over a relatively long time and persist (compare with acute).

Chronic rejection (of a transplantation): immunological rejection that occurs over months or years by a combination of cell-mediated and humoral immunity.

Chylomicron: large lipoprotein complexes that transport triacylglycerols, phospholipids and lipid-soluble vitamins from the intestines to the tissues.

Chyme: the watery mixture of gastric juice and partially digested food released from the stomach into the small intestine.

Cirrhosis: damage to the liver cause by the replacement of functioning tissues by fibrotic tissues.

CK-MB: a variant of CK (creatine kinase) found in cardiomyocytes; increased serum levels are found after myocardial necrosis, generally following myocardial infarction.

Classical pathway of complement activation: the pathway initiated when certain classes of antibody bind to an immunogen (compare alternative pathway).

Clinical audit: a process whereby practices and procedures are monitored and, if necessary, revised to provide a more effective, efficient and cost-effective service.

Clinical biochemistry: the science concerned with investigating the biochemical changes associated with diseases.

Clone: genetically identical cells derived from a single cell.

Cluster of differentiation: a system for classifying molecules on the surface of the cell.

Coagulopathy: a disease caused by over- or under-activity of the coagulation pathway.

Cobalamin: the chemical basis of vitamin B_{12} and related molecules.

Coefficient of variation: a measure of precision, and so the variability (and so reproducibility) of a particular test or process.

Coeliac disease: an autoimmune disease of the small intestine due to sensitivity of enterocytes to gliadin, resulting in atrophy.

Cold ischaemia time: time lapse between the harvest of an organ and its transplant into its recipient.

Colitis: an inflammation of the colon.

Colony forming unit: a stem cell in the bone marrow that ultimately gives rise to mature blood cells.

Colony-stimulating factor: a cytokine that regulates haemopoiesis but promotes the differentiation of stem and blast cells.

Compensation: in acid–base disorders, the physiological mechanisms that attempts to return the pH of the blood to values within the reference range.

Co-morbidities: two or more coexisting diseases.

Complement: a series of proteins that support defence against bacteria.

Confidence interval: a statistical term to denote the range of values over which there is 95% confidence that a particular number is accurate.

Congenital: apparent at birth or shortly thereafter.

Conjugate: refers to an antibody that has been conjugated to a detector molecule such as a fluorochrome, radioisotope or enzyme.

Conn's syndrome: a condition caused by the overproduction of aldosterone by the adrenal cortex, perhaps by a tumour.

Contact allergies: reactions produced by a number of chemicals that directly affect the skin, leading to a delayed type hypersensitivity.

Control samples: samples that are identical in composition to test samples except that they contain known concentrations of the test analyte.

Corneal arcus: opaque fatty deposits around the periphery of the cornea that occur in familial hyper-cholesterolaemia.

Coronary artery disease: a condition inevitably caused by atherosclerosis of the coronary arteries and resting in ischaemic heart disease.

Cortisol: the major gulcocorticosteroid; released from the adrenal cortex in response to stress. Active in glucose regulation and immunosuppression.

Corticosteroids: steroids secreted by the adrenal cortex; includes cortisol/cortisone and aldosterone.

COSHH: Control of Substances Hazardous to Health. A set of regulations that must be considered before a new process is introduced and adhered to thereafter.

Coumarin: a plant product that is the basis of a number of anticoagulants, notably warfarin.

C-reactive protein: liver-produced acute-phase reactant whose levels reflect infection, inflammation and general physiological shock.

Creatine kinase: an enzyme in muscle cells involved in the metabolism of glycolysis and the generation of ATP. Increased serum levels imply muscle damage.

Creatinine: a breakdown product of creatine phosphate in muscles; serum levels mark long-term renal function.

Cretinism: a congenital condition characterized by mental retardation and other symptoms, often due to failure of the thyroid gland and/or lack of iodine.

Crithidia lucilae: a microorganism whose kinetoblast contains double-stranded DNA, allowing quantification of anti-dsDNA antibodies in diseases such as systemic lupus erythematosus.

Crohn's disease: an inflammatory bowel disease; can affect any part of the intestines, but most commonly the terminal ileum.

Cross-match: a process in blood transfusion where donor blood packs are mixed with recipient (patient's) blood to determine if compatible.

Cryoglobulins: proteins in the plasma that become insoluble, and so pathogenic, in the cold.

Cushing's disease: an endocrine disorder caused by overproduction of ACTH by a pituitary or ectopic tumour.

Cushing's syndrome: product of excessive production of cortisol by the adrenal or a high intake of corticosteroids.

Cyanosis: blueish tinge to the lips and/or skin due to low levels of oxygen.

Cystic fibrosis: genetic condition caused by a mutation in a gene that regulates the components of sweat, digestive fluids and mucus.

Cytochemistry: the biochemistry of cells, mostly considering the contents of the cytoplasm. Often uses certain dyes and stains.

Cytochromes: a key group of enzymes involved in metabolism, perhaps the most important being cytochrome P450.

Cytogenetics: the microscopic study of chromosomes.

Cytokines: hormone-like proteins secreted by cells and stimulating activities in other cells after binding to receptors on their surfaces.

Cytotoxic cells: white blood cells that directly destroy other cells are infected with pathogens or that they consider to be non-self and so foreign.

Cytotoxic T lymphocyte: T lymphocyte that is programmed to recognize and destroy cells it regards as non-self, such as those expressing viral antigens.

D-dimers: breakdown product of thrombosis; high levels are present in thrombotic disease and (often) their risk factors. Low levels exclude venothromboembolism.

Deep vein thrombosis: a clot in a vein in the leg or groin.

Defaecation: expulsion of faeces from the body.

Dehydration: occurs when water loss from the body exceeds intake; can be due to insufficient drinking, or more commonly excessive loss in urine.

Deletion: when a section of a chromosome is missing, so that the remaining chromosome is shorter.

Dendritic cell: a highly modified cell that can present antigens to other cells (i.e. can be an antigen-presenting cell).

Densitometry: method for quantifying the proportions of protein or nucleic acids separated by electrophoresis.

Deoxyhaemoglobin: haemoglobin not carrying oxygen.

Diarrhoea: the frequent passage of faeces that are larger in volume and are more fluid than normal.

Diabetes insipidus: a condition caused by deficiency of antidiuretic hormone that in turn produces a large amount of watery urine.

Diabetes mellitus: a condition characterized by hyperglycaemia, itself due to lack of or failure of the action of insulin.

Diabetic ketoacidosis: a complication of diabetes mellitus where hyperglycaemia drives pathways resulting in high blood ketones and an acidosis.

Diamond–Blackfen anaemia: the principal congenital cause of pure red cell aplasia.

Diastole: relaxation of the heart's ventricles.

Differential: assessment of the different proportions of five major classes of white cells.

Differential diagnosis: two or more possible explanations for the cause(s) of the same disease or condition.

Differentiation: the process of cell maturity.

Digoxin: a drug that reduces the heart rate, thus giving the heart more time to beat effectively. Used in the treatment of atrial fibrillation, flutter and heart failure.

2-3-Diphosphoglycerate: a red cell metabolite that partially regulates oxygen carriage by haemoglobin.

Diploid: the presence of pairs of chromosomes in a cell. Normal human somatic (body) cells contain 23 pairs of chromosomes (compare haploid).

Direct antiglobulin test: a test used to determine the presence of an antibody of the surface of a red blood cell.

Direct thrombin inhibitor: an anticoagulant that acts directly on thrombin without the need for cofactors, unlike the need of antithrombin for heparin.

Disseminated intravascular coagulation: the most serious coagulopathy, associated with a high risk of haemorrhage and mortality.

Diurnal: refers to a marked variation in levels of an analyte during the day (e.g. higher or lower in the morning or evening).

Diuretic: a drug that promotes the production of urine and so reduces blood volume; consequently, a treatment for oedema and hypertension.

Dominant: genetically controlled factors that are expressed phenotypically in both the homozygous and heterozygous conditions (see recessive).

Dopamine: a catecholamine substrate for adrenaline and noradrenaline.

Down's syndrome: a common congenital condition caused by an extra copy of chromosome 21.

Dyslipidaemia: abnormalities in blood lipids and lipoproteins; may be high or low levels of individual molecules.

Ectoparasite: a parasite (such as a leech) on the outside of the body.

Ectopic: generally production of a molecule (often a hormone) outside its normal site, but also used to describe a pregnancy in the fallopian tubes.

EDTA: ethylenediaminetetraacetic acid, a laboratory anticoagulant.

Electrocardiogram: the record of the electrical potentials associated with the spread of depolarization and repolarization through the myocardium during its cycle.

Electrolyte: a fluid that promotes the passage of electricity, although in our setting refers to sodium and potassium.

Electronic issue: approval of blood to be transfused without a standard cell/plasma cross-match.

Electrophoresis: separation and analysis of plasma proteins according to their electrical change.

Elliptocytosis: refers to abnormally shaped red blood cells.

Embden–Meyerhoff glycolytic pathway: principal metabolic pathway by which energy (in the form of ATP and NADH) is obtained from glucose.

Embolus: a blood clot or part of a clot that migrates from a primary source (such as leg vein) to lodge in another vessel (such as of the pulmonary circulation).

Endocardium: the layer of smooth tissue lining the inside of the chambers of the heart.

Endocrine: referring to glands that pass their hormone products directly into the bloodstream.

Endocrine diseases: diseases that arise from the over- or under-production of hormones, or from resistance of a target tissue to a particular hormone.

Endoparasite: a parasite (such as a tapeworm) within the body.

Endothelial cell: flattened orthogonal cell that lines the inner surface of blood vessels and lymphatics.

Enterocyte: a cell lining the intestines and involved in the secretion of enzymes and the absorption of nutrients.

Enzyme immunoassay: the use of antibodies labelled with an enzyme that permits the measurement or identification of an antigen.

Eosinophil: one of the three types of polymorphonuclear leukocyte, characterized by a bilobed nucleus and red granules; others are the neutrophil and basophil.

Epidemiology: the study of disease in large populations.

Epistaxis: nosebleeds.

Epitopes: small regions of immunogenic molecules (antigens) which are specifically recognized by components of the immune system.

Erythroblast: a precursor cell in the lineage of red blood cell production.

Erythrocyte: red blood cell; an anucleate cell rich in haemoglobin that transports oxygen around the body.

Erythrocytosis: increase in the red cell count above the top of the reference range.

Erythropoiesis: the production of red cells in the bone marrow.

Erythropoietin: hormone and growth factor derived from the kidney that stimulates the production of red blood cells.

Erythrocyte sedimentation rate: a nonspecific marker of inflammation, infections, anaemia, cancer, and other pathology and physiological conditions.

Essential fatty acids: fatty acids that must be supplied in the diet since humans lack the enzymes necessary to produce them.

Essential: nutrients that the body cannot make from other compounds and must therefore be supplied in the diet.

Essential thrombocythaemia: an increased platelet count caused by a clonal proliferative state of megakaryocytes.

Euthyroid: literally 'true thyroid', refers to a normally functioning thyroid.

Excretion: expulsion of the waste products of metabolism from the body.

Exocrine: referring to glands that pass their products along ducts.

Exophthalmos: protruding eyeballs.

Extramedullary: haemopoiesis occurring outside the bone marrow, such as in the liver and spleen.

Extractable nuclear antigens: antigens that lead to the formation of autoantibodies in certain diseases such as rheumatoid arthritis.

Extravascular haemolysis: the uptake and subsequent destruction of erythrocytes in the spleen and liver.

Extrinsic pathway: part of the coagulation pathway culminating in the activation of factor X to factor Xa by a complex of factor VIIa, tissue factor and phospholipids.

Fab: part of the antibody molecule that binds antigen.

Factor V Leiden: genetically determined variant of coagulation factor V that is resistant to its inhibitor and, therefore, continuing to promote thrombosis.

Factor VIII: coagulation factor whose absence causes haemophilia – the leading genetic cause of haemorrhage.

Faeces: a mixture of the material in the diet that is undigestible and dead enterocytes.

Fainting: common term for syncope.

False negative: an incorrect negative result present when it should be positive.

False positive: an incorrect positive result present when it should be negative.

Familial hypercholesterolemia: an autosomal dominant condition associated with defective receptors on the liver cells. Leads to high serum cholesterol.

Fanconi's anaemia: the leading genetic cause of pancytopenia.

Favism: the haemolytic crisis brought on by the consumption of fava or broad beans (*Vicia fava*) in persons with G6PDH deficiency.

Fc: part of the antibody distant from the Fab that may bind white cells.

FcR: Fc receptor; molecule on the surface of leukocytes into which the Fc part of the antibody molecule can locate.

Ferritin: Liver-derived protein that stores iron largely within cells.

Fibrin: product of the cleavage of fibrin by thrombin: is stabilized by factor XIIIa and so forms a thrombus with platelets.

Fibrinogen: a liver-derived blood protein that is the substrate for the enzyme thrombin.

Fibrinolysis: the regulated digestion of fibrin clot by the enzyme plasmin; generates fragments called d-dimers.

Fibroblast: general purpose connective tissue cell present in the bone marrow and elsewhere that synthesizes molecules such as collagen.

Flow cytometry: a technique for determining the number and type of blood cell.

Fluorescence-activated cell scanning: a technique that uses antibody-linked fluorochromes to estimate numbers of specific cell populations.

Fluorescence *in-situ* hybridization: a technique in molecular genetics for determining the presence of a particular gene.

Fluorochrome: a molecule that fluoresces; often linked to an antibody or probe to determine levels or presence of an analyte.

Folate: an essential micronutrient required for the development of the red blood cell.

Fresh frozen plasma: a blood product rich in coagulation factor used in the treatment and prophylaxis of haemorrhage.

Fructosamine: glycation of serum proteins that may be used to indicate short-term glycaemic control.

Full blood count: the most frequently requested haematology test providing information of red blood cells, white blood cells and platelets.

Galactorrhea: an inappropriate secretion of milk.

Gallstones: bodies formed in the gall bladder from cholesterol, bile pigments and calcium salts. May obstruct the bile duct and so cause cholestasis.

Gamma glutamyl transferase: a metabolic enzyme found in many tissues; also a liver function test.

Gastritis: inflammation of the stomach.

Gating: the process of selecting a group of cells in FACS analysis for further analysis.

Gene: a segment of DNA that ultimately gives rise to a protein product.

Genetic diseases: diseases that arise due to defects in the genes or chromosomes.

Genetics: the biology of genes.

Genome: the entire DNA make-up of an individual.

Genotype: the genetic or hereditary constitution of an individual or a pair of alleles that an individual possesses at a specific locus on a chromosome.

Gilbert's syndrome: an inherited condition of hyper-bilirubinaemia that, under stress, precipitates jaundice.

Glandular fever: common term for infectious mononucleosis; a brief viral infection associated with a lymphocytosis and often lymphadenopathy of the neck.

Glanzmann's thrombasthaenia: haemorrhagic disorder caused by defects in the platelet surface molecules GPIIb/IIIa.

Gliadin: a glycoprotein component of gluten in certain cereals and which causes enteropathy or coeliac disease.

Globin: a globular molecule that makes up the protein part of haemoglobin, and that is abnormal in sickle cell disease and thalassaemia.

Glomerular filtration rate: leading method for assessing renal function, derived from an equation that inputs serum creatinine, age, sex and often other features.

Glomerulonephritis: inflammation of the glomerulus.

Glomerulus: part of the nephron that first interfaces with blood.

Glossitis: inflammation of the tongue.

Glucagon: a pancreatic hormone involved in the release of energy substrates such as glucose into the blood.

Glucocorticoid: a steroid hormone from the adrenal that affects protein, fat and glucose metabolism.

Glucose-6-phosphate dehydrogenase: an enzyme involved in the metabolism of the red cell; deficiency causes a haemolytic anaemia.

Gluconeogenesis: synthesis of glucose from non-carbohydrate sources such as amino acids.

Glycated haemoglobin: non-enzymatic modification of haemoglobin; a marker of general hyperglycaemia over the previous 3–4 months.

Glycogenesis: synthesis of glycogen in the liver.

Glycogenolysis: breakdown of glycogen in the liver to release free glucose.

Glycoprotein: protein to which a carbohydrate such as glucose is attached.

Goitre: swelling in the neck due to an enlarged thyroid gland.

Gonads: organs that generate gametes; testes in the male and ovaries in the female.

Gout: a painful metabolic disease caused by deposits of uric acid crystals in the synovial fluid and skin.

Graft-versus-host disease: condition caused by lymphocytes for a blood transfusion or bone marrow transplantation which attack the recipient.

Granule: an intracellular organelle that contain bioactive molecules such as enzymes.

Granulocyte: a leukocyte with a lobed nucleus and many granules in the cytoplasm; consists of neutrophils, eosinophils and basophils.

Granulocyte–monocyte colony stimulating factor: a growth factor that stimulates the generation and differentiation of granulocytes and monocytes.

Granulocytopoiesis: production of granulocytes in the bone marrow.

Graves' disease: an autoimmune condition characterized by overactivity of the thyroid gland.

Group and save: a key process in blood transfusion where patient's blood group is determined and then the sample saved for possible use in a cross-match.

Growth hormone: pituitary hormone acting on many cells and organs to induce their growth. Increased levels cause acromegaly.

Gynecomastia: the development of enlarged breasts in males.

Haem: a non-protein prosthetic group in haemoglobin comprising a porphyrin ring and an atom of iron.

Haematocrit: proportion of blood that is made up of blood cells, but in practice is the proportion made up by red cells.

Haemochromatosis: disease caused by the excessive deposition of iron in organs such as the liver, heart and brain; often hereditary.

Haemolytic disease of the newborn: present when maternal antibodies (often to red cells) cross the placenta and attack the foetus, causing haemolysis.

Haemopoiesis: production of blood cells; in the healthy adult, occurs only in the bone marrow.

Haemodialysis: removal of toxins and waste metabolites in those with failing renal function.

Haemoglobin: oxygen-carrying compound found in red cells.

Haemoglobin A1c: proportion of haemoglobin that is glycated; increased levels imply hyperglycaemia and possibly diabetes mellitus.

Haemoglobinopathy: clinical condition that results from mutations in the genes for globin molecules.

Haemolysis/haemolytic anaemia: the inappropriate destruction of red cells.

Haemolytic uraemic syndrome: a thrombotic disease characterized by acute renal failure and thrombocytopenia, often preceded by *Escherichia coli* infection.

Haemophilia A: (*the* haemophilia) haemorrhagic disease caused by lack of coagulation factor VIII.

Haemophilia B: haemorrhagic disease caused by lack of coagulation factor IX.

Haemophilia C: haemorrhagic disease caused by lack of coagulation factor XI.

Haematology: the study of blood and its disorders.

Haemoptysis: coughing up of blood-containing fluid from the lungs.

Haemosiderin: intracellular bodies storing iron, often of insoluble ferritin.

Haematuria: blood in the urine.

Haemopoiesis: the generation of blood cells in the bone marrow.

Haemoproliferative: refers to diseases where there are increased numbers of blood cells in the blood or tissues, such as leukaemia and lymphoma.

Haemorrhage: the inappropriate loss of blood as bleeding; can manifest as nose bleeds, bruising or haematuria, or can be into the body cavities.

Haemosiderin: a granular and storage form of iron present in cells of the liver, spleen and bone marrow.

Haemostasis: the healthy balance between overcoagulation (causing thrombosis) and undercoagulation (causing haemorrhage).

Heavy chain: major portion of the molecular make-up of the immunoglobulin molecule.

Half-life: time taken for a drug to fall to half of its original concentration.

Haptoglobin: a plasma protein that binds and removes haemoglobin from the circulation.

Hashimoto's thyroiditis: chronic inflammation of the thyroid gland caused by autoantibodies.

Health: the state of physical, mental and social well-being; not merely the absence of disease.

Health and safety: a set of rules and regulations designed to ensure the safety of laboratory workers.

Heart attack: see myocardial infarction.

Heart failure: present when the heart is unable to pump sufficient blood around the body, caused by disease of the left ventricle and a reduced ejection fraction.

Heinz body: inclusion body within red cells consisting of denatured haemoglobin.

Helminths: collective term for parasitic worms such as nematodes, cestodes and trematodes.

Helper T lymphocyte (T helper cell): lymphocyte required for the development of other lymphocytes that react specifically towards antigens.

Heparin: a natural anticoagulant that can also be used as a therapy to prevent inappropriate thrombus formation.

Heparin-induced thrombocytopenia: low platelet count caused by heparin; risk factor for haemorrhage.

Hepatitis: inflammation of the cells of the liver; has many causes.

Hepatocellular carcinoma: cancer of epithelial cells within the liver; the most common cause of a hepatoma.

Hepatocyte: liver cell, responsible for the anabolic and catabolic function of this organ.

Hepatoma: mass within the liver; can be primary or secondary, and is generally malignant.

Hepatomegaly: enlargement of the liver.

Hereditary: refers to a condition that is passed from parent to offspring (i.e. is genetic).

Heterophile antibody: an antibody that reacts with red cells of another species.

Heterozygous: the condition in which the two alleles of a gene are different (compare homozygous).

High-density lipoprotein cholesterol: species of cholesterol that has high density and confers protection against cardiovascular disease.

High-performance liquid chromatography: a key laboratory method for analysing molecules in fluids.

Hirsutism: the development of inappropriate increased body hair on the face, chest, upper back and abdomen in females.

Histology: branch of biomedical science that analyses sections of organs and tissues.

Hodgkin lymphoma: a form of lymphoma characterized by the presence of Reed–Sternberg cells within the affected lymph node.

Homeostatic diseases: diseases that arise when mechanisms for controlling homeostasis are disrupted or defective.

Homocysteine: an amino acid, high levels of which are due to metabolic disease which may in turn cause additional disease, such as mental retardation.

Homozygous: the condition in which the two alleles are identical (compare heterozygous).

Howell–Jolly body: red cell inclusion body consisting of remnants of DNA.

Human leukocyte antigen: molecules on the surface of nucleated cells that participate in immune recognition of self and non-self.

Humoral immunity: the production of antibodies to protect against pathogens and parasites (compare cell-mediated immunity).

Hyper: prefix attached to a wide range of molecules, markers and processes to denote high or increased levels or activity of a particular analyte or organ.

Hyper-acute rejection: failure of a graft within hours or minutes of transplantation due to the presence of pre-formed antibodies against graft antigens.

Hypercapnia: a high partial pressure of CO_2.

Hypertension: blood pressure higher than that regarded as normal.

Hypersensitivity: an excess or exaggerated immune response to an antigen that causes a pathological condition.

Hyperviscosity: refers to the increased viscosity of blood, perhaps due to abnormal levels of certain plasma proteins, possibly interacting with red cells.

Hypo: prefix attached to a wide range of molecules, markers and processes to denote low or reduced levels or activity of a particular analyte or organ.

Hypocapnia: a low partial pressure of CO_2 (compare hypercapnia).

Hypochromic: refers to red cells deficient in haemoglobin.

Hypogonadism: failure of gonads to generate gametes and sex hormones (testosterone in males, oestrogen and progesterone in females).

Hypovolemia: a reduction in the volume of blood in the body.

Hypoxemia: a low partial pressure of O_2.

Iatrogenic diseases: clinical conditions that arise as a consequence of treatment.

Icterus: another name for jaundice.

Idiopathic diseases: diseases of unknown cause.

Idiopathic thrombocytopenia purpura: a low platelet count and bruising whose aetiology is unknown.

Iliac crest: protruding part of the hips, often used to obtain bone marrow.

Immune complexes: combined super-molecule of an antibody and its antigen; capable of activating the complement pathway.

Immune response: a desirable collection of actions that defend the body from infections by micropathogens.

Immune system: the set of organs, tissues, cells and molecules that protect the body from diseases caused by pathogens.

Immune thrombocytopenia purpura: a low platelet count and bruising whose aetiology is an autoantibody to platelets.

Immunity: defence against defined microbial pathogens.

Immunoassay: the use of antibodies to quantify the amount of an analyte in a clinical sample.

Immunodeficiency: condition that occurs due to an inadequate immune system; inevitably associated with opportunistic infections.

Immunofluorescence: a type of immunoassay in which the antibody is labelled with a fluorescent dye.

Immunogen: any molecule or organism that stimulates a specific immune response. Synonymous with antigen.

Immunoglobulins (Igs): a collective name for the five classes of antibodies.

Immunohistochemical technique: refers to the use of labelled antibodies to detect antigens on or in cells.

Immunological memory: a molecular memory that allows a rapid protective response on second or subsequent contact with an immunogen.

Immunology: the study of the immune system and how it works.

Immunophenotyping: FACS technique for determining the identity of cells according to their surface molecules.

Immunosuppression: suppression of the immune system; can be deliberate or accidental due to chemotherapy, or by a pathogen such as HIV.

Impaired fasting glycaemia: fasting blood glucose between 6.1 and 6.9 mmol/L.

Impaired glucose tolerance: blood glucose between 7.8 and 11.1 mmol/L 2 h after ingestion of 75 g of glucose.

Incidence rate: the number of new cases of a disease in a population occurring within a specified period of time.

Inclusion body: a body found (mostly) within red blood cells in certain pathological states.

Incubation period: the time that must elapse before a disease becomes apparent following exposure to the aetiological agent.

Incompatible (blood transfusion): occurs if there is an adverse reaction between donor blood cells and the recipient because of a failure to match blood groups.

Indirect antiglobulin test: method for detecting antibodies present on the surface of red blood cells; often used in autoimmune haemolytic anaemia.

Infarction: the death (necrosis) of tissue following a disruption to its blood supply; consequence of long-term ischaemia.

Infection: the successful persistence and/or multiplication of a pathogen on or within the host.

Infectious diseases: diseases caused by pathogens, such as viruses, bacteria, fungi and parasites.

Infectious mononucleosis: self-limiting viral disease, often causing lymphocytosis and neck lymphadenopathy; also known as glandular fever.

Inflammation: the array of responses to infection and tissue damage, such as localized pain, redness, swelling and heat.

Innate immunity: nonspecific natural defence against pathogens that does not have to be learned (i.e. acquired).

Interleukin: hormone produced by a leukocyte that acts primarily on a leukocyte, but which may act on other cells.

International normalized ratio: prothrombin time on a vitamin K antagonist (such as warfarin) compared with the prothrombin time when free of that drug.

International sensitivity index: calibrated numerical value defining the sensitivity of a thromboplastin preparation in the determination of the INR.

Intramedullary: of haemopoiesis – within bone marrow.

Intravascular haemolysis: destruction of red cells within blood vessels.

Intrinsic factor: a glycoprotein secreted by the gastric parietal cells that binds vitamin B_{12} in the stomach and carries it to specific receptors on the ileum.

Intrinsic pathway: part of the coagulation cascade involving phospholipids, calcium, and factors XII, XI, IX and VIII, leading to the conversion of factor X to factor Xa.

Irradiation: the inactivation of donor leukocytes by gamma-radiation to prevent transfusion-associated graft-versus-host disease.

Ischaemia: pathological state of tissues when deprived of oxygen (as in ischaemic heart disease or ischaemic stroke). Closely related to hypoxia.

Islets of Langerhans: pancreatic cells with endocrine activity; alpha cells produce glucagon whilst beta cells generate insulin.

Isohemagglutinins: antibodies against blood group antigens that cause the agglutination of erythrocytes.

JAK2: a kinase enzyme important in promoting cell growth and differentiation. A mutation (*JAK2 V617F*) leads to haemoproliferative disease.

Jaundice: yellow coloration of the skin present in hyperbilirubinaemia; may imply liver disease or profound haemolytic anaemia.

Juxta-glomerular apparatus: cells close to the glomerulus that regulate the entry of blood into the nephron, and that also secrete renin to part-regulate blood pressure.

Karyotype: the characteristic number, size and shape of the chromosome complement of an individual or species.

Kernicterus: brain damage caused by the accumulation of bilirubin in the brain to concentrations of 200 μmol/L and above.

Ketoacidosis: condition following hyperglycaemia, characterized by ketone bodies and a low blood pH (acidosis).

Ketone bodies: acetoacetate, acetone, and β-hydroxybutyrate.

Ketonaemia: accumulation of ketone bodies (e.g. acetoacetate, 2-hydroxybutyrate) in the blood.

Ketonuria: excretion of ketone bodies in the urine.

Kinins: a group of low molecular weight molecules with roles in inflammation, haemostasis, blood pressure regulation and pain.

Knock-out: genetic manipulation of, for example, a mouse, where a particular gene is switched off, allowing its function to be dissected.

Kuppfer cell: specialized macrophage resident in the liver.

Kussmaul respiration: deep sighing breathing associated with the hyperventilation of metabolic acidosis.

Langerhans cell: a macrophage modified into a dendritic cell resident in the skin that can act as an antigen-presenting cell.

Latent period: the time when an illness is developing but overt signs and symptoms are not apparent.

Lectins: glycoproteins that bind to carbohydrate residues on macromolecules or cell membranes.

Lectin pathway: one of the three pathways by which the complement system is activated.

Left-shift: describes an increased proportion of immature neutrophils (such as band forms and metamyelocytes) in the circulating blood.

Lesions: structural or functional abnormalities.

Leukaemia: a proliferative neoplastic disease of white blood cells originating in the bone marrow.

Leukocyte/leucocyte: one of several types of white blood cell each with its own function.

Leukocytosis: a leukocyte count in excess of the reference range.

Leukodepletion: removal of leukocytes from blood by passing it through filters which trap the leukocytes but not the smaller erythrocytes or platelets.

Leukopenia: white blood cell count below the bottom of the reference range.

Leukopheresis: the removal of leukocytes from blood with the resulting plasma and erythrocytes being returned to the blood donor.

Leukopoiesis: generation of white blood cells; in the healthy adult, occurs only in the bone marrow.

Leukotrienes: inflammatory mediators, derived from arachadonic acid, with several functions, including smooth muscle cell contraction and roles in allergy.

Light chain: lower molecular weight part of the structure of the immunoglobulin molecule.

Linkage disequilibrium: present when two or more genes are co-inherited more frequently than can be accounted for by chance.

Lipids: a group of biological materials insoluble in water but a crucial component of cell membranes; major groups are cholesterol and triacylglycerols.

Low-density lipoprotein cholesterol: a form of cholesterol; high levels bring a risk of atherosclerosis.

Low molecular weight heparin: form of heparin that does not require monitoring by the APTT, and causes much less heparin-induced thrombocytopenia.

Lipoproteins: complexes of cholesterol, triacylglycerols and proteins involved in transport of these lipids in the blood.

Lupus anticoagulant: a type of antiphospholipid antibody specifically detected in specialized coagulation tests.

Lymphadenopathy: swollen lymph nodes.

Lymphoblast: a blast of the lymphocyte lineage that ideally develops into a mature lymphocyte. Abnormal lymphoblasts characterize certain leukaemias.

Lymphocyte: small mononuclear leukocyte, involved in antigen-specific immunity, whose nucleus occupies some 90% of the cell.

Lymphoid stem cells: haemopoietic stem cells found in bone marrow which can divide to produce the precursors of lymphocytes.

Lymphoma: a tumour that arises within a lymph node.

Lymphopenia: a lymphocyte count below the bottom of the reference range.

Lymphopoiesis: generation of lymphocytes.

Lysozyme: an antibacterial enzyme that can be secreted, but is also present in polymorphic leukocyte granules and contributes to phagocytosis.

Macrocyte: a mature red blood cells whose MCV exceeds the top of the reference range.

Macrocytic anaemia: anaemia when the MCV exceeds the top of the reference range; often caused by lack of vitamin B_{12}.

Macrophage: phagocytic immune cell that develops from monocytes in the tissues.

Margination: adhesion of leukocytes to the vascular endothelium that lines the blood vessels.

Major histocompatibility complex: a region of the chromosome that encodes membrane proteins, some of which are HLA molecules.

Malabsorption: failure of the intestines to absorb sufficient nutrients from an otherwise healthy diet.

Malaria: an infectious haemolytic disease of red blood cells caused by various *Plasmodium* species and transmitted by *Anopheles* mosquitoes.

Malnutrition: present when the diet is deficient in sufficient nutrients to maintain health.

Mass spectrometry: a technique that measures the mass/charge ratio of ions and used to determine chemical structures.

Mast cell: a type of granule-containing white blood cell that resembles a basophil but which is found in solid tissues.

Mean cell haemoglobin: the amount of haemoglobin inside the 'average' red cell.

Mean cell haemoglobin concentration: the concentration of haemoglobin inside the 'average' red cell.

Mean cell volume: the size of the average red cell; defines microcytes, normocytes and macrocytes.

Medulla: the inner part of an organ, such as the adrenal.

Megakaryocyte: stem cell in the bone marrow whose cytoplasm forms platelets.

Megaloblast: an unusually large red cell precursor such as an erythroblast, often a consequence of vitamin B_{12} deficiency.

Megaloblastic anaemia: anaemia characterized by hypersegmented neutrophils and often a consequence of vitamin B_{12} deficiency.

Membrane attack complex: a large hydrophobic structure constructed from complement proteins that inserts itself into a cell membrane, leading to cell lysis.

Menopause: stage of the female life cycle when the ovaries no longer produce ova.

Metabolic acid–base disorder: the occurrence of an abnormal blood pH because of a metabolic or renal dysfunction.

Metabolic syndrome: formed by a combination of obesity, insulin resistance or diabetes, hypertension, reduced HDL and raised triacylglycerols.

Metamyelocyte: in the development of polymorphonuclear leukocytes, the final immature stage before the differentiation into a mature cell.

Metastasis: the detachment of cells from a primary tumour allowing it to spread and form secondary tumours.

Metformin: a biguanide drug used by diabetics to improve insulin sensitivity and reduce hepatic glucose production (i.e. promote hypoglycaemia).

Methaemoglobin: a type of haemoglobin caused by the oxidation of ferrous (Fe(II)) iron in the haem group of haemoglobin to Fe(III) iron.

Microalbuminuria: the presence of small amounts (more than a trace) of albumin in the urine, inevitably an early sign of renal disease.

Microangiopathic haemolytic anaemia: present when red cells are lost or damaged because of pathological changes to small blood vessels.

Microarray technique: an assay system that allows simultaneous multiple analyses on a single sample.

Microcyte: a red blood cell with an MCV below the bottom of the reference range.

Microcytic anaemia: anaemia when the MCV is below the bottom of the reference range.

Microbiology: the science concerned with the detection and identification of pathogenic microorganisms.

Micronutrient: a nutrient required in very small amounts, such as trace elements and vitamins.

Mixed acid–base disorder: describes the situation in which a patient presents with more than one acid–base disorder.

Molecular genetics: the science concerned with the identification of genetic abnormalities.

Monoblast: a precursor of monocytes found in the bone marrow; differentiates into the promonocytes.

Monoclonal antibody: an antibody that recognizes only a single epitope on an antigen.

Monoclonal gammopathy: presence of an abnormal plasma protein, defined by electrophoresis and often present in myeloma and related conditions.

Monocyte: a type of mononuclear leukocyte that circulates in the blood for about 72 h before entering the tissues and developing into a macrophage.

Monocytosis: a monocyte count above the top of the reference range.

Mononuclear leukocytes: leukocytes characterized by a round rounded nucleus. They are subdivided into monocytes and lymphocytes.

Morbidity: the ill-effects of a disease on the patient.

Morphology: literally the study of shape, but in blood science referring to the morphology of different types of blood cells.

Mortality: describes the possibility of a disease causing death; usually expressed as a percentage.

Multiple endocrine neoplasia: present when two or more different endocrine cancers are present at the same time, such as of the pituitary, pancreas and thyroid.

Mutation: a change in the genotype, which occurs as a result of incorrect replication of DNA or from a random change to it caused by physical or chemical agents.

Myeloblast/myelocyte: precursor cells in the development of polymorphonuclear leukocytes.

Myelodysplasia: disease caused by failure of the correct function of myeloid stem cells, leading to anaemia and myeloid leukaemia.

Myelofibrosis: disease caused by the overgrowth of functioning haemopoietic cells by collagenous connective tissue cells such as fibroblasts.

Myeloma: malignancy of B lymphocytes that originates in the bone marrow.

Myelopoiesis: the production of myeloid cells (red cells, granulocytes, monocytes and platelets) in the bone marrow.

Myeloperoxidase: enzyme present in the granules of polymorphonuclear leukocytes; participates in the production of bacteriocidal reactive oxygen species.

Multisystem disease: present when more than one organ or organ system is affected, such as rheumatoid arthritis, which can affect the lungs, skin and joints.

Myocardial infarction ('heart attack'): possibly fatal condition that occurs when some or all of the blood supply to the heart muscle is restricted or cut off.

Myocardium: the cardiac muscle-rich wall of the heart.

Myxoedema: cutaneous and dermal oedema of the face that is caused by lack of thyroid hormones as occurs in hypothyroidism.

Natural killer cells: a type of lymphocyte that kills virus-infected cells and some tumour cells nonspecifically.

Natriuretic peptides: molecules that promote increased sodium and water loss by the kidneys and so reduce blood volume; isoforms ANP and BNP.

Near-patient testing: analysis of a patient's sample that occurs close to the point of collection, and far from the laboratory, such as in an outpatient clinic.

Necrosis: cell death due to injury or disease.

Neoplasm or neoplastic diseases: often used synonymously with cancer; literally means 'new tumour' or 'new mass'.

Nephelometry: measurement of light scatter through a fluid containing particulate matter. Often used to determine levels of plasma proteins using antibodies.

Nephritis: inflammation of the kidney.

Nephron: functional unit of the kidney; each is composed of a tuft of capillaries, called a glomerulus, and a tubule.

Nephropathy: disease of the kidney; common in advanced diabetes and inflammatory disease such as systemic lupus erythematosus.

Nephrotic syndrome: severe renal disease associated with hyperlipidaemia and the profound loss of protein in the urine, and so hypoalbuminaemia.

Neutropenia: present when the neutrophil count is below the bottom of the reference range.

Neutrophil: one of the three types of polymorphonuclear leukocyte (see basophil and eosinophil).

Neutrophilia: as neutrophil leukocytosis.

Neutrophil leukocytosis: present when the neutrophil count goes above the top of the reference range, often due to a bacterial infection.

Nitric oxide: a multifunction molecule that relaxes smooth muscle of the arterial wall, can act as a bacteriocide, inhibits platelet activation, and is also a neurotransmitter.

Nonessential: a nutrient that the body can make from other nutrients and need not be supplied in the diet (compare essential).

Non-self: expression used in immunology to signify macromolecules or cells that are foreign to the body, compared with self.

Nonspecific defences: the first line of immunological defence, including physical barriers and responses such as inflammation and the acute-phase response.

Normoblast: a red blood cell precursor in the bone marrow – same as an erythroblast.

Normocyte: red blood cell with an MCV in the reference range.

Normocytic anaemia: an anaemia when the MCV is in the reference range.

Nucleated red blood cell: a red blood cell in the bone marrow that still has a nucleus, which it sheds and becomes a reticulocyte.

Null allele: a mutant gene that lacks the function of its normal counterpart.

Nutrition: the study of food and nutrients and how they are used by the body.

Nutritional diseases: diseases that result from an inappropriate intake of nutrients.

Obesity: present when the body mass index exceeds $30 \, kg/m^2$.

Obstructive cholestasis: blockage of the bile duct by a gall stone, generally leading to jaundice and raised liver function tests.

Oedema: an excessive accumulation of water in the intercellular spaces.

Oestradiol: steroid hormone synthesized mostly by the ovaries (and responsible for female sexual function) and testes (being the substrate for testosterone).

Oliguria: passing less than 400 mL of urine per day.

Oncogene: a gene linked to (and possibly causing) cancer.

Oncology: the branch of medicine involved with the study of malignant tumours, their epidemiology, diagnosis and treatment.

Oncotic pressure: osmotic pressure due to plasma proteins (mostly albumin) that is required to keep fluid within blood vessels.

Opportunistic infections: present when immunodeficiency leaves the individual susceptible to infections by microorganisms that are normally non-pathogenic.

Opportunistic pathogen: a microorganism that causes opportunistic infections.

Opsonization: the process by which phagocytic cells bind to bacterial cells coated with activated complement proteins or antibodies, so promoting phagocytosis.

Oral glucose tolerance test: procedure used to make a diagnosis of diabetes.

Orthopaedic: referring broadly to the treatment of bone disease and injury, often by surgery.

Osmolality: a measure of the osmotic concentration, usually expressed as moles of solute per kilogram of solvent.

Osmolar gap: the difference between the measured and the calculated osmolality; a large difference implies an unknown substance in the blood.

Osteoblast: specialized fibroblast responsible for bone formation from osteoid and a mineralized connective tissue matrix.

Osteoclast: specialized macrophage responsible for breaking down (resorbing) bone.

Osteoarthritis: limited inflammation of the major joints most often brought on by the 'wear and tear' of prolonged overweight.

Osteomalacia: a bone disease characterized by poor bone structure and deformity; the result of lack of vitamin D.

Osteomyelitis: inflammation of the bone and/or bone marrow, generally caused by an infection.

Osteoporosis: a common bone disease characterized by loss of bone density, leading to deformity and fractures.

Oxyhaemoglobin: haemoglobin that is carrying oxygen.

Paget's disease: metabolic bone disease characterized by raised alkaline phosphatase with disordered bone resorption and construction leading to deformity.

Pallor: a pale coloration of the skin and sign of anaemia.

Palpitations: abnormal or irregular heartbeats of such severity that the patient is conscious of them.

Pappenheimer bodies: red cell inclusion bodies formed from iron, often present in sideroblastic anaemia and sickle cell disease.

Pancytopenia: reduced levels of all blood cells; indicative of profound damage to the bone marrow.

Paraprotein: an unusual plasma protein, part of an immunoglobulin and inevitably the product of a myeloma or related disease.

Parasite: an organism that lives at the expense of another; the terms pathogen and parasite are virtually interchangeable.

Parathyroid hormone: product of the parathyroid gland that acts to increase plasma calcium.

Parenteral: administration of a drug through the skin.

Parietal cell: gastric epithelial cell secreting hydrochloric acid and intrinsic factor, the latter required for the absorption of vitamin B_{12}.

Paroxysmal nocturnal haemoglobinuria: present when red cells lack molecules that normally give protection from complement lysis; leads to haemolytic anaemia.

Pathogen: an organism that causes disease.

Pathogenesis: the factors involved in the development of a disease.

Pelger–Huet anomaly: congenital or acquired condition characterized by hyposegmented neutrophils.

Penicillin: an antibiotic first discovered (in the mould *Penicillium notatum*) in 1928 but only used clinically in 1940.

Pericarditis: inflammation of the pericardium.

Periodic acid – Schiff: a dye that stains polysaccharides such as glycogen, and may be useful in certain malignancies such as erythroleukaemia.

Peristalsis: the rhythmic waves of smooth muscle contractions that propel food along the gastrointestinal tract.

Perls' Prussian blue: a stain used to estimate iron stores and in the diagnosis of sideroblastic anaemia and identification of sideroblasts and siderocytes.

Pernicious anaemia: the consequence of lack of vitamin B_{12} because of the failure of gastric parietal cells to secrete intrinsic factor.

Petechiae: red pin-point sized haemorrhagic spots that form as a result of blood leaking from capillaries.

Peyer's patches: lymphoid tissues within the wall of the intestines that may perform the same local functions as lymph nodes.

Phagocyte: literally an eating cell; neutrophils and macrophages are phagocytes that absorb and digest micropathogens (bacteria, fungi and viruses).

Phagolysosome: membrane-enclosed vesicle formed from the fusion of a lysosome with a phagosome, enabling digestion of the contents of the phagosome.

Pharmacodynamics: the action, effects and metabolic breakdown of drugs within the body.

Pharmacogenomics: a branch of pharmacology that considers the effects of genes on the response of an individual to a particular drug.

Pharmacology: the study of the effects of drugs that can be used for the prevention, diagnosis, prevention and, treatment or cure of disease.

Phenotype: the visible or measurable characteristics of an individual or, indeed, any observable biological trait.

Philadelphia chromosome: formed by the translocation between chromosomes 9 and 22, and which generates the fusion oncogene *BCR–ABL*.

Phospholipids: components of the cell membrane; in platelets, phospholipids support the coagulation pathway, but can also be the objective of autoantibodies.

Plasma: the protein-rich fluid remaining when erythrocytes, leukocytes and platelets have been removed from blood.

Plasma cell: antibody-secreting cell that develops from a B lymphocyte.

Plasma viscosity: describes the 'thick' or 'thin' quality of the plasma; it is increased in the acute-phase response and especially by a paraprotein.

Plasmapheresis: the collection of plasma, with the return of the erythrocytes and leukocytes to the blood donor.

Plasminogen: zymogen that is the substrate for tissue plasminogen activator, producing the enzyme plasmin that digests fibrin and so dissolves thrombi.

Plasminogen activator inhibitor: molecules that regulate the activity of tissue plasminogen activator and so plasmin and ultimately fibrinolysis.

Platelet: vesicle-like, subcellular fragment of the cytoplasm of megakaryocyte that is active in coagulation.

Platelet aggregation: occurs when platelets extend pseudopodia and become sticky and clump together when stimulated by an agonist such as by ADP.

Plateletpheresis: the collection of platelets from blood with all other components being returned to the blood donor.

Pluripotent: refers to a stem cell that has the ability to differentiate into any one of a number of unipotent stem cells.

Poikilocyte: an unusually shaped red blood cell: types include elliptocytes, acanthrocytes and stomatocytes.

Poikilocytosis: variation in the shape of red blood cells.

Point-of-care testing: as near-patient testing; an analysis performed closer to the patient than to the laboratory.

Polyacrylamide gel electrophoresis: technique for separating and quantifying proteins and nucleic acids.

Polychromasia: present when there is a marked variation in the degree of coloration of red cells, reflecting different levels of cellular haemoglobin.

Polyclonal antiserum: a collection of different antibodies each recognizing different epitopes on an antigen.

Polycythaemia: literally, many cells in the blood, but taken to be increased numbers of red cells as a result of proliferative disease polycythaemia rubra vera.

Polydipsia: intense thirst, a symptom of diabetes mellitus.

Polymerase chain reaction: an experimental, *in-vitro* method for duplicating in an exponential way short strands of specific DNA fragments.

Polymorphism: variations in the sequence of genes between individuals.

Polymorphonuclear leukocyte: one of the two main types of leukocytes that can be distinguished in having lobed nuclei and granular cytoplasm.

Polyuria: the production of abnormally large volumes of urine.

Porphyria: a group of diseases whose basis is abnormal production of the porphyrins that make up the haem group.

Post-transfusion purpura: an unexpected thrombocytopenia occurring after transfusion and caused by donor anti-platelet antibodies.

Precision: the ability of a test for an analyte to give the same result every time the same sample is analysed. Compare with accuracy.

Predisposing factors: conditions or situations that make an individual more susceptible to disease; they include age, sex, heredity and environmental factors.

Prevalence: the proportion of people in a population affected at a specific time.

Primary tumour: malignant tissue growing in the site where it initially started.

Primary care: refers to treatment not by a hospital but by a general practitioner.

Primary immunodeficiency: clinical condition resulting from a genetic defect leading to a failure of one or more components of the immune system.

Procoagulant: physiological, pathological or pharmacological process that promotes thrombosis.

Progesterone: steroid hormone produced by the ovary, adrenal gland and placenta. Supports female reproductive capacity.

Prognosis: the likely outcome of a disease.

Proinflammatory: description of a number of cytokines that promote (or a situation that promotes) inflammation.

Prolymphocyte: stage of differentiation of lymphopoiesis between the lymphoblast and the lymphocyte.

Promyelocyte: a precursor of the polymorphonuclear leukocyte lineage; in health, found only in the bone marrow.

Prophylaxis: the process of providing drugs to prevent or reduce the risk of suffering an adverse event, as in warfarin to prevent venous thrombosis.

Prostacyclin: a potent inhibitor of platelet function and vasodilator.

Prostaglandin: an active lipid-derived prostanoid metabolite that acts on various cells, such as platelets, mast cells, smooth muscle cells and endothelial cells.

Protamine sulphate: agent used to reverse the effects of heparin.

Protein C: zymogen of activated protein C; the natural inhibitor of coagulation factor Va and factor VIIa.

Protein S: a cofactor for activated protein C; deficiency leads to an increased risk of thrombosis via poor activated protein C function.

Proteinuria: protein in the urine – a general marker of renal failure, common in nephrotic syndrome

Proteolytic: enzymes that break down proteins.

Prothrombin time: principle coagulation test that assesses a number of coagulation factors, the extrinsic pathway and, as the INR, the effect of warfarin.

Prothrombotic: physiological, pathological or pharmacological process that promotes thrombosis.

Proto-oncogene: a normal gene that is at risk of becoming transformed into an oncogene.

Pulmonary circulation: the circuit which the blood follows between the right side of the heart, the lungs and the left atrium.

Pulmonary embolism: a potentially fatal thrombosis in the blood vessels of the lung; prevented and treated with anticoagulants.

Pulse: the bulging of the arteries that can be felt at a number of sites in the body during the cardiac cycle.

Purpura: red discoloration of the skin caused by subcutaneous haemorrhage often associated by conditions such as thrombocytopenia and DIC.

Pyrexia: a rise in body temperature; often a sign of an infection and/or an active acute-phase response.

Pyruvate kinase: an enzyme that participates in glycolysis; deficiency causes a haemolytic anaemia.

Quality assurance: where an external organization provides test samples to the host laboratory to enable performance comparison.

Quality control: analysis of samples of known analyte composition in each batch of analyses; used to ensure batch-to-batch reproducibility.

Radioimmunoassay: a type of immunoassay using radioisotopes. Has excellent technical sensitivity and specificity but carries health hazards.

Radiotherapy: treatment using ionizing radiation to treat or kill cancer cells.

Reactive oxygen species: ions such as superoxide and the hydroxyl radical that can be generated by phagocytes and which are bacteriocidal.

Recessive mutations: mutations that are only expressed in an individual who is homozygous for the condition.

Red blood cells: cells that carry oxygen.

Red cell distribution width: an assessment of the degree of variability in the size of population of red cells; a high RDW is synonymous with anisocytosis.

Reference range: numerical limits that determine the degree of concern regarding a particular analyte.

Reflux esophagitis: a back flow of gastric juice from the stomach into the lower end of the oesophagus.

Regurgitation: an imperfect closing of the heart valves leading to leakage and backflow.

Relapse: occurs when, following a period of apparent recovery, the symptoms of the disease return.

Remission: a period of good health with a reduction or disappearance of symptoms following a disease; may be permanent or temporary.

Renin: product of kidney in response to low sodium or low blood pressure; acts to increase aldosterone and so recover sodium from the urinary filtrate.

Respiratory acid–base disorder: the occurrence of an abnormal blood pH because of lung dysfunction.

Reticulocytes: the immediate immature precursors of erythrocytes.

Reticulo-endothelial system: a collection of cells of the blood, lymph node, liver and spleen that perform immunological and scavenging functions.

Retinoblastoma: a rapidly developing tumour of retinal cells of the eye. The most common inherited UK childhood cancer; caused by a mutation in the *RB1* gene.

Retinopathy: disease of the retina; a chronic complication of diabetes that ultimately leads to blindness.

Rheology: study of the physical properties of blood and plasma; principle measurements are erythrocyte sedimentation rate and viscosity.

Rhesus: refers to a blood group system more correctly designated as Rh; comprises antigens D, C, E, c and e.

Rheumatoid arthritis: a specific inflammatory connective tissues disease.

Rheumatoid factor: autoantibodies present in the plasma of patients with rheumatoid arthritis which are to be directed against their own IgG molecules.

Rickets: a juvenile form of osteomalacia caused by lack of vitamin D.

Risk assessment: legal requirement to assess the safety of a material or process before it is used or performed.

Risk factors: certain dietary, occupational or lifestyle conditions that increase the chances of developing a disease.

Ristocetin: once used as an antibiotic, now used to assess platelet aggregation and in the diagnosis of von Willebrand disease and Bernard–Soulier syndrome.

Romanowsky: a type of stain, sub-types of which are widely used to facilitate the identification of cells and tissues by light microscopy.

Rouleaux: term used to describe the close association of red cells, often because of high plasma protein levels as may be found in myeloma.

Sarcoma: a tumour of muscle cells.

Saturated: chemical denotation of the lack of double bonds (e.g. in a fatty acid).

Schistocyte: fragments of red blood cells, often found in haemolytic anaemia, and proof of a destructive pathological process.

Secondary: a term often used to refer to a condition that arises from an existing disease. It is also used to describe a stage of a disease.

Secondary care: refers to care in a hospital.

Secondary immunodeficiency: this arises as a consequence of other conditions, such as cytotoxic chemotherapy and HIV infection.

Secondary tumours: tumours formed when a primary tumour metastasizes and spreads to other tissues and organs. Often loosely referred to as 'secondaries'.

Secretors: individuals who express the A, B and H blood antigens on soluble proteins and secrete them into saliva and other body fluids.

Self: used in immunology to refer to the components of the body itself, in contrast to foreign material (such as a virus) that is non-self.

Senescence: the decline in the functions of almost all parts of the body and at all levels of organization, from cells to organ systems, seen on aging.

Sensitivity: ability of a method to correctly identify true positives (those who have the particular disease). See also specificity.

Sepsis: the presence of pathogen(s) in the blood likely to cause septicaemia (blood poisoning).

Serology: study of molecules and ions in the plasma serum.

Seropositive: identification of antibodies to a pathogen or tissue that defines the individual's status; examples are rheumatoid arthritis and HIV infection.

Serotonin: also known as 5-hydroxytryptamine (5HT); inflammatory mediator, neurotransmitter and vasoconstrictor.

Serum: the clear straw-coloured liquid remaining after the blood has been allowed to clot, and the clot and all the cells removed.

Severe combined immunodeficiency: a life-threatening clinical condition involving deficient numbers of both B and T lymphocytes.

Sickle cell anaemia/disease: a haemoglobinopathy characterized by the presence of crescent-shaped erythrocytes.

Sideroblast: nucleated red blood cell in the bone marrow with particles of iron within perinuclear mitochondria, often in a ring form; hence ring sideroblast.

Sideroblastic anaemia: anaemia characterized by sideroblasts; can be acquired (e.g. excess alcohol) or hereditary (e.g. mutations in haem-forming enzymes).

Siderocyte: red blood cells characterized by the presence of iron granules in the cytoplasm; a consequence of sideroblastic anaemia.

Sign: a physical phenomenon indicative of a particular disease or condition in a patient.

Sinoatrial node: one of the two 'pacemakers' that generates electrical impulses that regulate the rate of contraction of the heart. The other is the AV node.

Sinus rhythm: the natural pattern of a heart beat, generated by impulses from the sinoatrial node.

Smear cells: ruptured leukaemic cell found on a blood film; the cells are too fragile for the spreading of the film and break up, leaving a purple smear.

Sodium citrate: anticoagulant used in the processing of blood to permit coagulation tests on the plasma.

Specific immune response: the response that allows the development of true specific immunity against a particular infectious agent.

Specificity: ability of a method to correctly identify those persons who do not have a particular disease (i.e. true negatives).

Spectrin: parts of the cytoskeleton of the red blood cell.

Spherocyte: a spherical red blood cell; the shape is adopted because of the loss of function of the cytoskeleton.

Splenectomy: surgical removal of the spleen; often performed in certain cases of haemolytic anaemia, such as haemoglobinopathy.

Splenomegaly: enlarged spleen; present in many conditions, such as leukaemia, haemoglobinopathy, infectious mononucleosis and parasitic infections.

Standard operating procedure: a set of instructions enabling an intelligent practitioner to perform a task or procedure.

Stem cell: a precursor cell in the bone marrow that ultimately gives rise to mature cells in the blood.

Stenosis: a narrowing in a tube (blood vessel, duct of a gland), or a heart valve that restricts the free flow of fluid.

Stomatocyte: a red cell with a characteristic oblong area or central pallor; can be hereditary, due to a dysfunctional cytoskeleton, or acquired, such as in dehydration.

Streptokinase: fibrinolytic agent derived from a bacterium used to treat the symptoms of acute arterial thrombosis as in myocardial infarction or stroke.

Stroke or cerebrovascular accident: consequences of the restriction of the blood supply to the brain, leading to damage or death due to lack of oxygen.

Subclinical: a stage where a disease is established even though any overt signs and symptoms are not apparent.

Subendothelium: the connective tissue underneath the endothelium; comprises connective tissues such as elastin and collagen that can activate platelets.

Subcutaneous: pertaining to an injection; below the skin.

Sudan black: a dye used in cytochemistry to help in the diagnosis of diseases such as leukaemia.

Superoxide: a reactive oxygen species that is cytotoxic to cells and baceteria.

Surface membrane immunoglobulin: antibodies at the surface of B lymphocytes that define their lineage.

Symptom: some sort of indication that a disease is present and something of which the patient complains; for example, nausea, malaise or pain.

Symptomatic: to exhibit symptoms that may be linked to a disease; to be tired and lethargic is to be symptomatic of anaemia.

Syncope: fainting with a temporary loss of consciousness resulting from a temporarily inadequate supply of oxygen and nutrients to the brain.

Syndrome: a term applied to describe certain diseases that are characterized by multiple abnormalities and features that form a distinct clinical picture.

Syngeneic stem cell transplants: these involve donors and recipients who are genetically identical; for example, between identical twins.

Synovial fluid: the liquid lubricant secreted into the synovial cavity of joints. Increased volumes in disease of the joints such as arthritis.

Systemic circulation: the flow of blood from the left ventricle to all the tissues of the body with the exception of the lungs and back to the right atrium.

Systemic disease: one that affects the whole body.

Systemic lupus erythrematosus: a systemic inflammatory autoimmune connective tissue disease; a common finding is increased levels of antinuclear antibodies.

Systole: contraction of the heart ventricles.

T cell receptor: molecules at the surface of the T lymphocyte that recognize antigens.

T lymphocytes: lymphocytes that leave the bone marrow in an immature state and require the thymus to become mature; defined by surface expression of CD3.

Tachycardia: an inappropriately fast heartbeat, such as greater than 100 beats per minute.

T_C cells or cytotoxic T lymphocytes: T lymphocytes that, when appropriately stimulated, are capable of killing virus-infected cells.

Tertiary care: when a patient is passed from a hospital to a specialist unit in another hospital.

Testes: gonads that generate testosterone and sperm.

Testosterone: steroid sex hormone that stimulates male sexual characteristics.

T_H **cells or helper T lymphocytes:** blood cells that develop into cytokine-secreting T_H cells when stimulated by an immunogen, and assist B cells make antibodies.

Thalassaemia: a haemoglobinopathy of varying severity that originates from point mutations or deletions in globin genes.

Therapeutic drug monitoring: measurement of plasma levels of a drug in order to provide effective doses with minimal toxicity.

Thiazide diuretics: a class of drugs that decrease the reabsorption of sodium, and so the loss of water. Used to reduce blood pressure and excess tissue fluid.

Thrombin: a key enzyme in the coagulation pathway; cleaves fibrinogen into fibrin, but also stimulates platelets.

Thrombin time: coagulation test that mainly assesses fibrinogen activity.

Thrombocytopenia: platelet count below the bottom of the reference range.

Thrombocytosis: platelet count above the top of the reference range.

Thrombolytic therapy: use of drugs such as streptokinase and alteplase to digest clots.

Thrombophilia: literally, loving to clot; taken to be unusual propensity to inappropriate thrombosis – opposite of haemophilia (loves to bleed).

Thromboplastin: artificial platelet substitute rich in certain molecules such as tissue factor and phospholipids used in prothrombin and partial thromboplastin times.

Thrombopoiesis: the process of the development of platelets from megakaryocytes in the bone marrow.

Thrombopoietin: growth factor that stimulates thrombopoiesis.

Thromboxane: a vasoactive (inducing arterial constriction) eicosanoid product of platelet activation; can also stimulate the activation of other platelets.

Thrombus/thrombosis: the formation of a thrombus (clot).

Thymocyte: an immature T lymphocyte in the thymus.

Thyroid-stimulating hormone: product of the pituitary that induces the thyroid to generate and release thyroid hormones.

Thyroglobulin: globubular protein of the thyroid and substrate for thyroid hormones.

Thyrotoxicosis: overactivity of the thyroid gland and excessive release of thyroid hormones, and so hyperthyroidism.

Thyroxine: one of two thyroid hormones; has many effector functions. Also known as tetra-iodothyronine (hence T4); therapy for hypothyroidism.

Tissue factor: initiator of coagulation in partnership with factor VIIa; found at the surface of endothelial cells, monocytes and macrophages, and in the plasma.

Tissue factor pathway inhibitor: regulator of the tissue factor pathway of the coagulation cascade.

Translocation: movement of one section of a chromosome to another.

Transcobalamin: a blood protein that carries vitamin B_{12}.

Transferrin: a blood protein that carries iron.

Transplantation: moving tissue from one body or part of the body to a different body or another part of the same body.

Transfusion reaction: a series of signs and symptoms caused by an incompatible blood transfusion.

Transient ischaemic attack: short-lived loss of motor function due to shortage of oxygen in the brain; generally a precursor of stroke.

Trephine: obtaining a sample of bone marrow that retains the internal structure.

Triacylglycerol: (also known as triglycerides); a lipid composed of a molecule of glycerol and three long-chain fatty acids.

Tri-iodothyronine: a thyroid hormone with many effector functions.

Toxic diseases: diseases caused by the ingestion of a variety of poisons that may be encountered in the environment.

Toxicology: the study of the adverse effects of toxic chemicals on organisms.

Tumour necrosis factor: a pro-inflammatory cytokine secreted by monocytes and macrophages.

Turbidimetry: measurement of the loss of light transmittance through a suspension; used to quantify plasma proteins.

Ulcerative colitis: an inflammatory bowel disease; chronic inflammation of the colon.

Unfractionated heparin: a relatively crude heparin preparation that requires monitoring with the APTT, and a cause of thrombocytopenia.

Unsaturated: chemical notation for the presence of double bonds (e.g. in a fatty acid).

Urate: salt of uric acid; high levels accumulate in cool hypoxic tissues, become insoluble as crystals, and cause tissue damage as gout.

Urea: a liver product that removes nitrogen from the body; high levels imply renal disease.

Uraemia: levels of urea above the top of the reference range.

Uric acid: end product of the breakdown of purines nucleotides.

Urolithiasis: the formation of renal stones rich in uric acid; a consequence of hyperuricaemia.

Vaccination: the process of protecting people from infection by deliberately exposing them to parasite components that initiate a protective immune response.

Vacutainer: a glass tube under partial vacuum commonly used in venepuncture.

Vasculitis: inflammation of the blood vessels.

Vasoconstriction: a reduction of the lumen of a blood vessel; if arterial, results in increased blood pressure.

Vasopressin: a pituitary hormone, also known as antidiuretic hormone, acts on the nephron to increase the resorption of water, thus increasing blood volume.

Venepuncture: process of obtaining blood from a vein, most conveniently one on the inside of the elbow.

Venous thromboembolism: a clot in the veins; principally, deep vein thrombosis in the legs and groin, pulmonary embolism in the lungs.

Ventilation–perfusion scanning: use of inhaled and injected radioisotopes to investigate a possible pulmonary embolism.

Virchow's triad: three features that summarizes the pathology of thrombosis; the blood vessel wall, blood cells and blood molecules.

Virology: the study of viruses; a subdiscipline of microbiology.

Viscosity: a measure of how thin or thick a fluid is; can be of the blood or of plasma. In either case, increased viscosity is linked with disease.

Vitamin B_{12} and vitamin B_6: essential cofactors in the generation of haem; their absence leads to anaemia.

Vitamin K: essential cofactor for the production of certain coagulation factors. Given as an antidote for a dangerously high INR.

Von Willebrand disease: haemorrhagic disorder caused by qualitative or quantitative changes in von Willebrand factor.

Von Willebrand factor: large molecule derived from endothelial cells; functions as a cofactor for coagulation factor VIII and participant in thrombosis.

VTE: venous thromboembolism.

Waldenstrom's macroglobulinaemia: a myeloproliferative disease of B lymphocytes allied to myeloma and characterized by an abnormal IgM paraprotein.

Warfarin: a common oral coagulant used for the treatment and prophylaxis of venous thromboembolism and (in atrial fibrillation) stroke.

Weibel–Palade body: endothelial cell organelle that is a storage granule for von Willebrand factor.

Wells' score: clinical method to help determine the likelihood of a venous thromboembolism.

White blood cell: a blood cell involved in inflammation and/or immunity.

Wilson's disease: an inherited disorder of copper metabolism, the metal being deposited in organs such as the liver and brain.

Xanthoma: yellowish deposit on the tendons that occurs in familial hypercholesterolaemia.

Zymogen: a precursor molecule that turns into an active protease when subjected to a specific proteolytic cleavage.

Index

Blood Science: Principles and Pathology, First Edition. Andrew Blann and Nessar Ahmed.
© 2014 John Wiley & Sons, Ltd. Published 2014 by John Wiley & Sons, Ltd.